CHORDATE EMBRYOLOGY

Developmental Biology of Nonchordates and Chordates

S Chand And Company Limited

(ISO 9001 Certified Company)

Head Office: Block B-1, House No. D-1, Ground Floor, Mohan Co-operative Industrial Estate, New Delhi – 110 044 | Phone: 011-66672000

Registered Office: A-27, 2nd Floor, Mohan Co-operative Industrial Estate, New Delhi – 110 044 | Phone: 011-49731800

www.**schandpublishing.com**; e-mail: **info@schandpublishing.com**

Branches

Chennai	:	Ph: 23632120; chennai@schandpublishing.com
Guwahati	:	Ph: 2738811, 2735640; guwahati@schandpublishing.com
Hyderabad	:	Ph: 40186018; hyderabad@schandpublishing.com
Jalandhar	:	Ph: 4645630; jalandhar@schandpublishing.com
Kolkata	:	Ph: 23357458, 23353914; kolkata@schandpublishing.com
Lucknow	:	Ph: 4003633; lucknow@schandpublishing.com
Mumbai	:	Ph: 25000297; mumbai@schandpublishing.com
Patna	:	Ph: 2260011; patna@schandpublishing.com

First Edition 1975

Subsequent Editions and reprints 1976 (Twice), 78, 80, 83, 85, 87, 89, 91, 92, 93, 94, 95, 96, 97, 99, 2000, 2001, 2002, 2003, 2004, 2005, 2006 (Twice), 2007, 2009, 2010, 2011, 2014 (Twice), 2015, 2016 (Twice) , 2017, 2018, 2020 (Twice), 2021 (Twice)

Reprint 2022

ISBN : 978-81-219-0261-8 **Product Code :** H6ZEB68ZOOL10ENAJ0XR

PRINTED IN INDIA

By Vikas Publishing House Private Limited, Plot 20/4, Site-IV, Industrial Area Sahibabad, Ghaziabad – 201 010 and Published by S Chand And Company Limited, A-27, 2nd Floor, Mohan Co-operative Industrial Estate, New Delhi – 110 044.

PREFACE TO THE TENTH EDITION

It is quite heartening for the authors to release this thoroughly revised, enlarged and up-to-date edition of **Chordate Embryology.** In this edition every effort has been made to incorporate all the current information of embryology (Developmental Biology) yet retaining classical views to provide a complete picture of the subject to the students and teachers.

In this edition, several new chapters/topics have been added. These are as follows :

Cellular Basis of Development;

DNA, RNA and Protein Synthesis;

Parthenogenesis;

Mid-blastula Transition;

Nucleus and Cytoplasm in Development;

Cell Lineage;

Morphogenesis, Growth;

Metaplasia;

Teratogenesis;

Birth Control;

Impotency, Sterility, Artificial Insemination, Test-Tube Baby and GIFT.

Almost all the old chapters have been either rewritten or refashioned. More than one hundred new illustrations and many new tables have been added. Throughout the revision of the book, our main target has been to make it more acceptable and immensely useful for the students and teachers. And all this has been done keeping an eye on the lurking danger of the book attaining an encyclopaedic dimension. However, the authors hope that their sincere effort to educate their readers about various embryological problems will be appreciated by them.

Despite their most careful and conscious efforts, the readers may come across a few errors/shortcomings in the present book. It is hoped that the readers will point them out so that such errors may be eradicated from the next edition. Suggestions from students and teachers for further improvement of the book will be highly appreciated and incorporated in the future edition.

The authors are deeply grateful to Shri Rajendra Kumar Gupta, Shri Ravindra Kumar Gupta and Shri T.N. Goel of M/s S. Chand & Company Ltd., for the pain they have taken in the publication of this book.

AUTHORS

PREFACE TO THE FIRST EDITION

Recently many Indian universities have incorporated 'Chordate Embryology' in their syllabii of B.Sc., B.Sc. (Hons.) and M.Sc. classes. Keeping in view this change, we have written this textbook for B.Sc., B.Sc. (Hons.), and M.Sc. students of all Indian universities. During the writing of this book we always remained conscious of the difficulties of students and have tried to present a simple, well-illustrated and up-to-date account of the subject. No originality is claimed in the account of this book. This is merely a compilation from different sources, *viz.*, several classical and modern textbooks, monographs of embryology, developmental biology and cell biology.

Though we have taken special care to give a correct account, yet we are fully aware of our limitations. If the readers come across any inconsistencies, the authors extend their due apology for that .

The authors are thankful to Dr. A.P. Tyagi, D.A.V. College, Muzaffarnagar; Prof. Kuldeepak Kumar, S.D. College, Muzaffarnagar; Dr. Mahesh Chandra Sharma, M.S. College, Saharanpur; Prof. N.M. Malhotra, Meerut College, Meerut, and Prof. Smt. Sadhna Kaushik, R.G. College, Meerut for rendering various types of help during the preparation of this book. The authors are also grateful to Shri Shyam Lal Gupta ''Padam Shri'' and Shri T.N. Goel, M/s S. Chand and Company Ltd., New Delhi for their whole hearted cooperation in the publication of this book.

Suggestions for the improvement of the book shall be highly appreciated from the teachers and students.

AUTHORS

A Review On 'Chordate Embryology'

Rapid developments in the field of genetics, cytology, biochemistry, embryology, etc., in the last few decades made us to realize the utmost significance of embryology in understanding various problems of the life. The old descriptive embryology, therefore, has been combined with dynamic experimental physiological approach or more correctly an altogether a new branch called, the developmental biology, has come forth.

The book under review is intended for beginners who wish to understand the fundamentals of developmental biology. The whole gamut of subject has been described in a simple and systematic manner suited for readers who have not been exposed sufficiently to the subject. The authors have tried to put more emphasis on morphological approach but, nevertheless, a few aspects of cell biology, biochemistry and endocrinology are freely discussed at appropriate instances. These instances are enough to keep the arguments down to earth and so reduce the confusion in grasping the subject. The five new chapters in this second edition further strengthen and enhance the integrated spectrum of descriptive-morphological and experimental-physiological approach of the text subject matter.

The text starts with a good introductory chapter which briefly describes the history of embryology including its basic laws and required techniques. Six chapters are devoted to embryogenesis of individual chordates, and others to general account. Above all the text, a mere compilation of many standard sources, comprising of 20 chapters, is well linked.

In summary, the book may be recommended as a concise, lucid and simple textbook for beginners. The topics dealing with nonchordate embryology have been described in various chapters, especially in the five newly incorporated ones in this second edition.

SCIENCE REPORTER
July, 1977

CONTENTS

1

INTRODUCTION

1-1. DEFINITION AND SCOPE

The survival of each species of viruses, bacteria, fungi, plants or animals requires that its individual members multiply to produce new individuals to replace the one killed by the natural death in old age or by predators, parasites, environmental pollution or other ecological hazards such as shortage of water, oxygen, light, optimum temperature, food, etc. The multiplication of a species is achieved by two basic types of reproduction: asexual and sexual. In **asexual reproduction,** the progeny arises from a single existing organism which splits, buds or fragments to give rise to two or more offsprings, all of which have hereditary traits identical to those of the parental organisms. In **sexual reproduction,** the progeny arises from the fusion of two genetically distinct gametes (sperms and eggs of multicellular animals), each of which is derived from a different mating type or sex of organisms. The resultant cell, the **fertilized egg** or **zygote** has a genome different from that of either parent. The processes which transform a zygote or some body rudiment of parental somatic cells (*i.e.,* asexual reproductive bodies), into a multicellular and well-organized organism similar to adult animal of that species, constitute two different aspects of **ontogenetic development**. The ontogenetic development differs from the **phylogenetic development,** which is the historical or evolutionary development of a species. The ontogenetic development of a species, thus, includes **embryogenesis**, the development of a new individual from the fertilized egg or zygote, and **blastogenesis**, the development of a new individual from an asexual reproductive body, (*i.e.,* bud, body fragment, gemmule, etc).

The development of an organism involves all the changes it undergoes from its beginning until its death (*i.e.,* its entire life history). Most metazoan animals include two distinct periods in their entire life history: 1. **Embryonic** or **pre-natal period** which is passed within the egg or the womb of the mother, and 2. **Post-embryonic** or **post-natal period** which extends from hatching or birth up to the natural death of the organism due to aging. Thus, all the developmental events and changes which occur from the time of fertilization of the egg through formation of the embryo or foetus up to hatching or birth belong to the pre-natal or embryonic period of the life history of the animal, are included in embryogenesis. The study of embryogenesis is called **embryology.**

Embryological development does not end at birth but continues during the post-natal or post-embryonic period also. The newly born or hatched young animal is small in size and imperfectly developed both structurally and physiologically. During subsequent days, weeks, months, or years, depending upon the species, many changes of '*progressive nature*' occur in the young animal, as a result of which young one matures into adult animal which is capable of doing reproduction. Further, after the adult animal reaches a peak, a series of other kinds of developmental changes, this time *'degenerative in nature'*, set in which ultimately lead to its death.

The field of biology which deals with the study of those embryogenetic and blastogenetic processes by which organism undergoes progressive and orderly changes in structure and function during their entire life history is called **developmental biology**. In fact, developmental biology, unlike embryology, does not restrict itself to the study of the development of embryo (*i.e.,* embryogenesis) but deals with all aspects (genetic, biochemical, physiological and morphological) of the entire developmental period of the individual organism (**Goel,** 1984).

The developmental biology occupies a central position among various biological disciplines and offers a great challange for the future. It provides understandable explanations to various aspects of cell biology, genetics, physiology, ecology, evolution, histology and anatomy. For example, developmental biology and genetics have become closely related to form a distinct biological discipline, called **developmental genetics** which deals with the expression of genes during development and tries to understand the organization of chromosomes and mechanism underlying cell division. These problems bring developmental biology close to cell biology. Some major problems common to both of them are endocytosis (pinocytosis and phagocytosis), cellular secretions (acrosome, cortical granules), cell movements (morphogenetic and other movements), cell association and affinity, cytodifferentiation, cell growth, cell aging and cell death. Similarly, certain problems common to developmental biology and physiology comprise not only the basic life processes of the embryo but also the processes that permit the maintenance and normal development of an embryo in a harmonious and synchronized manner, *e.g.,* the physiology of foetal nutrition in amniota (*i.e.,* reptiles, birds and mammals), gastrulation and metamorphosis. Developmental biology has also provided convincing explanations to various problems of evolution and phylogeny. For instance, in last century, phylogeny and classification of the parasitic *Sacculina* as a cirripede crustacean arthropod or of the sessile ascidian as a chordate could be possible only due to careful study of their developmental biology. In fact, developmental biology allows the establishment of homologies and in the reconstruction of evolution. Ecology is also related to developmental biology. The study of nitrogen metabolism during development or the evolution of viviparity or metamorphosis provides crucial evidence of ecological adaptation of animals. To the medical student, the embryology is of primary importance because it provides a comprehensive and rational explanation of the intricate arrangement of human anatomy. It also helps in understanding the malformations, origin of vestigial structures and origin of various pathological conditions.

1-2. A PREVIEW OF CHORDATE EMBRYOLOGY (OR DEVELOPMENTAL BIOLOGY)

The embryogenesis of an animal species includes following stages or phases :

(i) Gametogenesis. The embryogenesis is started from the time of differentiation and specialization of haploid male and female sex cells, *viz.*, **sperm** and **ova** respectively, from diploid somatic cells of each parent during a process called **gametogenesis**. The gametogenesis includes **spermatogenesis** and **oogenesis**. The spermatogenesis is a process which occurs in male gonads (testes) and produces small-sized, motile, haploid sex cells, the **sperms** or **spermatozoa**. A spermatozoon is devoid of excess cytoplasm and has an unspecialized nucleus, a generally differentiated middle piece and tail, and a highly specialized distal acrosome, the latter being the derivative of Golgi-apparatus, containing enzymes for the penetration of sperm into egg cell. The oogenesis occurs in female gonads (ovaries) and produces a large, non-motile, nutrient-filled, metabolically inert, haploid sex cell, the **egg** or **ovum** and three haploid, small-sized **polar bodies** or **polocytes**. In both spermatogenesis and oogenesis, meiosis by discarding half of the chromo- somes, singles out the sets of genes which are to operate in the development of a particular individual.

(ii) Fertilization. The second phase of embryogenesis is fertilization, which is the union of a spermatozoon and an ovum and 'activation' of egg to begin the embryogenesis of a new individual. It includes the (i) encountering of spermatozoa and ova by the act of copulation of male and female parents; (ii) species-specific 'chemical lock' formation between **fertilizin** mucopolysaccharide molecule of ovum's plasma membrane and **antifertilizin** protein molecule of sperm's plasma membrane; (iii) acrosomal reaction in the formation of clear passage through the envelopes of ovum and formation of a acrosomal tubule from fused membranes of acrosome and sperm head ; (iv) migration of contents of sperm head and mid-piece inside ovum through acrosomal tubule ; (v) activation of egg to form fertilization cone, to elevate fertilization membrane outside the egg plasma membrane from the contents of cortical granules, to start replication of DNA and initiation of mitosis for cleavage and to initiate metabolism ; (vi) conversion of sperm nucleus into pronucleus and migration of both (male and female) pronuclei to the site of **amphimixis** or fusion, where both nuclei are fused to form zygote nucleus.

(iii) Cleavage. During third phase of embryogenesis, the **cellulation, segmentation** or **cleavage**, no growth of egg-cytoplasm **(ooplasm)** takes place, but rate of synthesis of some macromolecules such as DNA and proteins is increased at the expense of reserve food substances of egg (*viz.*, glycogen, fats and yolk); and the fertilized egg undergoes repeated and successive mitotic cell divisions and produces a compact heap of cells or **blastomeres**, called **morula**. The blastomeres of morula soon get arranged in a hollow spherical body, the **blastula**, with a layer of blastomeres, called **blastoderm**, surrounding a fluid-filled central space or cavity, called **blastocoel**. The process of conversion of a fertilized egg into a multicellular blastula is called **blastulation**.

(iv) Gastrulation. Following the cleavage phase, there occurs an extensive movement and rearrangement of the cells of blastoderm of blastula, transforming the one-layered thick embryo (blastula) into a two or three-layered thick embryo, the **gastrula**. These three layers of gastrula are complex rudiments from which are derived the various organs of animal body, so, they are called **primary germ layers**. The outermost germ layer is called **ectoderm**, the innermost **endoderm**, and the middle one is called **mesoderm**. The movements (called **morphogenetic movements**) by which these germ layers are brought into position, collectively comprise **gastrulation**. The fully formed gastrula has a central cavity called **archenteron**, which is lined by endoderm and communicates to the exterior by an opening called **blastopore**. During later developmental stages, the archenteron becomes the cavity of alimentary canal, while blastopore either becomes **mouth opening** in **protostomia** (*i.e.,* all invertebrates except echinoderms and hemichordates) or **anal opening** (anus) in **deuterostomia** (*i.e.,* echinoderms, hemichordates and chordates).

During gastrulation, in typical, there is a decrease in the rate of cell division and the first evidence of gene transcription (RNA synthesis) becomes apparent. This transcription in part specifies the proteins required for gastrulation. Transcription later in development specifies the proteins characteristic of the differentiated state of cells (*e.g.,* acetylcholine esterase in nerve cells, myosin in muscle cells and melanin pigment in epidermal cells) (**David Epel**, 1973).

(v) Organogenesis. During the fifth phase of embryogenesis, the **organogenesis** or organ formation, the continuous masses of cells of the three germ layers split up into smaller groups of cells, called the **primary organ rudiments,** each of which is destined to produce a certain organ or part of the adult animal body. The primary organ rudiments, further, subdivide into **secondary organ rudiments** which are rudiments of the subordinate and simpler organs and parts. With the appearance of primary and secondary organ rudiments the embryo begins to show some similarity to the adult animal, or to the larva, if the development includes a larval stage.

(vi) Growth. The sixth phase of embryogenesis is the period of **growth**. Growth, by definition, simply means developmental increase in mass. The increase in mass of embryo is achieved by synthesis of new nuclear material and cytoplasm and by subsequent cell multiplication. Thus, during growth period the organ rudiments of the embryo begin to grow and greatly increase their volume. In this way the embryo gradually achieves the size of its parents.

(vii) Differentiation. The differentiation refers to the events by which parts become different from one another and also different from what they were originally. During development differentiation includes following kinds:

(a) Morphological differentiation. As the cells of organ rudiments multiply, the individual cells and groups of cells become structurally different from one another **(cytodifferentiation)**. For example, from a common starting point in generalized ectoderm, nerve cells and epidermal cells acquire distinguishing features of size, shape and internal architecture.

(b) Behavioural differentiation or physiological differentiation. Although all cells exhibit common basic attributes such as metabolism (biosynthesis of DNA, RNA, proteins, lipids, carbohydrates, etc., and oxidation of foodstuffs) and irritability, special functional capabilities are eventually superimposed on these general properties. Thus, nerve cells come to transport electric disturbances, muscles to contract, gland cells to secrete special products, and so on.

(c) Chemical differentiation. The cytodifferentiation and physiological differentiation of cells are the products (or outcomes) of chemical differentiation or chemodifferentiation, which in its turn, is based on enzymes which direct the synthesis of organic compounds that give the cells its uniqueness. Therefore, it has been said that *"the differentiation is the production of unique enzymatic patterns,"* (**Spiegelman**, 1948). But, because all enzymes are proteins, the foregoing aphorism may properly be altered to read *"differentiation is the production of unique protein patterns,"* (**Balinsky**, 1970). Since cell constituents such as carbohydrates, lipids and organic acids are synthesized under the direction and control of enzymes, a given pattern of enzymes will dictate the pattern of these organic compounds. Not all the proteins of a cell are enzymes, however, some are so called **structural proteins**, which are the integral parts of biochemical architecture of the cell. It is now known, that the synthesis of different enzymatic and structural proteins takes place under the guidance of genetic material (*viz.*, DNA, RNA) of cell and chromosomal histone proteins, ribonucleic acids, enzymes, hormones, vitamins, certain inorganic ions, etc., have specific roles in regulation of protein synthesis during embryogenesis.

(viii) Morphogenesis. The acquisition of the body form of an organism during its embryonic development is called **morphogenesis**. It involves those developmental processes that lead to the characteristic size and shape of the tissue or organs that make up an organism (**Maclean** and **Hall**, 1987). The **form** of an organism depends on two main factors, the form of individual cells (essentially differentiation) and the relative position of the different cells. The process of morphogenesis rests on cell communication and cell interaction (*i.e.*, the capacity for recognition between cells). In fact, embryonic cells of animals are experimentally found to be motile and if, disaggregated, they can ultimately segregate out again, so that the cells of same type become reassociated together. Basically, the process of morphogenesis involves the following types of **morphogenetic movements** : (1) **Cell migration** ; (2) **Cell aggregation**, forming masses, cords or sheets ; (3) **Localised growth** resulting in enlargements or constrictions (separation of single sheet into separate layers) or cavitation ; and (4) **Folding** such as evaginations, invaginations and bending (due to unequal growth).

Communication between embryonic cells falls into two main categories: **short-range communication** that is provided by morphogens and long-range communication that is provided by hormones (**Graham** and **Wareing** , 1976).

Morphogens are chemicals produced during development to determine the morphogenesis and / or differentiation of cells, tissues or organs. Morphogens are of two types : 1. **Intra-cellular morphogens** are found within a oocyte and are called **morphogenetic determinants** because of their widely studied role in the determination of cell fate in the early embryo. Their presence leads to mosaic development and to the establishment of programmed lines of cells (**cell lineages**) found in groups such as insects, nematodes and tunicates. 2. **Extra-cellular morphogens** are found in particular regions of the post-oocyte embryos. They become operational once the embryo has become multicellular (*i.e.*, onwards 2-cell stage of egg). These morphogens are analogous chemicals to hormones. Thus, while **hormones** are produced at one site and act, usually at a single concentration, at a site distant from their place of synthesis, being carried there by the blood vascular system. **Morphogens**, by contrast, are produced at one site (the **source**) and diffuse through the developing cells or tissues to a distant site (the **sink**) where they are broken down. A concentration gradient is, therefore, set up between the source and the sink. Cells lying along that gradient will respond to the particular concentration of the morphogen that they encounter and furthermore, their differentiative or morphogenetic response will differ depending on the particular threshold concentration of morphogen that they encounter. There are developmental systems such as **primary embryonic induction** and **regional organisation of the nervous system**, where the existence of two opposing concentration gradients of different morphogens, *i.e.,* neuralising morphogen and mesodermalising morphogen, provide the simplest explanation for the differential developmental events observed.

Hormones bind to specific receptors either on the cell surface or within the cytoplasm so as to form complexes that are carried to the nucleus. There they bind to specific sites on the chromatin and selectively regulate gene activity. Ecdysone in insects, thyroxine in amphibians and sex hormones in vertebrates are the best-understood examples of hormone's action in development (**Maclean** and **Hall**, 1987).

(ix) Metamorphosis. The growth, differentiation and morphogenesis transform the embryo into a young animal which can have its independent existence—an existence in which it has to procure food from the surrounding environment. The young animal which has emerged (**hatched**) from the egg either has morphological resemblances with adult animal except in being of miniature size and sexually immature (being so called **juvenile stage**) or has no resemblance with adult animal, *i.e.,* it has different types of organs which serve the young animal to lead a different mode of existence than the adult. Such young stages of animals are called **larval stages**. The young larva later on passes through a most important developmental process, the **metamorphosis**, which causes drastic morphological and physiological changes in larva and transforms it into an animal similar to the adult. The metamorphosis is regulated by hormones.

1-3. HISTORICAL REVIEW

Our ancient records distinctly reveal the human's natural curiosity in the sexual reproduction and development of man and domestic animals from the time of Egyptian civilization (3000–1500 B.C.). They cited speculations about the formation and growth of human embryos and gave prescriptions for abortion and prevention of pregnancy, etc. Certain holy books of Hindus of about 600 B.C. have distinguished three ways of origin of new individuals—from eggs, from living beings and from germ. The ''*Susrutā Samhita*'', a monumental Indian medical book, that was written during second or third century A.D., describes the development of a human child in the mother's womb. **Anaximander** (600 B.C.) probably was the first Western philosopher who conceived that ''living creatures arose from the moist element, as it was evaporated by the sun. Man was like another animal, namely fish in the beginning.'' Another, Greek philosopher, **Empedocles** (500 B.C.) believed that the foetus arose partly from male and partly from female semen. He knew that there was a regular sequence of events in development, and he spoke of the heart as formed first in development, the nail last. The Greek biologist, **Hippocrates** (460–377 B.C.) observed the development of hen's egg.

The earliest treatise on embryology which has been preserved is **Aristotle's** (384–322 B.C.), entitled "*De Generatione Animalium*"—concerning the generation of animals. This work describes the reproduction and development of many kinds of animals. Another treatise of Aristotle, "*De Historia Animalium*", contains the first account of the development of the hen's egg. He compared the different types of reproduction and elaborated a kind of classification of animals according to their mode of reproduction. Aristotle placed the mammals first, for, being unable to discover the egg, he thought that their young arose from a mixture of male and female fluids and were "born alive". Sharks, on the other hand, arose from eggs which were retained within the body of the mother and so were also born alive. Next he placed the type of reproduction shown by reptiles and birds in which the egg is "complete", that is to say, furnished with albumen and a shell. Lowest among the vertebrates were the amphibians and bony fish with "incomplete" eggs. His account of development showed great power of observation, keen skill in comparison, and imagination in interpretation. Notable among his speculations is one which has been given the name "**epigenesis**". From his observations on the development of the hen's egg Aristotle concluded that development always proceeded from a simple formless beginning to the complex organization of the adult (**Shumway**, 1942). Aristotle also speculated that there was a kind of 'soul', a non-material purposive principle which guided development. He called it "**entelechy**". He imagined that the female contributed only the 'substance', 'matter', 'material cause' or 'soil' in which embryo grew, while the 'soul' or 'form' along with entelechy was provided by father. The mystical influence of the male semen has been called *aura seminalis* by **Harvey** (**Strickberger**, 1985). Aristotle established embryology as an independent biological science. He is, thus, regarded as the **founder of the science of embryology**. His doctrines about development were accepted as gospel truth through most of the medieval ages. No one tried to confirm them by actual observation or test. Consequently no significant embryological study was made during the medieval ages. This long time is called **dark age** of science.

Aristotle's doctrines about development appeared in the writings of **Fabricus** (1537–1619) and **William Harvey** (1578–1657). Harvey's book "*Exercitationes de Generatione Animalium*", published in 1651, is based largely on the development of the chick, which he described in great detail. He also carefully studied the development of deer and compared it with that of chick. From purely theoretical considerations he came to the conclusion that mammals also formed eggs. Harvey popularized the dictum "*Ex Ovo Omnia*"—all animals arise from eggs.

The use of the word "ovary" was introduced by **Steno** in 1667. **R. de Graaf** (1641–1673) in 1672 discovered follicles in human ovary and considered them to be eggs which developed into foetus by some mysterious influence of the male's semen. From the time of de Graaf, the word "testis" has been restricted to designate the male organ essential to reproduction.

With the advent of microscope in seventeenth century, the sperm and egg have been discovered, but, their exact significance in development could not be appreciated. During this period another doctrine called "**preformation theory**" concerning the nature of development gained dominance.

1-3.1. Preformation Theories

After the invention of the microscope, **Marcello Malpighi** (1628–1694) published an account of the development of hen's egg, "*De Ova Incubato*", illustrated with fine figures of development from the 24-hour stage of incubation on. His work (1672) was responsible for the theory of **preformation** as opposed to Aristotle's epigenesis. On theoretical grounds, he held that the various parts of the embryo were contained in the egg and become visible as they increased in size. Later on, a Dutch scientist **Swammerdam** (1637–1680) theorized that sex cells contained preformed embryos or miniatures of the adult. **A. van Leeuwenhoek** (1632–1723), another Dutch physician, discovered the human spermatozoa in 1675 in the seminal fluid. In 1677, he discovered the sperms of some animals but mistook them at first for "wild animalcules" arising in the seminal fluid by some sort of putrefaction (**Maheshwari**, 1950). Leeuwenhoek believed that every spermatic animalcule of the ram contained a lamb, but it did not assume the external appearance of a lamb, until it had been nourished and grown in the uterus of the female. He expressed this view to criticise the views of **N. Hartsoeker** (1656–1725). Hartsoeker published in 1695 certain drawings of human spermatozoa each containing the preformed miniature body of the human individual called **homunculus** or "little man" in its head (Fig.1.1).

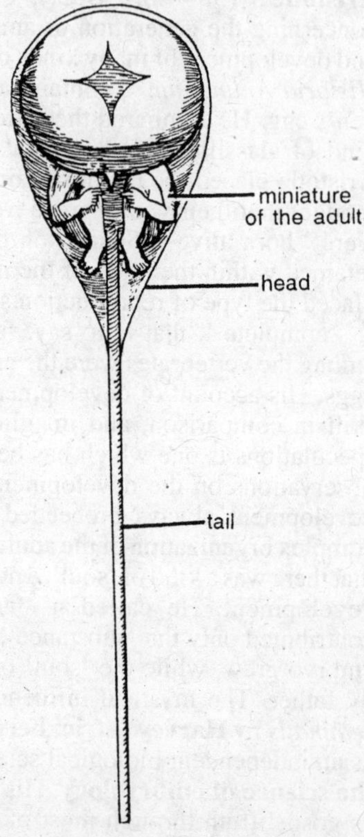

Fig. 1.1. Homunculus, "little man in a sperm cell". Drawing published by N. Hartsoeker in the *Jornal des Scavans*, Feb. 7, 1695 (after Burns, 1983).

Thus, two distinct schools of preformationists emerged : 1. **Animalculists** or **spermatists** such as **Leeuwenhoek** and **Hartsoeker** believed that preformed miniature organism was present in the head of the sperm. They theorized sperm as the seed and egg as the soil in which the seed was planted. 2. **Ovists** such as **Malpighi, Swammerdam**, **Haller** (1708–1777), **Bonnet** (1720–1793) and **Spallanzani** (1729–1799) postulated that the ovum (egg) contained a transparent, highly folded, small and unobservable preformed miniature of the adult, which was in some way stimulated to growth by the seminal fluid.

Some ovists namely **Swammerdam** and **Bonnet** advocated an extreme form of preformation theory called **encasement** or ''**emboitement**'' **theory**. This theory holds that successive generations of individual organisms pre-existed one inside the other in the germ cells of the mother. It was estimated that, as many as, 200 million years of human beings were present, already delineated in the ovaries of Eve.

Such theories of preformation persisted well in the eighteenth century, by which time (in 1759) the German investigator **Caspar Friedrich Wolff** (1733–1794) offered experimental evidence that no preformed embryo existed in egg of the chicken. He suggested that during embryonic development the organs formed successively in an epigenetic manner.

1-3.2. Epigenetic Theory

Wolff advocated that the future embryonic regions of an egg first consist of granules or ''globules'' (*viz.,* cells or their nuclei) lacking in any arrangement, *i.e.*, these globules do not reveal any resemblance with the form or structure of the future embryo. Only gradually did these ''globules'' organize into rudiments (germ layers) which, in turn, took on the characteristics of the various organs of the embryo. This method of progressive development from the simpler to the more complex, through the utilization of building units (globules or cells) is called **epigenesis**. For example, Wolff demonstrated that the tubular intestine of chick arises from the folding of a flat layer in an earlier stage of incubation. In fact, Wolff believed that embryonic organs arose completely *de novo* through mysterious vital forces while his famous successor **Karl Ernst von Baer** (1792–1876) proposed the more accepted view that they arose through a gradual transformation of increasingly specialized tissue (**Strickberger**, 1985).

1-3.3. Baer's Law

K.E. von Baer, the **father of modern embryology**, was the first embryologist who first of all, presented the embryological data in a coherent form, made various landmark embryological investigations and made certain very important generalizations. He forwarded the **germ layer theory** which states that ''various structures of the body arise from the same germ layers in different species of animals''. His most important generalization is known as **Baer's law** which states that ''more general features that are common to all the members of a group of animals are, in the embryo, developed earlier than the more special features which distinguish the various members of the group''. Thus, the features that characterize all vertebrates (*viz.*, brain, spinal cord, axial skeleton in the form of notochord, segmental muscle, arotic arches) are developed earlier than the features distinguishing the various classes of vertebrates (*viz.*, limbs in quadrupeds, hair in mammals, feathers in birds, etc.). The characters distinguishing the families, genera and species come last in development of the individual. The early embryo, thus, has a structure common to all members of a large group (phylum) of the animal kingdom and may be said to represent the basic plan of organization of that particular group.

Baer's law was formulated before the recognition of evolutionary theory, therefore, later on it is reinterpreted in the light of evolutionary theory by **Muller** and **Haeckel** (1864) and named as **biogenetic law**.

1-3.4. Biogenetic Law

The biogenetic law of **Muller** and **Haeckel** states that the features of ancient origin (characters of phylum) develop early in development, while the features of newer origin (characters of that particular animal species) develop late, hence, the ontogenetic development presents the various features of the animal's organization in the same sequence as they evolved during the phylogenetic development. Ontogeny, thus, is a recapitulation of phylogeny.

Following the principles of Baer's law and the biogenetic law, embryologists have systematically investigated the development of animals belonging to all the major groups of the animal kingdom. As a result of very extensive and painstaking investigations, the growth of **descriptive embryology** and **comparative embryology** have taken place. For example, **von Baer**, in 1827, discovered the mammalian ovum and studied embryonic development of many animals (1828 and 1837). The cellular nature of sperms and ova was established with the formulation of the **cell theory** in 1838–1839 by **M.J. Schleiden** (1804–1881) and **T. Schwann** (1810–1882). **George Newport** (1854) first of all observed the fertilization of the frog egg by the actual entry of the sperm into egg. In 1876, **O. Hertwig** (1849–1922) studied reproduction in sea urchin and concluded that fertilization involves the union of sperm and egg nuclei. **H. Fol** (1879) observed the penetration of single sperm in the egg during fertilization of sea urchin. **Hertwig** (1890) published a work on the spermatogenesis and oogenesis of the worm, *Ascaris megalocephala*, in which he showed, for the first time, the parallelism in the changes which occur during the ripening of both kinds of germ cells and proved that the polar bodies were the degenerate sisters of the egg (**E.W. MacBride**, 1914). In 1878, the American zoologist **Charles Otis Whitman** (1842–1910) traced for the first time the detailed history of the cells formed by the dividing egg (in the leech, *Clepsine*), thereby initiating the study of **cell lineage**. His pupil, **E.B. Wilson** (1892) applied the cell lineage technique to the study of development of *Nereis*. In 1899, **Jacques Loeb** (1859–1924) discovered a method of inducing development in unfertilized eggs (**artificial parthenogenesis**) which has led to extensive research on the nature of fertilization.

During the last days of 19th century and early days of 20th century, various embryologists, such as, **Weismann** (1883), **Roux** (1888), **Driesch** (1891), **Endres** (1895), **Spemann** (1901, 1903), **Morgan** (1908), etc., made experimental and analytical investigations on the developing eggs to understand developmental phenomena more clearly. Their work originated a new branch of embryology, the **experimental embryology**. Experimental embryologists have forwarded certain most spectacular theories to explain different embryological riddles. Some of their theories are following :

1-4 SOME LANDMARK THEORIES OF EMBRYOLOGY

1-4.1. Germplasm Theory

In 1883, **A. Weismann** (1834–1913) proposed that early development involves the orderly unpacking of an embryo already reorganized, if not actually preformed, in the chromosomes of the nucleus. He spoke of his theory as "the **architecture of the germplasm**". He postulated that there are units of heredity and development, which he called "**ids**" or "**determinants**." According to him, the first cell divisions are **differential**, that is, they first separate the determinants for the right side from those for the left, then the determinants for the anterior end separate from the posterior, and so on for several divisions of the egg. After that, so he thought, interactions make epigenetic development possible. He further showed in 1893 that a child in no way inherited its characters from the bodies of the parents, but, rather from their sex-cells alone. These germ cells, in turn, acquired their characters directly from pre-existing germ cells of the same kind. He pictured the "germ plasm" as a self-perpetuating, cellular legacy which has existed as an unbroken stream through the ages. At each new generation a temporary body (soma) is built up around it, to serve as carrier of the germ cells and to hold them in trust for the forthcoming offspring.

1-4.2. Mosaic Theory

Wilhelm Roux (1850–1924) in 1881, performed a classical experiment which may be viewed as marking the beginning of the science of experimental embryology. He took a frog's egg at the two cells stage of cleavage and touched one of the two cells with a hot needle, thus, destroying the nucleus. He observed that the uninjured cell continued dividing and developed into what he interpreted to be a one-half blastula, a one-half gastrula, and ultimately a one- half embryo. He thus, concluded that certain areas of the egg are already destined in the ovary to develop into special region. Thus, the pigmented cytoplasm of animal pole of frog's unfertilized egg chiefly develops into the head region of the animal, while the yolky cytoplasm of vegetal pole of egg forms the posterior region. Before fertilization, the egg is radially symmetrical, but after fertilization it becomes bilaterally symmetrical. As cleavage continues the blastomeres gradually become smaller and smaller and more areas for specific tissues and organs are delimited in the blastula, so that, finally it comes to resemble a mosaic with regard to the potencies of its blastomeres. The mosaic theory of **Roux**, thus, holds that the *ontogenetic development consists of differentiation of cells from a more or less undifferentiated and homogeneous egg with crude and simple prelocalization of materials to greater differentiation and specialization of cells as development goes on.*

1-4.3. Regulative Theory

In 1891, the German Scientist **Hans Driesch** performed experiment on sea urchin eggs, similar to Roux. Blastomeres of the two cell stage of the sea urchin, each of which normally gives rise to one-half of the embryo and larva, behave differently when separated by him, from each other, and each gives rise to a whole larva, though of somewhat reduced size. Blastomeres isolated at the four cell stage showed the same results. Together they form one whole : separately each becomes a whole. Visualizing these peculiar results, he reasoned that the early cleavages of the egg are equal (equational), "a quantitative division of homogeneous material". Hence, at an early stage the blastomeres have equal potencies. Their fate, he said, is determined by their position in the whole. Thus, he regarded a cleaving egg, as a harmonious, equipotent or equipotential system, in which every part has potentially the properties of the whole.

Development as observed by Driesch in sea urchin eggs was to be called **regulative development** and the eggs which were capable of performing such regulative development were called **regulative eggs**.

1-4.4. Concept of Potency and Totipotency

Potency is the capacity or ability of a cell to give origin to a specific kind of cell or structure or to various kinds of cells and structures. The word **totipotent** was introduced by **Wilhelm Roux** and it refers to the power or abiltiy of an early blastomere or blastomeres of a particular animal species to give origin to the many different types of cells and structures characteristic of the individual species. Speculation concerning the meaning of totipotency of a single blastomere received encouragement from the discovery by **Hans Driesch** (1891) that an isolated blastomere of the two- or four-cell stage of the cleaving sea urchin's egg could give origin to a ''perfect larva''. Driesch described this condition as constituting an **equipotential state**, while **Roux** referred to it as a **totipotential** condition. Totipotency may be of two kinds :

1. **Totipotency** is the ability of a cell or cell group to give origin to all or nearly all the different cells and tissues of the particular species to which it belongs, but it is lacking in the ability to organize them into an harmonious organism.
2. **Harmonious totipotency** is the ability to produce the various types of tissues of the species but possesses, in addition, the power to develop a perfectly organized body.

The fertilized egg or the naturally parthenogenetic egg constitutes an harmonious totipotential system. This condition is true also of isolated blastomeres of the two- or four-blastomere stage of the sea urchin development, of the two-cell stage of *Amphioxus*, or of the first two blastomere stage of the frog's egg when the first cleavage plane bisects the gray crescent. However, in the eight-cell stage in these animals, potency becomes more limited in the respective cells of the embryo. **Restriction of potency**, therefore, is indicated by a restriction of power to develop into a variety of cells and tissues, and potency restriction is a characteristic of cleavage and blastulation. When a stage is reached in which the cells of a particular area are limited in potency value to the expression of one type of cell or tissue, the condition is known as **unipotency**. A **pluripotent state**, on the other hand, is a condition in which the potency is not so limited and two or more types of tissues may be derived from the cell or cells.

1-4.5. Gradient Theory

Since the time of Driesch's experiments, the sea urchin egg has been chosen for immense number of investigations. This is because the eggs of sea urchin are readily obtained in large quantities, are easily fertilized artificially, and are easy to maintain through the process of gastrulation to the pluteus larval stage. Various operative and chemical procedures have been employed in attempts to analyze the developmental processes leading to / or involved in the formation of the blastula, the gastrula, and the actively swimming larva.

Such an experimental approach of **T. Boveri** (1910), **J. Runnstrom** (1928), **S. Horstadius** (1928) and **C.M. Child** (1936) on sea urchin eggs has contributed a most important theory, the **gradient theory**. The gradient concept was introduced in 1910 by **T. Boveri**, who suggested that the vegetal region of the sea urchin egg exerted a dominant control over more animal regions, the influence decreasing towards the animal pole. However, he could not give any plausible reason for his concept and they were **J. Runnstrom** (1928) and **S. Horstadius** (1928) who developed the **double gradient theory**. According to this theory, in the sea urchin's egg two factors or principles exist which are mutually antagonistic and interact with each other at the same time, and normal development is dependent on a certain equilibrium between the two principles. Each has its centre of activity at one of the poles of the egg. The activity diminishes away from the centre, producing a **gradient** of activity. The two gradients of activity decline in opposite directions : the one from the animal pole, the other from the vegetal pole. Considering this type of distribution of activities, the factors or principles themselves have been called **gradients**. The two gradients are, therefore, the **animal gradient**, with a centre of activity at the animal pole, and the **vegetal gradient**, with a centre of activity at the vegetal pole.

Normal morphogenesis and differentiation in the early development of sea urchin depends on the presence of both gradients and on an equilibrium between them. If the animal gradient is weakened, the vegetal gradient becomes dominant and the embryo is **vegetalized** ; *i.e.*, it develops, in excess, parts pertaining to the vegetal gradient, such as the gut. If the vegetal gradient is weakened, the animal gradient becomes dominant, and the embryo is **animalized** ; *i.e.*, it develops, in excess, parts pertaining to the animal gradient, such as the tuft of stiff cilia at the animal pole. Other structures of the embryo, such as the ciliary bands and the skeleton, can develop only if both gradients are active. The development is the more nearly normal the more the two gradients approach a correct equilibrium.

C.M. Child (1936), while recognizing the physico-chemical nature of the two gradients, proposed the existence of a single physiological or oxidative metabolic gradient in sea urchin egg. Using the rate of reduction of vital dyes, particularly Janus green, as an indicator of the rate of oxidative metabolism, **Child** showed that such a **reduction gradient** is present in the oocytes, cleavage stages and blastulae of sea urchin and starfish. There exist differences in rate of general metabolism in the ooplasm of animal and vegetal hemispheres of oocytes of sea urchin. Such differences in rate of oxidative metabolism are found to be extended in the plane of animal-vegetal axis, so Child's concept is also named as a **metabolic axial gradient**. It is found that cytoplasm of animal pole of egg divides with high frequency and this frequency gradually becomes low towards the vegetal pole. This fact indicates that ooplasm of animal pole has high metabolic rate which gradually decreases towards vegetal pole.

It was in 1969 and 1972 that **Horstadius** and **Josefsson** succeeded in isolating animalizing and vegetalizing substances from the mature unfertilized egg and early cleavage stages of sea urchin. Chemically animalizing agent is found to contain the amino acid tryptophan and vegetalizing agent is found to contain a nucleotide. The general indications are that the animal factor promotes synthesis of ribonucleic acid and proteins, while the vegetative factor seems to inhibit the onset of these processes for a while, so that the time pattern of early protein synthesis will be different in the two regions. Animal and vegetal gradients are also known as **morphogenetic gradients** (see **Grant**, 1978) or **morphogens** (**Maclean** and **Hall**, 1987). Now, it is generally believed the doube gradient system is located in the egg cortex which includes the fluid-mosâic plasma membrane and associated cytoskeleton of microtubules and microfilaments of gel-like ectoplasm. By experimenting with the squid's eggs, **Arnold** (1976) has suggested that the egg cortex by controlling the displacement of membrane receptors and enzyme systems, modulates metabolism in growth, division and cell surface interaction.

Gradients are also found in the early development of Amphibia and other vertebrates, *e.g.*, dorsoventral gradient exists in both ectodermal and mesodermal layers of amphibian gastrula. In certain invertebrates such as planarians and the hydroids, the ability to regenerate lost parts shows a regular decrease along the main axis of their bodies (**Child**, 1929,1941).

1-4.6. Spemann's Theory of Organizers

In 1901, **Spemann** performed certain constriction experiments on some amphibian eggs. These investigations into the behaviour of isolated blastomeres of the two-celled newt egg revealed that halves separated along the median plane developed into complete embryos, whereas, of the two blastomeres separated along a plane at right angles to the median plane, only the dorsal half formed an embryo. Spemann concluded that as early as the two-celled stage the dorsal half of the egg differs from the ventral half and it alone possesses the ability to form a complete embryo. Between 1901 and 1912 Spemann demonstrated the dependence of lens formation upon the presence of an optic vesicle. Thus, he found that when the prospective eye rudiment was removed from grass frog embryos, a lens did not form ; and conversely, an optic vesicle transplanted under belly ectoderm caused formation of a lens instead of belly skin. In 1918, Spemann reported that small pieces of presumptive neural plate and skin ectoderm exchanged between early salamander gastrulae developed according to their new location, whereas a piece of the dorsal lip of the blastopore developed independently of its new surroundings into an embryo-like body. He suggested that the dorsal lip might represent a ''centre of differentiation'' from which spreads an influence determining the development of the gastrular ectoderm. He further concluded that the organizing effect of the dorsal lip might be related to an effect of archenteron roof upon the overlying ectoderm comparable with that of the optic vesicle upon the overlying ''lens'' ectoderm. In 1924 **Spemann** and his student, **Hilde Mangold**, published a classical paper providing definitive proof of the organizing action of transplanted dorsal lip in the production of secondary embryos, establishing firmly the concept of induction as a basic mechanism in embryonic development. Later on, other workers found such organizing capacity in the chorda mesoderm cells of many other animals. Thus, we can define an embryonic organizer in the following manner : ''*when one embryonic tissue transmits a stimulus that influences another tissue to produce a structure that otherwise would not come into being, then the former tissue is called* **inductor** or **organizer** *and its morphogenetic effect is called* **induction**''. Spemann, thus, has recognized a primary organizer in the form of dorsal lip or roof of archenteron in amphibian gastrula and got Nobel prize of 1935, for such a landmark discovery in experimental embryology.

In modern terms induction can be defined as a type of intercellular communication which is required for differentiation, morphogenesis and maintenance (**Graham** and **Wareing**, 1976). It is also found that during induction some chemical is transmitted from one tissue to the other and this chemical acts on the genes of the cells being induced to develop into a particular manner. What the substance is has not been still determined, but it appears to be a relatively larger molecule (**Pai**, 1985).

1-5. BRANCHES OF EMBRYOLOGY

Modern embryology or developmental biology is truly a scientific cross road, a meeting place of such sciences as physics, chemistry, zoology, botany, microbiology, genetics, mathematics and psychology. However, modern embryology has following branches :

1. Descriptive embryology. For centuries observation and description of different embryonic stages of the ontogenetic development of a species have remained the chief concern of early embryologists. This type of embryological knowledge constitutes the descriptive embryology.

2. Comparative embryology. In comparative embryology, a comparative study is made between the embryology of most animal types and the latter are classified, compared and common trends and principles are sought.

3. Experimental embryology. The field of embryology which attempts to understand those fundamental developmental mechanisms which are involved in different animals is called, **developmental mechanics, casual embryology, analytical embryology** or **experimental embryology.** This is youngest and most vigorously investigated field of embryology and it has provided clear cut understanding regarding the concept of gradients, fertilization, cleavage, gastrulation, embryonic induction, determination and differentiation.

4. Chemical embryology. The field of embryology in which various biochemical, molecular biological, biophysical and physiological techniques are employed to the developing embryo to understand various developmental phenomena in molecular terms, is called **chemical embryology** (**Needham**,1931 ; **Brachet**, 1950) or **physiological embryology** (**Lehmann**, 1945).

1-6. MATERIAL AND METHODS IN EMBRYOLOGY

1-6.1. Selection of Material

An animal for embryological studies is generally chosen by considering that (1) the chosen animal should be ideally adopted to demonstrate the developmental process which it is expected to illustrate, and (2) the chosen animal should be readily available and easy to handle. Embryologists have found amphibian eggs good for study of cleavage and germ layer formations. Further, their eggs can be obtained in all the seasons by normal injections of pituitary hormones. The formation of the embryonic body and the establishing of the organ systems in their simple primordial form can be best studied in chick embryos, than in any other form. Like amphibian eggs, the hen's egg can be secured in any desired numbers. For the study of the development of the various organ systems in mammals, pigs, rabbits, guinea pigs and men have been found as best materials. For experimental and chemical embryology, most classical and modern experimental and chemical embryologists have chosen the eggs of echinoderms (mainly sea urchins), amphibians (*e.g.*, frogs, salamanders, etc.) and ascidians.

1-6.2. Certain Techniques of Descriptive and Comparative Embryology

The orderly progress of development is ascertained partly by observing entire embryos and dissections in proper age, sequence, and even more profitably by the microscopical study of stained sections. When sections of an embryo are arranged in the same progressive order as cut, they furnish a related series that permits the changing appearances from level to level to be traced and reconstructed either as mental images or physical models. The modern light microscopical and electron microscopical techniques have greatly revolutionized the descriptive and comparative embryology. Further, the initial methods for tracing cell movements during gastrulation include mainly carbon marking and vital staining. Recent more sophisticated embryological procedures include antibodies, radiotracers and genetic markers.

1-6.3. Certain Experimental Embryological Methods

The experimental embryology consists of altering some conditions under which a state exists or a process proceeds and then studying the results. These results provide, bit by bit, the elements upon which synthesis and comprehensive understanding are based. Certain new methods of experimentation include microsurgery, transplantation of parts on embryos or of nuclei from one cell type to another cell type, explantation of organs and tissues in culture or in other suitable locations such as the chorio-allantois, vitelline membrane, eye cavity, coelom, etc. The techniques of cell dissociation, cell reaggregation and cell hybridization have provided deep insight into the causes of morphogenetic movements and gene expression. Some of these techniques which are widely used in experimental embryology can be considered in detail as follows :

A. Determination of effect of external factors on the egg. One line of experiment tests the dependence of the developing egg or embryo on its environment. The several physical and chemical environmental agents are called **external factors**. The effects of external factors are naturally seen best in those organisms that are directly exposed to their influences ; the embryos of viviparous animals, and especially mammals are less subject to variations of the environment. The most important external factors are following :

1. **Mechanical factors**, such as pressure, gravity and centrifugal forces. 2. **Physico-chemical factors**, such as pH and osmotic pressure. 3. **Radiational factors**, such as, heat, ultraviolet rays and X-rays. 4. **Chemical factors**, such as, ionic effects.

These agents also become useful as tools in conducting experiments. For example, the centrifuge disarranges the constituents of the egg and redistributes them into layers according to their specific gravities; heat, ultraviolet rays or X-rays can be used to kill local regions of the cytoplasm ; X-rays or radium discharges can destroy the nucleus within a cell ; excess sodium chloride can prevent the closure of the spinal cord; calcium free sea water or sea water containing lithium ions causes cleavage stages to separate into their component blastomeres.

B. Determination of effect of internal factors on developing egg. A second line of attack tests the dependent relations existing between the embryos as a whole and its parts, and the influence of component parts on one another. These interactions and influences are called **internal factors**. Their effects are tested by subtracting parts, by altering relationships through interchange or substitution, and by adding parts or wholes. Few important procedures to test the effect of internal factors are following :

1. Isolation. This is carried out by removing a part or region and allowing it to develop in its natural medium (*e.g.*, sea water) or, as an **explant**, in an aseptic, nutrient medium (*i.e.*, tissue culture). When that embryonic part is introduced into a reasonably indifferent environment provided by another embryo (*e.g.*, belly cavity, chorio-allantoic membrane) it is called an **interplant**.

2. Recombination. Blastomeres can be displaced into strange positions. The grafting of an excised part into the place left by the removal of another part is called **transplantation**. Because, transplantation technique has been used most vigorously by classical experimental embryologists, such as **Spemann**, in his classical embryological experiments, therefore, it demands somewhat more detailed discussion.

Transplantation. During experimental work in developmental biology, small pieces may be cut out of embryos in various stages of development and inserted into suitably prepared wounds of the same or another embryo. In the case of transplantation of piece of the same embryo to another place, the transplantation is said to be **autoplastic**. If the transplantation is from one individual to another of the same species, the transplantation is called **homoplastic**. If the transplantation is to an individual of another species belonging to the same genus, the transplantation is **heteroplastic**. A transplantation to an individual more distally related than species of one genus is called **xenoplastic**. The animal (embryo) from which a part is taken is referred to as the **donor** ; the animal to which the part is transplanted is called the **host**.

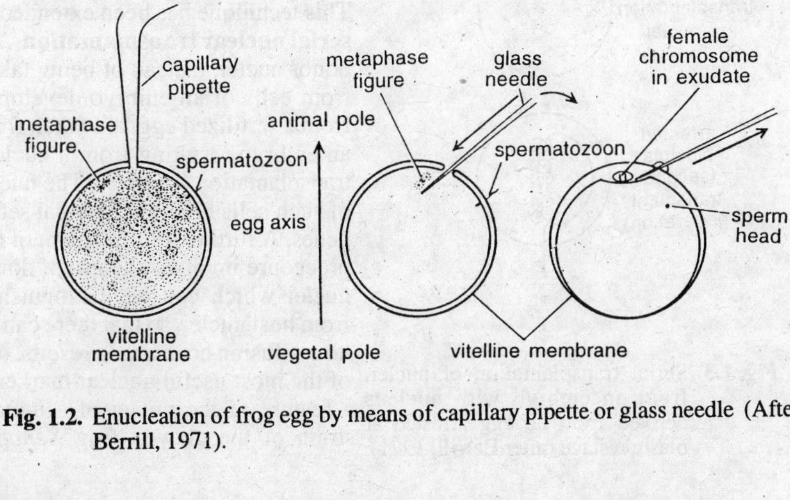

Fig. 1.2. Enucleation of frog egg by means of capillary pipette or glass needle (After Berrill, 1971).

In adult animals, especially highly organized animals such as vertebrates, transplantation is not easy, only autoplastic and homoplastic transplantations usually being successful. In embryos, however, and in lower invertebrates (such as coelenterates), the grafts may heal successfully even after xenoplastic transplantations. Successful transplantations have been carried out between embryos of frogs and salamanders and mammals and birds.

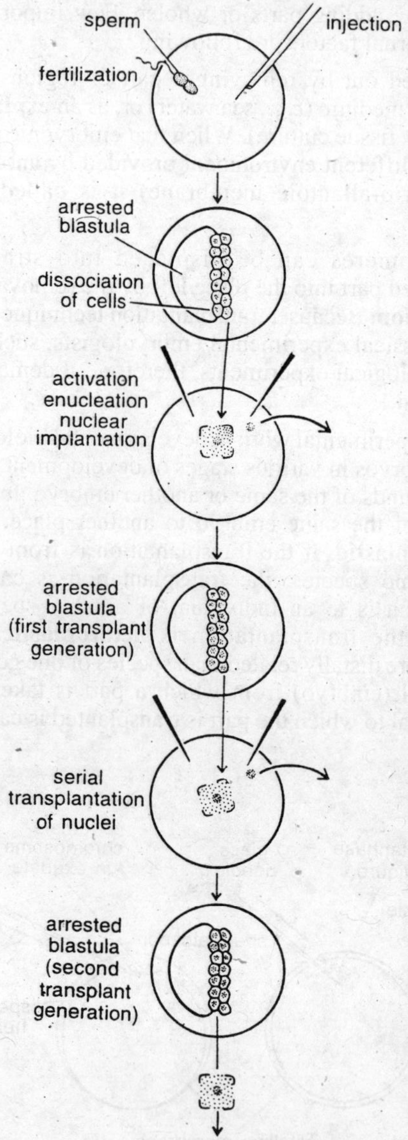

Fig. 1.3. Serial transplantation of nuclei, from an embryo with nucleus derived from an egg arrested at blastula stage (after Berrill, 1971).

Nuclear transplantations. **Briggs** and **King** (1952) have evolved a nuclear transplantation technique for frog eggs and embroys (*Rana pipiens*). The nucleus of an unfertilized egg, which is conveniently located just beneath the animal pole, is effectively removed with a needle or inactivated by ultraviolet light (Fig.1.2). A substitute nucleus is obtained by using a micropipette with a diameter less than that of the particular donor cell, so that the cell is ruptured and the isolated nucleus is obtained. The donor cell may be in the early or late, blastula stage, the gastrula stage, the neurula stage, or even from mature tissues of the functioning tadpole or frog. The last and most difficult step in the procedure involves the insertion of the donor-cell nucleus into the uncleaved enucleated egg. This technique has been extended to **serial nuclear transplantation**, *i.e.*, donor nuclei, instead of being taken from cells of an embryo developed from a fertilized egg, are taken from an embryo resulting from a nuclear transplantation (Fig.1.3). The nuclei of such cells have an identical set of genes. A further improvement of the procedure involves the use of donor nuclei which can be distinguished from host nuclei, so that there can be no confusion concerning results, one of the most useful nuclear marker is a feature of the nuclei of a mutant strain of the clawed frog *Xenopus* *laevis*, which contains a single nucleolus instead of the two nucleoli typical of most frogs.

3. Defect. An egg or embryo, after the excision or destruction of a local region, becomes a defect experiment. An egg nucleus can be, thus, destroyed by exposure to X-rays or ultraviolet radiation or sucked out with a micropipette. Cytoplasmic areas can be destroyed by local pricking, heat or ultraviolet treatment. A fine hair can be employed to cut or constrict an egg or cleavage group. Tiny knives, glass needles and scissors are used for excising the parts of embryos. Some regions can be eliminated by their differential susceptibility to toxic substances.

4. Addition. When a part is added to an already completed embryo, as a supernumerary structure, is called **implantation**. By surgical method called **parabiosis**, fusions of whole eggs or early cleavage stages can be accomplished.

1-6.4. Certain Chemical Embryological Techniques

The molecular changes which take place during development have been followed in various ways. For example, the cells of an embryo may be ''homogenized'' by grinding them with crushed glass or stirring them with blender, or subjecting them to high frequency vibrations. Next the constituent molecules of the protoplasm can be separated and characterized on the basis of molecular size (**dialysis** and **filtration**), density (**centrifugation** and **ultracentrifugation**), electric charge (**electrophoresis** and **isoelectric focussing** commonly used for proteins), and absorptive and solubility properties (**chromatography**). Finally the fractions thus obtained can be analyzed chemically. In other studies the molecules are first ''labelled'' with radioactive isotopes, so that, their history within the embryonic cells can be traced. The proteins and other antigens are recognised by their immunological properties.

1-7. DEVELOPMENTAL BIOLOGY AND HUMAN WELFARE

Developmental biology plays a key role in human welfare. It helps in understanding the causes of congenital malformations, cancer, aging and in improving the breeds of domestic animals, in controlling pests and vectors of diseases and in the formation of test-tube babies. Some important applied aspects of developmental biology are the following :

1-7.1. Teratogenesis

About 2 to 3 per cent of the human children born have congenital malformations or **terata**. In ancient times it was believed that the human malformations were due to factors like curses, gods, devils, or celestial bodies, human-animal matings, maternal imagination during pregnancy or disease. The real basis of teratogenesis is being investigated by developmental biologists. Many congenital abnormalities have been traced to exposure of the pregnant mother and, therefore, the foetus to certain **teratogens** (*viz.*, destructive environmental prenatal influences) such as drugs, diseases (diabetes, malnutrition and infectious diseases) and ecological factors (radiation, extreme pressure on uterus, wide fluctuations in temperature, too much carbon dioxide or too little oxygen). For example, the tranquilizer

thalidomide was found to be a teratogen since its user pregnant mothers of West Germany in 1960s gave births to babies either lacking limbs (**ectromelia**) or having grossly deformed limbs (**phocomelia** ; Fig. 1.4) or having mental retardation. Other chemicals or drugs which are found to be teratogens, include pregnancy-testing oral pills, cigarette-smoking, alcohol, morphine and heroin. Certain infectious diseases of the pregnant mothers such as rubella or German measles, mumps, syphilis and polio, may reach the foetus, since they are caused by viruses small enough to pass through the placental filter. They may be responsible for abnormalities including blindness and deafness. Even the consumption of blighted potatoes, containing an anti-fungal chemical, during the first month of pregnancy, has been isolated as a cause of **spina bifida** (a crippling and sometimes fatal gap in one of the spinal vertebrae which allows the spinal cord to protrude, Fig.1.5) and **anencephaly** (a fatal defect in which the top of the head, brain included, does not develop). In modern times teratogenesis has assumed greater importance because of introduction of several new man-made teratogens such as drugs, radiations and various pollutants in the human environment. It is, therefore, advisable that prior to the release of any new drug or chemical in the environment, proper testing on different animal systems should be undertaken. Even it is important that pregnant women working in industrial establishments should be kept under keen observation because they are exposed to various chemical and physical teratogens.

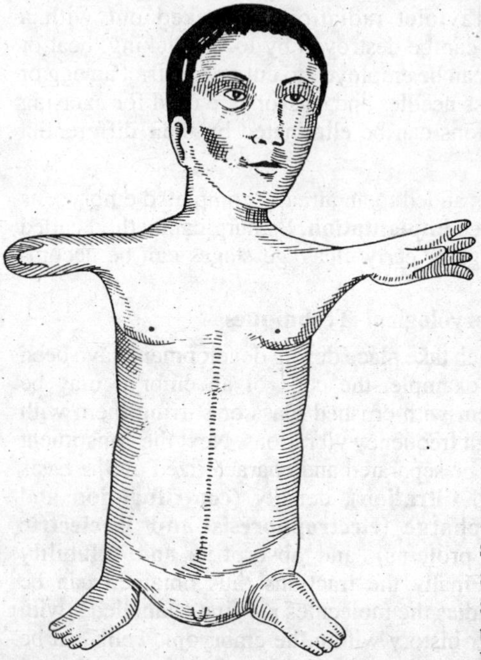

Fig. 1.4. A case of phocomelia (*i.e.*, the presence of one or more limbs like seal's flippers). Here no development of long bones of limbs takes place and hands and feet are attached directly to trunk.

1-7.2. Cancer

Developmental biologists have also enlightened us about the most dreaded often fatal human disease, the cancer. In the normal human body the constituent parts, except for the brain cells, are continuously renewed, for example, bone cells are replaced every 15 to 25 years, scalp hairs every 2 to 4 years, liver cells every 8 months, kidney cells every 3 weeks, intestinal cells are renewed every 5 days and scavenging white blood cells live no more than about 30 hours. In fact, there is a fine balance between the rate of cell multiplication and cell death in order to maintain the size of any organ or tissue. If the cell multiplication rate exceeds cell death rate the organ begins to grow beyond its normal size and a tumour is formed which is called **cancer**. How a normal cell loses control over its multiplication rate and becomes a cancer cell is still not well understood. Many external factors (biological, chemical and physical) can cause cancer. All cancers can be cured if we can regulate the multiplication rate of these cells.

1-7.3. Animal Breeding

In animal breeding programmes, the rate of growth during different developmental periods plays an important role. Developmental biologists can, thus, help in breeding larger animals with minimum input and economize the operation. Another successful animal breeding technique involving reproduction in mammals including humans is that of being able to store not only sperm but also embryos at extremely low temperatures (–196°C in liquid nitrogen). Careful thawing and reimplantation into natural or foster mothers has resulted in the development and birth of normal progeny. Such frozen embryos of domestic animals can be shipped or air lifted in vials over great distances. Domestic animals can be exchanged in this manner between breeders of different states and different countries with speed and economy. Certain modern techniques involve the injection of hormones to stimulate the ovulation of more eggs than occur normally referred to as **superovulation**. In United States, in 1979, a prize Simmental cow was superovulated, mated, and the embryos were removed. The embryos which appeared to be developing normally were implanted into foster mothers. In that year, this one cow could have produced 89 calves by this method (**Seidel, Jr.,** 1981). By adopting similar technique, in 1983 a successful implantation of a previously frozen human embryo was reported in Australia. The normal baby that developed was born in 1984 (**Pai**, 1985).

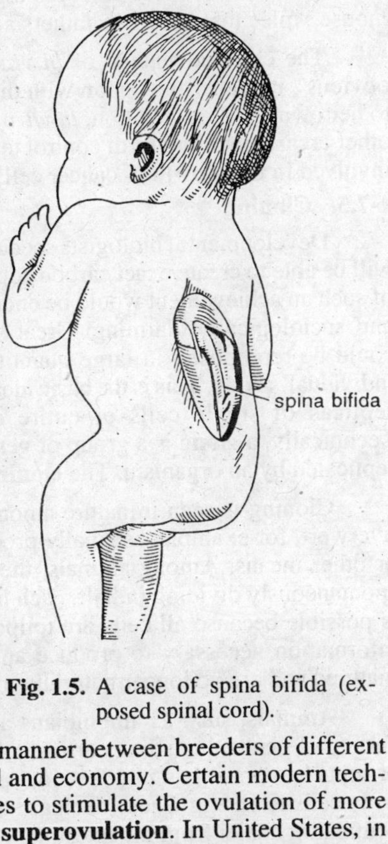

Fig. 1.5. A case of spina bifida (exposed spinal cord).

1-7.4. Test-Tube Babies

Test-tube babies are the result of *in vitro* fertilization. An ovum is removed from a woman, maintained in a culture medium (calf serum or human serum) and fertilized by semen in a petridish. The zygote then undergoes cleavage and forms blastocyst (blastula). The developing blastocyst is transferred to the uterus of either the natural mother or a foster mother, where implantation occurs. The only difference between a test-tube baby implanted in its natural mother and a baby conceived in the usual way is that fertilization of the egg occurs outside the body. There is no effect on its genome, unless the cells are somehow damaged and mutation is induced (**Pai**, 1985).

It is also possible to recover early embryos from two different mothers, mix the cells from these embryos together, reimplant them and get a normal mouse ; mice that have two fathers and two mothers are known as **chimeras**.

The chief advantage of *in vitro* fertilization (*i.e.*, test-tube babies) is obvious : it enables a woman with diseased or missing oviducts to give birth to her own child. In addition, *in vitro* research is yielding vital information in other areas, including birth control teachniques and possibly the mechanisms involved in the growth of cancer cells (**Goldsby**, 1976).

1-7.5. Cloning

Developmental biologists speculate that, in the not too distant future, we will be able to create exact carbon copies of ourselves. The possible outcomes of such an achievement would be enormous, at once scientifically remarkable and sociologically alarming. Great scientists, statesmen, artists and others could be reproduced in large quantities from single cells scraped from the individual's body. This is the basic idea behind cloning, by which exact genetic replicas of single cells or entire organisms are reproduced asexually. Technically, a **clone** is a group of genetically identical individuals asexually replicated by an organism. The **cloning** refers to the production of a clone.

Cloning is seen in nature among many plants and some animals. The liverwort, for example, asexually produces exact copies of itself by budding or other means. Among animals, the flatworm can reproduce asexually by spontaneously dividing in half, each half forming a complete worm. Cloning is possible because all cells are totipotent, carrying within them the genetic information necessary to produce an entire organism even after they have matured and are no longer naturally dividing.

Animals, such as amphibians and mice, have been cloned, as have a variety of plants, including carrots, tobacco and sugarcane. By the technique of renucleation, it has now become possible to transplant the nucleus from body cells to an egg whose nucleus has been removed. Such enucleated eggs containing nuclei from the intestinal cells of tadpole or the skin cells of adult frog can develop into normal frogs (**Gurdon**, 1962). By this method thousands of genetically identical frogs with a desirable genotype can be produced. The same is theoretically possible in man but there are many technical problems yet to be overcome.

The advantages of cloning plants and animals are numerous. High-yield food plants such as wheat, corn, and rice can be selected and abundantly reproduced. Cloning would give animal breeders a tool for exactly reproducing highly desirable animals; for example, cloning would make it possible to create 1,000 copies of prize dairy cows to help feed growing populations. Endangered species might be saved by cloning numerous replicas of the best of the few remaining individuals.

The great disadvantage of cloning is the loss of genetic variability. Cloning eliminates the chance of producing superior individuals by sexual reproduction, thus, seriously weakening the species' chances of survival.

1-7.6. Pest Control

Developmental studies can help in controlling pests and vectors of diseases with minimum damage to the ecosystem. For example, the progress in malaria control was possible only after the knowledge of the life cycle of the malarial parasite was obtained and the mosquito vector was controlled. This is especially useful because embryos are generally more susceptible than adults to various controlling agents.

REVISION QUESTIONS

1. Explain the relationship between embryology and developmental biology.
2. How is developmental biology related with other branches of biology and different scientific disciplines ?
3. Write an essay on the historical growth of developmental biology.
4. Write short notes on the following:
   (*i*) Aristotle as founder of embryology;
   (*ii*) Preformation theory; (*iii*) Baer's law;
   (*iv*) Mosaic theory; (*v*) Germplasm theory;
   (*vi*) Epigenesis theory; (*vii*) Gradient theory; and
   (*viii*) Spemann's organizer.
5. Is developmental biology a socially relevant and applied science ? Explain it with examples.
6. Describe various techniques of experimental embryology. How can nuclear trasplantation techniques be exploited for the benefit of human kind ?
7. Discuss in brief the following :
   (*i*) Totipotency ; (*ii*) Transplantation ;
   (*iii*) Teratogenesis ; (*iv*) Test-tube babies ; and
   (*v*) Cloning.
8. Write down the main contribution of the following embryologists : **C.M. Child**; **von Baer** ; **Wilhelm Roux** ; and **Hans Driesch**.

2

CELLULAR BASIS OF DEVELOPMENT (Cell, Cell Surface and Cell Division)

The spermatozoa, ova, zygote, blastomeres and differentiated body cells of animals are basically the eukaryotic cells, so the following brief discussion of biology of a typical or composite animal cell will enhance our understanding of various developmental phenomena such as the role of cell surface in early development ; changes in shape and morphogenetic movements (gastrulation); cytodifferentiation, etc.

To recall, the **cell theory** states that cells are the fundamental units of both structure and function in all living things ; that all forms of life (animal, plant and microbial) are composed of cells and their secretions ; and that cells arise only from pre-existing cells, each cell having a life of its own in addition to its integrated role in multicellular organism. In fact, all kinds of cells share the following three basic characteristics : 1. A **set of genes** which form the blue-prints for regulating cellular activities and making new cells. 2. A **limiting plasma membrane** that allows controlled exchange of matter and energy with the outside world. 3. A **metabolic machinery** for sustaining life activities such as digestion, respiration, secretion, growth, reproduction and repair of parts.

2-1. ULTRASTRUCTURE OF ANIMAL CELL

Animal cells vary considerably in size, shape, organelle composition and physiological roles. The ultrastructure of the typical or composite animal cell can be studied under three heads : plasma membrane, cytoplasm and its organelles and nucleus :

2-1.1. Plasma Membrane

It is outermost limiting membrane of the animal cell. It is a very thin, elastic, dynamic membrane of molecules of lipids (phospholipids and cholesterols), proteins and carbohydrates (*i.e.*, oligosaccharides of glucose, galactose, mannose, fucose, sialic acid, acetylglucosamine or acetylgalactosamine). Plasma membrane has fluid mosaic configuration — in it the molecules of proteins are studded in a mosaic pattern in a basic bimolecular fluid membrane of phospholipids. These protein molecules exist either at the outer or inner surface (the **extrinsic proteins**) of the bimolecular lipid layer or some of them, called **intrinsic proteins**, span the lipid layer. All of these protein molecules are capable of movement both vertically and laterally. To the external surface of some protein molecules and also of lipid molecules are attached with the polysaccharides (oligosaccharides), some of them giving immunological properties to the cell and others forming the surface coat or **glycocalyx**. Some protein molecules of inner side of membrane are also attached with microfilaments of the **cytoskeleton**. The cellular cytoskeleton is found to exist in the form of **microtrabecular lattice** which comprises three types of unit fibres — **microfilaments** (of actin protein), **intermediate filaments** (of proteins such as desmin, keratin, vimentin, etc.) and **microtubules** (of tubulin protein). The plasma membrane gives out finger-like cytoplasmic outgrowths called **microvilli** and has a variety of intercellular connections (or **cell surface differentiations**) such as **basement membrane, tight junctions** (zonula occludens), **demosomes** and **gap junctions** (**nexus**).

Besides its selective permeability shown by **diffusion**, **osmosis**, **facilitated transport** and **active transport** (by ATP - driven $Na^+ - K^+$ exchange pump, Ca^{2+} – pump, proton pump, etc.), plasma membrane performs two very important functions for the cell : endocytosis and exocytosis. **Endocytosis** refers to taking in or ingestion of foreign substances including viruses, pathogenic bacteria, or food particles by the plasma membrane. When the engulfed food is in the fluid form, such endocytic activity is called **pinocytosis** and resultant liquid-filled membrane bounded vesicles as pinocytic vesicles or simply, **pinosomes**. And when the engulfed food particles are solid in nature, then such an endocytic activity is called **phagocytosis** and resultant vesicles are referred to as **phagocytic vesicles** or **phagosomes**. The contents of pinosomes and phagosomes are ultimately subjected to digestion by enzymes of lysosomes.

Ingestion of specific macromolecules or substances (called **ligands**) is performed by **receptor-mediated endocytosis** in which the ligand molecules bind with the receptors of the plasma membrane which travel in the membrane. Such receptors with or without ligands are concentrated in **clathrin protein-coated pits** which invaginate to form **clathrin-coated vesicles**. As these vesicles move deeper into the cytosol, the clathrin coat is lost and recycled to the plasma membrane and vesicles eventually form the **endosomes**. In endosomes, the ligands and receptors are uncoupled and receptors too are recycled to the plasma membrane. Endosomes may fuse with lysosomes whose hydrolytic enzymes degrade the ligands. Receptor-mediated endocytosis is performed during uptake of extracellular yolk (the vitellogenin) by oocytes of chick (**Roth**, 1984) and insects, of maternal antibodies or immunoglobulins (IgG) through the placenta by the foetus, and of macromolecules (*i.e.*, protein hormones and growth factors such as insulin and low-density lipoprotein or LDL and protein-bound nutrients, *e.g.*, cholesterol, iron, vitamin B_{12} ; **De Robertis** and **De Robertis, Jr., 1987).**

Exocytosis involves extra-cellular transport of cellular secretions. All the proetins which are biosynthesized by the cell for the extra-cellular use are synthesized either on the free ribosomes (*e.g.*, milk proteins in the cells of mammary glands or thyroglobin in the cells of thyroid endocrine gland) or on the attached ribosomes of rough endoplasmic reticulum (RER) (*e.g.*, pancreatic enzymes, hormones, mucus, etc.) are ultimately processed, transformed, conjugated with oligosaccharides (if needed) by the Golgi apparatus. Finally, these secretions are packed into tiny circular membrane bounded packets, called **secretory vesicles** which pinch off from the lamellae

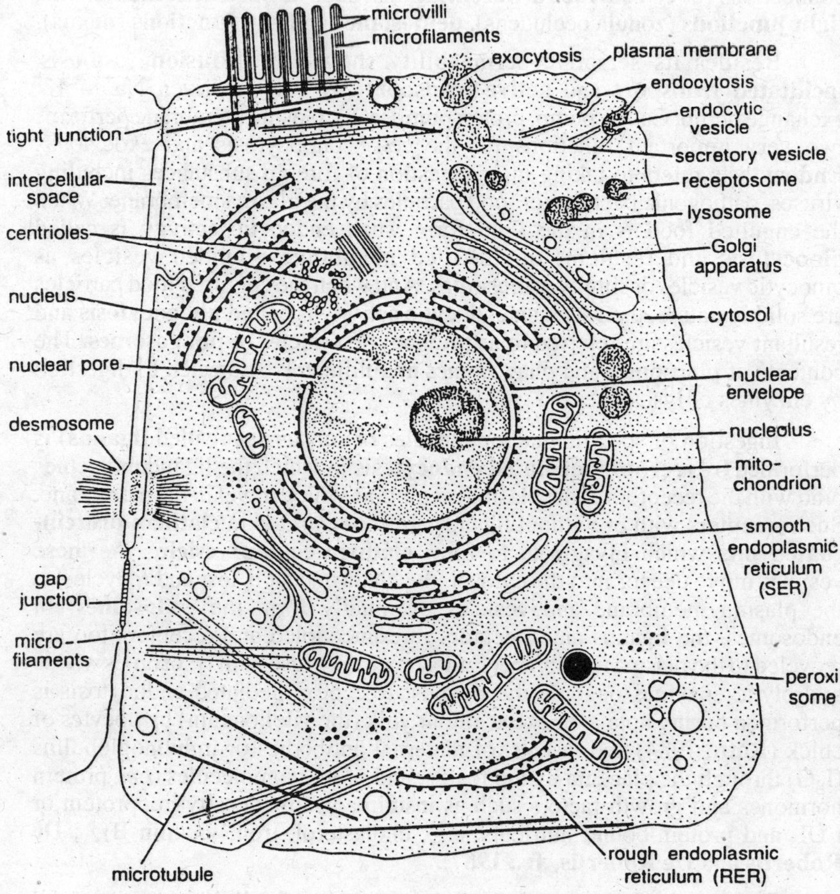

Fig. 2.1. A typical or composite animal cell (after Sheeler and Bianchi, 1987).

of **maturing** or **trans face** of the Golgi apparatus.They migrate to cell cortex (ectoplasm) and on arrival there, the membrane of each vesicle fuses with the plasma membrane to pour out their secretions to cell exterior. This last step needs Ca^{2+} and energy (ATP) similar to the process of endocytosis. Some types of secretory granules, such as **cortical granules** of unfertilized ova of sea urchin, amphibians and other animals, are produced by Golgi apparatus during oogenesis and they get arranged in egg cortex just beneath the oolemma. These cortical granules burst open only after the entry of first sperm inside the ovum during fertilization and elevate the **fertilization membrane** to check the polyspermy.

2-1.2. Cytoplasm and its Inclusions and Organelles

Plasma membrane is followed by the colloidal organic fluid called **cytoplasmic matrix, hyaloplasm** or **cytosol**. Cytosol contains dissolved organic substances of enormous variety and quantity. It is the seat of various biochemical or metabolic reactions, so it has a variety of substrates, enzymes, coenzymes, factors, end products and required raw materials for various life activities. Thus, cytosol is rich in water, oxygen, ions, glucose, amino acids, minerals, enzymes, coenzymes, vitamins, hormones, lipids, polysaccharides, proteins, nucleotides of DNA and RNA, RNA molecules and so on. In cytosol are also suspended a variety of cellular inclusions (such as glycogen, pigment granules and yolk (the yolk forming the **deutoplasm**) and organelles. Some very significant cellular organelles are the following :

1. Endoplasmic reticulum or ER. The endoplasm of the cell is traversed by an anastomose reticulum or network of membrane-bounded flattened tubules called **endoplasmic reticulum** (**ER**). More often the membranes of ER remain in continuation with that of plasma membrane and nuclear envelope. In fact, there occurs an active exchange of membranes between various cellular organelles such as ER, Golgi apparatus, nuclear envelope and plasma membrane. Certain protein components of membrane are synthesized at the ER-attached ribosomes. ER too is of two types : rough ER that has attached ribosomes and smooth ER that do not have any ribosomes.

Rough ER is found to contain certain ribosome specific receptor glycoproteins, called **ribophoreins I** and **II** to which ribosomes are attached during protein biosynthesis. The polypeptide chains which are synthesized on the attached ribosomes according to the genetic codes of mRNA molecules are entered inside the lumen of RER and undergo tailoring, maturation and folding to form functional secondary or tertiary molecules. RER pinches off certain protein filled vesicles which ultimately get attached with forming or cis face of Golgi apparatus. RER manufacture proteins for lysosomes, plasma membrane and extracellular transport.

Smooth ER is specialized in lipid metabolism (both catabolism and anabolism). It synthesizes a variety of phospholipids, cholesterol and steroids. In some cells (*e.g.*, liver cells) smooth ER performs **glycogenolysis** (degradation of glycogen) and **drug detoxification**.

2. Golgi apparatus. The size and number of Golgi apparatus in a cell indicate active metabolic, mainly synthetic, state of that cell. Golgi apparatus is a cup-shaped organelle consisting of a set of smooth **cisternae** (*i.e.*, closed fluid-filled flattened membranous sacs or vesicles) which are stacked together in parallel rows. These cisternae fall into three types : **cis Golgi**, **median Golgi** and **trans Golgi**, each of which has distinct enzymatic activities. The forming or cis face of Golgi apparatus is convex and close to the nucleus and it witnesses fusion of various secretion-filled vesicles of RER and SER. Golgi apparatus has a variety of enzymes for the synthesis of polysaccharides and for the formation of **glycoproteins** and **glycolipids**. It also contains enzymes for the processing of proteins, *i.e*, for glycosylation, phosphorylation, sulphation and selective proteolysis. Being a packaging station of the cellular secretions, its tubules of maturation or trans face (which is concave) continuously pinch-off either the lysosomes, secretory vesicles or cortical granules.

3. Mitochondria. Mitochondria are oxygen-consuming ribbon-shaped semi-autonomous organelles containing its own single, small, circular and double stranded DNA molecule and tiny 55S ribosomes, called **mitoribosomes**. Each mitochondrion is bounded by two membranes. The outer mitochondrial membrane resembles more closely with the plasma membrane in structure and chemical composition. Inner mitochondrial membrane is rich in many enzymes, coenzymes, pumps (*e.g.*, proton pump) and other components of electron transport chain for the energy metabolism. It gives out finger-like outgrowths (**cristae**) towards the central lumen of mitochondrion and bear tennis-racket shaped **F_1 particles** which have ATP-ase enzyme for ATP synthesis. The mitochondrial lumen is filled with colloidal **matrix** which contains the soluble enzymes of Krebs cycle which completely oxidize the acetyl-CoA (an end product of cytoplasmic glycolysis and mitochondrial oxidative decarboxylation) to produce CO_2, H_2O and hydrogen ion (H^+). Hydrogen ions reduce the molecules of NAD and FAD, both of which pass on H^+ to respiratory or electron transport chain where oxidative phosphorylation takes place to generate energy-rich ATP molecules. Since mitochondria act as the ''power houses'' of the cell, they are abundantly found on those sites where energy is earnestly required such as sperm tail, muscle cells, oocytes, microvilli, etc. Thus, each of liver cells may contain up to 1600 mitochondria and each of some types of ova may contain more than 300,000 mitochondria.

Mitochondria also have some other functions. They may accumulate iron-containing pigment (ferritin), ions of Ca^{2+} and HPO_4^{2-} (phosphate) or protein molecules, thereby forming yolk bodies in the eggs of molluscs or crystalline yolk platelets of amphibian oocytes.

4. Lysosomes. Lysosomes are about 0.1 to 0.8 μm in diameter, bound by a single membrane having proton pumps, contain about 50 hydrolytic enzymes (*e.g.*, **acid phosphatase** which is cytochemically identified) and have an acidic medium (pH 5). They are formed by Golgi apparatus and carry on useful scavenger-like activities in killing and destroying pathogens, digesting food and autophagy, *i.e.*, removal of worn-out old cellular organelles of cells and making their constituent molecules available to the cell for the reuse. The acrosome of the spermatozoa is a special lysosome and during development lysosomes are active in the shedding or remodelling of tissues ; for example, during amphibian metamorphosis, the degeneration of tadpole tail is performed by the lysosomal proteolytic enzymes called **cathepsins.** Yolk digestion during development is also done by lysosomes.

5. Peroxisomes. They are tiny circular membrane-bounded organelles containing a crystal-core of enzymes (urate oxidase, peroxidase, D-amino oxidase and catalase). These enzymes are needed by peroxisomes in detoxification activity, *i.e.*, in the metabolism (*i.e.*, production and decomposition) of H_2O_2 molecules which are produced during neutralization of certain superoxides—the end products of mitochondrial or cytosolic reactions). Peroxisomes are also related with β-oxidation of fatty acids and thermogenesis similar to the mitochondria and also in degradation of the amino acids.

6. Ribosomes. They are tiny spheroidal dense particles of 150 to 200 Å diameter that contain approximately equal amounts of RNA and proteins. They are primarily found in all cells and serve as a scaffold for the ordered interaction of the numerous molecules involved in protein synthesis. Ribosomes have a sedimentation coefficient of about 80S and are composed of two subunits, namely 40S and 60S. The small 40S ribosomal subunit contains one molecule of **18S rRNA** (M.W. 0.7×10^6 ; 1900 nucleotides) and 30 proteins (S_1, S_2, S_3, and so on). The large 60S ribosomal subunit contains three ribosomal RNA molecules, namely, **28 S rRNA** (M.W. 1.7×10^6 ; 4800 nucleotides), **5.8 S rRNA** (**M**.W. 5×10^4 ; 160 nucleotides) and **5S rRNA** (M.W. 3.6×10^4 ; 120 **nucleotides**) and 40 proteins (L_1, L_2, L_3 and so on). The 18S, 28S and 5.8 S rRNAs are the transcriptional products of closely linked genes (*i.e.*, ribosomal DNA or rDNA) in the chromosome called **nucleolar organizer** (NO) of the cell nucleus. The genes for 5 S rRNA are located outside the NO region.

7. Microtubules and microtubular organelles. With few exceptions such as human erythrocytes, the cytoplasm of most cells contains certain ultrafine (25 nm diameter) tubes of indefinite length, called **microtubules**. The microtubules are made up of two kinds of protein subunits : **α - tubulin** (tubulin A) and **β - tubulin** (tubulin B), each of M.W. 55,000 daltons. The wall of microtubules comprises helical array of repeating α and β tubulin subunits. Assembly studies have indicated that the structural unit of the microtubule is an α β dimer 8 nm in length. Thus, in a microtubule, there are 13 protofibrils or protofilaments, each composed of α β dimers that run parallel to the long axis of the tubule. In fact, the repeating unit is an α β heterodimer and within the microtubule all heterodimers are arranged 'head to tail', that is α β →α β→ α β and so on. Due to this arrangement all the microtubules have a defined polarity : their two ends are not structurally equivalent. Microtubules undergo reversible assembly — disassembly (*i.e.*, polymerization — depolymerization), depending on the need of the cell or organelles. This process requires energy (GTP) and calcium ions.

The microtubules form the cytoskeleton of the cell and also constitute the unit structures of various cellular organelles such as centrioles, mitotic or meiotic spindles, basal bodies, cilia and flagella.

2-1.3. Nucleus

Nucleus is the heart of every type of eukaryotic cell. It contains **genes**, the hereditary units. Genes are locted on the chromosomes which exist as **chromatin network** in the non-dividing cell, *i.e.*, during interphase. Chromatin has two forms : 1. **Euchromatin** is the well-dispersed form of chromatin which takes lighter DNA-stain and is genetically active, *i.e.*, it is involved in gene duplication, gene transcription (DNA dependent RNA synthesis) and **phenogenesis** or phenotypic expression of a gene through the biosynthesis of a specific type of protein. 2. **Heterochromatin** is the highly condensed form of chromatin which takes dark DNA-stain and is genetically inert. Such types of chromatin exist both in the region of centromere (the **constitutive heterochromatin**) and in the sex chromatin (called **facultative heterochromatin**) and is late-replicating one.

Chemically, the chromatin contains a single DNA molecule, equal amount of five basic types of histone proteins, some RNA molecules and variable amounts of different types of acidic proteins. In fact, the chromatin has its unit structures in the form of **nucleosomes**. Chromatin binds strongly to the inner part of **nuclear lamina**, a 50 to 80 nm thick fibrous sheath (or lamina) lining the inner side of the **nuclear envelope**. Nuclear lamina is made up of three types of proteins, namely **lamin A**, **B** and **C**, having M.W. of 74,000, 72,000 and 62,000, respectively. Lamins are found to play a role in the assembly of interphase nuclei after each mitosis. For example, during amphibian development, in the first few cleavages the nuclear envelope initially forms around individual chromosomes, forming several vesicles that then fuse together to form a single nucleus. This suggests that chromatin is the nucleating centre for the deposition of a nuclear lamina and envelope.

The nuclear envelope, which comprises two nuclear membranes — an **inner nuclear membrane** and **outer nuclear membrane**, contains many **pores** that have octagonal **pore complexes** to regulate exchange between the nucleus and cytoplasm. The number of nuclear pores in a nucleus seems to be correlated with the transcriptional activity of the cell. For example, in the frog *Xenopus laevis* oocytes (which are very active in transcription), there are 60 nuclear pores/μm^2 and up to 30 million pores and pore complexes per nucleus; whereas, mature erythrocytes (which are inactive in transcription) have only about 3 pores/μm^2 and a total of only 150 to 300 pores per nucleus (**Scheer**, 1973). The nuclear envelope binds the **nucleoplasm** which is rich in those molecules which are required for DNA replication, transcription, regulation of gene actions and processing of various types of newly-transcribed RNA molecules.

Nucleus also has a circular suborganelle the **nucleolus**. Nucleolus lacks any limiting membrane and is formed during interphase by rDNA of nucleolar organizer. It is the site where ribosomes are manufactured. It is in the nucleolus where ribosomal DNA transcribes most of rRNA molecules and these molecules undergo processing before their stepwise addition to 70 types of ribosomal proteins to form the ribosomal subunits. Any increase in the size or number of nucleolus indicates the active transcription of rRNAs and fast pace of ribosome biogenesis. For example, the germinal vesicle or nucleus of growing primary oocyte of amphibians have several nucleoli (up to 1500 nucleoli in *Xenopus*).

2-2. CELL SURFACE

The cell surface includes the extracellular materials secreted outside or adherent to the plasma membrane (about 100 A° beyond the outer leaflet) that make up the glycocalyx and extracellular matrix or ECM. It may also include the underlying microfilaments and microtubules attached to proteins floating in the lipid bilayer of plasma membrane (**Grant,** 1978). ECM includes the products that lie in the immediate vicinity of the plasma membrane and that have been synthesized and deposited by the cells themselves. The surfaces of animal cells are coated with a **glycocalyx** which is made of structural **glycoproteins** and of a class of enzymes called **glycosyltransferases** that play important roles in adhesion of cells to and detachment of cells from one another and from substrates (**Shur**, 1982). These enzymes are involved in cell migration, morphogenetic movements, interaction of determined cells prior to their cytodifferentiation, or the binding of cells to inductively active substrates (**Roth** *et al.*, 1977).

The **extracellular matrix** beyond the cell surface and glycocalyx is typically made up of two types of molecules : 1. **glycoproteins** which include **collagen**, **fibronectin** (or **mucin**) and **laminin** ; and 2. **proteoglycans** or **glycosaminoglycans** or **GAGs**, which are glcoproteins in which carbohydrates are in excess (*e.g.*, **hyaluronic acid**, **chondroitin**, **chondroitin sulphate**, **heparin sulphate** and **keratan sulphate**). Collagen fibres are known to determine normal cell form and movement of loose mesenchyme cells (*e.g.*, fibroblasts). Fibronectins are involved in cell contact behaviour such as spreading and flattening of cells on the substratum during their migration or movement (**Trinkaus**, 1984). Laminins play a role in cell spreading, locomotion, contact inhibition, integration of cells in tissues and cell transformation (**Timple** *et al.*, 1983). GAGs regulate the motile behaviour of embryonic cells. They are involved in cell adhesion and in providing direction to cell movement during development (see **Trinkaus**, 1984).

2-3. CELL CYCLE AND CELL DIVISION

All cells are produced by divisions of pre-existing cells. Continuity of life depends on cell division. A cell born after a division, proceeds to grow by macromolecular synthesis, reaches a species-determined division size and replicates to repeat the cycle. This cycle acts as a unit of biological time and defines the life history of a cell.

2-3.1. The Cell Cycle

Each cell capable of division passes through a cycle called **interphase cycle** (**Burns** and **Bottino**, 1989) or **cell cycle** which can be defined as the entire sequence of events happening from the close of one nuclear division to the beginning of the next. The total time varies with the species, maturation, tissue and temperature, among other factors. **Howard** and **Pelc** (1953) have divided the cell cycle into the following four stages or phases, of which G_1, S and G_2 phases forming the interphase and M phase forming the mitotic phase:

1. G_1 phase. It is the **first gap** or **growth stage** of interphase in which nucleus and cytoplasm are enlarging towards mature size. This phase begins immediately following cell division. During G_1 phase, chromatin is fully extended and not distinguishable as discrete chromosomes with the light microscope. This is a time of active synthesis of RNA and protein and of the resuming of normal cell metabolism that has slowed during preceding cell division. RNA synthesis (or transcription) is indicated by the appearance of nucleolus in the interphase nucleus. Proteins synthesized during G_1 phase are (1) enzymes necessary for DNA synthesis or replication of the next stage ; (2) possibly a protein that acts to trigger nuclear division, and (3) tubulin and other mitotic apparatus proteins. G_1 is most variable stage as to duration ; it may occupy 30 to 50 per cent of the total time of the cell cycle or be entirely lacking in rapidly dividing cells (*e.g.*, those of early mammalian embryo). Differentiated somatic cells that no longer divide (*e.g.*, neurons) are arrested usually in the G_1 stage which is often referred to as **G_0 phase**.

2. S phase. During **S phase** or **synthetic phase**, replication of DNA and synthesis of histones occur. DNA molecule of each chromosome becomes precisely doubled in its amount, so that each chromosome now carries a duplicate set of genes or two DNA molecules. By the end of S phase, each chromosome is composed of two sister **chromatids** sharing a common **centromere** ; thus, the cell retains the original diploid (2n) chromosome number but now has a duplicate set of genes. In fact, early in S phase, there is a rapid rise in **DNA polymerase enzyme** (this requires its transport from cytoplasm to nucleus) and in a type of RNA (which acts as an enzyme, called **ribozyme** and is needed for degeneration of nuclear membrane during a later stage of cell division) along with other RNAs. S phase occupies roughly 35 to 40 per cent time of the cell cycle.

3. G_2 phase. This is a second **gap** or **growth phase**, not as well understood as G_1 phase. During G_2 phase new DNA is rapidly complexed with chromosomal proteins and synthesis of RNA and protein continues. It may occupy 10 to 20 per cent time of the cell cycle.

As the G_2 stage draws to a close, the cell gradually enters mitotic division or M phase.

2-3.2. Cell Division

Growth and development of every organism depend in large part on multiplication, enlargement and differentiation of its cells, beginning with the zygote. Most sexually reproducing forms also depend directly on cell division for the formation of sex cells or gametes.

Division of nucleate cells consists of two distinct but integrated activities, nuclear division (**karyokinesis**) and cytoplasmic division (**cytokinesis**). In general, cytokinesis begins after nuclear division is well underway, but in many instances it may be deferred or entirely lacking, *e.g.*, early steps of cleavage of insect eggs include only karyokinesis, the cytokinesis occurs at a much later stage (see Chapter 10).

Two types of nuclear divisions, **mitosis** and **meiosis** are the characteristic of most animal cells. Mitosis is regularly associated with nuclear division of somatic cells ; meiosis occurs in conjunction with the formation of reproductive cells (gametes) in sexually reproducing cells.

2-3.3. Mitosis

Mitosis (Gr., *mitos* = fibre, thread) starts at the culmination point of the interphase. It is a short period of chromosome condensation, segregation and cytoplasmic division. Mitosis is important for growth, for replacement of cells lost to natural friction (attrition) and wear and tear and for wound healing. It is especially rapid during the early development of the organism, then it slows down with age. Different cell populations divide at widely different rates. The S and G_2 phases are relatively constant in all species, but the G_1 phase shows the greatest variability in response to physiological conditions. For example, cells in early embryos are almost entirely in S phase since their generation times are quite short, often equivalent to the 15 to 20 minute cycles of prokaryotes. And like the prokaryotes, they may have no G_1 or G_2 phases. For instance, in the early embryo of *Xenopus* the total cell cycle is of 20 to 25 minutes with no G_1 phase, a 17-minute S phase, a 4-minute G_2 and a mitotic time of 4 minutes. But this pattern changes during development ; a substantial G_1 and G_2 phases become apparent as the cell cycle gradually lengthens. Since cell division rates decline during development with increase in length of S phase to several hours, this means total genome DNA is replicated 100 times faster in early embryos than in late embryos or adult tissues.

Further, early in development, all cells divide, but later, as the embryo grows, DNA replication slows down and cells sepcialize. Some cells irreversibly lose the capacity to divide, as if division genes are permanently shut off (*e.g.*, nerve cells and skeletal or striated muscle cells). Other cells, though functionally differentiated and non-dividing, never lose mitotic potential. They can be induced to proliferate when isolated in a nutrient culture medium or exposed to strong stimuli such as wounding (*e.g.*, liver cells). Another class of adult somatic cells remains mitotically active throughout the life span of the organism, responding to functional demand for new cells in such tissues as bone marrow and epithelium of skin and gastro-intestinal tract.

Lastly, since mitotic mechanisms are similar in most eukaryotic cells, we assume a first approximation that a common control system is shared by all. Cells escaping these controls, either by mutation or by accident, proliferate into a **neoplastic** or **cancerous** growth.

A. Four steps of karyokinesis of mitosis. Mitotic process is a smoothly continuous process and is divided into following four stages or phases for convenient reference :

1. Prophase. (Gr., *pro* = before). The appearance of the thin-thread-like condensing chromosomes marks the beginning of mitosis. At the same time, the cell becomes spheroid, more refractile and viscous.

Each prophase chromosome is composed of two coiled filaments, the **chromatids** which are a result of the replication of the DNA and synthesis of histones during the S phase. As prophase progresses, the chromatids become shorter and thicker and the two sister chromatids of each chromosome are held together by a specialized region called **primary constriction** or **centromere**. The centromere of each sister chromatid includes a **kinetochore** to which about 34 microtubules of the spindle are attached. Kinetochores appear to be organized in prophase only after chromosome contraction has taken place.

During early prophase, the chromosomes are evenly distributed in the nuclear cavity ; as prophase progresses, the chromosomes approach the nuclear envelope, causing the space of nucleus to become empty. During prophase, the nucleolus gradually disintegrates.

In the cytoplasm the most conspicuous change is the formation of the spindle or **mitotic apparatus**. In early prophase there are two pairs of centrioles (which are duplicated earlier in interphase), each one surrounded by the so-called **aster**, composed of microtubules that radiate in all directions. The two pairs of centrioles migrate to opposite poles of the cell along with the asters and become situated in antipodal positions. The centrioles remain connected by numerous longitudinal microtubules, collectively forming the **spindle** ; the spindle tubules elongate as the centrioles move apart. The spindle is made up of three types of fibres : (1) **chromosomal fibres** or **kinetochoric fibres** which extend from one pole or the other to each kinetochore to which they are attached ; (2) **continuous** or **polar fibres** which extend pole to pole or centrioles to centrioles ; and (3) **interzonal** or **free fibres** which are formed between the daughter chromosomes and nuclei during anaphase and telophase.

Dissolution (degeneration and disappearance) of the nuclear envelope marks the end of prophase. This process is incompletely understood, but the aggregation of mitochondria suggests enzymatic processes. Physical stresses exerted by microtubules, which are attached to the nuclear membrane, may also play a role in membrane degeneration (see **Burns** and **Bottino**, 1989).

2. Metaphase. (Gr., *meta* = between). During metaphase the chromosomes are shortest and thickest. Their centromeres occupy the plane of the equator of the mitotic apparatus (a region known as the **equatorial** or **metaphase plate**), although the chromosomal arms may extend in any direction. At this stage the sister chromatids are still held together by connecting chromatin fibres at the centromere region. Electron microscopy shows that the kinetochores of the two sister chromatids face opposite poles ; this would permit proper separation in the next phase (anaphase).

3. Anaphase. (Gr., *ana* = back). At anaphase the equilibrium of forces that characterizes metaphase is broken by the separation of the kinetochores — a process that is carried out simultaneously in all the chromosomes. The kinetochores move apart, and the chromatids separate and begin their migration towards the poles. The kinetochore always leads the rest of the chromatid or **daughter chromosome**, as if it is being pulled by the chromosomal fibres of the spindle. During anaphase the microtubules of the chromosomal spindle fibres shorten one-third or one-fifth of the original length. Simultaneously, the microtubules of the continuous spindle fibres increase in length. Some of these stretched spindle fibres now constitute the so-called interzonal fibres.

4. Telophase. (Gr., *telo* = end). The end of the polar migration of the daughter chromosomes marks the beginning of the telophase ; it is in turn, terminated by the reorganization of two new nuclei and their entry into the G_1 stage of interphase. In general terms, the events of prophase occur in reverse sequence during this phase. New nuclear membranes are constructed from materials that may be remnants of the original membrane, or derived from the endoplasmic reticulum. The mitotic apparatus except the centrioles disappears, high viscosity of the cytoplasm decreases, the nucleoli are reformed at the nucleolar organizing sites of specific chromosomes and the chromosomes resume their long, slender and extended form as their coils relax.

B. Cytokinesis. If cytokinesis or cytoplasmic division occurs, the process takes place during telophase. When an egg or cell divides, a constriction appears in the cell surface, usually perpendicular to the spindle axis, deepens into a furrow, and slowly cuts the cell into two daughter cells.

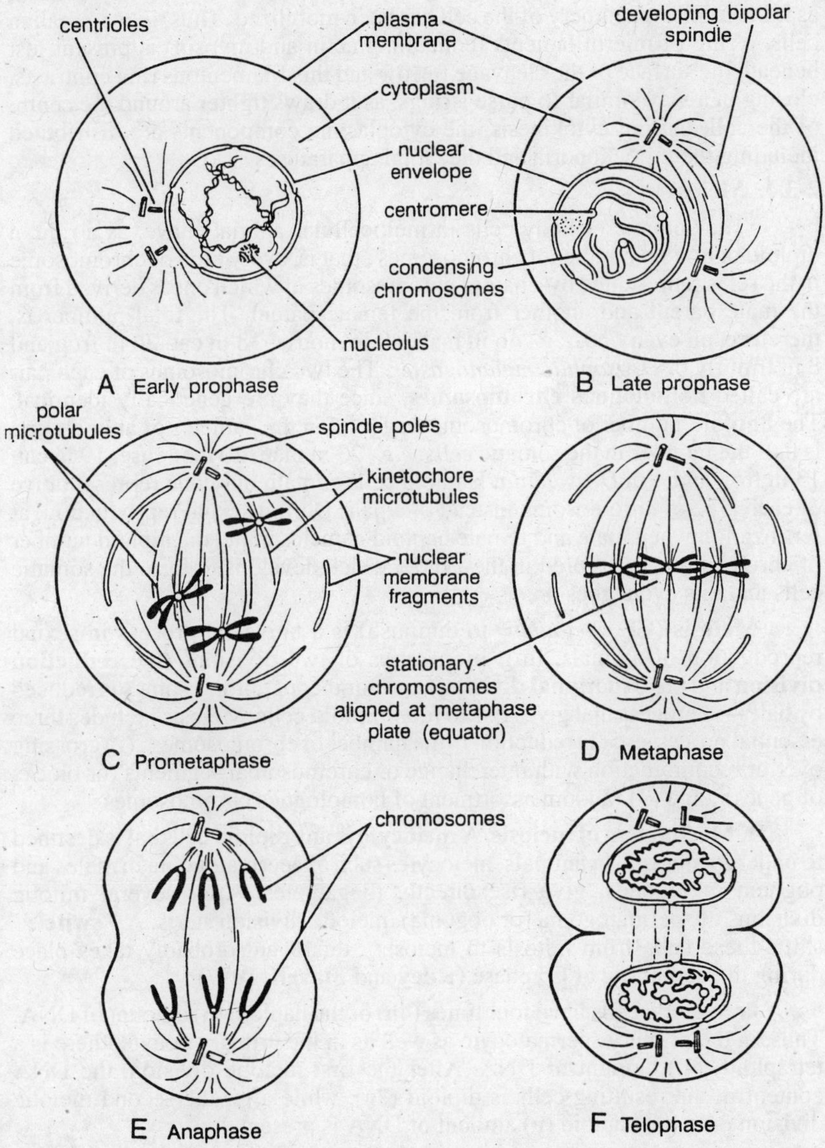

Fig. 2.2. Diagrammatic summary of mitosis in the animal cell. A—**Interphase** : synthesis of DNA, RNA and proteins such as histones, enzymes and tubulins. B—**Prophase** : chromatid coiling, nucleolar disintegration and spindle formation. C—**Prometaphase**. D —**Metaphase** : chromosomal orientation at the equatorial plate. E—**Anaphase** : movement of the daughter chromosomes toward the poles. F—**Telophase** : reconstruction of the daughter nuclei.

The entire process involves change of cell shape and contraction of cytoplasm, as if the motile machinery of the cell has been mobilized. Thus, in mammalian cells, a ring of microfilaments (containing actin and myosin) is present just beneath the surface of the cleavage furrow and this filamentous ring contracts, closing in a way similar to purse strings, as it draws tighter around the centre of the cell. During cytokinesis, the cytoplasmic components are distributed including the mitochondria and the Golgi apparatuses.

2-3.4. Meiosis

The somatic or body cells in multicellular animals have, as a rule, a diploid or double number of chromosomes ; that is, every kind of chromosome in the set is represented by a pair of chromosomes of which one is derived from the male parent and another from the female parent. The total number is, therefore, an even one, *e.g.*, 46 in man, 40 in mouse, 38 in cat, 26 in frog and 8 in fruit fly or *Drosophila melanogaster*. The two chromosomes of each pair are called **homologous chromosomes**, since they are genetically identical. The **haploid** number of chromosomes is found in the gametes of animals and is half the number in the somatic cells, *e.g.*, 23 in man, 20 in mouse, 19 in cat, 13 in frog and 4 in *Drosophila*. Haploid cells contain only one representative of each type of chromosome instead of a pair. During sexual reproduction, at fertilization when male and female haploid gametes unite, the diploid number of chromosome is restored in the zygote which develops to form the somatic cells and this cycle goes on.

Meiosis (Gr., *meioum* = to diminish) is a universal process in sexual reproduction of animals. In it by a series of two divisions (the **reduction division** and the **equational division**), the number of chromosomes is reduced by half — a process that gives rise to four **haploid** cells. Meiosis includes three essential processes : (1) reduction in the number of chromosomes ; (2) crossing over or recombination with interchange of chromosomal segments (or blocks of genes) ; and (3) random assortment of homologous chromosomes.

1. Mechanism of meiosis. A **meiocyte** is any diploid cell that is destined to undergo meiosis. In animals, meiocytes such as spermatogonia in males and oogonia in females, give rise directly to gametes. After several mitotic divisions of spermatogonia (or oogonia), meiotic division starts. A "**switch**" shifts these cells from mitosis to meiosis ; this event probably takes place during the beginning of G_2 phase (**Riley** and **Flavell**, 1977).

Premeiotic cells have four times (4n) of the haploid (n) amount of DNA. Thus, in the primary spermatocyte as well as in the primary oocyte, there is a tetraploid (4n) amount of DNA. After the first meiotic division, the DNA content in the resulting cells is diploid (2n), while after the second meiotic division only the haploid (n) amount of DNA is present.

Thus, like the mitosis, meiosis is preceded by the duplication of the chromosomes during interphase but in contrast to mitosis, this duplication is followed by two successive cell divisions, designated as **meiosis I** and **meiosis II**, each with its own prophase, metaphase, anaphase and telophase (Fig. 2.3).

A. Meiosis I. It includes the following four phases :

1. Prophase-I. It is longest phase of meiosis and may be measured in weeks or months or even longer. In the human female it persists in each primary oocyte from a foetal age of 12 to 16 weeks until ovulation, generally once each 28 days, after sexual maturity. Some human oocytes, therefore, persist in arrested prophase-I until the end of reproductive cycle at menopause, which is about 45 years. Prophase-I too includes the following five substages :

(i) Leptotene or Leptonema. (Gr., *leptos* = thin ; *nema* = thread). The nucleus enlarges. The chromatin material condenses to form distinct, long, thin, thread-like chromosomes. All chromosomes are longitudinally single and their number is equal to the number found in other somatic cells. (Thus, although chromosome duplication has already occurred and each of them has the material of two chromatids, leptotene chromosomes look single rather than double and these chromosomes have bead-like chromomeres at irregular intervals along their length). The centrioles duplicate and start moving towards opposite poles.

(ii) Zygotene or Zygonema. (Gr., *zygon* = adjoining or a yoke). Chromosomes inherited from each parent begin to pair or **synapse**. This pairing or synapsis (Gr., *syn* = together ; *apsis* = a joining) occurs between two homologous chromosomes, of which one comes from mother via the ova and the other from the father via the spermatozoa. The paired homologous chromosomes are called **bivalents** or **dyads**.

The synapsis is remarkably exact and specific, taking place point by point, with two homologues usually somewhat twined about each other. Between the paired homologous chromosomes are formed ultramicroscopic scaffoldings or frameworks of proteins and DNA, called **synaptonemal complexes** or **SCs**. A SC helps to stabilize the pairing of homologous chromosomes and to facilitate recombination by crossing over. In fact, synaptonemal complexes are not found in organisms in which crossing over does not occur (*e.g.*, the male fruit fly, *Drosophila* ; See **Burns** and **Bottino**, 1989). Synapsed chromosomes continue to shorten and thicken, and in the process, their longitudinal doubleness becomes apparent in light microscopy.

(iii) Pachytene or Pachynema. (Gr., *pachus* = thick). During this stage the formation of synaptonemal complex is completed and each homologous chromosome of each bivalent or dyad splits up longitudinally into two chromatids attached at the centromere. Each bivalent now has four strands (chromatids) and is called **quadrivalent** or **tetrad**. The two chromatids of each homologue are called **sister chromatids**.

During late pachynema, two of the non-sister chromatids (one paternal and other maternal) of the homologous chromosomes exchange some genes or portions of chromosomes (*i.e.*, recombine). Transverse breaks occur at the same level on each of the non-sister chromatids of the tetrad and this event is followed by interchange or exchange and final fusion of the chromatid segments. This is a cytogenetic activity and is called **crossing over** or **recombination.** The process of crossing over involves a variety of molecules and enzymes such as a 125- nucleotide long RNA molecule (**Stern** and **Hotta**, 1980), endonuclease, ATP-dependent DNA unwinding protein, R-protein and β-DNA polymerase (see **De Robertis** and **De Robertis, Jr.,** 1987).

Pachynema may last for days, weeks, or even years, whereas leptonema and zygonema may last only a few hours.

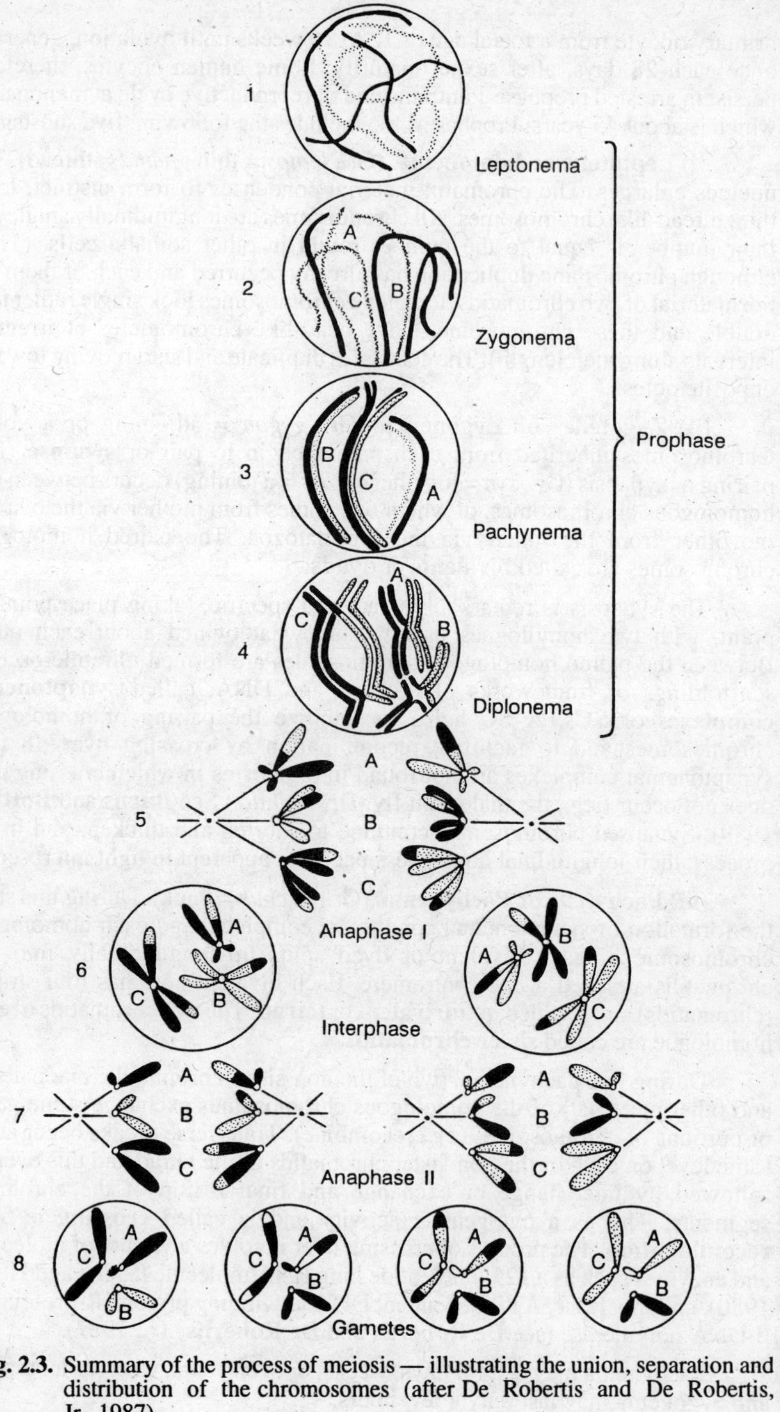

Fig. 2.3. Summary of the process of meiosis — illustrating the union, separation and distribution of the chromosomes (after De Robertis and De Robertis, Jr., 1987).

(iv) Diplotene or Diplonema. (Gr., *diploos* = double). At the diplonema stage the SC is disappeared and intimately paired homologous chromosomes repel each other and begin to separate. However, this separation is not complete, because the homologous chromosomes remain united by their points of interchange or **chiasmata** (plural for **chiasma**) (Gr., *chiasma*=cross-piece). Chiasmata represent the sites or points, at which exchange of genes or chromosomal segments has occurred during crossing over.

Diplonema is a long-lasting period. In the fifth month of prenatal life, for example, human oocytes have reached the stage of diplonema and remain in it until many years later, when ovulation occurs. In fish, amphibian, and reptilian oocytes, there is a marked uncoiling of chromosomes during diplonema, so that the greatly enlarged nucleus assumes an interphase appearance. Such a nucleus is called **germinal vesicle**. In these cases, the bivalent chromosomes may attain a special configuration known as the **lampbrush chromosomes**, in which chromatin thread unwinds at many points along the axis to form numerous lateral loops. It will be shown later (See Chapter 5) that the presence of these lampbrush chromosomes is related to an intensive RNA synthesis and to the enormous growth of the oocyte.

(v) Diakinesis. (Gr., *dia* = across). During this final stage of prophase-I, chromosomes reach maximum contraction. At the outset of diakinesis the synapsed homologues become spaced out well in the nucleus, often occur near the nuclear membrane. Chiasmata gradually **terminalize**, that is, they appear to move toward the ends of the arm and finally "slip off", owing to the continued shortening of the chromosomes. The nucleolus then disappears, the nuclear membrane degenerates, centrioles reach to opposite poles of the cell and start forming spindle apparatus of microtubules.

2. Metaphase-I. The arrival of synapsed chromosome pairs at the equator of the spindle begins metaphase-I. The spindle microtubules become attached to the kinetochores ; each homologue is being attached to one of the poles by the homologous kinetochores ; thus, the two sister chromatids behave as a functional unit. Meiotic metaphase-I can be distinguished by the fact that the homologues are still attached by chiasmata at their ends, while the kinetochores are being pulled toward the poles.

3. Anaphase-I. During this phase actual **disjunction** occurs, *i.e.*, the partners of homologous chromosomes completely separate from each other and move towards opposite poles of the cell. The centromeres that hold chromatids together in each chromosome do not divide as they do in mitosis. So, the pair of chromatids of each chromosome move together and do not separate.

4. Telophase-I. The arrival of chromosomes at the poles of the spindle signals the end of anaphase-I and the beginning of telophase-I. During this phase the chromosomes may persist for a time in the condensed state, the nucleolus and nuclear membranes are also formed. Thus, at each pole a nucleus is reformed as in mitosis. But these nuclei are haploid (having a double amount of DNA and histones), each containing one chromosome from each homologous pair.

Nuclear division is usually followed by cytokinesis resulting in two daughter cells, which in animals are called **secondary spermatocytes** (in the male) and **secondary oocyte** plus the first **polocyte** or **polar body** (in the female). During the short interphase between meiosis I and meiosis II, there is no replication of chromosomes, a situation different from that of mitosis. Now the second meiotic division follows :

B. Meiosis-II. Each haploid daughter cell, resulting from the first meiotic division now divides mitotically to give rise to two haploid daughter cells. Meiosis II includes the following four steps :

1. Prophase-II. This stage is generally short and superficially resembles mitotic prophase, except that sister chromatids of each chromosome are usually widely divergent, exhibiting no relational coiling. So during prophase - II, the centrioles duplicate and migrate towards opposite poles. Each set of centrioles is surrounded by aster rays. Nuclear envelope and nucleolus disappear.

2. Metaphase-II. In both daughter cells, a spindle is formed which is generally oriented at right angles to the first division's spindle. The haploid number of chromosomes, each consisting of two chromatids joined at the centromeres, are arranged in the equatorial plane.

3. Anaphase-II. Kinetochores of each centromere separate and the sister chromatids of metaphase-II move poleward as daughter chromosomes, similar to their movements in mitosis. Their arrival at the opposite poles marks the end of this phase.

4. Telophase-II. Haploid sets of chromosomes at two different poles uncoil and form chromatin material. Nucleolus reappears. Nuclear envelope is formed around the chromatin and the nucleolus to form nucleus. Thus, each of four nuclei of telophase-II has one chromatid of a tetrad of pachynema stage and each nucleus has a haploid number of chromosomes.

Telophase-II is followed by cytokinesis and, thus, ultimately four haploid cells are formed.

REVISION QUESTIONS

1. What is an organelle ? By means of a labelled diagram, indicate the parts of a generalized animal cell.
2. Describe the role of cell surface in development.
3. Show in a drawing how mitosis would proceed in a cell that contains four pairs of chromosomes.
4. Describe the process of meiosis. Add a note about its genetical significance.
5. Write short notes on the following :

   (1) Cell cycle; (2) Cell surface; (3) Receptor-mediated endocytosis; (4) Lysosomes in development.

3

DNA, RNA AND PROTEIN SYNTHESIS

Life's dynamism and diversity reside in the catalytic and structural properties of proteins. Proteins carry out all vital activities and are building blocks for all levels of biological organization. They act as molecular switches and turn biochemical reactions on or off. An organism is but the functional expression of its own constellation of proteins. Protein diversity resides in linear sequences of amino acids, and these in turn are encoded in nucleic acid messages, or genes. Development, therefore, may be viewed (at the molecular level) as a spatiotemporal ordered appearance of new proteins, while new temporal regulation of expression of different sets of genes may be defined as a developmental change at cellular and supracellular levels. To understand various embryological phenomena at the molecular level, the following brief discussion of DNA, RNA and protein synthesis is earnestly required:

3-1. DEOXYRIBONUCLEIC ACID (DNA)

Now it is well established that genetic information of all living systems except some RNA viruses, is encoded in DNA molecule. DNA is present in the form of a long, linear molecule in each chromosome of the nucleus.

3-1.1. Molecular Structure of DNA

The DNA molecule is a double-stranded very long helical polymer or polynucleotide of four nucleotides. Each nucleotide consists of a deoxyribose pentose sugar, a nitrogenous base (*i.e.*, either a **purine**—adenine or A, or guanine or G; or a **pyrimidine**–cytosine or C, or thymine or T) and phosphate. The four nucleotides are deoxyadenylic acid (d AMP), deoxythymidylic acid (dTMP), deoxyguanylic acid (dGMP) and deoxycytidylic acid (dCMP). With molecular weights of the order of 10 to 20 million, DNA must contain thousands of nucleotide subunits in linear array. These are arranged as two interwoven helical chains held together by hydrogen bonds between nucleotides stacked like poker chips.

Because of their molecular configuration, nucleotides form complementary pairs: adenine (A) nucleotides on one chain of the double helix form hydrogen bonds only with thymine (T) nucleotides on the other chain, and guanine (G) nucleotides pair only with cytosine (C) nucleotides. This strict stereochemical pairing is the principal "grammar rule" of "communication" or information transfer between each strand of the double helix. Nucleotide sequences on one helix automatically specify the sequence on the other. For example, if one strand of DNA has the sequence ATCGAAG..., the opposite strand must have the sequence TAGCTTC... .DNA molecules are **polarized**; they have a beginning and an end; thus, whatever information is contained in the polynucleotide sequence of DNA is read in one direction. In fact, two strands of DNA have an opposite polarity with reference to 5′ and 3′ ends, based on the carbon atom of sugar moiety to which phosphate is attached. Of the two strands in a gene (DNA), only one is a **sense strand** and is transcribed to RNA; the other is an **antisense strand**. The same strand, in two adjoining genes, may be sense strand in one gene and antisense in the other. The sequence of bases in DNA provides the message for specificity of proteins (the genetic code). It is also instrumental in determining its conformation (shape) and consequent identification by different molecules, such as proteins.

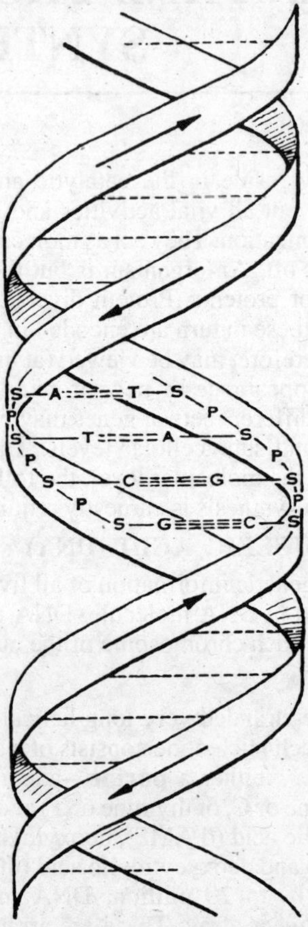

Fig. 3.1. The DNA double helix.

3-1.2. The Genetic Code

There is a species-specificity of nucleotide sequences in DNA. The analysis of human haemoglobins also suggests that a gene codes for the primary amino acid sequence of a protein. A nucleotide sequence in DNA encodes the amino acid sequences in polypeptide chains. Since there are four different nucleotides in DNA and 20 amino acids in proteins, how can an alphabet based on 4 letters be translated into an alphabet of 20 letters ? Obviously combinations of more than 1 nucleotide must specify a single amino acid.

More than a decade of genetic and biochemical studies have successfully unravelled the genetic code. A code word in DNA is a **triplet**; an arrangement of three nucleotides defines a single amino acid in a polypeptide chain. This furnishes 64 different code-word combinations, or **codons**, more than enough to specify 20 amino acids (Table 3-1). It implies code **degeneracy**; that is, several trinucleotide combinations code interchangeably for one amino acid. Thus, amino acid valine is coded by the codons GUU, GUC, GUA, GUG. The genetic tape, therefore, is a linear sequence of triplet codons, the instructions for ordering amino acid sequences in proteins. For the most part the genetic code is **universal** for all living organisms and for viruses. This means the existence of same triplets code for the same amino acids in a wide variety of organisms. Codons are **non-overlapping**, that is, each is contiguous with its neighbour. The code is read in one direction; it has no commas or spaces between each unit.

Table 3-1. The 'universal genetic code' of codon assignment in mRNA. The trinucleotide codons are written in the 5′ → 3′ direction.

First base (5′ end)	Second base U		Second base C		Second base A		Second base G		Third base (3′ end)
U	UUU	Phe	UCU	Ser	UAU	Tyr	UGU	Cys	U
	UUC		UCC		UAC		UGC		C
	UUA	Leu	UCA		UAA	Non-sense condon (stop)	UGA	Non-sense codon (stop)	A
	UUG		UCG		UAG		UGG	Trp	G
C	CUU	Leu	CCU	Pro	CAU	His	CGU	Arg	U
	CUC		CCC		CAC		CGC		C
	CUA		CCA		CAA	Gln	CGA		A
	CUG		CCG		CAG		CGG		G
A	AUU	Ileu	ACU	Thr	AAU	Asn	AGU	Ser	U
	AUC		ACC		AAC		AGC		C
	AUA		ACA		AAA	Lys	AGA	Arg	A
	AUG*	Met (start)	ACG		AAG		AGG		G
G	GUU	Val	GCU	Ala	GAU	Asp	GGU	Gly	U
	GUC		GCC		GAC		GGC		C
	GUA		GCA		GAA	Glu	GGA		A
	GUG*	(start)	GCG		GAG		GGG		G

Note : GUG is a codon for initiation but is very rare.

Recently, the following two important facts have become clear about DNA organization: 1. Only a small fraction of total DNA is represented in the cytoplasm in the form of rRNA, tRNA and mRNA; a large part is either not transcribed or its transcript is degraded. 2. The primary transcript for a mRNA molecule which codes for a protein, consists of non-coding **(introns)** and coding (**exons**) sequences, apart from non-translated sequences. The introns are removed by **gene splicing** before exteriorization of mRNA molecule in the cytoplasm; any slight (minor) mistake in splicing will make the protein molecules non-functional.

3-1.3. Replication of DNA

The replication of DNA occurs in the S phase of cell cycle. It is a very complex process involving unwinding and separation of strands; formation of origins (*i.e.*, points from where replication starts), of replication forks and of replicons; simultaneous replication of complementary strands (in both strands new DNA is synthesized in 5′ to 3′ direction); joining of newly synthesized DNA fragments, etc. Each polynucleotide strand of old DNA molecule serves as a **template** for the synthesis of a new DNA molecule. DNA replication follows the base pairing rules by which A pairs only with T, and G with C, and as a consequence of this, each daughter molecule is an exact replica of the parental molecule. DNA replication is **semiconservative**, *i.e.*, only one strand is synthesized; the other half of the original DNA is retained.

3-2. RIBONUCLEIC ACID (RNA)

There are three major classes of RNA molecules: ribosomal RNA (rRNA), transfer RNA (tRNA) and messenger RNA (mRNA). Each is transcribed from different genes and each has a distinct function in cells. Besides these RNAs, there are heterogeneous nuclear RNA (hn RNA) and the low molecular weight small nuclear RNA (Sn RNA). All RNAs are single stranded polymers of nucleotides (most RNA molecules assume a secondary structure by folding back on themselves and form hydrogen bonds in a double-stranded structure). However, they differ from each other by size, function and stability (Table 3-2). The RNAs differ from DNA in mainly the following respects: 1. The sugar moiety is **ribose** in RNA and not deoxyribose of DNA. 2. Of the four nitrogen bases in DNA thymine (a pyrimidine) is replaced by the **uracil** (U) (a pyrimidine) in RNA (Table 3-3). Like the proteins the tertiary structure of an RNA molecule is dependent on its primarv structure (base sequence) and has functional significance.

Table 3-2. Types of RNA molecules.

S.No.	Class	Principal location in cell	Molecular weight	Function	Stability
1.	rRNA	Ribosome	1.5×10^6 (28 S*) 7.5×10^5 (18 S) 4.5×10^4 (5.8 S) 3.5×10^4 (5 S)	Form ribosomes for protein synthesis	Very stable
2.	tRNA	Cytosol	2.5×10^3 (4 S)	Transfers specific amino acids to ribosomes	Fairly stable
3.	mRNA	Cytosol	Variable, generally $< 2\times 10^6$	Carries message for amino acid (aa) sequence in proteins	Variable
4.	hnRNA	Nucleoplasm	Heterogeneous, up to $> 10^7$	Yields mRNA molecules on processing	Short lived

* S represents Svedberg number (a unit of sedimentation during ultracentrifugation); the longer the RNA molecule, the larger is its S value and earlier it sinks on centrifugation.

Table 3-3. Ribonucleosides and ribonucleotides.

Base	Ribonucleoside (Base + Sugar)	Ribonucleotide (Base + Sugar + Phosphate)
1. Adenine	Adenosine	Adenylic acid
2. Uracil	Uridine	Uridylic acid
3. Cytosine	Cytidine	Cytidylic acid
4. Guanine	Guanosine	Guanylic acid

3-3. PROTEIN SYNTHESIS

The following central dogma of molecular biology depicts the main events of protein synthesis:

$$\text{DNA} \xrightarrow{\text{Transcription}} \text{RNA} \xrightarrow{\text{Translation}} \text{Protein.}$$

In the flow of information from DNA to RNA, the language (nucleotide sequence) remains the same. However, in the flow of information from RNA to proteins, the language is changed from a nucleotide sequence to an amino acid sequence. Just as in the process of translating from one language to another, this process of using information in RNA to make a protein is called **translation** (**Burns** and **Bottino**, 1989).

The process of protein synthesis involves the ribosomes, tRNA, mRNA, numerous enzymes, (*e.g.*, amino acid-activating system and peptide polymerase system), substrates (amino acids), energy sources (ATP and GTP), soluble protein initiation and transfer factors and various inorganic cations (for example K^+ or $NH_4{}^+$, and Mg^{2+}).

The process of protein synthesis includes the following three basic steps:

1. Initiation. Protein synthesis begins when non-translated 5′ end of mRNA binds to 18S rRNA in the 40S ribosomal subunit. This binding requires the presence of a protein molecule, the **eIF$_3$** (eukaryotic initiation factor 3; a 50,000 dalton molecule made of at least 10 polypeptides) and perhaps also the mRNA cap. Thereafter, the first codon (usually AUG) of mRNA molecule to be translated can receive, in the presence of **eIF$_1$**, a complex formed by **met-tRNA** (aminoacyl tRNA carrying the amino acid methionine), **eIF$_2$** and **GTP**. As soon as the codon in mRNA and anticodon in tRNA are base paired, all the three initiation factors (*i.e.*, eIF$_1$, eIF$_2$ and eIF$_3$) are released, the 60S ribosomal subunit is attached to the mRNA and GTP is hydrolyzed to GDP. Thus, a complete 80S ribosome is formed and the protein synthesis begins.

There are two sites in the ribosome to receive tRNA: (1) **P site** (or **peptidyl** or **donor site**) for p-tRNA (peptidyl tRNA, a molecule with a peptide chain attached to its amino acid), and (2) **A site** (or **aminoacyl** or **acceptor site**) for aa-tRNA (a tRNA molecule that has only its specific amino acid attached). However, the met-tRNA deceives the ribosome to occupy the P site and not A site, even though it has no peptide attached to it.

2. Elongation. For elongation of the peptide chain the next codon in mRNA is base paired by the appropriate aa-tRNA, at the A site. This requires the presence of **EF$_1$** (Elongation factor 1) and energy (one GTP is hydrolyzed to GDP). A peptide bond is then formed to join the amino group of aa-tRNA with the carboxyl group of p-tRNA. This is achieved by a **peptidyl transferase** present in the 60S ribosomal subunit. The tRNA in the P site is immediately discharged. There upon **EF2** (also called **translocase** or **G factor**) causes the translocation of newly formed p-tRNA and its codon from A to P site; again one GTP is hydrolyzed to GDP to provide the necessary energy. In this way the peptide chain continues to be elongated.

3. Termination. The termination of chain is achieved when a termination codon comes to occupy the A site. There is no tRNA for these codons. The releasing factor 1 to 2 with the help of peptidyl transferase releases the peptide from the p-tRNA. This causes also the release of tRNA from the P site and dissociation of 60S and 40S ribosomal subunits. There is evidence to show that eIF3 doubles up as a releasing factor.

Each mRNA molecule is simultaneously translated by several ribosomes, arranged at least 80 nucleotides apart; the mRNA-ribosome complex is called **polysome**. Each ribosome is capable of translating at a great speed of about 400 codons in 10 seconds. For example, in haemoglobin production, one polypeptide chain is nearly 150 amino acid residues long and

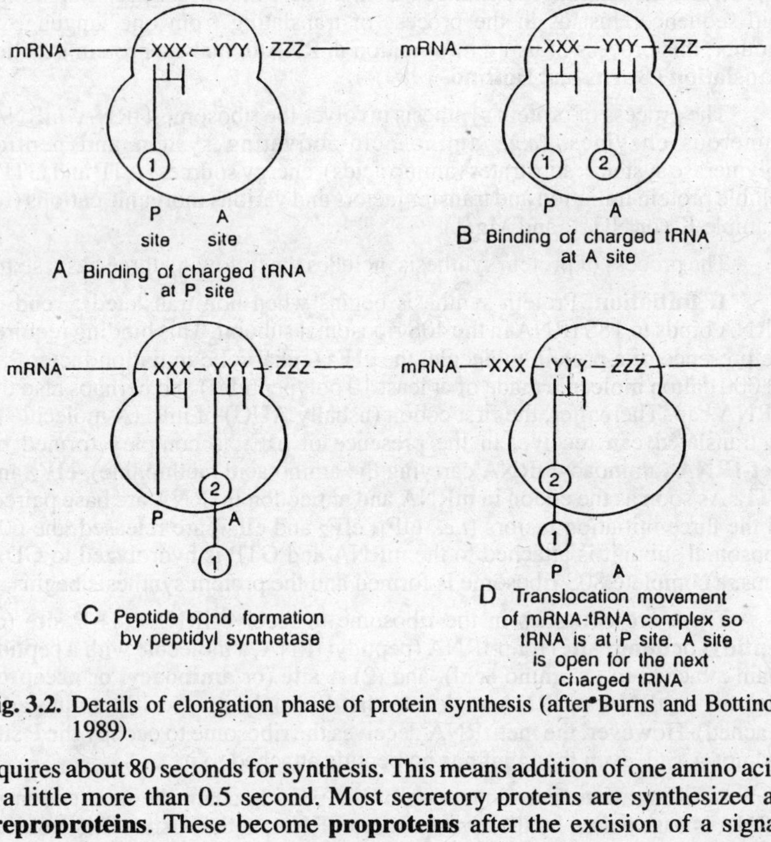

Fig. 3.2. Details of elongation phase of protein synthesis (after Burns and Bottino, 1989).

requires about 80 seconds for synthesis. This means addition of one amino acid is a little more than 0.5 second. Most secretory proteins are synthesized as **preproproteins**. These become **proproteins** after the excision of a signal peptide (15 to 30 amino acid long on NH_2 end of proprotein). Further cleavage at specific points leads to formation of final protein molecule. In some cases (*e.g.*, ACTH or adrenocorticotropic hormone and β-lipotropin) it is formed by a further excision of the protein molecule. In eukaryotes a unit of translation, the **translon**, normally comprises mRNA for one polypeptide chain only.

The life of any mRNA molecule is finite. It continues to function in the manner described as long as ribosomes attach to its 5′-end, but this terminus is vulnerable to **ribonucleases**. These are enzymes that hydrolyze RNA, and degrade it into its constituent nucleotides which may then be used to construct wholly new and different mRNAs.

REVISION QUESTIONS

1. How much of total DNA is transcribed, and how much of it codes for protein ?
2. Differentiate between the terms replication, transcription and translation.
3. Describe in brief the process of protein synthesis.
4. Write short notes on the following:
   1. DNA molecule; 2. RNA molecule ; 3. Genetic code.

4 MALE GONADS AND SPERMATOGENESIS

4-1. DEFINITION OF SEX

The embryogenesis (embryonic development) of a sexually reproducing multicellular animal is prefaced by the **gametogenesis**, *i.e.*, the formation and ripening of two highly dissimilar and specialized **sex-cells** or **gametes**, namely a large-sized, non-motile, nutrient-filled cell the **ovum** or **egg**, and a small-sized, motile, sex-cell, the **spermatozoon** or **sperm**, both of which unite and give origin to a diploid zygote. In some animals both kinds of gametes (*viz.*, spermatozoa and ova) are produced by a single individual, which is called **monoecious** or **hermaphrodite**. But in majority of cases, both kinds of gametes are produced by different individuals of an animal species, so that, the species is bifurcated into two **sects, sections** or **sexes** (L. *sexes*=section or separation) —a section of species includes those individuals which produce ova and are called **female sexes**, and other section of species, includes those individuals which produce sperms and are called **male sexes**. This kind of separation of sexes is called **gonochorism**. The organisms in which both male and female sex cells or gametes (*i.e.*, sperms and ova) are produced by different individuals are called **dioecious**.

The kind of sex-cells (spermatozoa and ova) and those organs which are immediately concerned with their production, *i.e.*, the **male gonads**, the **testes** and the **female gonads**, the **ovaries** form the **primary sexual characters** of each sex. Besides these primary sexual characters the male and female sexes differ from each other in many somatic characters, which are called **secondary sexual characters**. In man, for example, the secondary sex differences include the accessory sex organs (such as, penis, epididymis, vas deferens, prostate and seminal vesicles of human male, and clitoris, vagina, uterus and oviduct of human female) and various other morphological and behavioural differences. Thus, the breasts or mammary glands, well developed in female and rudimentary in male; the voice, pitched higher in female than in male; and body hair, sparse in female and relatively abundant in male. Further, such secondary sexual differences are well evident in the distribution of subcutaneous fat, which is generally more abundant in human female than man; and also in details of muscular and skeletal structures. Such a phenomenon of morphological differentiation or specialization in secondary sexual characters between male and female individuals of a species, is called **sexual dimorphism**. Recent studies have revealed that sexual dimorphism has certain genetical and hormonal basis.

4-2. ORIGIN OF PRIMORDIAL GERM CELLS

Both types of gametes—egg and sperm—originate as descendants of the **primordial germ cells (PGC)**. There exists much controversy among embryologists regarding the place of origin of primordial germ cells. Briefly, the main point of controversy is that whether, as **August Weismann** originally suggested, a germinal plasm set aside in embryonic development, later gives rise to all the primordial germ cells that migrate into the developing gonad, or, there are functional germ cells developed in close association with or in the gonad at later various stages of life. Now, it is well established that in many invertebrates and vertebrates, the primordial germ cells (PGC) are set aside from all the other cells of the body, the somatic cells, very early in development. Such early segregation of PGC comes about through the occurrence of special regions in the cytoplasm of the egg (the **germ plasms**) which cause the cells that come to include some of this cytoplasm to differentiate as germ cells or, in some cases, accessory cells connected with them (**Ede**, 1978). Pattern of origin of primordial germ cells in different animal groups can be studied as follows:

4-2.1. Origin of Primordial Germ Cells in Invertebrates

Among the invertebrates the early origin or segregation of the germ cells remains obscure in coelenterates and annelids. In other invertebrates, notably in the ascarid worm and dipterous insects, there is early segregation of the germ cells. In both cases segregation of germ plasm takes place by different methods and at an early stage of development.

1. Early segregation of germ plasm by chromosome diminution in *Ascaris*. The fertilized ovum or egg of the nematode *Ascaris megalocephala univalens* has two large chromosomes (which is a diploid chromosome number). The painstaking studies of **Boveri** (1899) on *Ascaris* development, showed that the future germ cells can be recognised because the chromosomes remain intact in them, while somatic cells undergo a process called **chromosome diminution**. *Ascaris* germ (gamete-forming) cells have only two large chromosomes which have a diffuse centromere (**holocentric chromosomes** with multiple microtubule attachement sites). In the cells that will become somatic (body-forming), the central part of the chromosomes (those having repetitive sequences of satellite DNA) breaks into multiple small chromosomes and the heterochromatic ends (which do not have microtubular attachment sites) remain as visible chromatin masses in the cytoplasm of the daughter cells of the **animal cell** or **S_1 AB blastomere** (Fig. 4.1). These heterochromatic ends of chromosomes are eventually lost from the animal cells. The germ line (*i.e.*, **vegetal cell** or **P_1 blastomere**) does not undergo such a process of chromatin diminution. At each cleavage division a single cell retains its chromosomes intact and is called **germ line stem cell**. For example, at the fifth cleavage the embryo of *Ascaris* has 31 somatic cells and one germ cell (**De Robertis** and **De Roberties, Jr.**, 1987). According to one study, in *Ascaris lumbricoides* also about 27 per cent DNA is lost during the process of chromosome diminution (see **Goel**, 1984).

2. Determination of germ cells in insects. Insect eggs have been shown to contain cytoplasmic determinants (germ plasm) whose function is to segregate germ cells from somatic cells early in embryogenesis. For the first few hours after fertilization, insect eggs undergo nuclear but not cytoplasmic division, producing many (up to 3500) nuclei in a common pool of cytoplasm, that is, a syncytium. The nuclei accumulate around the periphery of the zygote to form a cellular blastoderm; each nucleus eventually becomes surrounded by a cell membrane. The determination of primordial germ cells in insects takes place by the following two methods:

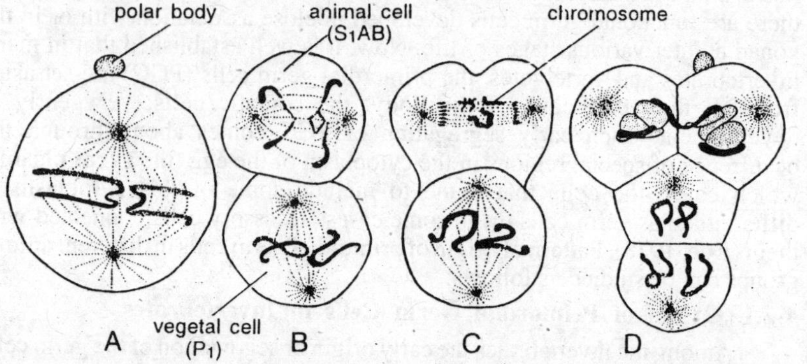

Fig. 4.1. Chromosomal diminution and determination of primordial germ cells during cleavage of *Ascaris*. A—Egg at metaphase of first cleavage with polar body at animal pole; B—Two-celled stage, with chromosome diminution occurring in animal cell or S_1AB blastomere but not in vegetal cell or P_1 blastomere; C—Dividing animal cell; D—T-shaped four-celled stage (After Berrill, 1971).

A. Determination of primordial germ cells by chromosomal diminution. The embryos of two species of the gall midges or fungus gnats (*viz.*, *Mayetiola destructor* and *Wachtiella persicariae*) undergo a process of chromosomal diminution similar to *Ascaris* embryos. In each of these species of gall wasps, 32 of the original 40 chromosomes of the oocyte are eliminated from somatic cells but are retained in germ cells.

At the 16-cell stage of *Mayetiola* (Hessian fly), 2 nuclei from the syncytium enter the cytoplasm at the posterior end of the egg where they are protected from undergoing fifth nuclear division. These nuclei retain all their chromosomes. The remaining 14 nuclei divide and in so doing lose 32 of their 40 chromosomes. These 32 heterochromatic chromosomes fail to move toward the spindle poles during the first cleavage division, implying defective kinetochore function. Instead, they accumulate at the equator of the spindle where they disintegrate. Those cells that have undergone chromosomal diminution become somatic cells. Those that retain a full set of chromosomes become germ cells.

The presumption that the cytoplasm at the posterior pole of the egg protects chromosomes from undergoing diminution has been tested by ligating eggs so that nuclei cannot enter the posterior pole cytoplasm. Chromosomes of such nuclei do undergo diminution, with the result that the adult midges are sterile. In another experiment, centrally placed nuclei are protected from chromosomal diminution if posterior pole cytoplasm is centrifuged in the centre of egg. Therefore, the posterior pole cytoplasm is germ plasm and acts to commit cells as germ cells by ensuring that all their chromosomes are preserved intact.

B. Determination of primordial germ cells by determinants of polar plasm. In fruit fly, *Drosophila*, chromosome elimination does not occur in somatic nuclei. In it, the egg is torpedo-shaped having zygote nucleus situated

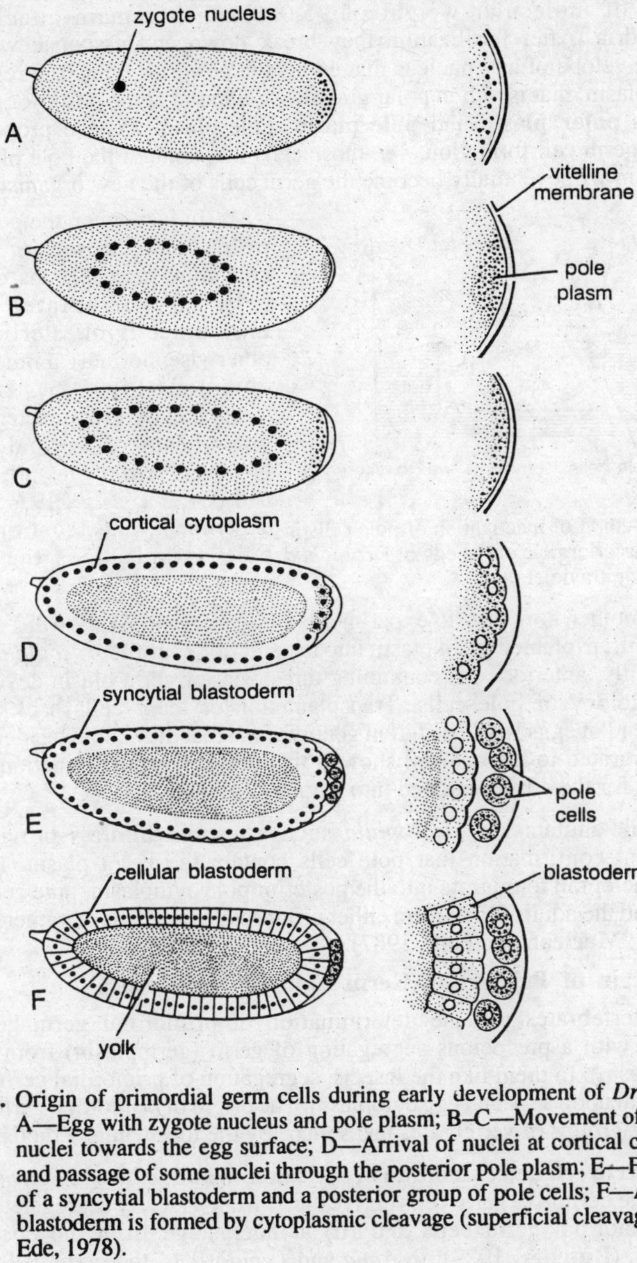

Fig. 4.2. Origin of primordial germ cells during early development of *Drosophila.* A—Egg with zygote nucleus and pole plasm; B–C—Movement of dividing nuclei towards the egg surface; D—Arrival of nuclei at cortical cytoplasm and passage of some nuclei through the posterior pole plasm; E—Formation of a syncytial blastoderm and a posterior group of pole cells; F—A cellular blastoderm is formed by cytoplasmic cleavage (superficial cleavage) (After Ede, 1978).

towards the anterior end (Fig.4.2). At its posterior pole, the *Drosophila* egg has a specialised region of cytoplasm which contains special granules called **polar granules** or **oosomes** (or **germ cell determinants**). Polar granules are meshes of fibrous elements about 0.5 to 1.0 μm in diameter and RNA-rich proteins of molecular weight of 95,000 which remain attached to mitochondria. After fertilization they break down and associate with the nuclear envelope of any nucleus that enters the posterior pole. The region of egg cytoplasm, that is rich in polar granules and found at the posterior pole, is known as **polar plasm** and **pole plasm**. Pole plasm has the property of inducing germ cell formation, *i.e.*, those cells that contain the pole plasm or germ plasm will eventually become the germ cells of the new organism.

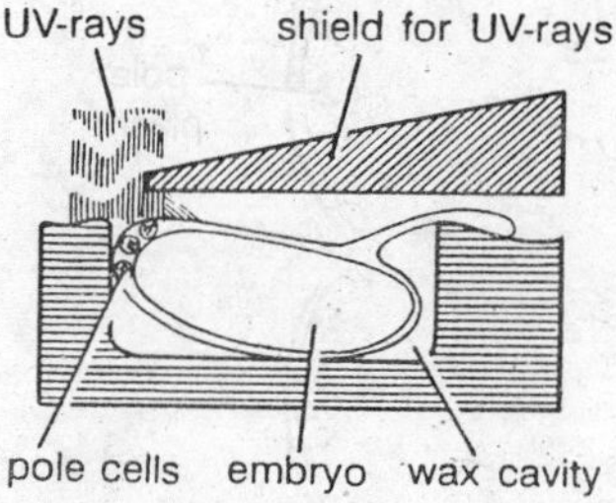

Fig. 4.3. Method of inactivation of pole cells in posterior pole of the egg of *Drosophila* by ultraviolet rays.

In fact, when the posterior poles of *Drosophila* eggs containing pole plasms or pole cells are irradiated with ultraviolet light, sterile (but otherwise normal) animals are obtained (Fig. 4.3). If UV-treated eggs are injected with pole-plasm of normal eggs, however, fertile flies are obtained (**Okada** *et al.*, 1974). This experiment, thus, confirms that pole cytoplasm is the germ plasm.

In another now classic experiment, **Illmensee** and **Mahowald** (1974) injected or transplanted germ plasm into the anterior ends of *Drosophila* eggs. Subsequently, anterior cells containing this posterior germ plasm developed the morphology of pole cells. Transplantation of these cells back to the posterior end of eggs with a different genotype (so that they could be identified as they migrated to the gonads) showed that they are indeed functioning as germ cells for they differentiated into gametes.

Sterile mutants of *Drosophila* such as **grandchildless** provide an independent confirmation that pole cells contain the germ plasm. In this mutant, nuclei fail to migrate into the posterior pole cytoplasm, pole cells fail to form, and the adults fail to form either primordial germ cells or gametes (see **Ede**, 1978; **Maclean** and **Hall**, 1987).

4-2.2. Origin of Primordial Germ Cells in Vertebrates

In vertebrates too the determination of primordial germ cells is correlated with a precocious segregation of germ (germplasm) from soma (somatoplasm). In them like the insects, segregation of primordial germ cells is dependent upon a special egg cytoplasm. The origin of primordial germ cells in different groups of vertebrates takes place by the following methods:

1. Germ cell determination in cyclostomes and fishes. In *Petromyzon marinus*, the definitive germ cells derive either from–(i) early segregated cells, called primordial germ cells and (ii) at later stage from the coelomic epithelium, (**Butcher**, 1929). In *Amia* and *Lepidosteus*, the germ cells arise early in development from the endoderm and migrate from there through the endoderm and overlying mesoderm to their final location within the developing gonad (**Allen**, 1911).

2. Germ cell determination in anuran amphibians. Like the insects, the cytoplasm at the vegetal pole of fertilized eggs of anuran amphibians, such as frogs and toads, is cytologically distinctive. In frog eggs, this unique cytoplasm (germinal cytoplasm) stains differently from the surrounding vegetal yolk and can be followed during early cleavage. If the vegetal pole of the egg of the frog (*Rana pipiens*) is irradiated with ultraviolet light shortly after fertilization, it renders the frogs sterile. Gonads are formed, but they lack primordial germ cells. Replacement of irradiated cytoplasm with vegetal pole cytoplasm from unirradiated eggs allows such animals to produce functional primordial germ cells (**L.D. Smith**, 1966). Thus, in frogs and toads, vegetal pole cytoplasm contains germ plasm.

In anurans, the blastomeres or cells bearing germ plasm are larger than most other blastomeres, but their mitotic behaviour is like the somatic cells, at least in early stages. A cytoplasmic component, sensitive to UV radiation and necessary for normal gamete development is segregated during the first cleavage divisions into presumptive germ cells. Later, large **primordial germ cells** carrying the stainable germinal plasm can be localised in the vegetal ''belly'' region of the embryo (Fig. 4.4). In young tadpoles they can be seen migrating in single file along the border of intestine. Their identity as germ cells has been shown by the following experiment: A belly region from an

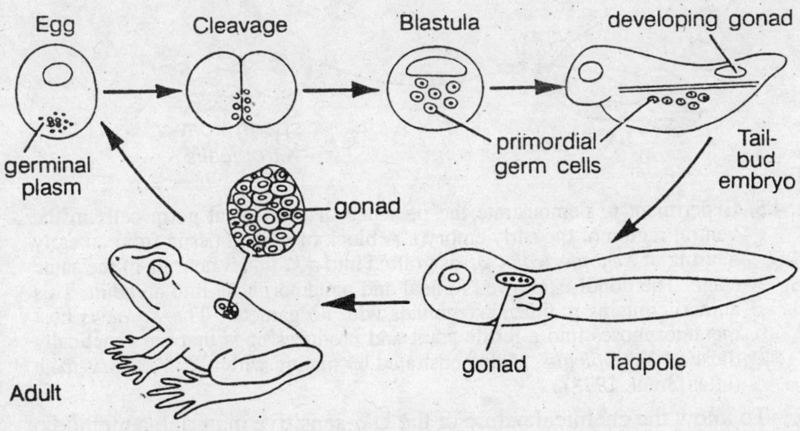

Fig. 4.4. Continuity of germ plasm in frog. Germ plasm, localised in the vegetal pole of eggs, enters several blastomeres during cleavage. These become primordial germ cells which migrate in the wall of the larval intestine to the developing gonad. They invade the gonad and multiply as oogonia or spermatogonia, respectively. Maturation is completed in the adult and during oogenesis additional germ plasm is synthesised and stored in the oocytes (after Grant, 1978).

embryo of one *Xenopus* strain is replaced with a block of ventral tissue from another strain bearing a nucleolar marker gene (Fig. 4.5). Cells containing this gene have only one nucleolus rather than two and can be easily identified in the embryo. The embryos develop into animals whose gonads contain gametes genetically identical to the donor strain and not the host; that is, they carry the nucleolar marker. Primordial germ cells in the graft populate the **host gonad** and generate gametes characteristic of the donor strain. Most importantly, when these animals are mated their progeny resembles the strain from which the graft was derived and not that of the host (see **Grant**, 1978).

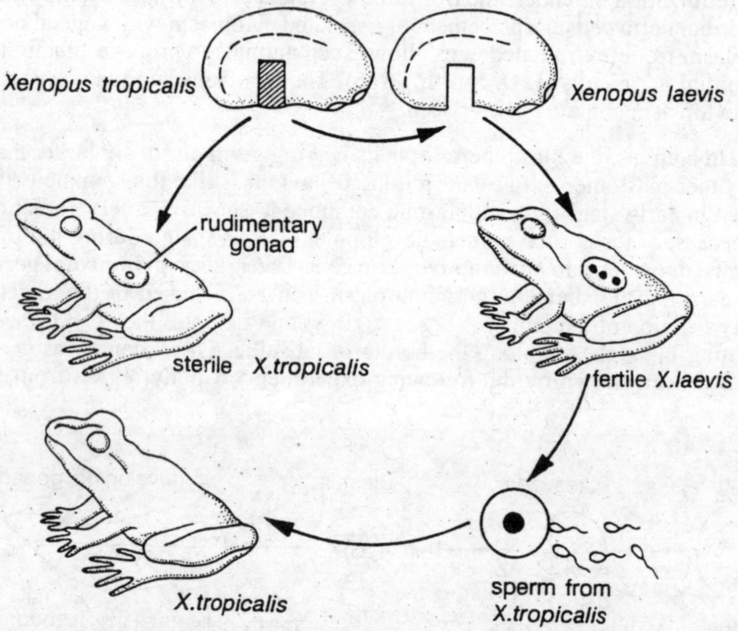

Fig. 4.5. Experiment to demonstrate the presence of primordial germ cells in the ventral region of the early embryo. A block of ventral tissue from an early neurula of *Xenopus tropicalis* is grafted into a *X. laevis* neurula of the same stage. The donor is allowed to heal and metamorphose into an adult. This animal contains rudimentary gonads with no gametes. The *X. laevis* host metamorphoses into a fertile adult and produces eggs that are genetically those of *X. tropicalis*, as demonstrated by mating with a *X. tropicalis* male (after Grant, 1978).

To know the chemical nature of the UV-sensitive material, a method of **action spectrum analysis** was employed, *e.g.*, eggs were irradiated at several different UV wavelengths and the extent of inhibition of germ cell formation was measured for each wavelength. Peak inhibition was obtained at 2537 A^o which is the absorption peak for the nucleic acids. Since the vegetal pole region does stain positively for RNA, the active germ plasm may be RNA, a ribonucleoprotein, or even DNA. An electron micrographic study of the vegetal pole reveals small clusters of dense granules that resemble polar granules in insect eggs. These stain for RNA and do migrate into cleaving blastomeres as predicted by the ultra-violet studies.

Thus, during frog development, the germ plasm is incorporated into the germ cells that lie within the presumptive endoderm. These cells then migrate into the floor of the blastocoele and from there into the mesoderm, and ultimately into the developing gonads.

3. Germ cell determination in urodele amphibians. In contrast to anurans, the germ cells of urodeles arise in the mesoderm and they arise from presumptive somatic cells after an inductive interaction with embryonic endoderm. Experiments involving separation and reassociation of various regions of urodele blastulae have shown that the ventral endoderm provides the necessary stimulus for germ cells to differentiate, a finding that has been established in at least four species (*Ambystoma mexicanum*, *Pleurodeles waltlii*, *Triturus alpestris* and *T. cristatus*) (**Nieuwkoop** and **Sutasurya**, 1979; **Maufroid** and **Capuron**, 1984).

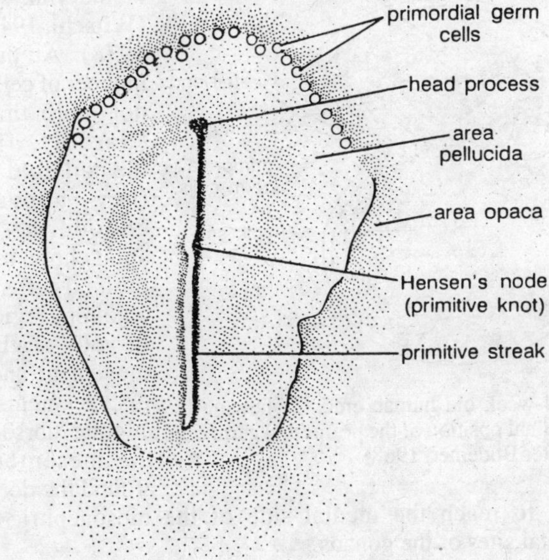

Fig. 4.6. Original position of the primordial germ cells on the margin between the area pellucida and the area opaca of chick embryo in the head process stage (after Balinsky, 1976).

4. Origin of primordial germ cells in reptiles and birds. In reptiles and birds, the germ cells are found to have an extragonadal origin. For example, in the chick embryo the primordial germ cells arise in the posterior margin of the disc of embryonic cells which appear even before incubation, and from there migrate to a position in a crescent-shaped area of extra-embryonic cells situated anterior to the head-fold of the early embryo (Fig. 4.6). Then they make their way between the endothelial cells of the extra-embryonic blood-vessel walls and are carried with the circulating blood until they reach the neighbourhood of the developing gonads. There PGC leave the blood vessels and move into the gonads, probably by active amoeboid movement and possibly oriented towards them by chemotaxis (**Ede**, 1978).

5. Origin of primordial germ cells in mammals. Studies concerning the site of origin of primordial germ cells have been performed in various mammals and it has been found that in most cases they have an extra-gonadal origin. Earlier it was maintained that the germ cells of the mouse originated from the germinal epithelium of the gonad; but recent studies suggest that in mouse, as in other vertebrates, the germ cells originate into endoderm before migrating into the gonad. By utilizing the marking capacity of high alkaline phosphatase enzyme contents of the germ cells, it was possible to determine their location within the caudal endoderm and their subsequent migration into the developing gonad. The primordial germ cells can be identified in human embryos at the 13-somite stage (Fourth week), in the endoderm at the caudal end and in the adjacent part of the yolk sac (Fig. 4.7) (**Witschi**, 1948; **Fujimoto**, 1978). At this stage the number of cells is probably not more than 20 to 30 (**Hardisty**, 1967). When the tail fold has formed they appear in the endoderm and the splanchnic mesoderm of the hind gut as well as in the adjoining region of the wall of yolk sac. By amoeboid movement and by growth displacement, these PGC migrate dorso-cranially in the mesentery, passing around the dorsal angles of the coelom to reach the medial side of. the mesonephric ridges—the developmental sites of the gonads.

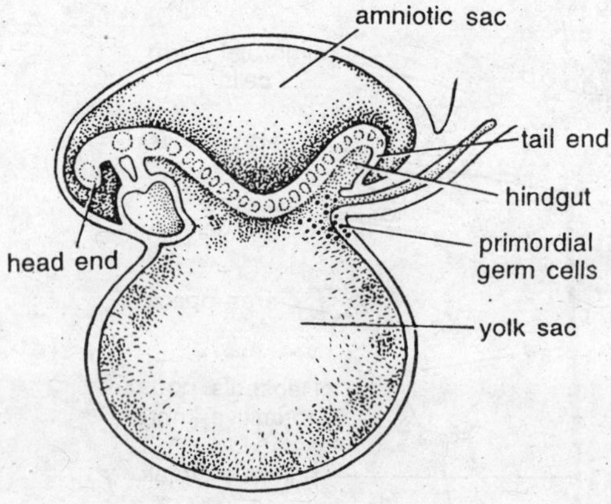

Fig. 4.7. A 4-week old human embryo showing the original position of the primordial germ cells (After Bodemer, 1968).

In most vertebrates, mitosis in the PGC is arrested after their early segregation, to be renewed only when they reach the genital primordia. However, in mammals there is no such arrest, and the primordial germ cells proliferate both during and after migration to the mesonephric ridges; cells which do not complete this migration degenerate. After segregation the PGC are often termed as **primary gonocytes**, which in turn divide to form **secondary gonocytes**.

The early association of PGC with the endoderm suggests that they are in fact derived from primitive endodermal cells or perhaps from common stem cells (**Franchi** *et al*., 1961). Alternatively, the endoderm may exert an inductive effect on their development and differentiation without being their actual source (**Carlon** and **Stahl**, 1973; **Williams** and **Warwick**, 1980).

4-3. MORPHOLOGY OF SPERMATOZOON

The male gamete or spermatozoon (Gr., *sperma* = seed + *zoon* = animal) of most animals is a highly specialised cell (Fig. 4.10). The main functions of a sperm during sexual reproduction are (1) to swim to the egg; (2) to fuse with the egg surface; (3) to cause the egg to start its development; and (4) to contribute a haploid set of chromosomes having paternal genes to the diploid offspring. In addition, the sperm typically provides a pair of centrioles to the fertilized egg, which is used in the mitotic divisions (cleavage) that follow fertilization. **Brachet** (1974) has compared a spermatozoon with a bacteriophage. Both of them inject DNA, the genetic material into the recipient cell.

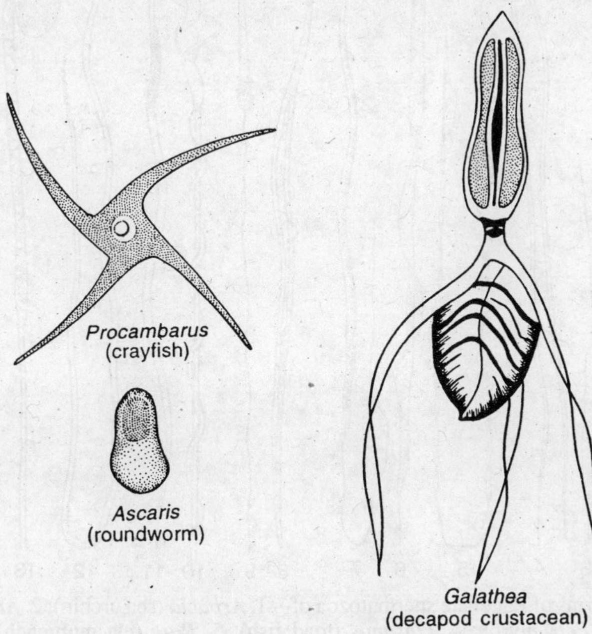

Fig. 4.8. Some non-flagellate spermatozoa (after Berrill, 1971).

The spermatozoon shows a remarkable uniformity of structure from lower invertebrates to mammals, with some notable exceptions (Fig. 4.8). The main structural subdivisions of a typical sperm are the head, mid-piece and tail (Fig. 4.10), all of these are contained by a continuous plasma membrane, like any other living cell:

1. Head. The shape of the sperm head varies considerably in different species (Fig. 4.9). It may be ovoid and flattened from the sides (in man, bull and rabbit), drawn into a scimitar shape with a pointed tip (in rodents and amphibians) or spirally twisted like a corkscrew (in birds and some molluscs; Fig.4.9 and Fig.4.11). The head contains two main parts, the nucleus and the acrosome, both performing two basic functions of the sperm—**genetic** and **activating**, respectively.

The **nucleus** occupies most of the available space of the sperm head and its shape ultimately determines the shape of the head of the sperm (Fig.4.10). It has its typical double-membrane envelope. However, the posteriormost part of outer nuclear membrane is sometimes thickened to form a **basal plate**. The nuclear envelope of sperm lacks the nuclear pores for the most part, except on the basal plate of several mammals. In some animal species, the redundant portion of the nuclear envelope forms membranous scrolls (the lamellar bodies) in the sperm neck.

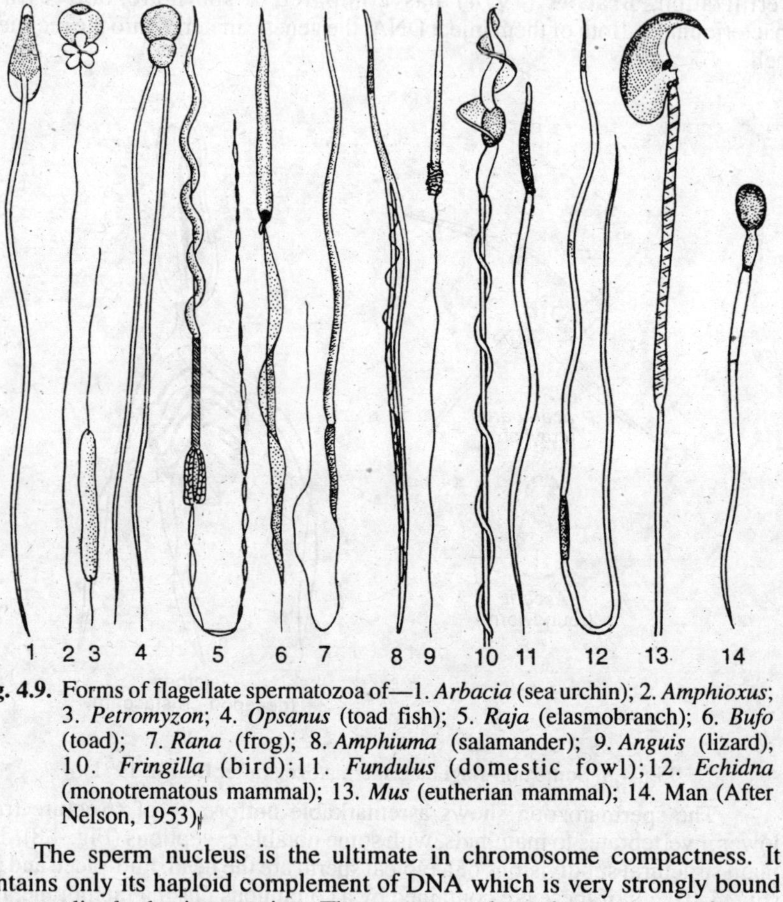

Fig. 4.9. Forms of flagellate spermatozoa of—1. *Arbacia* (sea urchin); 2. *Amphioxus*; 3. *Petromyzon*; 4. *Opsanus* (toad fish); 5. *Raja* (elasmobranch); 6. *Bufo* (toad); 7. *Rana* (frog); 8. *Amphiuma* (salamander); 9. *Anguis* (lizard), 10. *Fringilla* (bird); 11. *Fundulus* (domestic fowl); 12. *Echidna* (monotrematous mammal); 13. *Mus* (eutherian mammal); 14. Man (After Nelson, 1953).

The sperm nucleus is the ultimate in chromosome compactness. It contains only its haploid complement of DNA which is very strongly bound to exceedingly basic proteins. The sperm nucleus does not contain any nucleolus, RNAs and fluid contents. Instead of usual histone proteins of the nucleosomes, there are either 'very arginine-rich' histones or **protamines** (polypeptides with a very high content of arginine) (**Brachet**, 1974). The protamines are small, basic, DNA-binding proteins that serve to bring the DNA helices into closest possible proximity (*i.e.*, the protamines cause the high degree of condensation of sperm genome; **Warrant** and **Kim**, 1978). Thus, in a mature sperm, DNA-basic protein complex exists almost in a crystalline state and genetically such a DNA is completely repressed and inactive (*i.e.*, it neither replicates nor transcribes).

At the anterior extremity of the sperm is a special granule, called **acrosome** (Gr. *akron* = extremity; *soma* = body). The acrosome occupies the main space between anterior half of the nucleus and the plasma membrane of sperm tip. The shape and size of the acrosome vary among different species (Fig.4.11). The acrosome is bounded by a unit membrane and is formed by the Golgi apparatus. It contains a number of **acid hydrolases** (hydrolytic enzymes; also called **sperm lysins,** see **J. Brachet**, 1974) such as **acid phosphatase, cathepsin** (a proteolytic enzyme), **hyaluronidase** (which attacks the mucoproteins) and so on. The acrosome in mammals contains **acrosomin**, a lipoglycoprotein complex made of hyaluronidase and acrosin (zona lysin). The **hyaluronidase** enzyme in ram (male sheep) is a mucoprotein of a molecular weight of about 60,000 daltons and has pH optimum at 3.5—6.8. The **acrosin** enzyme is a lipoglycoprotein complex having trypsin-like property, a molecular weight of about 30,000 daltons and in ram a pH optimum at 8.0—8.5. The acrosome of sea urchin, *Strongylocentrotus*, contains an insoluble protein of molecular weight of 30,500 daltons. It is called **bindin**, since it helps in binding sperm to egg. The acrosome, in view of its enzyme contents and also of its origin from the Golgi apparatus, can be regarded as a specialized lysosome (**Brachet**, 1974; **De Robertis** and **De Robertis**, 1980).

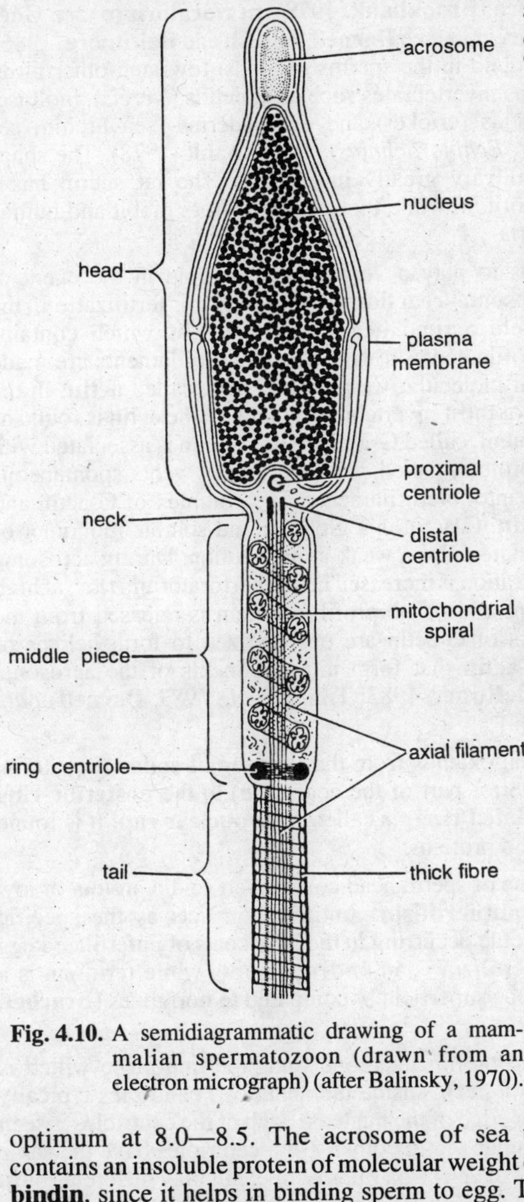

Fig. 4.10. A semidiagrammatic drawing of a mammalian spermatozoon (drawn from an electron micrograph) (after Balinsky, 1970).

Various synonyms are often used for the fluid-filled space that exists between the acrosome and the nucleus. These are **axial body** (**Galtsoff** and **Philpott**, 1960), **acrosomal cone** (**Kaye**, 1962), **acrosome granule** (**Brachet**, 1974), **subacrosomal space** (**Brookbank**, 1978), **perforatorium** (see **Goel**, 1978) and **periacrosomal region** (see **Darnell, Lodish** and **Baltimore**, 1986). Such a perforatorium is found in the sperms of toads, fowl and other birds, most mammals and various invertebrates such as annelids (*Nereis*), molluscs (*Spisula*, oyster), arthropods (cricket) and echinoderms (sea urchin, sea cucumber and sand dollar, *Echinarachnius*) (**Brookbank**, 1978). The shape and size of perforatorium vary greatly in animals. The rat sperm has a particularly large perforatorium, while the sperms of frog, rabbit and human have no distinct perforatoria.

A perforatorium has to play a very significant role in the event of fertilization. During **acrosome reaction** at the time of fertilization, the perforatorium develops into a rigid **acrosomal filament** which contains fibrous material, the microfilaments. In fact, these microfilaments are made up of a globular protein of molecular weight of 42,000, called **actin**. In the perforatoria of the sperms of most invertebrates, much of the actin is found as unpolymerized protein subunit, called **G-actin**. The G-actin is associated with a low-molecular weight protein, called **profilin**, that prevents spontaneous polymerization of G-actin into microfilaments. The complex of G-actin and profilin is called **profilactin**. G-actin is a globular and soluble monomer of actin and it is held in this state only in weak ionic solution. During acrosome reaction, when salt concentration is increased inside perforatorium (*i.e.*, at high Mg^{2+}, K^+ and Na^+ concentration), the profilin protein is released from the G-actin and the monomers of G-actin are polymerized to form helices of insoluble fibres, called **F-actin** that form microfilaments of the acrosome filament (**Engelman** and **DeRosier**, 1983; **Tilney** *et al., 1983;* **Darnell** *et al.*, 1986).

The narrow space that extends from the **acrosomal collar** (equatorial segment or thickened posterior part of the acrosome) to the **posterior-ring** (also called **nuclear** or **striated ring**) is called **post-nuclear cap**. It is found to contain phospholipids and proteins.

The plasma membrane of sperm head contains an acidic protein of low molecular weight, called **antifertilizin**. Antifertilizin acts as the specific receptor for **fertilizin** molecule occurring in the jelly coats of unfertilized egg of the same species. Antifertilizin is an **androgamone**, while fertilizin is a **gynogamone** and both can be superficially compared to hormones (**Brachet**, 1974).

2. Mid-piece. Behind the nucleus, one finds the mid-piece which is connected to the head with a **neck**. Inside the neck, two centrioles typically pressed against the posterior edge of the nucleus. Each of the centrioles is seen in a pair with one at right angles to the other. Both centrioles have the same complicated structure under the electron microscope, but play different roles. The **anterior** or **proximal centriole** of the sperm typically is donated to the egg and helps in forming the first mitotic spindle after fertilization. It lies at a right angle to the sperm axis in a depression in the posterior surface of nucleus. (Histologically, therefore, proximal centriole should be included in the domain of sperm head; **Balinsky**, 1981). The **posterior centriole** or **distal centriole** is responsible for the formation of the microtubules (axoneme) of the sperm

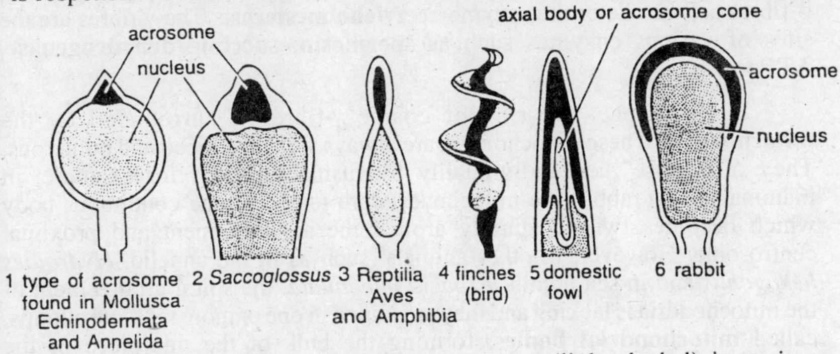

Fig. 4.11. Acrosomes (dark-shaded) and nucleus (light-shaded) in various spermatozoa (After Berrill, 1971).

tail (or flagellum). It lies at the basis of the flagellum and acts like the **basal body**. Sometimes, both centrioles are completely surrounded, or arched over, by the cephalic ends of γ-fibres (gamma or coarse fibres) of sperm tail (Fig.4.12).

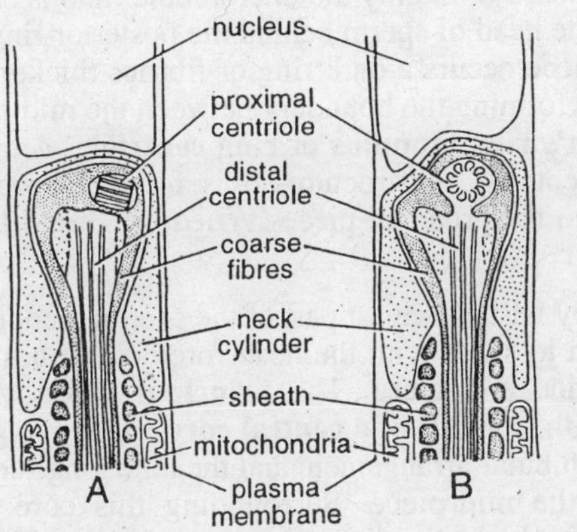

Fig. 4.12. Vertical sections of snake spermatozoon, A—parallel and; B—at right angles to the long axis of the proximal centriole (After Berrill, 1971).

The distal centriole and the proximal part of the axoneme or axial filament lie in the mid-piece (or middle-piece) of the spermatozoon. The axial filament of the sperm tail has the same organization as the axial filaments of flagella, cilia and kinocilia. Thus, it has a pair of longitudinal fibres (called **beta-** or **β-fibres**) along its middle and a ring of nine pairs of longitudinal fibres (called **alpha** or **α-fibres**) surrounding it. These fibres are anchored in the distal centriole in the same way that the fibres of a cilium or a flagellum are connected to their basal body.

In mammals, the α-fibres of axial filament are accompanied on the outside by nine much thicker fibres, called **γ-fibres** or **gamma fibres** or **coarse fibres**. The γ-fibres are wedge-shaped in cross section and they start in the middle piece but do not reach up to the tip of the sperm flagellum. The γ-fibres are also present in some other animals such as some birds, snakes, molluscs (*Helix*) and insects (honeybee, grasshopper and *Drosophila*) but are rudimentary in snail, squid and rooster. They are altogether absent in marine invertebrate's sperm.

The microtubular α-, β- and γ-fibres of sperm tail are the sites of various enzymes. For example, α-fibres are the site of **ATPase enzyme**, while, β-fibres are the site of the enzyme **acetylcholinesterase**. The γ-fibres are the sites of various enzymes such as **spermosin**, **succinic dehydrogenase**, **ATPase**, etc.

In mid-piece, the ring of coarse γ-fibres is surrounded by the **mitochondria**. These mitochondria are always well developed and numerous. They often lose their individuality by fusing together, for example, in mammals (*e.g.*, rabbit) the mitochondria are joined in one continuous body which becomes twisted spirally around the axial filament and proximal centrosome. However, in other animals, such as in the annelid, *Hydroides hexagonus* and in sea urchin, *Arbacia punctulata*, the spiral arrangement of the mitochondria is lacking and they are joined in one or more massive clumps, called **mitochondrial bodies**, forming the bulk of the mid-piece of the spermatozoon. The mitochondria of sperm contain all the respiratory enzymes and are extremely active in oxidative phosphorylations. They actively oxidize either endogenous (*e.g.*, phospholipids) or exogenous (*e.g.*, fructose) substrates and supply the sperm flagellum the ample energy that is used for the propulsion of the spermatozoon.

Around the periphery of mid-piece of the sperm is found a condensed layer of cytoplasm that is composed of mainly the microtubules and is called **manchette**. It also surrounds the head of sperm behind the posterior ring. At the posterior end of the mid-piece occurs a dark ring or fibrous thickenings beneath the plasma membrane, forming the boundary between the mid-piece and the tail. It is called **Jensen's ring**, **annulus** or **ring centriole**. Annulus does not resemble the true centriole in structure. It is found to prevent mitochondria from slipping into the principal piece (**Wheater**, **Burkitt** and **Daniels**, 1979).

3. Tail. The tail usually by far the longest part of the spermatozoon, by its movement causes the sperm to swim with the head foremost. It has two main parts—principal piece and end piece. The **principal piece** which constitutes most of the tail length, consists of a **central core**, comprising the axial filament with 9 + 2 microtubular arrangement and the nine longitudinal or γ-fibres continuing from the mid-piece. Surrounding this core is a microtubular fibrous **tail sheath** which sometime appear as semicircular ribs oriented perpendicular to the long axis of the filament, or as helical coils. In human sperms, two of the γ-fibres are fused with the surrounding ribs so as to form **anterior** and **posterior columns** extending throughout the length of the principal piece. Such an arrangement divides the principal piece longitudinally into two functional compartments, one containing three γ-fibres and the other containing four. This asymmetry is thought to somehow cause more powerful stroke of the tail in one direction, the so-called "power stroke" (**Wheater** *et al.*, 1979).

The **end piece** is merely a short tapering portion of the tail containing only the axial filament covered with cytoplasm and plasma membrane. The sperm tail of most invertebrates is not so well differentiated and has only axial filament.

The sperm cell is totally devoid of stored food (yolk) and protective envelope, which characterize the female gamete, the egg. It is also devoid of most cytoplasmic organelles that sustain cell viability, such as ribosomes and endoplasmic reticulum.

4-3.1. Variations in Sperm Structure

Spermatozoa of certain turbellarians and *Ascaris* (nematode) and microgametes of sporozoans (protozoans) lack the acrosome. Likewise sperms in several species do not possess flagella (tail) and are called **non-flagellate spermatozoa**. Such non-flagellate sperms are found in certain invertebrates such as *Ascaris*, *Procambarus* (crayfish), *Callinectes* (blue crab) and other crustaceans, myriapods and arachnids. The flagellate spermatozoa contain one or two flagella. Most animals contain monoflagellate spermatozoa, but, some animals, such as, *Opsanus* (toad fish) contains biflagellate spermatozoa.

The non-flagellate spermatozoa of the decapod crustaceans (*e.g.*, *Procambarus clarkii*) do not even possess centrioles and mitochondria and are incapable of swimming. Sperms of *Ascaris* and *Polyphemus* (crustacean) are amoeboid. Certain insects such as *Sciara* have only one centriole in their flagellate sperms. In sperms of some fishes and amphibia (*e.g.*, toad and salamander), there is an undulating membrane which stretches along most of the length of the tail (Fig. 4.9).

4-4. SPERM METABOLISM

Spermatozoa are most active and motile cells. Adenosine triphosphate (ATP) is the source of energy for the flagellar movement and possibly also for the basal metabolism necessary for maintaining the effectiveness of a sperm. Therefore, the primary role of the sperm metabolism is to generate and maintain an adequate ATP level. The spermatozoon produces ATP by catabolic processes such as glycolysis and respiration in anaerobic and aerobic conditions, respectively (Fig. 4.13). While respiration in sperm is common, glycolysis is associated mainly with internal fertilization; both the processes are dependent on sperm concentration (density) and sperm movement (motility). In fact, the energy which is supplied by sperm mitochondria is limited in amount and once expended, it obviously cannot be renewed. Spermatozoa are launched, so to speak, like torpedoes with limited range. If

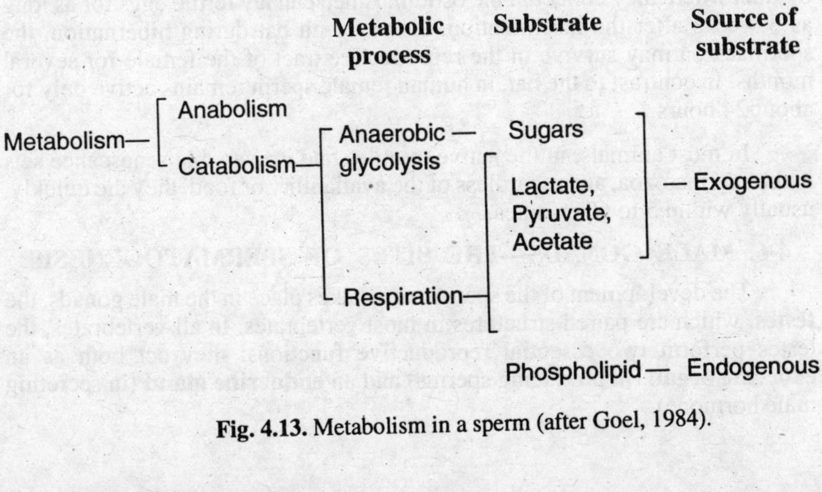

Fig. 4.13. Metabolism in a sperm (after Goel, 1984).

they do not reach the egg within the allotted time, they exhaust their energy supplies and die. In mammals, they are aided in covering the long distance through the female reproductive tubes by the contractions of the wall of the tube and also they get exogenous substrates for their energy metabolism for the prolonged time.

For respiration, the complete cytochrome system and related enzymes are present in the cytoplasm and mitochondria of sperm mid-piece. The sperms consume oxygen to oxidize either (i) the fatty acids (which are liberated from intracellular lipid reserves) or glycogen (sperm of gastropod, *Paracentrotus*, teleost fish and *Rana* (frog) contain large reserves of glycogen) in **endogenous respiration**; or (ii) lactic acid, glycerol, etc., (which are taken directly from outside) in **exogenous respiration**. However, in the absence of oxygen, for example, when the sperms are densely packed in a fresh mammalian ejaculate, the sperm uses glycolysis for ATP production. In glycolysis, extracellular glycolyzable sugars (such as fructose) are used. There are also recent evidences of anabolism, that is, RNA and protein synthesis, but only in mitochondria. No anabolism occurs in rest of the sperm cell, *i.e.*, sperm nucleus and sperm cytoplasm.

Both α-fibres and γ-fibres of sperm tail have ATPase enzyme for the hydrolysis of ATP molecules to generate energy for the flagellar movement. γ-fibres contain **actin** and **myosin** proteins and so form the contractile elements of mammalian sperm. In sperms, where γ-fibres are missing, α-fibres are involved in flagellar movement. The function of central β-fibres is not certain, however, these are believed to be involved in conduction and coordination of contraction or in providing rigidity to the flagellum.

4-5. SPERM LONGEVITY

The fact that sperm is so tiny, contains little or no energy stores, and expends energy as it moves about, definitely limits its functional life. Therefore, in dormant, non-motile state in the male or female reproductive tracts of honeybees, birds and bats and also in cold storage, some sperm can live for very long periods. For example, the sperms in honeybees (queens) can live up to five years. Likewise, in hen, the sperms migrate the length of the oviduct where they collect in the ostium. A hen can lay fertile eggs for as long as 3 weeks after the insemination. However, in bat during hibernation, the spermatozoa may survive in the reproductive tract of the female for several months. In contrast to the bat, in human female sperm remains active only for about 24 hours.

In most animals, in the active, motile state irreversible senescence sets in the spermatozoa, and regardless of the availability of food, they die quickly, usually within 5 to 60 minutes.

4-6. MALE GONADS—THE SITES OF SPERMATOGENESIS

The development of the spermatozoa takes place in the male gonads, the **testes**, which are paired structures in most vertebrates. In all vertebrates, the testes perform two essential reproductive functions: they act both as an **exocrine organ** (in producing sperms) and an **endocrine gland** (in secreting male hormone).

The testes of most vertebrates are located in the upper part of the abdominal cavity lateral to or just anterior to kidneys (*e.g.*, frog). A double fold of peritoneal mesentery, the **mesorchium**, usually suspends them from the dorsal wall. In the majority of mammals, however, the testes do not remain in this position within the abdominal cavity, but descend posteriorly into the **scrotal sacs** (which are small, posteroventral pouch-like extensions of abdominal wall and cavity), *e.g.*, rabbit, cat, bull, man, etc. The testes are lodged in the scrota to keep them at a temperature lower (*viz.*, 1 to 8°C lower) than the abdominal temperature. This is essential as the mature spermatozoa survive best at a temperature few degrees lower than the abdominal temperature. In fact, while spermatogonia are resistant to high temperature and die only after prolonged exposure to heat, mature spermatozoa are sensitive to the high abdominal temperature. Scrotal sacs are not present in those mammals which have lowest body temperature among the mammals. These mammals are monotremes, whale, elephant, seals and rhinoceros. In them, the testes exist in the abdominal cavity throughout life. In certain mammals which have moderate body temperature, *i.e.*, bats, moles and rodents, testes descend into the scrotal sacs where they stay temporarily during the breeding season. The descent of the testes in human males normally occurs about the seventh or ninth month of gestation. If the testes persist within the abdominal cavity and do not descend into the scrota, the condition is called **cryptorchidism**. Birds too lack the scrotal sacs like certain mammals. Although they have a higher body temperature than mammals, even then they are able to function spermatogenically. Possibly, in these animals, the air sacs wrap around the testes in the abdominal cavity and keep them cooler than the other visceral organs (**Phillips**, 1975).

Histologically, each testis is covered by a thin layer of peritoneal epithelium under which lies a tough connective tissue capsule, the **tunica albuginea**. In some animals, septa extend radially to the capsule, thus, partitioning the gland into smaller areas; in other cases the testis is lobular or segmented. Two types of structures, namely **seminiferous** (L. *semen* = seed + *fero* = to carry) **tubules**, which produce the sex cells, and **interstitial tissue**, which secretes androgen (the general term for the male hormone), compose the testis regardless of its shape, size or location.

4-6.1. Seminiferous Tubules

The seminiferous tubules may be short (ampulla) or long and tortuous, but in either case, their walls usually consist of an outer connective tissue capsule, a thin basement membrane and a lining of stratified **germinal epithelium**. Spermatogenesis occurs in these seminiferous tubules. By a series of cell divisions, the **sperm mother cells**, **undifferentiated sex cells**, or **spermatogonia** that are located near the basement membrane, transform into mature spermatozoa and ultimately pass into lumen of the tubules. A cross section through the tubule (Fig. 4.14), reveals many stages of differentiating sperm (*viz.*, spermatogonia, primary spermatocytes, secondary spermatocytes, spermatids and mature spermatozoa), usually in more or less distinct layers, located between the basement membrane and the cavity of the tubule. All these cells of germinal epithelium are connected to one another by narrow cytoplasmic bridges which only break down upon the release of spermatozoa into the lumen of the seminiferous tubule.

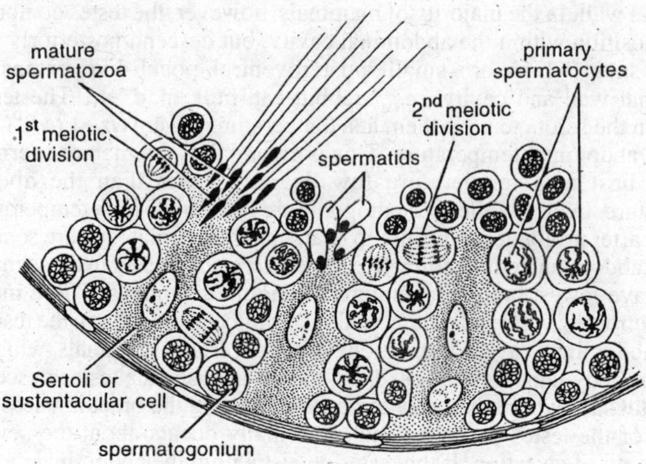

Fig. 4.14. Transverse section through a mammalian seminiferous tubule showing stages of differentiating spermatozoa located in layers between basement membrane and the cavity of the tubule (after Balinsky, 1970).

Interspersed among the spermatogonia of the seminiferous tubules are the second type of cells, the **sustentacular** or **Sertoli cells** (named after the name of discoverer, Italian histologist, **Enrico Sertoli**, 1842–1910). Each Sertoli cell is a non-spermatogenic cell. It is a tall, columnar, phagocytic cell. The Sertoli cell rests on the basement membrane and its cytoplasm extends to the lumen of the tubule, thereby filling all the narrow spaces between the cells of the spermatogenic series. Consequently, the cytoplasmic outline of the Sertoli cell is highly irregular and constantly changing to permit the progressive movement of developing spermatozoa toward the luminal surface. There also exist junctional complexes between cytoplasmic extensions of Sertoli cells. The nucleus of the Sertoli cell is often triangular or ovoid in shape with a very prominent nucleolus and relatively homogeneous chromatin. The cytoplasm of Sertoli cell contains a moderate number of mitochondria, many lipid droplets and relatively little rough endoplasmic reticulum. The presence of highly ordered smooth endoplasmic reticulum suggests that Sertoli cells are highly active in lipid metabolism (see **Wheater** *et al.*, 1979).

The Sertoli cells impart structure to the seminiferous tubule and function as nurse cells for the sperm. The mature spermatozoa have no way of manufacturing their food and they have to depend upon some external source for their nutritive materials. The heads of several mature spermatozoa remain embedded in membrane folds (or cytoplasmic pockets) of the Sertoli cells, where they receive nourishment and persist until released into the genital tract of the male. Further, during spermiogenesis, Sertoli cells phagocytose excess cytoplasm cast off by spermatids.

In fact, the outer wall of the tubule serves as an effective barrier to separate the developing germ cells from the other body tissues and to provide the proper microenvironment for spermatogenesis to occur. For example, in the seminiferous tubule of sheep there is less Na^+, more K^+ and Cl^-, much more glutamic acid, more aspartic acid, glycine and alanine, but less of all other amino acids than in the plasma of blood (see **Berrill** and **Karp**, 1978).

4-6.2. Interstitial Cells

In the testes also occur clusters of endocrine gland cells, called **interstitial cells** or **Leydig cells** (discovered by German anatomist, **Franz von Leydig**, 1821–1908). These cells exist in the connective tissue existing between the seminiferous tubules. Leydig cells secrete an androgen (the masculinizing hormone), the **testosterone** which is a steroid hormone, responsible for the development of male secondary sexual characteristics at puberty (*e.g.*, horns of sheep, plumage of birds, claws of turtle, anal fins of fish and hair patterns and voice timbre of human beings). Testosterone is also essential for the continued function of the seminiferous epithelium. The secretory activity of Leydig cells is controlled by the pituitary gonadotropic hormone, **luteinising hormone**, often referred to as **interstitial cell stimulating hormone (ICSH)** in the male.

In general, most male vertebrates have mature sperm only during breeding seasons, whereas during periods of sexual quiescence the seminiferous tubules undergo changes in which large amounts of lipid material accumulate in them. However, some vertebrates have mature sperm the year round. For example, amphibia and certain mammals such as man stay spermatogenically active throughout the year **(Phillips,** 1975).

The seminiferous tubules usually anastomose (or join together) at their distal ends and empty into straight tubules. These tubules empty into a network of sinuses or channels known as the **rete tubules**. In man, the rete tubules converge to form 12 to 14 **efferent ductules (vasa efferentia)** which open into **epididymis** which is a long coiled tube applied to the posterior surface of the testis. The epididymis leaves the testes as **vas deferens (ductus deferens)** to open into the urethera at a place which is inside the prostate gland.

4-7. THE PROCESS OF SPERMATOGENESIS

The spermatogenesis is a continuous process, but for the sake of convenience, it can be divided into following two stages:

1. Formation of spermatids ;
2. Spermiogenesis.

4-7.1. Formation of Spermatids

The cells of germinal epithelium which produce the spermatozoa are called **primary germinal cells** or **primordial germ cells**. For the formation of spermatids, each primary germinal cell undergoes following phases or stages:

A. The multiplication phase. An undifferentiated germ cell or primordial germ cell contains a large-sized and chromatin rich nucleus. It divides by repeated mitotic cell divisions and produces the cells which ultimately become the **sperm mother cells** or **spermatogonia** (Gr., *sperma*=sperm or seed; *gone*=offspring). The spermatogonia are diploid cells and found in vertebrates next to the basal membrane of the seminiferous tubules.

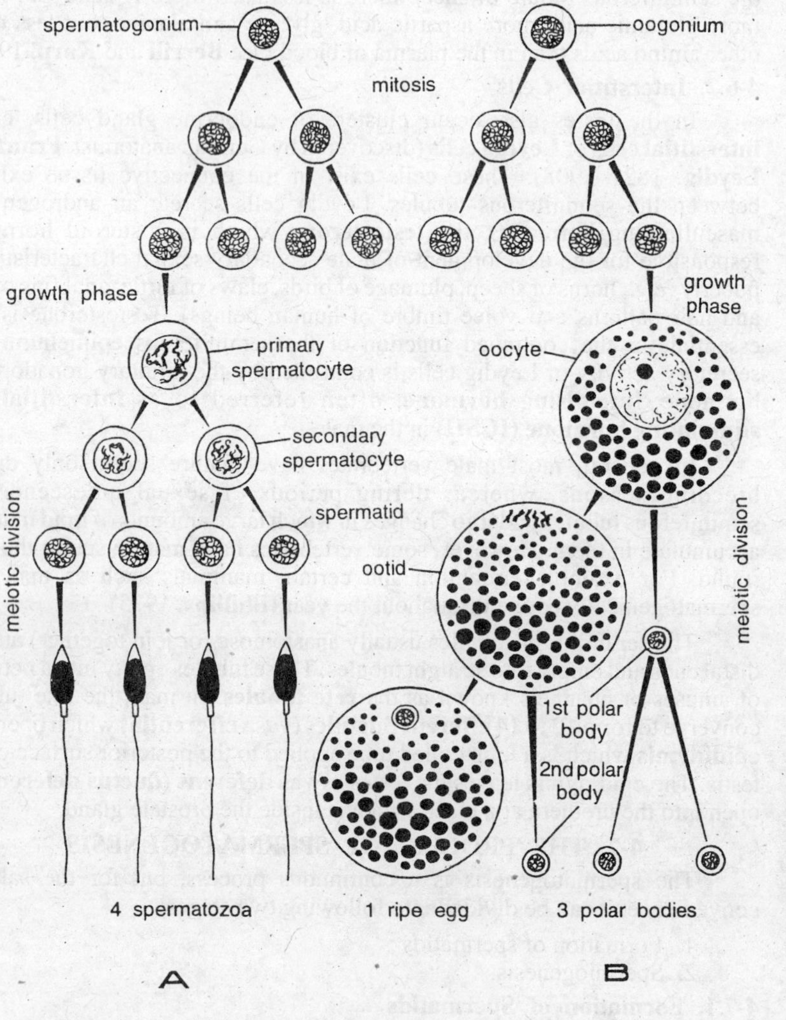

Fig. 4.15. A—Spermatogenesis; B—Oogenesis (After Balinsky, 1970).

In mammals, spermatogenesis typically proceeds in cycles of activity and the spermatogonia in one region of a tubule are typically at the same stage in this cycle. Adjacent regions are often in the next stage and waves of spermatogenesis can be found along the length of the tubules of many mammals. At the start of the process in the rat, the spermatogonia can be identified as **A_1 spermatogonia.** This type of spermatogonia divides twice to form four **A_2 spermatogonia**. One of these four cells becomes a cell that remains dormant and does not divide again until another cycle of spermatogenesis begins. Thus, in a way, this cell acts as a replacement for the cell that originated the cycle. The process of spermatogenesis usually continues throughout the adult life of the animal. Replacements must be set aside, otherwise all the potential germ cells will rapidly be exhausted and no new primordial germ cells will arise during the animal's life time.

The other three A_2 spermatogonial cells divide mitotically into six intermediate cells, which in turn divide mitotically into twelve **B_1 cells** and then into twenty-four B_2 cells. Thus, in man, there are found seven distinct classes of spermatogonia. These different spermatogonia can be identified by their cytological appearance and their diminishing size after each division. (In fact, doubling of cell volume before mitosis does not occur as in most cell lines).

B. Growth phase. After the last spermatogonial division, the nature of the cell changes; instead of preparing for another mitosis, it grows by accumulating nourishing material obtained from germinal cells to become double in volume. Such cells are called **primary spermatocytes** and these diploid cells have to develop into haploid spermatozoa, therefore, at this stage, they undergo **meiosis**, which constitutes the next phase of spermatogenesis, called the maturation phase.

C. The maturation phase. The primary spermatocytes enter into the prophase of **first meiotic** or **maturation division** (See Section 2-3.4). The first meiotic nuclear division is followed by the division of cytoplasm or **cytokinesis**, which divides each dividing primary spermatocyte into two haploid, **secondary spermatocytes**. Both secondary spermatocytes undergo **meiosis II** or **second maturation division** which is mitosis and produce four haploid **spermatids**. Thus, during maturation phase of spermatogenesis, each diploid primary spermatocyte produces four haploid spermatids. These spermatids, though having a haploid set of chromosomes, are still not capable of functioning as male gametes. To become functional spermatozoa, they have to undergo a process of differentiation or specialization called, **spermioteliosis** or **spermiogenesis**.

Role of Y chromosome during maturation division. In *Drosophila* the genes for sex phenotype of male individuals are uniformly distributed on the autosomes, Y chromosome contains genes for male fertility (see **Burns** and **Bottino**, 1989). In *D. heydei*, the Y chromosome is heterochromatic and inactive in all cells of the male including the primordial germ cells and proliferating spermatogonia in the testes. As meiosis begins, during diplotene stage, in primary spermatocyte the Y chromosome undergoes an explosive change in morphology and opens out into loops—the "ribbons" or "nooses" that resemble loops of lampbrush chromosomes of oocytes. Six loop pairs can be distinguished, each probably representing one fertility gene carried by the Y chromosome (Fig. 4.16). As loops emerge, RNA synthesis begins, which is a sign of gene activation. RNA synthesis or transcription persists for a short period into secondary spermatocyte but shuts down during spermatid differentiation (*i.e.*, spermiogenesis). Most RNA messages are synthesized premeiotically and are probably translated (into proteins) post-meiotically when spermatid differentiate. All loop-forming regions are essential for differentiation of mature sperm. If portions of the Y chromosome are deleted, certain loops are lost and sperm cell differentiate abnormally (**Grant**, 1978).

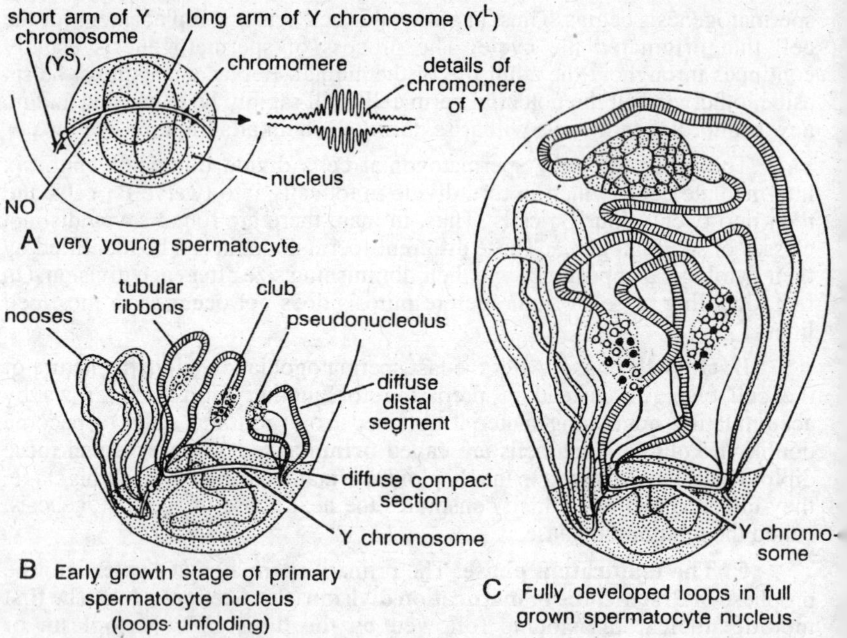

Fig. 4.16. Y chromosome of *Drosophila heydei* transforms from a condensed heterochromatic state to euchromatic lampbrush chromosome during spermatogenesis (after Grant, 1978).

4-7.2. Spermiogenesis or Spermioteliosis

Because, a sperm or spermatozoon is a very active and mobile cell, so, to provide great amount of mobility to the sperm, the superfluous materials of the developing spermatozoon are discarded and a high degree of specialization takes place in the sperm cell. During spermiogenesis (Fig. 4.17), the two major parts of sperm, the head and tail are formed by following methods:

(i) Formation of head of spermatozoon. The two major parts of sperm head, *i. e.*, the nucleus and acrosome, undergo following changes to form a sperm head:

(a) Changes in nucleus. In going from the spermatid to the spermatozoon, the nucleus loses its entire fluid content, all its RNA, nucleolus and most of its proteins. Only one macromolecule remains untouched, *i.e.*, its haploid amount of DNA. In some cases the proteins associated with the DNA (histones) are replaced by an unusual class of small, basic proteins, called **protamines**. The general shape of the sperm nucleus also changes—instead of the usual spherical form the nucleus becomes elongated and narrow. In different animals the nucleus assumes different shapes which ultimately determine their prospective shapes. Some of these shapes are an obvious adaptation for propulsion in water.

(b) Acrosome formation. The acrosome of a spermatozoon is derived from the Golgi apparatus of a spermatid (Fig. 4.18). The Golgi apparatus of an early spermatid consists of a series of cisternae arranged concentrically around an aggregation of small vacuoles. During acrosome formation, one or

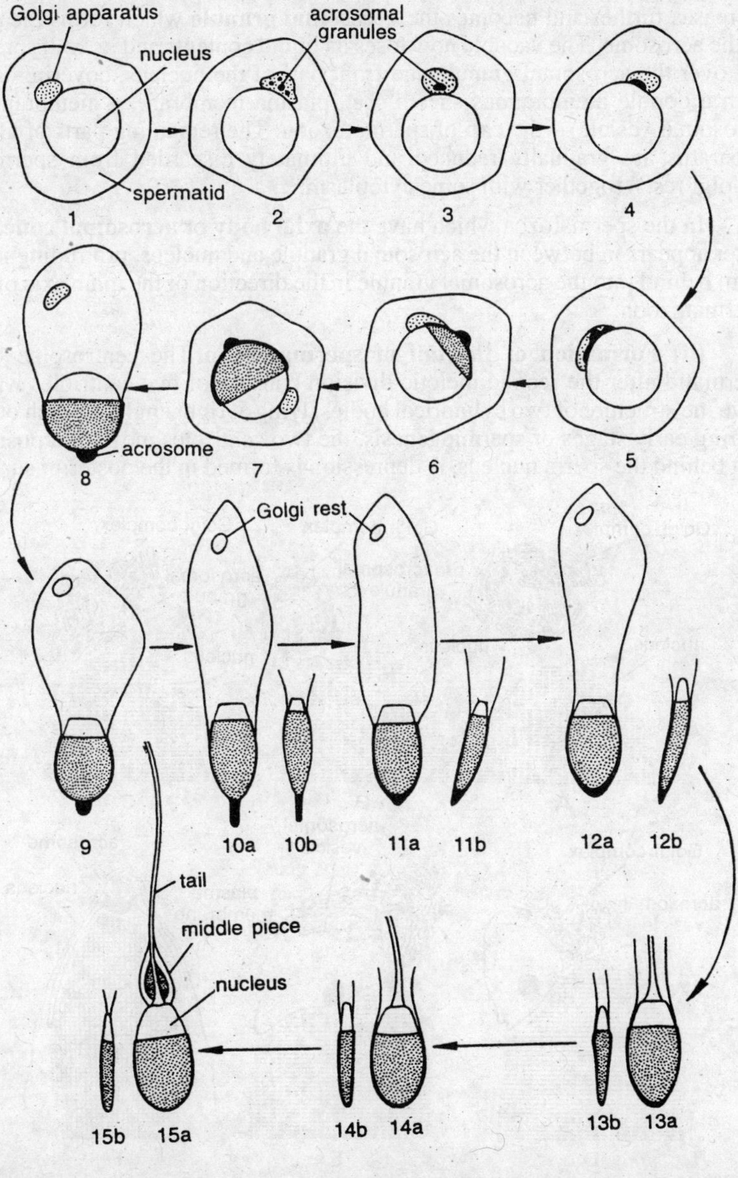

Fig. 4.17. 1 to 15 stages of metamorphosis of the spermatid into the spermatozoon (spermiogenesis) (After Balinsky, 1970).

more vacuoles start enlarging and inside the vacuole appears a small dense body the **proacrosomal granule**. The contents of vacuole and granule give a positive staining reaction for mucopolysaccharides (**Clermont** and **Leblond**, 1955). The vacuole or **acroblast** (see **Grant**, 1978) enlarges in its volume by fusing with other small vacuoles containing proacrosomal granule and become closely applied to the forwarding tip of the elongating nucleus. The granule increases further and becomes the **acrosomal granule** which forms the core of the acrosome. The vacuole now loses its liquid contents and its wall spreads out over the acrosomal granule and front half of the nucleus, covering them with a double membranous sheath (*i.e.*, plasma membrane + membrane of acrosomal vesicle)—the **cap** of spermatozoon. The remaining parts of Golgi apparatus are gradually reduced and ultimately discarded from sperm as "Golgi rest" together with some cytoplasm.

In the spermatozoa which have the **axial body** or **acrosomal cone**, the latter appears in between the acrosomal granule and nucleus, protruding itself from behind into the acrosomal granule in the direction of the main axis of the spermatozoon.

(ii) Formation of the tail of spermatozoon. The centrosome of a spermatid after the second meiotic division consists of two centrioles which have the structure of two cylindrical bodies, lying at right angles to each other. During early stages of spermiogenesis, the two centrioles move to a position just behind the sperm nucleus. A depression is formed in the posterior surface

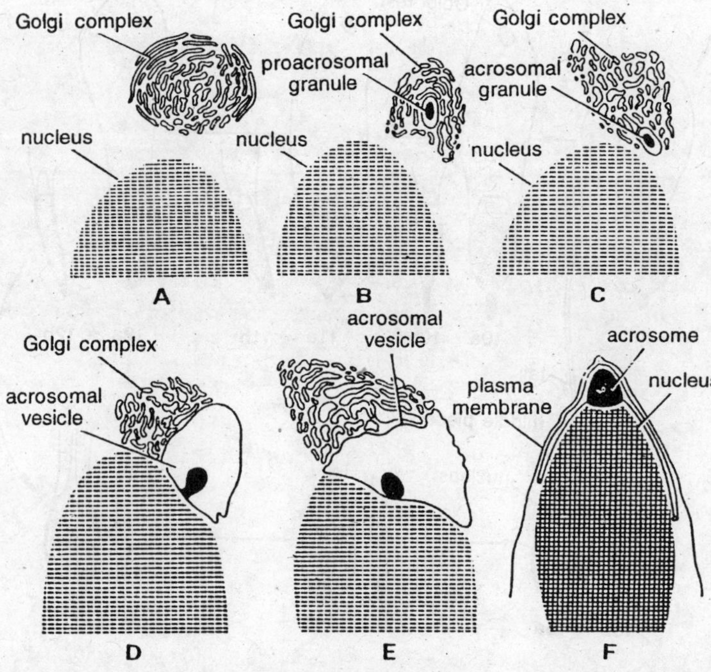

Fig. 4.18. Formation of acrosome of a spermatozoon (After Balinsky, 1970).

of the nucleus and one of the two centrioles is placed in the depression with its axis approximately at right angles to the main axis of the spermatozoon. This is **proximal centriole** of the spermatozoon; the other centriole, the **distal centriole**, takes up a position behind the proximal centriole with its axis coinciding with the longitudinal axis of the spermatozoon. The distal centriole now, gives rise to the **axial filament** of the flagellum of the spermatozoon for which it serves as basal granule.

Most of the mitochondria of spermatids concentrate around the distal centriole and proximal (upper) part of the axial filament and form the neck and mid-piece of the tail of spermatozoon. In the mid-piece of sperm, the mitochondria lose their individuality by fusing to become either spirally arranged or form large mitochondrial bodies. The tail of sperm comprises an axial filament which remains differentiated into a principal piece and a tail piece, as described before.

Thus, much of the cytoplasm of the spermatid becomes redundant in the spermatozoon, and it is simply discarded. With the formation of the acrosome at the anterior end of the spermatid nucleus, the cytoplasm flows away from it in the opposite direction, leaving only a very thin layer with the plasma membrane covering the acrosome and the nucleus. The bulk of the cytoplasm is then attached to prospective middle piece of the spermatozoon, while the tail is growing out at the posterior end (Fig. 4.17). After the mitochondria have arranged themselves around the base of the axial filament of the flagellum, the remainder of cytoplasm (containing the Golgi rest) is simply pinched off from the spermatozoon, leaving a thin film of cytoplasm around the mitochondria of middle piece. The detached part of the cytoplasm disintegrates.

4-8. FACTORS CONTROLLING SPERMATOGENESIS

In animals the production and liberation of spermatozoa are precisely controlled, either environmentally or physiologically. Temperature, light, hormones and psychological state play part in some degree depending on organism. For example, the onset and completion of spermatogenesis are regulated by certain gonadotropin hormones which are secreted by the anterior lobe or **adenohypophysis** of pituitary (See Chapter 7).

REVISION QUESTIONS

1. Write an essay on the spermatogenesis. Add a note on the significance of the process of spermatogenesis.
2. Discuss in brief about the origin and determination of primordial germ cells in various animal groups. Explain how does it explain the classical concept of germplasm and somatoplasm.
3. Give an illustrated account of the ultrastructure of the sperm of a mammal.
4. Define the term spermiogenesis. Discuss the major events involved in the metamorphosis of spermatid into sperm. Why are these changes essential ?
5. Trace the future of the following cellular organelles during spermatogenesis :
   (i) Golgi apparatus; (ii) Mitochondria;
   (iii) Cytoplasm; (iv) Nucleus;
   (v) Centriole.
6. Write short notes on the following:—
   (i) Acrosome; (ii) Axial body;
   (iii) Sertoli cells; (iv) Spermiogenesis.

5

FEMALE GONADS AND OOGENESIS

The female gamete, the egg or ovum is a unique cell; it is the cell fated to segregate from the organism and develop into a new individual. Larger than most somatic cells, the ovum is a non-motile and food-laden cell. It derives its developmental potential, not only from a rich dowry of organelles but also from the organisation of its outer cortex and plasma membrane. An ovum or egg acquires its developmental potential during **oogenesis**, a period of growth and maturation occurring in the female gonads, the **ovaries.**

5-1. OVARIES

The vertebrate ovary functions both as an **exocrine** and an **endocrine** organ. It produces ova or eggs and the steroid sex hormones—estrogen and progesterone, which regulate the reproductive tract, secondary sex characteristics and mating behaviour of the females. For example, in human female estrogen is responsible for the broadening of the pelvis, development and lactation of mammary glands (breasts), growth of the uterus and vagina, growth of pubic hairs, change in voice quality and the onset of the menstrual cycle.

In most vertebrates, the ovary is suspended from the dorsal body wall just behind the kidneys by a section of peritoneum, called the **mesovarium**. The shape, size and morphological types of ovaries of animals are as follows:

1. Number of ovaries. Most vertebrates have a pair of ovaries, but a single structure is found in many fishes, probably representing a fusing of the two original ovaries, since, it has a typical vertebrate structure. In some animals, birds, for example, only one side of the reproductive tract develops and the ovary on the other side remains inactive or degenerates. This is one of the most significant aspects of flight adaptation of birds, *i.e.*, it helps in the reduction of body weight of birds.

2. Shape of vertebrate ovaries. The external shape of the ovary varies according to the species and to the reproductive state of the individual. In teleost and ganoid fishes they are elongated structures extending over a considerable area of the body. In amphibians, they are composed of a series of lobes, each of which is a mass of eggs during the breeding seasons. Thus, in frog, during the breeding months the multilobed ovaries fill the body cavity, actually stretching the body wall. After ovulation, the ovary involutes and looks like a ruffled piece of mesentery. In reptiles and birds, they have the general form of a bunch of grapes. For example, the ovary of birds is an enormous structure because of both the accumulation of yolk in individual eggs and of the fact that the female bird ovulates one egg per day for several months. In mammals, they are flattened ovoid structures in the resting condition, but during the reproductive or breeding phase, they may assume rounded appearances containing mound-like protrusions.

3. Size of vertebrate ovaries. In teleost fishes the size of the ovaries is enormous compared to the body of the female, while in human, cow, etc., they are small structures in comparison to the adult body. The human ovary is relatively small, walnut-shaped body measuring 4 × 2 × 1 cm. Further, a ovary may contain millions of mature eggs in the codfish and conger fish during each breeding season, whereas, only a single egg is commonly matured at a time in the cow, elephant or human. During the reproductive season the ovary may assume a condition of striking colour effects as in the shark, frog, reptiles and birds, only to recede into an appearance of drab (light brown colour), shrunken and dishevelled in the non-breeding season.

4. Morphological types of vertebrate ovaries. Regardless of their many shapes and sizes, the ovaries of vertebrates may be divided morphologically into following two main types:

1. Compact ovaries. The solid or compact type ovaries are found in cyclostomes, elasmobranchs, osteichthyes fishes, reptiles, birds and mammals (Fig. 5.1). Each ovary has following regions: 1. An outer region, the **cortex**, containing many ova in various stages of development, and, 2. An inner **medulla**, consisting of irregularly arranged connective tissue, blood and lymph vessels. A dense connective tissue layer the **tunica albuginea**, covers the cortex. External to the tunic, the peritoneal epithelium (or coelomic

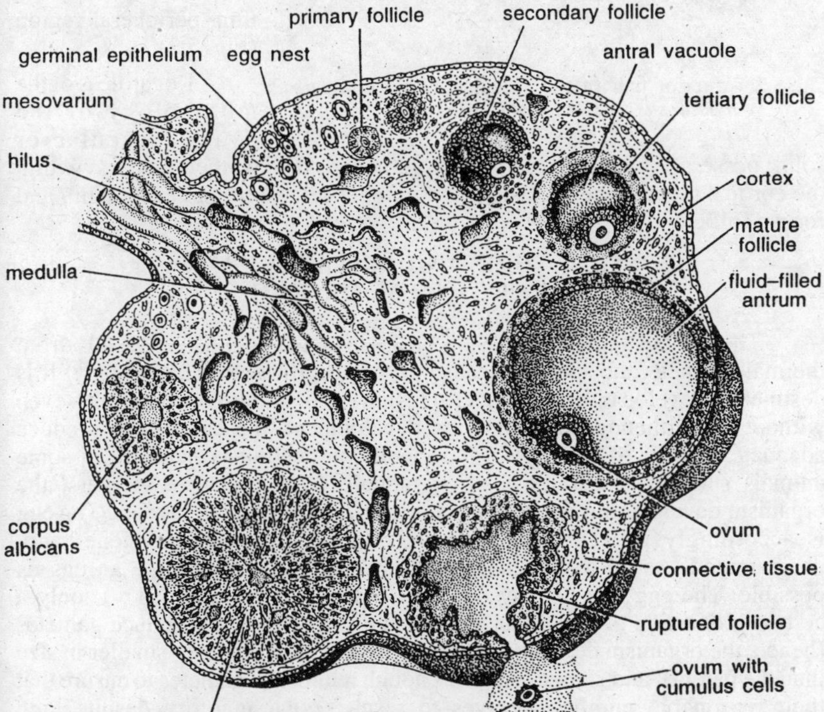

Fig. 5.1. A three dimensional representation of a mammalian ovary showing its morphology and cyclic changes (After Nelson, 1953).

epithelium) surrounds the ovary. This epithelium was unfortunately named the **germinal epithelium** by the early biologists who believed that these cells were the source of the primordial germ cells, the future eggs. But as we have already discussed in last chapter, the primordial germ cells originate in other areas of the body, outside the gonad and migrate into it during development.

2. Saccular ovaries. The saccular ovaries are found only in amphibians among the vertebrates. An ovary has a cortex and germinal epithelium, somewhat similar to the compact ovary but, the area which forms the medulla in the compact ovary is here represented by a large space. During early development, the amphibian ovary is a compact structure, but, later there is a hollowing out and disappearance of the compact medullary portion and the cortical area of the ovary remains as a relatively thin, peripheral region (Fig. 5.2).

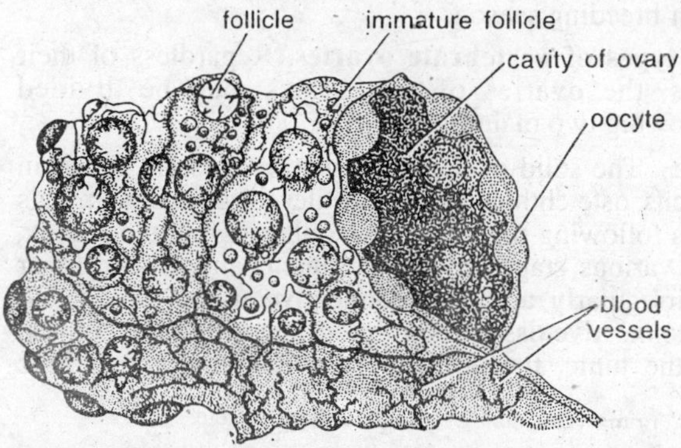

Fig. 5.2. Anterior half of the saccular ovary of *Necturus maculosus* (After Nelson, 1953).

Regardless of the morphology of the ovary (saccular or solid) possessed by the vertebrate, the germ cells take up their position within the cortical region where clusters of epithelial cells surround each ovum and form a **follicle**.

5-2. STRUCTURE AND METABOLISM OF MATURE OVUM OR EGG

The female gamete, the ovum or the ovulated egg differs greatly from the male gamete, the sperm, because of the specific role it has to play. It is destined to develop either after sperm fusion (fertilization) or sometimes even without it (parthenogenesis). During development, the egg has to produce adequate number of cells to form a viable multicellular organism; in some animals such as insect larvae, frog tadpoles and honeybee workers, the organism developed from the egg is quite different from the parents. The egg is accordingly packed with food and other substances and genetically programmed. As a consequence, most species produce as large an egg as possible. The egg size usually becomes limited for two reasons : 1. only a certain proportion of an animal's substance can be used to produce gametes (hence, the organism developed from the zygote is necessarily smaller in size than the parent); and 2. there must be enough number of gametes to ensure that their reasonable number survives to reach sexual maturity despite high mortality rate; the progeny has genetic variation to cope with the various ecological factors; and the progeny has a fair chance of dispersal over a territory.

The mature ovum (or ovulated and unfertilized egg) is typically more or less spherical and large-sized. It is like any other eukaryotic animal cell, so has a plasma membrane (plasmalemma or oolemma) inside which are situated the cytoplasm, nucleus, yolk (deutoplasm) and other organelles and inclusions.

In most types of eggs, the egg cytoplasm (also called **ooplasm**), remains differentiated into outer **ectoplasm** or **egg cortex** and inner **endoplasm**. The ectoplasm occurs just beneath the egg's plasma membrane and it is a transparent, viscous layer and is not displaced by centrifugation. These properties of egg cortex are due to the presence of some cytoskeletal elements such as microtubules and microfilaments in it (**Balinsky**, 1981; **De Robertis** and **De Robertis , Jr.,** 1987). The egg cortex also contains pigment granules and numerous tiny membrane-bounded vesicles or granules, called **cortical granules**, which contain mucopolysaccharides.

The endoplasm occurs inside the ectoplasm. It contains many mitochondria, Golgi apparatuses, ribosome-bearing endoplasmic reticulum (*i.e.*, rough ER), annulate lamellae (*i.e.*, stored preassembled pore complexes in stacks of cisternae of ER membranes) and a wide variety of molecules such as late-translating mRNAs (informosomes), tRNAs, histones, tubulin, actin and enzymes. Egg endoplasm is also rich in various types of inclusions such as species-specific amount of yolk, glycogen granules and fat droplets or lipochondria all of them meant for food of developing embryo.

The nucleus of ovum is greatly enlarged having large amount of nucleoplasm and a prominent nucleolus. It is called **germinal vesicle** and is located in that part of endoplasm which is clear and without yolk. In frog's egg and other telolecithal eggs this side of ovum is called **animal pole**. Actually, in all types of eggs or ova of animals, the animal pole can be recognised by the fact that during maturation or meiotic division of oogenesis, the **polar bodies** or **polocytes** are always given off from the animal pole of the ovum.

Outside the plasma membrane of the egg are found various types of egg membranes about which we would have a discussion after a short while. The egg size and yolk contents are inter-dependent variables. The small eggs of many marine invertebrates usually produced in enormous numbers and set free in the water, range from about 50 μm in diameter (*e.g.*, in many polychaete worms) to about 150μm (*e.g.*, ascidians), while others such as echinoderms are in between. Eggs of larger dimensions characteristically contain proportionately more yolk reserves and range all the way to the extremely large and yolky eggs of birds, reptiles, and some fishes, particularly sharks and rays. Remarkably in mammals, which are descendants of reptiles, the eggs are minute, almost yolkless and measure about 100 μm in diameter.

An animal egg-cell has to perform the following functions : (i) it has to supply a nucleus containing half of the chromosomal complement (*i.e.*, haploid set of chromosomes) to the future embryo; (ii) it has to supply almost all cytoplasm to the future embryo upon union with spermatozoon; and (iii) it has to supply food reserves that will enable the embryo to develop to a stage where it can begin to feed upon exogenous food materials. Therefore, for these future needs, an egg cell becomes immotile, large-sized, and well specialized.

5-2.1. Metabolism of Egg

The mature ovum or ovulated egg, irrespective of the stage of meiosis, is usually metabolically inert. There is no detectable protein synthesis, despite everything necessary for it, such as, mRNAs, tRNAs, ribosomes, amino acids, enzymes, etc., being present in the ovum. The respiratory rate of the mature ovum is also very low; this is perhaps partly because of the enormous change in the ratio of surface area to cell volume.

5-3. OOGENESIS

An egg acquires its developmental potential during oogenesis—a period of growth and maturation in the ovary. During oogenesis, an egg has to be programmed and packaged. The **programme** is in the form of molecular information somehow coded into the structure of the egg. It provides a potential for that cell to unfold—through a predictable sequence of changes—from a quite simple, homogeneous-appearing cell into the complex pre-adult form. Since it is from the egg that this transformation emerges, it must be within the organisation of the egg that the directions are laid. The **packaging** refers to the presence within the egg of all the materials that will be needed to build the structures and to provide the energy for the embryo up to the point where it can feed and, therefore, take its own supporting nutrition (**Berrill** and **Karp**, 1976).

Since various events of the process of oogenesis hang on the scaffolding of meiosis, so, oogenesis or manufacturing of ova, involves in it the following two important phases: 1. Premeiotic or proliferative phase; and 2. Meiotic phase which includes various substages such as growth of oocyte, yolk synthesis or vitellogenesis and maturation, egg membrane deposition, etc.

5-3.1. Premeiotic Phase (Proliferative Phase)

As in spermatogenesis, the growing egg cells or oocytes arise from the primordial germ cells that are set aside very early in development and multiply by mitosis to form a population of **oogonia,** from which the oocytes will arise. Thus, premeiotic or proliferative phase is a period of oogonial proliferation within the ovary, usually during embryonic, larval or foetal life. This occurs in all vertebrates, but its timing differs in major groups. In mammals and birds, primordial germ cells proliferate into a limited number of oogonia; the premeiotic or proliferative phase is completed once in the life of an organism, long before sexual maturity. No proliferating oogonia are seen in an adult ovary. In fish and amphibians, the premeiotic proliferative phase is seasonal throughout the adult life of the organism. New oogonia are replenished each mating season from a reserve population of mitotically active gonocytes in mature ovaries.

All oogonia are not matured into functional eggs since many do not survive into the adult. Surviving oogonia usually begin meiosis in larval and foetal stages and when they do, they become primary oocytes. This marks the transition between premeiotic and meiotic phases.

In a way, egg development in vertebrates is a wasteful process. Most oogonia fail to mature and degenerate. Such wholesale abrasion of oogonia accompanies gonadal development in human females. Counts made early in development indicate that approximately 1,700 primordial germ cells migrate and invade the undistinguished female gonad at one month of foetal development. These proliferate actively so that at two months of gestation, the gonad contains approximately 600,000 oogonia, and by the fifth month, when the gonad assumes its female characteristics, there are 7 million oogonia. Mitosis usually ceases at this point, and many oogonia enter the first meiotic prophase and become oocytes. Many oogonia undergo degeneration so that at birth only 2 million oocytes remain. After birth there is a continuing decline so that by puberty only 3 lac oocytes remain and these too continue to undergo degeneration. Each month, several oocytes are stimulated to mature, but only one usually completes meiosis and ovulates; all others degenerate. This means that during the roughly 40 years of reproductive life of a women (*i.e.*, from **menarche** or first menses to **menopause** or the last menses), only 400 to 500 eggs or ova will have been ovulated. All others degenerate since no oocytes can be detected in the ovaries of menopausal women. According to **Grant** (1978), there may be a selective advantage of such reproductive wastage that it insures that some mature eggs are always available for fertilization. In humans, some mature eggs will have been in arrested meiosis for 40 to 50 years, which may explain the higher incidence of congenital abnormalities (*i.e.*, Down syndrome) among children from older women. Delayed ovulation may lead to overly ripe eggs that are more susceptible to chromosomal aberration.

5-3.2. Meiotic Phase

Before oogonia enter meiosis and are to be known as the **primary oocyte**, their chromosomes replicate as if preparing for another mitotic division. The amount of DNA doubles from the diploid 2C value to a 4C or tetraploid value (C-value stands for characteristic DNA content of the cells of a given species). Shortly thereafter, prophase begins and replicated chromosomes condense, but in contrast to mitosis, prophase is markedly prolonged, lasting for months and even years. Thus, *arrested prophase is the most peculiar feature of oocyte meiosis*. During meiotic prophase of oogenesis, the oocyte grows and acquires its developmental information. The dual functions of an egg, its role as a sexual recombinant and its morphogenetic potential are acquired during this long prophase. The meiotic phase can be conveniently subdivided into the following phases :

1. Growth phase of primary oocyte. During vegetative growth, primary oocyte's stores of gene products accumulate in the cytoplasm. These products are coded in the oocyte genome and in the genome of accessory maternal cells surrounding the oocyte. Thus, growth phase of primary oocyte involves the following events :

A. Growth and differentiation of oocyte. The processes of growth and differentiation of primary oocyte involve not only the accumulation of nutrients and other reserves necessary for embryonic development, but also genetic programming for development. For example, in *Xenopus* eggs compared to somatic cells, there are 100,000 times more RNA polymerase enzymes, histone proteins, mitochondria, ribosomes, etc. Such a hoarding is required for rapid cell division during cleavage since the essential components become distributed into the daughter cells or blastomeres, without the necessity of their at the spot synthesis.

B. Magnitude of growth. The oocyte grows to enormous proportions, normally to become the largest cell of the animal body. Most of the primordial germ cells are approximately 10 microns (0.01 millimeter) in diameter, which is the size of an average body cell. In mammals, such cells increase to a size that, in some species may reach a diameter of 200 microns. In the frog, the increase is up to a diameter of 2000 microns, in birds the diameter of the ovum is as large as 40,000 microns. Thus, volume increases × 43 in mouse, × 27,000 in *Rana pipiens*, × 90,000 in *Drosophila*. This growth is due not so much to an increase in the active cytoplasm, as to an increase in the reserves such as yolk.

C. Rate of oocyte growth. The time required to achieve the maximum size is not related to the increase that takes place. The rate of growth of primary oocyte may be slow or higher. Thus, *Drosophila* needs only three days for oocyte growth, whereas the mouse needs sixteen days and the frog needs three years. In *Rana pipiens*, for example, the oocyte increases in its volume by a factor of 27,000, so that, a mature egg has 1500 microns diameter. In it, this growth takes place over a period of three years (Fig. 5.3). The young oocyte starts growing after the tadpoles metamorphose into young froglets. One-year old and two-year old frogs do not yet have mature eggs, but by the third year the eggs are ready and the frogs may spawn for the first time. Thus, in it eggs take at least three years for its growth. In the hen the last rapid growth of oocyte occurs in the 6 to 14 days preceding ovulation, and during this time the volume of oocyte increases 200 fold.

D. Site of oocyte growth. The site of growth of primary oocyte may be different in different animals. In many annelid worms and echiurans early primary oocytes may develop as independent cells, floating freely in body fluids usually after detaching from the coelomic epithelium. They absorb nutrients directly from body fluids. The oocyte is liberated either alone (*e.g.*, the echiuran *Urechis*) or with attached nurse cells (some annelids). In most instances, however, oocytes develop in a localised organ, an ovary. They may mature independently of surrounding ovarian cells or may be directly associated with follicle and /or nurse cells.

E. Mode of growth of oocyte. During growth phase of primary oocyte qualitative and quantitative changes take place both in the nucleus and the cytoplasm.

(1) Growth of nuclear substances. Among oocytes developing independently of nurse cells, the oocyte nucleus or germinal vesicle serves as the principal source of developmental information. It swells to a volume several hundred times that of the original oogonial nucleus, and its newly extended lampbrush chromosomes are metabolically active. By contrast, oocytes surrounded by the nurse cells generally lack a germinal vesicle or lampbrush chromosomes. Instead, nurse cell nuclei enlarge, become metabolically active and synthesize gene products for storage in oocyte cytoplasm. Thus, growth of nuclear substances can be studied under the following headings:

(i) The lampbrush chromosomes (*i.e.*,Bumper synthesis of mRNAs from the oocyte genome). Lampbrush chromosomes are found in oocytes of diverse groups, including humans, but their behaviour has been most thoroughly studied in amphibian oocytes. In fact, they are best visualised in salamander's oocytes because they have a high DNA content and, therefore, very large chromosomes.

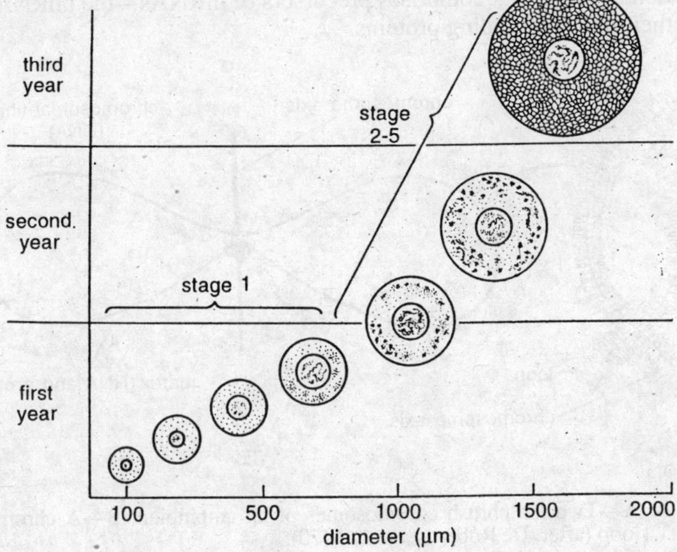

Fig. 5.3. Growth of amphibian oocyte showing enormous accumulation of yolk granules proceeding from peripheral region inward. (After Berrill, 1971).

In the frog, *Xenopus*, diplotene begins as chromosomes elongate into the lampbrush condition. Vitellogenesis (*i.e.*, the period of yolk synthesis and storage) also begins; cytoplasm becomes opaque as yolk platelets accumulate. The height of growth and metabolic activity, however, is reached later when lampbrush chromosomes are in their most extended and active condition. The lampbrush stage, though it persists for 4 to 8 months in *Xenopus*, is transient stage. While the oocyte is still growing and yolk is deposited, chromosomes lose their lateral loops and begin to condense, a process that is completed in mature oocytes. The germinal vesicle enlarges to accompany cytoplasmic growth, but the chromosomes contract to their ultimate compact configuration in preparation for meiotic metaphase.

Lampbrush chromosomes (Fig. 5.4) are homologous **bivalents** in which the maternal and paternal chromosomes are held together by chiasmata at those sites where crossing-over has previously occurred. The long axis of each chromosome (homologue), made up of two chromatids, consists of a series of chromomeres—the tight coils of the principal axial thread. Lampbrush **loops** are regions where chromomeres have uncoiled and extended laterally and symmetrically on either side; each chromosome has two of them, one for each chromatid. There are about 10,000 loops per chromosome set of the salamander, *Triturus*. Each lateral loop has **axis** formed by a single DNA molecule that is unfolded from the chromosome for doing intense transcription or RNA synthesis. Only about 5 to 10 per cent of the DNA is in lateral loops, the rest being tightly condensed in the chromomeres of the chromosome axis, which are transcriptionally inactive. Each loop is covered by a **matrix** of ribonucleoprotein that comprises precursors of mRNAs—the hnRNAs along with their hnRNA-binding proteins.

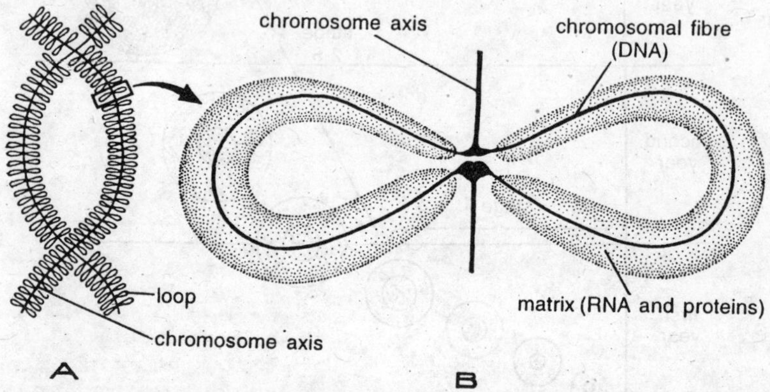

Fig. 5.4. A—Two lampbrush chromosomes of an amphibian; B—A chromosomal loop (after. De Robertis, *et.al.*, 1970).

Many of the loops correspond to a single transcriptional unit and other loops contain several units of transcription. Some transcriptional units in lampbrush chromosomes are extremely long, for example, in newt one such unit is exceptionally long measuring over 100 μm in length (1 μm of DNA = 3000 base pairs). **Joseph Gall** and collaborators (1981) have reported that in the American newt, *Notophthalmus viridescens*, such extremely long loops contain clusters of the five histone genes (*i.e.*, genes for five types of histone proteins of the nucleosome namely H1, H2A, H2B, H3 and H4). From each of such transcriptional unit of the loop are transcribed very long precursor molecules, each containing mRNA molecules for the five types of histone proteins.

The mRNA molecules which are synthesized during oogenesis are of three types:

(a) Immediately translating mRNAs. These are those mRNAs which are translated immediately to produce proteins of storage value such as ribosomal proteins, histones, enzymes, etc.

(b) Masked mRNAs. These mRNA molecules are made late translating by 'masking' them with proteins and are called **informosomes** (**Spirin**, 1965). Translation of these 'masked' mRNAs of mature ova is activated only after fertilization. For example, sea urchin eggs increase ten-fold their protein synthesis after fertilization and this is due to stored maternal 'masked' mRNAs.

(c) Heat-shocked mRNAs. Recently it has been reported that certain amphibian oocytes contain some special types of stored and late-translating mRNAs, called **heat-shocked mRNAs**. These mRNAs are induced to translate into proteins at high temperatures. For example, when oocytes of *Xenopus* are exposed to high temperature, the synthesis of a major heat-shock protein of M.W. of 70,000 daltons is induced, while the synthesis of other proteins is repressed (**Bienz** and **Gurdon**, 1982).

(ii) Gene amplification, nucleolus and synthesis and storage of ribosomes. No other cell, even the most metabolically active liver cells, can match the oocytes in its rate of ribosome production and storage. The ribosome reserve sustains the embryo through most of its development until it begins to synthesize its own ribosome after hatching.

During the early stages of amphibian oogenesis, before the lampbrush stage, RNA synthesis is one of the principal synthetic activities. Most of the RNA made during this period is **transfer 4S RNA** (*i.e.*, 4S tRNA) and **5S ribosomal RNA** (*i.e.*, 5S rRNA); very little 28S, 5.8S and 18S rRNAs are synthesized at this time. Oocytes are found to meet the increased demand for ribosomes by amplifying the ribosomal genes during which the number of genes for 28S, 5.8S and 18S rRNAs is greatly increased. However, the matching increased production of 5S rRNA is not done by gene amplification but by activating all 40,000 5S rRNA genes which are located on telomeres or tips of most chromosomes.

Gene amplification is the process by which one set of genes is replicated selectively, while the rest of the genome remains constant. Such gene amplification can be seen more prominently in ribosomal DNA (rDNA) of **nucleolar organizer** of oocytes of amphibians, some beetles and spiders.

The nucleolus is the principal organelle for ribosome biosynthesis and assembly. Species with small, yolk-poor eggs, for example, sponges, hydroids and echinoderms, usually contain one large metabolically active nucleolus, while those with yolk-rich eggs, such as amphibians, may contain thousands of uniform nucleoli distributed at the periphery of germinal vesicle. The gigantic nucleoli and /or the multiple nucleolar systems in many oocytes provide the best case of redundancy of a gene locus. Genes coding for rRNA are copied thousands of times within oocyte nucleoli, on templates in the nucleolar organizer region.

For example, in *Xenopus*, eggs are very large cells (1.3 mm in diameter) that accumulate large numbers of ribosomes (10^{12}) for use during early development. Anucleolate embryos can survive throughout the first week of development using these maternal ribosomes. Oocytes must synthesize all these ribosomes using the genes contained in a single nucleus (oocyte nuclei have a tetraploid amount of DNA). In contrast, a somatic cell has 10^7 ribosomes; if it divides every 24 hours, it must be making 10^7 new ribosomes per day. At the same rate, a somatic cell nucleus would need 10^5 days to make the 10^{12} ribosomes present in an oocyte. But since an oocyte nucleus has twice the amount of DNA as a diploid cell, the somatic cell, it would still require more than 100 years to synthesize 10^{12} ribosomes from a single nucleus.

Oocytes achieve this synthesis by amplifying the number of rDNA genes 1000-fold (see Table 5-1).

Table 5-1. Number of ribosomal genes in somatic cells and oocytes in *Xenopus laevis* (after **De Robertis** and **De Robertis,** 1987).

Cell Type	Nuclear DNA (pg)	Number of chromosome sets per nucleus	Chromosomal DNA (pg per nucleus)	Ribosomal DNA (pg per nucleus)	Number of rRNA genes	Number of nucleoli
Somatic	6	2	6	0.012	900	2
Oocyte (after amplification)	37	4	12	25	2,000,000 (2 million)	1000–1500

The process of gene amplification takes place in very small oocytes in the early stages of meiotic prophase. During pachytene, excess DNA begins to accumulate on one side of the nucleus, forming a Feulgen-positive cap. By the end of amplification process this newly synthesized DNA contains about 2 million rRNA genes in *Xenopus*. As oocytes grow, the extra DNA is accommodated in 1000 to 1500 extra-chromosomal nucleoli. The rDNA in amplified nucleoli consists of circles of about 450 tandem gene repeats. The circles result from the way in which amplified DNA is replicated (''rolling circles''). Amplified rDNA is used only to make ribosomes during oogenesis; it is not inherited by the embryo and it gradually disappears during the first embryonic cleavage.

(2) Stockpiling of membranous organelles. The membranous organelles of the oocyte such as mitochondria, endoplasmic reticulum and Golgi apparatus make up an essential portion of egg's developmental organization. Though some of these organelles are necessary for oocyte metabolism, the bulk is stockpiled in oocyte cytoplasm to sustain the developing embryo to hatching.

Mitochondria are the most numerous of all membranous organelles. The structure of oocyte mitochondria is typical that is found in most somatic cells; however, in some oocytes they are more spherical, smaller and have fewer cristae. Large number of mitochondria accumulate in the oocyte, more than are needed for respiratory metabolism. There are approximately 2×10^6 mitochondria in a mature *Rana pipiens* oocyte compared with the 200 to 1000 found in most somatic cells. In very young oocytes, they are localised at the periphery of nuclei, but later as the oocyte grows, they are dispersed randomly in the cytoplasm. In growing oocytes of amphibians and birds, the mitochondria aggregate to form large ''**mitochondrial clouds**''. During the period of maximum yolk deposition, mitochondria are often seen at the periphery of the cytoplasm in areas of active micropinocytosis, a process requiring large amount of energy. Mitochondria that are stockpiled during oocyte growth probably arise from autonomous replication of their circular DNA. In fact, in many mature oocytes, for example, mitochondrial DNA makes up the bulk of total DNA; *i.e.*, up to 1000 haploid equivalent in *Xenopus*.

In some oocytes, mitochondria come from nurse cells, migrating down nutritive cords into oocyte cytoplasm.

Metabolically, oocyte mitochondria resemble those found in somatic cells; they carry on oxidative phosphorylation. Some mitochondria are found to be more active than the others. Recently, it has been suggested that egg mitochondria are metabolically immature and during the course of embryogenesis become mature, assembling batteries of mitochondrial enzymes that might be characteristic of each tissue type. For instance, sea urchin egg mitochondria lack **cytochrome C** but acquire it during development (**Grant**, 1978).

Most oocytes contain a **smooth endoplasmic reticulum** (**SER**) of small vesicles which are dispersed evenly through the cytoplasm. **Rough endoplasmic reticulum** (**RER**) is often not found in young or mature primary oocyte; since the proteins are being synthesized for storage rather than for secretion. The storage proteins are normally synthesized on free ribosomes. The growing oocytes also contain **annulated lamellae** in their cytoplasm. These are membranous structures which appear as stacks of cisternae of flattened sacs with an intrasaccular space 100 to 200 A° wide. They are formed by the budding of the nuclear envelope and are found to contain octagonal pore

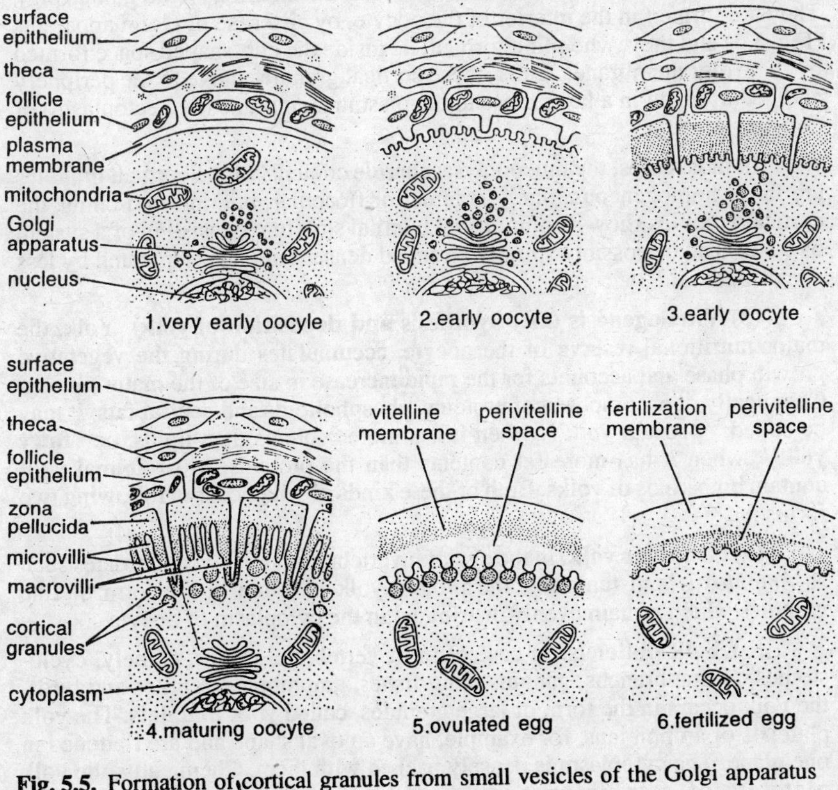

Fig. 5.5. Formation of cortical granules from small vesicles of the Golgi apparatus (After Berrill, 1971).

complexes similar to those of the nuclear envelope (**Maul**, 1977). Annulate lammellae are usually associated with rapidly proliferating cells and are considered as a storage form of preassembled pore complexes (**De Robertis** and **De Robertis, Jr.,** 1987).

A **Golgi apparatus** is also present in most oocytes, performing a variety of secretory, storage and transport functions. The stacked lamellae vesicles around the centrosome form a typical feature of young oocytes in the process of vitellogenesis, implicating them in yolk synthesis. Golgi also might be concerned with the elaboration of membranes of endoplasmic reticulum. In mature oocyte, it either becomes well developed and forming large spherical mass in some mammals, or becomes located in the subcortical cytoplasm of frog and chick or sometimes may disappear completely. The Golgi apparatus of oocyte besides performing its normal function, manufactures the **cortical granules**.

The cortical granules. The cortical granules are membrane bound spherical bodies of diameter from 0.8 micron (in sea urchins) to 2 microns (in frogs) containing acid mucopolysaccharides for the synthesis of fertilization membrane during fertilization. They are found in bivalve molluscs, some annelids, fishes, frogs and some mammals (rabbit, man, etc.), but are found to be absent in gastropods, insects, urodels, birds and mammals (rat, guinea pig). They are formed in the interior of the oocyte, by cisternae of Golgi apparatus (Fig. 5.5) and they, when first formed, lie inside the cup-shaped space formed by the Golgi membranes. Later, these cortical granules move to the periphery and get arranged in a layer close to the plasma membrane (*i.e.*, ectoplasm) of oocyte.

In most cases, the mucopolysaccharide contents of the cortical granules are fairly homogeneous, finely granular or floccular, but, in sea urchins, the cortical granules show a complicated internal structure consisting of a system of concentric or possibly spirally arranged denser lamellae, separated by less dense material.

(3) Vitellogenesis (*i.e.*, Synthesis and deposition of yolk). Yolk, the major nutritional reserve of the oocyte, accumulates during the vegetative growth phase and accounts for the rapid increase in size of the maturing egg. Chemically, it is composed of proteins, phospholipids and neutral fats. It may be called "**protein yolk**", when it has more proteins than lipids, or "**fatty yolk**", when it has more fat contents than the proteins. Most animal eggs contain both kinds of yolks. Both of these kinds of yolk occur in following two forms :

(a) Granular yolk. In the egg of invertebrates and lower chordates such as *Amphioxus* and tunicates, the protein yolk is found in the form of fine granules which remain evenly distributed in the cytoplasm of oocyte.

(b) Yolk platelets. In case of most vertebrate groups, namely, cyclostomes, elasmobranchs, ganoids, lung-fishes, amphibians, reptiles and birds, the yolk occurs in the form of large granules, called **yolk platelets**. The yolk platelets of amphibians, for example, have an oval shape and are flattened in one plane. The cytoplasm is densely packed with them. Chemically, the yolk platelets of amphibian oocytes contain two main proteinaceous substances—**phosvitin** and **lipovitellin**. Phosvitin is a highly phosphorylated protein (phosphorus content, 8.4 per cent) with a molecular weight of 35,000. The lipovitellin is a protein with a very much larger molecule (molecular weight 40,000) and containing a very considerable amount of bound lipid (17.5 per cent). In the yolk platelets two molecules of phosvitin are associated with one molecule of lipovitellin in a structural unit. The units are being arranged in the platelet in a crystalline lattice with regular hexagonal packing. Other animals have yolk platelets similar to amphibians, however, the yolk of some bony fishes has fat droplets inside its mass.

The yolk is accumulated inside the oocyte by diverse means. In some coelenterates, echinoderms and bivalve molluscs (*i.e.*, animals with microlecithal eggs) the oocyte, unaided by any other cells, takes up the raw materials (amino acids and the like) and synthesized the yolk (**autosynthesis** of yolk) by the membrane system of the rough and smooth ER and the Golgi apparatus. In some animals such as *Hydra*, several grown oocytes fuse into one to produce a large oocyte and all the nuclei except one degenerate. In most species, however, vitellogenesis involves the surrounding cells (*i.e.*, follicle and nurse cells) and the extra-ovarian cells of the mother (**heterosynthesis** of yolk).

Yolk proteins are synthesized in the amphibian liver (or vertebrate liver) and transported to the oocyte via the circulation. A yolk protein precursor, a lipo-phosphoprotein, called **vitellogenin** is detected in the serum of vitellogenic females during the period of oocyte growth. Its synthesis is under hormonal control (*i.e.*, estrogen such as 17B-estradiol). Serum vitellogenin disappears rapidly in vitellogenic females because it is selectively taken up by

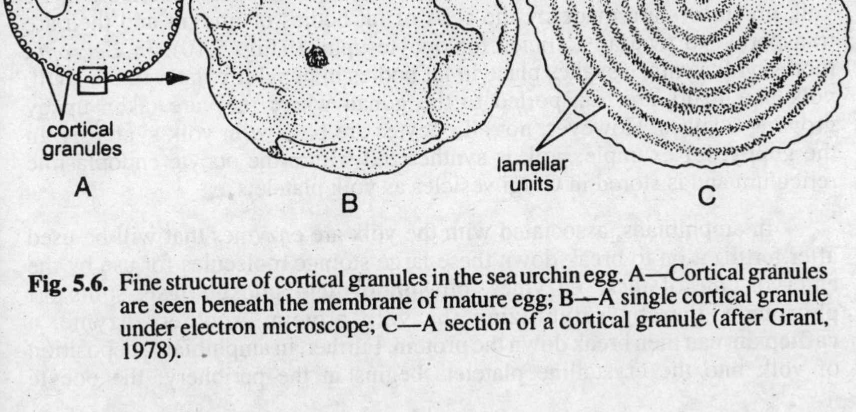

Fig. 5.6. Fine structure of cortical granules in the sea urchin egg. A—Cortical granules are seen beneath the membrane of mature egg; B—A single cortical granule under electron microscope; C—A section of a cortical granule (after Grant, 1978).

the ovaries. From ovaries, the yolk protein precursors are taken by oocyte membrane by **micropinocytosis**. The formed coated vesicles in micro-

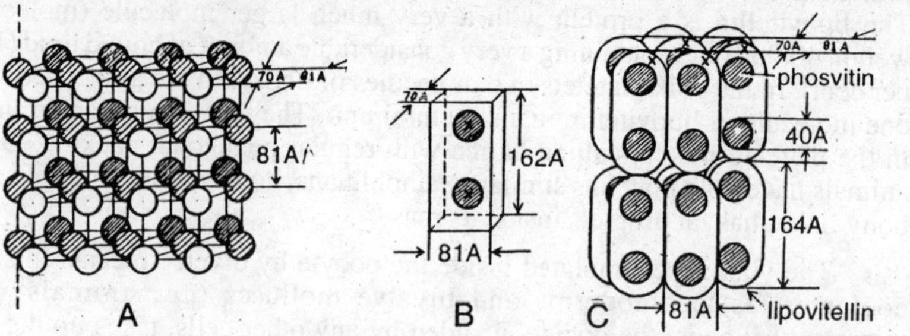

Fig. 5.7. Diagram of crystalline structure in the main body of an amphibian yolk platelet. A—Arrangement of phosvitin molecules in the lattice; B—A structural unit consisting of two molecules of phosvitin and one molecule of lipovitellin; C—Arrangement of phosvitin and lipovitellin molecules in a simple hexagonal lattice (After Balinsky, 1970).

pinocytosis are much smaller, usually lined with small granules and appear in great numbers near the surface of oocytes showing rapid yolk deposition. Presumably a hormonal signal, pituitary gonadotropin, acts directly on the cell surface and stimulates micropinocytic uptake of serum vitellogenin. The appearance of these vesicles on the surface of growing oocytes may be the first stage in yolk platelet formation (Fig. 5.8). Since vitellogenin is not detected in mature yolk, it must be precursor of phosvitin and lipovitellin. During platelet formation, the precursor is converted to yolk proteins by proteolysis. The yolk platelet begins at the cell periphery as a small droplet and gradually increases in size as it migrates into the interior, where it crystallizes into a definite pattern. Most yolk platelets are assembled within the cytoplasm as micropinocytic vesicles enlarge with crystalline yolk. In *Rana pipiens* oocytes, crystalline yolk platelets are sometimes (1 to 2 per cent) seen within the mitochondrial matrix, suggesting a direct role of mitochondria in yolk platelet formation (Fig. 5.9). In molluscan eggs too mitochondria are transformed into yolk bodies (see **De Robertis** and **De Robertis, Jr.**, 1987).

Likewise, in insects, micropinocytotic uptake of yolk precursors from extracellular fluid is a common mechanism of yolk accumulation. In them, the yolk precursors are made by the cells of fat bodies or midgut and then are transported to ovarian follicles for oocyte uptake (Fig. 5.10). In *Cecropia* moths, yolk synthesis takes place in fat bodies in pupal life; later in the adult, yolk precursors are transported to the gonad where they are taken up by growing oocytes. However, not all oocytes are passive in yolk synthesis. In the guppy, for example, yolk is synthesized within the oocyte endoplasmic reticulum and is stored in Golgi vesicles as yolk platelets.

In amphibians, associated with the yolk are enzymes that will be used after fertilization to break down these large storage molecules for use by the embryo. One of these enzymes, **phosphoprotein phosphatase**, splits off phosphate, thereby solubilizing the yolk protein. Another enzyme, a **cathepsin**, can then break down the protein. Further, in amphibians, deposition of yolk into the crystalline platelets begins in the periphery; the oocyte becomes filled from the outside to near the edge of the nucleus, where the process stops. At the end of oogenesis, the yolk platelets are larger and more tightly packed at one end of the egg, the vegetal pole. However, in those vertebrates that possess very large yolky eggs, such as reptiles and birds, yolk deposition continues far beyond that of amphibians, until the active cytoplasm is only a very small fraction of the finished product.

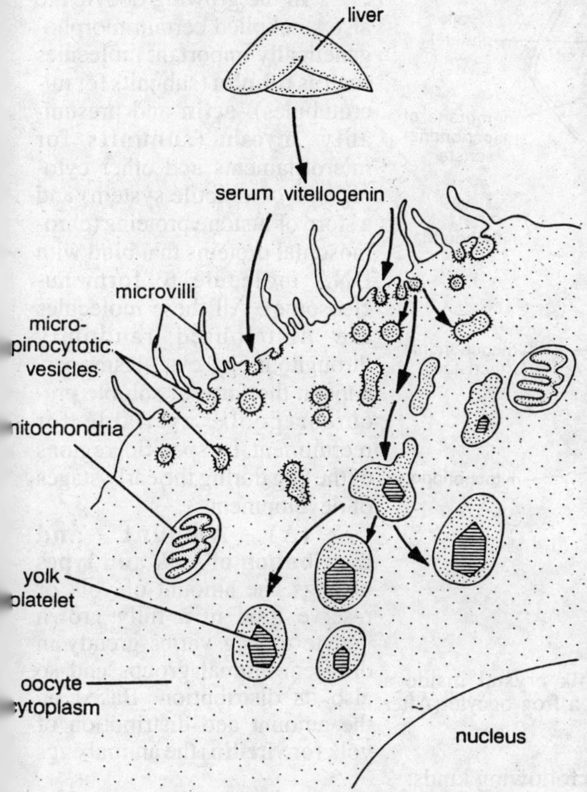

Fig. 5.8. Formation of yolk platelets in the amphibian oocyte (after Grant, 1978).

For long time, it was believed that yolk is synthesized inside the oocyte by a special body, called **yolk nucleus** or **Balbiani body**. The yolk nucleus has been observed in the young oocytes of many animals (*i.e.*, *Spirorbis*) lying next to nucleus, nearer to vegetal pole. Electron microscopic observation of yolk nucleus revealed the fact that yolk nucleus may consist of numerous stacked membranes with mitochondria and vesicular structures (spiders), a mitochondrial cloud with numerous vesicles (*Xenopus laevis*) or Golgi apparatus (mammals) or mixture of them all (chick), so, its exact role in yolk synthesis and accumulation have been questioned (**Goel**, 1984).

In addition to yolk, other nutritive substances such as glycogen, lipids and oil droplets are stored in the egg. Glycogen granules are probably one of the most important energy reserves and in the frog egg, they are more numerous than ribosomes. Various kinds of fat droplets and lipids also accumulate in the ooplasm. Lipids pile up with a thin, surrounding protein-coat and these are termed **lipochondria**. As a rule, lipids made elsewhere in the organism are stored as droplets in oocyte cytoplasm prior to yolk platelet synthesis. Carbohydrates and lipids act as nutrient reserves for energy production during embryogenesis. A lipid store of yolk supplies the phospholipid precursors for plasma membrane synthesis during cleavage, when new cells are rapidly produced.

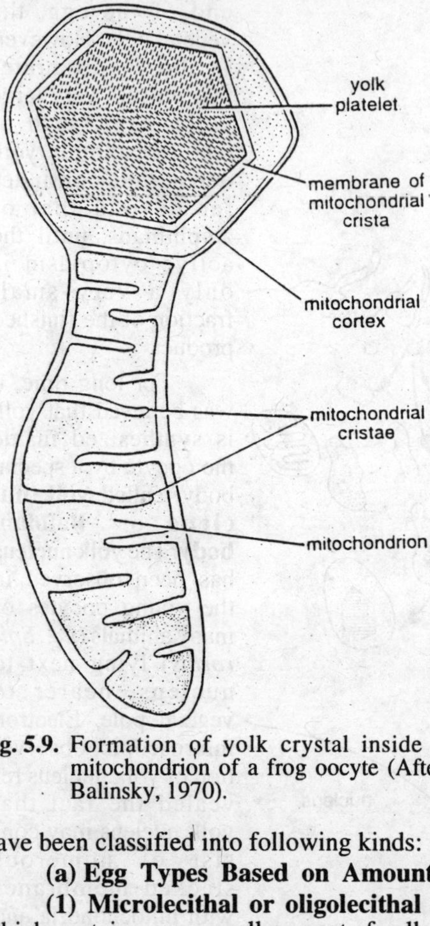

Fig. 5.9. Formation of yolk crystal inside a mitochondrion of a frog oocyte (After Balinsky, 1970).

In the growing oocyte are also stockpiled certain morphogenetically important molecules such as **tubulin** (subunits for microtubules), **actin** and presumably **myosin** (subunits for microfilaments and other cytoplasmic contractile system) and a store of histone proteins (chromosomal proteins that bind with DNA molecule to form nucleosome). All these molecules are distributed randomly throughout egg cytoplasm, perhaps in the form of soluble precursor pools, available for recruitment in specific regions of the egg during the early stages of development.

(A) Amount and distribution of yolk and types of eggs. The amount of yolk or reserve food in a fully grown oocyte or egg varies greatly in different animal groups and so also its distribution. Based on the amount and distribution of yolk (or **vitellin**) the animal eggs have been classified into following kinds:

(a) Egg Types Based on Amount of Yolk

(1) Microlecithal or oligolecithal eggs. These are small-sized eggs which contain a very small amount of yolk. Such microlecithal eggs are found in certain marine invertebrates such as *Hydra* and sea urchin and in various chordates such as *Amphioxus* and other cephalochordates, tunicates and marsupial and eutherian mammals. **Kent** (1969) had described microlecithal eggs as **alecithal** eggs, but this term is not appropriate as there are no egg without yolk.

(2) Mesolecithal eggs. These types of eggs contain moderate amount of yolk and are found in annelid worms, molluscs, Petromyzontia (*Petromyzon*), Dipnoi (lung-fishes) and Amphibia.

(3) Megalecithal, macrolecithal or polylecithal eggs. These types of eggs contain enormous amount of yolk and are found in insects, Myxinoidea (*Myxine*), Chondrichthyes (elasmobranch fishes), reptiles, birds and Monotremata (prototherian mammals).

(b) Egg Types Based on the Distribution of Yolk

(1) Homolecithal or isolecithal eggs. In microlecithal eggs, the amount of yolk is so little that it is found scattered almost uniformly throughout the egg cytoplasm. Such eggs with evenly distributed yolk in ooplasm are described as **homolecithal** or **isolecithal** eggs (Fig. 5.11A).

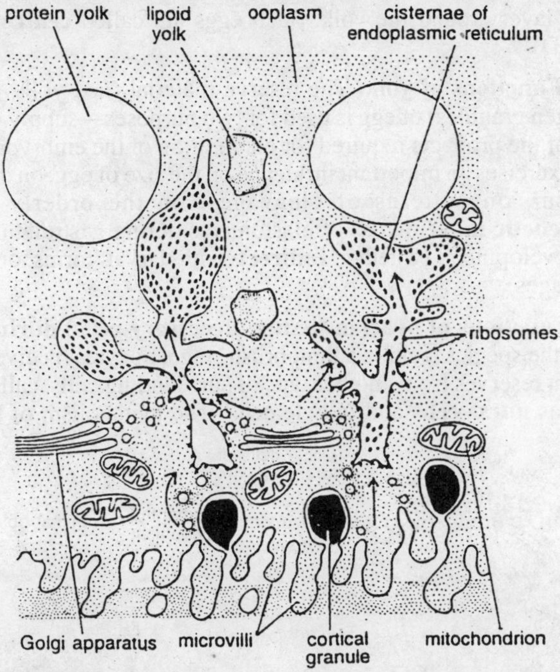

Fig. 5.10. Vitellogenesis in an insect. The lipid component of yolk first accumulates into follicle cells and then is transferred from them to the oocyte. The protein component of yolk is either manufactured in midgut and then transferred to ovaries by blood or synthesized in follicle cells. It is either taken directly from the haemocoel or indirectly through the follicle cells by the oocyte by the act of pinocytosis. Pinocytic vesicles from the periphery fuse with the endoplasmic reticulum and these are joined by vesicles from the Golgi apparatus. Materials accumulate in the cisternae of endoplasmic reticulum and eventually form the protein yolk spheres (After Berrill, 1971).

(2) Telolecithal eggs. These types of eggs have a polarised distribution of yolk in the ooplasm and are found in the mesolecithal and macrolecithal eggs. Thus, in telolecithal eggs, the yolk, due to its gravity, is concentrated more in one hemisphere than in the other. Because of the uneven distribution of yolk, such an egg is said to have a **vegetal pole**, where the concentration of yolk is greatest and an **animal pole**, where the concentration of yolk is smallest. In fact, in macrolecithal eggs, the amount of yolk is so massive that it almost occupies most of the space of animal pole, and the active cytoplasm and germinal vesicle (nucleus) remain confined to a small cap at the animal pole.

The telolecithal eggs may be either **moderately telolecithal** (*e.g.*, eggs of Amphibia, Petromyzontia and Dipnoi; Fig. 5.11 B) or **highly telolecithal** (*i.e.*, cartilaginous and bony fishes, reptiles and birds, Fig. 5.11 C).

(3) Centrolecithal eggs. In insects and some hydrozoa, the yolk is concentrated in the centre of the egg and the active cytoplasm forms a thin peripheral layer around the yolk. Such eggs are called **centrolecithal eggs** (Fig. 5.11 D).

(B) Functions of yolk. The reserve material (*i.e.*, yolk, lipid droplets and glycogen granules) of egg is used for two purposes—supply of energy and synthesis of the products required for elaboration of the embryonic body. The yolk also exercises an important influence on the size of egg, on differentiation of ooplasm, on patterns of cleavages, on the orderly movements (**morphogenetic movements**) of blastomeres during gastrulation and on the type of development (whether **indirect** with larval forms or **direct** with juvenile stages).

F. Functions of ovarian tissues (accessory cells) during growth of oocyte. In the species lacking a lampbrush stage, oocytes receive information and nutrient reserves from another type of surrounding cells, called **accessory cells** such as **nurse cells** and **follicle cells**. The metabolism of both types of

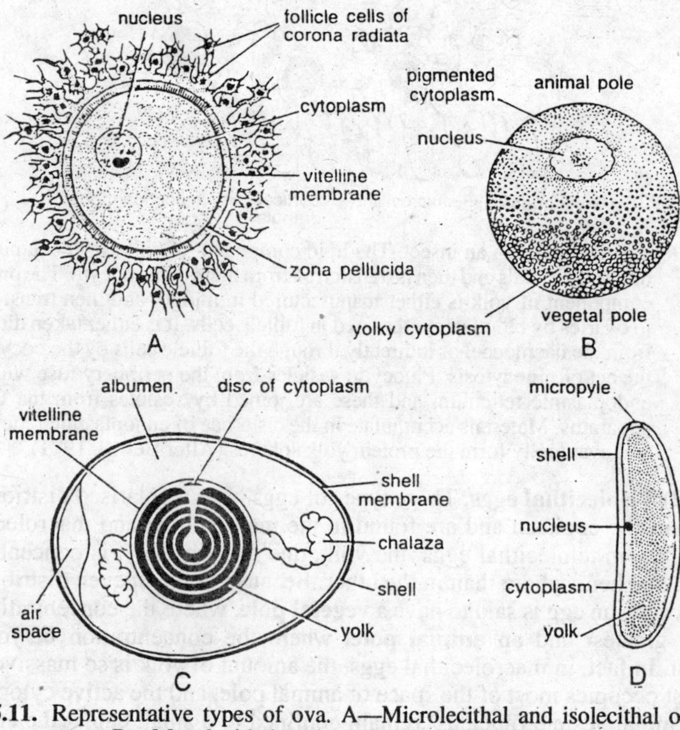

Fig. 5.11. Representative types of ova. A—Microlecithal and isolecithal ovum of man; B—Mesolecithal and moderately telolecithal ovum of the frog; C—Macrolecithal and highly telolecithal egg of the hen; D—Macrolecithal and centrolecithal egg of an insect, the fly (After Arey, 1965; and Balinsky 1970).

accessory cells is coupled to oocyte growth. The intimacy is so great that there exists a direct cytoplasmic connection between nurse cells and/or follicle cells and the growing oocyte. The principal role of these accessory cells is to (1) supply oocyte nutrient reserves during its growth; (2) act as selective barrier between the vascular system and the oocyte, transporting essential molecules (yolk precursors) into oocyte cytoplasm, and (3) synthesize accessory egg membranes.

1. The **nurse cells** are found only in invertebrates such as coelenterates, annelids, molluscs (snail, *Helix*) and insects. They are derived from the same oogonium that give rise to the oocyte. During oogonial divisions in the ovary, at some point a differential division separates cells destined to become oocytes from their sisters which develop into nurse cells. For example, in insects sometimes this division correlates with visible differences between oocyte and nurse cell nuclei. In water beetle *Dytiscus*, a large Feulgen-positive mass (*i.e.*, DNA) in an oogonial nucleus is segregated into only one of the two cells following an oogonial division. This process continues into subsequent divisions, so that only one cell out of a total of 16 retains this extra DNA. This cell is destined to become the oocyte, whereas the others, without the Feulgen positive body, become nurse cells. In *Dytiscus*, this extra DNA represents a mass of ribosomal RNA genes, a mechanism of gene amplification similar to nucleolar amplification in amphibian oocytes.

The oocytes of *Drosophila* lack lampbrush chromosomes and do not synthesize RNA. In it the nurse cells are the chief source of maternal genetic information and ribosomes. Gene amplification mechanisms for RNA synthesis are found in the genomes of nurse cells, due to which the chromosomes of nurse cells become polytenic (giant chromosomes) and metabolically active in RNA synthesis. Thus, during oocyte vitellogenesis of *Drosophila*, the volume of a nurse cell nucleus, nucleoli and cytoplasm doubles once every 4 to 5 hours. As *Drosophila* nurse cells develop, they retain cytoplasmic connections with the oocyte forming **ring canals**, **bridges** or **fusomes** (**Goel**, 1984). Oocytes and nurse cells develop from stem cells within a germarium, a constricted chamber in the ovariole (Fig. 5.12). A single stem cell undergoes a differential division; one daughter cell remains as a stem cell, while the other, called **cystoblast** or **oogonium**, completes four mitotic divisions to form a cluster of 16 **cystocytes**, one of which becomes an **oocyte** and rest 15 become nurse cells. The follicle cells surround all the cystocytes. Nurse cells are connected to the oocyte by direct cytoplasmic bridges or ring canals. The cell retaining the maximum number of cytoplasmic bridges is the future oocyte. The presumptive oocyte usually appears at the first division, while all subsequent divisions produce nurse cells (**Ede**, 1984).

The ovaries of *Drosophila* are **polytrophic**, *i.e.*, only a few nurse cells connect directly with each growing oocyte, which means that products from other nurse cells must funnel through these before they enter oocyte cytoplasm. In other type of insect ovaries, called **telotrophic** ovaries, all nurse cells connect with the developing oocyte, emptying nutrients directly into it. In both types of ovaries, informational macromolecules (RNA messages) from nurse cell nuclei enter the oocyte via ring canals. In the fruitfly this has been demonstrated by showing that RNA labelled with 3H-cytidine in the nurse cell nuclei is found 5 hours later pouring into the oocyte cytoplasm (Fig. 5.13).

2. The **follicle cells** are the accessory cells of vertebrates and other groups. In vertebrates, these cells are derived from the germinal epithelium and become organized as a single layer of epithelial cells surrounding the developing oocyte. Though they may synthesize some substances for oocyte storage, they act more as a selective barrier, mediating transfer of materials from the blood into oocyte cytoplasm, *i.e.*, the materials (*e.g.*, yolk proteins) entering the oocyte from blood stream, have to pass through the membranes of the follicle cells on the way. These cells, therefore, must in some way act as a screening influence to determine what materials will be allowed through. To facilitate this transfer process, there are thousands fine cytoplasmic processes, the **microvilli**, reaching out from the surfaces of both the oocyte and the follicle cells. These microvilli interdigitate to some extent, thereby greatly increasing the surface area over which contacts and, thus, exchange, can occur. In contrast with nurse cells, follicle cells do not arise from oogonia, are not in direct cytoplasmic communication with the oocyte and generally transport rather than synthesize oocyte-bound materials.

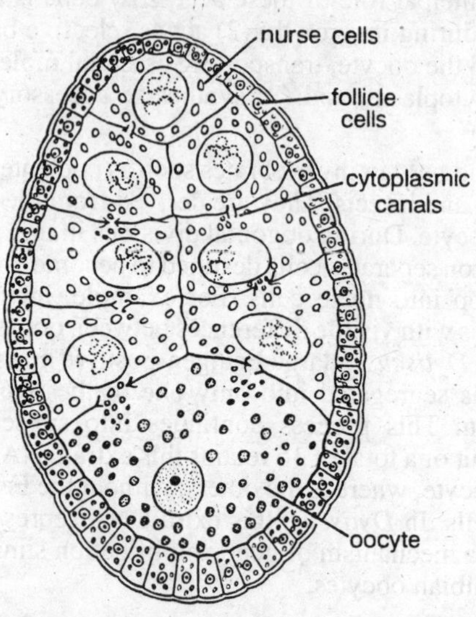

Fig. 5.12. Intercellular cytoplasmic ring canals between the nurse cells and oocyte in a stage 8–9 egg chamber of *Drosophila melanogaster* (after Goel, 1984).

For example, mammalian oocyte (100 μm in diameter) develops within connective structure called a **follicle** (Fig. 5.14). As a given oocyte proceeds towards maturation, this follicle becomes larger and is filled with fluid, called **liquor folliculi** and oocyte becomes surrounded by more and more follicle cells. At the mature stage, the follicle is called **Graafian follicle** having an eccentric, fluid-filled cavity called **antrum** and oocyte remains suspended within the liquor folliculi. A thick gelatinous membrane, the **zona pellucida** (or vitelline membrane) separates the oocyte from a layer of follicle cells, the **corona radiata**, which is several cell thick. Follicle cells also surround the antrum (fluid cavity) and are responsible for the secretion of latter's fluid.

2. Establishment of egg organization. The egg is not just the sum total of molecules contained within it, but is highly complex and well-ordered or organized. This orderliness is reflected in the distribution of various contained substances, particularly yolk, cortical granules, pigments, enzymes, etc. Thus, as the oocyte matures, an organization is set up in it that will affect the future development of the embryo. Such an organization of the egg occurs at both

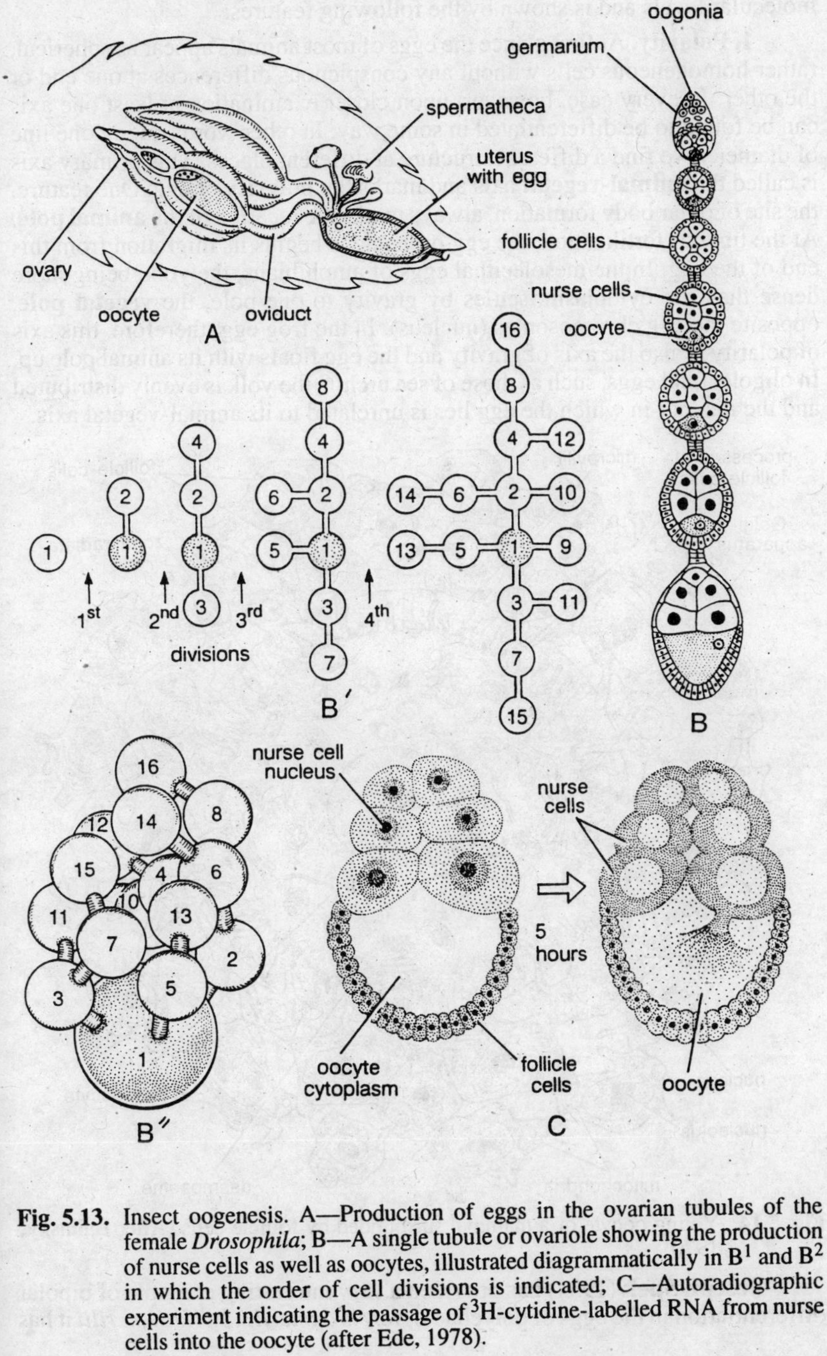

Fig. 5.13. Insect oogenesis. A—Production of eggs in the ovarian tubules of the female *Drosophila*; B—A single tubule or ovariole showing the production of nurse cells as well as oocytes, illustrated diagrammatically in B^1 and B^2 in which the order of cell divisions is indicated; C—Autoradiographic experiment indicating the passage of ^{3}H-cytidine-labelled RNA from nurse cells into the oocyte (after Ede, 1978).

the macroscopic level and on the microscopic and submicroscopic or molecular levels and is shown by the following features:

1. Polarity. At first glance the eggs of most animals appear as spherical, rather homogeneous cells without any conspicuous differences at one end or the other. In every case, however, upon closer examination at least one axis can be found to be differentiated in some way. In other words, along one line of diameter we find a different structure at different places. This primary axis is called the **animal-vegetal axis** and marks the **polarity** of egg. One feature, the site of polar body formation, always marks the location of the **animal pole**. At the time of fertilization, the egg pronucleus begins its migration from this end of the egg. In the mesolecithal eggs of amphibians, the yolk, being more dense than the cytoplasm, settles by gravity to one pole, the **vegetal pole**, opposite the egg chromosomes (nucleus). In the frog egg, therefore, this axis of polarity is also the axis of gravity and the egg floats with its animal pole up. In oligolecithal eggs, such as those of sea urchin, the yolk is evenly distributed and the manner in which the egg lies is unrelated to its animal-vegetal axis.

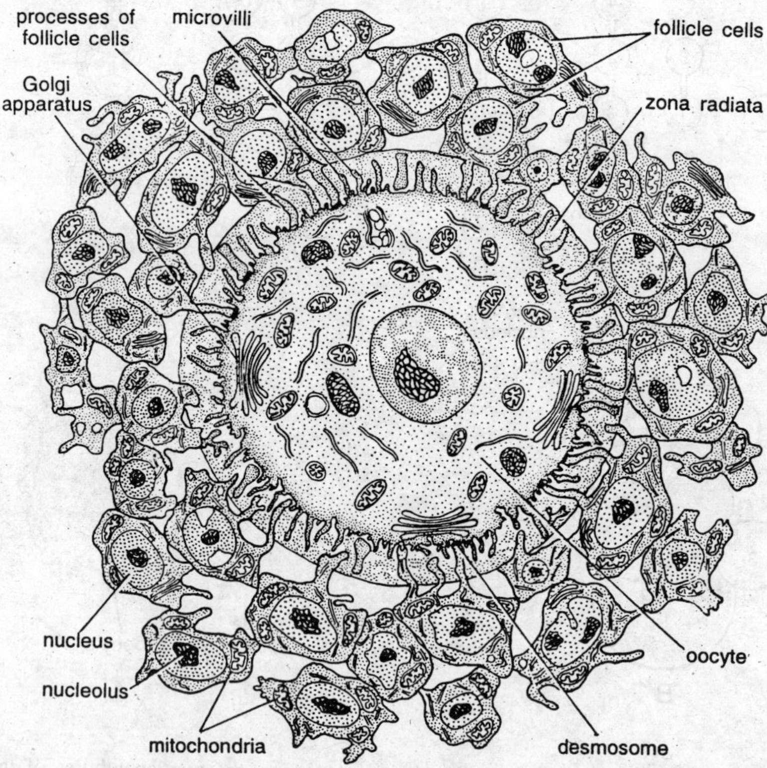

Fig. 5.14. Young oocyte of a mammal surrounded by follicle cells (After Balinsky, 1970).

Joseph Spek (1934) has provided a very interesting example of bipolar differentiation in the eggs of polychaete *Nereis dumerilii.* In *N. dumerilii* it has

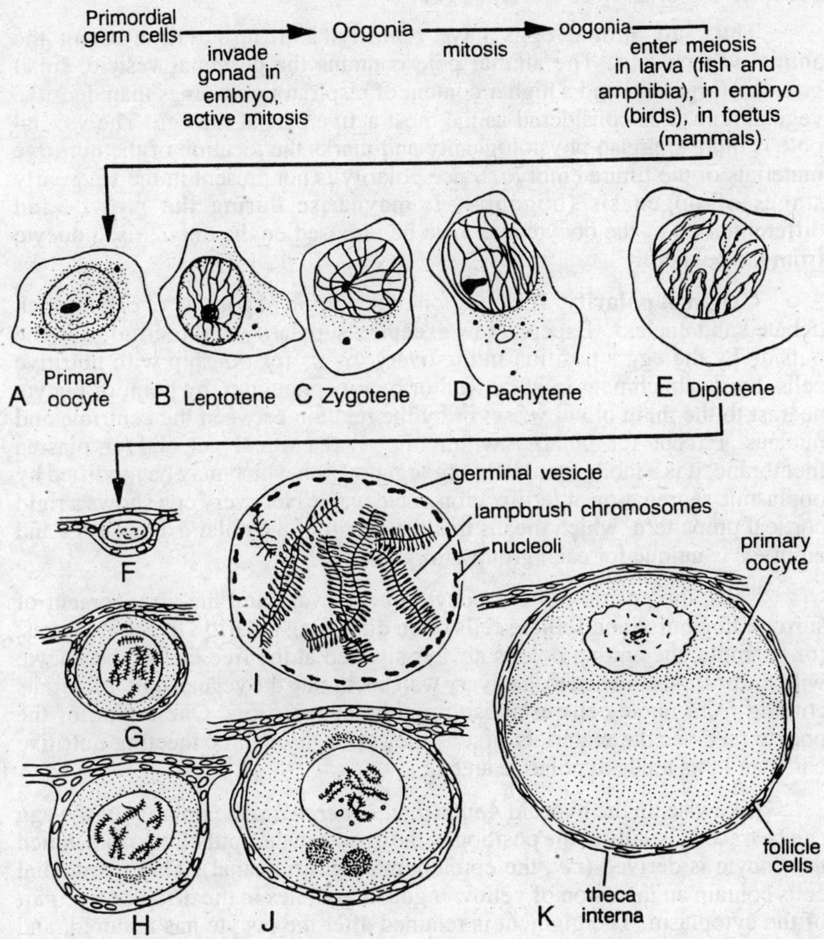

Fig. 5.15. Oogenesis and meiosis in vertebrates. A—Primary oocyte of cat at the beginning of meiosis; B to E—Meiotic prophase in cat, the leptotene to diplotene (at this stage meiosis is arrested for many months or years). F to K—Vegetative growth phase in the amphibian oocyte (after Grant, 1978).

been found that during bipolar differentiation of the eggs, the cellular or ooplasmic substances become distributed in such a way that substances of high pH migrated towards the animal pole, while the substances of low pH migrated towards the vegetal pole. In the egg cells of *N. dumerilii* there is a homogeneously dissolved yellow pigment which is a natural indicator changing colour at about pH 5.4 from yellow to violet from the alkaline towards the acid side. During the bipolar differentiation the pigment is accumulated in the vegetative half of the egg cells, while the animal half becomes colourless (see **Berrill**, 1971).

Thus, all animal eggs have polarity, a differentiation along the animal-vegetal axis. The animal pole contains the germinal vesicle, small granules (pigment) and a higher content of respiratory enzymes than does the vegetal pole. It is considered as the most active part of the egg. The vegetal pole is more sluggish physiologically and marks the location of the nutritive materials of the future embryo. Since polarity is not present in the very early stages of oogenesis (oogonia), it may arise during the growth and differentiation of the oocyte and must be imposed on the unpolarised oocyte from the outside.

Origin of polarity. The question of the origin of polarity of egg is much debated and unclear. Except a few exceptions, polarity is conferred from the outside by the egg's position in the ovary, by its relationship with nutritive cells, by its attachment to body wall or ovary epithelium, by point of oocyte nearest to the main blood vessel or by the relation between the centriole and nucleus. Except for polarity within the rigid cortical gel and /or plasma membrane, it is a labile feature of egg organization which may be modified by ooplasmic segregation at fertilization or cleavage. Not every egg shows a rigid cortical prepattern, which means that the origin of the polar axes in eggs and embryos is unique for each group of organisms.

Thus, the position of the oocyte in the ovary and the arrangement of surrounding nurse and follicle cells often direct egg polarity. In echinoderms, for example, the animal pole is always situated at the free end of the oocyte with the portion attached to the ovary wall becoming the vegetal half. Materials coming from nurse cells establish a unidirectional flow. One region of the oocyte receives the materials first as they pour through connecting nutritive cords and this tends to polarize an egg.

Likewise, in the hydroid *Amphisbetia operculata* the polarity of the egg can be traced visually to the position of the oocyte in the epithelium from which the oocyte is derived (*i.e.*, the epithelium of gastric canal). These epithelial cells contain an inclusion of yellow pigment granules in the distal or free part of the cytoplasm. The pigment is retained after the oocyte has matured, and even after the eggs have started to do cleavage. The part of the developing embryo which contains the yellow pigment becomes the vegetal pole, *i.e.*, side in which endoderm is formed by ingression of cells into the blastocoele (Fig. 5.16).

In hydroid coelenterate such as *Hydractinia echinata* and *Phialidium gregarium* the polar bodies are produced at the free, distal end of the oocyte (the end furthest from the attachment at the base of the gamete-producing epithelium). After one end of the egg, where the polar bodies have been formed, was marked by a vital stain by **Freeman** (1979) and the egg was fertilized and allowed to develop, it was found that the first cleavage furrow started within the stained part of the egg and subsequently this part of the egg gave rise to the posterior end of larva (planula) which after attachment to the substratum gave rise to the oral end of the polyp (**Balinsky**, 1981).

Another mechanism of polarity determination is found in amphibians and mammals in which the growing primary oocytes are surrounded by follicle cells from all sides. It has been suggested that the course of the nearest blood vessel supplying parts of the ovary with nourishment might cause the parts of the oocyte nearest to the vessel to develop into the vegetal pole. But according to **Child** (1941), the animal pole, as the more active one, should develop from that part of the oocyte which has a better oxygen supply and on this principle the part of the oocyte nearest to a blood vessel should become the animal pole.

According to **Balinsky** (1970), the position of animal and vegetal poles seems to have no connection with the course of the blood vessels, because, polarity is a phenomenon which is found not only in egg cells, but in other cells as well. For example, in epithelial cells there is a distinct difference between their proximal ends (the ends resting on the underlying basement membrane) and their distal ends (which form the free surface of the epithelium). Likewise, in nerve cell the polarity of the cell takes the form of the opposite differentiations of axons and dendrites. Further, in the interior of any cell, the position of the centrosome in respect to the nucleus establishes a general form of polarity, the main axis of which is the line drawn through the centrosome and the nucleus. This polarity not only affects the distribution of cytoplasmic inclusions (Golgi apparatus is often found grouped around the centrosome), but, may influence the nuclear structures. For example, in oocyte meiotic prophase, the bivalent chromosomes show a well marked orientation in respect to main axis of cell.

In some animals (*i.e.*, *Limnaea*), the egg polarity seems to be fixed in cortical structures. For example, fat droplets appear first at the vegetal side of the *Limnaea* oocyte, forming a dense cap. If ovarian oocytes are stratified by centrifugation, the droplets are driven centripetally, but within six hours after centrifugation they return to their original location at the vegetal pole, perhaps attracted there by the cortical field (**Grant**, 1978).

However, there are evidences which show that in some animals such as in *Xenopus*, the animal-vegetal polarity could be achieved as a result of intrinsic factors contained in the oocyte itself, quite independently of the maternal body. **Wallace** and **Misulovin** (1978) grow the oocyte of the frog, *Xenopus laevis*, in a culture medium in complete isolation from the female body. An essential ingredient of the culture medium has been vitellogenin (the precursor of the yolk accumulated in the oocyte). This is necessary because the oocyte does not synthesize yolk in its cytoplasm, but draws on a supply provided from the blood. The smallest oocytes used in the experiments had a diameter of 0.66 mm, and were barely pigmented. In the course of cultivation they became fully pigmented over their entire surface, and subsequently the pigment migrated to the animal hemisphere of the oocyte. The oocyte in this way acquired the typical heteropolar structure, with a pigmented animal hemisphere and a vegetal hemisphere largely devoid of pigment. When these oocytes reached to a size of 1.1 mm in diameter, they were treated with progesterone hormone, then, these oocytes immediately were converted into mature eggs by rupturing the germinal vesicles and forming first polar bodies.

In fact, in the oocyte of *Xenopus* it has been found that morphological polarity is the end product of physiological processes. By applying specially adapted electrodes to the oocytes of *Xenopus* (which have a diameter of 0.5 mm and in which pigment is just started to be formed), it was found by **Robinson** (1979) that there is a continuous current 1 μ A/cm^2 in strength passing through the oocyte, taken as the movement of positive charges. The positive current is maximal at the animal pole, is reduced to zero just below the equator and becomes negative in the vegetal hemisphere with a negative maximum at the vegetal pole. However, with maturation of oocyte, this current stops within a short time.

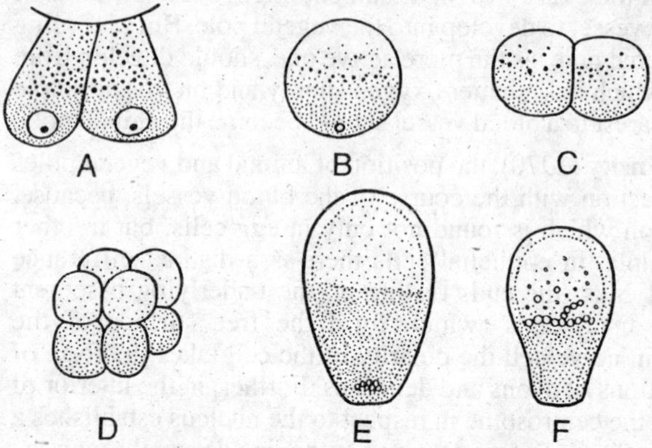

Fig. 5.16. Origin of polarity in the egg and embryo of the hydroid *Amphisbetia operculata.* A—Oocytes as epithelial cells; B—Egg; C, D—Cleavage; E, F—Formation of endoderm at vegetal pole (After Balinsky, 1981).

The actual carrier of the charges was found to be Cl^- ions moving through the oocyte from the vegetal to the animal pole, with the permeability of the plasma membrane to Ca^{++} ions being an essential factor. The current automatically produces a potential difference along the animal-vegetal axis, and this in turn could cause a directed movement of charged particles inside the oocyte, establishing as it were an electrophoresis system in the cell. Such an 'electrophoresis-like' system could well contribute to the morphological animal-vegetal polarization in the oocyte (**Balinsky**, 1981).

Significance of egg polarity. The main significance of egg polarity lies in the fact that this property is often translated into the major axes in the embryo. For example, the animal-vegetal polarity of the amphibian egg becomes the anterior-posterior axis of the embryo, with the head appearing at the animal pole and the anus developing near the vegetal pole. In general, the animal pole is differentiated into the external covering (ectoderm) of embryo, while the vegetal pole typically is pushed into the interior of the egg at gastrulation and differentiates into the endoderm and some time into the mesoderm. The same rule applies to axes of symmetry of the embryo, that is, dorsal-ventral and left-right. These too are often "preformed" in the egg. For example, in the macrolecithal eggs (*e.g.*, chick), the vegetal pole becomes the ventral side of the embryo.

The cleavage planes are also formed in relation to the axis of polarity. For example, the first and second planes of the sea urchin and frog coincide with the animal-vegetal axis, *i.e.*, they are meridional; the third is perpendicular to it, *i.e.*, the equatorial. In the sea urchin the axis of polarity becomes the axis of radial symmetry during cleavage, while in the frog it comes to lie within the plane of bilateral symmetry (**Berrill** and **Karp**, 1978).

2. Symmetry. In animals with a symmetrical body plan that is roughly bilateral such as vertebrates, the body axes can be defined as the antero-posterior, the dorso-ventral and the left-right. We have just described how the antero-posterior axis is typically related to the axis of polarity and as such is established in the ovary. The other two axes become established simultaneously, though in different manner among various animals. In some species such as cephalopods, molluscs, some insects and chordates (*e.g.*, tunicates, cephalochordates and amphibians), the unfertilized egg already possesses these three identifiable axes as a result of oogenesis. In some insects, these axes become established within the oviduct. In other cases, the point of fertilization is clearly related to the plane of bilateral symmetry, as in many molluscs, annelids and amphibians; in other animals such as birds, reptiles and mammals, these axes are not established until the end of cleavage (**Berrill** and **Karp**, 1978).

3. Polarplasm and ooplasmic segregation. In the eggs of molluscs, annelids and ascidians, the ooplasm displays some definite organization. For example, in *Dentalium*, the egg is released from the ovary with the germinal vesicle still intact and it proceeds to the meiotic divisions only afterwards. After release from the ovary, the egg partially rounds up but remains slightly flattened from animal to vegetal pole. In a vertical section through the egg, the light area at the vegetal pole (the pole that was attached to the wall of ovary) is seen to be made up of clear yolk-free cytoplasm. This cytoplasm reaches inward to the nucleus and surrounds it, at the outer edges it is continuous with a very thin cortical layer covering the whole of the egg surface. There is also a small patch of clear cytoplasm at the opposite pole; this is the place from where the polar bodies will be given off. The rest of the cytoplasm is filled by rather densely packed yolk granules. The yolk-free, unpigmented cytoplasm of vegetal pole is called **vegetal polarplasm**. While, the pigment free **animal polarplasm** is only partly free from yolk. Both types of polarplasm have great significance in development.

In certain eggs (*e.g.*, gastropods), in a primary oocyte or egg, various components of cytoplasm move in such a fashion that there establishes a condition in which different parts of the egg cytoplasm (ooplasm) differ more or less in their chemical composition. This is called **ooplasmic segregation**. Thus, glycogen tends to shift and concentrate towards the animal pole of the egg; RNA concentrates in both polarplasms; and vitamin C, presumably bound to the Golgi-apparatus, tends to concentrate into an annular band near the equator of the egg. The displacement and selective accumulation of cytoplasmic components during ooplasmic segregation have been observed among chordates, in ascidians and mammals.

4. Ooplasmic pigments and their significance in development. During differentiation phase of oogenesis of certain animals (*e.g.*, sea urchins, ascidians, frog, etc.), there appear some kind of pigment granules in the cytoplasm of primary oocyte. These ooplasmic pigment granules occur in ectoplasm of egg and are found to have no developmental significance, because many animals (*e.g.*, *Amphioxus, Triturus cristatus,* etc.) lack the ooplasmic pigment granules but have normal development. However, in the eggs in which these pigment granules (yellow in ascidian *Cynthia* (*Styla*) *partita*; red in sea urchin *Paracentrotus lividus* and dark brown in amphibians) occur, there arises a distinct pattern and organization in the ooplasm in respect to them. For example, in frog, the ooplasmic pigments make bipolar differentiation of oocyte well distinct and after fertilization they help in the establishment of bilateral symmetry of fertilized egg (Fig. 5.17).

5. Cortical differentiation. The cytoplasm of growing oocytes and unfertilized eggs of some animals such as echinoderms, molluscs, annelids and amphibians, becomes differentiated into following two regions:

(a) Cortical layer or egg cortex. Developmental information is also stored in the egg cortex, that is, the plasma membrane and underlying cortical gel. The cortical layer of cytoplasm or ectoplasm exits in a viscous state and its components are not displaced either by cyclosis (cytoplasmic streaming) or centrifugation. In its ultrastructure, egg cortex is found to contain network of **microfilaments** which are composed of bundles of **actin** protein or **stress fibres**. Certain cytoplasmic components such as cortical granules, pigment granules or vitamin C containing granules of mollusc, *Aplysia limacina*, are either embedded in or attached to the egg cortex and so remain in fixed position. Centrifugation experiments have suggested that the egg cortex is the actual carrier of the polarity of the eggs of sea urchin (*Arbacia*) and mollusc, *Ilyanassa*. In normal development, the cytoplasmic substances in the egg are distributed in relation to local differences in the egg cortex, and, thus, it is the egg cortex that foreshadows in some way, the pattern of future development of the embryo.

(b) Endoplasm. The interior of egg cytoplasm is occupied by yolk, other inclusions and cellular organelles such as nucleus, mitochondria, Golgi apparatus, endoplasmic reticulum, etc., and is called **endoplasm**. All the endoplasmic components display a graded distribution along an axis (*i.e.*, animal-vegetal axis) and the nucleus. Further, they are freely movable in natural conditions owing to cytoplasmic streaming or cyclosis.

5-3.3. Maturation of Egg (Completion of Reduction Division)

Maturation of egg involves completion of meiotic cell division, *i.e.*, formation of three tiny-sized, yolkless haploid polar bodies and one large-sized, yolk-rich haploid ootid, ovum or egg due to equal meiotic karyokinesis (nuclear division) and unequal cytokinesis (cytoplasmic division). The germinal vesicle breakdown marks the end of the vegetative growth phase of oogenesis and initiation of meiotic metaphase.

It has been mentioned earlier, that during the growth and differentiation periods of oogenesis, the nucleus of primary oocytes remains in a prolonged meiotic prophase during which stage various genetical events such as synapsis, duplication, chiasma formation, and crossing over take place between the homologous chromosomes. After differentiation phase of oogenesis the oocyte nucleus or germinal vesicle resumes meiosis and it reaches the diakinesis stage. Due to some type of stimulation the nuclear membranes of oocyte nucleus break up and the contents of the nucleus become intermingled with the surrounding cytoplasm. The rupture of the nuclear membrane, apart from liberating the chromosomes for the subsequent division, is of great importance because substances such as rRNA, tRNA, mRNA, riboproteins and also certain stimulatory chemical molecules which initiate the formation of microtubules, all of which are accumulated in the nucleoplasm during growth phase of oogenesis, are released and become mixed with the rest of the cytoplasm. After the breakdown of the nuclear membrane, the chromosomes which have become greatly contracted and concentrated towards the centre of unbroken germinal vesicle, are carried to the periphery of animal pole of the oocyte. Here, centrioles form a spindle of microtubules which takes up a position perpendicular to the surface of the primary oocyte. Typically, the microtubules of the meiotic apparatus become attached to a narrow region around the animal pole which remains predetermined for the expulsion of polar bodies. Each synapsed homologous chromosome or tetrad is ultimately separated into two component chromosomes. A bulge now appears at the surface of the animal pole of the oocyte and the outer centriole of spindle with half of the chromosomes (haploid set of chromosomes) enters into this cytoplasmic bulge during the anaphase of first meiosis. The part of the cytoplasm which is bulged out and having a centriole and haploid set of chromosome is pinched off from the rest of the oocyte. Due to this equal nuclear division but unequal cytoplasmic division (cytokinesis) there forms a small-sized, haploid **polar body** or **polocyte** (the name polocyte has been given to it, because, it pinches

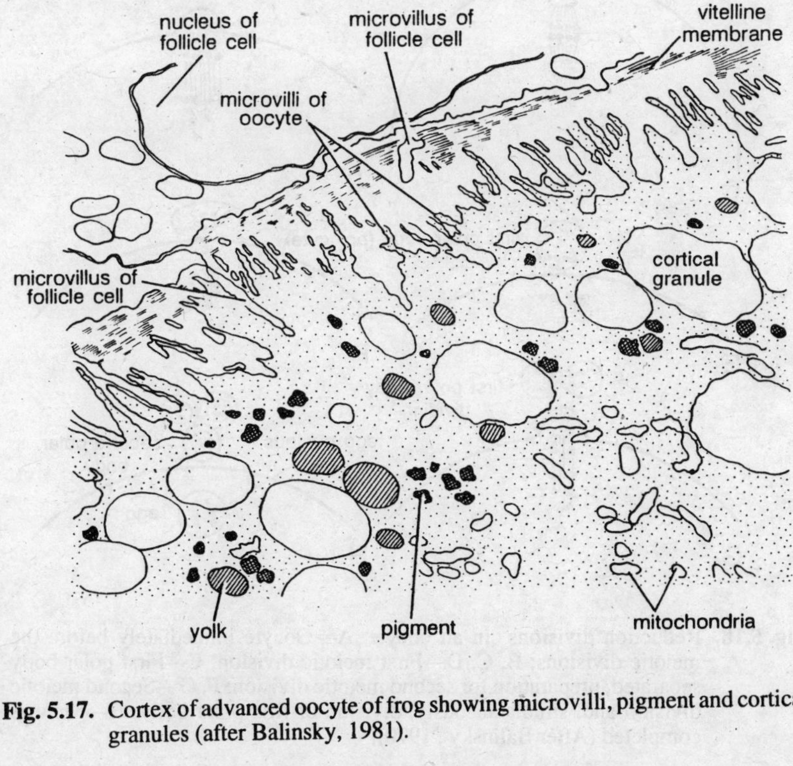

Fig. 5.17. Cortex of advanced oocyte of frog showing microvilli, pigment and cortical granules (after Balinsky, 1981).

off at the "animal pole" of the egg), and a large-sized cell, called **secondary oocyte** or **ootid**, of the same size as the original cell, the primary oocyte.

The ootid undergoes an equal nuclear division by meiosis II, which is followed by an unequal cytoplasmic division and, thus, a haploid **secondary polocyte** and a haploid large-sized **ovum** are formed. The first formed polar body (**primary polocyte**) also divides by meiosis II and two secondary polocytes are formed. Out of these four meiotic products, only the ovum or mature egg remains functional as female gamete and has the capacity to develop into an adult animal, while, three polocytes because have very little

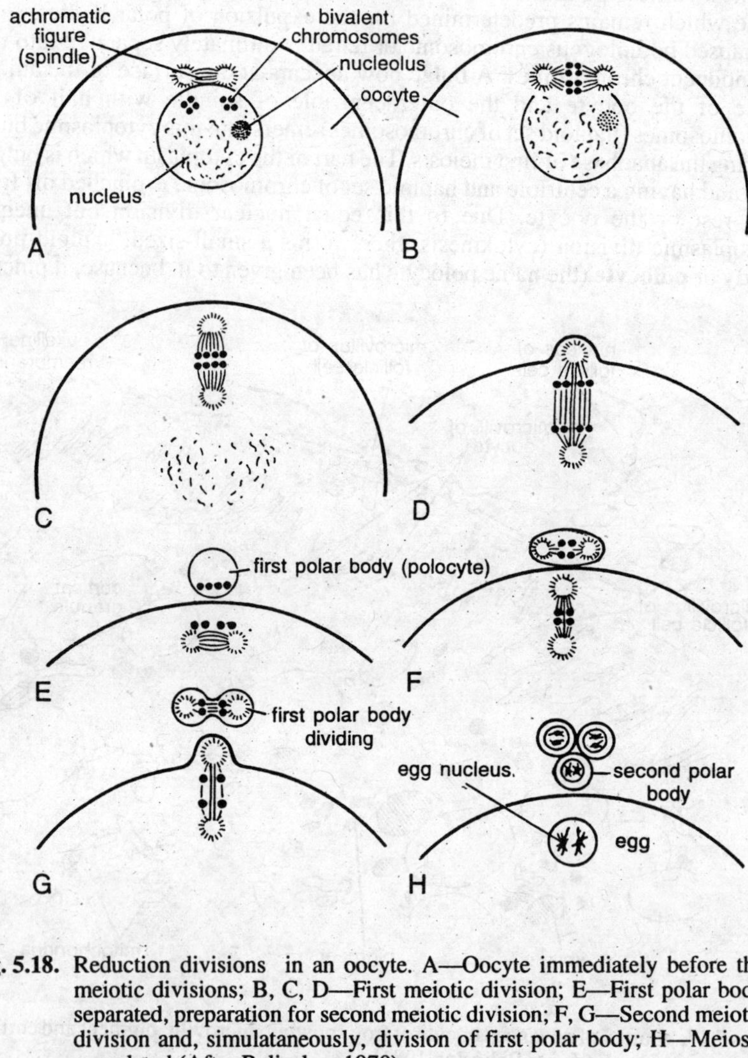

Fig. 5.18. Reduction divisions in an oocyte. A—Oocyte immediately before the meiotic divisions; B, C, D—First meiotic division; E—First polar body separated, preparation for second meiotic division; F, G—Second meiotic division and, simulataneously, division of first polar body; H—Meiosis completed (After Balinsky , 1970).

cytoplasm, with no food reserve (yolk), lack the capacity to develop further, so later die off.

In fact, eggs need not complete meiosis before they are released from the ovary. They may be ready for fertilization at various stages in meiosis as long as vegetative growth is completed. Depending on species, meiosis may be completed within the ovary, in the oviduct, or outside the body before or even after fertilization. For example, in the sea urchin, germinal vesicle breakdown and polar body formation are completed within the ovary before eggs are shed. In related species such as starfish, eggs are shed into the sea before germinal vesicle breakdown, while in many worms germinal vesicle breakdown is induced only by after sperm entry (For more details See Chapter 8).

1. Causes of meiotic arrest during oogenesis. The question may be raised as to why meiosis is delayed until after oocyte differentiation in the female, unlike the male. According to **Berrill** and **Karp** (1978), by delaying meiosis until after oocyte growth and differentiation, the tetraploid amount of DNA can be utilized as a template for RNA synthesis rather being restricted to one-fourth of this amount. Another factor may relate to the DNA/ cytoplasmic ratio. From several different examples, it has been understood that there is a significant relationship between the size of a cell and its DNA content. Under experimental conditions, amphibian embryos can be raised with haploid, diploid, triploid, tetraploid or pentaploid amount of DNA per cell. In these embryos there is a striking relation of DNA content to cell size. Tissues of a pentaploid embryo contain cells approximately five times larger than a haploid embryo, but approximately one-fifth the number. It is conceivable that the higher premeiotic DNA content of the germinal vesicle facilitates the greater cytoplasmic volume.

2. Significance of unequal cytokinesis during oogenesis. The asymmetrical fractioning of the primary oocyte into three small-sized, haploid polocytes and one large-sized, yolky haploid egg, has great embryological significance. If the equal cytoplasmic divisions of primary oocyte might have been occurred, the reserved food amount (yolk, etc.) would have been distributed equally to the four daughter cells which might prove quite insufficient for the developing embryos. Therefore, unequal cytokinesis allows one cell out of four oogenetic meiotic products to contain most of the cytoplasm and reserve food material which may be quite sufficient for the developing embryo.

5-3.4. Formation of Egg Membranes

In most animals, except the sponges and coelenterates, oocyte maturation is not completed until additional structures, called **egg envelopes** or **egg membranes**, are produced outside the plasma membrane of egg. These egg membranes vary in different animal groups and often reflect the adaptations made by the animal in order to ensure development of the young in its particular environment. There are several ways of classifying these membranes, but the simplest way is to group them according to their origin such as follows:

1. Primary egg membranes. These egg membranes are formed in the ovary and laid down between the egg plasma membrane and follicle cells.

They are formed either by ovum or follicle cells and sometimes by the cooperative effort of both. Primary egg membranes are of the following types:

(a) Vitelline membrane. In fact, this is not a membrane, but a noncellular, transparent layer of mucoprotein, forming an extracellular matrix which is both fibrous and amorphous. It is often much thicker and stronger than the underlying fine plasma membrane of egg and has tensile and elastic properties. The vitelline membrane has been given different names in different animals. For example, in amphibians and birds it is in close contact with the ooplasmic surface until the egg is fertilized, at which time it separates from this surface, forming a tough fertilization membrane. In fish this membrane is called **chorion**, and in reptiles and mammals it is given the name **zona pellucida**. It usually remains intact through several stages of cleavage where it appears to help hold the dividing cells together.

(b) Zona radiata. The primary envelope of shark, some bony fishes, some amphibians and some reptiles have striated appearance and is called **zona radiata**. It represents the degraded microvilli of the growing oocyte.

(c) Jelly envelope. In echinoderms and many other eggs of marine invertebrates, the primary egg envelope is much thicker structure of jelly coat.

All these primary egg membranes, usually adhere closely to the surface of the oocyte, but at a later stage a space filled with fluid may appear between the egg plasma membrane and primary egg membrane. This space is called **perivitelline space**.

Micropyle. In some eggs a small opening, a **micropyle**, is found in the primary egg membrane. The location of the micropyle always bears a definite relation to the animal-vegetal polarity of the egg, being situated either at the presumptive animal pole, (*e.g.*, Echinoidea, Pisces) or at the presumptive vegetal pole (*e.g.*, certain Mollusca). Sperm usually makes its entry inside egg during fertilization through the micropyle. Micropyles arise through inhibition of the growth of egg membrane either because of the attachment to the ovarian wall (vegetal pole) or because a follicle cell establishes contact with the egg surface (animal pole).

2. Secondary egg membranes. The secondary egg membrane is secreted outside the primary egg membrane by a layer of follicle cell that surrounds the oocyte. It occurs in the form of a chitinous shell surrounding the egg in insects, ascidians and cyclostomes and is called **chorion**. (In teleost fishes, the chorion is formed by oocyte and contains proteins and polysaccharides). In many insects, the surface of chorion shows a complicated sculpture, which is typical of each species.

No secondary egg membranes are found in amphibians, reptiles, birds and even in mammalian eggs. However, in mammals, when a Graafian follicle ruptures and releases the mature ovum, the egg carries with it, for some distance down the fallopian tube, a layer of follicle cells. These cells are columnar in shape and arranged radially in a single layer, called **corona radiata**, around the ovum outside the zona pellucida. The corona radiata does not form a true membrane, since, its cells are peeled off as the egg descends the oviduct.

3. Tertiary egg membranes. These are secreted by the cells of the oviduct as the egg travels down the duct toward the cloaca. The tertiary egg membranes involve albumen layers, jelly coats, shell membranes and the shell itself.

(a) In oviparous sharks and rays (elasmobranchs), the egg is surrounded by albumen and hard horny capsule of a complicated shape. The shell of capsule is drawn out into long twisted horns which serve to entangle the eggs among sea weeds. The horny egg capsule is secreted by **shell glands** of the oviduct.

(b) As the egg of the amphibian spirals down the oviduct, three rather uniform layers of albumen (jelly) are deposited around it. The inner layer is a thin one and most viscous; the outer, wide and more fluid; the middle layer, intermediate. When the amphibian egg is deposited in water, the jelly absorbs water and swells. These jelly envelopes hold the egg together in masses or strands, protect the eggs from infection, insulate the eggs, make them unappetising to predators since they are tasteless, and anchor the eggs to twigs or plants. Besides these physical functions, the jelly coats apparently also help to make the frog's egg fertilizable, since up to now no egg taken directly from the body cavity has been induced to develop into embryo.

(c) In reptiles, birds and chick five tertiary membranes make up the envelopes of the egg. External to vitelline membrane are added the white of egg (albumen), two shell membranes and a porous shell (Fig. 5.11C).

Significance of egg membranes. The extraneous envelopes of eggs of different oviparous vertebrates basically provide the protection to the contents of eggs of developing embryos from different ecological hazards (variable pH, temperature variations, radiations, pollutions, danger of desiccation etc.) and also from mechanical injuries; secondarily, they may prevent self-fertilization as in ascidians, or may provide buoyancy to the eggs as in amphibians and other chordates. Even in viviparous mammals, the extraneous coat such as zona pellucida is found (1) to check polyspermy, *i.e.*, fertilization by more than one spermatozoon; (2) as a means of preventing egg fusion and (3) as a necessary mean of maintaining normal cleavage of the egg following fertilization. Likewise, the primary egg membranes may be protective, osmoregulative and may even be of great morphogenetic significance.

5-4. COMPARISON OF SPERMATOGENESIS AND OOGENESIS

Although one may distinguish three phases, **proliferation (multiplication)**, **growth** and **nuclear maturation** (meiosis), leading to the maturation of the ova or sperm, these events do not occur in the same order in two kinds of gametogenesis, *i.e.*, spermatogenesis and oogenesis. Both types of gametogenesis are distinguished by the following features:

1. Spermatogenesis occurs in the testes of males, while oogenesis takes place in ovary.

2. The reserve food material of spermatocyte never is yolk, while the reserve food material of the growing oocyte is always yolk.

3. In the male, each of four spermatids which are produced from a primary spermatocyte after meiotic division, becomes a functional

spermatozoon, while in the female, out of four cells produced by meiotic division of primary oocyte, only one cell, the egg, is the functional gamete and the other three cells, the polar bodies which have very little cytoplasm and no food reserve (yolk), so later die off.

4. In male, after maturation (reduction) division, the spermatids must still go through the process of differentiation before they become spermatozoa, that is functional male gametes. On the other hand, the egg, after reduction divisions, does not undergo any significant change (*i.e.*, differentiation) and it is quite ready to start development.

5. A further distinction between the male and female gametes is that the male gametes, the spermatozoa, become capable of fertilizing the egg sometimes after they have completed the reduction divisions and subsequently have become differentiated. The eggs become capable of being fertilized sometimes before they have completed reduction divisions and before they have given off the polar bodies.

REVISION QUESTIONS

1. Discuss in brief the process of oogenesis.
2. What is yolk ? Describe its physico-chemical structure and role in determining various types of eggs.
3. Define the term vitellogenesis. Describe the process of yolk synthesis and packaging inside the growing oocyte.
4. Why do lampbrush chromosomes are observed during oogenesis of amphibians ? Describe the structure and function of these chromosomes.
5. What do you understand by polarity in eggs ? Describe how polarity and symmetry are established during oogenesis and what are their significance ?
6. Describe the role of nurse and follicle cells in the growth of primary oocyte.
7. Describe the peculiar features of meiosis involved in oogenesis.
8. What is egg cortex ? Describe its structure and significance in animal development.
9. What are egg membranes ? Describe the structure, types and function of egg membranes.
10. Compare and contrast the process of gametogenesis in male and female animals.
11. In the growth phase of oogenesis, why does the egg accumulate the following:

    (1) Yolk; (2) Mitochondria; (3) mRNAs; (4) Ribosomes; (5) Proteins; (6) Cortical granules.
12. Write short notes on the following:

    (i) Gene amplification; (ii) Yolk nucleus; (iii) Cortical granules; (iv) Germinal vesicle; (v) Vitelline membrane; (vi) Animal pole.
13. Differentiate between the following:

    (a) Telolecithal and centrolecithal eggs; (b) Zona radiata and corona radiata; (c) Nurse cells and follicle cells.
14. Draw a well labelled diagram of electron microscopic view of a growing primary oocyte of a mammal.

6

SEMINATION, OVULATION AND TRANSPORTATION OF GAMETES

When the spermatozoa and ova become mature and capable for fertilization they are released from their respective gonads, *viz.*, testes and ovaries, and are transported from the place of their origin to the area or site, where conditions are optimum for their union or **fertilization**. The process of releasing of mature spermatozoa from the testes is called **semination**, while, the process of setting free of ova from the ovaries is called **ovulution**.

6-1. SEMINATION AND TRANSPORTATION OF SPERMATOZOA

The processes of semination and transportation of spermatozoa, from the seminiferous tubules of testes to the site of fertilization, are performed by testes and accessory reproductive organs of male vertebrates. The male accessory reproductive organs are well adapted for two fundamental kinds of fertilization, *i.e.*, **external fertilization** and **internal fertilization**.

6-1.1. Male Accessory Reproductive Structures of Forms Utilizing External Fertilization

The animals in which external fertilization occurs, the male accessory reproductive organs are of primitive category and are fairly simple in organization. For example, the **cyclostomes** lack the male reproductive ducts, therefore, the mature spermatozoa are shed into the coelom and pass towards the posterior side to emerge externally be means of two **abdominal pores**, each of which empties into the urogenital sinus. In some **teleost fishes** (*e.g.*, *Amia*, *Polypterus*) and **anuran amphibians** (frogs and toads), the mature spermatozoa are collected by short, fine tubules called **vasa efferentia** from the seminiferous tubules of testes and then are conveyed to the mesonepheric ducts, which fundamentally act as urinary ducts to transport urine from the kidneys. The mesonephric ducts are also called **Wolffian ducts**, **opisthonephric ducts** or **ductus deferens**. The mesonephric ducts of both sides open into **cloaca** in anuran amphibians or **urinogenital sinus** in *Amia, Polypterus* and other fishes. In caudate amphibians (*e.g.*, *Necturus* and *Ambystoma*), the mesonephric ducts transport no urine but only spermatozoa, therefore, are called **sperm ducts**, **vasa deferentia** or **ductus deferens**. In frog and salamander a dilation or lateral pocket forms from the sperm duct near the cloaca. These pockets are known as the **seminal vesicles** (Fig. 6.1). Sperms collect here until released.

6-1.2. Male Accessory Reproductive Structures of Forms Utilizing Internal Fertilization

Entirely different, more complex, male accessory reproductive organs are found in those vertebrates where gametic union occurs within the protective structures of the maternal body (*e.g.*, elasmobranch fishes, certain amphibians, reptiles, birds, and mammals). In such animals the mature spermatozoa are collected by vasa efferentia which convey them to the main reproductive duct, the vas deferens. The anterior portion of vas deferens becomes greatly lengthened, twisted or contorted, and highly coiled to form a compact, oval-shaped structure, called **epididymis**. The epididymis remains situated at the cephalic end of a testis and it is a place of physiological maturation of sperms and also a place of storage of mature sperms.

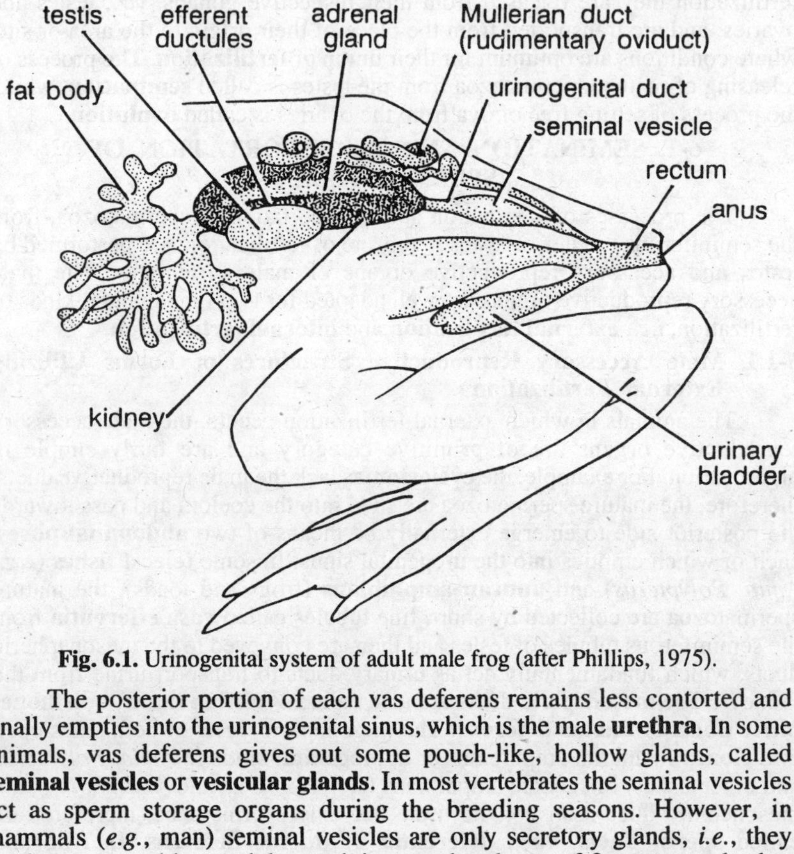

Fig. 6.1. Urinogenital system of adult male frog (after Phillips, 1975).

The posterior portion of each vas deferens remains less contorted and finally empties into the urinogenital sinus, which is the male **urethra**. In some animals, vas deferens gives out some pouch-like, hollow glands, called **seminal vesicles** or **vesicular glands**. In most vertebrates the seminal vesicles act as sperm storage organs during the breeding seasons. However, in mammals (*e.g.*, man) seminal vesicles are only secretory glands, *i.e.*, they produce a mucoid material containing an abundance of **fructose** and other nutrients, as well as, large quantities of **prostaglandins** and **fibrinogen**.

In metatherian and eutherian mammals, there are found some more glands such as prostate glands and mucous glands—such as Cowper's glands or bulbo-urethral glands. In man, the **prostate gland** secretes a thin, milky, alkaline fluid containing **citric acid**, **calcium**, **acid phosphate**, a **clotting enzyme** and **profibrinolysin**.

Intromittent organs. When fertilization is internal, the male vertebrate usually develops intromittent (copulatory) organs for introducing sperm into the reproductive tract of the female. Intromittent organs are particularly characteristic of reptiles and mammals. They are absent in most fishes, amphibia (except gymnophionans which have a definite protrusible copulatory organ of modified cloaca), since fertilization is external in these groups. In urodels, the internal fertilization takes place by taking in of spermatophores by the cloaca. The intromittent organs are also absent in most birds.

The intromittent organs of elasmobranchs are modifications of the pelvic fins and are known as **claspers**. These are grooved, digitiform appendages that are inserted into the cloaca of the female. In many teleost fishes, the anal fin is modified for sperm transport and, thus, becomes a **gonopodium**.

The intromittent organs of reptiles and mammals are of two types—paired **hemipenes** and **penis**. Male snakes and lizards have a pair of stubby, grooved, sac-like, **hemipenes** lying in pockets under the skin besides the cloaca. These can be everted for insertion into the cloaca of the female. After sperm transfer, they are retracted by modified body wall muscles.

Male turtles, crocodiles, a very few birds (drakes, ganders, ostriches and so forth), and male mammals exhibit an unpaired erectile **penis**.

Seminal fluid. Semen or **seminal fluid** is the all important substance which the male contributes during the reproductive event, the copulation. It is the product of the entire male reproductive system and is composed of **spermatozoa** and the **seminal plasma**. The seminal plasma is a fluid and lymph-like substance which contains many enzymes, nourishment in the form of fructose and proteins, and certain those chemical molecules which protect the spermatozoa from other environmental hazards. In many mammalian species, the semen tends to coagulate after its discharge from the male system. In the mouse, rat, guinea pigs, opossum, etc., the semen coagulates into a solid mass, the **vaginal plug**, once it reaches the vagina of the female. The probable function of the vaginal plug is to prevent the semen from seeping out of the vagina. The formation of vaginal plug is due to a protein present in semen and which comes in contact with **vesiculase** enzyme of semen. Coagulation of the semen also occurs in man, boar, etc., but not in dog, bull and many other mammalian and non-mammalian animals. Human semen coagulates immediately after discharge but liquefies a short time afterward due to two enzymes namely, **fibrinogenase** and **fibrinolysin**, both of which are synthesized in prostate glands and occur in semen of man.

Prostaglandins of human semen are cyclopentane-ring containing lipids which act as **parahormones** (or **local** or **tissue hormones**) to regulate or modulate cell metabolism through activating cyclic AMP molecules. They are produced by all types of nucleated body cells and are found to affect a broad-range physiological functions such as reproduction, smooth muscle contraction, nerve impulse transmission, blood flow, etc., (**Tortora** and **Anagnostakos**, 1984). The prostaglandins are believed to aid fertilization in two ways: (1) by reacting with the cervical mucus of uterus to make it more receptive for the sperms, and, (2) possibly by causing reverse peristaltic contractions in the uterus and Fallopian tubes to move the sperms towards the ovaries; in fact, a few sperms reach the upper end of the Fallopian tubes within five minutes (**Guyton**, 1986).

6-1.3. Mechanism of Semination

In most higher vertebrates (reptiles, birds and mammals), the oozing of sperm and seminal fluid from the seminiferous tubules through the rete tubules into the efferent ductules of the epididymis, possibly may be the result of accumulated pressure within the seminiferous tubules themselves. This pressure may arise from secretions of the Sertoli cells, the infiltration of fluids from the interstitial tissues and by the addition of sperm to the contents of the tubules. As the seminiferous tubules are blind at their distal ends, increased pressure of this kind would serve efficiently to push the contained substance forward towards the efferent ductules connecting the testis with the sperm duct (vas deferens). The accumulated pressure of semen, contraction of walls of efferent ductules of epididymis, and contraction of muscular walls of vas deferens, drive forward the semen. The nervous stimulation, psychological state, environmental factor and hormones have active influence on semination.

In the species lacking epididymis, (*e.g.*, fishes, and frogs and toads, etc.), the mature sperms are stored in seminiferous tubules. The pressure within the seminiferous tubules of the testes associated with contractions of the reproductive ducts (vasa efferentia and vas deferens), serve to move the sperm along the reproductive duct. There is some evidence that in frog the fluid accumulates because of the conversion of an acid mucopolysaccharide, **hyaluronic acid**, into its component part. **Van Oordt** *et al.*, (1958) showed that when this process occurred, water is taken up in the seminiferous tubule from the surrounding capillaries. The force of the current is great enough to loosen the sperm bundles from the Sertoli cells and move them toward the efferent ducts.

1. Copulation. In most aquatic animals the eggs and spermatozoa are discharged externally at about the same time and place. Their meeting depends largely upon a chance, enhanced by the production of immense numbers of sex cells. Some animals increase the certainty of such cell union by **pseudocopulation**. For example, during pseudocopulation the male frog clasps the female and pours his spermatic (seminal) fluid over the eggs as they are extruded. On the other hand, many invertebrates and most vertebrates including all reptiles, birds and mammals, have their sex-cells unite within the genital tract of the female. This is brought about by the sexual embrace called **copulation** or **coitus**, during which the penis or some intromittent organ of male introduces the spermatic fluid or semen, in the vagina or some female reproductive organ.

2. Sperm discharge. At the male climax or **orgasm** during coitus, **ejaculation** occurs. During ejaculation, involuntary muscular contraction forcibly eject the older spermatozoa along with seminal plasma into the external environment or inside the body of female.

6-2. OVULATION AND TRANSPORTATION OF OVA

The ovulation and transportation of ova are performed by ovaries and accessory reproductive organs of female vertebrates. The **female accessory reproductive organs** include those organs which assist in the transportation of mature eggs from the place of their origin (ovary) either to the exterior in case of **oviparous** or egg-laying animals (*e.g.*, cyclostomes, most fishes, amphibians, reptiles and birds) or to a female reproductive organ, such as

uterus which retains the eggs for **ovoviviparous** or **viviparous** development. The accessory reproductive structures of female vertebrates fall into following three categories:

(i) The total absence of or the presence of a pair of short funnel-like structures which convey the eggs from the peritoneal cavity through an opening into the urinogenital sinus and then to the outside as in **cyclostomes**.

(ii) A short sinus-like tube remains attached to each ovary and to the urinogenital sinus or to a separate body opening as in many **teleost** fishes.

(iii) In most animals, however, there occur two elongated tubes called **oviducts** or **Mullerian ducts**. The oviducal ducts are variously modified in different animals. Most vertebrates have two oviducts, but reptiles and birds have single oviduct in adult animal. Each oviduct is placed near the ovary and is generally open at its anterior or cephalic end by an orifice. In some vertebrates the anterior orifice of the oviduct may be located at a considerable distance from the ovary, as in frogs, toads and salamanders (Fig. 6.2).

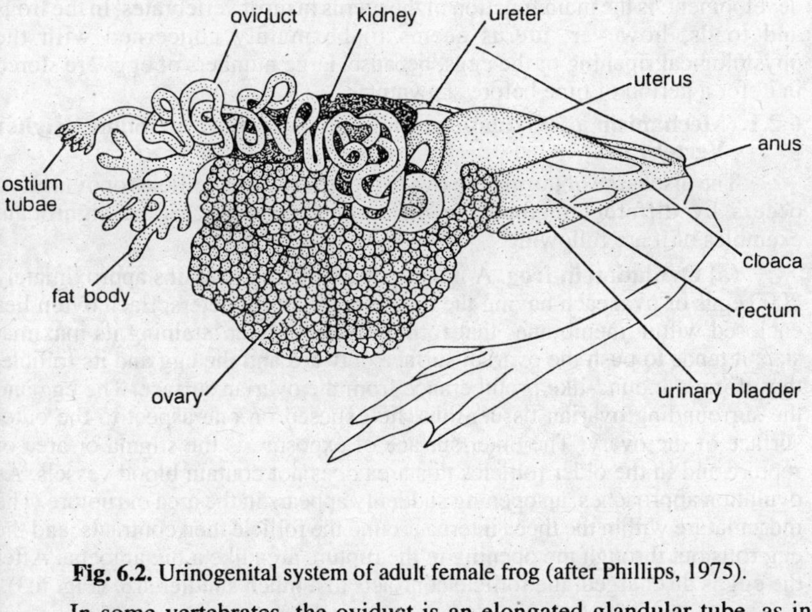

Fig. 6.2. Urinogenital system of adult female frog (after Phillips, 1975).

In some vertebrates, the oviduct is an elongated glandular tube, as in certain urodele amphibians and in ganoid fishes; in others, such as frogs, birds or mammals, it is composed of two main parts, an anterior glandular structure and a more caudally placed uterine portion. The latter may unite either directly with cloaca, as in the frog, or with **vaginal canal** or **vagina**, a structure located either between the uterus and the cloaca in elasmobranch fishes, reptiles and birds, or sinus in mammals. The vaginal canal may be single, as in eutherian mammals, or double, as in metatherian mammals. One of the main functions of vaginal canal or vagina is to receive the intromittent organ of the male during copulation.

The anteriormost opening of each oviduct is called **ostium tubae abdominale**. The anterior end of each oviduct is a fringed, ciliated funnel, called **infundibulum** which is located near the ovary. In the transport of the egg from the ovary to the oviduct, the infundibulum actually engulfs and swallows the egg in many species.

The portion of oviduct anterior to the uterus is often called the **convoluted glandular part**, it is highly twisted and convoluted in many species. In amphibians, reptiles, birds and in some mammals the glandular portion functions to secrete an albuminous coating which is applied to the egg during its passage through this region. It forms the major portion of the oviduct in amphibians, reptiles and birds, but in mammals the glandular part is much reduced in size and extent and is called **uterine** or **Fallopian tube**.

The **uterus** is muscular, posterior segment of the oviduct. Like the convoluted glandular part of oviduct, it also has glandular functions but these are accessory to its more particular property of expanding into an enlarged compartment, where the egg or developing embryo may be retained. The protection and care of the egg or of the embryo during a part or all of its development, is the main function of the uterus in most vertebrates. In the frogs and toads, however, uterus seems to be mainly concerned with the physiological ripening of the eggs, because, large numbers of eggs are stored in it, for a period of time before spawning.

6-2.1. Mechanism of Ovulation and Egg-Transport in Some Higher Vertebrates

The discharge of the mature ova from their ovarian follicles or **ovulation** occurs by different methods in different vertebrates. Some significant examples of it are following:

(a) Ovulation in frog. A mature ovary of frog contains approximately 2000 eggs or ova, each having the diameter of 2 millimeters. Each ovum lies enclosed within membranes that form a follicle. After attaining its maximal size, it tends to push the ovarian surface outward and the egg and its follicle, thus, form a mound-like protuberance from the ovarian surface. The egg and the surrounding ovarian tissue, thus, lie exposed on one aspect to the outer surface of the ovary. The outer surface of exposure is the stigma or area of rupture and in the older follicles this area does not contain blood vessels. As ovulation approaches, an opening suddenly appears in the area of rupture. The musculature within the theca interna around the follicle then contracts, and the egg rolls out through the opening in the rupture area like a big amoeba. After the egg is discharged, the follicle contracts to a much smaller size (Fig. 6.3).

Egg transport in frog. Thus, during ovulation, many mature ova or eggs fall simultaneously into the body cavity or coelom. The coelom is lined with peritoneum; a thin cellular and ciliated layer covering its surface. A rhythmic, synchronized, beating action of the cilia literally sweeps the ova forward from the vicinity of the ovary to the opening of the oviduct, which lies dorsal to lung. The eggs, thus, traverse a large section of the general body cavity before entering the opening (ostium) of the oviduct (Fig. 6.4). As ova pass through the oviducts, they are covered with various coats of jelly finally coming to rest in the enlargements of the oviducts called **uteri**. When about 2000 ova become accumulated in the uteri, the entire mass of egg is released at the time, the male librates its sperms during pseudocopulation and fertilization occurs externally in the water.

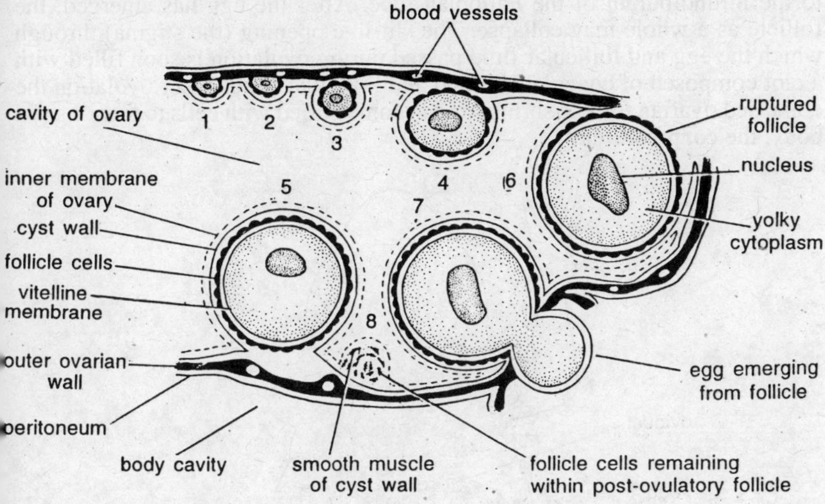

Fig. 6.3. Ovulation in frog (After Berrill, 1971).

(b) Ovulation in hen. As soon as an ovum develops, it gradually pushes the ovarian surface outward and becomes suspended from the general surface of the ovary of hen by means of a narrow stalk, the **pedicle**. When the ovulatory changes are initiated, the musculature of the ovarian wall overlying the outer surface of the eggs appears to contract, and an elongated narrow area along its outer surface becomes **avascular**. This avascular area represents the place where the ovarian surface eventually ruptures to permit the egg to leave the ovary; it is called variously, the **rupture area**, **stigma** or **cicatrix**. Gradually cicatrix widens and finally a slit-like opening is formed by a tearing apart of tissues in the central region of the cicatrix. Contractions of the smooth muscle fibres appear to be responsible for this tearing procedure. The egg is eventually expelled through the opening and in many instances it rolls into the infundibular funnel of the oviduct, which at this time is actively engaged in an attempt to engulf or "swallow" the egg.

Transportation of egg in hen. During ovulation the egg falls into the ovarian pocket, *i.e.*, the space formed around the ovary by the contiguous body organs or in body cavity, it is engulfed by infundibulum of the oviduct. Peristaltic movements of oviduct aid the process of engulfing and further transport of egg inside the oviduct.

(c) Ovulation in mammals. In mammals, as the Graafian follicle enlarges and matures, it moves closer to the ovarian surface. The surface of the ovary, over the ripening follicle, bulges outward, forming a mound-like protuberance. In the rabbit, the central part of the original protuberance pushes out still further and forms a papilla-like swelling (Fig. 6.5). As the papilla develops, a small area of the follicular wall becomes deprived of its vascular supply, so that the underlying tissues become thin and greatly distended. As the distended papillary area continues to grow thinner, a small amount of blood followed by some of the follicular fluid containing the egg emerges from the follicle and passes into the surrounding area (body cavity) in close proximity to the infundibulum of the Fallopian tube. After the egg has emerged, the follicle as a whole may collapse. The slit-like opening (the stigma) through which the egg and follicular fluid passed during ovulation is soon filled with a clot composed of coagulated blood and follicular fluid. After ovulation the collapsed ovarian follicle shrinks and becomes filled with cells to form a solid body, the **corpus luteum**.

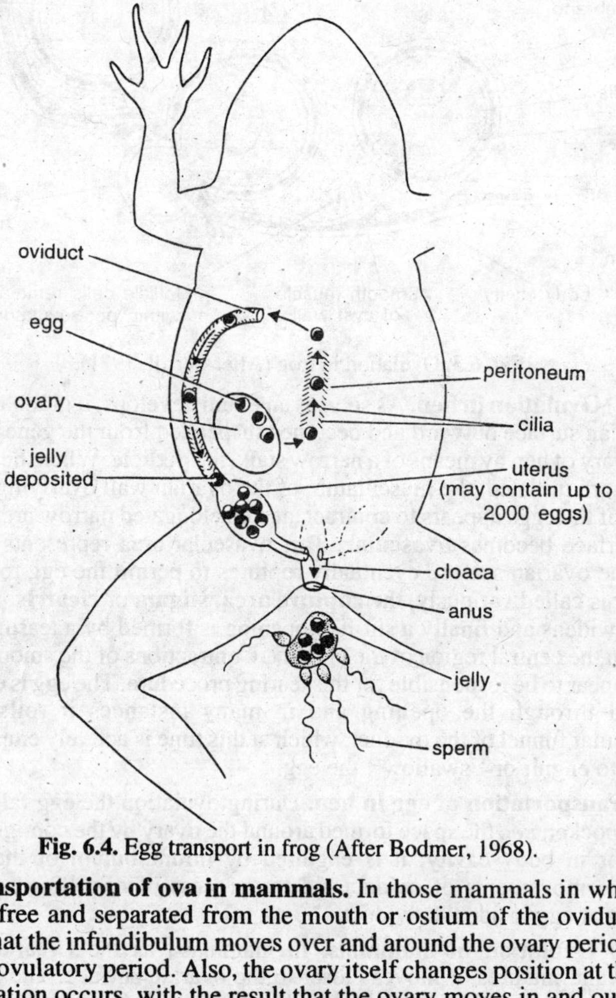

Fig. 6.4. Egg transport in frog (After Bodmer, 1968).

Transportation of ova in mammals. In those mammals in which the ovary lies free and separated from the mouth or ostium of the oviduct, it is probable that the infundibulum moves over and around the ovary periodically during the ovulatory period. Also, the ovary itself changes position at the time when ovulation occurs, with the result that the ovary moves in and out of the infundibular opening of the uterine tube. In the **Monotremata** (prototherian mammals) during the breeding season, the enlarged membranous funnel (infundibulum) of the oviduct engulf the ovary, and a thick mucous-like fluid lies in the area between the ovary and the funnel. At ovulation the relatively large egg passes into this fluid and then into the Fallopian tube. In the rat and the mouse which have a relatively closed ovarian sac, the **bursa ovarica** around the ovary, contractions of the Fallopian tube similar to those of other mammals tend to move the fluid and contained eggs away from the ovary and into the tube. Thus, the activities of the mouth and upper portion of the oviduct serve to move the eggs from the ovarian surface into the reproductive duct at the time of ovulation in the mammal (also in birds, reptiles and elasmobranchs). The intra-oviducal transport of egg is also facilitated by the activity of cilia lining the fimbriae, mouth, and ampullary portions of Fallopian tube.

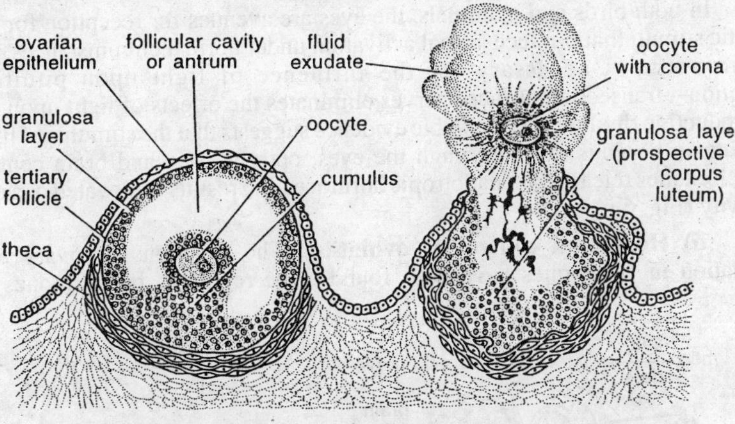

Fig. 6.5. Ovulation in a mammal (After Torrey, 1971).

6-2.2. Time and Regulation of Ovulation

The time of ovulation is directly related to the breeding cycle of animal involved, thus, the timing and control of ovulation are the important biological problems. In animals following three methods of controls of ovulation are known: (i) Environmental control, (ii) Hormonal control, and (iii) Neural control.

(i) Environmental control. Various environmental factors such as duration and intensity of light, temperature, visual, olfactory, auditory and copulatory stimuli, rain fall and lunar tides have been reported to have some definite effect on the process of ovulation. For example, most animals time their breeding (also ovulation) with respect to the seasons in order to take advantage of favourable temperature and food supply. In some marine forms the ovulation has some correlations with the tides and phases of moon. Further, it has been observed that changes in climatic conditions result in variations in the time of ovulation of domestic animals. Animals transported across the equator from one hemisphere to the other, adjust their ovulatory cycle to the new conditions, that is, the reversal of summer and winter. Similar effects have been observed in relation to light. For example, when rats and mice were illuminated all night and kept in dark all day, they changed their reproductive habits and instead of ovulating near midnight, they ovulated near noon. Continuous exposure to light resulted in an inhibition of ovulation. Such kind of phenomenon in which periodicity of light controls the ovulatory or sexual cycle of female is called **sexual photoperiodicity**. Sexual periodicity has also been reported in male animals.

Mechanism of sexual photoperiodicity. During an experiment with the bird, ferret, it is observed that the inactive sexual state of the female ferret during the months from September to February is most easily changed to an active sexual state, if the birds are exposed to 16 hours of light per day of wavelengths between 3650 and 6500 angstroms. Further investigations on this bird suggest that light does not affect the ovary directly but affect it in an indirect manner, that is, via the pituitary gland. Removal of the pituitary gland prevents any effect of increased light upon the gonads. The eyes are also important in this process.

In both birds and mammals, the eyes are avenues for reception for the photic stimuli that produce sexual activation under normal circumstances, the optic nerve is necessary for the influence of light upon pituitary function—transection of these nerves eliminates the effects of light upon the reproductive rhythm. The available evidence suggests that the stimulus of light initiates a nervous reflex through the eyes, optic nerves, and brain centres which excites release of gonadotropic hormones and results in elevated ovarian activity (Fig. 6.6).

(ii) Hormonal control of ovulation. The functions of ovary and ovulation in vertebrates have been found to be regulated by hormones of

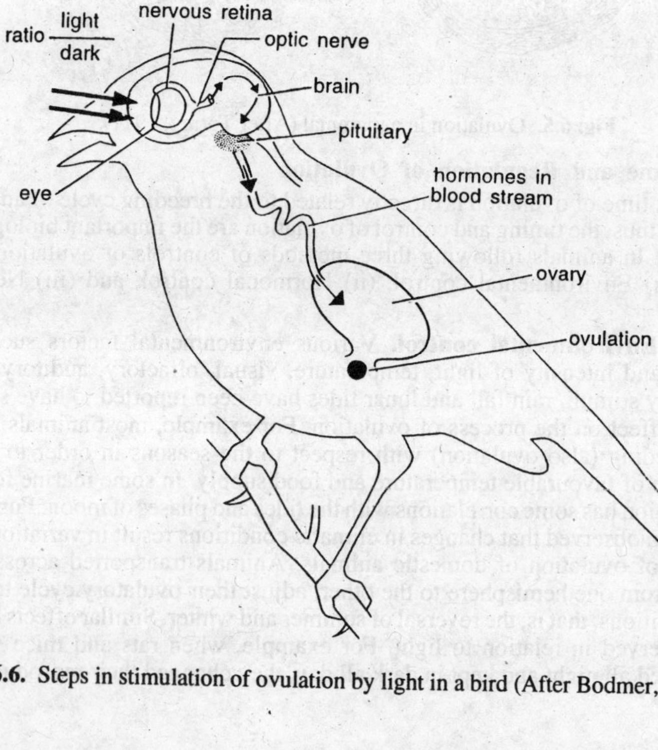

Fig. 6.6. Steps in stimulation of ovulation by light in a bird (After Bodmer, 1968).

pituitary gland. It has been found that the removal of the pituitary gland of the frog prevents ovulation, while, reimplantation of the pituitary gland into a hypophysectomized animal (one in which the hypophysis or pituitary has been removed) will induce ovulation. This is true for many vertebrates.

In mammals, the two processes of ovary, namely the growth of follicle, culminating in ovulation and the formation of the corpus luteum, are controlled by three gonadotropic hormones of pituitary. One hormone, the **follicle stimulating hormone (FSH)**, induces ovarian follicular growth in immature or hypophysectomized female mammal. A second pituitary hormone, the **luteinizing hormone (LH)**, acting alone, has little or no effect on ovarian size or follicular growth, but, when acting synergistically with a small dose of FSH, it elicits follicular maturation, ovulation and the formation of corpora lutea. A third pituitary hormone, the **prolactin** or **lactogenic** or **luteotropic hormone (LTH)**, is responsible for milk secretion and for maintaining the corpus luteum in a functional secretory state.

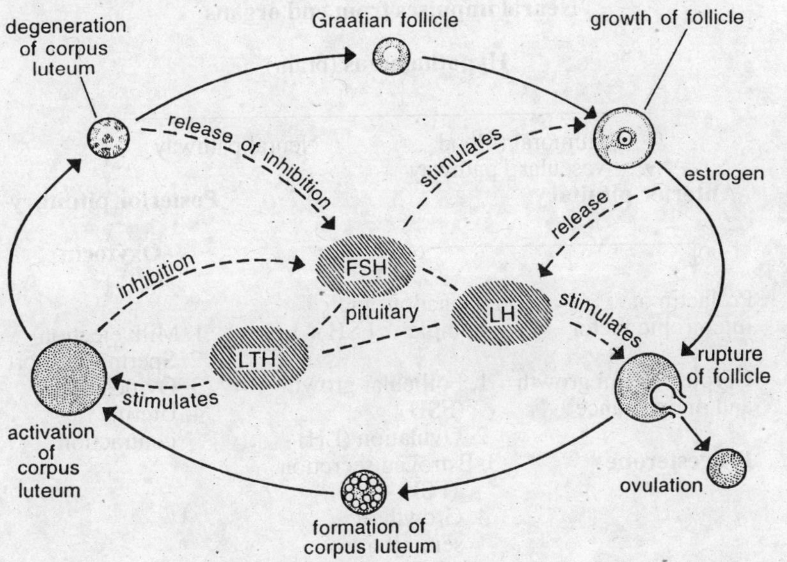

Fig. 6.7. Cyclic role of hormones in the regulation of oogenesis and ovulation in mammals (After Bodmer, 1968).

The pituitary, in turn, is influenced by two ovarian hormones, namely the **estrogen** and **progesterone**. There exist interdependent chain reactions in between the pituitary and ovarian hormones for the regulation of ovarian and other sexual activities of female mammal. Thus, the pituitary hormone **FSH**, with the synergistic effect of a little **LH**, stimulates follicular growth and the production of estrogen in the ovary. The estrogen acts upon the pituitary to stimulate secretion of increased amounts of **LH** and **LTH**, and to cause a decreased secretion of **FSH**. Under such conditions, ovulation and formation of corpora lutea occur. The corpora lutea, which produce the progesterone, are activated and maintained by **LTH**. When the corpora lutea begin to regress, the amounts of ovarian hormones are reduced, with the result that the pituitary again produces **FSH** to initiate repitition of the ovarian cycle (Fig. 6.7). During pregnancy, when the corpus luteum persists because of the presence of an embryo, there is no follicular growth or ovulation, and placental hormones replace the ovarian hormones.

(iii) Neural control of ovulation. The nervous system of vertebrates has an indirect control over ovulation via the pituitary gland. The activities of **adenohypophysis** (anterior lobe of pituitary) and **neurohypophysis** (posterior pituitary lobe) are controlled by neurohormones secreted by neurosecretory cells of hypothalamus of brain. The hypothalamus is supposed to synthesize certain "**releasing factors**" which ultimately release various reproductive hormones of anterior lobe of pituitary, such as, luteinizing hormones (LH), follicle stimulating hormone (FSH) and a variety of other non-gonadotropic hormones. The action of nervous system and pituitary over ovulation and semination can be summarized as follows :

Neural impulses from end organs
↓
Hypothalamus (brain)
↓

- Humoral (blood vascular) pathway → **Anterior pituitary**
  - Prolactin or luteotropic factor
    ↓
    Corpus luteum growth and maintenance
    ↓
    Progesterone
  - Gonadotropic complex : FSH + LH
    ↓
    1. Follicular growth (FSH)
    2. Ovulation (LH)
    3. Estrogen secretion (FSH + LH)
    4. Growth of seminiferous tubules (FSH)
    5. Spermatogenesis (FSH + LH)
    6. Androgen secretion (LH)
- Neural pathway → **Posterior pituitary**
  ↓
  Oxytocin
  ↓
  1. Milk ejection
  2. Sperm transport
  3. Oviposition
  4. Uterine contraction

Comparison of ovulation in invertebrates and vertebrates. Different types of hormones regulate oocyte maturation, ovulation and egg-shedding in invertebrates and vertebrates (Fig. 6.8). For example, in starfish, *Asterias*, at the end of oogenesis, oocytes are still in the germinal vesicle stage. Its ovulation is induced by a gonad-stimulating hormone secreted from the radial nerve and eggs are shed. Hormone secretion of *Asterias*, too, is probably regulated by some environmental stimulus such as temperature or light. The radial nerve factor (*i.e.*, hormone) is a small peptide (2100 mol. wt.) that is isolated directly from nervous tissue or from the coelomic fluid of naturally spawning females. It has been found that this hormone acts directly on ovarian tissue, which is stimulated to secrete another substance, **1-methyladenine**, which acts on the surface of oocyte of *Asterias* and starfishes and ultimately results in germinal vesicle breakdown, *i.e.*, completion of meiosis (maturation).

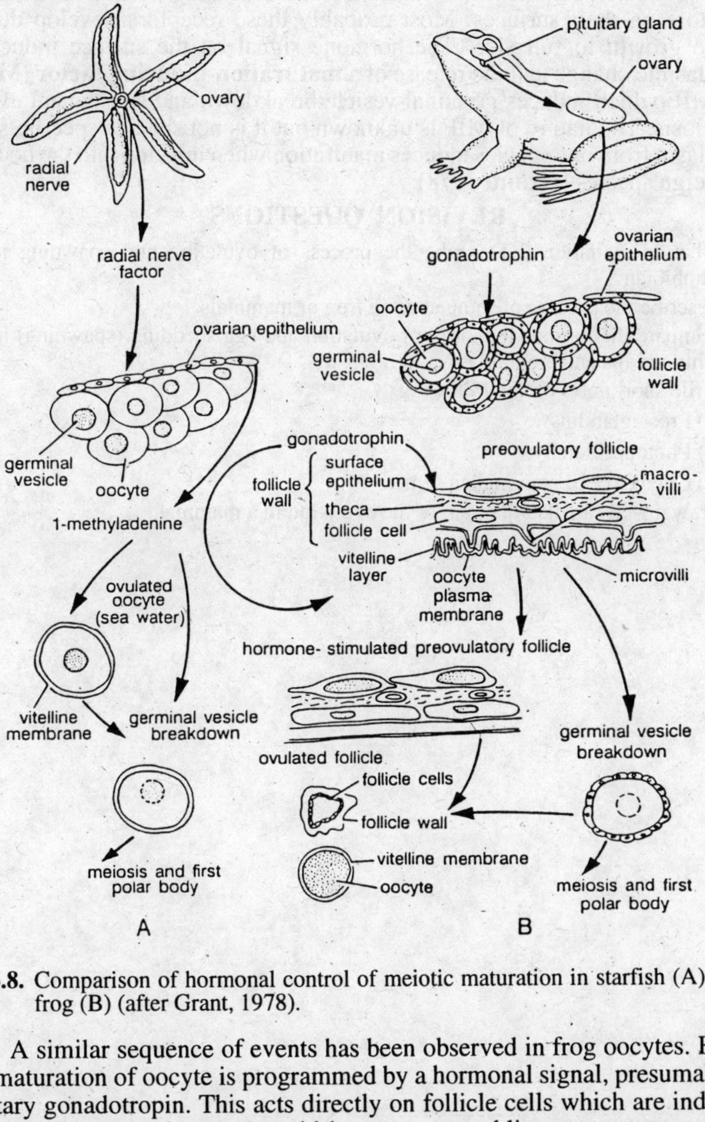

Fig. 6.8. Comparison of hormonal control of meiotic maturation in starfish (A) and frog (B) (after Grant, 1978).

A similar sequence of events has been observed in frog oocytes. Here, too, maturation of oocyte is programmed by a hormonal signal, presumably a pituitary gonadotropin. This acts directly on follicle cells which are induced to synthesize and release a steroid hormone resembling progesterone. This hormone, in its turn, acts on the oocyte surface which triggers germinal vesicle breakdown, completion of the first meiotic metaphase and polar body formation.

In fact, full-sized frog oocytes (about 1mm in diameter) will respond to exogenous progesterone *in vitro*, even when dissected from their follicle membranes. In consequence, the germinal vesicle breaks down and the first polar body appears. However, generally oocytes smaller than this (0.8mm) do not respond to progesterone, presumably because they do not have steroid receptors on their surfaces. Most probably these receptors develop during oocyte growth to full size. The hormone signal on the surface induces a cytoplasmic change and the release of a **maturation-inducing factor** (**MIF**). It is MIF which induces germinal vesicle breakdown and subsequent events of meiosis. The nature of MIF is unknown but it is not species-specific since cytoplasm from one species induces maturation when injected into the oocytes of foreign species (**Grant**, 1978)

REVISION QUESTIONS

1. What is ovulation ? Describe the process of ovulation and spawning in an amphibian.
2. Describe the process of semination in frog or mammals.
3. Compare the hormonal control of ovulation and egg shedding (spawning) in the echinoderms and amphibians.
4. Write short notes on the following:
   (1) Prostaglandins;
   (2) Photoperiodicity;
   (3) Chemistry of mammalian semen.
5. Draw a well-labelled diagram of the ovulation in a mammal.

7

REPRODUCTIVE CYCLES (ESTROUS AND MENSTRUAL CYCLES)

7-1. BREEDING AND NON-BREEDING SEASONS

The female reproductive or breeding activities of almost all animals are cyclic. Usually **reproductive cycles** are seasonal and coordinated so that the young are born at a time of year when conditions for growth are most favourable. Thus, the gonads of vertebrates are not independent structures and do not function continuously, but rather they fluctuate cyclically during the reproductive period of the individual's life. In most vertebrates, the reproductive cycle is a yearly one and in the temperate zone, many breed in the spring (when the ecological conditions are best for hatching or birth). In tropical zone, the breeding season falls either in autumn (*e.g.*, sheep) or in spring and autumn (*e.g.*, bitches). This **breeding** or **reproductive season** alternates with a sexually quiescent one, the **non-breeding season**, in which the ovaries and testes regress. In many male mammals, the testes descend into the scrotum only during the breeding season (See Chapter 4). Thus, the individual during non-breeding season is anatomically and physiologically incapable of reproducing. Domestication and selection modify this basic pattern (*i.e.*, cycle of breeding and non-breeding seasons) and some domestic birds (*e.g.*, hen) and mammals (*e.g.*, cow, buffalo) carry on reproduction throughout the year. Higher primates including man, can breed at any time. Generally the reproductive cycle of the male vertebrate parallels that of the female.

In general, during their reproductive season most poikilothermic (cold-blooded) vertebrates enter into a period of courtship that terminates in mating and spawning. The fertilized eggs are then abandoned by the adults who become reproductively inactive. The reproductive cycles of birds and mammals (homeothermic or warm-blooded animals) are more complex, involving periods of incubation or gestation that call for corresponding changes in adult behaviour.

7-1.1. Clutch Size and Incubation Period

Most American songbirds lay from three to six eggs, but the number of eggs laid in a clutch varies from one or two (some hawks and pigeons) to 18 to 20 (quail). Some birds are **determinate layers** and lay only a fixed number (clutch) of eggs in a season. If any of the eggs of a set are removed, the deficit is not made up by additional laying (*e.g.*, herring gull). **Indeterminate layers**, however, will continue to lay additional eggs for a long time if some of the first-laid eggs are continually removed (*e.g.*, flickers, ducks, domestic poultry).

Further, most birds incubate their eggs with the heat of their bodies by covering them and show considerable concern for their young until they are able to fend for themselves. Many birds, such as ducks, start incubating after the last egg has been laid. Some species such as owl, however, begin incubating as soon as the first egg is laid.

Incubation is usually done by both parents of a monogamous pair (**biparental incubation**), *e.g.*, sparrow and pigeon, or may be done either by female alone (*e.g.*, ducks, peahen, baya weaver bird) or by male alone (*e.g.*, jacana) (**uniparental incubation**). The length of incubation (*i.e.*, the time when the birds sit on the eggs until they hatch) varies. In small birds, it is from 12 to 15 days, whereas, in large birds, it is about 20 days.

Moreover, when birds hatch (*i.e.*, come out from the egg shell), they are of two types–**precocial** and **altricial**. The precocial young such as quail, fowl, ducks and most water birds, are covered with down feathers when hatched and can run or swim as soon as their plumage is dry. The altricial ones on the other hand, are naked and helpless at birth or hatching and remain in the nest as **nestlings** for a week or more.

7-1.2. Gestation Period and Litter-Size

The period of pregnancy or of development within the body of the mother (in viviparous species) varies with the species and is referred to as the **gestation period**. In mammals, the gestation period may be anywhere from 20 to 22 days for the rat to 607 to 641 days for the elephant (Table 7-1). The human embryo takes 270 to 295 days to complete its development.

In fact, viviparity requires structural modification of the reproductive tract. The establishment of an intimate association between mother and embryo allows the young to receive food and oxygen and to excrete wastes. The formation of this intimate relationship is referred to as **placentation**. The structure formed is called the **placenta** which develops partly from the maternal tissues derived from the female reproductive tract (the uterine wall) and partly from extra-embryonic tissues associated with the embryo.

The marsupials (opossum) have a very short gestation period of 13 days; at the end of that time the tiny young leave the vaginal orifice and make their way to the marsupial pouch where they attach themselves to nipples. Here they remain for more than 2 months before emerging.

The **number of young** produced by mammals in a season depends on many factors. Usually the larger the animal, the smaller the number of young in a litter. Perhaps one of the greatest factors involved is the number of enemies a species has. Small rodents can reveal astonishing **fecundity** (*viz.*, animal's potential capability to produce eggs). For example, lemmings in captivity may produce as many as 16 litters per year, each with four to eight young (*i.e.*, up to 128 youngs in a year). Most carnivores have but one litter of three to five young a year. Large mammals, such as elephants and horses, usually have only one young.

Among mammals, the condition of the young at birth also varies. Those young with hair and open eyes and ability to move around are called **precocial** (*i.e.*, ungulates and jackrabbits); those that are naked, blind and helpless (*e.g.*, carnivores and rodents are known as **altricial**.

7-2. TYPES OF REPRODUCTIVE CYCLES OF MAMMALS

In mammals, the reproduction is a well orchestrated process. It needs the coordinated preparation of many tissues in the female under the direction of hormones, starting from maturation of ovum, its release, transportation and fertilization and also the extensive preparation of uterine wall to receive developing egg and ultimately its implantation. All these preparations occur in a cyclic manner in female as a prelude to reproduction. In the male mammals, too, many cyclic changes take place in structure and function of reproductive organs.

Female mammals have the following two major types of reproductive cycles: 1. Estrous cycle (characteristic of most mammals); 2. Menstrual cycle (characteristic of primates only)

Table. 7-1. Duration of gestation (Source : **J.B. Phillips**, 1975).

	Animal	Days
1.	Dogfish (*Mustelus canis*)	300
2.	Opossum (Marsupial)	12—13
3.	Mouse	19—20
4.	Rat	20—22
5.	Red-necked wallaby (Marsupial)	30
6.	Rabbit	28—32
7.	Hare	30—36
8.	Squirrel	30—40
9.	Giant Kangaroo (Marsupial)	39
10.	Fox	52—63
11.	Guinea pig	62—70
12.	Dog	60—65
13.	Cat	52—65
14.	Lion	105—115
15.	Pig	112—120
16.	Sheep	145—155
17.	Rhesus monkey	150—170
18.	Chimpanzee	250
19.	Man	270—295
20.	Cow	275—290
21.	Horse	330—345
22.	Whale	334—365
23.	Giraffe	400—480
24.	Elephant	607—641

7-2.1. Estrous Cycle

A female mammal that follows an annual reproductive season, passes through several shorter sexual cycles during her breeding season when she is more receptive of the males and when ovulation occurs spontaneously. During these short sexual cycles, called **estrous cycles**, many anatomical and physiological changes occur. The dominant event of each estrous cycle is the maturation and release of the ova which usually occurs at the time, the female is most receptive to the male (period of **heat** or **estrus**). In other words, at regular intervals in her breeding season the female mammal comes into a state of maximum sex urge or heat, when she receives the male animal for mating; the state of heat exists for a relatively brief period, extending generally over only several hours (*e.g.*, 18 hours in cow). In mammals the length of estrous cycle varies with the species (Table 7-2). Thus, it ranges from 4 to 5 days in the rat, 21 days in cow and 3 to 4 months in the dog.

Table. 7-2. Duration of estrous cycle in mammals (Source : **Phillips**, 1975)

	Animal	Days
1.	Rat	4–5
2.	Mouse	5
3.	Guinea pig	16
4.	Sheep	16
5.	Asian elephant	21
6.	Sow	21
7.	Cow	21
8.	Mare	21
9.	Rhesus monkey	24–26
10.	Woman	28
11.	Opossum	28
12.	Chimpanzee	35–37
13.	Dog	90–120

The estrous cycle is divided into the following stages marked by characteristic changes in the ovary, uterus and vagina :

1. Estrus phase. A short period when the male and female are fully prepared for sexual reproduction and have strong sexual urge or desire.

2. Post-estrus or metestrus phase (Metestrum). A period during which preparations for pregnancy are completed and if pregnancy does not take place, a regression of these preparations takes place.

3. Diestrus phase (Diestrum). A period of rest during which the uterus becomes small and anaemic.

4. Proestrus phase (Proestrum). A period of active preparation for next estrus. During this period new ovarian follicles grow. Actually, the variations in length of estrous cycle among mammals are largely reflections of variations in the length of the diestrus. In season breeders, throughout the non-breeding season, the estrous cycle remains suspended in the female animal; this suspension of the estrous cycle is called the state of **anestrum**.

How often females are in heat varies greatly among the different mammals. Those animals that have only a single estrus (heat) during the breeding season are called **monoestrous**; those that have a recurrence of estrus during the breeding season are called **polyestrous**. Dogs, foxes, deer, wild sheep, and bat are the monoestrous; while the field mice and squirrels are polyestrous, as are many mammals living in the more tropical regions of the earth.

Rut cycle. Testicular activity in the majority of mammals is a seasonal event, as is the associated period of sexual excitement known as **rut**. For example, in some stags (deer family), there is a brief period of intense sexual activity, called **rutting season**, at particular season of the year and then a period of cessation of spermatogenesis. During this inactive period, in some mammals testes may shrink and may be withdrawn from the scrotal sac to the anterior part of body. The rut cycle corresponds with the female's period of mating impulse (period of heat). In man and most domesticated and semi-domesticated mammals, the rut cycles are not apparent.

7-2.2. Morphological and Physiological Changes during Estrous Cycle

A. Ovarian changes during the estrous cycle. The maturation and growth of a follicle of mammalian ovary proceeds during **proestrus phase** of estrous cycle. Growth of the follicle culminates in its rupture and the release of the ovum, ovulation. This event is timed to coincide with the onset of estrus. In other words, the ovum is made available for fertilization at the very time the female arrives at a stage of readiness for mating (*i.e.*, heat or estrus), and thus, when spermatazoa are most likely to be present. Yet there may be considerable variability in the time of ovulation. Sometimes, it occurs several hours before or after heat. Further, the ovulation is not a spontaneous event in all mammals. In some, notably the rabbit, the cat, and the mink, ovulation takes place only after nervous stimulation deriving from the act of copulation.

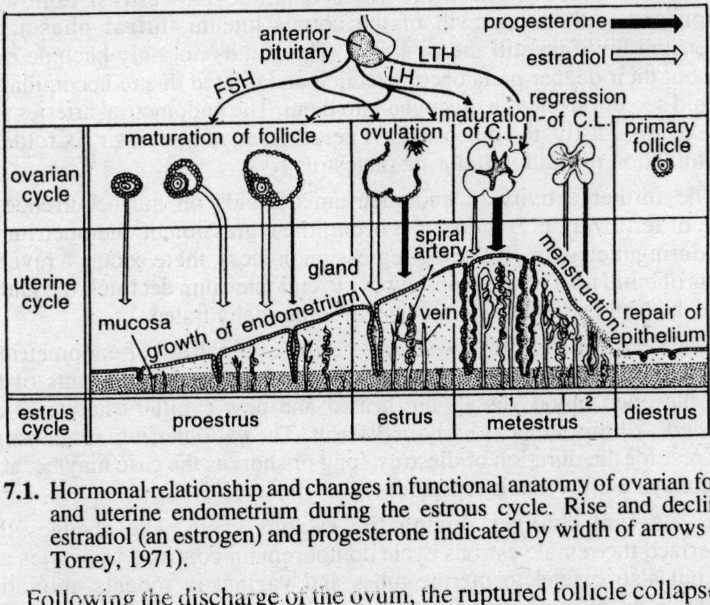

Fig. 7.1. Hormonal relationship and changes in functional anatomy of ovarian follicle and uterine endometrium during the estrous cycle. Rise and decline of estradiol (an estrogen) and progesterone indicated by width of arrows (after Torrey, 1971).

Following the discharge of the ovum, the ruptured follicle collapses and its wall becomes folded. The cells of the granulosa layer multiply, increase in size, frequently develop a yellow pigment and become a vascularized structure called **corpus luteum**. Development of corpus luteum occurs during early part of metestrus. If the liberated ovum is now fertilized, the growth and function of corpus luteum are prolonged throughout the duration of pregnancy, toward the end of which the corpus luteum degenerates. By the same token, metestrus I is maintained until the end of gestation. If, however fertilization does not occur, the ovum and corpus luteum, both degenerate. The degeneration of corpus luteum occurs during metestrus II and it includes the breakdown of the cellular part of the corpus with a concomitant invasion by fibrous connective tissue from the ovarian theca (or stroma). Eventually, a white scar of tissue called **corpus albicans**, is formed at the place of corpus luteum.

In the mammals with long estrous cycles, the diestrus phase represents for the ovary a phase of relative inactivity, during which oogonia and neighbouring ovarian cells remain quiescent, until a new cycle of maturation is inaugurated in proestrus.

B. Uterine changes during the estrous cycle. During the estrous cycle, the mammalian uterus undergoes profound morphological and physiological changes. The part of the uterus primarily involved is the **endometrium** which consists of a thin uterine **epithelium** bordering the uterine cavity and the highly vascular **mucosa** beneath the epithelium. It also includes numerous **uterine glands** which developmentally are tubular invaginations of the uterine epithelium projected into the mucosa.

At the beginning of proestrus, the endometrium remains relatively thin and its glands are short and stubby. As an accompaniment of follicular maturation during proestrus, the uterine mucosa gradually thickens and the uterine glands deepen. These processes continue into estrus. Following ovulation and with the growth of the corpus luteum **(luteal phase),** the endometrium build up still more rapidly. The glands not only become even longer, but their deeper parts become coiled and dilated due to accumulation of mucoid secretions rich in glycogen into them. The endometrial arteries also become spiral. Lastly, the mucosa itself gets thickened by five or six fold due to accumulation of fluid in it during metestrus I.

The further activity of endometrium depends on the occurrence or absence of fertilization. If it does not occur, the regression of endometrium is started during metestrus II. During regression process, there occurs a reversal of events of build up, *i.e.*, the blood supply to endometrium declines, the glands become short and straight, and mucosa becomes dehydrated.

The metestrus II is followed by a period of repairing of endometrium. Regeneration of the uterine epithelium proceeds from the remnants of the glands; the open blood vessels are healed and new capillary networks are established. All this happens early in diestrus. The endometrium remains in a state of rest for the duration of diestrus, long or short as the case may be, after which the cycle of events is repeated.

C. Vaginal changes during the estrous cycle. The changes that characterized the female estrous cycle do not remain confined to ovaries and uterus, but also extend to uterine tubes and vagina. In rodents, migrating leucocytes present at diestrus disappear just before estrus and the epithelial cells proliferate and begin to specialize. During estrous, these epithelial cells become cornified and shed in great number. In primates the vaginal changes are less extensive and distinct.

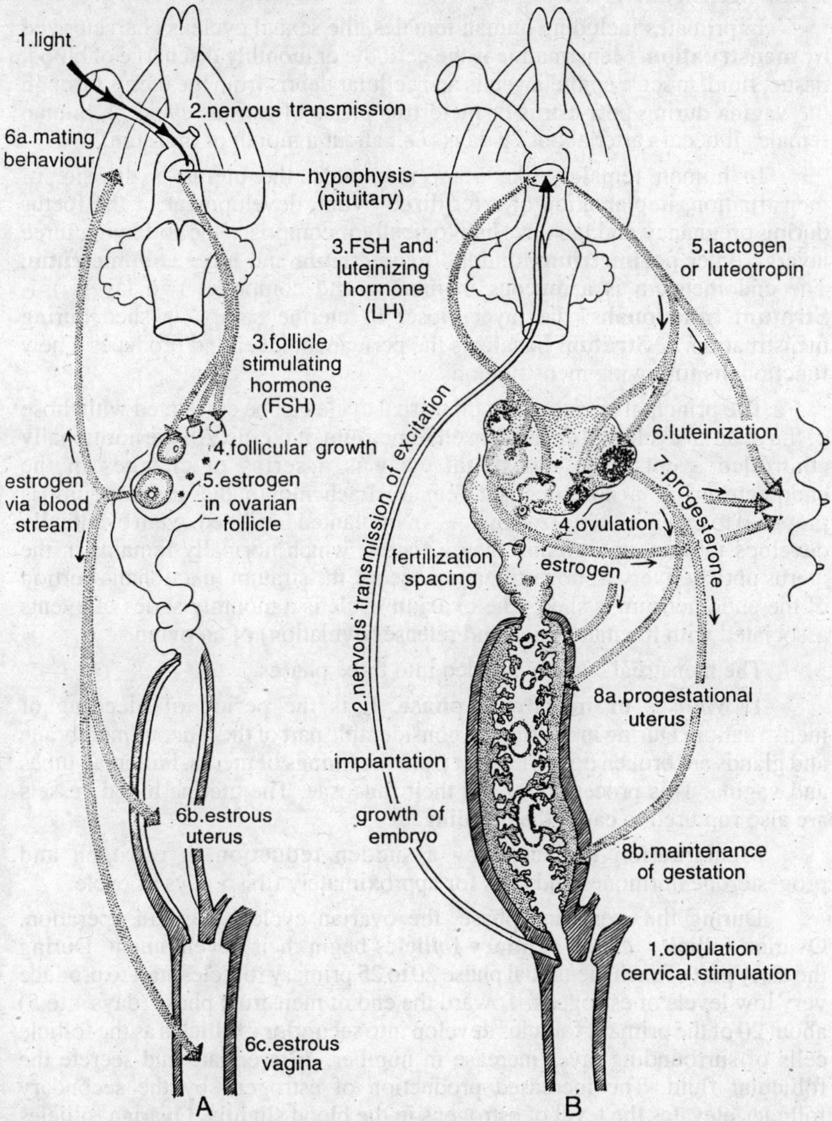

Fig. 7.2. Sequence of events in the reproductive (estrous) cycle of the rabbit. A—Reactions leading to the estrus state; B—Sequence of events following mating (after Nelsen, 1953).

D. Changes in mammary glands during the estrous cycle. In monkey, as in lower mammals, growth and regression take place in mammary glands in co-ordination with the ovarian cycle. In human female, the breasts do tend to become larger and firmer preceding menstruation. They remarkably develop during pregnancy but regress after the nursing period.

7-2.3. Menstrual Cycle

In primates including human females, the sexual cycle is characterized by **menstruation**. Menstruation is the periodic or monthly discharge of blood, tissue, fluid, mucus, epithelial cells and cellular debris from the uterus through the vagina during post-estrus or metestrus phase of estrous cycle. In human females it occurs after about 28 days (*i.e.*, about a month or mensum).

In human females, like other mammals, the uterus is the site of menstruation, implantation of a fertilized ovum, development of the foetus during pregnancy and labour. Histologically it comprises the following three layers : outer **perimetrium**, middle **myometrium** and inner **endometrium**. The endometrium is a mucous membrane and comprises two layers : 1. **Stratum functionalis**, the layer closer to uterine cavity, is shed during menstruation. 2. **Stratum basalis** is the permanent layer and produces a new functionalis following menstruation.

The principal events of the menstrual cycle can be correlated with those of the ovarian cycle and changes in endometrium of uterus. All are hormonally controlled events. The **menstrual cycle** is a series of changes in the endometrium of a non-pregnant female. Each month the endometrium is prepared to receive a fertilized ovum. An implanted fertilized ovum eventually develops into an embryo and then a foetus, which normally remains in the uterus until delivery. If no fertilization occurs, the stratum functionalis portion of the endometrium is shed. The **ovarian cycle** is a monthly series of events associated with the maturation and release (ovulation) of an ovum.

The menstrual cycle is divided into three phases :

1. Menses or menstrual phase. It is the period of bleeding or menstruation. During menstruation considerable part of the mucous membrane and glands are broken down and lost from the linings of uterus, Fallopian tubes and vagina. This process thins out their mucosae. The uterine blood vessels are also ruptured to cause the bleeding.

Menstruation is caused by a sudden reduction in estrogen and progesterone hormones and lasts for approximately first 5 days of cycle.

During the menstrual phase, the ovarian cycle is also in operation. Ovarian follicles, called **primary follicles** begin their development. During the early part of each menstrual phase 20 to 25 primary follicles start to produce very low levels of estrogens. Toward the end of menstrual phase (days 4 to 5) about 20 of the primary follicles develop into **secondary follicles** as the follicle cells of surrounding layer increase in number, differentiate and secrete the follicular fluid. The increased production of estrogens by the secondary follicles elevates the level of estrogens in the blood slightly. Ovarian follicles development is the result of secretion of a releasing hormone (RH) of FSH by hypothalamus, which in turn stimulates FSH production by the anterior pituitary. Although a number of follicles begin development in each menstrual cycle, **only one attains maturity**, the others undergo atresia (death).

2. Follicular phase or proliferative phase. During this second phase of the menstrual cycle, the uterine endometrium regenerates and becomes thickened under the influence of estrogens that are secreted by the matured ovarian follicle under the stimulation of FSH and LH of anterior pituitary.

Thus, during proliferative phase, in a couple of days, the broken mucous membranes of uterus and Fallopian tubes undergo some repairs such as formation of a continuous epithelial lining on the mucous membrane and repair of ruptured blood vessels in it. Then a secondary ovarian follicle starts growing and maturing into mature Graafian follicle. An ovum gradually matures in the follicle and estrogens are released from the maturing follicle. Estrogens reach the secondary sex organs and stimulate growth and proliferation in their mucosae. The uterine mucosa or endometrium thickens greatly, uterine glands elongate and ultimately become cork-screw-shaped, contractions of uterine muscles are greatly increased to enhance the uterine movement, the Fallopian tube epithelium gets thickened and its cilia and their movements are increased. While these changes in the Fallopian tubes are intended for conveying the ovum down those tubes, the uterine changes are preparations for the expected pregnancy (Fig. 20.4).

The proliferative phase extends for about 10–12 days (it lasts from days 6 to 13 in a 28-day cycle) and near its end, the ovum is ejected (**ovulation**) on day 14 from the Graafian follicle of ovary in the abdominal cavity (*i.e.*, pelvic cavity). Ovulation occurs due to decreased levels of FSH and LH in the blood.

It has been found that LH is necessary for final follicular growth and ovulation. Approximately two days before ovulation, the rate of secretion of LH by anterior pituitary gland increases six to ten-fold and peaking about 18 hours before ovulation. This is called **ovulatory surge of LH** (**Guyton**, 1986). FSH also increases about two fold at the same time, and two of these hormones act synergistically to cause rapid swelling of follicle. LH also stimulates the granulosa to secrete less estrogen but progressively increasing amount of progesterone.

3. Secretory or luteal phase. This phase lasts from days 15 to 28 in a 28-day cycle and represents the time between ovulation and onset of next menses. During this phase the ruptured follicle changes into a corpus luteum (yellow body) in the ovary. The corpus luteum secretes the increasing quantities of the estrogens and progesterone. Progesterone produces changes in the secondary sex organs to further the preparations for the anticipated pregnancy. So, the uterine endometrium thickens further, uterine glands become more cork-screw-shaped and begin secreting a juice in the uterus and uterine movements are reduced to keep the uterus quiet. All these uterine changes help in the implantation of the fertilized ovum on the uterine wall. Progesterone also inhibits further follicular maturation and ovulation from the ovary in this phase.

If the ovum is not fertilized and implanted, the corpus luteum undergoes degeneration in the ovary towards the end of luteal phase; it changes into **corpus albicans** and progesterone secretion declines. The overgrown endometrium cannot be maintained in the uterus and Fallopian tubes. They begin to break down, blood vessels in the uterine endometrium start bleeding and menstruation starts. Thus, new menstrual and ovarian cycle are initiated.

Menarche and menopause. The menstrual cycle normally occurs once each month from menarche, the first menses to **menopause**, the last menses.

The advent of menopause is signalled by the **climacteric**, *i.e.*, menstrual cycle becomes less frequent between ages 40 to 50. In the postmenopausal woman there will be some atrophy of the ovaries, uterine tubes, uterus, vagina, external genitalia and breasts.

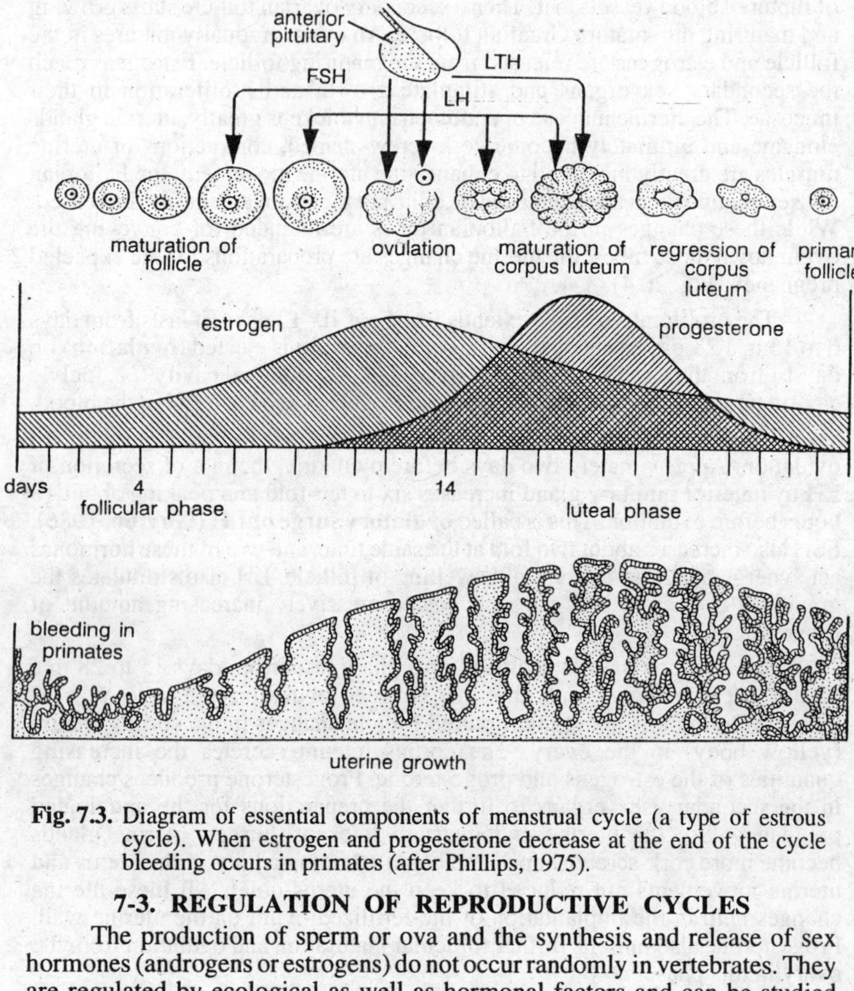

Fig. 7.3. Diagram of essential components of menstrual cycle (a type of estrous cycle). When estrogen and progesterone decrease at the end of the cycle bleeding occurs in primates (after Phillips, 1975).

7-3. REGULATION OF REPRODUCTIVE CYCLES

The production of sperm or ova and the synthesis and release of sex hormones (androgens or estrogens) do not occur randomly in vertebrates. They are regulated by ecological as well as hormonal factors and can be studied under the garb of the estrous cycle as follows :

7-3.1. Ecological Control of Estrous Cycle

Environmental conditions have a definite regulative effect on the estrous cycle of many mammals. As many mammals, time their estrous cycle with respect to seasons in order to take advantage of favourable temperature and food supply. Various ecological factors such as duration and intensity of light, temperature, visual, olfactory, auditory stimuli, rainfall, and lunar tides are found to effect estrous cycle of different mammals in one way or other.

7-3.2. Hormonal Control of Estrous Cycle

It has been already noted in the beginning of this chapter that the cyclical events in the ovary and endometrium (due to estrous cycle) are neatly co-ordinated. Each developmental circumstance in one is timed to correspond with a developmental circumstance in the other. It has long been established that this timing is controlled by chemical agents circulating in the blood stream, *viz.*, **hormones** and also by peculiar devices called **biological clocks**. (**Villee**, 1976, has described biological clocks as the means by which activities of plants or animals are adopted to the regularly recurring changes in the external physical conditions, and perhaps to change in internal milieu as well).

Gonadotropic (gonad-stimulating) hormones produced by the anterior lobe of pituitary gland (hypophysis), control the functions of the gonads. These hypophyseal hormones are released into the blood-stream and carried to the ovary or testis where they produce their effect. The gonadotropic hormones are believed to be the same in both male and female, differing quantitatively rather than qualitatively. The target organ itself determines the effect of their stimulation. The names given to these hormones refer to their functions in the female.

We can study the roles of different hormones involved in maintaining the reproductive cycles of vertebrates, especially of mammals, under the following headings :

1. Hormones of anterior lobe of pituitary (adenohypophysis). Anterior lobe of pituitary secretes the following three gonadotropins : **follicle-stimulating hormone (FSH)**, **luteinizing hormone (LH)** and **prolactin (PRL).** Of these FSH and LH affect both the ovaries and testes, while, PRL affect only the ovary, especially the mammary glands. Secretion of these gonadotropins is subject to the regulating factors or neurohormones produced by hypothalamus of brain, called **releasing hormones** (RH) such as FSH-RH, LH-RH and PRL-RH.

(1) Follicle-stimulating hormones (FSH). In female, FSH is transported from the adenohypophysis by the blood to the ovaries, where it initiates the development of ova each month. It also stimulates ovarian follicles to secrete the female sex hormone, called **estrogens**.

In male, FSH acts on the seminiferous tubules to initiate spermatogenesis and stimulate sustentacular cells (Sertoli cells) to produce **inhibin** hormone (**Guyton**, 1986).

(2) Luteinizing hormone (LH). In the female, LH together with estrogens, stimulates the ovary (*i.e.*, Graafian follicle) to release an ovum (ovulation) and prepares the uterus for implantation of a fertilized ovum. It also stimulates the formation of the corpus luteum in the ovary which secretes another female sex hormone called **progesterone**. In fact, LH acts on granulosa cells of ruptured Graafian follicle and stimulates them to become lutein cells which seems to be programmed to go through a predetermined sequence of (a) proliferation, (b) enlargement, (c) progesterone secretion and (d) degeneration.

In the male, LH stimulates the interstitial cells of Leydig of testes to secrete the androgen or male sex hormone, called **testosterone**; this is why LH is also known as **interstitial cell stimulating hormone (ICSH)**.

(3) Prolactin (PRL). PRL or **lactogenic hormone** (LTH), together with other hormones (estrogens, progesterone, cortico-steroids, growth hormones, thyroxine and insulin), initiates and maintains milk secretion by the mammary glands after the birth of child. PRL acts directly on tissues and by itself has little effect on lactation.

2. Hormones of posterior lobe of pituitary (neurohypophysis). The posterior lobe of pituitary also secretes a hormone, called **oxytocin (OT)** that has a role to play in reproduction. OT is secreted into the blood when the hypothalamic neurons are stimulated either due to the distention of uterus by the full-term foetus or due to the sucking of the breast by the infant. Oxytocin stimulates contraction of the smooth muscles of uterus and mammary glands. Uterine contraction, stimulated by OT at the end of pregnancy, help in child-birth. The OT-induced contractions of the mammary gland muscles help in the ejection of stored milk from the mammary glands to the mouth of the suckling infant. Due to these specific functions, OT is also referred to as **birth hormone** or ''**milk ejection hormone**''.

3. Hormones of ovaries. The mammalian ovary secretes the following different hormones which regulates different stages of reproductive cycles :

(1) Estrogens. At least six different estrogens have been isolated from the plasma of human females. However, only three are present in significant quantities. These are **β-estradiol, estrone** and **estriol**. Of these β-estradiol exert the major effect. It is synthesized from cholesterol or acetyl coenzyme A in the ovaries (*i.e.*, ovarian follicles).

Estrogens are hormones of growth and have three main functions : 1. Development and maintenance of female reproductive structures, especially the endometrial lining of the uterus, secondary sex characteristics and breasts. 2. The control of fluid and electrolyte balance. 3. Increase in protein anabolism.

(2) Progesterone. This hormone is synthesized from cholesterol or acetyl coenzyme A in the ovaries by the corpus luteum. Progesterone is a hormone of maturation and works with estrogens to prepare the endometrium of uterus for implantation of a fertilized ovum and the mammary glands for milk secretion.

(3) Relaxin. This hormone is produced by both corpus luteum and placenta near the end of pregnancy. It stimulates relaxation of pelvic ligaments (the symphysis pubis) and helps dilate the uterine cervix to facilitate delivery.

4. Hormones of testes. In males, testes secrete two types of sex hormones :

(1) Testosterone. It is synthesized from cholesterol or acetyl coenzyme A in the testes by interstitial cells of Leydig. Testosterone is the principal male sex hormone (androgen) and has a number of effects on the body. It controls the development, growth and maintenance of the male sex organs. It also stimulates bone growth, protein anabolism, sexual behaviour, final maturation of sperm and the development of male secondary sexual characters at puberty. Testosterone also stimulates descent of the testes in the scrotal sacs just prior to birth.

(2) Inhibin. It is a protein hormone that is secreted by sustentacular or Sertoli cells in response to FSH of adenohypophysis.

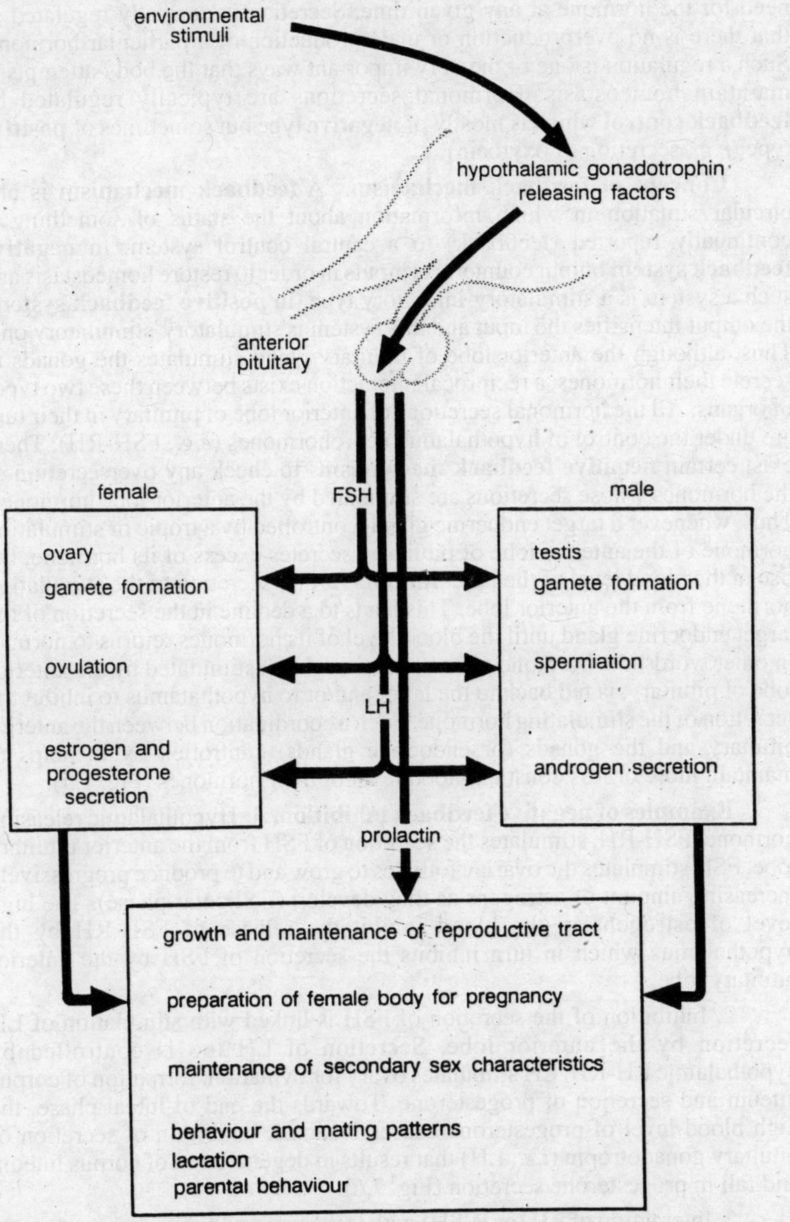

Fig. 7.4. Control of seasonal reproductive cycles by the pituitary hormones and hypothalamic releasing hormones (after Phillips, 1975).

5. Control of hormonal secretion : feedback control. The amount of hormone released by an endocrine gland or tissue is determined by the body's need for the hormone at any given time. Secretion is normally regulated so that there is no overproduction or underproduction of a particular hormone. Such a regulation is one of the very important ways that the body attempts to maintain homeostasis. Hormonal secretions are typically regulated by **feedback control** which is mostly of **negative** type but sometimes of **positive** type (*e.g.*, secretion of oxytocin).

Concept of feedback mechanism. A **feedback mechanism** is any circular situation in which information about the status of something is continually reported (feedback) to a central control system. In **negative feedback system** output counteracts inputs in order to restore homeostasis and such a system is a stimulatory-inhibitory type. In **positive feedback system**, the output intensifies the input and this system is stimulatory-stimulatory one. Thus, although the anterior lobe of pituitary gland stimulates the gonads to secrete their hormones, a reciprocal interaction exists between these two types of organs. All the hormonal secretions of anterior lobe of pituitary in their turn are under the control of hypothalamic neurohormones (*e.g.*, FSH-RH). There exist certain **negative feedback mechanisms** to check any oversecretion of the hormones whose secretions are stimulated by the anterior lobe hormones. Thus, whenever a target endocrine gland, controlled by a tropic or stimulating hormone of the anterior lobe of pituitary, secretes excess of its hormone, the rise in the blood level of the latter **inhibits** further secretion of the stimulating hormone from the anterior lobe. This leads to a decline in the secretion of the target endocrine gland until the blood level of its hormones returns to normal. In other words, the hormone of an endocrine gland, stimulated by the anterior lobe of pituitary, is fed back to the latter and/or to hypothalamus to inhibit the secretion of the stimulating hormone. Such a coordination between the anterior pituitary and the gonads (or endocrine glands) controlled by it, helps to maintain more or less constant blood level of their hormones (Fig. 7.5).

Examples of negative feedback inhibition. 1. Hypothalamic releasing hormone, FSH-RH, stimulates the secretion of FSH from the anterior pituitary lobe. FSH stimulates the ovarian follicles to grow and to produce progressively increasing amount of estrogens as they develop (Follicular phase). The high level of estrogens in the blood inhibit the release of FSH-RH by the hypothalamus which in turn inhibits the secretion of FSH by the anterior pituitary lobe.

2. Inhibition of the secretion of FSH is linked with stimulation of LH secretion by the anterior lobe. Secretion of LH too is controlled by hypothalamic LH-RH. LH stimulates ovary for ovulation, formation of corpus luteum and secretion of progesterone. Towards the end of luteal phase, the high blood level of progesterone causes feedback inhibition of secretion of pituitary gonadotropin (*i.e.*, LH) that results in degeneration of corpus luteum and fall in progesterone secretion (Fig. 7.6).

3. Interaction of LH (or ICSH) with testosterone illustrates the operation of another negative feedback system. In male, ICSH stimulates the production of testosterone, but once the concentration of testosterone in blood reaches a certain level, it inhibits the release of hypothalamic releasing hormone for LH

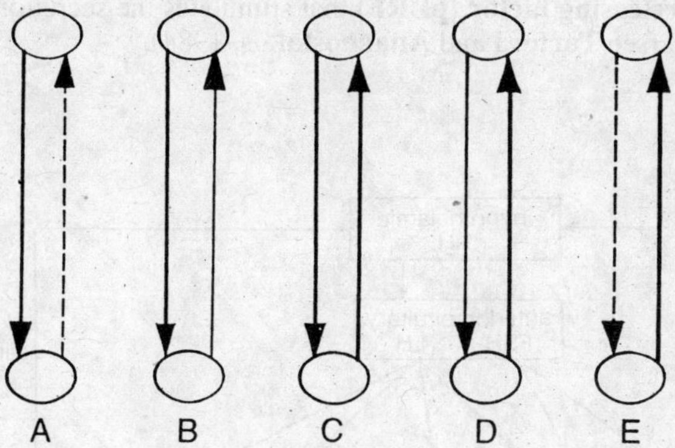

Fig. 7.5. Diagram of a simple negative-feedback mechanism operating in endocrine control. A and B—At first the pituitary stimulates the gonads to secrete their hormones; C to E—As the secretion of gonadal hormone increases, they in turn inhibit the secretion of tropic hormones by the pituitary. Ultimately, the concentration of gonadal hormone falls (A) and pituitary, released from its inhibition once more secretes its gonadotropins (after Phillips, 1975)

or ICSH, *i.e.*, ICSH-RH. This, in its turn, inhibits the release of ICSH by the anterior lobe of pituitary. Thus, the production of testosterone is decreased. However, once the testosterone concentration in the blood decreases to certain level, ICSH-RH is released by the hypothalamus which stimulates the release of ICSH by the anterior lobe. ICSH stimulates testosterone production by testes. Thus, the testosterone-cycle is completed (Fig. 7.7).

4. Inhibin has direct effect on the anterior pituitary by inhibiting the secretion of FSH. FSH brings about spermatogenesis and stimulates sustentacular cells. Once the degree of spermatogenesis required for male reproductive functions has been achieved, the sustentacular cells secrete inhibin. The inhibin feeds back negatively to the anterior pituitary to inhibit FSH and, thus, to decrease spermatogenesis.

6. Hormones of pregnancy and parturition. The corpus luteum is maintained throughout pregnancy. For about 8 to 10 weeks after fertilization, it continues to secrete **estrogens** and **progesterone**. Both of these hormones maintain the lining of the uterus during pregnancy and prepare the mammary glands to secrete milk. The amount secreted by the corpus luteum, however, is only slightly more than that produced after ovulation in a normal menstrual cycle. So the high levels of estrogens and progesterone that are needed to maintain pregnancy and develop the mammary glands for lactation are provided by the placenta.

(1) Hormones of placenta. Placenta produces the following hormones that help in pregnancy and parturition :

A. Human chorionic gonadotropin (HCG). The chorion of placenta secretes **human chorionic gonadotropin** (**HCG**). The primary role of HCG is to maintain the activity of the corpus luteum, especially with regard to continuous progesterone secretion—an activity necessary for the continued attachment of the foetus to the lining of the uterus. In male foetuses, HCG stimulates testosterone production by testes.

Recent evidences have shown that the placenta produces a hypothalamic releasing hormone for LH-like (*i.e.*, LH-RH-like) regulating factor called **placental luteotropic releasing factor (pLRF)** that stimulates the secretion of HCG by the placenta (see **Tortora** and **Anagnostakos**, 1984).

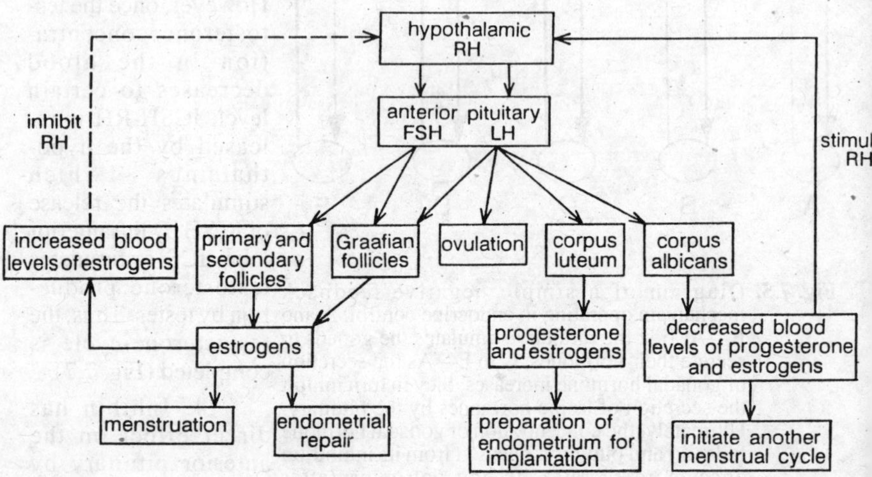

Fig. 7.6. Summary of hormonal interactions of the menstrual and ovarian cycles (after Tortora and Anagnostakos, 1984).

2. Estrogens and progesterone. The placenta begins to secrete estrogens and progesterone no later than the sixtieth day of pregnancy. Both of these hormones are secreted in increasing quantities until the time of birth. By the fourth month, when the placenta is established, the secretion of HCG is greatly reduced because the secretions of the corpus luteum are no longer essential. The placenta supplies the levels of estrogens and progesterone needed to maintain the pregnancy. Thus, the foetal hormones take over management of the mother's body in preparation for parturition (birth) and lactation. Following delivery, estrogens and progesterone in the blood decreases to normal levels.

3. Human chorionic somatomammotropin (HCS). Another hormone produced by the chorion of placenta is **HCS** that stimulates breast development (*i.e.*, protein deposition for lactation), protein anabolism and glucose and fatty acid catabolism.

4. Relaxin. This hormone is produced by the placenta and ovaries. The main physiological roles of relaxin are to relax the ligaments that bind the pelvic bones and to dilate the cervix of uterus.

B. Hormones involved in parturition. The term **parturition** refers to birth. After about 9 months, the human embryo and placenta are ejected from the mother's uterus by forceful muscular contractions called labour. Peristaltic

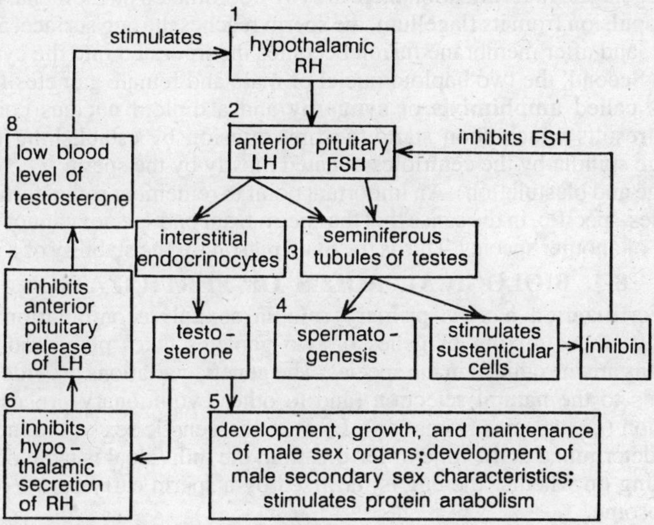

Fig. 7.7. Secretion, physiological effects and control of testosterone and inhibition (after Tortora and Anagnostakos, 1984)

muscle contractions sweep from the top of the uterus downward. Labour comprises three stages : 1. stage of dilation of cervix ; 2. stage of expulsion and 3. placental stage. The initiation and control of labour is regulated by the following hormones : 1. There is a shift in the relative amount of progesterone and estrogens at the end of pregnancy. Thus, in the final phases of gestation, estrogen concentrations continue to increase and estrogen is known to promote uterine contraction. 2. Oxytocin is released from the posterior pituitary during labour. This hormone stimulates contraction of the uterus. 3. Relaxin allows birth to proceed through an enlarged birth canal.

REVISION QUESTIONS

1. What is reproductive cycle ? Describe the estrous and menstrual cycles of vertebrates.
2. Differentiate between the ovarian follicles and the corpus luteum.
3. Discuss how hormones regulate reproduction in mammals.
4. Give an account of hormones involved in growth of ovum, maintenance of pregnancy, parturition and lactation in mammals.
5. Write short notes on the following :

   1. Negative feed-back regulation ; 2. Sex hormones ; 3. Corpus luteum ; 4. Rutting; and 5. Gestation.

8

FERTILIZATION

Fertilization of the egg by the sperm is the first step in the process of embryogenesis. In fertilization there are two coordinated phenomena. First, by self-propulsion from its flagellum, the sperm reaches the egg surface, achieves contact, and after membrane fusion becomes incorporated into the cytoplasm of egg. Second, the two haploid nuclei of male and female gametes fuse in a process called **amphimixis** or **syngamy** and a diploid nucleus containing zygote results which soon starts the first division by establishing the first cleavage spindle by the centrioles donated mostly by the sperm (*i.e.*, onset of cleavage and blastulation). An important point to remember is that fertilization is species-specific, in the sense that the sperm from one species cannot fertilize the egg of another species. This is the main reason for the stability of a species.

8-1. BIOLOGICAL ROLES OF FERTILIZATION

Fertilization has its primary role in sexual recombination during syngamy where mixing of genes of two gametes takes place and genetic variations are introduced in the species. The genetic variations provide the raw materials to the natural selection (and to other evolutionary processes) for speciation (*i.e.*, species formation). Further, the genetic sex of the individual is also determined at the time of fertilization; the individual is male or female depending on whether the egg is fertilized by a sperm carrying a Y or an X chromosome.

Fertilization as its secondary role, triggers the developmental programme in a rapid burst of metabolic activity. In a sense, this moment of activation "saves" the life of the egg, delicately balanced between quick extinction or growth and morphogenesis. Because of their enormous size and unbalanced nucleocytoplasmic volumes, mature unfertilized eggs have a limited life after ovulation, often a matter of a few hours. They "ripen" or age, gradually losing developmental information as their nuclei and cytoplasm become disorganized. If overripe eggs are fertilized, they develop abnormalities, including chromosomal aberrations whose magnitude is directly related to the egg's age before fertilization. Many known congenital abnormalities in humans (*e.g.*, Down's syndrome) have been attributed to the development of overripe eggs. In fact, ripening induces irreversible changes in egg organization and unless an egg is rescued by fertilization, it soon dies. Hence, the process of fertilization accomplishes three purposes: (1) sexual recombination and determination of the genetic sex; (2) activation of developmental programme (*i.e.*, onset of mitosis); and (3) sustaining egg survival (**Grant**, 1978).

The process of fertilization has been studied mostly in marine invertebrates, such as, sea urchins, polychaetes, echiuroids and balanglossids, whose eggs are typically translucent, and without much yolk. Among vertebrates, most fertilization concerning investigations have been done in amphibians and mammals. Attention has been focussed on amphibians because they are available and suitable for analytical and experimental procedures, and on mammals because man himself is a mammal and the need to know what happens is great. Fertilization studies of mammals also help in improving breed of domestic animals such as cattle, horse, sheep, goat, donkey, etc.

8-2. BASIC REQUIREMENTS OF FERTILIZATION

In most animals the act of fertilization requires following fundamental requirements for its fulfilment :

1. In most animals the fertilization requires a fluid medium which may be seawater in marine forms, freshwater in freshwater forms or some body-fluid in viviparous animals.

2. The life-spans of gametes are limited. Mature eggs quickly become "overripe", they must be activated promptly if they are to be activated at all. Eggs that are shed into the water, like those of most invertebrates, fish, and amphibians, must be fertilized immediately or within a few minutes. Eggs that are fertilized within the body of the female generally have a longer life-span; for example, the human egg can be fertilized for at least 24 hours after ovulation.

Sperms are also short-lived, but there are more exceptions to the rule in nature. For example, in bat viable sperm may persist in the female reproductive tract over the winter, during hibernation. Further, the life time of the human sperm in the female tract is of the order of 24 hours.

3. To increase the probability of fertilization, the number of sperms must exceed the number of eggs. A foolproof method that will allow the perfect union of one egg and one sperm without wasting millions of sperms has not been devised. Thus, in mammals, a single ejaculation may range from 0.05 ml., in the bat to 2 ml., in the rat to 500 ml., in the boar, containing 6 million, 2 to 5 million, and 100,000 sperms per microliter, respectively. In man, the ejaculate averages 3 ml. of which about 20 per cent by volume is sperm (100,000 cells per microliter).

8-3. MECHANISM OF FERTILIZATION

The process of fertilization completes in following stages :

8-3.1. Encounter of Spermatozoa and Ova

A major problem in sexual reproduction is how to bring together the spermatozoa and ova. It is important that ripe eggs and sperms be brought together in the same general locality and that individual sperm may reach the surface of ova. In general, most invertebrates and vertebrates accomplish the close approximation of spermatozoa and eggs through special devices or particular forms of behaviour. According to place and nature of fluid media following two kinds of fertilization have been reported :

A. External fertilization; B. Internal fertilization.

A. External fertilization. When the fertilization occurs in the aquatic medium outside the bodies of male and female parents, it is called **external fertilization**. The aquatic medium for the external fertilization may be either seawater or freshwater.

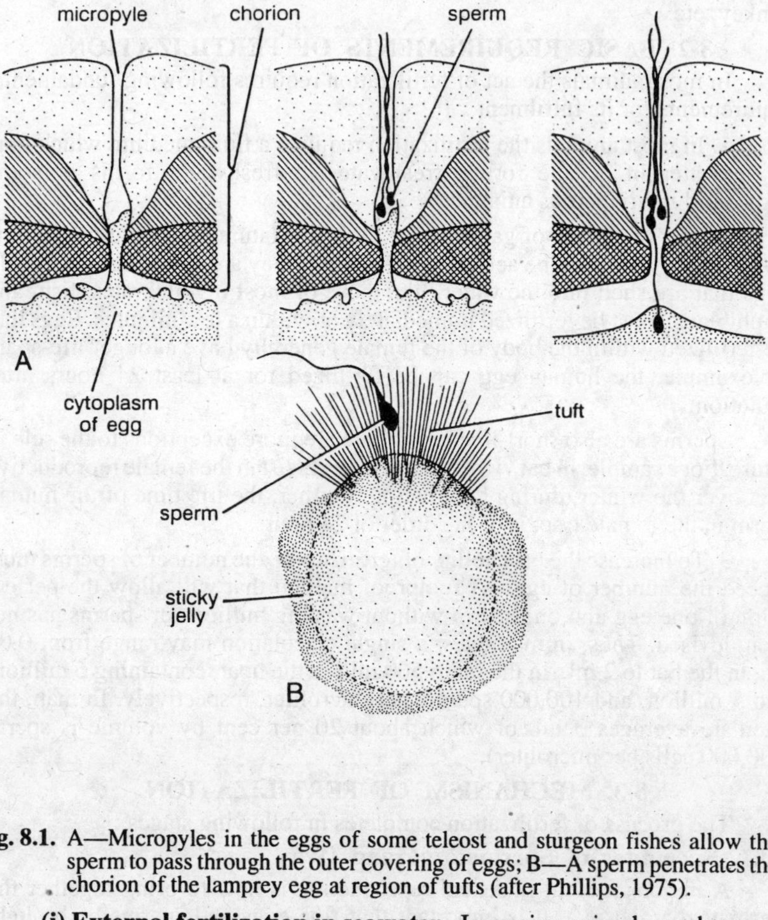

Fig. 8.1. A—Micropyles in the eggs of some teleost and sturgeon fishes allow the sperm to pass through the outer covering of eggs; B—A sperm penetrates the chorion of the lamprey egg at region of tufts (after Phillips, 1975).

(i) External fertilization in seawater. In marine animals, such as, sea urchins, the communities of sexually mature adults shed eggs and sperms freely into the surrounding water. In such cases, because the developmental hazards are greatest, astronomical numbers of eggs as well as sperms are spawned during any one spwaning period. The adult members of a species, usually become sexually mature depending on the environmental temperatures of the preceding weeks or months, and as a rule, one or more ripe female spawn at dawn or dusk, stimulated by the rapid change in light intensity. Substances liberated with the oviducal fluid at the time of shedding of eggs in turn stimulate other ripe females and also the ripe males in the vicinity. Consequently clouds of eggs and sperms are formed in the seawater at the same time, and mass fertilization occurs. Even so, to reach the eggs, spermatozoa have to travel long distance and their chances of encounter with the eggs may greatly be scarced.

The importance of timing (life span of gametes) relates to two factors—(1) the extent of diffusion of the spermatozoa, increasing with the distance they have to travel, and (2) the limited life with regard to fertilizability of the reproductive cells, especially, the sperm. In the sea, where the complex salt water solution itself is essentially a primeval physiological medium, the eggs remain fertilizable for less than a day, while spermatozoa expand all their potential energy for motility in a few hours at the most.

(ii) External fertilization in freshwater. Among the freshwater animals, the timing of spawning of eggs and shedding of sperms by female and male parents, respectively, becomes more crucial and specific, because, their spermatozoa remain active in freshwater usually for minutes rather than hours. Hence, the common mating procedures of fish, amphibians and freshwater invertebrates, are to deliver sperms directly to the eggs of an individual female at the moment of laying, if not before.

B. Internal fertilization. In terrestrial forms, particularly where eggs are completely enclosed in impermeable envelops before being laid (*e.g.*, oviparous animals such as reptiles and birds), or where they are retained within the maternal body throughout development (*e.g.*, ovoviviparous and viviparous animals such as elasmobranchs and mammals), the spermatozoa are delivered internally (in the body of female) by some type of copulatory mechanism or device (intromittent organ) of male. In such forms the fertilization may occur either in lower portion of the oviduct near or at the external orifice (*e.g.*, Urodela); in the upper portion of oviduct (*e.g.*, salamanders, reptiles, aves and most mammals); or in the ovarian follicles (*e.g.*, viviparous fishes such as *Gambusia affinis* and *Heterandria formosa* and certain eutherian mammals such as *Ericulus*). On land, in terrestrial animals, there remains no problem of timing of spawning of eggs and shedding of sperms because the mature sperms are commonly stored in a physiological medium capable of maintaining their life and potential activity for days, weeks, or months, either in moisture conserving capsules or in compartments of the male or female body to be picked up, transferred, utilized in one way or another.

The mere deposition of eggs and sperms in the general vicinity of each other within a liquid medium (*viz.*, sea or freshwater in case of external fertilization; fluid medium of female reproductive tract in case of internal fertilization) does not insure their meeting, even though spermatozoa possess a propelling tail for active motility. The movements of the spermatozoa are entirely at random and the spermatozoa collide with the eggs as a matter of pure chance. Such chance encounters occur regularly in nature partly due to enormous number of spermatozoa produced by the male gland, and partly the result of the eggs being a relatively very large target so that it can be hit fairly easily. In vertebrates, where the fertilization is internal, the movement of the spermatozoa from the site of deposition to the site of fertilization usually depends little on the active swimming of the spermatozoa themselves. Rather, spermatozoa tend to be transported passively by muscular contractions of the female tract, and also by the counter currents in the cilia-propelled, backward flowing liquid content of the tract.

8-3.2. Approach of the Spermatozoon to the Egg

The sperm reaches near the ovum by adopting one of the following two strategies :

A. Chemotaxis. Chemotaxis is the ability of the sperm to detect a difference in concentration of some substance released into the water by the egg and to move from an area of lower to one of greater concentration of that substance. Apparently, such a chemical attraction of male gamete towards the ovum operates only in the plant, mosses and ferns. However, chemotaxis has also been suspected to be operative in some animals such as coelenterates and fishes. In them spermatozoa appear to be guided to the egg by chemical substances. For example, in the hydroid, *Campanularia*, the eggs are produced by reduced female gonangia, which are enclosed in a theca with an opening at the distal end. During fertilization spermatozoa have been observed to converge toward the opening of theca. The attractant chemical substance extracted from female gonangia by **Miller** (1966) was found to be a small molecule having molecular weight less than 5000 daltons.

A somewhat similar mechanism has been observed in fish eggs (*e.g.*, sturgeon) and lamprey eggs, where each of the eggs is enclosed in a hard chorion. The chorion is perforated by narrow canals called **micropyles**, through which the sperm enters to reach the egg cytoplasm (Fig. 8.1 A). The spermatozoa tend to accumulate at the opening of the micropyle, as if they are attracted to it (Fig. 8.1). It is noteworthy that in both cases (*i.e.*, hydroids and fishes), the egg is not accessible from every angle and the spermatozoa are directed to a certain opening which leads to the egg (**Balinsky**, 1981).

B. Fertilizin-antifertilizin interactions. We have known for many years that the eggs and sperms of some animals such as sea urchin and other echinoderms, polychaetes, molluscs, tunicates, cyclostomes, fishes, amphibians and some mammals, are chemically attracted to one another. These eggs release into the surrounding medium substances that bring about agglutination of the sperm and the adherence of the sperm to the egg. These substances are considered to be more of a trapping agent for the sperm than a chemotaxic factor.

Factors that mediate sperm-egg interaction even before they make contact were identified in 1914 by **F.R. Lillie**, who proposed the first theory of the physiology of fertilization, called **fertilization theory**. He observed that **egg water** (seawater surrounding unfertilized sea urchin eggs) **agglutinated** the sperms and activated their motility. The reaction was specific since some sperms from related species were unaffected. The factor called **fertilizin**, came from the egg jelly coat which slowly dissolved as eggs stood in seawater. **Lillie** postulated that fertilizin was continuously secreted by the egg before fertilization, but it is now clear that fertilizin is the constituent of both jelly coats and egg membranes such as vitelline membrane and plasma membrane (**Grant**, 1978).

For sperms to be agglutinated by fertilizin, they must have receptor sites on their surface (*i.e.*, on their plasma membrane). These receptor sites or

antifertilizins interact with fertilizin molecules (Fig. 8.2). The antifertilizins can be extracted from the spermatozoa by heating, freezing or thawing or acidifying the water.

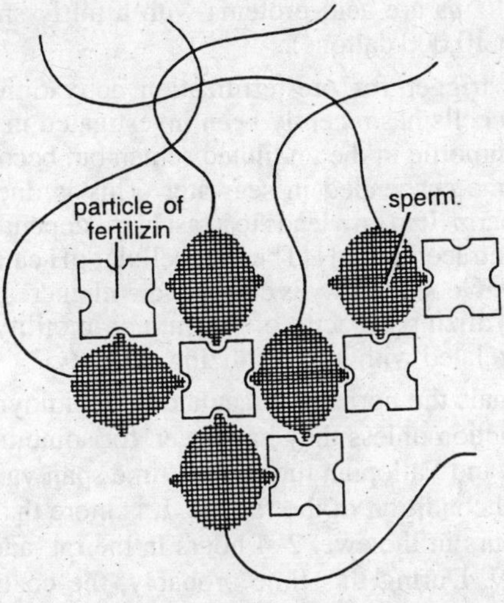

Fig. 8.2. Mode of binding of sea urchin spermatozoa by particles of fertilizin (After Balinsky, 1970).

Regarding the mode of interaction of fertilizin and antifertilizin, **Balinsky** (1981) has forwarded the following two analogous mechanisms : 1. **Antigen-antibody type interaction.** The interaction between the sperm and the egg resembles an antigen-antibody reaction; fertilizin in egg membranes may behave as an antibody combining with antigenic site (*i.e.*, antifertilizin) on the sperm surface. In both cases, the bond between the two complementary substances is supposed to depend on the spatial arrangement of the atoms in certain parts of the molecules in such a way that the shape of the surface of one molecule fits closely onto the surface of other. The two bodies correspond to each other as a template to the model. **2. Key and lock type interaction.** Another often-used analogy is the correspondence of a key to a lock. Such an interaction is currently called **adhesive recognition** which proposes that such an adhesion depends on bonding between complementary macromolecules (*i.e.*, 'locks' on one cell might fit 'keys' on another (**Trinkaus**, 1984). However, in case of this analogy it may be relevant to remember that sometimes a lock may be opened by a wrong key, if it is adequately similar to the right one. Such a situation can be observed in the cross reaction of fertilizins and antifertilizins of very closely related species (*e.g.*, donkey and horse).

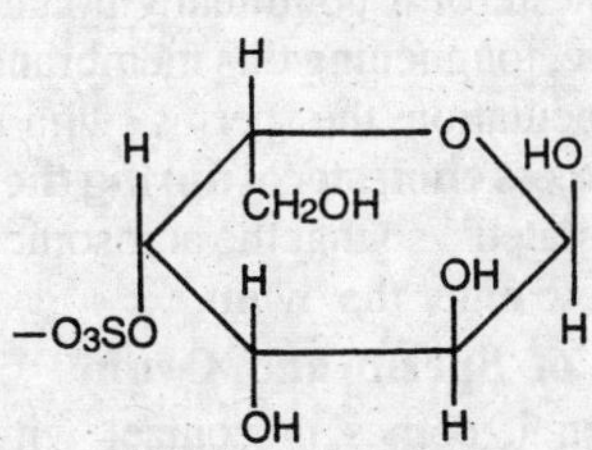

Fig. 8.3. Chemical formula of monosaccharide of fertilizin esterfied by sulphuric acid.

Chemically, a fertilizin is a glycoprotein or mucopolysaccharide, now called **proteoglycan (Trinkaus**, 1984). As a protein, it contains a number of amino acids and as a polysaccharide it includes sialic acid residues as well as one or more monosaccharides such as glucose, fucose, fructose or galactose, each of which is esterified by sulphuric acids as shown in the chemical formula (Fig. 8.3).

Both the monosaccharides and amino acids of fertilizin vary from one species to another, so that, each species possesses its specific type of fertilizin. The molecular weight of fertilizin is 300,000 daltons. Each molecule of fertilizin has more than one "active group", so that one fertilizin particle may attach to two or more spermatozoa, thus, binding them together (Fig. 8.2). In contrast to fertilizin, the antifertilizins are acid proteins with a fairly small molecule (molecular weight about 10,000 daltons).

C. Sperm motility. The trigger for pre-fertilization activation of metabolism and motility in sperm cells has recently been investigated in the sea urchins. The sperms remain immotile in the undiluted semen but become active immediately as they become suspended in seawater. This is due to spontaneous influx of Na^+ in the sperm. It in turn leads to a fast Na^+ - dependent release of acid and an increase of intracellular pH. The intracellular pH causes increased metabolism and motility. Good motility is correlated with increased metabolism, irrespective of the fertilizing capacity of sperm; non-motility in most flagellate sperm types is correlated with non-fertilizing capacity.

D. Capacitation. In mammals the normally ejaculated or epididymal sperms do not show acrosome reaction unless they have been for some time in the female genital duct (vagina and Fallopian tube). This time span varies with the species and physiological condition of the female. It is more than 1 hour in the mouse, at least 1.5 hours in the ewe, 2–4 hours in the rat, and 6 hours in the rabbit, man and sow. During this time probably, the coating substances on the sperm surface particularly those on the acrosome, are gradually removed. The receptor sites on the acrosome are, thus, exposed to enable the sperm to recognize signals emanating from the ovulated egg. This is called **capacitation**. Rabbit sperm can be capacitated by *in vitro* treatment with β-amylase. The factors that induce capacitation in the female genital tract do not appear to be strictly species-specific.

Significance of capacitation. According to **Guyton** (1986) the process of capacitation of human beings and other mammals results from the following effects and has the following significance : The acrosomes of the human sperm contain extremely powerful hydrolytic and proteolytic enzymes that could literally destroy the male genital tract if all the sperms should release these enzymes prematurely. Fortunately, the fluids in seminiferous tubules, in the epididymis and in the vas deferens all contain minute floating vesicles filled with large amounts of cholesterol. This cholesterol is continually donated to the cellular membrane covering the acrosome, toughening this membrane and preventing release of the enzymes. After ejaculation, the sperms swim away from the vesicles and **gradually lose their excess cholesterol during the next few hours**. In so doing they become "capacitated" so that the acrosomes can now release the enzymes that allow a sperm to enter the ovum.

8-3.3. Acrosome Reaction and Contact of Sperm and Ovum

The first reaction of the sperm when it comes in contact with the constituents of the egg surface, involves the acrosome. Onset of acrosome reaction takes place when there are optimal physiological conditions such as optimum pH, Ca^{++}, Mg^{++} ions concentration and temperature. The presence of Ca^{++} is an essential condition for acrosome reaction.

The acrosome reaction involves breakdown of membrane of acrosomal vesicle, release of acrosomal enzymes, formation of acrosomal tubule and fusion of plasma membranes of sperm and egg. Acrosome reaction has been studied in a variety of animals such as echinoderms, annelids, *Saccoglossus* and mammals.

1. Trigger mechanisms for breakdown of acrosome. The following two trigger mechanisms have been observed among the animals : 1. In annelids (*Hydroides*), molluscs (*Mytilus*), echinoderms (*Arbacia, Asterias*), decapod crustacean (*Procambarus*) and hemichordate *Saccoglossus*, the apical part of the sperm plasma membrane and underlying outer acrosomal membrane undergo dehiscence. The severed edges of the two membranes fuse to form an opening through which the contents of the acrosomal vesicle are released. Inner acrosomal membrane grows into one or many **acrosomal tubules** which in some cases may be almost as long as the entire spermatozoon. It is important to note that the membrane surrounding the acrosomal tubule is always derived from an inner membrane of the acrosomal vesicle. It is this membrane that makes first contact with egg jelly, vitelline membrane and plasma membrane, in that order (Fig. 8.4). 2. In mammals the plasma membrane and the outer acrosomal membrane break and fuse to give rise to extensive vesiculation (Fig. 8.7). The inner acrosomal membrane joins with the plasma membrane at the level of the acrosomal collar. Unlike *Mytilus*, *Hydroides* and echinoderms, in mammals, the inner acrosomal membrane does not participate in gametic fusion. Moreover, the sperm is possibly phagocytosed by the egg and the sperm plasma membrane does not form part of the zygote plasma membrane.

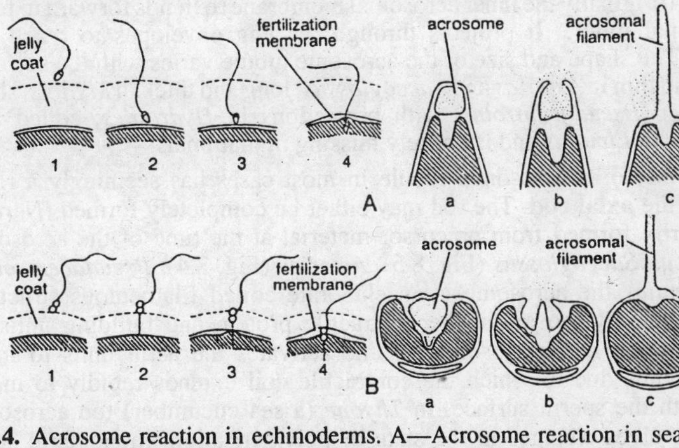

Fig. 8.4. Acrosome reaction in echinoderms. A—Acrosome reaction in sea urchin, *Arbacia* as seen in the light microscope (A′) and in electron microscope (A"); B—Acrosome reaction in starfish, *Asterias* as seen in light microscope (B′) and in electron microscope (B") (After Grant, 1978).

2. Release of acrosomal contents. The trigger mechanism releases lytic enzymes or lysins located in the acrosome. The lysins are of wide occurrence and help the sperm penetrate the egg envelopes by liquefying them locally, without affecting the egg plasma membrane or oolemma. In sea urchins (*Strongylocentrotus purpuratus* and *S. franciscanus*) the acrosomal reaction results in the liberation of a species-specific egg-binding protein, **bindin** (**Vacquier** and **Moy**, 1977) and a protease or lysin called **acrosomin**. The bindin has a molecular weight of 35,000 and plays a role in causing adhesion of sperms to eggs of the same species (the vitelline membrane of the egg contains specific glycopeptide receptors for bindin, **Trinkaus**, 1984). Acrosomin is responsible for the initial digestion of the vitelline membrane that covers the unfertilized egg (**De Robertis** and **De Robertis**, Jr., 1987).

In human females and other mammals, fertilization of ovum takes place soon after the ovum enters the Fallopian tube. However, before a sperm can enter the ovum, it must first penetrate the multiple layers of granulosa (follicular) cells which are attached to the outer side of the ovum, forming the corona radiata. The follicular cells of corona radiata are held together by an adhesive cementing substance called **hyaluronic acid** (a glycosaminoglycan). The sperm entry through the corona radiata and zona pellucida is believed to require the **hyaluronidase** and **proteolytic** enzymes released by the acrosome of the sperm. The hyaluronidase is supposed to dissolve the cement between the cells of corona radiata. In fact, even though only one sperm is required to fertilize the ovum, if an insufficient number of sperms are released in the semen, the male donor of the semen is likely to be **infertile**. This is believed to result from the fact that many sperms are required to attack the corona radiata before one sperm can succeed in penetrating the barrier. In addition to the enzymes from the sperm, bicarbonate ions in the Fallopian tube's secretions probably also help in dispersing the granulosa cells. The **zona lysins** (or proteolytic acrosomal enzymes) are held responsible for the passage of the sperm through the zona pellucida.

3. Formation of acrosomal tubule. The apical part of sperm plasma membrane (originally the inner acrosomal membrane) extends forward to form an acrosomal tubule. It projects through the egg envelopes to reach the oolemma. The shape and size of the acrosome tubule varies with the species; it is long and thin in *Asterias* and *Saccoglossus*; long and thick in *Mytilus*; short and thick in *Pseudocentrotus*, with projections in *Hydroides*; coiled and filamentous in *Limulus* and is entirely missing in mammals.

The inside of acrosomal tubule in most cases has seemingly a rigid substance, the **axial rod**. The rod may either be completely formed (*Nereis*) or only partly formed from precursor material at the time of the acrosome reaction, *e.g.*, *Saccoglossus* (Fig. 8.5), *Asterias* (Fig. 8.4), *Pseudocentrotus*, etc. In *Limulus* the acrosome is an elaborate coiled filamentous structure containing mainly the **actin**, the contractile protein and building units of **microfilaments**. Discharge of acrosome activates the actin units to form microfilaments, due to which the contractile coil extends rapidly to make contact with the sperm surface. In *Thyone* (a sea cucumber) too acrosome reaction involves changes in the state of polymerisation: monomeric actin precursors polymerise into long microfilaments which contract to pull sperm toward the egg surface (**Grant**, 1978).

4. Fusion of sperm and egg plasma membrane. The tip of the acrosomal tubule fuses with the plasma membrane of the egg and a continuity between the cytoplasms of the egg and the sperm is established. In sea urchins, it involves only 0.002 per cent of the egg surface. In mammals, in the absence of acrosomal tubule, the spermatozoon appears to contact the egg surface by its lateral aspects. Following this action, the plasma membranes of the egg and the spermatozoon dissolve at the point of contact, and the nucleus and

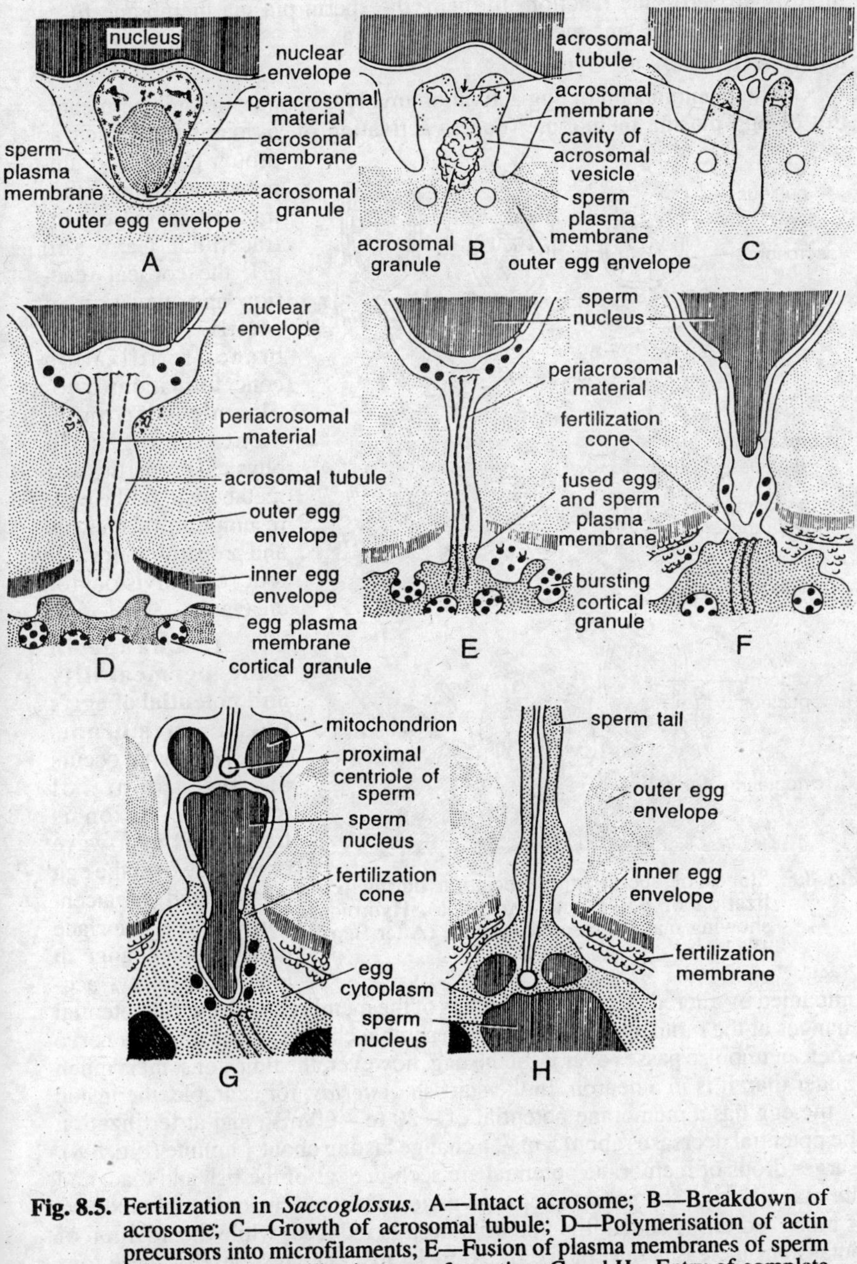

Fig. 8.5. Fertilization in *Saccoglossus*. A—Intact acrosome; B—Breakdown of acrosome; C—Growth of acrosomal tubule; D—Polymerisation of actin precursors into microfilaments; E—Fusion of plasma membranes of sperm and ovum; F—Fertilization cone formation; G and H—Entry of complete sperm inside egg cytoplasm (After Torrey, 1971).

cytoplasmic components of the spermatozoon are drawn into the interior of the egg. Thus, a single cell, called **zygote**, is formed.

In organisms lacking an acrosome in their sperm (*e.g.*, teleost fishes), there is no acrosome reaction. In them, the sperm plasma membrane fuses directly with the plasma membrane of the ovum.

8-3.4. Activation of Ovum

The phenomenon of egg activation involves change in ionic permeability of egg plasma membrane (due to activation of various pumps such as proton pump, sodium-potassium pump, etc.); transient increase in intracellular Ca^{++} and pH; the cortical reaction and elevation of fertilization membrane; fertilization cone formation; cytoplasmic rearrangement; changes in physical properties and metabolism of the egg; resumption of meiosis and preparation for mitotic cell division for cleavage.

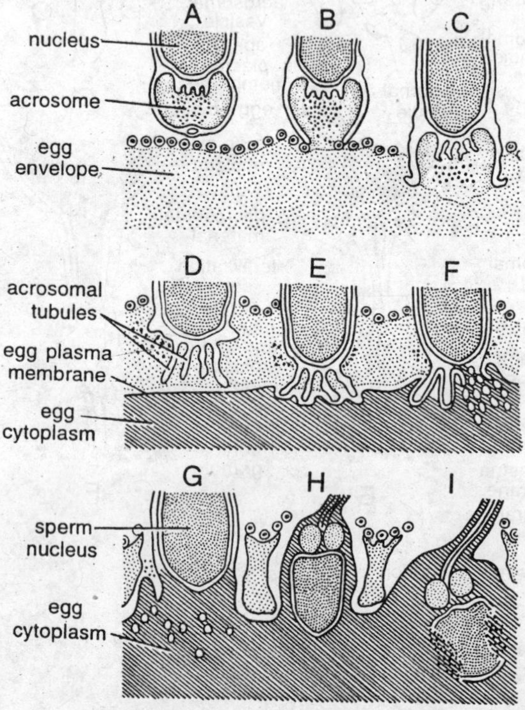

Fig. 8.6. Stages of sperm-egg association during fertilization of polychaete (Annelida), Hydroides, showing many acrosomal tubules. (After Berrill, 1971).

1. Change in ionic permeability and potential of egg's plasma membrane. The change that occurs at the point of membrane fusion is transmitted as a wave or impulse over the egg surface and subjacent cortex. The egg surface undergoes a change in ion permeability, accompanied by altered electrical properties of the membrane. Electrical potential changes at the moment of fertilization resemble changes occurring in a nerve when an impulse passes over it. In the egg, however, the time constant is much longer than it is in a neuron. In the starfish, *Asterias*, for example, the inside of the egg has a membrane potential of – 30 to – 60 mV, and at fertilization the potential decreases about 5 mV, a change lasting about 1 minute (Fig. 8.8). Larger drops of membrane potential are seen in eggs of the fish and toad, and the magnitude of the potential change varies with the ion content of medium. If more sodium is added, the potential drop is reduced, while the addition of potassium increases the change by 10 to 15 mV. The response of these ions

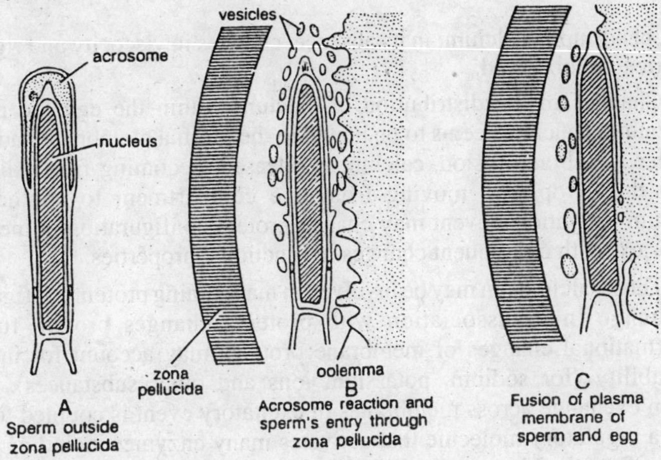

Fig. 8.7. Acrosome reaction and sperm entry in ovum of rat (After Goel, 1984).

suggests that the primary activation event of the egg is a change in ion permeability at the site of sperm contact. Sodium, potassium and calcium ions diffuse rapidly into the frog egg, reaching a peak within a minute after sperm attachment. The permeability change is then propagated as a wave over the membrane within a matter of seconds and involves a variety of transport proteins or pumps of membrane.

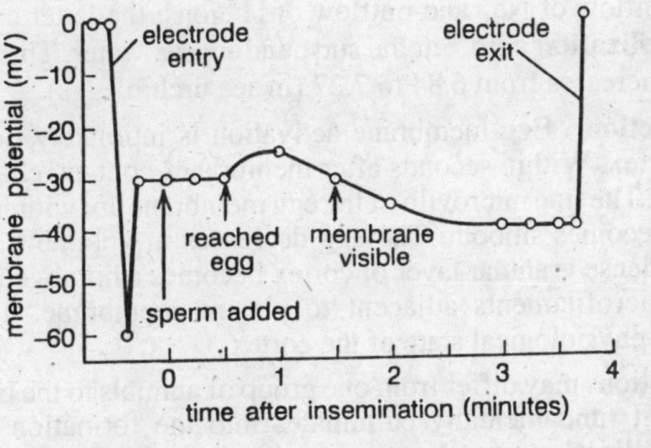

Fig. 8.8. Change in membrane potential of the starfish (*Asterias*) egg at fertilization (after Grant, 1978).

2. Transient intracellular rise in calcium ions. The first directly detectable intracellular change is a transient rise in Ca^{++}, released in the egg cortex from intracellular calcium store, which can be recharged after 40 minutes. The rise begins 20–25 seconds after sperm fusion and lasts for 2–3 minutes, when the basal levels are again attained. The transitory Ca^{++} increase is sufficient to cause the cortical reaction not only in sea urchins but also in fish eggs.

Role of calcium ions in fertilization. Of all the ions involved, calcium seems to play a major role in egg activation. In the absence of calcium, fertilization does not occur. Sperm viability, motility and fertilizing capacity are also diminished in calcium-free media. Furthermore, the acrosome reaction is calcium dependent; without it, no acrosome filament explodes. Egg stability also depends upon calcium; in its absence, changes in viscocity and rigidity of the cortex are observed.

Changes in the distribution of calcium within the egg are noted at fertilization. Calcium seems to be bound to the plasma membrane and cortex, but after membrane fusion, calcium is released, becoming freely diffusible within the cytoplasm, moving from one compartment to another. This "calcium activation" event may change protein configuration in membrane and cortex, with consequent changes in functional properties.

Thus, calcium ion may be involved in maintaining protein configuration; any change in its association with proteins changes protein function. Conformational changes of membrane proteins may account for increased permeability (for sodium, potassium ions and other substances). Finally calcium exchange across membranes in excitatory event is coupled to cyclic AMP, a regulatory molecule that activates many enzymes including protein kinases. **Protein kinases** are enzymes that phosphorylate proteins such as **histones** and may also phosphorylate **microtubular proteins** that function in cell division. Since histones have a structural and possibly a regulatory role in chromosome function, we can see that activation of histones and microtubular proteins, among others, might be essential to further development of the egg (**Grant**, 1978).

3. Transient intracellular increase in pH. The second important transient change beginning at the same time (*i.e.*, increase in Ca^{++}) but lasting for about 10 minutes is the intracellular increase in pH by 0.3 to 0.4 units. The increase is due to an inflow of Na^+ and outflow of H^+ ions; the latter causes the formation of "**fertilization acid**" in the surrounding sea water. Thus, the pH of the cytoplasm increases from 6.84 to 7.27 (in sea urchin eggs).

4. Cortical reaction. Egg membrane activation is intimately tied to changes in the egg cortex. Within seconds after membranes contact, changes occur in the egg cortex. The fine microvilli of the egg membrane are withdrawn and the egg surface becomes smooth. The egg decreases in volume (*i.e.*, it slightly shrinks). The dense granular layer of cortex becomes a homogeneous region of oriented microfilaments adjacent to plasma membrane. These changes signal altered physiological state of the cortex.

The cortical reactions may differ from one group of animals to the other, but, in most groups, it fundamentally culminates into the formation of a membrane, called **fertilization membrane**, outside the egg plasma membrane. The fertilization membrane blocks the entrance of late arriving spermatozoa. The process of cortical reactions and fertilization membrane formation have been studied more extensively only in certain echinoderms (sea urchins and starfishes).

A. Fertilization membrane formation in sea urchins. In sea urchins, as soon as the apical end of acrosomal tubule touches the surface of egg, from the site of contact, a wave-like colour change from yellow to white (under dark field microscopy) travels rapidly around the egg cortex and is shortly followed by the elevation of fertilization cone from the egg surface and the formation of fertilization membrane around the egg plasma membrane.

Electron micrographs of sea urchin's unfertilized eggs show that the egg cortex is bounded by two membranes (i) an outer 30A° thick, vitelline membrane and (ii) an inner, 60 A° thick plasma membrane. Beneath the plasma membrane occurs a layer of cortical granules (about 15,000 cortical granules) (Fig. 8.9). A fertilization membrane is formed in following stages:

(i) The outer vitelline membrane separated from the plasma membrane, undergoes expansion and becomes the outer layer of the fertilization membrane.

(ii) The cortical granules explode and release following three components: (a) Dark, denser, **lamellar** and **folded parts** of the granules unfold and fuse with the inner side of the already elevating vitelline membrane. (b) **Globules**, which fuse together and build up a new surface of viscous **hyaline layer**, just at the outer side of the egg plasma membrane. The hyaline layer adheres closely to the surface of the egg and during cleavage, it helps to keep the blastomeres together. (c) The **liquefied component** of the cortical granules fills the perivitelline space. It contains mucopolysaccharides and abundant water. The contents of cortical granules also include a trypsin-like protease enzyme.

(iii) All these structures, namely, vitelline membrane and contents of cortical granules, thus, form a fertilization membrane.

Ultimately, the membrane of cortical granules fuses with egg plasma membrane to form a mosaic. Such an altered membrane may contribute to the change in permeability. **Grant** (1978) has speculated that cortical granule membranes of an egg are already differentiated as special transport systems, so, when they are fused with the plasma membrane, they are activated to alter the permeability of membrane.

The sequence of cortical granule fusions with the egg membrane resembles an enzymatic activation travelling as a wave over the egg surface from the point of sperm attachment. It is in sea urchin's eggs propagated over the surface in 20 to 30 seconds. One hypothesis for impulse spread is that a self-propagating chain reaction is initiated by enzymes formed at the site of sperm attachment, which diffuse over the egg surface and initiates cortical granules breakdown. For example, minutes after fertilization, **lipases** and **proteases** become active, and they may digest cortical granule membranes to initiate breakdown. Another enzyme activated at fertilization is a calcium-dependent **adenosine triphosphatase**, which could explain the requirement for calcium at fertilization. Splitting the ATP may furnish energy for explosive release of the granules or for contractions of the egg surface (**Grant**, 1978).

Among other invertebrates, cortical granule reaction and fertilization membrane formation have been observed only in certain animals such as *Nereis* (an annelid worm). In fact, there are conflicting opinions as to the significance of the cortical granule reaction at fertilization. In some eggs, such as the barnacle *Barnea*, cortical granules explode independently of fertilization. In the annelid worm, *Chaetopterus*, cortical granules do not breakdown at fertilization but are retained during early cleavage.

B. Fertilization membrane formation in vertebrates. Comparable changes occur in the cortex of different species similar to sea urchins. In bony fishes and frogs, the cortical granules contain mucopolysaccharides. In the eggs of these animals, the cortical granules are broken down after sperm's penetration into the egg cytoplasm; their contents become liquefied and

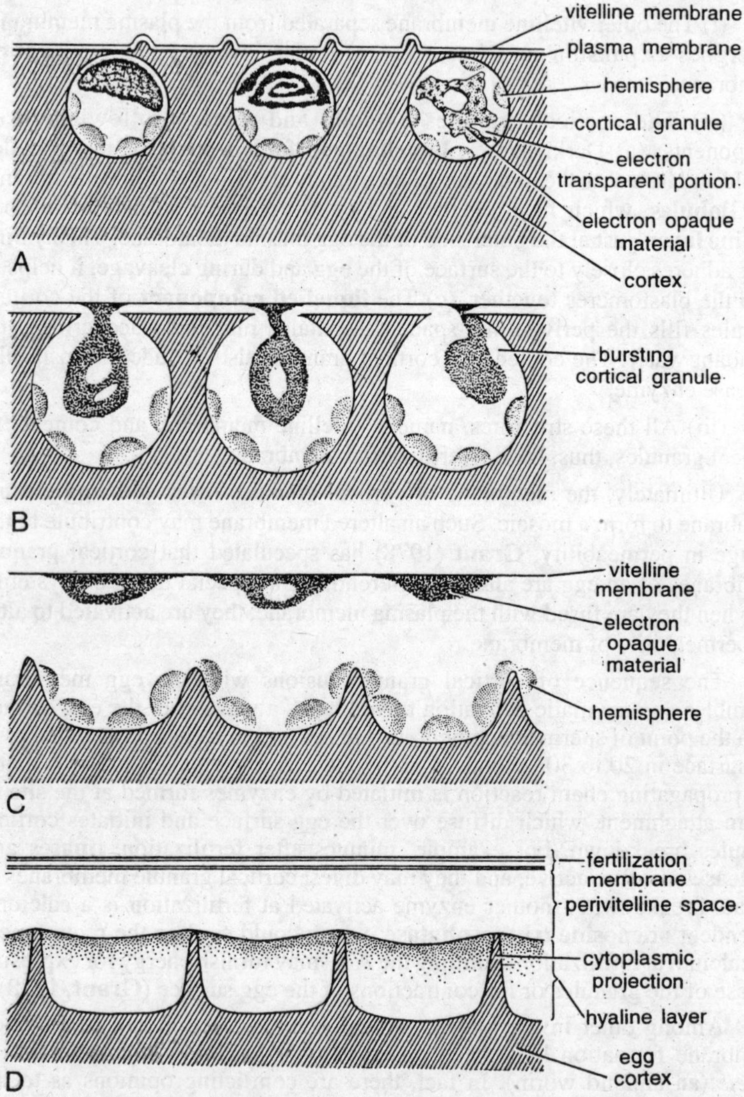

Fig. 8.9. Changes of the egg cortex of the sea urchin *Clypeaster japonicus* following fertilization. A—Unfertilized egg; B—The explosion of the cortical granules; C—Adhesion of electron-opaque material to the vitelline membrane now lifted up and formation of fertilization membrane. The hemispheric bodies remain at the surface of the egg and give rise to the hyaline layer; D—The egg surface upon completion of these changes (After Berrill, 1971).

extruded on the surface of the plasma membrane of the egg. They gradually fill up the perivitelline space which is already occurring in between the chorion and egg plasma membrane in case of bony fishes or the space lies in between the vitelline membrane and egg plasma membrane in case of frogs. In both cases, the vitelline membrane or chorion itself does not transform into the fertilization membrane like the sea urchins.

The unfertilized eggs of some mammals (*e.g.*, man, rabbit and hamster), have cortical granules in their cortical regions. In them, the sperm penetration is not followed by the formation of fertilization membrane but, the cortical granules burst open and release their contents into the perivitelline space, *i.e.*, the space between the egg plasma membrane and zona-pellucida. They ultimately disappear.

Lastly, there are many vertebrates, such as caudate or urodel amphibians, mammals, etc., which lack cortical granules. In these animals, no fertilization membrane formation occurs, however, in caudate amphibians a hyaline perivitelline space is formed in fertilized egg (see **Lovtrup**, 1974).

5. Monospermic and polyspermic fertilization. In most species successful fertilization occurs when one and only one sperm enters the egg. This is called **monospermic fertilization**. When more than one sperm enters, a genetic imbalance is created and development is abnormal. In 1902, **Boveri** noted that dispermic sea urchin eggs develop into bizarre forms. He could show that the first division of the egg was abnormal because a multipolar mitotic spindle formed by two sets of centrioles donated by two sperms. The triploid chromosome complement was randomly distributed to three cells at first division and subsequent cells were aneuploid and developed abnormally (see **Grant**, 1978).

When many spermatozoa penetrate the single ovum, this is called **polyspermic fertilization**. It is of two types : 1. **Pathological polyspermy.** Bispermy or trispermy occurs in some mammals and sea urchins if the immature or aging or overripe egg is put in a dense sperm suspension. This is called **pathological polyspermy**. 2. **Physiological polyspermy.** In some animals, particularly those with much yolk in their eggs, *i.e.*, in polylecithal eggs (*e.g.*, some insects, elasmobranchs, urodeles, reptiles and birds) and also in the microlecithal eggs of bryozoans, the egg is naturally entered by several sperms. This is called **physiological polyspermy.** In these cases, genetic material from only one sperm is incorporated in the zygote nucleus and other sperm nuclei are degenerated.

Mechanism to block polyspermy in monospermic forms. The mechanisms involved in checking of polyspermy is best studied in some sea urchins (*Lytechinus pictus* and *Strongylocentrotus purpuratus*). There are two main mechanisms to prevent polyspermy: (i) A **fast block** to polyspermy, operative from 3 to 45 seconds after the addition of sperm, is due to early influx of ions and membrane depolymerization. (ii) A **slow block** to polyspermy, starts about 30 seconds after sperm addition and lasts till hatching. It is due to (a) a steric effect because the cortical reaction elevates a fertilization membrane which physically prevents the sperm to reach to egg surface; and (b) an enzymatic effect of trypsin-like protease released from the cortical granules, which degrades bindin receptor from the vitelline membrane and detaches the bound sperm. All these mechanisms together are indeed very effective blocks to polyspermy.

In other species the block to polyspermy is only partially understood and is also either to membrane depolarisation (fish), or elevation of fertilization membrane (*Xenopus laevis*, hamster), or both (rabbit). In some fishes the micropyle is plugged. Ascidian eggs rely on the chorion or tough outer membrane to prevent multiple sperm entry; in mammals, the zona pellucida prevents polyspermy. In the latter case, penetration of one sperm changes the organisation of zona so that no other sperm can penetrate.

6. Metabolic activation. In animals, in which the egg, before fertilization, becomes metabolically inactive, the fertilization leads to a spurt of metabolic activity. Metabolic activation of the fertilized egg is depicted in the following changes :

(1) Respiratory changes. In a fertilized egg, the rate of respiration either increases (*e.g.*, sea urchins), or decreases (*e.g.*, *Chaetopterus* and molluscs, *Cumingia*) or remains static (*e.g.*, *Bufo* and *Fundulus*). There appears to be a relation between the post-fertilization oxygen consumption and the stage of maturation of the egg at fertilization. Because, at the time of fertilization the sea urchin egg has completed maturation, the egg of *Bufo* is at second maturation division stage, and the egg of *Chaetopterus* is at the first maturation division stage.

The increased oxygen consumption is related with the oxidation of glycogen and other foodstuffs of the egg and synthesis of numerous ATP molecules.

(2) Coenzyme changes. In a fertilized egg, occurs a inter-conversion of pyridine nucleotide coenzyme, NAD into another coenzyme NADP and also NADPH due to phosphorylation of the NAD in the presence of an enzyme **NAD kinase**.

$$\text{NAD} + \text{ATP} \xrightarrow{\text{NAD kinase}} \text{NADP} + \text{ADP}$$

There are ample evidences that NAD kinase enzyme, though, present in the unfertilized egg, it exists in an inactivated state. It is made activated only at the time of fertilization (**Epel** and **Iverson**, 1965).

Since, the coenzymes NAD and NADP are involved in numerous enzyme mediated oxidation and reduction reactions of the cell, and because of their obligatory role in accepting or donating electrons for these reactions, any change in concentration or availability of these coenzymes may have profound metabolic consequences. The increased NADP and NADPH contents may initiate many synthetic pathways of fertilized egg (**Epel**, *et al.*, 1969).

(3) Change in rate of protein synthesis. The cytoplasm of mature unfertilized egg, though, contains complete machinery of protein synthesis, such as DNA molecule, tRNA, mRNA, ribosomes and enzymes required during protein synthesis, none or virtually very little protein synthesis occurs in it. There are evidences that during later phase of oogenesis, some inhibitor or repressor proteins are manufactured in sea urchin's eggs and these proteins inactivate chromosomal genes, mRNA molecules, ribosomes, etc., (**Metafora**, *et al.*, 1971). During fertilization some proteolytic enzymes remove these inhibitor proteins from them and activate protein synthesis.

Thus, before fertilization, the ribosomes in sea urchin egg cytoplasm are single. After fertilization, increasing quantities of ribosomes are joined in groups, which shows they are actively engaged in protein synthesis (**Balinsky**, 1981). After five minutes of fertilization, the rate of protein synthesis increases three to twelve-fold: about 20 minutes post-fertilization, the DNA synthesis in the pronuclei begins. The protein and DNA synthesis herald cleavage and development.

7. Resumption of meiosis. In most animals meiosis is arrested at a particular stage and resumes only after fertilization when the polar bodies extruded. The time of fertilization varies from species to species. It has been found that spermatozoon may enter the egg at different stages of maturation in different animals as has been tabulated in Table 8-1.

Table. 8-1. States of egg at which fertilization occurs in various animals (after **Berrill**, 1971).

Young primary oocyte	Fully grown primary oocyte	First metaphase	Metaphase second	Female pronucleus
Brachicoelium	*Ascaris*	*Aphryotrocha*	*Amphioxus*	Coelenterates
Dinophilus	*Dicyema*	*Cerebratulus*	Most	Echinoids
Histriobdella	Dog	*Chaetopterus*	mammals	
Otomesostoma	Fox	*Dentalium*	*Siredon*	
Peripatopsis	*Grantia*	Many insects		
Saccocirrus	*Myzostoma*	*Petinaria*		
	Neries			
	Spisula			
	Thalassema			

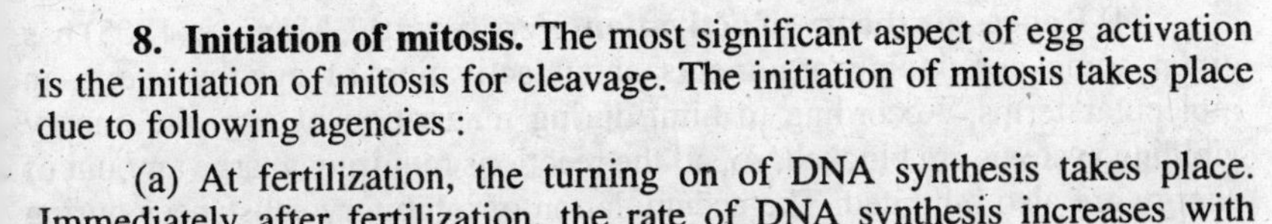

8. Initiation of mitosis. The most significant aspect of egg activation is the initiation of mitosis for cleavage. The initiation of mitosis takes place due to following agencies :

(a) At fertilization, the turning on of DNA synthesis takes place. Immediately after fertilization, the rate of DNA synthesis increases with greatest pace. It has been observed that in most of the animal eggs, the fertilization is followed by selective accumulation or uptake of cytoplasmic **DNA polymerase** enzyme (an enzyme used in DNA synthesis) by the nucleus (**Loeb** and **Fansler**, 1970).

(b) Recent investigations have shown that a spermatozoon must introduce its own centriole, if the egg is to form a mitotic spindle and divide in normal way. The unfertilized egg cytoplasm though has centriole but this centriole is incapable to divide and also to form a mitotic spindle of microtubules.

9. Theories concerning the activation of eggs by sperm. It is still not very clear that in what manner sperm activates the egg during fertilization. The following theories have been forwarded by different embryologists, to explain the mode of activation of egg :

(1) Bimolecular theory of activation. R.S. Lillie (1941) proposed that an activating substance (A) is formed from the union of two substances (B and S) that may be initially present in low concentration in the egg. Substance B is considered to be a product of hydrolytic processes in the egg and its formation to be stimulated by the action of such agents as acids and heat, which can act under anaerobic conditions. Substance S is considered to be formed from synthetic processes that may be stimulated by agents such as hypertonicity acting only in aerobic conditions. Since some substance of each type is present, the threshold concentration of A can be reached by increase in either B or S.

(2) "Sensitization to calcium" theory. The significance of calcium in egg activation has been emphasized in two rather independently developed theories, namely the "**sensitization to calcium**" theory of **Dlacq, Pasteels,** and **Brachet** (1936) and "**calcium release protoplasmic clotting**" theory of **Heilbrunn** (1930–40). Both of these theories based on experiments showing that (i) isotonic calcium chloride solutions or calcium-rich solutions are effective activators in many species of animals; (ii) activation of other chemical agents is dependent upon the presence of calcium ions; and (iii) depriving the eggs of calcium will sensitize them to the subsequent action of calcium solution. Artificial parthenogenetic studies also show a significant role of calcium in activation of egg.

(3) Loeb's theory. Loeb (1964) believed that in normal fertilization the sperm brings in a laytic principle which brings about cortical cytolysis and a second substance called "corrective factor" which regulates oxidation. **Lovtrup** (1974) has supported the Loeb's theory of activation. According to him, there may be some cytolytic agents which may function as activators by causing a partial dissolution and perforation of plasma membrane, opening of cortical granules and subsequent swelling of perivitelline space and further activation of fertilized egg.

(4) Repressor theory of activation. Very recently, **Monroy** (1965) has given some convincing arguments about activation of egg by sperm in molecular terms. According to him, during maturation of egg, the energy yielding systems are blocked, so, all the reactions requiring a large amount of energy are also inhibited. The accumulation of inhibitory substances during maturation of egg, seems to be the most likely explanation and indeed, he recognized one such inhibitor.

Monroy's repressor theory has been supported by many modern embryologists, such as **Tomkins**, *et al.*, (1969), **Metafora**, *et al.*, (1971), **Berrill** (1971) and **D. Epel** (1973), etc. Now, there are large amount of evidence that the repressor substances manufactured by the egg in the later phases of maturation, inhibit both the metabolic activities residing in the cytoplasm and the genetic activities of the nucleus. The key reaction of fertilization is, thus, the removal of the repressors, releasing at the same time the cytoplasmic metabolic activities and the activity of the nuclear genetic system.

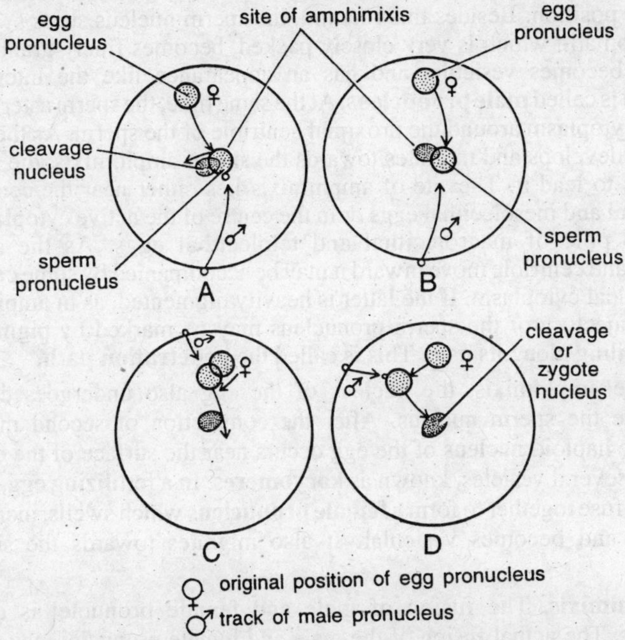

Fig. 8.10. Amphimixis in sea urchin. Tracks of the pronuclei of the sea urchin in their movements toward each other and to the site of the cleavage nucleus with sperm entering at opposite side of egg from site of egg nucleus and from same side respectively (after Balinsky, 1970).

Recently studies using video microscopic data and microelectrodes to record bioelectrical reponses have shed some light on the initial phenomenon that activates the egg. It appears that after acrosome reaction there is an elevation of Ca^{++} and an increase in pH inside the sperm and that localized increase in cytoplasmic Ca^{++} (at the place of sperm and egg contact), is sufficient to trigger the exocytosis of the cortical granules and elevation of the fertilization membrane. This is followed by a propagating wave of Ca^{++} released from intracellular stores and by an increase in cytoplasmic pH. These two components (Ca^{++} and pH) appear to be the regulators that trigger the various cytoskeletal changes occurring in fertilization. In addition to protons and Ca^{++}, calmodulin proteins and possibly cyclic nucleotides are also involved in these complex molecular mechanisms (**De Robertis** and **De Robertis Jr.**, 1987).

8-3.5. Migration of Pronuclei and Amphimixis

At the time of penetration of spermatozoon inside the egg cytoplasm, the sperm nucleus remains compact and its mitochondria and centriole remain situated behind it (Fig. 8.10). To perform the act of amphimixis, the sperm nucleus has to perform two activities : (i) it has to become pronuclei and (ii) it has to migrate from the site of fertilization to the site of amphimixis. As the sperm nucleus moves inward from the site of the fertilization cone, it soon rotates through the angle of 180°, so that its mitochondria and centriole assume the leading position. Besides this rotation, the sperm nucleus starts swelling and its chromatin which is very closely packed, becomes finely granular. It ultimately becomes vesicular and has an appearance like the interphase nucleus and is called **male pronucleus.** At the same time, the sperm aster forms in the egg cytoplasm around the proximal centriole of the sperm. As the male pronucleus develops and migrates towards the site of amphimixis, the sperm aster seems to lead it. The site of amphimixis lies either near the centre of microlecithal and mesolecithal eggs or in the centre of the active cytoplasm at the animal pole of macrolecithal and telolecithal eggs. As the sperm pronucleus and centriole move inward it may be accompanied by some cortical and subcortical cytoplasm. If the latter is heavily pigmented, as in amphibian eggs, the trajectory of the sperm pronucleus may be marked by pigmented granules trailing along its path. This is called the **penetration path.**

Before amphimixis, the nucleus of the egg also undergoes certain changes like the sperm nucleus. After the completion of second meiotic division, the haploid nucleus of the egg occurs near the surface of the egg in the form of several vesicles, known as **karyomeres**. In a fertilizing egg, these karyomeres fuse together to form a **female pronucleus** which swells, increases in volume, and becomes vesicular. It also migrates towards the site of amphimixis.

Amphimixis. The fusion of male and female pronuclei is called **amphimixis**. The actual fusion of the male and female pronuclei may differ in detail in different animals. In most animals, the two pronuclei actually fuse together, that is, the nuclear membranes of both pronuclei are broken at the point of their contact and the contents of both pronuclei unite into one mass, which is ultimately bounded by a common nuclear envelope. In other cases, (*e.g.*, *Ascaris*, some annelids and molluscs), however, the male and female pronuclei do not fuse as such, but the nuclear membranes in both dissolve and the chromosomes are released. In the meantime, the centrosome of the spermatozoon has divided in two and an achromatic spindle is formed and to this spindle, the chromosomes derived from the male (sperm) and female (egg) pronuclei become attached. Only after the completion of the first division of the fertilized egg, the paternal and maternal chromosomes become enclosed by common nuclear membranes to form the nuclei of two blastomeres.

8-4. OOPLASMIC SEGREGATION

One of the principal results of egg activation during fertilization is reorganization of egg cytoplasm. This, in turn, may change the original polarity of the egg and affect the future embryonic axis. Many cytoplasmic organelles move about the egg after fertilization to take up new positions, often within a matter of seconds after sperm entry. This cytoplasmic restructuring is called

ooplasmic segregation and it creates new axes of polarity and symmetry. During ooplasmic segregation morphogenetically active materials are redistributed in the cytoplasm and come to occupy specific regions. As a result, some morphogenetic substances are segregated into certain cells during the first few cleavage divisions. Sorting out morphogenetic substances in this way may determine cell fate early in development just as it does in the germ line.

To understand the phenomenon of ooplasmic segregation let us consider the following two typical examples :

1. Ooplasmic segregation in ascidians. In 1905 **Conklin** carried out the classical study on the ascidian *Styela partita* (Fig., 8.11). The ripe unfertilized egg of *Styela* is laid in the germinal vesicle stage and is full of gray yolk concentrated in the centre. Surrounding the yolk is a peripheral layer of protoplasm in which are embedded yellow pigment granules and minute mitochondria. A region of clear cytoplasm surrounds the germinal vesicle at the animal pole. As soon as the egg enters sea water, it begins to mature, the germinal vesicle breaks down, and the first metaphase spindle appears at the

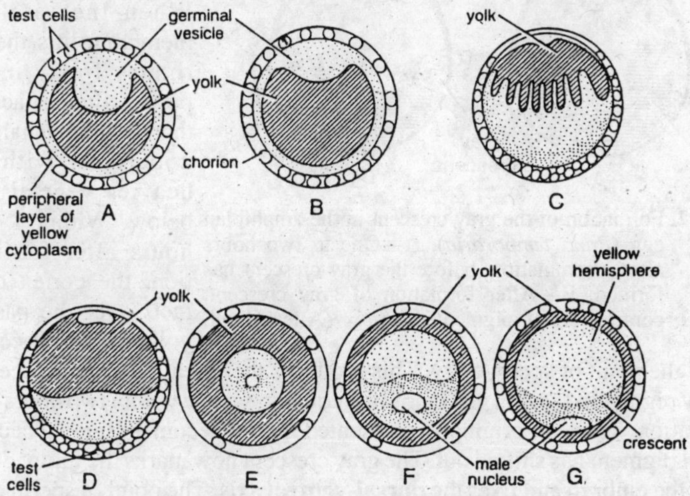

Fig. 8.11. Ooplasmic segregation after fertilization in the egg of *Styela partita* (after Grant, 1978).

animal pole. The egg remains in this state until fertilized. After sperm entry, the first and second polar bodies form; within 5 minutes, ooplasmic segregation takes place as the clear cytoplasm and the peripheral layer of yellow cytoplasm together stream violently toward the vegetal pole, forming a cap. They separate, with clear cytoplasm spread out over the yellow cytoplasm below. Meanwhile, gray yolk granules congregate near the animal half, at the maturation (*i.e.*, meiotic) spindle. The sperm, now a pronucleus, penetrates the vegetal region, migrates through the clear cytoplasm toward the animal hemisphere, and shifts toward the future **posterior side** of the egg. In so doing, it drags the clear and yellow plasm toward the posterior side near the equator, creating a clear layer of cytoplasm above a **yellow crescent** on one side of the egg, the future **dorsal-ventral plane** of the embryo. The gray yolk granules are likewise displaced into the vegetal pole region. This is the state of cytoplasmic localization before the first division. It displays an asymmetrical distribution of cytoplasm that will be irreversibly fixed after the first division, with each cell receiving different populations of egg organelles.

2. Ooplasmic segregation in amphibians. The ooplasmic segregation in amphibians involves two components, the cortex and the adjacent cytoplasm. Segregation results from contraction of the cortical gel over stationary yolk mass. The egg of *Rana temporaria* consists of a pigmented cortical region of variable thickness extending from animal half and surrounding a central yolk mass. Attachment of sperm to the egg surface initiates a surface reaction with cortical granule breakdown. The site of sperm entry is marked by a tail of pigment granules drawn in from the surface as the sperm penetrates the deeper cytoplasmic layers (Fig. 8.12). When the vitelline membrane lifts, the egg rotates freely in the perivitelline space so that the animal half lies uppermost with the heavier vegetal half below. Within 1 to 2 hours after fertilization, the cortex contracts over the passive yolk mass in the centre, toward the point of sperm entry. This produces the so-called **gray crescent**, a slightly pigmented region just below the equator, visible within two hours after fertilization. Thus, an originally pigmented region becomes "gray" because cortical pigment has shifted out. The gray crescent now marks the future dorsal side of the embryo and fixes the **dorsal-ventral** axis. The point of sperm entry corresponds to the future plane of bilateral symmetry of the embryo; that is, it usually coincides with the plane of first cleavage and occupies the ventral side of the embryo.

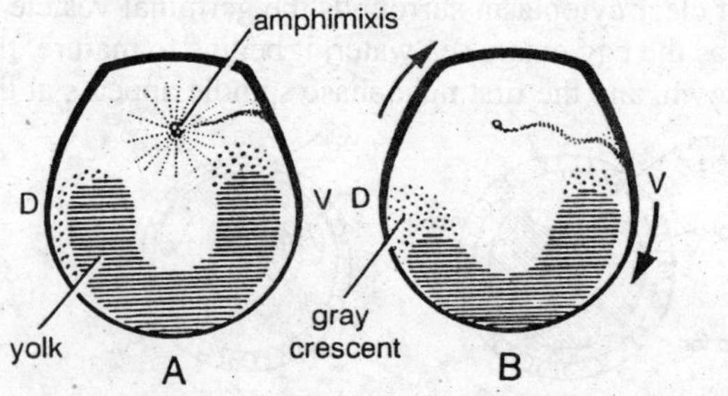

Fig. 8.12. Formation of the gray crescent in the amphibian egg (*Rana temporaria*). A—One to two hours after insemination, before the gray crescent has formed; B—After formation of gray crescent, contraction of pigmented cortex (After Grant,

3. Forces involved in ooplasmic segregation. What forces move cytoplasms or attract organelles into specific regions during ooplasmic segregation ? One hypothesis is that particulate components of the cytoplasm are **electrically charged** and fertilization by changing the charge distribution within the cell, induces the charge particles to move to one region or another.

Another hypothesis, first proposed by **Raven** in 1910, is that a "**cortical morphogenetic field**", a combination of membrane-cortical properties, "directs" the displacement of cytoplasmic substances during ooplasmic segregation. The localizing forces may indeed emanate from the cortex since it seems to orient microtubules and microfilaments, the organelles of cell motality and cytoplasmic flow. Microtubule assembly and microfilament contractility are both sensitive to changes in calcium ion concentration, one of the principal ion changes that occur at fertilization. In addition, both types of organelles are also responsible to changes in concentration of cyclic AMP, another regulatory moleçule that probably fluctuates at fertilization. The flux of microtubules and microfilaments after fertilization may orient cytoplasmic flow just as they displace pigment granules in fish melanocytes or induce axoplasm flow in neurons. Localization of specific cytoplasmic substances may reflect the fact that these are preferentially associated with microfilaments or microtubules, or both. The directed flow of cytoplasms could occur because microtubules and microfilaments are anchored in specific regions of the cortex so that a combination of cortical contractions with microtubular displacement could account for ooplasmic segregation.

4. Exceptions of ooplasmic segregation. In many eggs, ooplasmic segregation does not occur immediately after fertilization. Instead, it is a progressive process beginning at fertilization but completed after the first few divisions. Cytoplasmic localization is achieved by coupling ooplasmic segregation to cleavage, a situation found in the egg of *Nereis*.

The unfertilized egg of *Nereis* contains a large central germinal vesicle surrounded by a double row of oil droplets. After activation, large alveolar-like granules in the cortical layer are extended and swell into a thick jelly. The germinal vesicle ruptures and slow movements of cytoplasmic constituents begin. Most mitochondria aggregate into a small subcortical ring below the animal pole, and these are incorporated into only one of the two cells at the first division. Cytoplasmic movements continue well into the second division, and by the four-cell stage, ooplasmic segregation is completed with clear protoplasm in the animal half and the oil droplets and yolk spheres localized in the vegetal hemisphere. In the egg of *Nereis*, thus, cytoplasmic reorganization is stabilized by the cleavage furrowing process.

Thus, in a way, **timing of ooplasmic segregation** determines the cytoplasmic constituents in each cleavage blastomere. If cytoplasmic localization occurs early, immediately after fertilization as in *Styela*, then first cleavage produces qualitative differences in each blastomere as each receives different morphogenetic material. If it occurs late, after several cleavage divisions as in *Nereis*, then qualitative differences in blastomere cytoplasms are delayed. In this case, the cleavage process itself determines the localization of special cytoplasms. Finally, there are eggs, such as that of the sea urchin which exhibit no visible segregation except for a cortical shift in distribution of pigment granules. In this species, however, cytoplasmic localizations are completed precociously during oogenesis; the developmental blue print is programmed before fertilization. The division process alone partitions egg organization into cell populations with different developmental fates.

8-5. SIGNIFICANCE OF FERTILIZATION

The process of fertilization accomplishes following significant results :

1. The gametes or the germ cells constitute the only physical bridge between the successive generations. The gametes possess haploid (n) number of chromosomes. The fertilization restores the specific parental diploid chromosome configuration.

2. Fertilization brings together the chromosomes and genes from two different lines of ancestry and results in new genetic recombinations, which ensure improved adaptation to ever changing environment.

3. Fertilization activates the egg to start cleavage.

REVISION QUESTIONS

1. What is fertilization ? Write an essay on fertilization in animals.
2. Describe with well-illustrated diagram the major events that occur in the process of fertilization.
3. What is ooplasmic segregation ? Describe its significance.
4. What is egg activation ? What precisely is meant by activation of (a) egg and (b) sperm ? Discuss various theories regarding egg activation during fertilization.
5. What are the similarities between egg activation and nervous stimulation of a cell ?
6. Describe the process of cortical reaction in sea urchin eggs.
7. Give an account of metabolic changes occurring in a fertilized egg.
8. Discuss the significance of fertilization in the life cycle of living beings.
9. Write short notes on the following :

   (1) Fertilizin-antifertilizin reaction ; (2) Capacitation ; (3) Acrosome reaction ; (4) Hyaluronidase ; (5) Fertilization membrane ; and, (6) Amphimixis.

9

PARTHENOGENESIS

Parthenogenesis or **virginal reproduction** is a special type of sexual reproduction in which an egg (or ovum) develops without fertilization. It is generally agreed that a mature ovum contains within itself all the material potentiality to form a new being. This potentiality of the ovum to develop further without fertilization or by parthenogenesis either is released naturally (in many invertebrates) or can be released experimentally (in many invertebrates and chordates) by a variety of mechanical, chemical and physico-chemical means (**Williams** and **Warwick**, 1980).

Beatty (1967) has defined the phenomenon of parthenogenesis as the development of a female gamete into the embryo without any genetic contribution from the male gamete, with or without the attainment of adulthood. The word parthenogenesis (Gr., *parthenos* = virgin ; *genesis* = origin) literally means the birth of an individual from a virgin. The individuals reproducing by parthenogenesis are called **parthenotes**. Parthenogenesis frequently occurs, in its natural course, in certain orders of insects such as Hymenoptera (bees and wasps), Homoptera (aphids), Coleoptera (beetles); rotifers and lower crustaceans (*Apus* or tadpole fish, *Artemia* or brine shrimp, *Cypris* and *Daphnia*) and certain desert lizards. However, it has been successfully induced artificially in echinoderms, molluscs, annelids, amphibians and even mammals. Thus, the parthenogenesis is of two main kinds : 1. Natural or spontaneous parthenogenesis; 2. Artificial or induced parthenogenesis.

9-1. NATURAL PARTHENOGENESIS

Parthenogenesis occurs as a natural phenomenon in a number of species. In some organisms it is the usual (**facultative**) or exclusive (**obligatory**) mode of reproduction. Natural parthenogenesis falls into the following two broad classes :

9-1.1. Complete Parthenogenesis

In some platyhelminthes, rotifers and certain wasps parthenogenesis appears to be the only form of reproduction. This is called **complete parthenogenesis**. In complete parthenogenesis, every individual arises from an unfertilized egg; the species consists entirely of females, and males appear rarely, if at all. Sexuality is lost in these groups; the entire adult life is devoted to feeding and reproduction. The advantage of this system lies in the population's high reproductive capability, since each individual is a female, producing more offspring. On the other hand, the system loses the advantages offered by sexual recombination, which means that the only source of species variation is mutation.

9-1.2. Cyclic Parthenogenesis

In cyclic parthenogenesis which is found in aphids and some rotifers, several generations of parthenogenetic reproduction alternate with biparental reproduction in which the egg is fertilized. Thus, cyclic parthenogenesis retains the advantages of sexuality. These groups are benefited from both methods of reproduction since parthenogenesis and bisexuality are found in an animal's life cycle. Aphids, for example, increase rapidly during the parthenogenetic portion of the life cycle. In spring, egg hatch as females, which invade plants and reproduce many parthenogenetic offspring through the summer months. Gene recombination is achieved in the single sexual generation at the end of summer. Eggs from some winged females become sexual individuals which mate. From these fertilized eggs, female nymphs emerge in the spring the following year to repeat the cycle (see **Grant**, 1978).

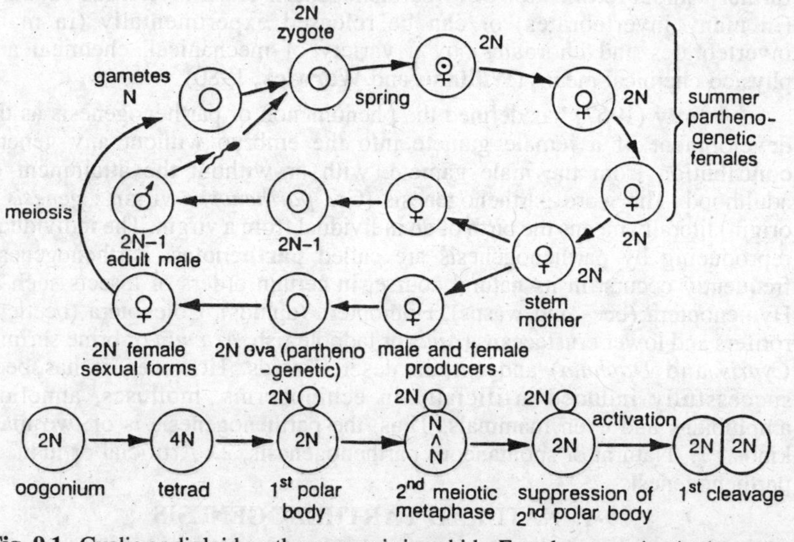

Fig. 9.1. Cyclic or diploid parthenogenesis in aphids. Females emerging in the spring produce many generations of females by diploid parthenogenesis resulting from suppression of first or second polar body (shown below). At summer's end some females produce sexual males and females by diploid parthenogenesis, males differing from females in lacking one sex chromosome. Both of these sexes, in turn, produce haploid gametes through normal meiosis which fuse to form diploid zygotes that emerge again in the spring as parthenogenetic females (after Grant, 1978).

Rotifers too have a curious type of reproduction. There are generally separate sexes. Females are of two types, **amictic** or **mictic** and each produces separate types of eggs. **Mictic eggs** are haploid and can be fertilized by males. A fertilized mictic egg becomes dormant and eventually develops into a diploid female. If mictic eggs are not fertilized, they develop into haploid males. **Amictic females** produce diploid eggs which develop directly into diploid females without fertilization, *i.e.*, by the process of parthenogenesis. The relative proportions of the sexes and the presence of dormant eggs appear to be controlled both by internal physiology and environmental conditions, but the nature of this relationship is obscure. Dormant eggs tend to be produced when extreme conditions of temperature or drying prevail. The eggs can survive for weeks and months in this state (see **Herreid II**, 1977).

Natural parthenogenesis has also been classified according to the following two ways, both of which are independent of each other :

1. Classification of natural parthenogenesis based on the behaviour of chromosomes. Based on the behaviour of the chromosomes during meiosis, the following three types of natural parthenogenesis have been recognised :

(1) Haplo-diploid parthenogenesis. (Haploid parthenogenetic males, diploid females). In this case normal meiosis occurs and parthenogenetic haploids develop into males. For example, queen honeybee is fertilized only once by a male (drone) or sometimes by more than one drone. She stores the sperm in her seminal receptacles and as she lays her eggs, she can either fertilize the eggs or allow them to pass unfertilized. The fertilized eggs (diploid with 32 chromosomes) develop into females (queens or workers); the unfertilized eggs (haploid with 16 chromosomes) become males (drone) (Fig. 9.2). According to **Grant** (1978), in this case, only the germ line remains haploid, since all somatic cells or tissues, as is true for many insects, become highly polyploid or polytenic during larval development. The male produces normal gametes because meiosis does not occur to prevent further reduction of chromosome set. Each spermatogonium undergoes one mitotic division

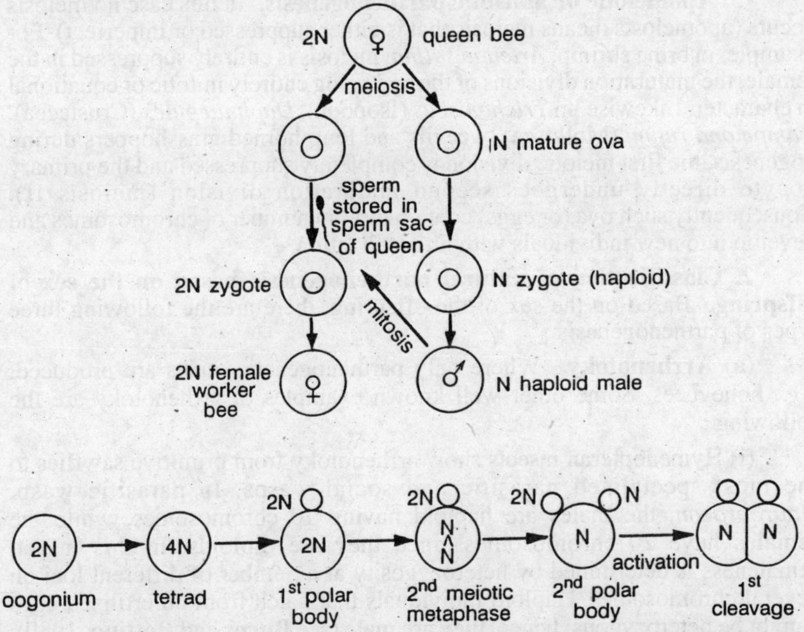

Fig. 9.2. Haploid parthenogenesis in bees. The diploid queen bee lays haploid eggs after meiosis. Most are fertilized by stored sperm as they pass out and develop as diploid female workers. At certain times some eggs emerge unfertilized; these develop as fertile haploid males which mate with another queen on a nuptial flight, and she starts the cycle again in a new hive. The mechanism of haploid parthenogenesis is shown below in which oocyte completes a normal meiosis but the haploid egg is activated without sperm and develops as a haploid male (after Grant, 1978).

without any crossing over and produces two haploid sperms. Meiosis in females is normal.

(2) Automictic, meiotic or diploid parthenogenesis. In this case, meiosis occurs but the chromosome number is doubled (**diplosis**) either before (*Moraba*, Orthoptera) or after (*Solenobia*, Lepidoptera) meiosis. Diplosis may take place by the following two methods :

A. By autofertilization. During this most common mechanism (Fig. 9.1), during oogenesis, the first meiotic division is normal and produces a first polar body or polocyte. However, the second part of meiotic division (*i.e.*, meiosis II) is suppressed and the second polar body's nucleus fuses with the egg nucleus (or nucleus of ootid) to produce the diploid cell.

B. By restitution. In insects of order Hymenoptera (*Nemertis conesceus*) and Lepidoptera, nuclear division or karyokinesis (of meiosis I) of primary oocyte forms a nucleus of secondary oocyte and a nucleus of the first polocyte. However, this karyokinesis is not followed by the cytokinesis. The chromosomes of both daughter nuclei get arranged on the equator and proceed for second meiotic division to form a diploid ootid (ovum) and a diploid polocyte. The diploid ovum develops into a parthenogenetic diploid individual.

(3) Apomeiotic or ameiotic parthenogenesis. In this case no meiosis occurs (apomeiosis means meiosis that is either suppressed or imperfect). For example, in brine shrimp, *Artemia salina*, meiosis is entirely suppressed in the female, the maturation divisions of the egg being entirely mitotic or equational in character. Likewise, in *Trichoniscus* (Isopoda), *Daphnia pulex* (Crustacea), *Campelona rufum* (Mollusca), weevils and long-horned grasshoppers during oogenesis, the first meiotic division is completely suppressed and the primary oocyte directly undergoes second maturation division (meiosis II). Consequently such ova (or eggs) contain diploid number of chromosomes and develop into new individuals without fertilization.

2. Classification of natural parthenogenesis based on the sex of offspring. Based on the sex of the offspring, there are the following three types of parthenogenesis :

(a) Arrhenotoky. Where only parthenogenetic males are produced, *e.g.*, honeybees. Some other well known examples of arrhenotoky are the following :

(i) Hymenopteran insects show arrhenotoky from primitive sawflies to the most specialized parasitic and social wasps. In parasitic wasp, *Habrobracon*, the males are haploid having 10 chromosomes, while the females have 20 chromosomes since they are diploids. In this insect, femaleness is determined by heterozygosity at a number of different loci on several chromosomes. Haploid individuals that hatch from unfertilized eggs cannot be heterozygous; hence, they are male (see **Burns** and **Bottino**, 1989).

(ii) In Coleoptera, *Micromalthus debilis*, males are haploid and develop from unfertilized eggs.

(iii) In Thysanoptera (*Anthothrips verbasi* and *Frankliniella insulasis*) virgin females produce all haploid males. Some embryologists have questioned the validity of these cases.

(iv) The rotifer *Asplanchna amphora* is an example of arrhenotoky, where the males are dwarf and develop from haploid eggs. In another rotifer, *Bolelloida* males are unknown and arrhenotoky is the only mode of reproduction.

(v) In Arachnida, arrhenotoky is mostly found in ticks and mites. In them, males are haploid and develop from the unfertilized eggs, whereas, the females are diploid and sexual. **Schrader** (1923) has reported male haploidy in the mite *Tetranynchus* and **Patau** (1936) has described male haploidy in *Pediculoides ventriconus.*

(b) Thelytoky. In this case only parthenogenetic females are produced. In thelytoky the reduction division is abortive and the young individuals develop from the unfertilized diploid eggs. It includes the already discussed categories of parthenogenesis such as meiotic and ameiotic parthenogenesis, *e.g., Solenobia, Ptinus, Lacerta saxicola armeniaca* (No males exist in nature of this species and its population is propagated exclusively by parthenogenetic females (**Balinsky**, 1981).

(c) Amphitoky. In this case parthenogenetic eggs may produce individuals of either sex, *e.g., Aphis.*

9-2. ARTIFICIAL PARTHENOGENESIS (Experimental Activation of the Egg)

The development of a ripe egg after certain treatments, but without fertilization is called **artificial parthenogenesis**. The study of artificial parthenogenesis is of interest because it throws light on the activation of the egg by the sperm at the time of fertilization.

Eggs of many animals, such as sea urchin, starfish, moth, fish, frog, turkey, hen, rabbit, etc., have been activated by the following ways and it has been found that the eggs which have been induced by artificial means seldom reach maturity and seldom develop into the functional adults :

1. Chemical treatment with hypo-and hypertonic solutions ; weak organic acids (butyric, lactic, oleic acids); alkalis (0.01 $N\,NH_3$); chloride salts of lithium, sodium, potassium, calcium and magnesium; strychnine (an alkaloid obtained primarily from the plant nux vomica or *Strychnos nuxvomica*); fat solvents (toulene, ether, alcohol, benzene and acetone); and other chemicals such as chloroform, urea, sucrose, corrosive sublimate, etc.

2. Treatment with physical agents such as heat or cold shocks (*e.g.*, when the egg is transferred from 30°C to –10°C), electric shock, shaking, pricking, etc.

3. Treatment with radiations such as fertilization by genetically inactivated (ultra-violet or UV-treated) sperm.

9-2. 1. Mode of Action of Agents of Artificial Parthenogenesis

The diversity of physico-chemical agents inducing parthenogenesis obscures the underlying activation mechanism. The one common feature is that all agents first act on the egg plasma membrane, which behaves like a coiled spring ready to be triggered into action at the slightest stimulus. Once the egg is activated, a developmental programme can unfold in the absence of sperm. Sublethal damage to egg cortex and Ca^{2+} ions and pH changes appear to have a role in the activation of the egg. To follow the action of some of the activation factors in artificial parthenogenesis, let us consider the following classic experiments :

1. Artificial parthenogenesis in sea urchin. Loeb (1913) was working on sea urchin eggs and discovered that they undergo activation when they are treated with hypertonic sea water. The fertilization membrane is elevated in the normal fashion, but only a single aster, a "**monaster**" (of course of maternal origin) can form. At best, the chromosomes can undergo a few cycles of replication around this monaster; since, there is only one aster, no typical mitotic apparatus with two asters and a spindle can form. The consequence is that the egg cannot divide and will not develop further. But **Loeb** found that if the activated eggs are submitted, after a few minutes, to a second treatment with butyric acid, this time a second aster will appear and the egg will develop parthenogenetically until the pluteus larval stage is reached.

2. Artificial parthenogenesis in frog. Bataillon (1910) worked on frog eggs and found that they can be activated by pricking them with a clean fine glass needle. They form a fertilization membrane and undergo a series of monasterial, abortive division cycles (to result the polyploidy). If, however, the needle is not clean and is smeared with freshly drawn frog blood, so that a nucleate cell is introduced into the egg at the time of pricking, a second aster will form around the injected cell. Mitosis will then be normal and development will proceed until the tadpole stage (see **Brachet**, 1974). The cell which has been introduced in the unfertilized eggs, thus, contains a "second factor" necessary for the formation of a second aster and further development. Since the introduction of an anucleate mammalian red blood cell had no effect, **Bataillon** concluded that the "second factor" must somehow be linked to the cell nucleus and that a **caryocatalysis** is required for parthenogenetic development.

The chemical nature of the "second factor" of parthenogenesis is not yet known with certainty. Recent work has shown that it is not located in the nucleus, but in cytoplasm. Most probably it is either centriole, mitochondria or Ca^{2+} ions.

Further, it is important to note that parthenogenetic activation even if it is not followed by further development, it induces the same metabolic changes in the egg as fertilization. Treatment with hypertonic sea water, even in the absence of a second treatment with butyric acid, is sufficient to increase both respiration and protein synthesis in sea urchin eggs. Thus, the part played by the spermatozoon in these metabolic events appears to be negligible.

3. Haploid syndrome. Parthenogenetic embryos are, as a rule, haploid and die during development. This is called **haploid syndrome (Brachet**, 1974). The haploid syndrome can very easily be demonstrated in frog eggs which have been fertilized with sperm previously irradiated for a short time (1 to 2 minutes) with UV lamp. The spermatozoa remain perfectly motile and the same percentage of fertilization can be obtained as with normal non-irradiated sperm. But the chromatin of the spermatozoon has been irreversibly damaged; the paternal chromosomes are eliminated into the cytoplasm, where their DNA undergoes complete degradation during the first cleavage of the egg. Development is at first normal; but a few eggs already fail to gastrulate properly. The others may reach at best a very abnormal tadpole stage. Such tadpoles exhibit strong **oedema** (*viz.*, an excessive accumulation of fluid in the cells, tissue spaces or body cavities due to a disturbance in the fluid exchange mechanism). They have a reduced head and very small gills and are unable to digest their yolk completely. This haploid syndrome is always lethal. A few eggs, however, may escape death and become adults. Cytological observation shows that they are always diploid (2n) or polyploid (4n, 6n or 8n). This regulation of the number of chromosomes is due to the fact that in a few eggs, the chromosomes undergo one or more monasteral cycles of replication before a second aster forms around the injected nucleate cell.

4. Artificial parthenogenesis by ionophores. Certain compounds, called **calcium ionophores** can activate (*i.e.*, cause parthenogenetic development) the eggs of a wide group of animals including sea urchins, tunicates, molluscs, amphibians and mammals (**Steinhardt** and **Epel**, 1974; **Epel**, 1977). Ionophores make cellular membranes permeable to certain ions—and not only the outer cell or plasma membrane but internal membranes as well. Thus, a calcium ionophore facilitates the release and flow of bound intracellular calcium in the sea urchin egg which is followed immediately by the action potential of ferilization, cortical granule breakdown, membrane elevation, and stimulation of metabolism. However, the activated egg does not cleave. This experiment, thus, proves that parthenogenetic agents may act by releasing calcium.

5. A case of egg-squeezing in *Habrobracon*. In the wasp *Habrobracon* egg activation seems to result from a mechanical stimulus imparted to the egg as it is squeezed through the ovipositor during laying. Eggs protected from this compression process do not develop (see **Grant**, 1978).

6. Artificial parthenogenesis in birds. Parthenogenesis can occur accidentally in many organisms by suppression of polar body formation in meiosis. This has been reproted in domestic turkeys and domestic hens. Eggs laid by female turkeys which had been isolated from males start developing in a fairly high percentage of cases (up to 41.7 per cent in experiments conducted by **Olsen** in 1960). Although there is still a high percentage of abnormal development, a small number of embryos reach the hatching stage. All the poults were males. A few of these reached maturity and one was used with success, to fertilize eggs of normal females.

More recently (**Sarvella**, 1973) similar results have been obtained with hens.

7. Artificial parthenogenesis in mammals. Among mammals, artificial parthenogenesis had been successfully induced in the eggs of rabbit by **Pincus** (1936). It was found that if the eggs of rabbit are simply kept for up to 48 hours in the ordinary tissue culture medium (blood plasma plus embryo extract), some of them become activated (*i.e.,* they complete their maturation division, produce polar bodies and even do cleavage) (see **Balinsky**, 1981).

9-3. SIGNIFICANCE OF PARTHENOGENESIS

Parthenogenesis has the following advantages and disadvantages :

9-3.1. Advantges of Parthenogenesis

The process of parthenogenesis has the following advantages :

1. Parthenogenesis is a well established means of reproduction. It is even simpler and easier than sexual reproduction.

2. Parthenogenesis represents a device of high multiplication. For example, aphids multiply very fast during their parthenogenetic phase of the life cycle.

3. Parthenogenesis permits the establishment of triploid (3n) and aneuploid races having three or more sets of chromosomes. In triploid forms, three sets of chromosomes fail to pair during meiosis at the time of gamete formation, thus, the retardation of the formation of gametes. This leads to sexual sterility. The parthenogenesis is, thus, a means for aneuploid individuals to propagate their own kind. It has been described by **Darlington** (1939) as an **escape from sterility.**

4. Parthenogenesis encourages advantageous combination of characters or genes originated by mutation to remain as such because these are not liable to segregate by crossing over as there is no meiosis.

5. Since all mutagenic changes and their combinations are not useful to the organisms, so, natural selection in the form of parthenogenesis tends to eliminate maladaptive combinations of genes.

6. Parthenogenesis is a means of sex determination in honeybee and other animals. It supports the chromosome theory of sex determination.

7. Parthenogenesis eliminates the need for mating and allows the whole energy of the adult life to be devoted to feeding and reproduction.

8. Finally, the evolution of parthenogenesis could have eliminated the need for an activation signal. As eggs mature in the ovary, they might develop immediately without going into a metabolically arrested state.

Except for its genetic and second-factor contribution, the importance of sperm in embryogenesis is relatively minor. The egg alone is repository of developmental information and can express this potential when stimulated either by sperm or by some artificial agent.

9-3.2. Disadvantages of Parthenogenesis

The main evolutionary disadvantage of parthenogenesis is that it tends to eliminate variation in a population. The offsprings developing parthenogenetically are more alike to and mostly the phenocopies of the parents, whereas, a sexually developed offspring exhibits marked differences from the parents.

REVISION QUESTIONS

1. Write a short essay on parthenogenesis.
2. What is parthenogenesis ? Explain how various methods of induced parthenogenesis help us in understanding the mechanism of fertilization.
3. Describe advantages and disadvantages of parthenogenesis.
4. Differentiate between the following :
   (i) Sexual development and parthenogenesis;
   (ii) Arrhenotoky and thelytoky;
   (iii) Natural and induced parthenogenesis;
   (iv) Complete parthenogenesis and cyclic parthenogenesis.
5. Write short notes on the following : (1) Haploid syndrome; (2) Ionophores.

10

CLEAVAGE AND BLASTULATION

This is the inherent nature of the sexually reproducing multicellular animals that their body originates from a large-sized, single-celled (unicellular), nutrient-filled, diploid and fertilized egg or zygote. In them, immediately after the activation of egg either through the intervention of a spermatozoon (fertilization) or through some parthenogenetic agent (parthenogenesis), the activated egg is passed through a phase of repeated mitotic cell divisions, which occur in rapid succession and produce an increasing number of cells. These cells build up the body of the metazoan embryo and eventually of the adult. The splitting or division of an activated egg by a series of mitotic cell divisions into a multitude of cells which become the building units of future organism, is called **segmentation, cellulation** or **cleavage**. In fact, a multicellular adult has different sets of cells performing different functions. Thus, the cleavage has two functions to perform. The first is to produce many cells. The second is to lay the foundation of differences, so that different cleavage cells follow different developmental pathways. In other words, cleavage prepares the ground for differentiation and morphogenesis.

The period of cleavage extends from fertilization to the formation of a characteristic developmental stage, the **blastula** and is generally characterized by rapid, successive divisions without intervening period of growth. Cleavage usually has a very well defined beginning. It starts with the mitotic division of the zygote into two cleavage cells or **blastomeres** (Gr., *blastos* = germ, *meros* = parts). The cessation of cleavage is not so well defined. Normally the beginning of gastrulation is regarded as the end of cleavage.

10-1. CHARACTERISTICS OF CLEAVAGE

The process of cleavage has the following peculiar features :

1. All divisions of cleavage are mitotic. They occur rapidly one after the other, but their rate depends upon the species concerned and on temperature. The intermitotic phase or interphase is of short duration and includes no growth of mitotic products, *i.e.*, cleavage cells or blastomeres. For example, during the cleavage of sea urchins, blastomeres may divide every 30 minutes. This is as fast as in bacteria. Indeed it is so fast that the S-period (DNA synthesis phase) for the next cleavage begins already in telophase of the preceding cleavage (**Trinkaus**, 1984).

2. Early cleavage divisions occur synchronously, *i.e.*, all blastomeres present divide simultaneously. During late cleavage, the cleavage divisions become asynchronous.

3. Cleavage divides the substances of the egg into an increasing number of cells of progressively decreasing size. The divisions continue until the average cell sizes are those characteristic of the differentiated cells of the parental organism. The number of successive divisions accordingly depends

on the volume of the egg compared with the volume of typical somatic cells and on the availability of intracellular or extracellular reserves, such as yolk.

4. Cleavage begins with a cell (*i.e.*, zygote) that has greatly bloated volume and a distorted nucleus-to-cytoplasm ratio in comparison with cells of the adult body. During cleavage, reduction of nucleo-cytoplasmic ratio in blastomeres is accomplished and this permits the cells to be more metabolically active because such nuclei have less cytoplasm to control. For example, according to **Brachet** (1950), in a mature sea urchin egg the nucleo-cytoplasmic ratio is :

$$\frac{\text{volume of the nucleus}}{\text{volume of the cytoplasm}} = \frac{1}{550}$$

The ratio of 1/550 is a drop from 1 : 6 during the previous oocyte stage that contains the large germinal vesicle. At the 4 cell stage the ratio is 1 : 18 ; it is 1 : 12 at the 64-cell stage and 1 : 6 in the blastula (see **Berrill** and **Karp**, 1976).

5. Gas exchange (oxygen into the blastomeres and carbon dioxide out of the cells) is more efficient because the surface/volume ratio is increased as the cells become smaller, permitting more ready diffusion of gases into and out of the cells.

6. During cleavage, there is no growth in the resulting blastomeres and the total size and volume of the embroy remains the same. This indicates that there is no quantitative change in the cytoplasm of blastomeres.

7. During cleavage, the blastomeres do not move, so that the general shape of the embryo does not change except for the formation of a cavity, called **blastocoele** in the interior.

8. Although there is no growth during the period of cleavage, chemical transformation (*i.e.*, qualitative change) does occur. Chemical conversion of reserve food materials (yolk, glycogen and ribonucleotides) into active cytoplasm takes place. An increase in the active cytoplasm is intimately related with the steady increase of respiration throughout cleavage.

9. The constituent parts of ooplasm remain comparatively unaltered during cleavage, *i.e.*, internal organization of ooplasm of a fertilized egg remains same in a cleaved egg.

10. The cleavage converts the egg into a compact mass of blastomeres called **morula** which ultimately transforms into **blastula**, having unilayered thick **blastoderm** around a central space, **blastocoele.**

In fact, cleavage does more than propagating multicellularity ; it also determines the fate of cells in the embryo. By segregating morphogenetic materials, it creates different cell lineages with unique developmental programmes. The timing of these events varies with each species, but it tends to occur when rapid division synchrony shifts to an asynchronous lengthening cell cycle (see **Grant**, 1978).

10-2. HISTORICAL

The process of cleavage could be noticed to be operative in animal eggs, only in the eighteenth century, when, in year 1738 first of all **Swammerdam** observed the first cleavage of the frog's egg. In 1780, **Spallanzani** described first two cleavage planes of toad's egg. **Prevost** and **Dumas** in 1824, made extensive observations on the cleavage of the eggs of frog. Their work is followed by other embryologists and cleavage studies of eggs of frogs, newts, coelenterates, nematodes, and starfishes were made. The first reported cleavage of the eggs of a rabbit was made in 1838–1839 ; a fish in 1842 and a bird in 1847. The modern cell biological techniques have made several aspects of cleavage more understandable.

10-3. PECULIARITY OF MITOSIS INVOLVED IN CLEAVAGE

Though, the mitosis involves in cleavage and the mitosis occurs in late embryos and in adult animals, is fundamentally same, but, both differ in following aspects :

1. In the later embryonic stages and in the adult, the mitosis is intimately connected with growth. After each mitotic cell division the daughter cells grow and when they are approximately doubled in size they divide again. The cells, thus, maintain an average size in every type of tissue. During cleavage this is not so. The consecutive divisions of the blastomeres are not separated by period of growth. A blastomere does not increase in size before the next division begins. Consequently, with each division the resulting blastomeres are only half the original size. Thus, cleavage begins with one very large cell and ends with a greater number of cells, each of which is no longer very much larger than the tissue cells of the adult animal.

2. The intermitotic interval or interphase remains very short in mitosis involved in cleavage.

3. During cleavage, the division of nucleus (karyokinesis) of fertilized egg or blastomere, usually is followed by division of cytoplasm (cytokinesis) like the mitosis of somatic cells, but, in certain animals (e.g., insects) each karyokinesis is not immediately followed by cytokinesis.

10-4. RATE OF CLEAVAGE

The rate of cleavage varies from species to species. For example, in the gold fish, divisions follow each other continuously at regular intervals of 20 minutes. The inter-cleavage interval of frog is one hour and of mouse is 10 to 12 hours. In most of these cases, the cleavage rate remains rapid (synchronous) during early developmental stages and becomes very slow (asynchronous) during completion of blastular stage.

The rate of cleavage is determined by the cytoplasm rather than by the nucleus, a fact long known but confirmed by the observations that haploid, diploid and triploid frog embryos, all undergo cleavage with the same rhythm. The rate of cleavage is affected by temperature. Slight increase or decrease of temperature of a cleaving egg may increase or decrease the time interval of cleavage. Importantly, temperature change affects the rate of respiratory or oxidative metabolism which appears to be base of all cellular activities. Respiration varies with the temperature and so does cleavage rate. In the absence of oxygen or the presence of excess of carbon dioxide, cleavage is inhibited.

10-5. SYNCHRONUS AND ASYNCHRONUS DIVISION

The first few cleavage divisions of most eggs, except those showing irregular cleavage, are **synchronus** ; every blastomere replicates its DNA and progress through each phase of the cell cycle at the same time. In a sense, cell cycle time or the rhythm of cell division acts as a clock measuring off the pace at which the developmental programme unfolds. Initially all cleaving blastomeres seem to possess the same **cell division clock** and are synchronized, but soon local cell populations establish their own mitotic rhythms. Morphogenetic events acquire regional properties, coordinated to the division cycles of distinct cell populations.

The length of cleavage synchrony varies in different species. Some maintain synchrony for long periods, possibly up to 15 cell cycles in the frog embryo ; in others, it is brief, only 2 to 3 cycles in spirally cleaving forms. In the sea urchin, division synchrony persists for three divisions but is lost at the fourth division as micromeres slow down and acquire their own mitotic rhythm, while macromeres and mesomeres divide synchronously for 6 to 8 more cycles. Likewise, pole cell nuclei in insect eggs are delayed at the fourth division, while all other cleavage nuclei replicate synchronously for 7 to 8 additional cycles.

Mitotic synchrony is not found in all eggs. In nematodes, rotifers and mammals, it does not extend the first division. Each daughter cell at the two-cell stage divides to its own rhythm, an asynchrony that becomes more pronounced in subsequent divisions. Early division asynchrony may be correlated with precocious determination of blastomere fate. The best example is in *Ascaris*, where the first division is differential, separating animal and vegetal hemispheres. Each then proceeds to divide independently and as we have seen in Chapter 4, the animal cell, the progenitor of the somatic cell line, undergoes chromosomal diminution, while the vegetal cell completes normal mitosis and gives rise to the germ line.

A similar correlation between division asynchrony and determination exists in other embryos, although the relationship is not as clear, since synchrony persists for many cell cycles and cell population is large when determination occurs. Nevertheless, this period is often marked by transition to asynchronus mitotic patterns with different regions dividing at different rates. A **mitotic gradient** is gradually established over the embryo, and mitotic waves pass periodically from one pole to the other. For example, in the sea urchin *Paracentrotus* the vegetal region acts as the pacemaker. Synchrony persists for six cycles, except for the micromeres which establish their own mitotic rhythm at the fourth cleavage. At the seventh cleavage, mitosis begins first in vegetal macromeres sitting directly over the micromeres and spreads toward the animal hemisphere. This mitotic gradient prevails through the next two cycles, but gradually disappears at the tenth division, when over 1000 cells are present. Gastrulation, the first sign of morphogenesis, usually begins at this time, and at the vegetal pole. In insects the transition occurs at the 9th division and the gradient progresses anterior to posterior. Transition to asynchrony occurs earlier in spirally cleaving eggs, at the third division in some cases.

In synchronus divisions, cell cycles are very short, 15 minutes to 1 hour. Most of this period is devoted to DNA replication in S-phase and in mitosis.

G_1 and G_2 phases of interphase are either absent or brief. Consequently, time for genome expression is limited. The cell cycle remains short and constant during the synchronus phase and lengthens after synchrony is lost and mitotic gradients are established. Though all phases lengthen, the G_1 phase exhibits the greatest change, and it is possible that some cells are committed to their fate at this time. (Let us recall here that G_1 phase or first gap is a growth stage during which takes place the synthesis of RNA such as rRNAs, mRNAs, etc., and proteins such as DNA polymerase enzyme, tubulin and cytoplasmic growth factors or the autocrine secretions ; see **Burns and Bottino**, 1989). Asynchrony is characterized by visible signs of nuclear activity. For instance, nucleoli appear in frog embryos at the asynchronous blastula stage of several thousand cells. Moreover, transcriptive activity of cell nuclei also becomes evident ; RNA is synthesized. By contrast, in mouse eggs which have longer cell cycles and no obvious synchronus phase, nucleoli appear early at the four-cell stage. Specific cells that display independent mitotic rhythms, such as insect pole cells and sea urchin micromeres seem to be determined precociously and assume an important morphogenetic role in the embryo.

Synchrony in divisions during cleavage may be contributed by the following two factors : (1) a large store of mitotic precursors in the egg, and (2) junctional communication between blastomeres that transmit pacemaker signals coordinating division cycles. We know that eggs contain large stores of ribo- and deoxyribonucleotides and the enzyme systems for their synthesis and interconversions. Some eggs, such as the brine shrimp *Artemia*, have rich supplies of strange nucleotides such as **guanine tetranucleotides**, which may serve as a nucleotide precursor store. These nucleotide reserves decrease during the synchronous phase and reach low levels at the time asynchrony begins (see **Grant**, 1978).

Cleaving cells also communicate with each other across gap junctions. Early blastomeres of the newt are in electric communication ; for instance , if a stimulating microelectrode is inserted in one blastomere, it generates a voltage that can be recorded in distant cells. Likewise cells of discoidally cleaving squid, chick and fish embryos are in intimate ionic exchange. In squid, junctional communication between embryos and yolk persists until larval stages, after which it disappears. These specialized juncations for cell communication may facilitate the flow of regulatory molecules that promote synchronus division behaviour. Perhaps synchrony is lost when the cell population becomes too large and diffusion rates cannot keep up with rates of signal initiation from pacemaker regions (see **Grant**, 1978).

10-5.1. The Mid-Blastula Transition

The first cleavage of the *Xenopus* egg takes place 90 minutes after fertilization (Fig. 10.1). The 11 subsequent cleavages take place synchronously every 35 minutes (**Newport** and **Kirschner**, 1982). This comapres with a doubling time of 24 hours for an adult frog somatic cell. For achieving such an extraordinarily short cell cycle, cleaving embryos increase the number of origins of replication in their DNA (which shorten the S-phase) and omit the G_1 and G_2 phases of the interphase of the cell cycle, so that the end of DNA synthesis is immediately followed by mitosis (**Graham** and **Morgan**, 1966). During this initial phase of development there is no transcription of DNA. After the 12th cleavage or 4096-cell stage, the cell cycles become longer and asynchronous, the cells become motile and RNA synthesis starts (Fig. 10.1). This turning point in development is called the **mid-blastula transition**.

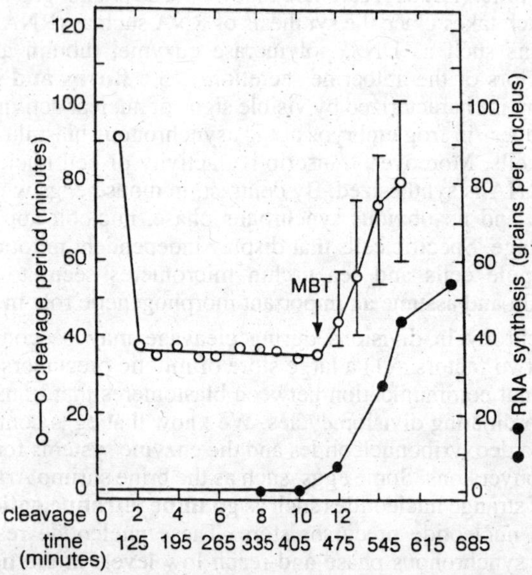

Fig. 10.1. Mid-blastula transition (MBT) during early *Xenopus* development (after De Robertis and De Robertis, Jr., 1987).

Since several types of RNA start to be expressed simultaneously at the mid-blastula transition (*i.e.*, mRNA, snRNA, tRNA and rRNA), so the question arises why they are not expressed earlier. The experimental evidence suggests that there is a critical ratio of nuclei to egg cytoplam before transcription can start. For example, polyspermic fertilization or the injection of large amounts of extra DNA will induce the mid-blastula transition earlier (**Newport** and **Kirschner**, 1982). Thus, it seems that *Xenopus* eggs may have a substance that binds to chromatin or DNA and turns it off transcriptionally. The amount of this substance would be sufficient to block up to 4096 nuclei, but then would become exhausted and expression of the genome would start. This is called **titration model** of mid-blastula transition (Fig. 10.2).

10-6. THE PLANES OF CLEAVAGES

During cleavages, different cleavage furrows may divide the egg from different planes. Few important cleavage planes are following :

1. Meridional plane. When cleavage furrow passes through the centre of animal-vegetal axis and bisects the both poles of the egg, then such plane of cleavage is called **meridional plane**. *e.g., Rana pipiens, Lepidosteus osseus* and first cleavage furrow of chick.

2. Vertical plane. The vertical plane of cleavage is a furrow which tends to pass in a direction from the animal pole toward the vegetal pole. But like meridional plane, it does not pass through the median axis of the egg, but courses to one side of this axis, *e.g.*, third cleavage furrows of *Amia calva, Lepidosteus osseus* and chick.

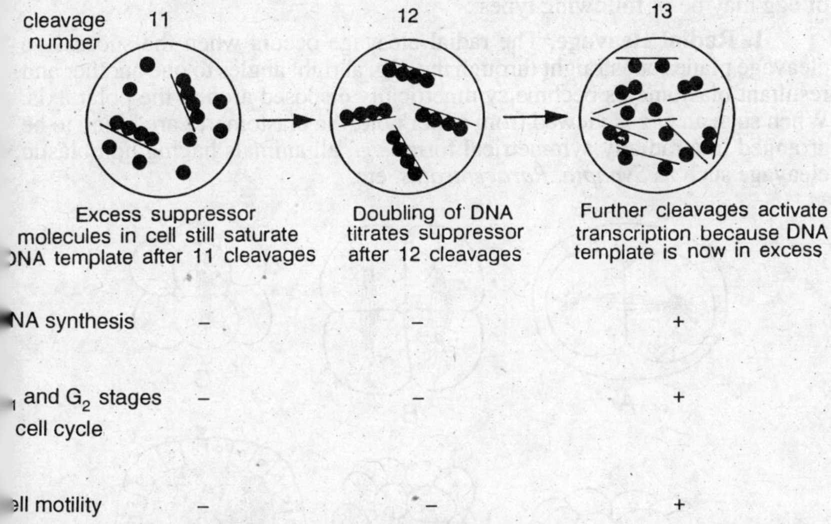

Fig. 10.2. A titration model suggested by **Newport** and **Kirschner** (1982) for the onset of transcription at the mid-blastula transition.

3. Equatorial plane. The equatorial plane of cleavage bisects the egg at right angles to the main axis and half way between the animal and vegetal poles, *e.g.*, fifth cleavage planes of the eggs of *Ambystoma maculatum*, and first cleavage plane of eggs of higher mammals.

4. Latitudinal plane. The latitudinal plane of cleavage is similar to the equatorial, but it courses through the cytoplasm on either side of the equatorial plane. It is also called **transvese** or **horizontal plane** *e.g.*, third cleavage planes of *Amphioxus* and frog.

Determination of plane of cleavage. The plane of cleavage is determined by the position of mitotic spindle of the dividing egg or blastomere, because, as a rule, the plane of cell division (cleavage) remains always at right angles to the long axis of the spindle. Therefore, the direction of the division plane in reference to the whole cell is determined by those forces responsible for the mitotic spindle. Ordinarily in a homogeneous (isolecithal) spheroidal egg, the mitotic spindle will be oriented transversely to the polar axis, and, thus, the first mitotic division will be vertical or meridional. The spindles for the second set of divisions will then line up parallel to the first division plane, so that the ensuing divisions are again vertical but at right angles to the first, with each quadrant as long as the entire egg, but no more than half its diameter, the spindles for the third set of divisions will now lie parallel to the polar axis, and the divisions themselves will be horizontal and, thus, at right angles to the previous two, *i.e.*, tranverse, horizontal or latitudinal.

10-7. PATTERNS OF CLEAVAGE

Most animal eggs incorporate an organization pattern and this pattern plays a far more important role in conditioning cleavage than does whatever amount of yolk may be present. The pattern of cleavage due to organization of egg may be of following types :

1. Radial cleavage. The radial cleavage occurs when the successive cleavage planes cut straight through the egg, at right angles to one another and resultant blastomeres become symmetrically disposed around the polar axis. When such an egg is viewed from either pole, the blastomeres are found to be arranged in a radially symmetrical form, *e.g.*, all animals having holoblastic cleavage such as *Synapta, Paracentrotus,* etc.

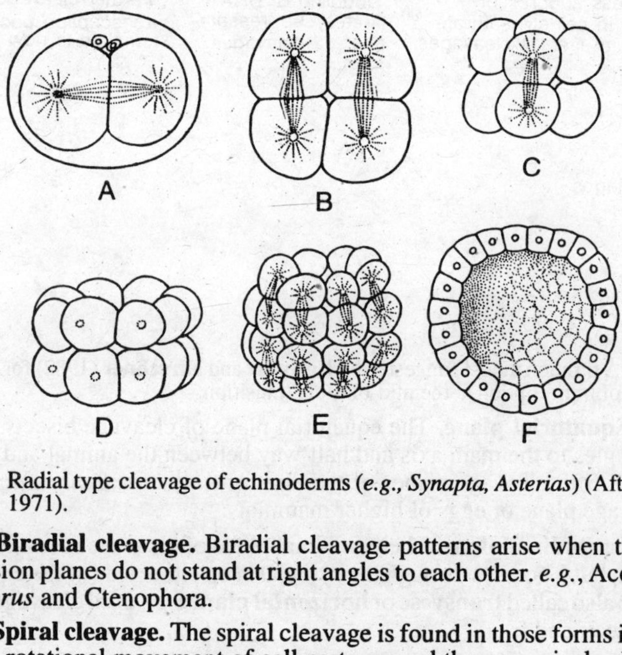

Fig. 10.3. Radial type cleavage of echinoderms (*e.g., Synapta, Asterias*) (After Berrill, 1971).

2. Biradial cleavage. Biradial cleavage patterns arise when the three first division planes do not stand at right angles to each other. *e.g.*, Acoela like *Polychoerus* and Ctenophora.

3. Spiral cleavage. The spiral cleavage is found in those forms in which there is a rotational movement of cell parts around the egg axis, leading to a displacement or inclination of the mitotic spindles with respect to the symmetrically disposed radii. The spiral cleavage because results due to oblique positions of mitotic spindles in the blastomeres, therefore, also called **oblique cleavage (Costello,** 1955). In successive cleavages, the rotational movements are characterized by a regular alternation in clockwise (right-handed) or anti-clockwise (left-handed) direction. In the first case the spiral cleavage is called **dexiotropic** or **dextral spiral cleavage** and in the second case, it is called **laeotropic** or **sinistral spiral cleavage**, *e.g.*, Turbellaria, Nematoda, Rotifera, Annelida, and all molluscs except cephalopods.

4. Bilateral cleavage. In bilateral cleavage the mitotic spindles and cleavage planes remain bilaterally arranged with reference to a plane of symmetry which coincides with the median plane of the embryo, *e.g.*, Tunicata, *Amphioxus*, Amphibia and higher mammals.

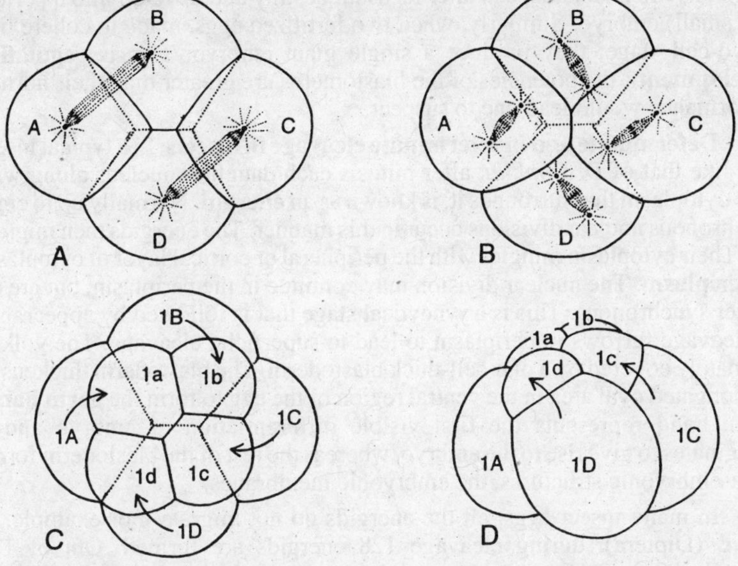

Fig. 10.4. Spiral cleavages of mollusc (*e.g., Trochus*) (After Balinsky, 1970).

10-8. DETERMINATE AND INDETERMINATE CLEAVAGE

According to the concept of **potency**, which refers to the total range of developmental possibilities, that an egg, blastomere or its parts is capable of realizing under any imposed condition either natural or experimental, following two types of cleavages have been recognized :

A. Determinate cleavage. In some animals (*e.g.*, tunicates, molluscs, annelids, etc.), cleavage follows a precise pattern and each blastomere has its characteristic position and unalterable fate. Such type of cleavage is called **determinate cleavage** and it transforms a fertilized egg into a blastula which is mosaic, because, its component blastomeres have received assignments according to an inflexible plan already completed at the time of fertilization. This parcelling-out process means that each blastomere at same period becomes the precursor of a definite part of the embryo. An isolated blastomere of the two-cell stage of a *Dentalium* egg, develops into a half-larva, while the destruction of even a part of a blastomere may result in a defective larva. Later blastomeres of this animal were found to have still more restricted possibilities. This is **mosaic development**, the potencies of the blastomeres just equal to the fates which they achieve, and blastomeres cannot be swerved (turned aside) away from those fates.

B. Indeterminate cleavage. In vertebrates, the plan of cleavage is less rigid. Although, their normal development demonstrates a general relation between blastomere position and fate, still the blastomeres possess more capabilities than they ordinarily show. For example, the destruction of a blastomere with a certain presumptive fate or its dislocation to a strange position is followed by readjustments and substitutions that produce a normal embryo. A first cleavage blastomere of a sea urchin or an amphibian or a mammal, when isolated, can alter its usual destiny and develop into a perfect (but small) embryo. Similarly, when two fertilized eggs, made to cohere like a two-cell stage, they produce a single giant embryo. This is **regulative development** ; the potencies of the blastomeres are greater than their normal performance would lead one to suspect.

Determinate and indeterminate cleavage in insects. In a typical insect egg, like that of *Drosophila*, after mitosis each daughter nucleus along with little cytoplasm that surrounds it, is known as an **energid**. Normally up to eight synchronous nuclear divisions occur in this manner. The energids then migrate out. Their cytoplasm mingles with the peripheral or cortical layer of cytoplasm or **periplasm**. The nuclear division may continue in the periplasm, but are no longer synchronous. This is a syncytical stage that is followed by appearance of cleavage furrows in periplasm to lead to superficial cleavage. The yolk is ultimately covered by a one-cell-thick blastoderm. The blastoderm thickens in an elongated oval area in the ventral region of the egg to form the **germ band**. Germ band represents the first visible differentiation of embryo and it invaginates to give rise to the embryo, whereas the rest of the blastoderm forms extra-embryonic structures, the embryonic membranes.

In many insect eggs, all the energids do not migrate. For example, in *Dacus* (Diptera), during cleavage 128 energids are formed. Out of 128 energids, 90 energids migrate to periplasm to form the blastoderm, and about 38 energids stay behind to form **primary vitellophages**. Later on, **secondary** and **tertiary vitellophages** arise from the blastoderm or other organ rudiments. The vitellophages help in the digestion of yolk and later may get incorporated in the midgut epithelium (see **Goel**, 1984).

Insects too exhibit the determinate and indeterminate or regulative cleavage. For example, in **holometabolous** insects (*i.e.*, insects having a drastic and complete metamorphosis such as higher Diptera, Hymenoptera and some Lepidoptera) the fate of different parts of the egg is already determined in the periplasm or egg cortex at the time of egg-laying before the egg nucleus has begun to divide. Since elimination of some of the migrating nuclei by ultraviolet irradiation at a sufficiently early stage of development does not affect morphogenesis, their place being taken by other nuclei arriving in the cortical region, the subsequent fate of the cleavage cells is determined by the regional nature of cortical plasm (which become chemically differentiated). The nuclei are clearly equivalent (see **Berrill** and **Karp**, 1976). In **hemimetabolous** insects (*i.e.*, insects with incomplete metamorphosis such as dragon flies), on the other hand, eggs are regulative. In them, the determinative chemodifferentiation of the egg cortex is not attained until some time after the cleavage nuclei have reached the cortical layer.

Further, there are some insects which show the phenomenon of **polyembryony**, in which one egg produces not one but many larvae (8 in *Platygaster*, a parasitic Hymenoptera and exceeding 1000 in *Litomastrix*, another parasitic Hymenoptera).

10-9. INFLUENCE OF YOLK ON CLEAVAGE

The cleavage occurs more readily in the active cytoplam than the yolk-laden cytoplam or **deutoplasm** of the egg. The yolk is a non-living component of the egg. It participates neither in cleavage nor in the formation of the embryo. It simply provides nourishment to the developing embryo. The yolk is present in all animal eggs in little (*e.g.*, microlecithal eggs) or larger (*e.g.*, mesolecithal and macrolecithal eggs) amounts and remains distributed evenly (*e.g.*, isolecithal eggs) or unevenly (*e.g.*, telolecithal eggs) in the ooplasm of different kinds of eggs. It has a profound deterring influence on cleavage and mechanics of moving the germ layers into their final positions. The amount of yolk influences the course of cleavage by following ways :

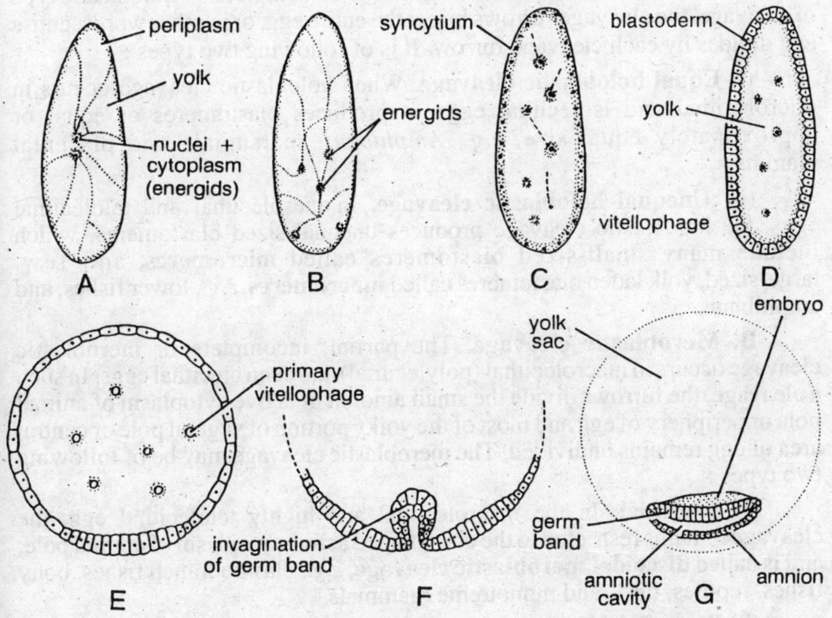

Fig. 10.5. Superficial cleavage in an insect (after Goel, 1984).

(1) When the amount of stored yolk increases, the amount of active cytoplasm gradually decreases, and the position of nucleus is variously affected. In mesolecithal eggs, such displacement of zygotic nucleus from the geometrical centre of the egg to the less yolky (active) cytoplasm is very common. The mitotic divisions of such a displaced nucleus result in unequal cytokinesis and unequal-sized blastomeres.

(2) Every mitosis of cleavage involves movements of the cell components, the chromosomes, parts of the cytoplasm constituting the achromatic figure, the mitochondria, and the surface layer of the cell, the

activity of which along the equator of the maternal cell leads to the eventual separation of the daugther cells. During these movements, the yolk granules or yolk platelets behave entirely passively and are passively distributed between the daughter blastomeres. When the yolk granules or platelets become very abundant, they tend to retard and even inhibit the process of cleavage. As a result, the blastomeres which are richer in yolk tend to divide at a slower rate and consequently remain larger than those which have less yolk. The yolk in the uncleaved egg is more concentrated toward the vegetal pole of the egg. It is, therefore, at the vegetal pole of the egg that the cleavage is most retarded by the presence of yolk, and where the blastomeres are of the larger size.

Thus, amount and distribution of yolk may determine the cleavage variously and following kinds of cleavage may be recognised :

A. Total or holoblastic cleavage. In total, complete, or holoblastic type of cleavage, the cleavage furrows bisect the entire egg, or in other words, entire egg divides by each cleavage furrow. It is of following two types :

(i) Equal holoblastic cleavage. When holoblastic cleavage occurs in microlecithal and isolecithal eggs, it produces blastomeres of equal or approximately equal size, *e.g., Amphioxus*, marsupials and placental mammals.

(ii) Unequal holoblastic cleavage. In mesolecithal and telolecithal eggs, the holoblastic cleavage produces unequal sized blastomeres, which include many small-sized blastomeres called **micromeres,** and few, large-sized, yolk laden blastomeres called **macromeres,** *e.g.*, lower fishes, and amphibians.

B. Meroblastic cleavage. The partial, incomplete or meroblastic cleavage occurs in macrolecithal (polylecithal) and centrolecithal eggs. In such a cleavage, the furrows divide the small amount of active cytoplasm of animal pole or periphery of egg and most of the yolky portion of vegetal pole or central area of egg remains undivided. The meroblastic cleavage may be of following two types :

(i) Discoidal. In the macrolecithal and highly telolecithal egg, the cleavage remains restricted to the disc-shaped active cytoplasm of animal pole, and is called **discoidal meroblastic cleavage,** *e.g.*, elasmobranch fishes, bony fishes, reptiles, birds and monotreme mammals.

(ii) Superficial. In centrolecithal eggs, the cleavage remains restricted to the peripheral cytoplasmic investment of egg, *e.g.*, arthropods, especially insects.

10-10. LAWS OF CLEAVAGE

By studying the cleavage in the eggs of different animal groups, it has been concluded that in all animal eggs, the cleavage seems to be governed by certain basic principles, rules or laws. Certain fundamental rules or laws of cleavage are following :

1. Sach's laws. In 1877, **Sach** proposed following two laws :

(a) Cells tend to divide into equal daughter cells.

(b) Each new division plane tends to intersect the preceding plane at right angles. (Acts to maintain the spheroidal shape of blastomeres).

2. Hertwig's laws. In 1881, **O. Hertwig** added following laws of cleavage in Sach's laws :

(i) The nucleus and achromatic figure (or mitotic spindle) occupy the 'centre of protoplasmic density' of the egg or blastomeres in which it lies. Hence, in the microlecithal and isolecithal eggs, the spindle is located centrally; in a telolecithal ovum it is nearer the animal pole.

Corollary. Blastomeres divide into two equal parts unless the yolk is unevenly stored in them.

(ii) The axis of a mitotic spindle occupies the longest axis of the protoplasmic mass in which it lies and division therefore tends to cut this axis transversely (evident in ovoid blastomeres).

Corollary. The ensuing plane of division cuts across the long axis, and the daughter cells revert to a more spheroidal shape.

3. Balfour's law. Balfour's law which was formulated by **Balfour** in 1885, states that the speed or rate of cleavage in any region of egg is inversely proportional to the amount of yolk or deutoplasm it contains. (In telolectithal eggs, blastomeres at the animal pole divide faster than those nearer the vegetal pole).

4. Pfluger's law. The mitotic spindle always elongates in the direction of least resistance.

10-11. MECHANISM OF CYTOKINESIS

The separation of the daughter nuclei (after each karyokinesis or nuclear division) and cytokinesis or **cell cleavage** may be two separate processes. For example, the eggs of most insects undergo division of the nucleus to form a multinucleate **plasmodium** without separation of the cytoplasmic territories. Cleavage of animal cells has been studied mainly in tissue culture and dividing eggs and is produced by furrowing of the cell.

There exists a controversial situation regarding the exact mechanism of cytokinesis and various contradictory views have been forwarded by different embryologists and cell biologists. Some important theories are the **spindle elongation theory (Dan,** 1947, 1958), **expanding surface theory (Mitchison,** 1952); **contractile ring theory (Marsland and Londau,** 1954) and **astral relaxation theory (Wolpert,** 1960). In recent years, one of these theories, the contractile ring theory has continued to gain favour of zoologists. According to this theory, in a cleaving egg, cytokinesis is performed by the formation of a **contractile ring** of actin and myosin in the following way :

The first visible changes consist of the appearance of a dense material around the microtubules at the equator of the spindle at either mid- or late anaphase. Then, while most spindle microtubules tend to become disorganized and disappear during telophase, some may persist and even increase in number at the equator, frequently being intermingled with a row of vesicles and dense material; the entire structure is called the **midbody.** Simultaneously, a depression appears on the cell surface—a **"constriction"** that deepens gradually until reaching the midbody. With the completion of the furrowing the separation of the cell is concluded.

An important point is that the position of the cleavage is apparently determined by the spindle poles. In fact, the cleavage furrow is positioned at the intersection of the sets of astral rays coming from the poles. This can be observed best in the initial divisions of sea urchin eggs in which microtubules from the two poles intercept at the surface where the contractile mechanism is activated. A signal that moves from the poles toward the surface, at a rate of 2 μm/min, has been postulated.

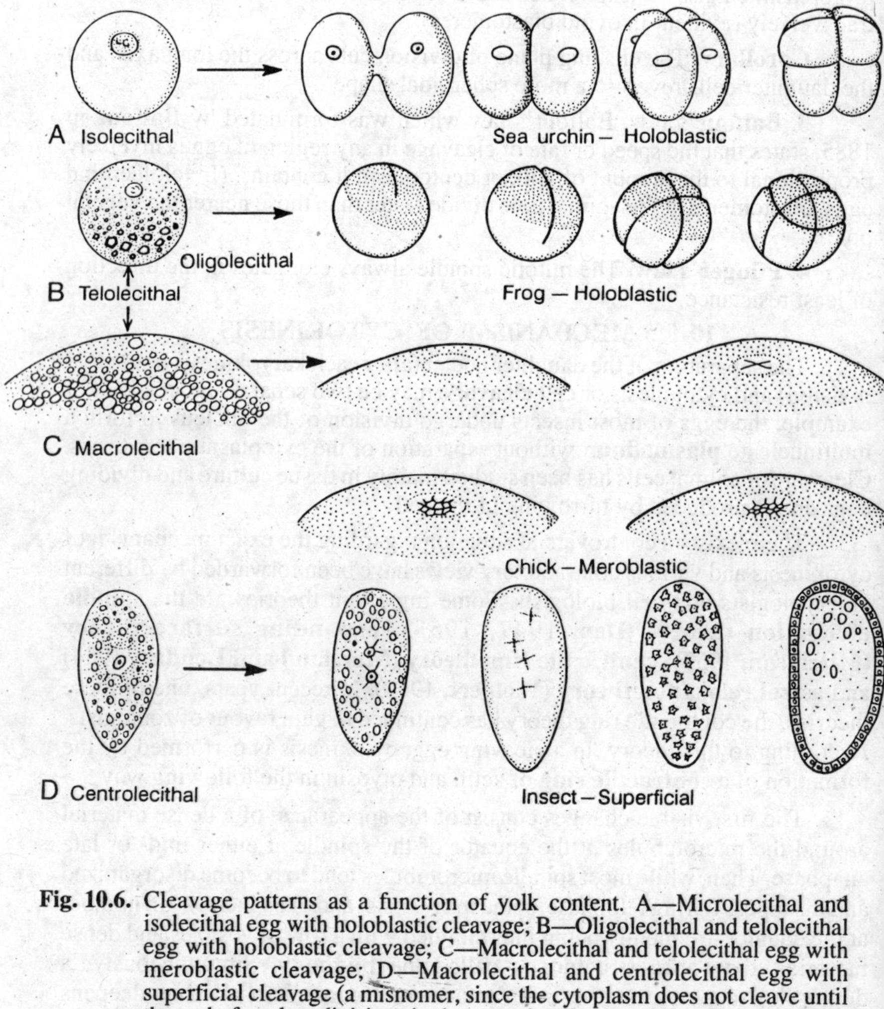

Fig. 10.6. Cleavage patterns as a function of yolk content. A—Microlecithal and isolecithal egg with holoblastic cleavage; B—Oligolecithal and telolecithal egg with holoblastic cleavage; C—Macrolecithal and telolecithal egg with meroblastic cleavage; D—Macrolecithal and centrolecithal egg with superficial cleavage (a misnomer, since the cytoplasm does not cleave until the end of nuclear divisions in the central yolk mass; (after Grant, 1978).

Cleavage by furrowing appears to be the result of a contractile ring present at the cell cortex and can continue in a cell lysed by detergents in the

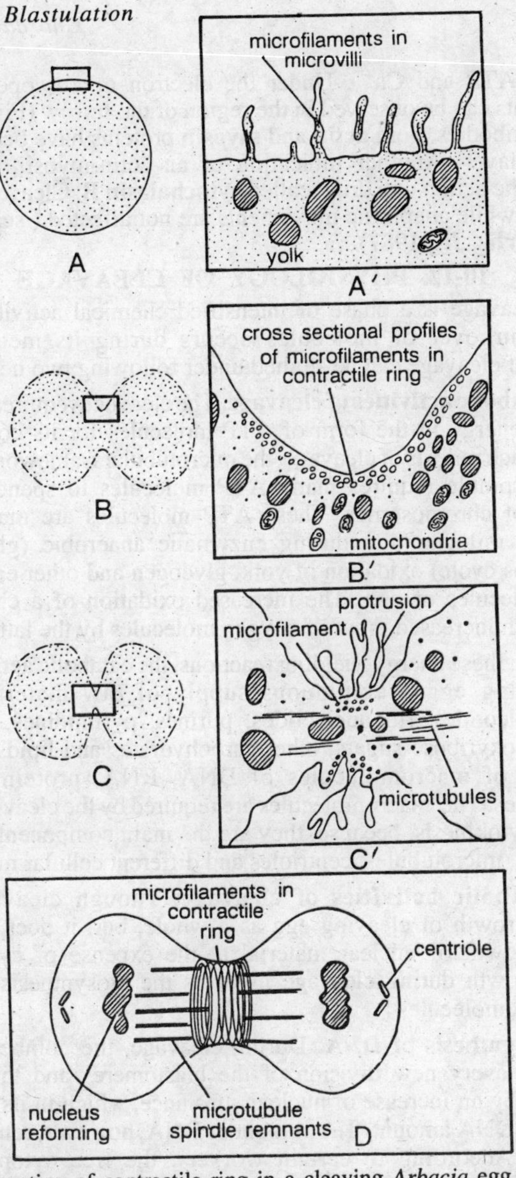

Fig. 10.7. Formation of contractile ring in a cleaving *Arbacia* egg. A—Meridional section of egg at late anaphase about 30 sec before furrowing begins; A′-Surface as seen in the electron microscope. No evidence of contractile ring is seen. Microfilaments are present only in microvilli; B—Meridional section 4 minutes after cleavage has begun; B′-Cleavage furrow enlarged. Note accumulation of cross-sectional profiles of microfilaments in contractile ring beneath plasma membrane in region of furrow; C—Meridional section of late cleavage. Daughter nuclei reformed; C′—Enlargement showing remnants of contractile ring in constricting cytoplasm; D—Three-dimensional view of dividing cell showing position of microfilaments in contractile ring and spindle remnants (after Grant, 1978).

presence of ATP and Ca^{2+}. Under the electron microscope, a system of microfilaments can be observed in the region of the furrow (Fig. 10.7) and by the use of antibodies, both **actin** and **myosin** proteins have been detected in the cortical layer. Another indication of an actin-myosin mechanism is provided by the action of a drug, called **cytochalasin B**. Cleavage is inhibited by this drug, while anaphasic movements are not affected (see **De Robertis** and **De Robertis, Jr.,** 1987).

10-12. PHYSIOLOGY OF CLEAVAGE

The cleavage is a phase of intensified chemical activity and a great amount of turnover of molecules occurs during its metabolism. The metabolism of cleavage can be studied under following two headings :

A. Catabolic activities of cleavage. The cleavage requires large amount of chemical energy in the form of ATP molecules for its completion. For example, in each mitosis of cleavage, the microtubular framework of centrioles and mitotic spindle requires many ATP molecules to spend in poleward movements of chromosomes. These ATP molecules are manufactured in ooplasm and mitochondira during enzymatic anaerobic (glycolysis) and aerobic (Krebs cycle) oxidation of yolk, glycogen and other energy yielding chemical molecules of egg. The increased oxidation of a cleaving egg is correlated with increased input of oxygen molecules by the latter.

Besides, these energy-yielding reactions, the catabolic activities provide to the dividing egg a continuous supply of raw materials (such as deoxyribonucleotides, ribonucleotides, purines, pyrimidines, amino acids, ribose and deoxyribose sugars, other carbohydrates and lipids, etc.) for the biosynthesis of macromolecules of DNA, RNA, proteins, lipids and polysaccharides. The macromolecules are required by the cleaving egg for the mitosis and cytokinesis, because, they are the main component molecules of chromosomes, microtubules, centrioles and different cellular membranes.

B. Anabolic activities of cleavage. Though cleavage involves quantitative growth of cleaving egg as a whole, but, it does include great qualitative growth of nuclear material at the expense of cytoplasm. The qualitative growth during cleavage involves the biosynthesis of following kinds of macromolecules :

(i) Biosynthesis of DNA. During cleavage, the number of nuclei is doubled with every new division of the blastomeres and this increase is accompanied by an increase of nuclear substance, which in its turn involves an increase of DNA amount. The amount of DNA, however, remains constant per nucleus. According to certain workers, the free cytoplasmic DNA molecules which have been reported to occur either freely in the ooplasm of Amphibia (**Brachet** and **Quertier,** 1963), and insects (**Muckenthaler** and **Mahowald,** 1966); or in the mitochondria (**Tyler,** 1967) or yolk platelets (**Brachet,** 1968), are also incorporated during nuclear DNA duplication. Some types of eggs (*e.g.,* sea urchin) have a specific enzyme called **ribonucleotide reductase** which converts ribonucleotides into deoxyribonucleotides (**Balinsky**, 1981).

(ii) RNA synthesis. During early cleavage, because the interphase is of very short duration, so, the chromosomal DNA remains hurriedly busy in its own duplication and exists in a highly coiled and, thus, in an inactive state. The result is that no or little transcription of any kind of RNA occurs during the period of early cleavage (Fig. 10.8.).

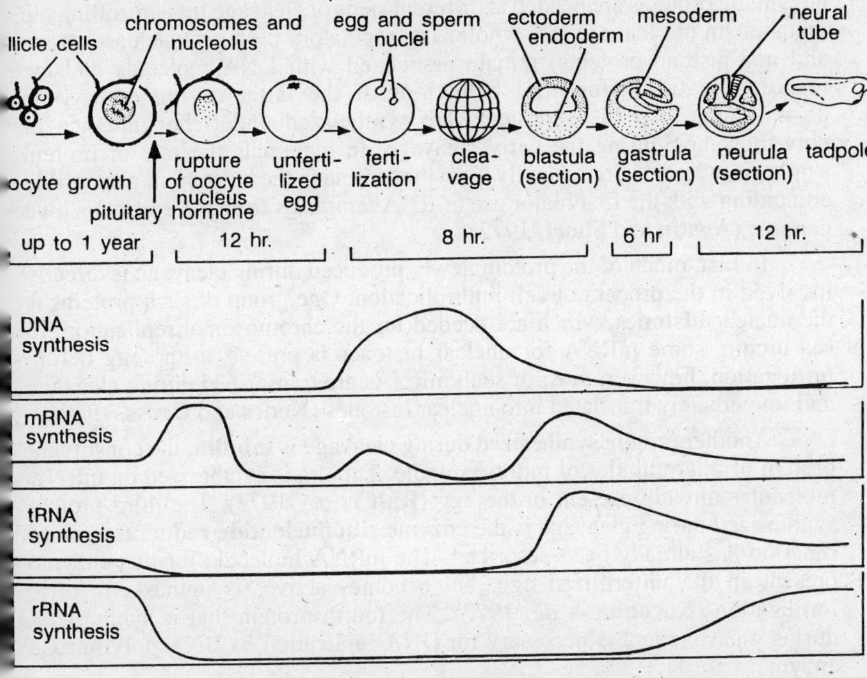

Fig. 10.8. Changes in the synthesis of nucleic acids in oogenesis, at fertilization and during early development (*i.e.*, during cleavage, blastulation, gastrulation, etc.) in the frog embryo (after Balinsky, 1981).

Messenger RNA (mRNA) and transfer RNA (tRNA), however, are synthesized during cleavage or at least in the later stages of cleavage (**Tyler** and **Tyler,** 1966; **Gurdon,** 1969). However, the synthesis of mRNA and tRNA molecules is not necessary for cleavage, because, when fertilized eggs are treated with actinomycin D which inhibit transcription, they continue to divide in a normal fashion. The newly synthesized mRNA molecules are not traslated in protein synthesis during cleavage and they are used during later developmental stages. Any protein synthesis, if, occurs during cleagave, it employs the mRNA molecules which are produced during oogenesis and these proteins, thus, have the characteristics of the maternal parent only.

Mammalian embryos differ strikingly from most other embryos (*e.g.*, those of amphibians, teleost fishes, echinoderms, insects) not only in requiring an external source for embryogenesis, but also in the precocious onset of RNA synthesis (*viz.*, transcription). In them transcription of different types of rRNAs and tRNAs is started from the 2-cell stage of the embryo. Synthesis of mRNA has not been conclusively demonstrated in early embryos (**Austin** and **Short,** 1972).

(iii) Protein synthesis. The cleaving egg requires different species of structural, enzymatic and regulatory proteins for its diverse mode of functions. For example, the **structural proteins** are required as building blocks for different cellular membranes, egg cortex, microtubules, ribosomes, etc. The enzymatic proteins are needed at different steps of cleavage for controlling the metabolism of cleavage as a whole. The regulatory proteins (such as histone and non-histone proteins) remain associated with DNA molecule and are responsible for differential behaviour of the latter during embryonic development. All these proteins are syntheiszed under the guidance of polyribosomes during the early cleavage. In mammals the rate of protein synthesis increases dramatically from the morula to the blastula stage possibly coinciding with the first major use of RNA templates from the embryo's own genome (**Austin** and **Short,** 1972).

In fact, much of the protein newly produced during cleavage is directly involved in the process of cell multiplication. One group of such proteins is the nuclear **histones,** which are needed for the chromosomal replication. In sea urchin, some mRNA for nuclear histones is present in the egg before fertilization; however, most of such mRNAs are transcribed during cleavage and immediately translated into nuclear histones (**Kedes** and **Gross**, 1969).

Another protein synthesized during cleavage is **tubulin,** the constituent protein of microtubules of mitotic spindle. Tubulin is synthesized on mRNA molecules already present in the egg (**Raff** *et al.,* 1972). The third protein synthesized during cleavage is the enzyme **ribonucleotide reductase** whose function has already been described. The mRNA molecule for this enzyme present in the unfertilized egg, but becomes active (is unmasked), after fertilization (**Noronha** *et al.,* 1972). The fourth protein that is synthesized during cleavage and is necessary for DNA replication, is **DNA polymerase** enzyme.

10-13. PRODUCTS OF CLEAVAGE

The cleavage is, thus, a fractionating process, which progressively splits fertilised egg into smaller cells, the **blastomeres.** By each cleavage furrow, the blastomere increases in the typical doubling sequence ad 2, 4, 8, 16, 32, etc. But, this type of geometrical regularity is existed only in early developmental stages, later on the process of cleavage becomes entirely irregular.

Further, during early cleavage stages, the blastomeres tend to assume a spherical shape and their mutual pressure flattens the surfaces of the blastomeres in contact with each other, but free surfaces of each blastomere remain spherical. Probably because of surface tension, they cluster into a spherical group with a minimal amount of surface area exposed. Cleavage process, thus, paves the way for a multicellular embryonic stage with loosely arranged blastomeres, called **morula**, the Latin word for mulberry, which they resemble. Later on, morula develops into next embryonic stage called **blastula.**

10-13.1. Morulation and Morula

Isolecithal (or hololecithal) eggs usually cleave to form a population of equal-sized cells in a tight cluster, the morula. The term morula is normally applied to most early cleavage stages, refers to a cluster of blastomeres tightly packed within the fertilization membrane; the cells have not organized into a tissue sheet as yet (see **Grant,** 1978).

The general shape of a cleaving cell population or morula depends on many factors. Primarily, yolk determines how newly formed blastomeres get arranged into a cell sheet. Secondarily, mechanical forces exerted by the surrounding fertilization membrane and hyaline layers (*e.g.*, echinoderms and amphibians), as well as, contact relationships established between the blastomeres such as tight junction and cytoplasmic bridges and intracellular cementing substances (*e.g.*, mucopolysaccharides) provide cohesion in blastomeres and contribute to the shape of embryo, *i.e.*, morula.

No morula is found in the centrolecithal eggs of insects. A true morula is regarded to be a solid ball of cells as found in coelenterates. In most morula, cells are arranged loosely and consist of narrow crevices between their round ends. A morula in marcolecithal and telolecithal eggs is composed of a small flattened disc of cells grouped at the animal pole.

10-13.2. Blastulation and Blastula

As cleavage proceeds, the number of morular blastomeres increases and the blastomeres undergo a profound rearrangement. The adhesion of the blastomeres to each other increases, and they arrange themselves into a true epithelium, which may be single cell thick, as in coelenterates, echinoderms, *Amphioxus,* etc., or many cell thick, as in most vertebrates. The resultant epithelium is called **blastoderm.** Due to rearrangements of blastomeres to form the blastoderm, a fluid-filled space or cavity called **blastocoele** appears in the centre of the blastoderm. This hollow, spherical and uniepithelial thick embryonic stage is called **blastula** and the process of its formation is called **blastulation.**

In a blastula, specialized cell adhesions appear in the most peripheral layer of cells and a true cell sheet is formed. As cells cleave, they are integrated into a physiological continuum by tight junctions, desmosomes and cytoplasmic bridges. Tight junctions appear at apical ends of epithelial cells and organize a coherent sheet. The epithelial layer surrounds the developing embryo and controls the exchange of materials between the environment and the blastocoele.

Desmosomes gradually appear in embryonic tissues as specializations of the cell surface. For example, in the early chick embryo, we find no desmosomes, but we do see tight junctions. The tissues are easily dissociated at these stages, but in older embryos, when tissues are more highly organized and distinct, desmosomes are found at numerous points in plasma membranes, and the tissues are much more difficult to disaggregate. When chick embryo tissues are disaggegated by trypsin, desmosomes seem to disappear, but as reaggregation begins, fully formed desmosomes reappear between contacting membranes and more common in fully organized aggregates. The high correlation between tightness of adhesive contact and high incidence of desmosomes suggests that these, above all other junctions, play key role in cell adhesion.

Desmosomes and tight junctions are uncommon in invertebrate tissues. In their palce are **septate desmosomes** with similar adhesive functions.

A. The process of blastulation. How does the first embryonic cell sheet, the **blastoderm,** develop ? This question has been studied in the hollow-sphered blastula of the echinoderms, where the hyaline layer at the cell periphery plays an important part in blastula formation. The hyaline layer is the extracellular matrix of the cleaving embryo. The following two theories have been suggested to explain the mechanism of blastulation :

1. Hydrostatic pressure theory. This theory was suggested by **K. Dan** in 1960 and according to it, the hydrostatic pressure within the blastocoele forces the cleaving cells to periphery. According to this view, fine cytoplasmic processes penetrate the enveloping hyaline layer, creating a cohesive expanding surface. As cell number increases and cells become smaller, they secrete fluids into extracellular spaces and these fluids accumulate in a central cavity. Hydrostatic pressure builds up in the cavity, impinges on outermost cells adhering to the hyaline layer, and generate a hollow sphere.

2. Surface tension theory. This theory was forwarded by **Gustafson** and **Wolpert** (1967). According to this theory the formation and expansion of blastocoele result from the manner in which the blastomeres are packed in a blastula. Just before each cell divides, the surface tension of the outer cell layer increases and the cells round up, divide and produce two rounded cells. Since cleavages are synchronus, this produces rhythmic tensions, and divisions are translated into pulsations of contraction and expansion. After a division, tension is reduced and cell contact is increased.

Further, when one spherical cell divides into two spherical cells, the combined diameter of the two daughter cells exceeds that of the original cell. As long as the dividing cells remain in one layer and their outer surface is attached to the hyaline layer, the layer of cells must inevitably push itself outward as a result of the increasing diameter of the cells. It is the radial pattern of cleavage (*e.g.*, sea urchin) that keeps the cells side by side rather than forming layers of cells. In case of the blastula of sea urchin, too, the volume is temporarily diminished but is expanded as new cells are added in the peripheral layer at the next cleavage.

The rounding up of cells prior to their division indicates that changes occur in the surface tension of the membrane and in the adhesiveness of one cell to its neighbours. The importance of these factors has been illustrated in the Figure 10.9, where transitions between A, B and C are possible simply by changes in surface tension and cell-to-cell adhesiveness. Case D in Figure 10.9 would require additional factors. Thus, in view of **Gustafson** and **Wolpert**, the blastulation of echinoderms can be explained as changes in the mechanical properties of the cells and geometrical packing of an increasing number of cells produce by radial cleavages that remain attached to an elastic hyaline layer. Because of the radial cleavage pattern and the attachment of cells to the hyaline layer, the elongating and constricting cells exert a tangential force upon each other leading to the increase in size of blastula and its blastocoele (see **Grant**, 1978).

B. Other blastular changes. The blastocoele, with its changing fluid environment, becomes part of new gradient system developing within the embryo. A spherical cell separates an external fluid environment from a different internal one in the blastocoele. The spatial property of ''inside-outsideness'' is now a reality as epithelial cells become polarized. The outer surfaces of cells of the peripheral layer are thrown into fine protoplasmic filaments embedded in a hyaline layer. Smooth inner surfaces are exposed to a blastocoele cavity of different ionic composition from the surrounding fluid environment. Cells are polarized, basal and apical ends exposed to different ionic environments which may in turn, lead to differential permeability with inner surfaces generally more permeable than outer surfaces.

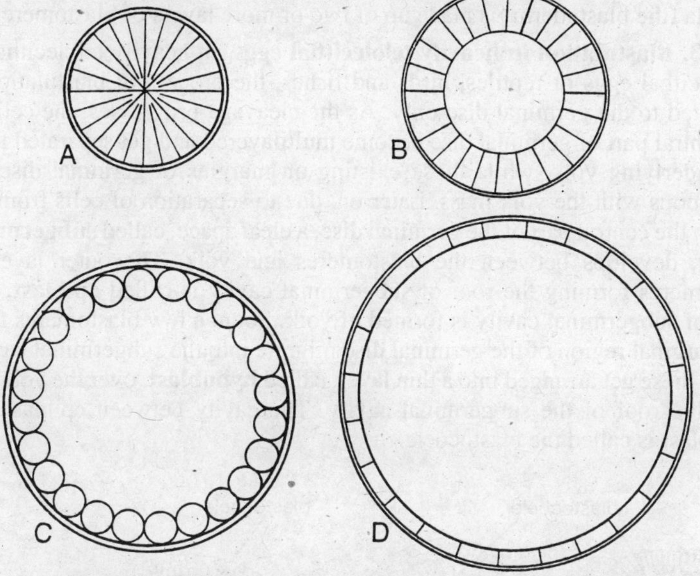

Fig. 10.9. Possible configurations of the cross section of a tube comprising 24 cells in a single layer attached to a supporting hyaline membrane, with no change in cell volume. A—Considerable contact between cells and no cavity; B—Moderate contact, and a cavity is present; C—Point contacts between the rounded-up cells; D—Considerable contact with the supporting membrane (after Berrill and Karp, 1976).

One of the first structural signs of this newly acquired polarity is the emergence of cilia. Most blastulae of marine embryos develop cilia and begin to swim about within the fertilization membrane. Cilia emerge on the outer surfaces of epithelial cells because basal bodies, from which cilia develop, localize in the cortex beneath the plasma membrane (see **Grant**, 1978).

C. Effect of yolk on blastulation. The amount of yolk present in the egg affects not only the cleavage but also the way a blastula is formed :

1. Blastulation in isolecithal eggs. The blastomeres produced by cleavage of an isolecithal egg tend to arrange in a single layer called **blastoderm,** on the periphery of the egg. This results in the appearance of a central cavity, called **blastocoele.** A true isolecithal egg is absent in chordates and, hence, the cells present toward the vegetal pole are comparatively large, *e.g. Amphioxus* and echinoderms.

2. Blastulation in moderately telolecithal eggs. In moderately telolecithal eggs such as frog and other amphibians, the yolk has an unequal distribution. This results in the displacement of the horizontal cleavage towards the animal pole and retardation of cleavage rate in the yolk-laden vegetal half. As a consequence the blastomeres of vegetal half (**macromeres**) are much larger than those of the animal half (**micromeres**). The small micromeres tend to form the roof of the blastocoele, while the large macromeres form the floor of blastocoele. In such a blastula, the roof of blastula (the blastoderm) is made up of two or more layers of blastomeres.

3. Blastulation in heavily telolecithal eggs. In heavily telolecithal or polylecithal eggs of reptiles, birds and fishes, the process of blastulation is restricted to the germinal disc only. As the cleavage progresses, the cells in the central part of germinal disc become multilayered and get separated from the underlying yolk, while those existing on margins of germinal disc are continuous with the yolk mass. Later on, due to separation of cells from the yolk in the central part of the germinal disc, a clear space, called **subgerminal cavity,** develops between the blastomeres and yolk. The outer layer of blastomeres forming the roof of subgerminal cavity is called **epiblast.** The floor of subgerminal cavity is formed of yolk. Soon, a few blastomeres from the marginal region of the germinal disc migrate into the subgerminal cavity, where these get arranged into a thin layer, called **hypoblast**, over the yolk and form the roof of the subgerminal cavity. The cavity between epiblast and hypoblast is called the blastocoele.

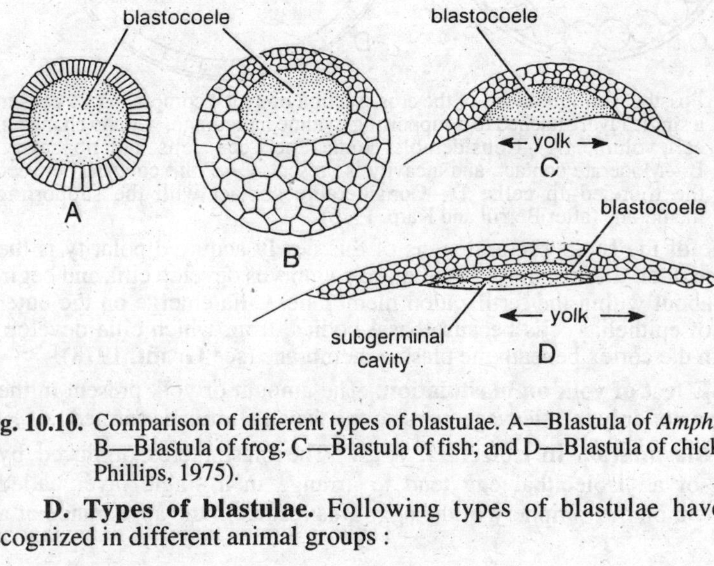

Fig. 10.10. Comparison of different types of blastulae. A—Blastula of *Amphioxus ;* B—Blastula of frog; C—Blastula of fish; and D—Blastula of chick (after Phillips, 1975).

D. Types of blastulae. Following types of blastulae have been recognized in different animal groups :

1. Coeloblastula. The blastula of echinoderms and *Amphioxus* is called **coeloblastula**. It is in the form of a hollow sphere. The blastocoele is filled with mucopolysaccharides and the blastoderm is formed of a single layer of cells.

2. Stereoblastula. In many spirally cleaving eggs of annelids, molluscs, nemertans and some of the planarians, no blastocoelic cavity appears in the blastula. In them the smaller micromeres accumulate as a cluster of cells over the larger vegetally placed macromeres. At the end of cleavage, this results in a solid blastula known as a **stereoblastula.**

3. Periblastula or superficial blastula. The superficially cleaving eggs of insects produce a periblastula without a blastocoele cavity as nuclei collect in the peripheral layer. In a sense, a cavity is present but filled with yolk from the beginning of cleavage.

4. Discoblastula. It is found in reptiles, birds and fishes which have large yolky eggs. It is called discoblastula because it appears at the animal pole as a small multilayered flat disc separated from the yolk by a narrow segmentation or subgerminal cavity.

5. Amphiblastula. Such a blastula is formed of two types of structurally different blastomeres. As for example, in *Sycon* the anterior half of blastula is formed of flagellated cells, while the posterior half is formed of large rounded granular cells. In amphibians, blastula contains two types of cells **micromeres** and **macromeres.**

6. Blastocyst. Such a blastula is found in mammals. Blastula formation in the mammalian embryogenesis is quite distinctive. Cleavage is regular and a small cavity appears inside the dividing cells. The cavity is called blastocoele and becomes progressively larger. The cluster of cells differentiates into two distinct groups, an epithelium-like layer of **trophoblast cells** or **nutritive** cells, surrounding the expanding cavity and an inner cell mass of **formative cells** of embryo which are displaced to one pole of the sphere. Though mammalian eggs have no yolk, they form a discoidal blastula just as in a chick embryo. The inner cell mass spreads inside the cavity as a flat disc, resembling the chick blastoderm. This stage of mammalian development is known as the **blastocyst,**

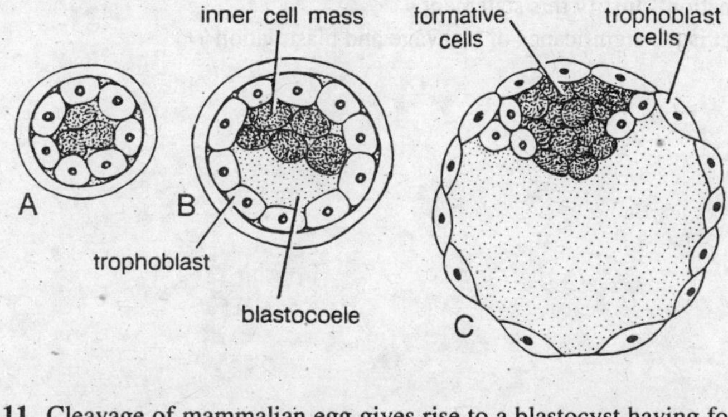

Fig. 10.11. Cleavage of mammalian egg gives rise to a blastocyst having formative cells and trophoblast or nutritive cells (after Phillips, 1975).

and it is at this stage that the embryo implants into the uterine wall. Trophoblast layer surrounding the blastocyst cavity, the blastocoele, is not part of the embryo proper but enventually becomes part of the placenta or extraembryonic membranes as the embryo implants.

10-14. SIGNIFICANCE OF CLEAVAGE, MORULATION AND BLASTULATION

The significance of cleavage, morulation and blastulation comprises three aspects : subdivision of the embryogenetic substrate into an array of cells, preparation for and occasionally even the initiation of the process of cell differentiation, or diversification, and the creation of a cell aggregate which, through the activity of morphogenetic processes, may be subjected to further morphological changes (gastrulation).

Further, due to cleavage and blastulation, the major presumptive organ forming areas of the future embryonic body are segregated into definite parts of the blastoderm. The blastocoele permits the migration and rearrangement of the major presumptive organ forming areas during gastrulation.

REVISION QUESTIONS

1. Describe the various cleavage patterns of eggs in the types of chordates studied by you.
2. Describe various types of blastulae and explain the mechanism of blastulation in echinoderms.
3. Describe the mechanism involved in the cytokinesis of cleavage.
4. Discuss various metabolic events involved in cleavage and blastulation.
5. What is the concept of masked mRNA ? How can it be related to cytoplasmic inheritance and phenomenon of determination ?
6. What are energids ? What are their fate ?
7. What is the role of yolk in determining holoblastic and meroblastic cleavage ?
8. In what ways the mitosis of cleavage and of adult somatic cell differ from each other ?
9. Why does mitotic synchrony of early cleavage change into asynchrony during late cleavage ? Explain.
10. Describe the phenomenon of mid-blastula transition.
11. 'The amount and disposition of yolk affects the cleavage patterns and blastula formation'. Justify this statement.
12. What is the significance of cleavage and blastulation ?

11

NUCLEUS AND CYTOPLASM IN DEVELOPMENT (Nucleocytoplasmic Interactions during Ontogenetic Development)

11-1. INFORMATION FLOW BETWEEN NUCLEAR AND CYTOPLASMIC COMPARTMENTS

Ever since it was recognised that most eukaryotic cells are divided into nuclear and cytoplasmic compartments, biologists have been fascinated by the interplay between the two, and there is still a tendency to exaggerate the role of one or the other compartment in ontogenetic development, for example, in the development and maintenance of differentiation. It is quite true to say that both cytoplasm and nucleus are entirely interdependent, one cannot survive without the other. The cytoplasm provides most of the energy for cell through oxidative phosphorylation (in mitochondria) and anaerobic glycolysis and the cytoplasmic ribosomes contain most of the mechanism for protein synthesis. Further, no molecular signal can reach the nucleus from outside the cell without relying on the conducting capacity of the cytoplasm and all proteins are made in the cytoplasm, even those that regulate gene activity or form the ribosomes. In turn, nucleus contains genes and provides mRNA molecules and other essential RNA molecules such as rRNAs and tRNA. Thus, there exists a balanced dynamic relationship between the two during cleavage, gastrulation, histogenesis and differentiation.

The classical experiments of **Goldstein** (1966) have shown that the exchange of molecules takes place between the nucleus and the cytoplasm. On entering the nucleus from the cytoplasm these molecules cause the selective activation of the nuclear genes. **Goldstein** cultured *Amoeba proteus* in a medium containing radioactive protein. He also fed *Amoeba* with labelled radioactive protein. Very soon the radioactive proteins entered the cytoplasm and were incorporated into the nucleus. He removed this radioactive nucleus and transplanted it to another *Amoeba* which had a normal nucleus. This *Amoeba* now had two nuclei, *viz.*, the transplanted radioactive nucleus and the normal host nucleus. In the meantime, the radioactive proteins from the transplanted nucleus slowly found their way to the host nucleus through the cytoplasm. Ultimately, both the nuclei reached to equilibrium and shared the equal amounts of radioactive proteins. This was verified by autoradiography or by a Geiger (– Muller) counter. This experiment clearly established the existent two-way traffic between the nucleus and the cytoplasm.

Thus, at a level of interaction between nuclear and cytoplasmic compartments, the cytoplasmic state has been shown to affect nuclear behaviour in cell functions. In a sense, at this level it is the cell's structural information that is involved since different cytoplasmic or nuclear "physiological states" are implicated. As cells adjust to an environment, they develop stable metabolic states that are transmissible to their progeny. Such metabolic states reflect networks of interactions between nuclear and cytoplasmic compartments. According to **Grant** (1978), the "informational" state of the cytoplasm or nucleus at any instant is an expression of spatial and temporal ordering of metabolic systems, that is, the cell's structural information.

In most cases these metabolic states are defined operationally, usually by one of the following methods : 1. One may remove a nucleus or some cytoplasm to see what effect its absence has on cell survival, motility, growth, or morphogenesis. Enucleation experiments reveal something about the role of the nucleus in cell behaviour and its effect on the cytoplasm. 2. One may transfer nuclei from one cytoplasmic state to another to see how nuclear behaviour changes in response to new sets of cytoplasmic signals. 3. One may fuse cells in different states to form a hybrid in which two different genomes interact in a common but mixed cytoplasm. The results of these experiments show whether nucleus and cytoplasm are mutually dependent and whether certain cytoplasmic states regulate nuclear behaviour.

11-2. NUCLEAR CONTROL OF CYTOPLASM

11-2.1. Enucleation Experiments (or Significance of Nucleus in Cell Survival)

Cell survival and reproduction are dependent on the nucleus. Removing a nucleus from a cell disrupts cytoplasmic activities and ultimately leads to cell death. But the nucleus exercises more indirect control over cellular activities as well.

Most investigations of nucleocytoplasmic interactions have been carried out on large single-celled organisms such as *Amoeba*, the large ciliate *Stentor*, and the unicellular alga *Acetabularia*. These cells can be cut into nucleated and enucleated fragments and because of their large size, nuclei are easily transplanted between the cells. Since most genetic directions come from the nucleus, so enucleation of a cell apparently eliminates a primary source of informational molecules and gene products. However, the removal of the nucleus does not always result in a sudden interruption of cell function. Many cell activities persist over long periods of time in the complete absence of a nucleus. In fact, photosynthesis, protein synthesis and even cell division continue in enucleated cells, suggesting that *nuclear control over cell function is not immediate but delayed.* This statement is well supported by following experimental evidences :

1. Effect of enucleation on nerve cells. The fact that the nucleus is necessary for cell survival was first found by **Waller** in 1852, who cut nerves in half and found that the nerve cells containing the nucleus survived, while the non-nucleated axons degenerated.

2. Effect of enucleation on *Amoeba*. One of the immediate effects of removal of a nucleus from an amoeba is cessation of motility. Pseudopodia do not appear and the cell rounds up into an inactive sphere. Since it does not form pseudopodia, it cannot feed and, thus, survives for less than two weeks. By contrast, an amoeba with its nucleus intact survives for three weeks even though it is starved. Enucleated halves are unable to metabolize glycogen and fat reserves and instead of them utilize a declining protein reserve. However, the most immediate change after enucleation is a dramatic loss in total cell RNA. Though protein synthesis continues, its rate declines as RNA is exhausted and is not replenished by the nucleus.

In fact, transfer of a nucleus from a normal amoeba into a quiescent or non-motile, enucleated amoeba cell restores normal function. The recipient cell begins movement and resumes feeding. Even the nuclear signal for reactivation is not species-specific since a nucleus from an unrelated species is equally capable of restoring function.

3. Effect of enucleation in *Acetabularia*. Certain cells such as *Acetabularia* survive for much longer periods without a nucleus. *Acetabularia* is a large unicellular green alga which is 5 cm long with a prominent nucleus in its **rhizoid**. Nucleus contains a giant RNA-rich nucleolus but very little chromatin (see **Brachet**, 1974). **Stalk** has many chloroplasts and bears at its apical end an umbrella or **cap** which will later serve for its sexual reproduction (*i.e.*, formation of cysts which produce gametes). The morphology of the cap is species-specific and allows the taxonomists to classify the different species of *Acetabularia*. The most important of these species are *A. mediterranea* and *A. crenulata* which can easily be cultivated, in large amounts, in the laboratory.

After enucleation, the enucleated stalk, if kept in the light, continues to photosynthesize and survives for periods up to seven months (**Hammerling**, 1930). It synthesizes cell wall, grows several millimetres in length, and in many cases will even regenerate a new cap. Survival time depends upon photosynthetic capacity; enucleated fragments kept in the dark survive for shorter periods.

Thus, survival of an enucleated cell is in part a function of feeding behaviour. An enucleated amoeba, which must move to feed, exhibits a shorter life than an enucleated *Acetabularia* stalk which is still able to synthesize its own food. Chloroplast turnover in *Acetabularia* is quite low; these organelles function without nuclear information for long periods of time, relying entirely on instructions and products from their own genomes.

4. Effect of enucleation in sea urchin eggs. The unfertilized eggs of sea urchin *Arbacia* can be cut into two halves by centrifugation in a density gradient. The heavy red pigment accumulates to the centripetal end of the egg, which successively elongates, takes a dumb-bell-shape and separates into two halves (Fig. 11.1). The light, transparent half contains the nucleus, fat droplets, mitochondria and a clear, ribosome rich layer, the hyaloplasm. The red heavy half is enucleated; it contains the red pigment granules and most of the yolk. However, electron microscopy shows that the light halves also possess some of the yolk platelets and likewise, heavy halves also contain mitochondria and lipid droplets.

Both kinds of fragments of the eggs can be fertilized or parthenogenetically activated by the classical method of **Loeb** (hypertonic sea water and butyric acid treatments). Nucleate halves develop very well after fertilization and give rise to normal plutei. However, the development of fertilized enucleate halves is less normal and plutei are not easy to obtain; this rather poor development is probably not so much due to haploidy as to the excess of yolk which makes cleavage difficult. Parthenogenetic development of the nucleate halves is almost as good as that of normal eggs. But the development of the enucleate halves, after parthenogenetic treatments, is completely abortive : there occur irregular fragmentation of the egg rather than a true cleavage (Fig. 11.1). Thus, asters can repeatedly form in the cytoplasm of the activated enucleate halves and irregular furrows can cleave the egg to form a kind of morula. But the typical cleavage patterns of sea urchin egg, with micromeres, mesomeres and macromeres is always missing. The blastula-like larvae neither hatch, nor form cilia. As one can see, there is no true morphogenesis. Only the formation of asters and cleavage furrows displays some autonomy in activated enucleate sea urchin eggs.

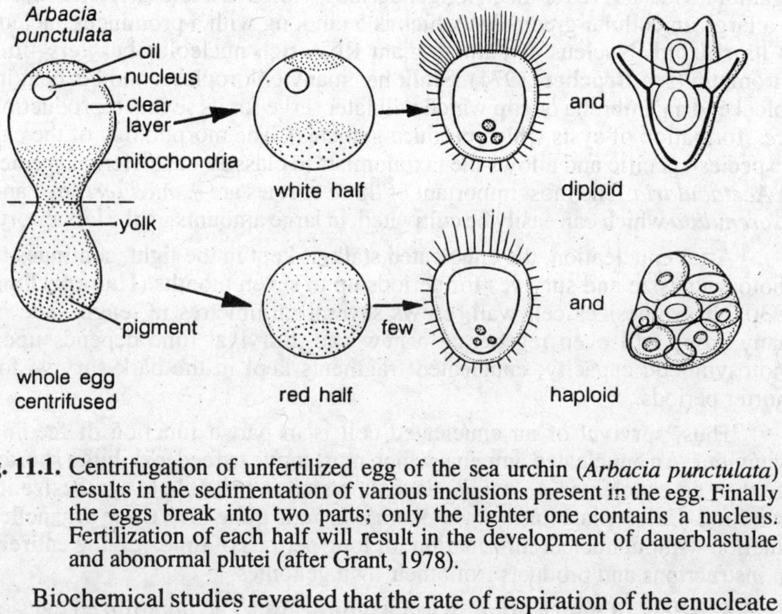

Fig. 11.1. Centrifugation of unfertilized egg of the sea urchin (*Arbacia punctulata*) results in the sedimentation of various inclusions present in the egg. Finally the eggs break into two parts; only the lighter one contains a nucleus. Fertilization of each half will result in the development of dauerblastulae and abonormal plutei (after Grant, 1978).

Biochemical studies revealed that the rate of respiration of the enucleate halves becomes higher than that of the nucleate ones. The higher oxygen consumption of the enucleate halves is probably related to the fact that the enucleate heavy halves contain a larger proportion of the reducing mitochondrial respiratory enzymes (*i.e.*, dehydrogenases) than their nucleate counterparts (see **Brachet**, 1974).

Further, though in cleaving sea urchin eggs, the protein synthesis remains independent of the presence of the nucleus; but, the enucleate halves of sea urchin eggs are unable to synthesize the proteins which appear at later stages only and can be detected, with immunological methods, in pluteus larva and in the adults. This indicates that the enucleate eggs can utilize the preformed mRNA's of maternal origin which have accumulated during oogenesis; but they are unable to synthesize proteins which are characteristic of more advanced differentiation; these proteins are coded for by new populations of mRNA's, synthesized by the nuclei at later stages than the blastula. One of the proteins that is synthesized in enucleate fragments of sea urchin eggs, is **tubulin**, the major protein of the mitotic apparatus (see **Brachet**, 1974).

5. Effect of enucleation in molluscan eggs. Another enucleate material that has been obtained from eggs and has been subject of biochemical studies is the polar lobe which can be separated, at the "trefoil" stage, from the eggs of the molluscs *Mytilus* and *Ilyanassa*. The isolated polar lobes never form asters and do not cleave; all one can see are localized contractions of the cortex. Such lobes of the *Mytilus* have a low respiratory rate. All that is known about protein synthesis in isolated polar lobes is that, in *Ilyanassa*, protein synthesis occurs both in cytoplasmic polysomes and in mitochondria.

6. Effect of enucleation in frog eggs. Enucleate frog eggs are also able to cleave and form partial, abnormal blastulae after parthenogenetic activation (*e.g.*, treatment of amphibian oocytes by progesterone); they synthesize the same proteins as normal fertilized eggs during early cleavage. But they are unable to increase the synthesis of certain specific proteins which normally occurs at the middle and late blastula stage. However, enucleate and normal frog oocytes have about the same respiratory rate.

All the above examples have clearly shown that in the absence of a nucleus, cytoplasmic system begin to degenerate, some more rapidly than others. These degenerative changes are reversible since early reintroduction of a nucleus into enucleated cytoplasm restores vitality. A point of no return is reached when degenerative changes become irreversible and nuclear transplants do not revive the cell. This indicates toward the possibility that nuclei synthesize informational molecules to balance their utilization by cytoplasmic organelles and metabolic machinery.

11-2.2. Nucleus during Early Development

Developmental biologists have been interested to investigate whether the nucleus undergoes any irreversible (permanent) change during cleavage and development; and whether there is any loss of genes or chromosomes in the nuclei of different blastomeres or differentiated cell types; or whether only the functional state of their genome (epigenotype) is different; and which factors bring about the changes in the nucleus. Application of certain recently developed techniques of the molecular biology to the developing embryo has attempted to answer some of above questions. Thus, microscopical observations, experimental embryological methods and most modern techniques of molecular hybridization, nuclear transplantation, cell hybridization and microinjection, etc., have established now that during development—1. Some cells lose certain chromosomes (*e.g.*, somatic cells of *Ascaris*); 2. Some cells retain full complement of chromosomes (*e.g.*, germ line cells of *Ascaris*); 3. Among the somatic cells, some nuclei undergo irreversible changes, some nuclei undergo partially irreversible changes and other nuclei bear completely reversible changes. In many animals, the nuclei of somatic cells retain their reversibility or potency throughout the life and the activity of genes of such nuclei is found to be regulated by certain cytoplasmic factors (*i.e.*, determinants). The dynamic states of nuclei or genes during the course of embryonic development have been shown by the following experimental evidences :

A. Morphological evidences for nuclear changes during development. There are evidences which show that in differentiated cells of an adult all nuclei are not identical and are in different functional states. Some of such evidences are the following :

1. In certain cells the nucleus divides without accompanying cytoplasmic division to produce **multinuclear** cells. For example, the ventral nephrocytes of the *Calliphora* larva or fat bodies in *Dacus*.

2. In certain cell types, the entire chromosomal complement of the cell is multiplied. Such **endopolyploidy** occurs in mammalian hepatocytes, human or rat Purkinje neurons in the cerebellum and in the larval tissues of endopterygote insects.

3. In salivary glands and some other tissues of dipteran larvae, the chromosomes undergo **polyteny**, that is partial replication of the haploid genome; the DNA strands remain attached side by side. In polyteny, the centromere does not divide, whereas in multicellularity, nuclear division occurs. The polytene chromosomes have thousands ($1–4 \times 10^3$) of copies of single genes, which are seen as characteristic band-interband complexes. While the euchromatin replicates through nine doublings; the rDNA (of nucleolar organizer) only through six or seven and the DNA of heterochromatin of centromere is not replicated at all.

Quite interestingly, there is a characteristic pattern of **puffs** (sites of RNA synthesis, *i.e.*, transcription) in the chromosomes in different tissues (such as salivary gland, Malpighian tubule and rectum) and at different stages of development. The puffing pattern of polytene chromosome is under genetic control and is also regulated by certain environmental factors. It provides the first evidence that eukaryotic gene activity is regulated at the level of RNA synthesis.

4. Similar to the puffs of polytene chromosomes, **lampbrush chromosomes** are formed during meiotic diplotene of oogenesis in the primary oocytes of most animals to do intense RNA synthesis (*i.e.*, transcription of mRNAs) of certain genes (see Chapter 5).

5. In some cells, there is a **differential gene amplification**; this amplification may be of thousandfold. For example, in amphibian and insect oocytes many copies of the rRNA genes (*i.e.*, ribosomal DNA of nucleolar organizer) are produced to facilitate rRNAs synthesis. Similarly, there is gene amplification in DNA puffs of the insect *Sciara* in late fourth instar larvae.

6. **Heterochromatization** of chromosomes is another type of nuclear change. The heterochromatin replicates late in the S-phase of the cell cycle, is genetically inactive and stains differently from euchromatin. For example, in mammalian females, during the dosage compensation mechanism of X-chromosomal genes, all X-chromosomes except one become heterochromatic to form the Barr body or sex chromatin. In male mealy bugs all paternal chromosomes become heterochromatic. In fact, in almost all cells small parts of the chromosomes may become heterochromatic, depending on tissues.

7. In *Ascaris megalocephala univalens* all non-germ line or somatic cells undergo **chromosome diminution** and **chromosome elimination** (see Chapter 4). Only the zygote and germ line cells have one pair of large

chromosomes (the full complement of diploid set of chromosomes). Chromosome elimination from somatic cells is seen also in *Sciara* (fungus fly), *Mayetiola* (Hessian fly), *Miastir* and several other animals. Since chromosome elimination always occurs in all somatic cells, it does not account for differences between them.

8. Some cells show the extreme phenomenon of nuclear disintegration or exclusion, for example, the lens fibre cells in the vertebrate eyes and the red blood cells in mammals.

B. Biochemical evidences for functional states of genome. The crucial biochemical evidence has accumulated only recently, from the **DNA-RNA hybridization technique** (a type of **molecular hybridization** method). This technique is quite simple in principle. It is based on the property of specific base pairing (adenine–thymine/uracil; guanine–cytosine) in the double stranded nucleic acid molecule. It means that a particular single DNA strand will base pair (or renature or reanneal) only with its own mRNA transcript (complementary mRNA) and not with any other mRNA.

In the technique of DNA-RNA hybridization, thin filters (called **membrane filters**) are used. These filters are made of nitrocellulose and are commercially available. These filters bind single-stranded DNA very tightly but fail to bind either double-stranded DNA or RNA. They provide an important technique for measuring hybridization. Renaturation or reannealing can be carried out on a filter in the following way — a sample of denatured DNA (*i.e.*, single-strands of DNA which are obtained by heating the double-stranded DNA or **native DNA**) is filtered. The single DNA strands bind tightly to the filter along the sugar-phosphate backbone, but their bases remain free or exposed. The filter is then placed in a vial containing a solution of small amount of radioactive RNA. After a period of renaturing, the filter is washed and hybridization is detected by the presence of radioactive RNA on the filter (see **Freifelder**, 1985).

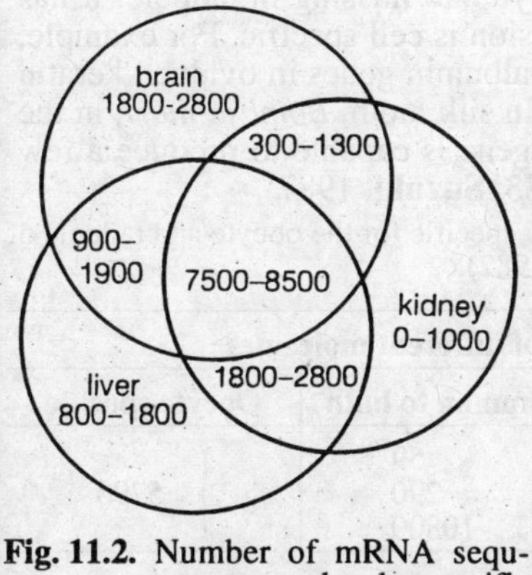

Fig. 11.2. Number of mRNA sequences or molecules specific to a tissue or common with other tissues in mouse (after Bishop, 1982).

In the earlier experiments a comparison of the total 'mRNA' from different stages of the development of frog was carried out. The results indicated that the mRNA from cleaving eggs was entirely different from the mRNA from the swimming tadpole. This technique, however, deals with the RNA transcripts from the repetitive DNA sequences and not with those from the non-repetitive (single-copy or unique) DNA.

The hybridization experiments indicate the existence of not more than three distinct abundance classes of mRNA. In each class the number of mRNA copies of each DNA sequence per cell varies in a narrow range (a few fold) and the average mRNA sequence concentration differs by more than ten-fold (Table 11-1). For example, a comparison of three mouse tissues—liver, kidney and brain, indicates that some mRNA sequences are tissue specific (*i.e.*, mRNAs for **cell specific** or **luxury proteins**), while others are shared by two or more tissues (*i.e.*, mRNAs for **housekeeping proteins**, **Maclean** and **Hall**, 1987) (Fig. 11.2). In other words, some genes are transcribed specifically in each tissue, while other genes are transcribed in common with other tissues, which means that the genome in the three tissues is in different functional states.

Table 11-1. Abundance classes and their properties in mouse liver mRNAS (Source **Bishop**, 1982)

Abundance classes	Number of different sequences (genes) in class	Average number of mRNA copies of each sequence/cell	Part of mRNA mass in cell
1	9	12400	0.22
2	750	280	0.41
3	12200	15	0.37

Notes : **Class 1** represents transcripts of moderately repetitive (around 3,000 copies per genome) DNA sequences.

Class 2 represents transcripts of low frequency repetitive (10 to 20 copies per genome) DNA sequences.

Class 3 represents transcripts of non-repetitive (single-copy or unique) DNA sequences.

It is with this background that the question of different gene expression during development has been explored. In *Xenopus laevis* an oocyte has been compared with the free swimming tadpole. The tadpole is formed in about 48 hours and the process of development involves cleavage, blastulation, gastrulation and organogenesis. Quite interestingly, the oocyte appears to have a greater number of mRNA sequences than the tadpole (Table 11-2), but there are only two and not three abundance classes. The important point is that about 6000 mRNA sequences present in the oocyte are missing in tadpole. It has become well evident now that gene expression is cell specific. For example, globin genes are active in reticulocytes, ovalbumin genes in oviduct, keratin genes in skin and crystallin genes in lens. In silk moth, *Bombyx mori*, in the cells of the posterior silk gland, only one gene is capable to produce a few million molecules of fibroin (**Gregory**, 1983; **Suzuki**, 1987).

Table 11-2. mRNA molecules common to and specific for the oocyte and tadpole of *Xenopus laevis* (Source **Bishop**, 1982).

Abundance class in tapole mRNA	Number of different molecules		
	Tadpole specific	Common to both	Oocyte specific
1	1	59	5790 (classes 1–3)
2	50	290	
3	1500	10800	

11-2.3. Nature of Changes in Nuclei during Development

During development nuclei of differentiated cells undergo the following type of changes :

1. Equivalence of cleavage nuclei. Weismann had postulated that the determinants of development are localized in the nucleus, particularly in chromosomes. According to him, diversified differentiation of the organism is the result of the (postulated) fact that as the egg divides into blastomeres and then into a large population of cells, the nuclei become non-equivalent, successively losing some of their determinants (or **idioplasms**) at each step of cell division. **Weismann** thought that in contrast to somatic cells, primordial germ cells retain complete set of determinants and are thereby equipped to transmit whole germ plasm to next generation.

We now know that as is the case for all the morphological and behavioural characteristics of the adult organism, the features of development, too, are determined by directives that originate in the genes. But we have also learned that with progressive cell differentiation the nuclei do not lose their equivalence (**J.F. Holtfreter**, 1988). This insight was derived in part from the classical experiments of **Driesch** (1892) and of **Spemann** (1928).

An early classical experiment of **Driesch** (1892) consisted of altering the distribution of nuclei into diverse sectors of an amphibian egg by forcing cleavage to occur under the pressure of a glass plate. Since the resulting embryo was normal, "equivalence" of the nuclei was indicated. In elegant experiment, **Spemann** (1928) constricted fertilized eggs of the newt *Triturus* (*Triton*) into two halves with a fine hair, just as they were about to begin to cleave. The constriction was not carried out completely, so that the two halves were still connected to each other by a narrow bridge of cytoplasm. The nucleus of the fertilized egg lay in one half, and the other half consisted of cytoplasm only (Fig. 11.3 A). When the egg nucleus began to divide, the cleavage was at first restricted to that half of the egg

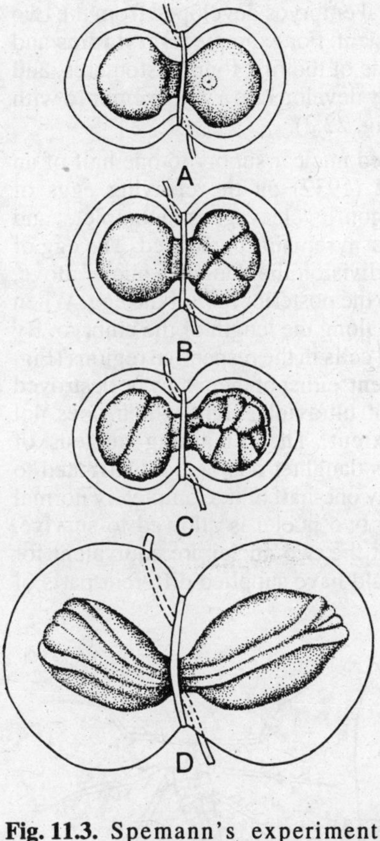

Fig. 11.3. Spemann's experiment of delayed supply of the nucleus to one of a newt's egg. A—Zygote immediately after constriction; B—One of the cleavage nuclei has penetrated into the uncleaved half; C—Fourth cleavage with first division of left part; D—Neurula stage, showing normal thin embryos but with difference in time of beginning of cleavage of the two halves still clearly seen (after Berrill, 1971).

which contained the nucleus. This half divided into 2, 4, 8 cells and soon, while the non-nucleated half remained uncleaved, at about the stage of 16 blastomeres, one of the daughter nuclei now much smaller than at the beginning of cleavage, passed through the cytoplasmic bridge into the half of the egg which had hitherto no nucleus (Fig. 11.3 B). Forthwith this half also began to cleave. After both halves of the egg were thus supplied with nuclei, Spemann drew the hair loop tighter and separated the two halves of the egg completely from each other. They were now allowed to develop into embryos. In a number of cases two completely normal embryos developed from the two halves of the egg as a result of this experiment. For example, **Horstadius** and **Wolsky** (1936) found in sea urchin that one of the first four blastomeres, and even occasionally one of the first eight, may develop into a whole embryo with all the normal parts but reduced in size (Fig. 22.3).

Spemann-like experiments of delayed nuclear supply to one half of an amphibian egg were repeated by **Seidel** (1932) on the cleaving eggs of dragonfly, *Platycnemis pennipes*. In a dragonfly, cleavage is incomplete, and only the nucleus divides at first the cytoplasm remains uncleaved. The egg of dragonfly is elongated and after the first division the daughter nuclei move, one into the anterior half and the other into the posterior half of the egg. When eight nuclei are available, they are spaced along the length of the embryo. By further divisions nuclei are provided for all cells in the respective regions (Fig. 11.4). In the stage when two nuclei are present, either of them may be destroyed by a short exposure to a narrow beam of ultraviolet light which does not damage the cytoplasm to any great extent. The remaining nucleus of UV-treated egg continues to divide and its daughter nuclei are distributed to all parts of the egg instead of supplying only one-half of it. Completely normal embryos develop no matter which of the two nuclei is allowed to survive. Seidel's experiment, thus, established that the two nuclei are equivalent for development, although normally they would have supplied different parts of the embryo.

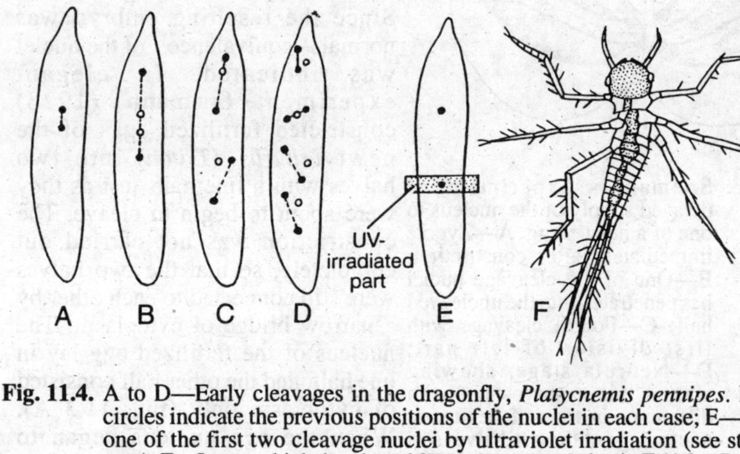

Fig. 11.4. A to D—Early cleavages in the dragonfly, *Platycnemis pennipes*. Hollow circles indicate the previous positions of the nuclei in each case; E—Killing one of the first two cleavage nuclei by ultraviolet irradiation (see strippled area); F—Larva which developed from an egg treated as in E (After Balinsky, 1981).

Further clear-cut evidences for the equivalence of the nuclei have been furnished by the nuclear transplantation experiments performed by **Briggs** and **King** and **Gurdon** and others.

2. Nuclear transplantation experiments. As already described in Chapter 1, **R. Briggs** and **T.J. King** (1952) in the United States, developed a simple technique of nuclear transplantation. In this procedure blastula nuclei are transplanted into enucleated, activated eggs and the capacity of surrogate nuclei to support a complete developmental programme is assessed. Thus, an intact egg organization with all its developmental potential is provided; if a nucleus has been irreversibly altered in development, it should not be capable of supporting normal development.

Such a technique begins with the newly laid unfertilized egg of *Rana pipiens*. After the egg is parthenogenetically activated by pricking it with a glass needle, a fertilization membrane is elevated and the female nucleus approaches the surface of cytoplasm in preparation of second maturation division, forming the second polar body. The site at which the second polar body forms contains the entire spindle apparatus which can be easily plucked out with a glass needle (second prick) since it appears close to the surface. The activated enucleated egg now serves as the recipient of donor nuclei. If nothing is done to the enucleated egg, it soon dies and cytolyzes. If, however, a nucleus is introduced, the fate of such an enucleated egg is entirely different. A donor cell from any region of the blastula is broken as it is sucked into a micropipette (its diameter is less than that of the cell) to release its nucleus and this broken cell (*i.e.*, the nucleus and debris) is injected deep into the recipient egg cytoplasm. When the pipette is withdrawn, the egg cytoplasm tends to escape through the hole in the egg's plasma membrane, forming an extraovate protrusion. This must be cut off by a pair of glass needles to prevent further loss of egg cytoplasm. More than 80 per cent of injected blastula nuclei induce normal cleavage to the blastula stage, and 50 to 80 per cent of these blastulae develop into normal tadpoles which can be reared to fertile adults. Thus, all blastula cell nuclei behave like zygote nuclei, they support a complete developmental programme even after 12 to 15 cell generations during which they have been exposed to different cytoplasms. This indicates that there is no genetic change in the nuclei during cleavage.

Later on, more elegant methods of nuclear transplantation were devised to test the totipotency of nuclei of cells which are even more advanced in the process of differentiation. By his classical experiments **J.B. Gurdon** has shown that the unfertilized eggs of frog or South African clawed toad, *Xenopus laevis*, can be irradiated with ultraviolet light to destroy the endogenous nucleus and can then be injected with a single *Xenopus* diploid nucleus. Diploid nuclei obtained from *Xenopus* tadpole intestinal cells (which are clearly differentiated cells, having a ''brush border'' of microvilli) are able to sustain development of normal adult frogs which are fertile (**Gurdon**, 1962). This experiment demonstrates that the intestine cells retained all the genes required for the complete life cycle of a frog, including the formation of germ cells. However, not all transplants are successful : the main limiting factor being the rapidity (90 minutes) with which the transplanted nuclei must replicate its DNA before first cleavage.

Subsequently, it is found that the success rate can be increased by using "partial" blastulae (which arises frequently in nuclear transplantation experiments) in which the nucleus escapes the first division and which can be recognized readily because only one-half of the egg cleaves. These nuclei take a shorter time to complete their DNA replication and have a much higher success rate when transplanted into new recipient egg. This technique is called **serial nuclear transplantation** (Fig. 1.3 and 11.5).

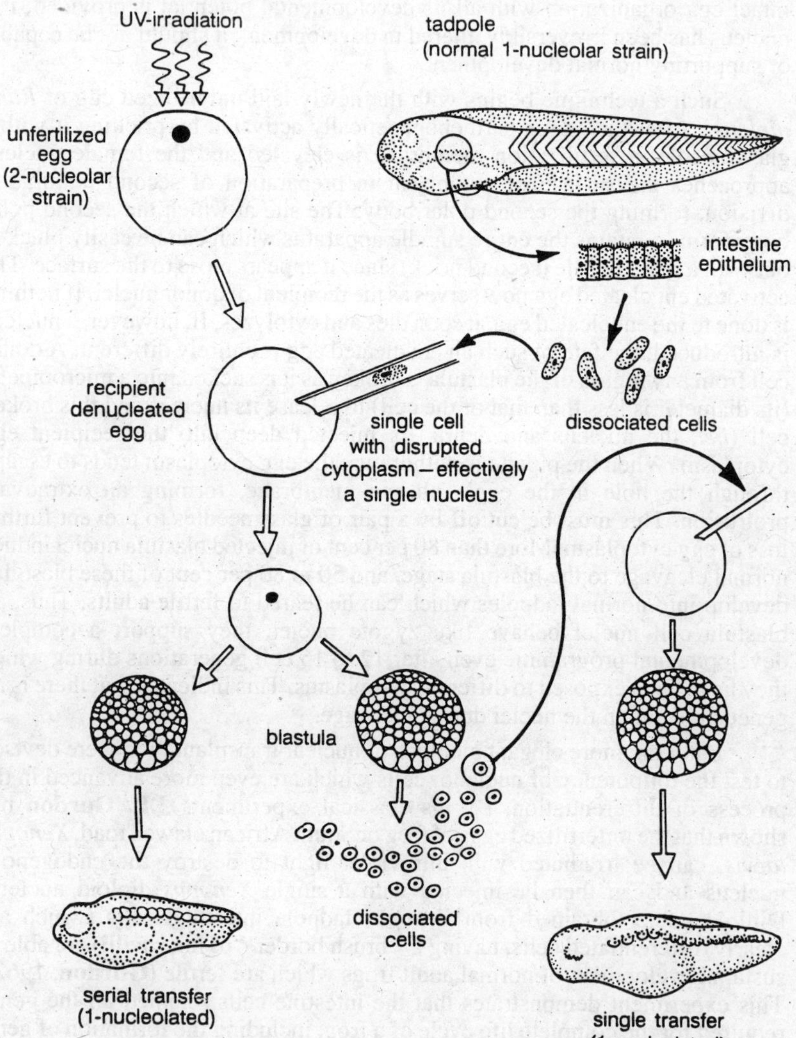

Fig. 11.5. Gurdon's technique of serial nuclear transplantation for the demonstration of the equivalence of somatic cell nuclei in *Xenopus*. Best results are obtained when the first blastula is dissociated and the blastomere's nuclei used for further transplantation into enucleated eggs (after Ede, 1978).

Development up to the stage at which the tadpole swims has been obtained by serial nuclear transplantation of a variety of adult tissues, such as keratinizing skin cells (**Gurdon**, **Laskey** and **Reeves**, 1975), lymphocytes (**Whabl**, **Brun** and **Du Pasquier**, 1975) and other tissues such as kidney and lung (see **Balinsky**, 1981). These results demonstrate that the genes required to make nerve, blood, muscle, cartilage and other tadpole tissues were not irreversibly inactivated in the donor nuclei. It has not yet been possible to obtain adult frogs using adult tissues and, thus, we cannot entirely exclude the possibility that some genes might in fact be lost in some cells (see **De Robertis** and **De Robertis, Jr.,** 1987).

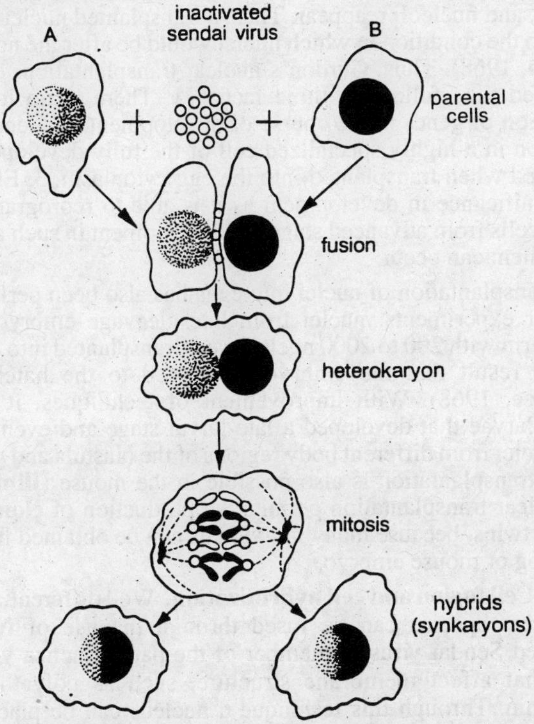

Fig. 11.6. Cell fusion with Sendai virus results first in the production of a heterokaryon with two nuclei, one from each parent. A hybrid cell line (synkaryon) arises when the two nuclei undergo mitosis synchronously. The cells of the hybrid cell line have one nucleus that contains chromosomes from both parents and are capable of multiplying in cell culture (after De Robertis and De Robertis, Jr., 1987).

The preceding discussion is not meant to imply that nuclei of differentiated cells are in every respect similar to the nuclei of fertilized eggs. In experiments with frog, nuclei transplanted in the egg cytoplasm undergo a change which makes them resemble early embryonic nuclei. The nuclei of newly fertilized eggs, although small in proportion to the egg cytoplasm, are much larger than the nuclei at later stages. Accordingly, transplanted nuclei increase up to 30-fold in volume during the first 40 minutes after transplantation. The functioning of the nuclei also changes drastically. Nuclei of advanced embryos do not divide rapidly and accordingly the synthesis of DNA (which is necessary for the replication of chromosomes) is slow. On the other hand, embryo's nuclei synthesize large quantities of ribosomal RNA and as this kind of RNA is produced in the nucleolus, they have prominent nucleoli. Nuclei of cleavage cells synthesize DNA rapidly but do not synthesize any ribosomal RNA and, therefore, do not show any nucleoli. Transplanted nuclei cease to synthesize ribosomal RNA and their nucleoli disappear. Instead they start to synthesize DNA rapidly as do normal cleavage nuclei. By the time the embryo reaches the stage of gastrulation, DNA synthesis in the descendants of the transplanted nucleus slows down and ribosomal RNA synthesis is renewed, and nucleoli reappear. Thus, a transplanted nucleus reverts in every respect to the condition in which nuclei would be after the normal fertilization (**Gurdon**, 1968). Thus, Gurdon's nuclear transplantation experiments have established the following three facts : 1. There is no loss or permanent inactivation of genes in the course of development. 2. Genes which have no expression in a highly specialized cell of the fully developed animal can be reactivated when transplanted into the egg cytoplasm. 3. Egg cytoplasm has great significance in development as it is able to reprogramme nuclei from somatic cells from advanced stages of development in such a way that normal development can occur.

Transplantation of nuclei into eggs has also been performed in insects. In earlier experiments, nuclei from late cleavage embryos of *Drosophila* (blastoderm with 200 to 2000 nuclei) were transplanted into unfertilized eggs, with the result that the embryo developed to the hatching larva stage (**Illmensee**, 1968). With improvement of techniques, it was possible to produce larvae that developed a late larval stage and even to pupate, using single nuclei from different body regions of the blastula and gastrula embryos. Nuclear transplantation is also possible in the mouse (**Illmensee**, 1981). In fact, nuclear transplantation permits the production of **clones** of genetically identical twins, because many individuals can be obtained from the cells of a single frog or mouse embryo.

3. Cell fusion and cell hybridization. Well differentiated cells of same or different species can be fused through the use of **fusogens** such as inactivated Sendai virus (a member of the parainfluenza viruses) and other agents that affect membrane structure, such as polyethyleneglycol and lysolecithin. Through this technique a nucleus can be placed in a different cytoplasmic environment. The initial product of the fusion of two different cells is a **heterokaryon** (*i.e.*, a single cell containing nuclei of two types) (Fig. 11.6). Eventually both nuclei might enter mitosis synchronously, form a single metaphase plate, divide and produce a **hybrid cell line** or **synkaryon**. The cells of a hybrid cell line have a single nucleus containing chromosomes from both parental nuclei.

The technique of cell fusion and cell hybridization was developed by **Harris** and **Watkins** in the 1960s. Cells which are cultured in the presence of Sendai virus may fuse together, even though they may be highly differentiated, *e.g.*, fibroblasts, macrophages, lymphocytes, nucleated erythrocytes and what is more astonishing even though they may be from quite different vertebrate species, *e.g.*, human, mouse, chick or *Xenopus*, providing material of enormous value in furthering our understanding of nucleocytoplasmic relationships, nuclear equivalence, control of transcription and other aspects of the molecular events underlying differentiation (**Ede**, 1978).

In 1965, **H. Harris** found that chick erythrocyte nuclei are reactivated when fused to HeLa cells (*i.e.*, an undifferentiated cell line so named because it was derived from the uterine carcinoma cells of a woman named *He*nrietta *La*cks). These heterokaryons are of immense interest because the erythrocyte nucleus does not normally synthesize RNA or DNA. Chick erythrocytes are terminally differentiated cells that have a highly condensed nucleus and are destined to die. (Mammalian erythrocytes normally eliminate their nuclei during red blood cell maturation, whereas red blood cells of birds, amphibians and reptiles retain their nuclei which become inactivated). When fused to HeLa cells, the chick erythrocyte nucleus increases 20 times in volume, disperses its chromatin, resumes RNA synthesis, develops nucleolus and eventually replicates its DNA. This process if accompanied by the uptake of large amounts of human nuclear proteins (**Ringertz** *et al.*, 1971) which are thought to reactivate the erythrocyte nucleus.

In fact, in a few cases it has been possible to activate latent genes coding for specific proteins by fusion of differentiated cells. For example, albumin secreting rat hepatoma cells fused with mouse lymphocytes (which do not produce albumins) will occasionally give rise to clones of hybrid cells that are able to produce both rat and mouse albumins (**Malawista** and **Weiss**, 1974). Thus, somatic cell hybridization experiments show that the patterns of nucleic acid synthesis and gene experssion by a nucleus can be modified by substances present in the cell cytoplasm.

4. Microinjection experiments. Frog oocytes are growing egg cells, obtained from the abdominal cavity of frogs, which due to their large size (1.2 mm) and tolerance for micromanipulation have been used in many microinjection experiments. Aside from their hardiness, oocytes interest developmental biologists becuase they are destined to become eggs and embryos and already contain most of the molecular machinery necessary for early development. Oocytes are active in RNA synthesis but do not synthesize DNA, and if somatic cell nuclei are injected into them, the transplanted nuclei also show this type of synthetic activity. The mature frog egg (which is laid into pond water), in contrast, does not synthesize RNA but actively synthesize DNA after penetration by sperm or a microinjection needle. Nuclei transplanted into mature eggs also mimic the behaviour of the host cell, actively replicating DNA but not transcribing any RNA.

The oocyte cytoplasm not only affects the pattern of macromolecular synthesis but is also able to **reprogramme** the expression of individual genes in transplanted nuclei. In an experiment of **Gurdon**, **De Robertis** and **Partington** (1976), *Xenopus* oocytes were microinjected with a suspension of about 200 HeLa nuclei. The nuclei survive for several weeks inside the oocytes and during that period synthesize considerable amounts of RNA. In the first few days (three days) after injection the somatic nuclei enlarge to about ten times their original volume and resemble the oocyte's own nucleus morphologically. This change is followed by chromatin dispersion, massive uptake of proteins from the surrounding cytoplasm and formation of lampbrush chromosomes. In fact, the oocyte cytoplasm reprogrammes the gene expression of the injected nuclei in such a way that only those genes that are normally active in oocytes are expressed. This can be determined by taking advantage of the fact that the RNA synthesized by the injected nuclei in the course of several days accumulate in the cytoplasm, where they code for new proteins that can be labelled with radioactive amino acids (**De Robertis** *et al.*, 1977). In one such experiment cultured kidney nuclei of *Xenopus* were injected into oocyte of a different amphibian species (the salamander *Pleurodeles*, whose proteins can be readily distinguished from those of *Xenopus* by two-dimensional electrophoresis. Those genes that are normally expressed in kidney cells but not in oocytes become inactive after injection into *Pleurodeles* oocytes. More importantly some oocyte-specific genes that were not expressed by the kidney cell nuclei were activated by the oocyte cytoplasm. The oocyte, therefore, contains components that can determine that a particular spectrum of protein-coding genes will be active and that others will be inactive.

11-3. CYTOPLASMIC CONTROL OF NUCLEUS DURING DEVELOPMENT

Many experiments have shown that the egg cytoplasm plays a major role in early development. Certain recent experiments on the sea urchin, *Anthocidaris* indicate that for cleavage to occur harmoniously, the occurrence of following two cycles becomes must : (1) Nuclear division, (2) cytoplasmic activity. The cytoplasm shows cyclic changes in the thickness of the hyaline layer and the amount of sulphydryl (SH) content in the cortex, etc. Heat shock (34°C) or ether treatment suppresses the SH-cycle (in the cortical plasm or egg cortex) and cleavage (cytokinesis), but allows the nuclear division to continue. The UV-irradiation of eggs, on the other hand, suppresses the nuclear division cycle and the cleavage but allows SH-cycle to take place. The control of a biological clock, maintained by the SH-cycles in the cortex and the nuclear division cycles, remains to be investigated.

Thus, if the information contained in the cleavage nuclei is identical, then the initial differences between cells should reside in the cytoplasm they inherit. Recently, two distinct concepts have been formulated to explain how the diversity of cell types is generated. 1. The egg already contains in its cytoplasm regionalized components called **intracellular morphogens** (**Maclean** and **Hall,** 1987) or **morphogenetic** or **cytoplasmic determinants** (**Jeffery**, 1988) which influence the fate of cells as a result of their differential segregation into various blastomeres during cleavage. 2. Certain embryonic cells, whose unique properties were established earlier by the action of localized cytoplasmic determinants influence the development of adjacent cells by **induction** through some substances called **inducing substance**, **evocator** or **extracellular morphogen** (**Maclean** and **Hall**, 1987).

11-3.1. Cytoplasmic Determinants

Cytoplasmic determinants are entities localized in specific cytoplasmic regions of the egg or embryonic cell that bias the cell containing them to assume particular properties or fates during embryogenesis. Roles of cytoplasmic determinants are evident in the formation of body axis and germ layers and in the expression of histospecific enzymes. Determinants are described as entities, rather than as particular kinds of molecules or organelles because they have yet to be isolated and characterized (**Jeffery**, 1988). The knowledge of the presence, localization, function and chemical properties of cytoplasmic determinants has been gained by the following indirect means :

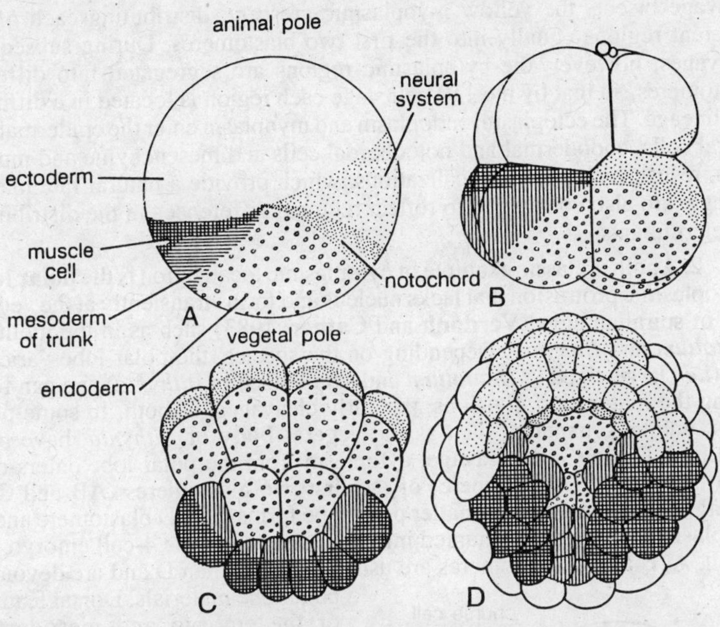

Fig. 11.7. Embryonic derivatives of four kinds of cytoplasm of the fertilized egg of *Styela partita* . A—Fertilized egg; B—8-celled stage; C and D — Early and late gastrulae (after Balinsky, 1981).

1. Egg cytoplasmic localizations. The visible regionalization of the egg cytoplasm known as **cytoplasmic localization**, has been described for a large number of animal eggs, particularly invertebrates. For many years, cytoplasmic localization was thought to signify the presence of determinants (**Wilson**, 1925 ; **Davidson**, 1986). 1. Possibly the most dramatic example of localization occurs in ascidian eggs. **Conklin** (1905) has described five differently coloured cytoplasmic regions in the unfertilized *Styela* egg. The three most obvious of these regions are the transparent **ectoplasm**, residing in the animal hemisphere ; the gray **endoplasm**, located mainly in the vegetal hemisphere ; and the yellow **myoplasm**, distributed uniformly in the egg periphery. The different cytoplasmic regions reflect an uneven distribution of variously pigmented granules (*e.g.*, yellow lipid granules complexed with mitochondria in the myoplasm) ; yolk particles (in endoplasm) and clear contents of the giant oocyte nucleus or germinal vesicle in ectoplasm. Between fertilization and cleavage, these cytoplasmic regions are rearranged by a process called **ooplasmic segregation** (**Conklin**, 1905 ; **Jeffery**, 1984). During ooplasmic segregation, the myoplasm and ectoplasm stream vegetally, while the endoplasm flows into the animal hemisphere. The ectoplasm subsequently moves back into the animal hemisphere, the endoplasm too returns to its original position in the vegetal hemisphere, and the myoplasm shifts upward from the vegetal pole, extending into a yellow crescent in the future vegetal-posterior region of the zygote (Fig. 11.7). The plane of the first cleavage bisects the yellow myoplasmic crescent, distributing each of the different regions equally into the first two blastomeres. During subsequent cleavages, however, the cytoplasmic regions are segregated into different blastomeres, so that by the 110-cell stage each region is located in a different cell lineage. The ectoplasm, endoplasm and myoplasm enter the epidermal and neural cells, endodermal and notochordal cells and mesenchyme and muscle cells, respectively. These localizations, which provide a natural fate map of the egg, are also considered to reflect regional differences in the distribution of determinants.

2. Another classic example of cytoplasmic localization is the **polar lobe**, a cytoplasmic protrusion that lacks nucleus and forms transiently at the vegetal pole of spiralian eggs (**Verdonk** and **Cather**, 1983) such as in the molluscs *Dentalium* and *Bithynia*. Depending on the species, the polar lobe varies in size (*i.e.*, large-sized in *Dentalium* and small-sized in *Bithynia*) and can form during the maturation divisions, the early cleavages, or both. In some polar lobes, unique organelles such as the **vegetal body** of *Bithynia*, have been identified. During first cleavage, cytoplasm from the polar lobe enters only one (the larger CD blastomere) of the first two blastomeres (AB and CD). During second cleavage, a smaller polar lobe forms the CD blastomere and its cytoplasmic contents are shunted into only one cell of the 4-cell embryo, the D-cell. A, B and C blastomeres are usually smaller than D and are devoid of polar lobe materials. Dorsal features of the embryo and mesodermal derivatives arise from the D cell. Thus, the polar lobe appears to be a mechanism used to shunt cytoplasmic determinants responsible for subsequent germ layers and embryonic axis formation into specific blastomeres during cleavage (See Fig. 12.9).

3. The eggs of some holometabolous insects also contain distinctive localizations at their anterior and posterior poles (Fig. 11.8). The cytoplasmic region at the anterior pole is thought to contain determinants responsible for the development of head segment (**Kalthoff**, 1979). A cytoplasmic region at the posterior pole the **oosome**, contains the **polar granules**, a localized group of electron-dense organelles believed to be inherited by the germ cells (**Mahowald**, 1968). After

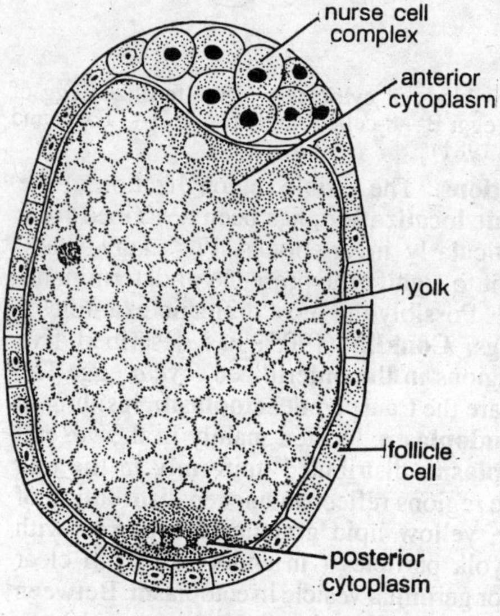

Fig. 11.8. The anterior and posterior (oosome) cytoplasmic regions in an oocyte of the dipteran insect, *Miastor americana* (After Jeffery, 1988).

fertilization and oviposition, the insect egg develops as syncytium ; the zygotic nucleus undergoes a series of divisions within the internal yolk-filled cytoplasm. Subsequently, many nuclei migrate to the periphery of the egg ; those that enter the oosome obtain special qualities : they cease division precociously and become the nuclei of **pole cells**, the first cells formed in the embryo. Pole cells eventually enter the interior of the embryo during gastrulation ; many of them migrate to the gonad and form germ cells. In insect species that exhibit chromosome elimination during early development, pole cell nuclei are protected from loss of chromosomes by their localization in the oosome (**Beams** and **Kessel**, 1974). Therefore, the oosome is considered a localization containing **germ cell determinant**.

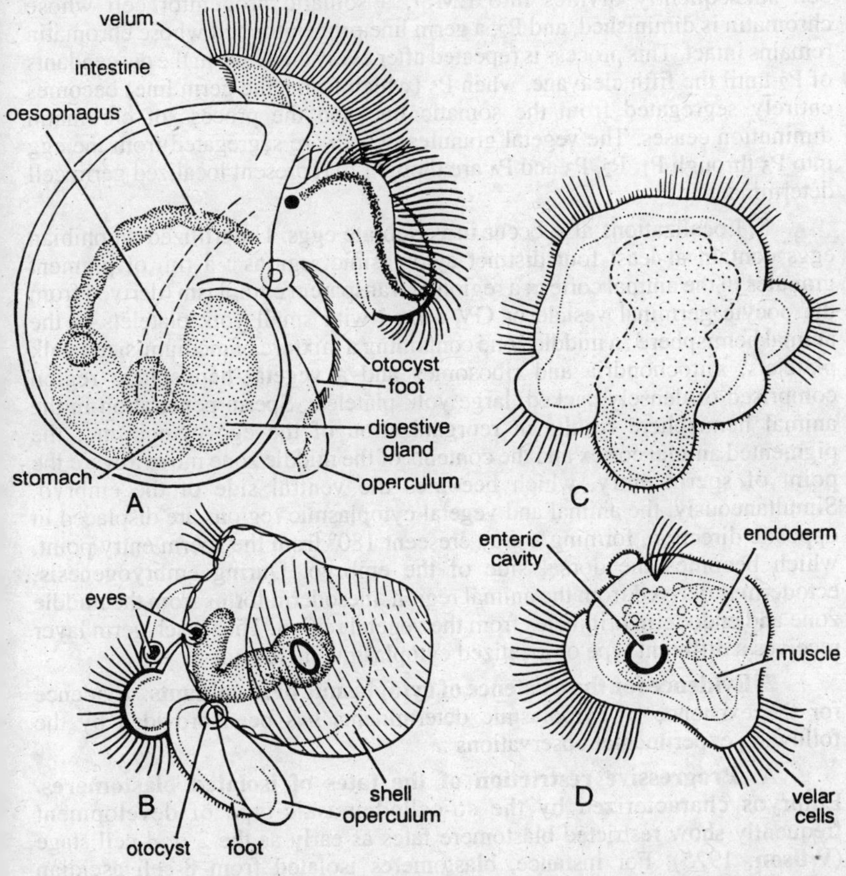

Fig. 11.9. A comparison of normal (A), lobeless (B) and partial (CD) *Ilyanassa* larvae. A—A normal larva; B—A lobeless larva; C—A partial larva that developed from an isolated CD cell; D—A partial larva that developed from an isolated AB cell (after Jeffery, 1988).

Likewise, the cytoplasmic determinants localized near the anterior pole of egg in chironomid (*Smittia*) are necessary for the development of the head and thorax ; if these determinants are removed a double abdomen without a head and thorax develops.

4. Among the nematodes, organelles known as **vegetal** (or P) **granules** are localized in the vegetal (posterior) hemisphere of *Caenorhabditis elegans* (**Strome** and **Wood**, 1982) and *Ascaris* eggs and are found to contain germ cell determinants which are segregated into the germ cell lineage during early development. The first cleavage of nematode eggs is equatorial, so that most of the vegetal cytoplasm is distributed to the vegetal (P_1) blastomere (Fig. 4.1). The animal (AB) blastomere gives rise to only somatic cells, while P_1 cell gives rise to both somatic and germ cells. *Ascaris* (but not *C. elegans*) undergoes chromatin diminution in somatic progenitors cell (*i.e.*, A B cell). P_1 cell subsequently divides into EMSt, a somatic progenitor cell whose chromatin is diminished, and P_2, a germ line progenitor cell whose chromatin remains intact. This process is repeated after every cleavage in the descendants of P_2 until the fifth cleavage, when P_5 (which forms the germ line) becomes entirely segregated from the somatic line and the process of chromatin diminution ceases. The vegetal granules, which are segregated from the egg into P_5 through P_1, P_2, P_3 and P_4 are thought to represent localized germ cell determinants.

5. Localizations also occur in vertebrate eggs. Unfertilized amphibian eggs contain at least four distinct cytoplasmic regions : a rim of pigment granules in the animal cortex ; a region of transparent cytoplasm (derived from the oocyte germinal vesicle or GV mixed with small yolk platelets in the animal hemisphere ; a middle zone containing a mixture of medium-sized yolk platelets, mitochondria and ribosomes and a vegetal hemisphere region comprised of densely packed, large yolk platelets. Sperm penetration in the animal hemisphere results in reorganization of the egg cytoplasm ; the pigmented animal cortex and the contents of the middle zone move toward the point of sperm entry, which becomes the ventral side of the embryo. Simultaneously, the animal and vegetal cytoplasmic regions are displaced in opposite direction, forming a gray crescent 180° from the sperm entry point, which becomes the dorsal side of the embryo. During embryogenesis, ectoderm is derived from the animal region, mesoderm forms from the middle zone and endoderm originates from the vegetal region. Thus, each germ layer receives a different type of localized cytoplasmic inclusion.

2. Evidence for the existence of cytoplasmic determinants. Evidence for the existence of cytoplasmic determinants has been provided by the following experimental observations :

A. Progressive restriction of the fates of isolated blastomeres. Embryos characterized by the so-called mosaic-type of development frequently show restricted blastomere fates as early as the 2-or 4-cell stage (**Wilson**, 1925). For instance, blastomeres isolated from 8-cell ascidian embryos, at the stage when the pigmented cytoplasmic regions have begun to be segregated into separate cells, each forms only the tissue that would be expected according to the fate map (**Reverberi**, 1971). Likewise, AB and CD blastomeres isolated from polar lobe-containing embryos of the gastropod *Ilyanassa* show different fates in culture (**Verdonk** and **Cather**, 1983). In this mollusc **partial embryos** derived from the CD cell form foot, shell and eye, whereas partial embryos derived from the AB cell form the velum, muscle, gut and pigment cells (**Clement**, 1956). Early restriction of the fates of blastomeres in these and other embryos suggests that developmental potential is unevenly segregated during the early cleavages.

B. Invariant segregation of developmental potential between specific blastomeres. Invariant segregation of developmental potential is illustrated by **cleavage-arrest experiments**. In these experiments, the spatial expression of a cell lineage or tissue-specific markers (*e.g.*, some histospecific enzyme or other protein) is examined in embryos arrested at one (or more) of the early cleavage stages by treatment with **cytochalasin B** (*i.e.*, an inhibitor of actin microfilaments). This treatment blocks cytokinesis, but cleavage-arrested cells carry on their mitotic cycles and DNA replication and ultimately express tissue-specific markers only in cells expected according to the fate map.

A clear example of this phenomenon occurs in the ascidian embryos, in which the spatial distribution of several histospecific enzymes and morphological markers has been thoroughly investigated (**Whittaker**, 1973, 1977, 1984). The enzyme **acetylcholinesterase** (**AChE**) (A tail-muscle specific enzyme) is a good marker of muscle differentiation. This enzyme appears in a maximum of two cells of cleavage-arrested 2-, 4-, or 8-cell embryos; four cells of cleavage-arrested 16-cell embryos ; six cells of cleavage-arrested 32-cell embryos and eight cells of cleavage-arrested 64-cell embryo. The cleavage-arrested cells that express ACh E at each developmental stage precisely match the position and maximum number of cells belonging to the muscle lineage. Other histospecific markers in cleavage-arrested ascidian embryos, including **alkaline phosphatase** in gut cells, **tryosinase** in brain melanocytes, a **colloidal substance** in notochord cells and extra-cellular secretions in epidermal cells, also exhibit pattern of gene expression that match their respective cell lineages.

C. Alteration of developmental fate after removal, destruction, or reorientation of egg cytoplasmic regions. When fertilized ascidian eggs are cut into fragments perpendicular to the animal-vegetal axis, only those fragments that contain a part of the vegetal cytoplasm develop into complete larvae (**Ortolani**, 1958). Thus, determinants important for larval development appear to be localized in the vegetal cytoplasm of ascidian egg.

Likewise, when the polar lobe is removed during the first cleavage, the lobeless embryo exhibits changes in the pattern and chronology of cell division, and later significant defects appear in larval morphology. **E.B. Wilson** (1904) removed the polar lobe at the two-cell stage of *Dentalium* by sucking up it by pipette. He found that larvae that develop from such lobeless embryos lacked the apical tuft (a ciliary organ), the mesodermal bands, and the post-trochal regions, from which parts of the mantle, foot and shell glands are formed. He, therefore, concluded that the polar lobe cytoplasm contains mesodermal determinants. Also in the gastropod *Ilyanassa*, larvae that develop from lobeless embryos lack the shell, foot, heart, eyes, operculum and tentacles (Fig. 11.9) (**Clement**, 1952).

In insect embryos, the oosome can be eliminated by ligature, cauterization or ultraviolet (UV) irradiation without affecting the nucleus (**Hegner**, 1911 ; **Geigy**, 1931 ; **Okada** *et al., 1974).* Embryos lacking the oosome never develop pole cells or functional germ cells, and the nuclei that enter the posterior regions of these embryos, are subsequently incorporated into general blastoderm cells.

When the nucleate animal hemisphere region of *Xenopus laevis* egg is separated from the enucleate vegetal hemisphere regions by ligation, the animal hemisphere develops into a ciliated structure lacking a dorsal-ventral axis (**Gurdon** *et al.*, 1985). By contrast, animal hemisphere fragments ligated so that they include a part of the vegetal hemisphere cytoplasm form ciliated embryos with partial axial structures. Nucleated vegetal or equatorial egg fragments also form axial structures but do not develop cilia, unless they include a part of the animal hemisphere cytoplasm. These experiments suggest that determinants specifying ectodermal ciliation are located in the animal cytoplasm of *Xenopus* eggs, whereas determinants required for dorsal-ventral axis specifications are concentrated in the vegetal cytoplasm. It was also found that determinants that control muscle actin gene expression are concentrated in the vegetal hemisphere cytoplasm near the future dorsal side of uncleaved *Xenopus* zygotes.

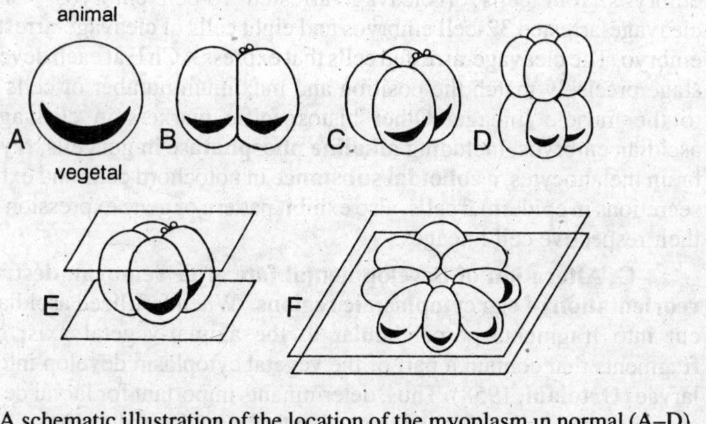

Fig. 11.10. A schematic illustration of the location of the myoplasm in normal (A–D) and compressed (E–F) *Styela* embryos. The myoplasm (shaded crescents) is distributed to 2 cells of the normal 8–cell embryo (D), but 4 cells of the compressed 8–cell embryo (F) (After Jeffery, 1988).

More information on the location of determinants has been obtained by reorientation of cytoplasmic regions in either the egg or embryo. The quality of egg cytoplasm received by different blastomeres during cleavage has been altered in two ways : by a change in the position of the spindle apparatus and cleavage furrow or by centrifugation. For example, when 4-cell *Styela* embryos are flattened by compression, the spindle apparatus is reoriented, causing the third cleavage to be meridional rather than equatorial (**Whittaker**, 1980). This causes the yellow myoplasm to enter four instead of two blastomeres. Some of the cells that obtain myoplasm in this way now become competent to express AChE (Fig. 11.10).

Centrifugation of uncleaved egg of *Ascaris* drives the vegetal granules to the centrifugal pole. The first cleavage furrow then forms at right angles to the plane of vegetal granule stratification, thereby separating the centrifugal zone of compacted granules into a spherical mass completely separate from each of two blastomeres. Consequently, the nucleus in each blastomere is diminished, providing further indication that the vegetal granules may contain the protectants of chromatin diminution (Fig. 11.11).

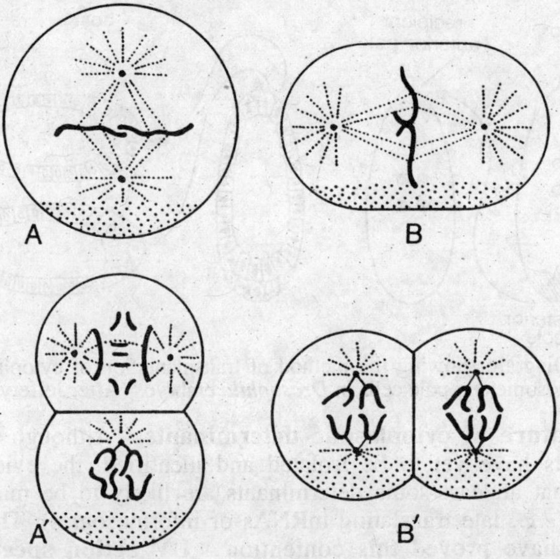

Fig. 11.11. Cytoplasmic control of chromosome diminution in *Ascaris*. A, A′ — Normal eggs; B, B′— centrifuged eggs. The shaded area indicates the distribution of a hypothetical cytoplasmic determinants. In blastomeres containing the determinants the two large chromosomes remain intact; in blastomeres lacking them the chromosomes undergo diminution (A′ – top cell) (After De Robertis and De Robertis, Jr., 1987).

D. Alteration of developmental fate by transplantation of egg cytoplasmic regions. The strongest evidence for the existence of cytoplasmic determinants is obtained when a portion of the egg cytoplasm can be shown to alter developmental fate after it is removed from one part of egg or embryo and transferred into another part of a different egg or embryo. As described in Section 4-2, **Illmensee** and **Mahowald** (1974, 1976) reported the first successful transplantation experiment of cytoplasmic determinants in *Drosophila*. The basic scheme for these transplantations has been illustrated in Fig. 11.12. Cytoplasm from the oosome of eggs or syncytial embryos was transplanted to the anterior or mid-ventral region of a precellular blastoderm embryo. The presence of transplanted cytoplasm led to the formation of histologically identifiable pole cells, which formed precociously at **ectopic sites** in the recipient. In a similar experiment, normal oosome cytoplasm was injected into the anterior-lateral region of UV-irradiated *Drosophila* eggs whose own oosome was inactivated. This transplant also resulted in the formation of ectopic anterior pole cells (**Okada** *et al.*, 1974). The ability of the ectopic pole cells produced in the experiments of **Illmensee** and **Mahowald** (1974, 1976) to form germ cells, was determined by a second transplantation experiment. The ectopically formed pole cells were transplanted to the posterior region of host embryos of the same age, but of different genotype. Some of the transplanted pole cells migrated to the gonads and become germ cells and the hosts produced progeny showing the genotypes of the eggs from which the oosome cytoplasm was originally derived.

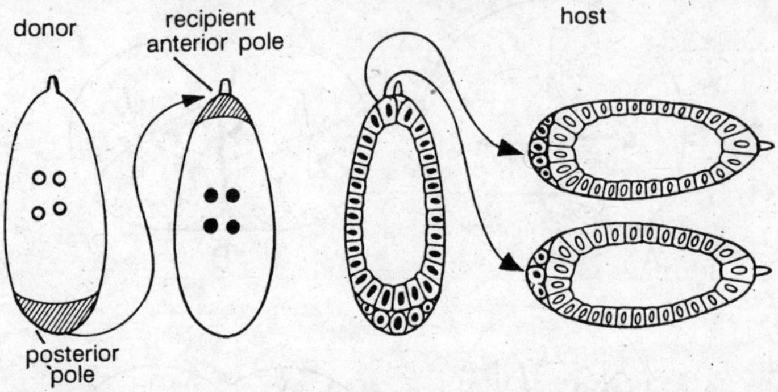

Fig. 11.12. Diagram showing the method of transplantation of cytoplasm from the oosome and pole cells in *Drosophila* embryos (After Jeffery, 1988).

3. Nature of cytoplasmic determinants. Although cytoplasmic determinants have yet to be isolated and identified, the evidences have indicated that at least some determinants are likely to be maternal RNA molecules (*e.g.*, late translating mRNAs or informosomes). The following evidences have proved this contention : **UV action spectrum data**, **photoreactivation** (*i.e.*, when pyrimidine dimers are formed in DNA or RNA by UV irradiation they can often be enzymatically excised and repaired in the presence of visible light), **RNase enzyme sensitivity**, and **rescue experiments** with purified RNA, especially mRNA (*i.e.*, to rescue an embryonic defect imposed by the modified determinants with microinjection of purified mRNA, **Togashi** and **Okada**, 1982). It should be stressed, however, that other determinants may be proteins, organelle systems or regulatory ions, such as Ca^{2+} (**Jaffe**, 1983).

Cytoplasmic determinants also appear to be bound/anchored to cellular structures, such as the **plasma membrane** (*e.g.*, molluscs such as *Dentalium*, *Crepidula* and *Nassarius* and Ascidians), the **egg cortex** (*e.g.*, the gray crescent cortex of uncleaved *Xenopus* egg), the **cytoskeleton** (*e.g.*, eggs of *Caenorhabditis elegans* and *Styela*), or a **complex** containing some or all of these structures.

4. Activation and function of determinants. How a determinant is activated depends on its identity or chemical nature. Proteins may be activated by a number of amino acid modifications, proteolytic cleavages, or higher order structural changes, whereas mRNAs may be activated by translation. The translational activation of mRNA determinants is part of the unresolved issue of how maternal mRNAs or informosomes are activated in general during early development. **Morgan** (1934) first proposed that cytoplasmic determinants may function by promoting differential gene expression in specific lineages and this idea has been a theme in subsequent hypotheses of how determinants function (**Davidson**, 1986). There appear to be at least two types of determinants based on their mode of action : those that do not require embryonic gene expression and those that promote specific gene activation in different regions of the early embryo.

11-3.2. Extracellular Morphogens

The extracellular morphogens obviously can only be operational once the embryo has become multicellular. In theory, once the oocyte has become a 2-cell zygote, the potential exists for the operation of such extracellular morphogens, although such early embryos are usually still under the control of the intracellular morphogens or cytoplasmic determinants. Extracellular morphogens are produced at one site (the **source**) and diffuse through the developing cells or tissues to a distant site (the **sink**) where they are broken down. A concentration gradient is, therefore, set up between the source and the sink. Cells lying along that gradient will respond to the particular concentration of the morphogen that they encounter and furthermore, their differentiation or morphogenetic response will depend on the particular threshold concentration or morphogen that they encounter. There are developmental systems, such as **primary embryonic induction** and **regional organization of the nervous system** where the existence of two opposing concentration gradients of two different morphogens, namely **neuralizing factors** and **mesodermalizing factors** provides the simplest explanation for the differential developmental event observed (see **Maclean** and **Hall**, 1987).

REVISION QUESTIONS

1. What sort of conclusions are drawn from enucleation experiments in *Acetabularia* and various animals ?
2. What are the main conclusions drawn by embryologists from the nuclear transplantation and cell hybridization experiments ?
3. Are the nuclei in amphibian blastomeres identical and/or totipotent ? If so, how do they have different prospective fates ?
4. What is the principle of molecular hybridization ? Explain how the technique of DNA-RNA hybridization is useful in developmental biology ?
5. Is it true that early development is under cytoplasmic control and later development under genetic control ? why ?
6. Give evidence from different species to show that the egg cytoplasm is highly organized. What is the causal/functional importance of this organization ?
7. 'Cytoplasm controls the activities of nucleus.' Justify this statement by giving suitable examples.
8. Write an essay on the concept of cytoplasmic determinants.
9. Write short notes on the following :
   (1) Equivalence of cleavage nuclei ; (2) Synkaryon ; (3) Cleavage-arrest experiments ; (4) Nuclear transplantation ; (5) Cytoplasmic transplantation ; (6) Andromerogons.

12

FATE MAPS AND CELL LINEAGE

In a blastula, the cells are located on or near the surface of the embryo and as yet do not exist in layers, so the area of the future germ layers during the blastula stage is referred to as **presumptive**. For example, that area of the blastula which is destined to form the neural ectoderm is called the **presumptive neural ectoderm**.

The early biologists constructed maps of the blastulae based on the trajectories of tagged or marked cells. They showed the fate of every region of the blastula under normal developmental conditions and were referred to as **fate maps**. According to these maps (Fig. 12.4), similar patterns exist in all vertebrates. Regardless of the specific shape of the blastula (round or disc-shaped), the organ forming areas are arranged in a pattern around the presumptive notochordal region.

12-1. METHODS OF PREPARATION OF FATE MAPS

The fate maps of different types of animal blastulae have been constructed by the following methods :

12-1.1. Construction of Fate Maps by Natural Marking

The cytoplasm of the fertilized eggs of some animals such as some ascidians (*e.g., Ciona, Styela (Cynthia) partita*), has natural colour differences in its various regions. For example, on the basis of pigmented differentiation, four general regions may be distinguished in the egg-cytoplasm of *Styela partita*: an upper hemisphere of light protoplasm, a yellow crescent postero-ventrally, a gray crescent antero-dorsally, and a vegetative area of dark gray yolky substance. Because of the natural differences in pigmentation, the developmental fate of these regions has been worked out with considerable precision. Knowing what these developmental prospects are, we may then view the areas as the components of a ''map'' of prospective organ regions. Accordingly, the upper clear plasm represents prospective epidermal ectoderm, the yellow crescent is prospective mesoderm, dark gray yolky plasm is prospective endoderm, and gray crescent is prospective neural ectoderm and notochord.

12-1.2. Construction of Fate Maps by Artificial Markings

Often natural–pigment–differentiation of ooplasm of the eggs of ascidians and amphibians is not well marked and presents ambiguity in determining the fate of various blastomeres of the blastula. So in them some artificial marking methods have been used to construct the fate maps. **Wather Vogt** (1925) is the pioneer embryologist who has discovered the artificial marking methods in the construction of fate maps of amphibians. Now following artificial marking methods are used in constructing the fate maps in different animals :

(a) Vital staining marking. This is done by the use of so called **vital stains** which are retained by cells for considerable periods and in no way

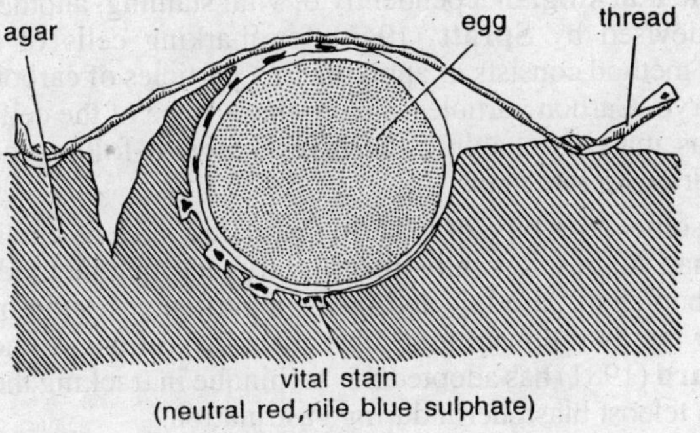

Fig. 12.1. Method of vital staining of a blastula (After Berrill, 1971).

interfere with normal processes, For example, for the construction of a fate map of an amphibian, a small piece of agar or cellophane (both are called stain **carriers**) is stained with a vital stain such as Nile blue sulphate, neutral red, Janus green or Bismark brown and is pressed against a chosen area of the blastula for a short period. The stain diffuses from the piece of agar or cellophane into the blastomeres, which will retain their distinctive colouration for several days. Only those cells receiving the dye retain the stain ; it usually does not diffuse into neighbouring cells. Hence, the vital dye or stain serves as an effective marker for the morphogenetic behaviour of small cell groups. By so marking several areas (see Fig. 12.2) simultaneously, it is possible to follow the movements of gastrulation by continuous observation. Moreover, the persistence of the stain makes it possible to establish the ultimate locations of the marked materials in the later embryo. This in turn, enables one to project the pattern of parts back onto the blastula and, thus, construct a map of prospective organ regions.

By vital staining marking methods, the prospective organ region maps or fate maps have been made for a number of vertebrates species (See Fig. 12.4).

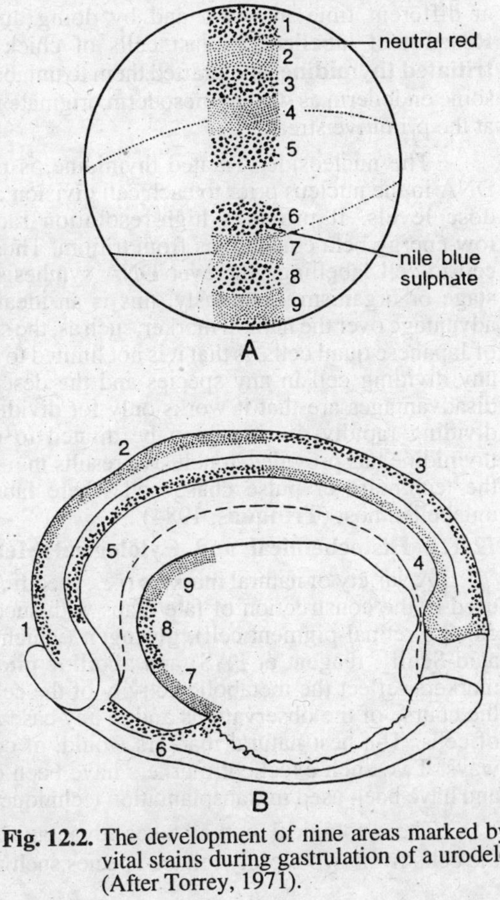

Fig. 12.2. The development of nine areas marked by vital stains during gastrulation of a urodele (After Torrey, 1971).

(b) Carbon particle marking. Independently of vital staining, another method has also been devised by **Spratt** (1946) for marking cells of a developing embryo. This method consists of applying tiny particles of carbon to the surface of the embryo. Carbon particles stick to the surface of the cells and can, thus, be used as markers enabling the investigator to follow the movements of cells and draw up fate maps.

Instead of carbon, some other type of adhering particles, such as chalk can also be used in fate map studies. Recently, in a slightly modified method, the marker particles (carbon/chalk) are injected inside a particular region of an embryo and then their fate is followed in a subsequent dissections of the developing embryo. **Ballard** (1981) has adopted this techinque in tracking the movement of cells within teleost blastoderm during gastrulation.

(c) Radioactive labelling method. In this method of fate map construction, the cells of an embryo are first labelled by providing the embryo with a suitable radioactive metabolite. A part of this embryo (**donor**) is then excised and aptly grafted on to a normal unlabelled embryo (**host**). The fate of the labelled primordium is easily found out by sectioning the host embryos at different time intervals and by doing autoradiography. For example, **Rosenquist** labelled epiblast cells of chick with radioactive thymidine, **tritiated thymidine**, and grafted them to unlabelled blastoderms to prove that some endoderm as well as mesoderm originate from the epiblast, invaginating at the primitive streak.

The nucleoside, tritiated thymidine, is permanently incorporated into DNA in the nucleus prior to each cell division and is not harmful at moderate dose levels. It provides high-resolution radioautographs as a result of low-energy beta emanations from tritium. Thus, tritiated thymidine can give cell-to-cell labelling wherever DNA synthesis occurs, regardless of tissue, stage or organism. Obviously, this is an ideal method of tagging cells. Its advantage over the natural marker, such as, the staining properties of the nuclei of Japanese quail cells, is that it is not limited to a particular organism. It labels any dividing cell in any species and the descendants of that cell. Its main disadvantages are that it works only for dividing cells and if these cells are dividing rapidly, it may soon be diluted to undetectable levels. Tritiated thymidine has provided conclusive results in many studies, however, both by the techinque of pulse chase and where labelled cells are grafted to an unlabelled host (**Trinquas**, 1984).

12-1.3. Histochemical and Cytological Methods

A variety of natural markers (*i.e.*, specific molecules of cells) have been used in the construction of fate maps with success : **melanin** granules (*e.g.*, chick's retinal pigment cell), **glycogen** (which can be stained with periodic acid-Schiff reagent or PAS) and **alkaline phosphatase**. However, all such markers reflect the metabolic activity of the cell which might change during the course of the observations and in any case are limited to particular kinds of cells. The best natural markers would, of course, not be limited in these ways. Two such excellent markers have been obtained for birds and frogs ; and have been used in transplantation techniques in preparing their fate maps.

The cytological method takes advantage of the fact that in certain cases the cells from two closely related species such as the chick and the quail, can be distinguished from each other under the light microscope (For example, only quail cells have a mass of nucleolar-associated heterochromatin) and the tissue from embryo of one species harmoniously cooperates with that of the other during development. Thus, it becomes possible to ascertain the fate of a xenoplastic graft (*i.e.*, transplanted or donor chick cells in a quail host or vice-versa) and to construct the fate map of a primordium. Similarly, in case of frog, *Xenopus laevis* there are two races which differ in their nucleolar structure. In *Xenopus laevis laevis* there is one nucleolus per nucleus whereas in *Xenopus laevis victorianus* there are two nucleoli per nucleus. Thus, graft from one race in another can be easily identified and followed.

12-1.4. Genetical Method

This method utilizes genetic mosaics (gyandromorphs) and was used by **Hotta** and **Benzer** in 1972 to construct fate maps for the early embryo of *Drosophila melanogaster*. In this method, advantage is taken of the fact that the ring-X chromosome (X_R) is unstable in the mutant strain (*i.e.*, In (1) w^{vc}) of *D. melanogaster* and is lost in 35 per cent of the female embryos (X_RX) during the first division of the zygote nucleus. One cleavage nucleus, therefore, becomes XO and the cells containing mitotic products of this nucleus will determine male phenotype and will express the recessive genes located on the X chromosome, the wild types of alleles were located on ring-X chromosome (X_R). The resultant gyandromorph flies have some genetically male cells (XO) and other genetically female cells (X_RX).

In such a case, it is known that the orientation of the mitotic spindle at first nuclear division is at random and there is almost no mixing between the clones of nuclei derived from the two cleavage nuclei, at the time they populate the blastoderm (*i.e.*, nuclei migrate to the periplasm). One half of the blastoderm is, thus, populated by the descendants of one cleavage nucleus (X_RX) and the other half from that of the other (XO) ; the dividing line corresponds to the orientation of the first spindle (Fig. 12.3). The probability of a randomly oriented dividing line in the mosaic passing between any two blastodermal sites is related to distance between the two sites. It is, thus, possible to measure the distance between any two presumptive sites by counting the percentage of mosaic flies in which two structures are of different phenotypes. This distance is measured in **strut units** (in memory of **Sturtevant**) : one strut represents "a probability of 1 per cent that among the entire set of adult mosaics (formed by this method), the two structures in question will be of different type." Such a mapping system is analogous to that of the genetic mapping. By considering three or more structures at the same time it becomes possible to form a two dimensional fate map of the blastoderm surface.

12-2. EXAMPLES OF FATE MAPS

The fate maps of various vertebrates reveal that there exists a similar pattern. The following general description of different types of vertebrate blastulae emphasizes the similarity of this basic plan :

12-2.1. Fate Map of Protochordate Blastula

In the blastula of *Amphioxus* (Fig. 12.4 D), the blastoderm forms a single layer around the blastocoele. The cells from the vegetal pole of the egg will

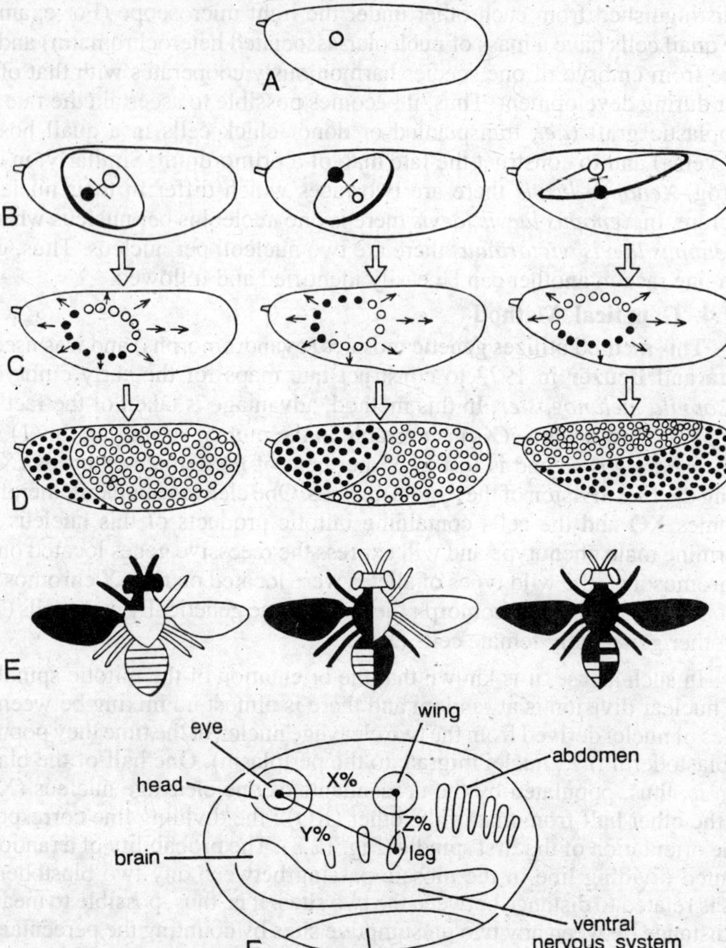

Fig. 12.3. Fate map of *Drosophila melanogaster*. A—Typical mosaic gynandromorphs of *D. melanogaster*. Body parts of each fly derived from female nucleus are shaded, while body parts derived from male nucleus are unshaded. The plane of spindle at first cleavage determines that which body part will be male or female; B—Mechanism of formation of a mosaic fly. 1. The zygote nucleus (female X_RX) divides into two. One daughter nucleus (open circle) loses its one X (*i.e.,* X_R) to become XO male; 2. Clones of nuclei derived from two cleavage nuclei; there is no mixing up; 3. Section of blastoderm; and 4. Surface view of blastoderm. Nuclei have migrated to egg cortex or periplasm to populate periplasm. Distribution of XX and XO nuclei corresponds to plane of first mitotic spindle; C—Fate map of blastoderm for external body parts of adult; distances between body parts projected over the blastoderm; D—Pictorial sketch of external parts of an adult fly on fate map of right half of blastoderm (after Goel, 1984).

become the floor of the blastocoele or **hypoblast**, and are the future **endoderm** of the embryo. Those cells from the region of the animal pole will form the roof over the hypoblast, the **epiblast** and will give rise to the **ectoderm** (*i.e.*, **epidermal** and **neural** cells) and **chordamesoderm** (*i.e.*, **notochordal** and **mesodermal** cells).

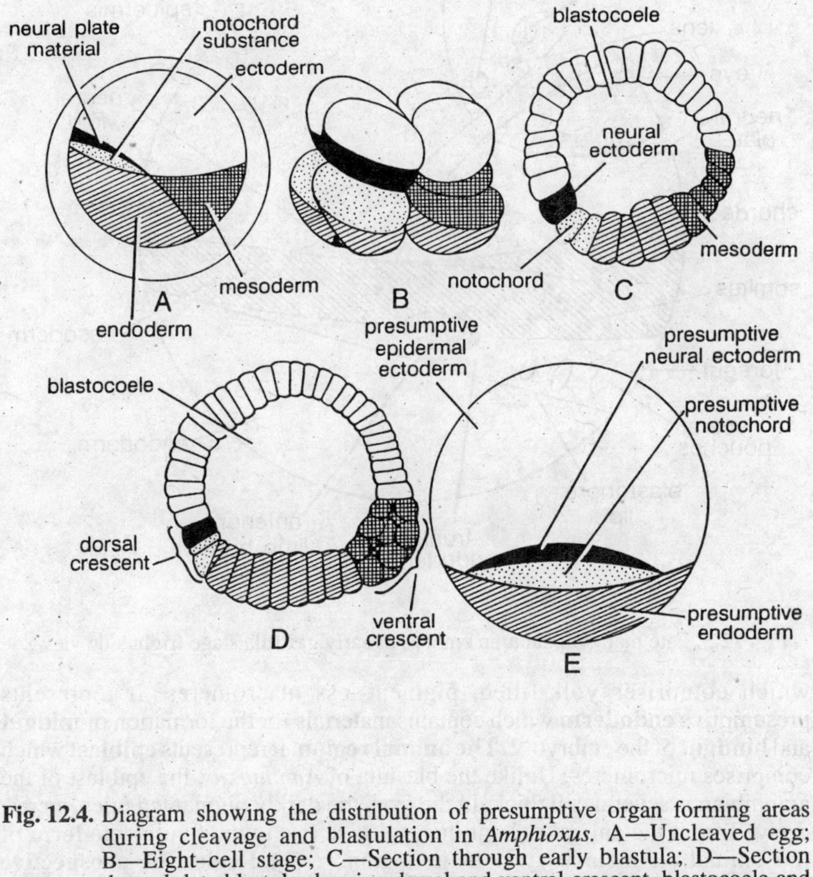

Fig. 12.4. Diagram showing the distribution of presumptive organ forming areas during cleavage and blastulation in *Amphioxus.* A—Uncleaved egg; B—Eight–cell stage; C—Section through early blastula; D—Section through late blastula showing dorsal and ventral crescent, blastocoele and elongation; E—Posterior view of dorsal crescent (after Phillips, 1974).

Most of the cells of the composite epiblast will give rise to epidermal ectoderm. Some small epiblast cells with special staining affinities located at the margins of the prospective endoderm layer, form a ring around the blastula (Fig. 12.4 D), shows that this ring of cells consists of two general areas, the **dorsal crescent** and **ventral crescent.** The cells of the dorsal crescent give rise to the presumptive notochord and neural plate cells (Fig. 12.4 E). The ventral crescent contains the cells that make up the presumptive mesoderm.

12-2.2. Fate Map of Amphibian Blastula

The blastula of the amphibian embryo is also a round structure that can be roughly divided into three regions (Fig. 12.5). They are similar but not identical to those of *Amphioxus*. **1. The vegetal region.** It includes hypoblast

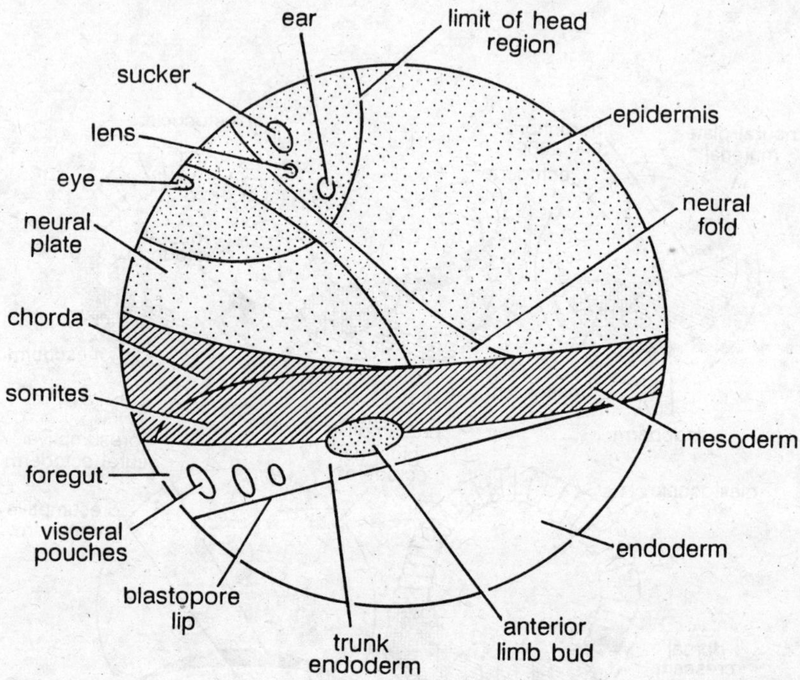

Fig. 12.5. Fate map of an anuran embryo at early gastrula stage (right side view).

which comprises yolk-filled, pigment-less macromeres. It represents presumptive **endoderm** which contains materials for the formation of **midgut** and **hindgut** of the embryo. **2. The animal region.** It represents **epiblast** which comprises micromeres. Unlike the blastula of *Amphioxus*, the epiblast of the amphibian is several cell thick. In the frog, the darkly pigmented micromeres, derived from the animal pole of the egg, give rise to the future **ectoderm** of the animal. It includes two main regions : (1) **Region of prospective epidermal ectoderm** which develops into the epidermis of the skin of the embryo. (2) **Region of prospective central nervous system** which contains material for brain, spinal cord and sub-areas for the sense organs such as eyes, ears and nose. **3. The intermediate or marginal region.** The third area is the grey-crescent area which in blastula represents a ring of cells around the equator. Most of these cells are presumptive **mesodermal** cells. Unlike the blastula of *Amphioxus* in which the presumptive mesoderm cells occupy approximately half of this marginal zone (ventral crescent), the presumptive mesodermal cells of the amphibians almost encircle the blastula. It contains the following sub-regions : **(1) Notochordal region** which is present on the dorsal side of blastula and gives rise to notochord. **(2) Area for anterior part of alimentary canal** that is situated below the notochordal area and gives rise to the endodermal lining of the mouth, gill region and pharynx. **(3) Region of somites or segmental muscles** that is present on both sides of the notochordal area and develops into trunk muscles or somites ; and **(4) Ventro-lateral mesoderm area** which lies on lateral and ventral parts of marginal zone and gives rise to the mesodermal lining of the body cavity, the kidney and reproductive organs.

12-2.3. Fate Map of Avian Blastula

The area pellucida, consisting of the central cells freed from yolk, contributes to the formation of the embryo. It is the part of the chick blastoderm that is included in the fate map (Fig. 12.6.). The area pellucida consists of two layers of cells — those cells exposed to the surface, the **epiblast**, and those cells located below the cleft, the **hypoblast**. Instead of having a rounded appearance, the bilaminar blastula of birds may be thought of as being flattened against the hypoblast, changing the blastocoele into slit-like cavity. Such a flattened blastula or **blastodisc** rests on a bed of undivided yolk.

Fate maps of birds too revealed the fact that the cells of the epiblast are organised around the notochord (Fig. 12.6.). The cells in the anterior part of the area pellucida form the **ectoderm** (*i.e.*, presumptive **epidermal ectoderm** and presumptive **neural plate ectoderm**) and embryonic membranes. The posterior half of the area pellucida gives rise to **mesoderm** of the body proper and to extra-embryonic mesoderm of the membranes. In the middle of the area pellucida the presumptive **neural plate** cells take the form of a transverse crescent-shaped mass. Posterior to the presumptive neural plate cells, near the middle of the area pellucida, lies the presumptive **notochord** material. The cells of the hypoblast area destined to form the **endoderm**. The gut incorporates some of these cells into its wall, but most of them will form the lining of the yolk sac.

12-2.4. Fate Map of Mammalian Blastula

It is impossible to mark or stain the cells of the mammalian embryo and follow their path through the sequence of developmental phases. The rat, mouse, rabbit, pig and monkey are the eutherian mammals most studied and much of our knowledge is based on these forms.

Despite the fact that the mammalian egg undergoes holoblastic cleavage, the cells composing the inner cell mass spread out and form a plate or disc resembling the blastodisc of the bird and reptiles. The mammalian blastodisc too becomes divided into epiblast and hypoblast as in birds and reptiles, but there is no consensus as to how this occurs. Regardless of its origin, the **hypoblast** will give rise to the **endoderm** as in other animals.

The cells that remain in the inner cell mass are considered to be the **epiblast** and will give rise to the **epithelium**, **neural ectoderm**, **notochord**, and **mesodermal** components of the embryo. Thus, the mammalian fate map resembles that of the reptile and bird.

12-3. USES OF FATE MAPS

The fate maps have their great uses in clearly following the mechanism of gastrulation. In fact, mapping techniques are helpful in following the movements and ultimate position of local tissues, but they tell us nothing about tissue developmental potentialities during morphogenesis (**Grant**, 1978).

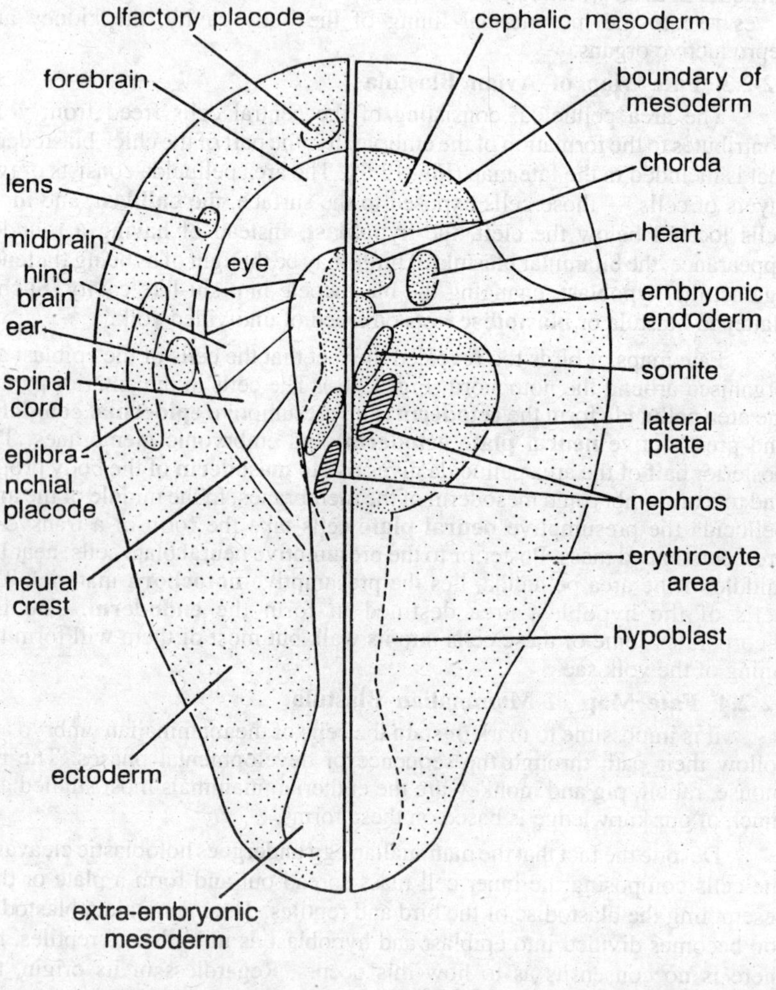

Fig. 12.6. Fate map of chick embryo at definitive primitive streak stage (stage 4).

12-4. CELL LINEAGE STUDY (CYTOGENY)

The analysis or study of the fate of each blastomere after first and second cleavage, is called **cell lineage study** or **cytogeny**. A cell lineage study is easier in spirally cleaving mosaic eggs of flatworms, annelids and molluscs than in radially cleaving regulative eggs of other animals. This is so, because, spirally cleaving eggs exhibit determinate cleavage ; that is, differentiation of each cleavage product (or blastomere) and its progeny is dependent on its prior division history.

12-4.1. Method of Study of Cell Lineage

In cell lineage study, to trace the fate of particular blastomeres, or in other words, to find out which part of the embryo or larva is formed by which cell of the blastula, it becomes necessary to identify the early blastomeres and their progeny by giving them certain names. Such names are given by the letters of the English alphabet and by numbers put as exponentials above the letters (Fig. 12.7).

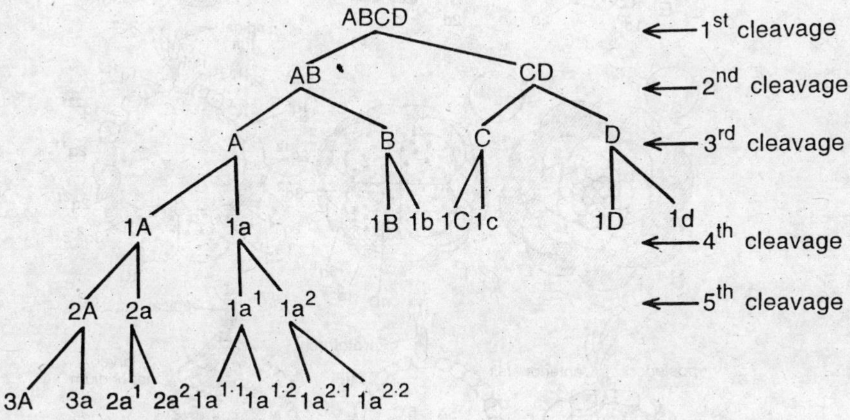

Similarly the progeny of B will be :
3B, 3b, $2b^1$, $2b^2$, $1b^{1\cdot 1}$, $1b^{1\cdot 2}$, $1b^{2\cdot 1}$, $1b^{2\cdot 2}$
The progeny of C will be :
3C, 3c, $2c^1$, $2c^2$, $1c^{1\cdot 1}$ $1c^{1\cdot 2}$, $1c^{2\cdot 1}$, $1c^{2\cdot 2}$
The progeny of D will be :
3D, 3d, $2d^1$, $2d^2$, $1d^{1\cdot 1}$, $1d^{1\cdot 2}$, $1d^{2\cdot 1}$, $1d^{2\cdot 2}$

Fig. 12.7. Method of nomenclature of blastomeres in cell lineage studies (after Majumdar, 1985).

12-4.2. Examples of Cell Lineage

1. An example of cell lineage study is given for the annelid, *Nereis*, a classic work done by **E.B. Wilson** in 1892. In *Nereis*, first four cleavages are synchronous. The **first cleavage** is unequal, giving rise to a small AB blastomere and a large CD. Sixty minutes later, the **second cleavage** furrow produces two equal-sized A, B blastomeres, while the larger CD cell divides

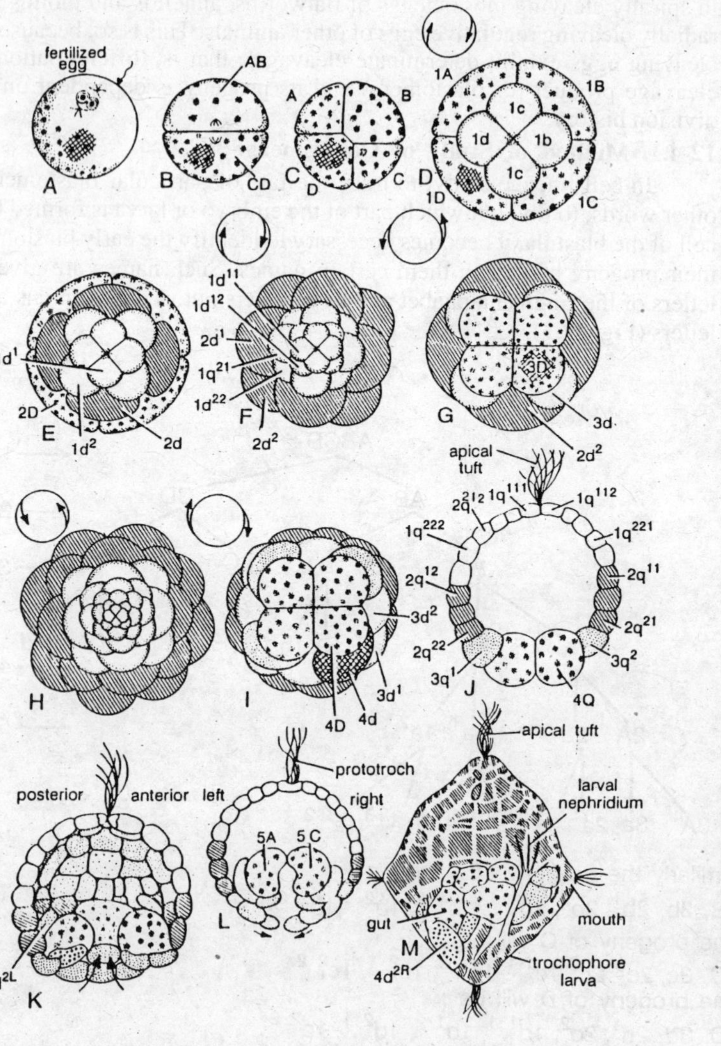

Fig. 12.8. Cell lineage pattern and spiral cleavage stages in *Nereis*. A—Fertilized egg with arrow indicating point of sperm entry; B—2-cell stage with a large CD and a small AB blastomeres; C—4-cell stage; D—8-cell stage (viewed from the animal pole); E—16-cell stage from the animal pole (cell lineage of only D cell has been labelled); F—32-cell stage viewed from the animal pole; G—32-cell stage viewed from the vegetal pole; H—64-cell stage from the animal pole; J—64-cell stage from the vegetal pole; K–Vertical section of a 64-cell stage (from left to right); L—Vertical half (anterior to posterior) of a later stage showing invagination of the micromeres to form gut, *i.e.*, section from B–D quadrant; M—Vertical section of a later stage (taken at right angles to the plane of section in L) showing how the floor of gut is formed, *i.e.*, from A quadrant to C quadrant; N—Trochophore larva with mouth at the site of the front of the 'blastopore' and anus perforated at the opposite end (after Cohen, 1967).

into a smaller C and a larger D cell. Cleavage is spiral, alternating dexiotropic to leiotropic with each successive division. At the **third cleavage**, the first quartet is designated 1a–1d, marking the micromere derivative of each A-D macromere. With each succeeding macromere division, the second through fourth micromere quartets are formed, identified as 2a–2d through 4a–4d, respectively. It should be noted that the third cleavage division, in producing the first micromere quartet, segregates yolk and oil droplets into the macromeres, which remain there until the end of cleavage.

At the **fifth cleavage**, the divisions become asynchronous and the 32-cell stage is reached in steps as cell groups divide at different times. The first micromere quartet at the animal pole gives rise to the most apical structures in the larva. It divides into four **central cells** at the animal pole (designated $1a^1$–$1d^1$) and four **trochoblast** cells, $1a^2$–$1d^2$. The latter cells develop into a band of powerful cilia, encircling the larva, known as the **prototroch**. This band develops after only two more divisions, when 12 of the 16 cells produced immediately differentiate cilia and never divide again. The apical tuft of the larva (Fig. 12.8), made up of large cilia extending from the former animal pole, is derived from the four central cells $1a^1$–$1d^1$. These divide unequally into four small cells at the animal pole, designated $1a^{11}$–$1d^{11}$ and four larger cells $1a^{12}$–$1d^{12}$. The former (*i.e.*, $1a^{11}$–$1d^{11}$) immediately differentiate into apical tuft cells with long cilia, while the latter contribute to larval epidermis. In this way, each cell can be followed into the larval organs such as gut, eye spots, ciliary bands and coelomic sacs.

2. When cleavage takes place in the fertilized mosaic egg of *Dentalium* (a marine mollusc), just before the first cleavage, a small part of the cytoplasm bulges out from the vegetal hemisphere of the egg as a **polar lobe**. In the tune of cell lineage studies, the fertilized egg of *Dentalium* can be named as A B C D. The first cleavage furrow forms two blastomeres A B and C D. At this stage the polar lobe hangs from lower side of C D blastomere. However, this two-cell stage embryo has the appearance of three cells, so this stage is called **trefoil** (**Berrill** and **Karp**, 1978).

Before the second cleavage starts, the entire contents of the polar lobe goes into CD blastomere. When the second cleavage starts, the polar lobe again appears. When the second cleavage is completed, four blastomeres are formed, two A and B from AB cell and C and D from CD cell. At the four-cell stage, the polar lobe remains sticking to the D blastomere. If at this stage, the polar lobe is removed, the resulting embryo does not possess any mesoderm. Such an experiment (called **defect experiment**) tends to show that there is specific substance in the polar lobe which is responsible for the formation of the mesoderm in the embryo.

Clement (1962) has obtained similar results during his cell lineage study of the snail *Ilyanassa*.

12-4.3. Outcomes of Cell Lineage Study

The cleavage patterns and lineages in three major phyla—flatworms, annelids and molluscs, are similar. Both annelids and molluscs develop a **trochophore larva** with a common body plan, a sign of close phylogenetic relationship. In each, larval **ectoderm** comes from the first three micromere quartets, the remaining quartets contributing to **mesodermal** derivatives, while the **endoderm** comes from the macromeres. The ciliated **trochophore bands** in these larvae come from the $1a^2$-$1d^2$ cells just described. Two micromere derivatives, the 2d cell of the second quartet and 4d cell of fourth quartet (both derivative of the D macromere) are called **first** and **second somatoblasts**, respectively. Both of these cells assume a special role in the development of larvae structures in both phyla. Both give rise to such mesodermal derivatives such as **heart**, **muscle bands** and **ventral nerve cord**, as well as other main tissues. The development of each phylum diverges from this common plan later on, as unique structural adaptations appear, such as **shell glands** in mollusc larvae and segmented or **metameric** body plan in annelids.

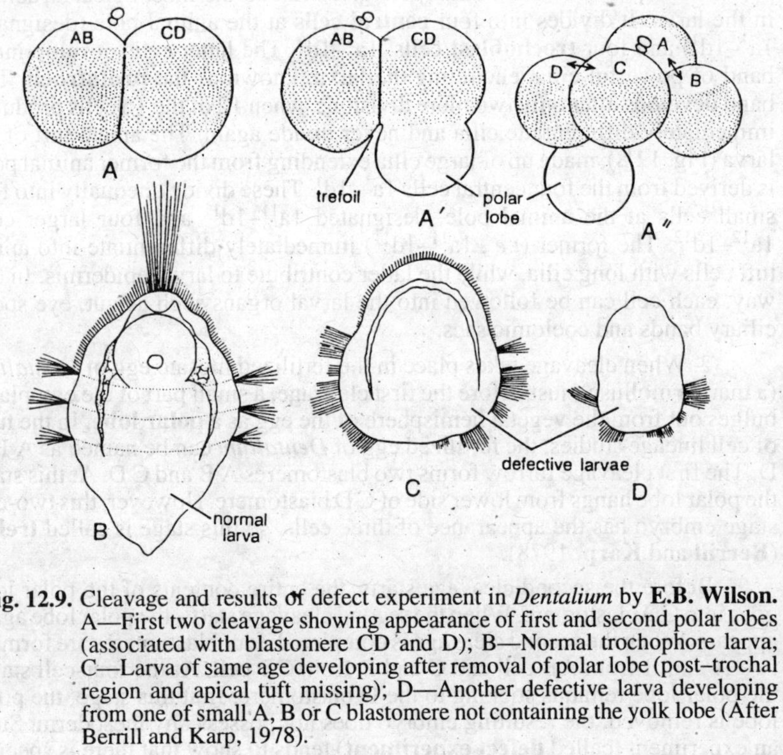

Fig. 12.9. Cleavage and results of defect experiment in *Dentalium* by **E.B. Wilson.** A—First two cleavage showing appearance of first and second polar lobes (associated with blastomere CD and D); B—Normal trochophore larva; C—Larva of same age developing after removal of polar lobe (post–trochal region and apical tuft missing); D—Another defective larva developing from one of small A, B or C blastomere not containing the yolk lobe (After Berrill and Karp, 1978).

Cell lineage studies also reveal that each blastomere carries out a prescribed number of divisions before it differentiates into a larval structure. Both its position and division programme seem to be set at the beginning of development, at the first two cleavages. If any of the four macromeres is separated after second cleavage, each divides exactly as it would have in the intact embryo, both in terms of number of divisions and position of progeny.

The cleavage partitions the egg organization according to a predetermined plan and special cytoplasms and morphogenetic substances are precociously separated into different cell lines (*e.g.*, polar plasm in *Dentalium*). More than in radially or bilaterally cleaving eggs, minor perturbations of spirally cleaving eggs upset the developmental fate of blastomeres, and abnormal embryo result. The spiral pattern is a rigid cleavage programme in which blastomere fate may be set as early as the first division. For this reason it has been called **mosaic** ; the progeny of each blastomere contributes one small piece of the "jigsaw puzzle" of the larval body plan. Loss of an early blastomere or a shift in its position produces abnormal larvae.

REVISION QUESTIONS

1. What is the fate map ? Describe different methods of fate map construction.
2. Give an illustrated account of fate map of different chordates.
3. Write short notes on the following : 1. Fate map ; 2. Vital staining ;
4. Write an essay on 'cell lineage'.
5. What is the significance of fate maps and cell lineage study . Discuss.

13
GASTRULATION

Gastrulatioin is the embryonic process that brings about the reorganization and displacement of organ-forming areas of a blastula. It converts a simple one-layered (**monodermic**) blastula into a two-layered (**didermic**, *e.g.*, *Amphioxus*) or often a three-layered (**tridermic**, *e.g.*, all vertebrata) embryonic stage, called **gastrula** (Gr., *gaster* = stomach or gut). The single layer of blastula is known as **blastoderm**, **ectoblast** or **protoderm**. The resultant three layers of gastrula—the outer **ectoderm**, the middle **mesoderm** and inner **endoderm**—are commonly known as **germ layers**. During gastrulation, the cavity of the blastula (blastocoele) is obliterated and a new cavity, called **archenteron** (a precursor of enteron or gut) or **gastrocoel** (cavity of gastrula) enclosed by endoderm is formed. It is also during the gastrulation that the embryo acquires its clear-cut antero-posterior polarity and bilateral symmetry.

At gross level, though the process of gastrulation varies greatly in different groups, but in most cases basically it involves the same cellular mechanisms : large-scale movements, called **morphogenetic movements** (*e.g.*, invagination, involution, ingression, epiboly, etc.) and sorting out of the cells and cell groups, usually accompanied by changes in the shapes of migratory cells. Thus, important and irreversible changes occur during gastrulation.

13-1. DEFINITION OF GASTRULATION

A fundamental difficulty in the study of gastrulation is defining just what we mean by it and when it begins and ends; especially since it differs in various important ways throughout the animal kingdom. For example, in amphibians there is invagination accompanied by involution in the formation of archenteron. In birds (and probably also in mammals) there is ingression accompanied by involution and there is no archenteron. In teleost fishes, there is neither invagination nor ingression, nor formation of an archenteron. And in echinoderms, there is invagination to form an archenteron, but apparently no involution. Visualizing these differences in the process of gastrulation among animals, quite recently **Ballard** (1976, 1981) has pointed out that the term gastrulation has become quite antique and inadequate in incorporating all types of gastrulations of animals. According to him, therefore, this term should be abandoned and be replaced by some more adequate and new term such as ''**phase II**''. But, **Trinquas** (1984) opined that we should stick to the 'Global usage' of the term gastrulation. He has redefined the term gastrulation as *an embryonic process that ensembles of processes during post-cleavage early development which when complete has brought the cells that will form the various organs to the places where those organs are to be formed*. In a word, gastrulation is the **embryo's way of laying down its body plan** (see **Trinquas**, 1984).

13-2. MAJOR EVENTS OF GASTRULATION

The process of gastrulation involves the following main events :

1. A rearrangement of blastular cells or blastomeres by means of **formative** or **morphogenetic movements** occurs.

2. The rhythm of cellular division (cellulation or cleavage) is slowed down or almost is inhibited.

3. Growth, if any, is insignificant.

4. The type of metabolism changes and the rate of oxidation is intensified.

5. The nuclei become more active in controlling the activities of the embryonic cells. The influence of the paternal chromosomes (genes) becomes evident during gastrulation.

6. During gastrulation **chemodifferentiation** is started by the synthesis of new and different kinds of protein molecules.

Thus, egg cleaves rapidly into a blastula, an epithelial sheet surrounding a central cavity. This sheet then transforms into three primary germ layers of the embryo in the bilateria (flatworms to vertebrates) and two germ layers in the radiata (sponges and coelenterata). New sheets (*i.e.*, germ layers) arise from blastula epithelium by a programme of tissue displacements, the gastrulation. The blastula epithelium expands or invaginates or both, so that organ-forming areas that were outside are now inside. The outermost sheet, the ectoderm, surrounds a mesodermal sheet which covers an innermost endoderm. By the end of gastrulation, a single blastular sheet has become three concentric cylinders making up the body plan of the organism. Each germ layer derivative, occupying a different position in the embryo, is committed to develop into a different set of organ systems.

A **cell sheet** may be defined as a monolayer of cells displaying integrated behaviour. Though it may consist of more than one cell layer, the several layers behave as one. Sheet cohesiveness depends on cell communication mediated by **gap** and **tight junctions**, extra-cellular matrices and **desmosome** connections. All these develop during cleavage ; as a result sheet behaviour arises **epigenetically** as cell motility is translated into an expanding or invaginating epithelium. Most movements take place on substrata, such as a basement membrane or the **extra-cellular matrix (ECM)** between sheets. Some substrata may favour directed migration. Guided by surface features of the ground mat, the leading edge of a sheet expands in directions of preferred adhesive contacts, usually outward into unoccupied areas, covering all available surfaces. Marginal edge cells display vigorous activity at their free surfaces ; they exhibit ruffling movements, while interior cells are dragged along passively. When leading edges of two epithelial sheets come into contact, they show a form of **contact inhibition**. Ruffled membrane activity at the marginal zone ceases and sheets fuse into a continuous monolayer.

Further, some cell sheets tend to spread more readily than others. If two sheets are combined, the one with greater spreading tendencies envelops the other, which may explain the inside-outside relationships of tissues within the embryo. Ectoderm has the greatest spreading tendencies ; it always envelops all others and occupies an outside position. In their turn endoderm and mesoderm sheets tend to invaginate or spread inside into cavities beneath the ectoderm (Fig. 13.1).

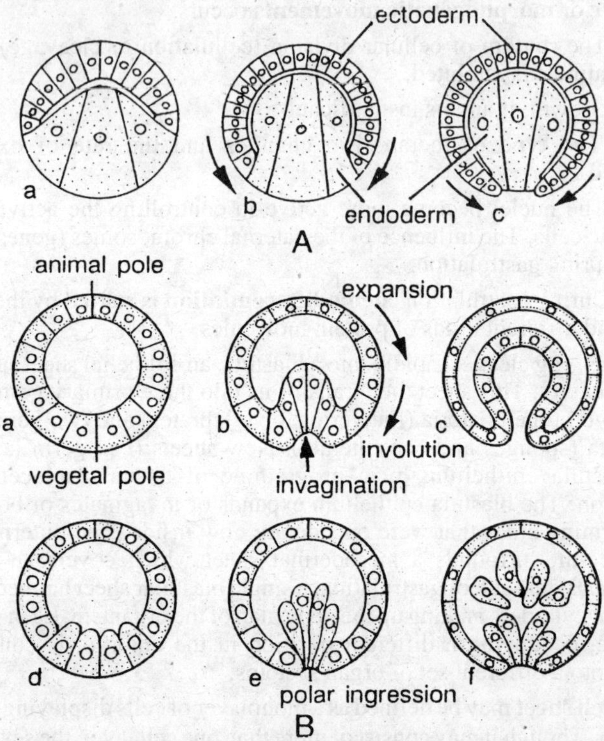

Fig. 13.1. Types of gastrular morphogenetic movements. A—Gastrulation by epibolic expansion of ectoderm over large endodermal cells, forming a blastopore; B—More complex embolic movements during gastrulation. Heavy line indicates the original cortex of the egg. In (a) through (c) the invaginating cells retain contact with the original cortex and pull it into the embryo as the blastopore forms. This is brought about by a combination of epibolic expansion, involution, and invagination. In (d) through (f) during ingression attachment to the cortex is broken (after, Grant, 1978).

13-3. GASTRULAR MORPHOGENETIC MOVEMENTS

Gastrulation is a nicely integrated, dynamic process which is controlled largely by intrinsic forces bound up in the specific physico-chemical conditions of the various presumptive organ forming areas of the late blastula and early gastrula. These internal forces in turn are correlated with external conditions. In most vertebrates, it includes different kinds of morphogenetic movements.

The movement of the blastomeres from one place of the embryo to another to establish a particular form or structure is a common embryological procedure. This type of cell movement is described as **formative** or

morphogenetic movement, because, it results in the generation of a particular form or structural arrangement (**Vogt,** 1925, **Trinkaus,** 1969 and **Berrill**, 1971). Such morphogenetic movements occur in the embryo during blastulation, gastrulation, tubulation and organogenesis. They do not occur only during the embryonic stages, but also in an adult animal. But, there is a fundamental difference in the two, that the morphogenetic movements of an adult animal are of a **reversible nature**, while, the morphogenetic movements of gastrula, etc., are of **irreversible nature**, *i.e.*, each part remains in the position into which it has been brought by the preceding movements.

The following two basic types of morphogenetic movement patterns are involved in gastrulation :

1. Epiboly ; 2. Emboly.

13-3.1. Epiboly

The word epiboly is derived from the Greek, meaning a "*throwing on*" or "*extending upon*". It represents the natural tendency of epithelial sheet to expand into free areas over a substratum. Epibolic sheets are usually outside, or **ectodermal**, and tend to envelope and surround inner sheets (Fig. 13.1). Sometimes sheets expand toward a single point on the surface. This type of epibolic movement is called **convergence** (see **Grant**, 1978).

Epiauxesis. In Ctenophora, in various spirally cleaving eggs (*e.g.*, Annelida and Mollusca), and in some vertebrata (Pisces), a modified type of epibolic movement called **epiauxesis** occurs (**Lovtrup**, 1974). Four conditions must be fulfilled for epiauxesis to occur, namely, absence of a hyaline membrane, presence of a surface to which the blastomeres adhere and on which they may spread, a certain intercellular adhesion and that they are mitotically active. In Annelida and Mollusca epiauxesis involves overgrowth of ectodermal (micromeres) over yolk-laden vegetal pole cells (macromeres). Overgrowth of micromeres not only depends upon the subdivision of the existing micromeres, but also upon the addition of new ones budding off from macromeres.

13-3.2. Emboly

This is just opposite to epiboly. In emboly, sheets fold in or invaginate; that is, a sheet migrates inside a cell aggregrate or beneath another layer because of local changes in cell shape or motility. Embolic movements include **invagination**, **involution** and **ingression**, movements displacing future or prospective endoderm and mesoderm into the interior of the embryo.

1. Invagination. It is the sinking in (inpocketing) of a layer of cells and is similar to punching in of an expanding layer at a specific point. The reverse of invagination is **evagination**

2. Involution. It is turning in of an expanding layer at a specific point. In involution, a surface layer of cells turns inside itself and continues to expand along the inner surface of the superficial layer. Thus, in involution, cells which are at some distance from the point of invagination, move to the margin of the site of invagination (*e.g.*, at a lip of blastopore during amphibian gastrulation), flow over it and move inside.

3. Ingression or polyinvagination. In this type of emboly, small groups of independent cells separate from a primary epithelial layer and migrate into the blastocoele cavity or other embryonic spaces as a new layer. For example, in primitive streak of amniote embryos, cells may sink inside as individuals, changing their relations to each other in the process.

Delamination is a type of ingression that forms an inside sheet by mass separation of a layer of cells from an epithelium. Sometimes, this process is accomplished by division of cells in a plane parallel to a tissue space.

13-3.3. Concrescence

The movement of masses of cells toward each other and their fusion into one cell mass is called **concrescence**. It occurs during the development of the feathers of birds.

13-3.4. Cell Proliferation

The gastrulation of *Amphioxus* and frog, somehow depends on increase in cell number due to mitosis.

13-3.5. Divergence

It is opposite of convergence. For example, after cells have involuted over the blastoporal lips during gastrulation, they migrate and diverge to their future positions within the developing embryo. Such divergence occurs in lateral plate and ventral mesoderm in frog ; and in lateral plate and extra-embryonic mesoderm in reptiles, birds and mammals.

13-3.6. Extension

During extension entire gastrula begins to elongate in the antero-posterior axis. The elongation of the presumptive neural and epidermal areas externally and of the notochordal, mesodermal and endodermal materials after they have moved inward beneath the neural plate and epidermal material are the examples of extension. The extension of cellular masses is a prominent factor in gastrulation in all Chordata from Tunicata to Mammalia.

13-4. EXAMPLES OF GASTRULAR MORPHOGENETIC MOVEMENTS

Gastrulation usually includes combination of embolic and epibolic movements as have been shown by the following examples from the Animal Kingdom :

13-4.1. Gastrulation by Epiboly in Mollusca

The spirally cleaving eggs of gastropod snail *Crepidula* and other molluscs produce the **solid** or **stereoblastula** which lack a segmentation cavity or blastocoele. In them **epiboly** is the main method of gastrulation. Upper micromere layers of a stereoblastula multiply and migrate as an enveloping cell sheet over the larger macromeres. Sometimes, this epibolic movement coincides with **ingression** (emboly) of larger macromeres toward the centre. Both of these types of gastrular movements ultimately result in a **stereogastrula** with two layers of cells : an inner cell mass, the future meso-endoderm and an outer sheet of cells developing into all ectodermal structures (Fig. 13.1).

The epibolic expansion of a micromere sheet over the macromeres below indicates some change in motility or adhesive relationships or both. Ectoderm glides over macromere surfaces until it fuses at the vegetal pole, closing up the free surface available for furthere expansion. Before fusion, however, the expanding ectoderm forms a circular outline at the vegetal pole. This is called **blastopore**, through which macromeres are seen. It develops into future mouth of larva. Meanwhile, multiplying endodermal cells create a cavity, the **archenteron** by a combination of embolic movements and displacements.

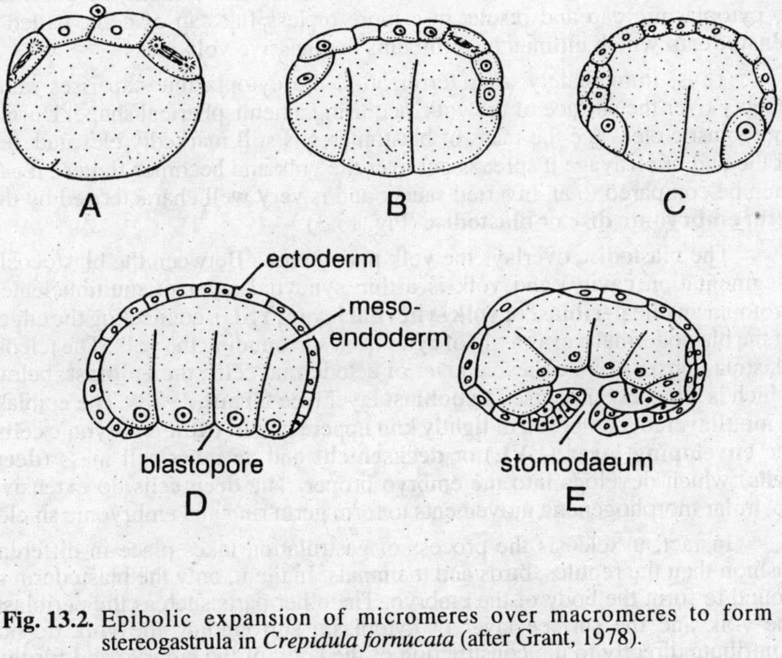

Fig. 13.2. Epibolic expansion of micromeres over macromeres to form a stereogastrula in *Crepidula fornucata* (after Grant, 1978).

13-4.2. Gastrulation by Epiboly in Teleost Fishes

Teleost eggs have attracted the attention of embryologists for the analysis of gastrulation because of their transparency and remarkable epiboly. Research on mechanism of epiboly in teleost (bony-fishes) eggs has been pursued mainly in three laboratories throughout the world :

1. Warren L. Lewis is the pioneer who worked mainly on the eggs of the zebra fish, *Brachydanio rerio* in Philadelphia.

2. Charles Devillers worked on the eggs of trout in Paris. He confirmed and extended Lewis's idea regarding the role of surface layer of yolk (now is called **yolk cytoplasmic layer** or **YCL**) in epiboly.

3. John Philip Trinkaus worked on the eggs of *Fundulus* in the laboratories of New Haven and Woods Hole. He has done a remarkable job in exploring all possible mechanisms involved in the epiboly of teleosts.

In an unfertilized teleost egg, the surface is formed by yolk-free cytoplasm containing numerous cortical granules. After fertilization, the cytoplasm moves toward the animal pole of the egg and there it forms a cytoplasmic cap. Active cytoplasm, in fact, is withdrawn both from the surface and from the interior of the egg. As a result, yolk platelets are pressed closer together. So, a fertilized egg of the bony fish consists of a cytoplasmic cap on the animal pole containing the nucleus or nuclei, the central mass of yolk and a thin layer of cytoplasm surrounding the yolk.

In fact, in the eggs of teleost fishes (*e.g.*, *Amia*, *Lepidosteus*, *Fundulus*, etc.), reptiles and birds, the amount of yolk is so enormous that it does not participate in cleavage. Thus, in them the **meroblastic cleavage** is confined to cytoplasmic cap and results in a more or less flat cap of cells, called a **blastoderm** which ultimately surmounts the massive yolk mass.

Thus, immediately after fertilization the cytoplasmic cap rises quite steeply over the surface of the yolk, acquiring a hemispherical shape. During meroblastic cleavage the mass of blastomeres is still markedly elevated, but at the end of cleavage it spreads out over the yolk and becomes flatter ; it can then be compared to an inverted saucer and is very well characterised by the term **embryonic disc** or **blastodisc** (Fig. 13.3).

The blastodisc overlays the yolk mass below. Between the blastocoele (segmentation cavity) and yolk is a thin syncytial layer of multinucleated protoplasm, the **periblast** or **yolk syncytial layer** (**YSL**), connecting the edges of the blastodermal cap as a rim of cytoplasm surrounding the yolk. The teleost blastula consists of a cohesive layer of ectodermal cells, the **epiblast**, below which is a loosely organised **hypoblast** layer (see **Grant**, 1978). The epiblast is multilayered, composed of tightly knit upper layer of extra-embryonic cells, the **enveloping layer** (**EVL**) or **deckschicht** and an inner cell mass (**deep cells**) which develops into the embryo proper. The deep cells do extensive gastrular morphogenetic movements to form germ ring and embryonic shield.

In fact, in teleosts the process of gastrulation takes place in different fashion than the reptiles, birds and mammals. In them, only the blastoderm is bound to form the body of the embryo. The other parts such as the periblast, the yolk and the surface layer of cytoplasm surrounding the yolk do not contribute directly to the construction of the body of the embryo and for this reason are designated as **extra-embryonic**. Further, three germinal layers are formed within the balstoderm. How this occurs is a subject of long-standing controversy. It has been claimed (**Pasteels**, 1940 ; **Oppenheimer**, 1947) that most fishes gastrulate basically like the amphibians, the cyclostomes and the more primitive fishes, with involution occurring at the "dorsal lip" of the blastopore (*i.e.*, endomesoderm arises from the edge of the blastodisc that corresponds to the lips of blastopore). However, careful experiments of **Ballard** (1966, 1976) have furnished evidence against this view. He has shown that there is no significant sinking in of the superficial material into the interior and that the endomesodermal layer is formed by a rearrangement of the deep-lying cells of the blastodisc.

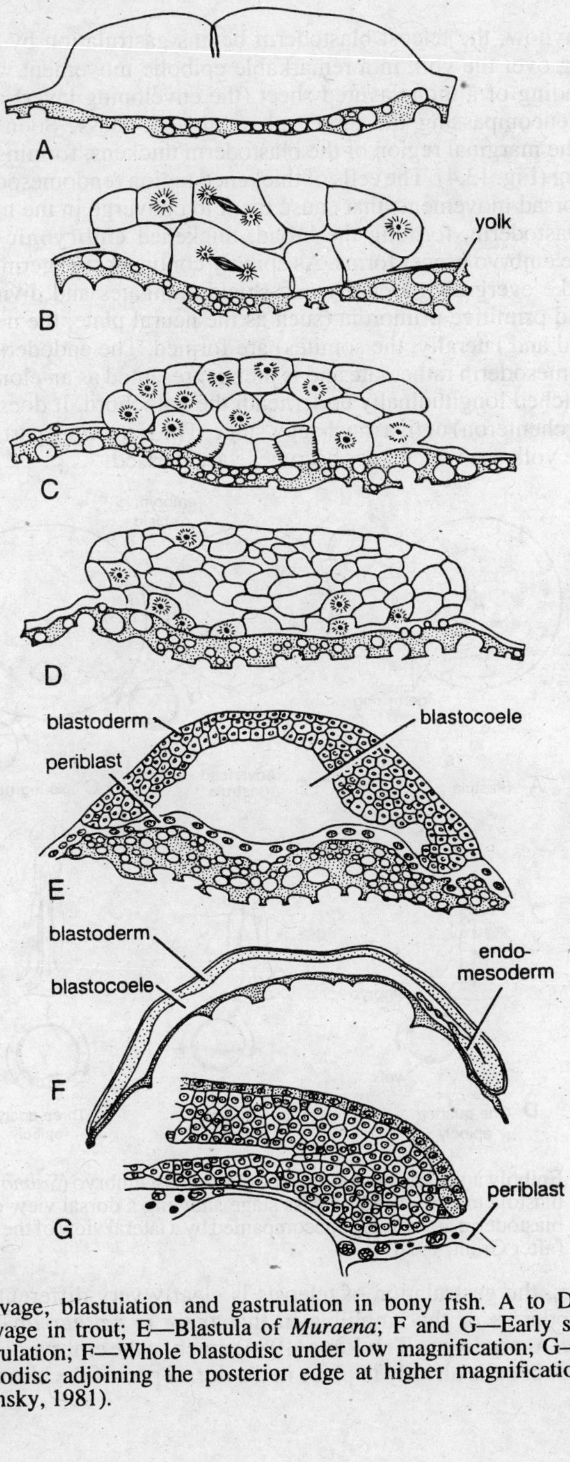

Fig. 13.3. Cleavage, blastulation and gastrulation in bony fish. A to D—Early cleavage in trout; E—Blastula of *Muraena*; F and G—Early stages of gastrulation; F—Whole blastodisc under low magnification; G—Part of blastodisc adjoining the posterior edge at higher magnification (after Balinsky, 1981).

Anyhow, the teleost blastoderm begins gastrulation by flattening and spreading over the yolk in a remarkable epibolic movement which involves the spreading of a monolayered sheet (the enveloping layer) and ultimately entirely encompassing the large spherical yolk mass. Soon after epiboly begins, the marginal region of the blastoderm thickens, forming the so-called **germ ring** (Fig. 13.4). The cells of thickened region (endomesoderm) undergo strong dorsad movements that cause them to converge in the mid-dorsal part of the blastoderm, forming the broad, thickened **embryonic shield** out of which the embryo proper forms. As epiboly continues, the germ ring thins and tongue-like evergrowing embryonic shield elongates and divides into germ layers and primitive primordia (such as the neural plate, the neural tube, the notochord and laterally, the somites) are formed. The endoderm is separated from the mesoderm rather late and is then represented as an elongated plate of cells stretched longitudinally underneath the notochord. It does not contain a cavity (archenteron) until a much later stage. The remaining part of blastoderm forms the yolk sac which later becomes vascularised.

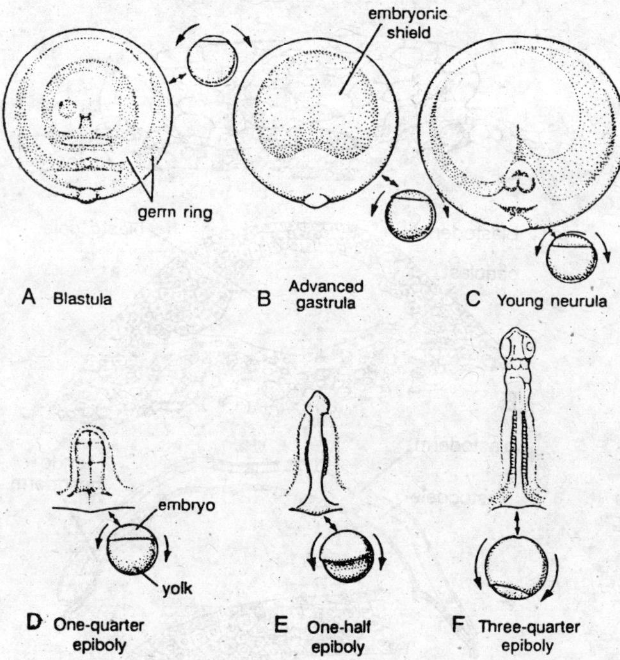

Fig. 13.4. Epiboly and early development of the teleost embryo (*Salmo iredeus*) from blastula to early head process stage showing a dorsal view of the isolated blastoderm at each stage accompanied by a lateral view of the entire embryo (after Grant, 1978).

Thus, the gastrulation of teleosts is clearly very different from that of the echinoderms or the amphibia in that *there is neither invagination nor involution*. According to **Trinkaus** (1984) much of the embryonic shield forms directly out of cells already in place in the dorsal part of the blastoderm before convergence begins. And the central nervous tube (*i.e.*, neural tube) forms by secondary hollowing out of a solid rod rather than by the rolling together of raised neural folds.

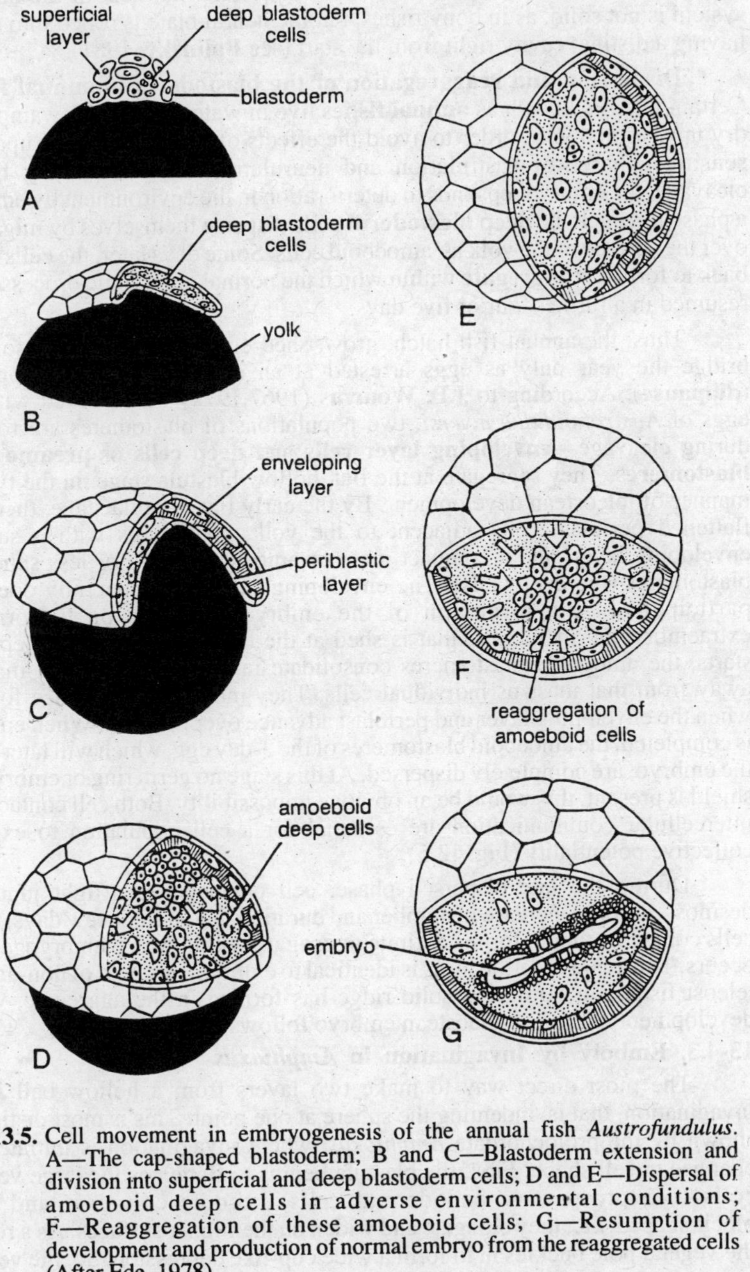

Fig. 13.5. Cell movement in embryogenesis of the annual fish *Austrofundulus*. A—The cap-shaped blastoderm; B and C—Blastoderm extension and division into superficial and deep blastoderm cells; D and E—Dispersal of amoeboid deep cells in adverse environmental conditions; F—Reaggregation of these amoeboid cells; G—Resumption of development and production of normal embryo from the reaggregated cells (After Ede, 1978).

Gastrulation in elasmobranch fishes. Gastrulation and the formation of the primary organ rudiments in elasmobranchs follow the same pattern as in bony fishes—with the difference, however, there an archenteron is formed at the posterior edge of the blastodisc. Further, the rudiment of the nervous system is not solid, as in bony fishes, but the neural plate is rolled into a tube having a distinct cavity right from the start (see **Balinsky**, 1981).

Dispersion and reaggregation of the blastoderm in annual fishes. Certain teleosts known as **annual fishes** live in water holes and swamps that dry up seasonally. In order to avoid the effects of this dehydration upon the sensitive phases of gastrulation and neurulation which normally follow cleavage, the embryo repsonds to deterioration in the environment by entering a phase in which the **deep blastoderm cells** disperse themselves by migration over the surface of the yolk as amoeboid cells. Some days later, the cells move back to form a **reaggregate** within which the normal embryonic processes are resumed in a further four or five days.

Thus, the annual fish hatch, grow, shed eggs and die in a season and bridge the year only as eggs arrested at an early stage of development (**diapause**). According to **J.P. Wourms** (1967,1972) who worked with the eggs of *Austrofundulus myersi*, two populations of blastomeres are formed during cleavage—**enveloping layer cells** and **deep cells** or **preamoeboid blastomeres**. They segregate at the flat, hollow blastula stage, in the typical manner of teleostean development. By the early flat blastula stage, there are flattened periblast cells adjacent to the yolk, continuous with a surface enveloping layer, and a compact mass of amoeboid, more or less spherical blastomeres (the deep cells). The enveloping cell layer apparently does not participate in the formation of the embryo but goes on to form an extraembryonic membrane that is shed at the time of hatching. As epiboly starts, the amoeboid blastomeres consolidate into a mass and then migrate away from that mass as individual cells. They move into the space formed when the enveloping layer and periblast advance over the yolk. When epiboly is completed, the amoeboid blastomeres of the 3-day egg, which will later form the embryo, are completely dispersed. At this stage no germ ring or embryonic shield is present, this would be an obvious impossibility. Both cell contact and intercellular communication are essential for a cell population to express collective potentiality (Fig. 13.5).

During the late dispersed phase, cell contacts (*e.g.*, tight junction, desmosome, etc.) increase in number and duration, until after a few days, these cells come together to form a definite **aggregate** within which embryogenesis occurs. This reaggregation mass is identical to embryonic shield of non-annual teleost fishes. In 10 days a solid ridge has formed in the aggregate, while development of typical teleostean embryo follows (**Berrill** and **Karp**, 1976).

13-4.3. Emboly by Invagination in *Amphioxus*

The most direct way to make two layers from a hollow ball is by invagination, that is, indenting the sphere at one point. This is most distinctly shown by the protochordate *Amphioxus*, where invagination is initiated by changes in cell shape. First, the blastula begins to flatten as the large ventral cells (*i.e.*, prospective endodermal cells) become columnar and then block-shaped, *i.e.*, they elongate and widen on their inner surfaces. As a result, the vegetal pole buckles in to form a wide cup-like depression. So, the ventral layer of blastula expands into the blastocoele cavity toward the animal pole until it has obliterated the space of blastocoelic cavity. The cavity of the cup is the future **archenteron** which communicates with the exterior through the **blastopore**. The outside ectoderm and inside endoderm remain continuous at rims or lips of the blastopore. The invagination of endoderm is assisted by **epibolic** expansion of the ectodermal sheet in the animal half, which closes the lips of the blastopore and lengthens the embryo in an anteroposterior direction (Fig. 13.6).

13-4.4. Emboly by Invagination Combined with Ingression in Echinoderms

Gastrulation of echinoderms involves the morphogenetic movements in the following sequence : **ingression → invagination → ingression**. The echinoderm blastula is spherical and made up of about 1000 ciliated cells attached to a distinct surface hyaline plasma layer. After about 24 hours of free-swimming life, the cells resorb their cilia and the blastula becomes flattened at its vegetal pole (Fig. 13.7). Thus, at this stage the process of morphogenesis begins, during which cells of vegetal pole change their shape from elongate wedges to cylinders, so that the ventral region of blastula thickens into a plate, called **vegetal plate**. To do **ingression**, in sea urchin, *Psammechinus*, about 40 cells of vegetal plate show pulsatory activity and round up (Fig. 13.7). They break contact with the surface layer and their neighbours and squeeze into the blastocoele. Here, they move around the inner surface of the blastula cells, making apparently random exploratory movements but eventually taking up a position in a doughnut-shaped arrangement around the edge of the vegetal plate ; possibly there is some matching of points of attachment to particular attachment sites. Locomotion of these **primary mesenchyme cells** occurs by extension of several long branched filopodia, which attach by their tips to the junctions between the blastula cells and subsequent contraction of these filopodia. **Tilney** and **Gibbons** (1969) have demonstrated the presence of orientated microtubules in these filopodia. The primary mesenchyme cells are destined to differentiate into skeletal spicules of the **pluteus larva**. **Okazaki** showed in 1960 that filopodia from several cells fuse to form a syncytial cable in which the spicules of the skeleton are secreted.

While the primary mesenchyme cells are taking up their position, the remaining cells of the vegetal plate begin pulsating and rounded up on their borders, but this time without detaching from each other or from the surface layer, so that the mechanical effect is to produce a hemispherical impushing or **invagination** of the whole plate into the blastocoele. Thus, the remaining cells of vegetal plate retain strong lateral connections via **tight junctions** and these cells fill the gap left by escaping primary mesenchymal (mesodermal) cells. Soon these cells elongate, buckle in and invaginate, pulling the sheet in as a unit because of lateral cell adhesions. Since their inner borders are motile than their outer surfaces attached to the hyaline layer, the direction of invagination is inward, into the blastocoele. This tendency for invagination has been experimentally found to be intrinsic to the cells of ventral plate (**Grant**, 1978). Microfilaments and microtubules are found to be involved in cell movements and change of cell shape (Fig. 13.8).

A **second ingression phase** follows or is concurrent with invagination. After a slight halt, the tip of invaginating sheet begins to show renewed

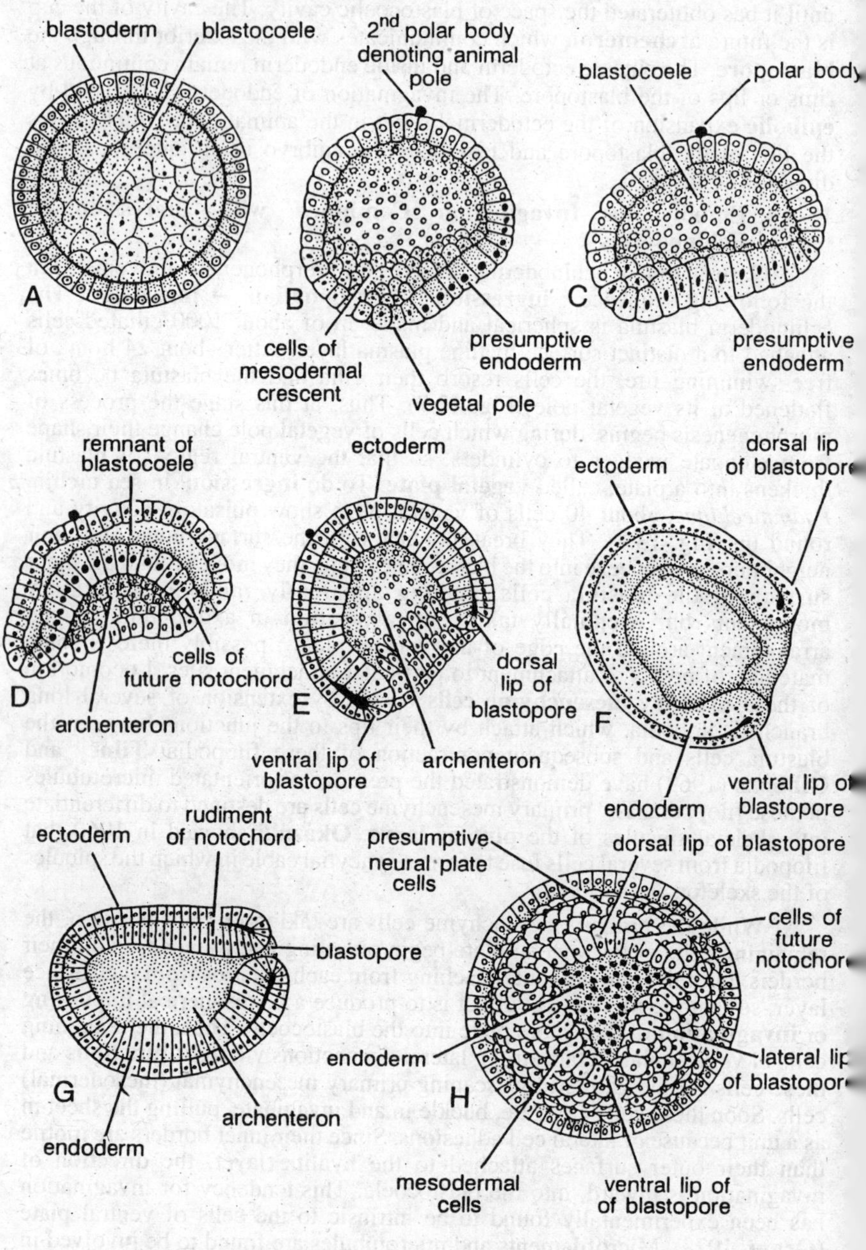

Fig. 13.6. Stages of gastrulation in *Amphioxus*. The embryos A to G are represented as cut in the median plane. A—Blastula; B and C—Beginning of invagination; D—Invagination advanced; E and F—Constriction of blastopore; G—Completed gastrula; H—A middle gastrula viewed from the side of blastopore (after Balinsky, 1981).

pulsatory activity. New mesenchyme cells or the **secondary mesenchyme**, larger than primary mesenchyme cells, squeeze from the sheet and congregate into the blastocoele cavity. Fine microtubule-containing filopodia are spun out from the secondary mesenchyme cells into the blastocoele cavity, their tips extending as far as the inner wall of the animal half. Attached as they are by filopodia, secondary mesenchyme cells exert a tension on the invaginating tip, pulling it further into the blastocoele.

Thus, contraction of filopodia in secondary mesenchyme probably contributes to invagination of the endodermal sheet in the second and final phase of gastrulation. Later on, these secondary mesenchymal cells separate from invaginating endoderm and form larval musculature.

Gastrulation in sea urchin produces an inner endodermal sheet surrounding an archenteron. The oral opening or mouth breaks through where archenteron tip contacts the inner wall of the animal half, while the blastopore becomes the future larval **anus**. A secondary mesenchyme containing outpocketing or evagination just behind the tip of archenteron ultimately produces the coelomic vesicles of the larva (the enterocoelomic coelom). The basic body plan of the bilaterally symmetrical **pluteus larva** is thereby established.

13-4.5. Emboly by Invagination Combined with Involution in Amphibians

Amphibian gastrulation has been studied in both urodeles (with tailed adults – newts and axolotles) and anurans (with tail-less adults – frogs and toads). Although the basic plans of gastrulation are similar, there are sufficient differences between the two which suggest that one cannot safely conclude from one species and assume it as the pattern for all. Each species has evolved special modifications of a common plan of gastrulation to adapt to the peculiarities and rhythms of its own developmental processes.

However, the pattern of gastrulation we are going to describe is more typical of anurans than of urodeles because it is simpler. Further, the various regions of the blastula are more clearly distinguished by pigment differences in anurans. Thus, a black pigmented animal region of micromeres (the **presumptive ectoderm**) ; an unpigmented vegetal region of macromeres (the **presumptive endoderm**), and a roughly equatorial band between them which is grey and approximates to the **presumptive mesoderm** ; within this band lies the grey crescent of the fertilized egg.

Like the echinoderms, in amphibians local changes in cell adhesion and motility initiate **invagination** in blastula. **Epibolic** spreading of an epidermal sheet and **involution** or the rolling in of tissue at the primary invagination site (the dorsal lip), contribute to invagination. In amphibians the pattern of tissue displacement (called gastrular morphogenetic movements) is also affected by the multilayered sheet of blastoderm, large yolk-filled cells in the ventral or vegetal half and eccentrically placed blastocoele.

The fate map of an anuran differs significantly from the urodele map in that the presumptive mesoderm is already inside the embryo as an inner layer beneath the ectoderm. Only the future notochordal region is in the outer layer in the marginal region. This along with endoderm is displaced into the interior by gastrulation, where it acquires its place in the embryo body plan. Various

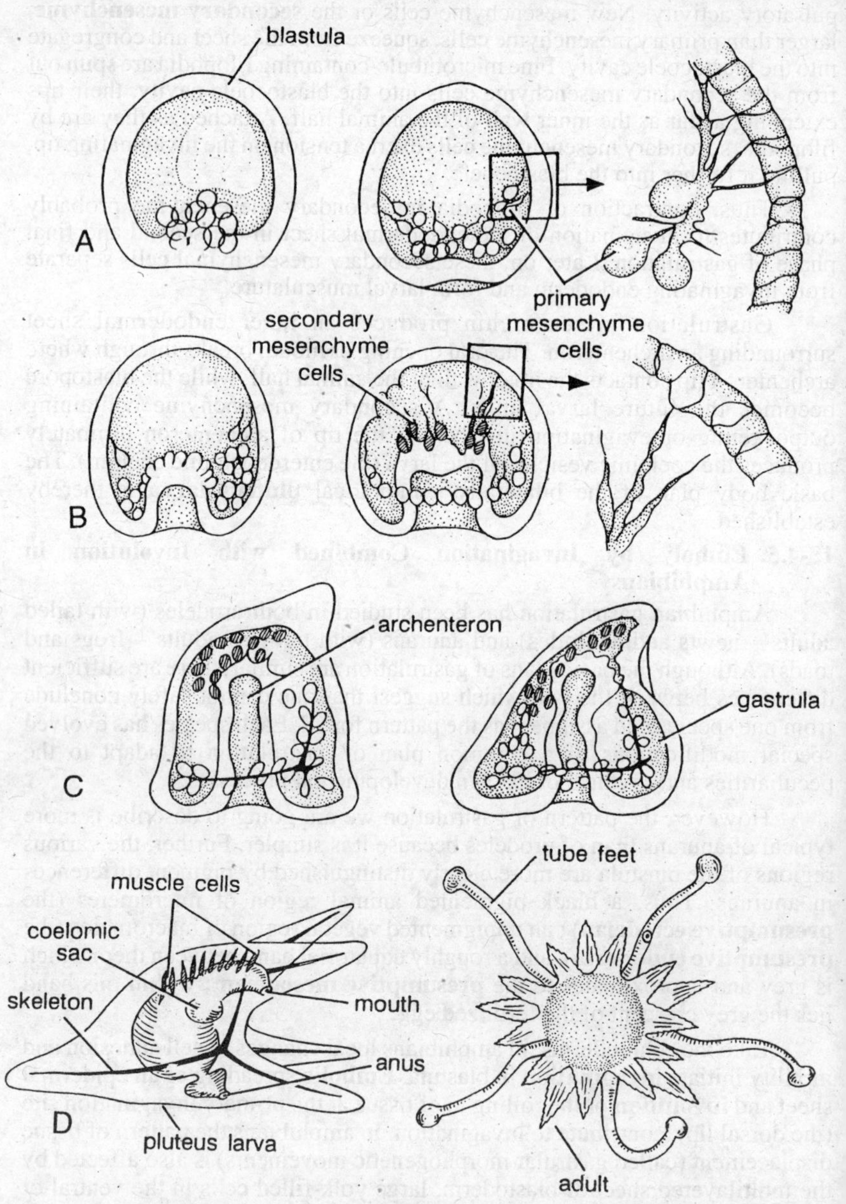

Fig. 13.7. Gastrulation and development of the pluteus larva in the sea urchin *Psammechinus.* A—Appearance and migration of the primary mesenchyme cells, which later secrete the skeleton of the pluteus; B—Appearance and behaviour of the secondary mesenchyme cells (coloured black), which first drag the archenteron up to the animal pole, then produce the larval muscles; C and D—Stages in formation of the bilaterally symmetrical pluteus larva and its transformation into the radially symmetrical adult form (after Ede, 1978).

events involved in the gastrulation of amphibians take place in the following steps :

1. Certain cortical events precede gastrulation of the amphibians. In late blastula, cortical tension in the vegetal pole increases and it contracts, more so on the future dorsal side (*i.e.*, in the region of grey crescent). This displaces, by epibolic expansion, the smaller pigmented animal-half cells (micromeres) in the marginal zone well below the equator. These initial changes are probably assisted by microfilaments. Cells in the dorsal marginal zone are smaller than ventral marginal cells, a factor that probably reduces cortical tension in this region. This leads to localised changes in cell surface behaviour, particularly in the region of the future dorsal lip of the blastopore.

In fact, epibolic expansion cannot continue indefinitely ; spatial considerations dictate that cells must be displaced somewhere, and the best place turns out to be the blastocoele cavity. It is the only "free space" into which epithelia may expand.

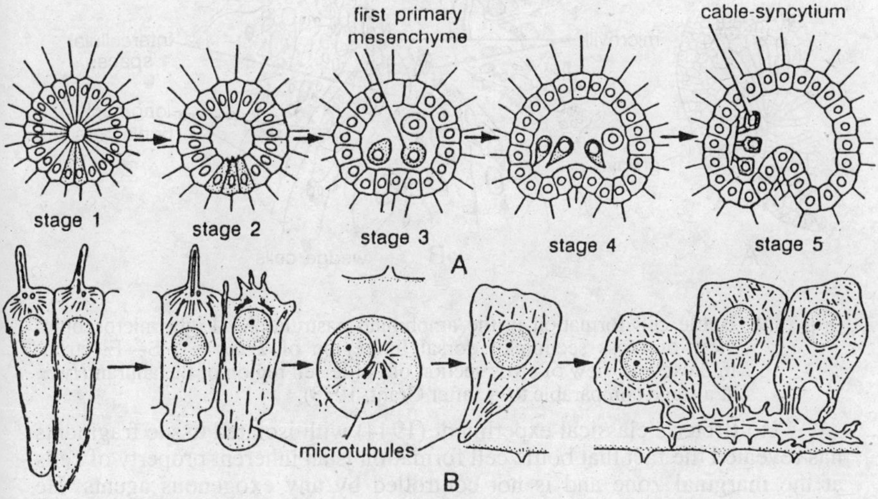

Fig. 13.8. Role of microtubules in regulation of cell shape changes in primary mesenchyme cells during ingression. A—Primary mesenchyme cells (black) exhibiting shape changes at different stages of ingression; B—Mode of distribution of microtubules at each stage of A is shown (after Grant, 1978).

2. The first visible sign of gastrulation is a sinking in (**invagination**) of a few presumptive endodermal cells at the original grey crescent or dorsal side of the embryo (Fig. 13.9). Like the ventral plate cells in the echinoderm embryo, these cells become motile and less adhesive, stretching into elongate bottle shapes, wider at inner surfaces than at the periphery. Here also the cell-shape change correlates with **microtubules** oriented parallel to the long axis. Contracting microfilaments too seem to transform cuboidal epithelial cells (*i.e.*, endodermal cells) into **bottle cells**, their distal tip remains attached

to the outer surface by **microvilli** interdigitating with neighbouring cell surfaces (Fig. 13.9). Ultimately, these cells sink into the interior as motility of their wide inner portion increases and a groove, called the **dorsal lip** of blastopore, appears. Pigment granules tend to collect in the fixed distal tip of each bottle cell as it invaginates, thereby marking the site of the dorsal lip. Adjacent endodermal and chordamesodermal cells, attached to bottle cells by tight junctions, seem to be drawn in the interior. As a result, the depression lengthens into a crescent-shaped or horse-shoe-shaped groove or pigmented slit.

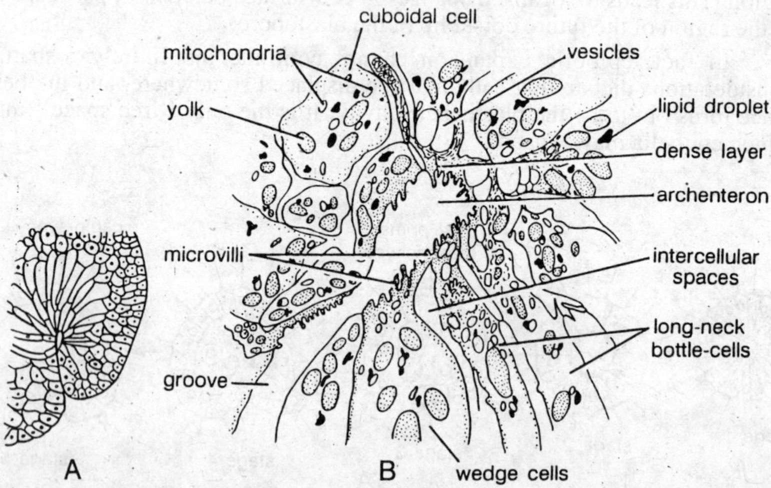

Fig.13.9. Bottle cell formation in the amphibian gastrula. A—Light microscopic view of cross section of dorsal lip region of *Triturus* ; B—Electron microscopic view of cross section of bottle cell region of the anuran *Hyla* at a stage comparable to A (after Grant, 1978).

Holtfreter's classical experiment (1944) with isolated tissue fragments has revealed the fact that bottle cell formation is an inherent property of cells at the marginal zone and is not controlled by any exogenous agents. He believed that the so-called surface coat of the egg plays a major role in invagination. He claimed that the dynamic tendencies of the embryo can be traced back to properties of the unfertilized eggs. However, the nature of egg-coat is still not clear. No specialised outer extracellular layer is seen in the electron microscope except for clear hyaline material. Surface coat properties are probably due to the cortical region of egg, which contain contractile microfilaments. As in the teleost egg after cleavage, cells at the surface are joined into a cohesive contractile layer by lateral membrane junctions (**tight junctions / desmosomes**). This layer acts as a unit, has spreading tendencies and communicates the pull of invaginating bottle cells to other cells in the region.

3. The entire animal half of the blastula undergoes epibolic expansion. The dorsal lip enlarges and becomes crescent-shaped because animal-half cells stretch, **converge** toward the blastopore, and **involute** over the lip, spreading

underneath the superficial layer. **Epiboly** is most clear in the dorsal region but gradually extends laterally and ventrally as the dorsal lip enlarges. More cells invaginate at a ventral blastopore lip, and the expanding blastopore becomes a circular opening. The blastopore is slowly reduced to **yolk plug stage**, while epibolic movements sweep superficial layers of cells into the interior. Inside, the invading chordamesoderm pushes and pulls yolky endoderm forward to line the cavity of the archenteron. The roof of the archenteron comes from endoderm pushed in by the chordamesoderm, plus any superficial ectoderm that has invaginated over the dorsal lip. The floor of the archenteron consists of yolk-laden cells pulled forward from the blastocoele floor.

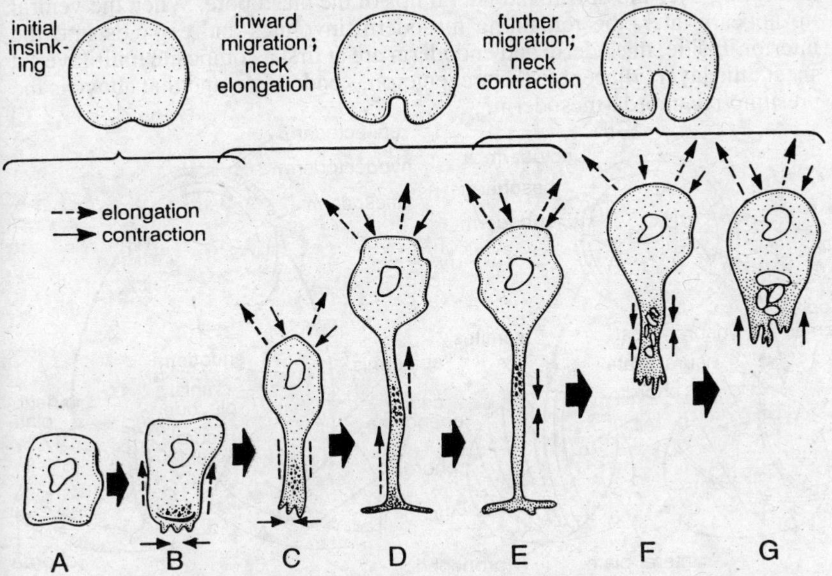

Fig. 13.10. Proposed scheme for transformation of invaginating bottle cells during gastrulation based on hypothesis of contraction and expansion layer. A—The cell on the surface before gastrulation; B—The distal surface contracts causing initial sinking in of blastoporal pit; C and D—The proximal cell surfrace begins to migrate inward, deepening the groove; the distal surface continues to contract and the neck elongates; E and G—The neck contracts after it is maximally elongated. This shortening, together with the pull exerted by the migrating inner ends, furthers invagination and pulls adjacent cells into the groove (after Grant, 1978).

Thus, the **ectoderm** has a greater spreading tendency than **endoderm**. Endoderm has a dual property ; it invaginates and spreads. In fact, only a superficial layer of endoderm surrounding the archenteron is capable of expansion ; the remainder, made up of large yolk-filled cells, is more cohesive and passive, usually carried along by other expanding sheets. Spreading tendencies of endoderm also change during development. For example, endoderm of young gastrula spreads less than the endoderm of neurula, indicating that motile and adhesive properties change with age.

4. In amphibians, the **mesodermal sheet** has a complicated origin. According to Vogt's fate maps, urodele presumptive mesoderm and notochord lie superficially at the marginal zone as a wide crescent in the blastula, just above the endoderm (Fig. 13.11). However, presumptive mesoderm in an anuran gastrula occupies an inner position just beneath the most superficial ectoderm layer. Thus, *all anuran mesoderm is derived from internal cells that do not invaginate* (see **Grant**, 1978). The combined endo-mesodermal sheet advances as a unit under the ectoderm, and the mesodermal sheet separates, probably by **delamination**, at the end of gastrulation.

In case of urodeles, some mesoderm is also inside but most occupies a peripheral position. This mesoderm invaginates behind the endoderm cells, involuting over the dorsal and lateral lips of the blastopore. When the ventral lip appears later, the remaining mesoderm involutes and expands into the interior. Inside, mesoderm and endoderm are at first a composite multilayered sheet ; the layer adjacent to the archenteron is endoderm, and that above is the presumptive chorda-mesoderm.

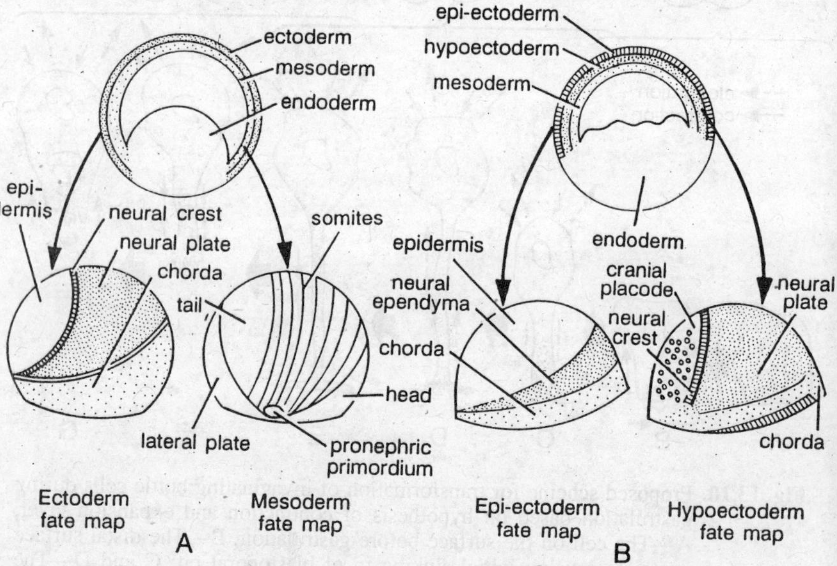

Fig. 13.11. Origin of mesoderm and fate maps of germ layers in early amphibian gastrulae. A—Urodeles; B—Anurans. Note in both cases that the mesoderm even at the early gastrula stage is a layer beneath the ectoderm. The fate map for anuran mesoderm is probably similar to that shown by urodeles (after Grant, 1978).

13-4.6. Emboly by Ingression of Endoderm and Mesoderm in Chick

To recall, the **epiblast** of avian blastula is the source of skin epidermis, neural tissue, future notochord, somites and lateral mesoderm tissue. It also contains the presumptive endoderm. **Hypoblast** is the source of some endoderm (the primordial germ cells) and notochord, but most of it develops into yolk sac membranes.

Gastrulation of the discoblastula of chick overlying a mass of yolk occurs in a very different manner compared to the coeloblastula (*Branchiostoma* or

Amphioxus) because yolky part of the egg is not involved in the process. Further, the most fundamental difference between amphibian and avian gastrulation is that no archenteron is formed in bird embryos ; the origin of the avian endoderm too has long been a matter of confusion and uncertainty (see **Ede**, 1978 ; **Grant**, 1978).

A short, broad and longitudinal band, called the **primitive streak**, develops in the epiblast about eight hours of incubation in the chick (Fig. 13.12). It initiates the conclusive **embolic** movements of gastrulation. The streak indicates the future antero-posterior axis of the embryo. Its formation involves the migration of the eipblast cells from the sides to the central area. The direction and position of the primitive streak is influenced by the underlying hypoblast which serves as an **inductor** for the formation of the primitive streak in the epiblast. It appears that contribution of cells from the margin of area pellucida to the hypoblast is instrumental in latter's capacity to induce the primitive streak. Up to this stage the development is entirely **regulative**, *i.e.*, even one-eighth of the area pellucida can form a complete embryo, provided the explant includes the margins of the area pellucida.

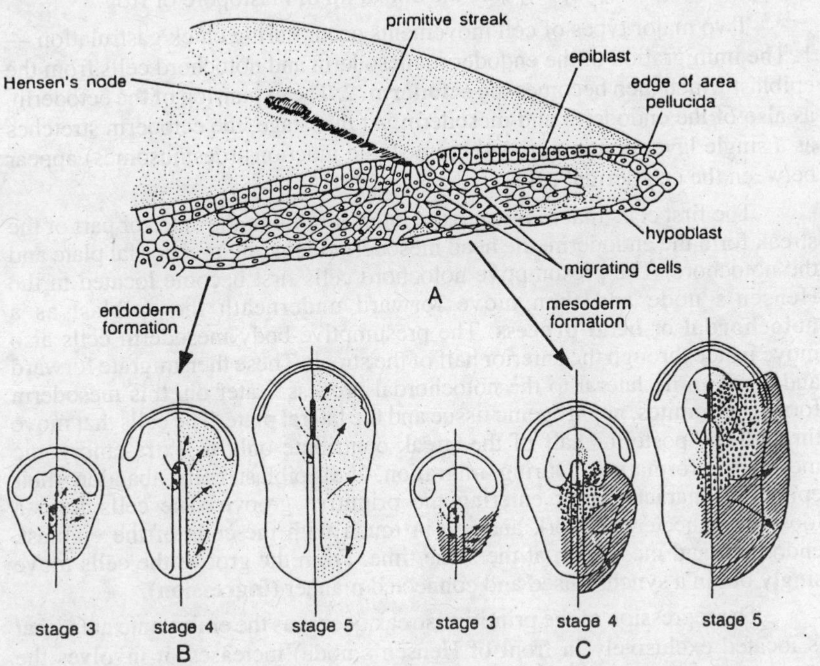

Fig. 13.12. Gastrulation in chick. A—Hemisected view of the primitive streak showing the invaginating and ingressing cells from the epiblast entering the segmentation cavity and spreading out to form the endoderm and mesoderm; B—Pattern of endoderm formation; C—Pattern of mesoderm formation and expansion of mesoderm (after Grant, 1978).

The primitive streak elongates and occupies the posterior half of the area pellucida and the intermediate primitive stage. Even the area pellucida soon begins to change its shape : first to an oval and then to pear-shape. The fully formed streak or **definitive primitive streak** has a characteristic morphology. Its anterior end forms **Hensen's node** (or primitive knot) in which there is a primitive pit. The streak length is formed by the **primitive groove** flanked by the **primitive ridges**. This longitudinal groove appears in the middle of streak as epiblast cells become flask-shaped and invaginate into the blastocoele. The basal ends of these cells are highly motile, whereas their apical ends remain fixed in the surface by junctional contacts (tight junctions). In this way they pull neighbouring cells into the groove so that they migrate in as a cohesive sheet.

The formation of the streak is considered to mark the end of gastrulation. The primitive streak is comparable to the amphibian blastopore, since the migration of the cells into the interior occurs through it. Hensen's node is comparable to the dorsal lip of the blastopore, since implantation of Hensen's node into a competent area pellucida induces a secondary embryo, *i.e.*, it contains **induction** property like the dorsal lip of blastopore of frog.

Two major types of cell movements occur during chick gastrulation — 1. The immigration of the endoderm, mesoderm and notochord cells from the epiblast which then becomes the **ectoderm**. 2. The expansion of the ectoderm, as also of the endoderm and mesoderm. At this stage, the ectoderm stretches as a single layer continuum and more stable junctions (desmosomes) appear between the ectodermal cells.

The first cells that move into the interior through the anterior part of the streak form the endoderm, the head mesoderm including prechordal plate and the notochord. The presumptive notochord cells first become located in the Hensen's node and then move forward underneath the epiblast as a **notochordal** or **head process**. The presumptive body mesoderm cells also move inside through the anterior half of the streak. These then migrate forward and come to lie lateral to the notochordal process. Later on, this mesoderm forms the somites, nephrogenic tissue and the lateral plate. The cells that move through the posterior half of the streak contribute only to extra-embryonic mesoderm formation. During migration, the epiblast cells abandon their epithelial character after entering the primitive groove. The cells form a loosely connected network and are in touch with the cells of the epiblast, endoderm and mesoderm at the same time. From the groove the cells move singly but in a synchronised and connected manner (**ingression**).

The regression of the primitive streak occurs as the embryonic area (that is located exclusively in front of Hensen's node) increases. It involves the shortening of the streak and the posterior shifting of Hensen's node. As the node shifts backwards and the notochord process elongates in its wake, the presumptive endoderm of middle and posterior part of the gut, located just behind the Hensen's node moves inside as an endodermal strip underneath the notochord. Ultimately Hensen's node and remnant anterior part of the primitive streak are included into the tail bud of the embryo.

After the streak disappears, continuity between three germ layers is also lost and they become independent expanding sheets. The endoderm, thus, has dual origin — the original hypoblast at the floor of blastocoele form the floor of archenteron, while, epiblast cells that invaginate through the streak form the roof and walls of developing archenteron.

Lastly, in chick, unlike the *Amphioxus* and the frog, the notochord or mesoderm never lines the archenteron ; because, the gut cavity develops between the yolk and endoderm, by the lifting up of yolk and endoderm from the yolk (see **Goel**, 1984).

13-5. THEORIES / VIEWS ABOUT THE MECHANISM OF INVAGINATION

Many attempts have been made throughout the present century to explain how and why the invaginative movement is accomplished. Certain important views regarding the cause of invagination are following :

13-5.1. Certain Classical Views about the Physiology of Invagination

Most of the classical embryologists fancied that some mechanical factors were operative in the process of invagination. Some of the early workers' well studied mechanical factors involved in invagination are following :

1. It has been assumed generally that continued increase in the number of cells by division and growth in the wall of blastula (*viz.*, blastoderm) results in a lateral pressure among the cells, which causes an infolding of the weakest part of the blastula sphere. However, by this hypothesis, it is difficult to explain why the thicker vegetal cells should first invaginate. Moreover, it has also been noted that invagination could take place without the increase in cell number. For example, **J. Cooke** (1973) showed that gastrulation in the amphibian *Xenopus laevis* proceeds normally when mitosis is blocked by colchicine or mitomycine C. Thus, all of the expansion of cell sheets without a free edge that occurs during amphibian gastrulation, *i.e.*, epiboly of prospective ectoderm, convergence and involution of the prospective endoderm of the marginal zone, and divergence of the endodermal lining of the archenteron, does not depend at all on cell division.

2. Rhumbler (1902) proposed that the formation of bottle cells is the main cause of invagination ; the great expansion of the membrane of these cells on their internal aspects is due to a gradient in surface tension between the inner and outer milieu of the late blastula. A decrease in surface tension within the blastula was thought to transform cell shape into flask form, with consequent invagination.

3. Butschli (1915) and **Spek** (1920) postulated a differential swelling between the inside and outside of the endodermal layer. The differential swelling was supposed by **Spek**, to be due to a special arrangement of lyophilic and lyophobic colloids (**lyophilic colloid** is a colloid that is readily dispersed in a suitable medium ; **lyophobic colloid** is a colloid that is not readily dispersed in a medium) in the endodermal plate and it was the main cause of the invagination of the endoderm.

4. **Assheton** suggested that the invagination might be brought about, if there were an attraction between cell of the blastula with the greatest force being exerted along a line passing through the peripherally located nuclei. **Moore** (1930) tested Assheton's hypothesis on late blastula of sand dollar and sea urchins and found no correlation between the tendency of invagination and position of nuclei in blastomeres.

5. **Waddington** (1940) has slightly modified the idea of Assheton to suggest **fibrilization theory** for invagination. According to his theory, fibrils, passing through cell boundaries produce the effect, necessary for invagination. No definite evidence in favour of this theory, however, is as yet available.

6. **Morgan** (1927) suggested that the blastula is ''water-tight'' and an absorption of blastocoele fluid by certain cells might cause an invagination because of hydrostatic pressure.

13-5.2. Certain Modern Views about the Physiology of Invagination

Recently many new interpretations have been forwarded regarding the invagination by modern embryologists. The most important theories of invagination are following :

(i) Tension theory; (ii) Inward cell migration theory.

(i) Tension theory of invagination. Moore and **Burt** (1939) and **Moore** (1941) attempted to analyze by operative means the question whether the entire blastula takes part in the gastrulation process, or whether the forces of gastrulation are localized in a special region. When the upper half or two-thirds of the early gastrula is cut away leaving an intact vegetal or endodermal plate, gastrulation is not damaged. This experiment disposes of the idea that differences between the blastocoele fluid and the external medium play a part during invagination (Fig. 13.13). In the cases where the endodermal plate was isolated by cutting away the remainder of embryo, the plate continued to roll up its rim, eventually closing to form a small spherical invaginated gastrula. These experiments led **Moore** and **Burt** to the conclusion that the forces of gastrulation do not act in a median plane through the poles of the embryo, but in the plane of the endodermal plate at right angles to the animal-vegetal axis. Thus, tension theory suggests that *the forces having to do with invagination exist in the endodermal plate and in intensity and direction are disposed from the centre outward in the form of a radial gradient.* It also suggests that the differential cohesion of cells may be an important factor in producing pocketing.

The tension theory is supported in various ways — one is that the hyaline layer of the late blastula exerts a tensile force which causes the vegetal cells to buckle in. In starfish embryos particularly, this layer is about 2 μm thick and is present at all stages ; when it is removed at an early stage by trypsin digestion, gastrulation (due to invagination) still takes place, even though it is not entirely normal.

(ii) Inward cell migration theory of invagination. According to **Trinkaus** (1984), Rhumblers's classical views have been supported by some Dutch workers who showed that there is a pH gradient in amphibian embryos, the blastocoele fluid being more alkaline (pH 9) than the cells (pH 6.7–7.0). **Trinkaus** believed that if these determinations are accurate, they would be of great interest, because *alkalinity has a profound effect on the cell surface, causing the formation of protrusions and greater cell motility*.

On the similar lines, **Holtfreter** (1943, 1944) proposed that the lower surface tension of the blastocoele fluid causes the inner surface of endoderm cells to expand and move into the interior. However, he emphasized that the connection of these cells to the surface coat is essential for invagination. Otherwise, they would slip into the interior as individuals.

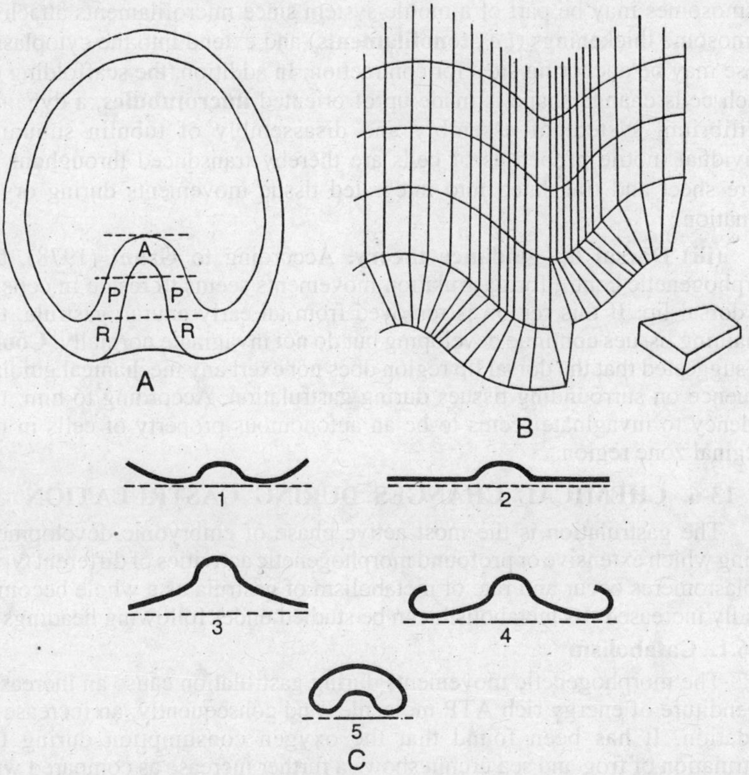

Fig. 13.13. A—Outline of normally invaginating gastrula. A, the originally active zone derived from the central spot of the gastral plate, P, the active zone derived from the plastic zone of the plate, which formed the periphery of the place surrounding **A R**, the new active ring; B—Change in cell shape associated with folding of columnar epithelium; C—Horizontal view of the positions assumed successively by the gastral plate of starfish *Patiria miniata* during hour following excision (After Moore, 1941).

Modern cell biologists have agreed with **Holtfreter** only about the inward migration of endodermal and mesodermal cells due to change in their shapes. In recent years, it is commonly accepted that these cells migrate inside by changing their respective shapes and they move in by performing ameboid movements. For example, **Ruffini** (1925) remarked that bottle cells are also associated with invagination of many kinds such as optic vesicle, nasal placode, mouth opening and so on.

Quite recently it has been found that machinery for the morphogenetic changes (*i.e.*, change in cell shape) resides in the contractile system of motile cells which is composed of **microfilaments** (see **Grant**, 1978 ; **Trinkaus**, 1984 ; **De Robertis** and **De Robertis**, **Jr.**, 1987). Contraction of these filaments may lead to shape changes or may be directly involved in movement of cells over substrata. The behaviour of cell sheets is probably due to adhesions between cells via specialised junctions such as **desmosomes**. Desmosomes may be part of a motile system since microfilaments attach to desmosome thickenings (*i.e.*, **tonofilaments**) and extend into the cytoplasm. These may be anchoring sites for contraction. In addition, the scaffolding on which cells change shape is made up of oriented **microtubules**, a dynamic equilibrium system of assembly and disassembly of tubulin subunits. Individual motile properties of cells are thereby transduced throughout an entire sheet and translated into integrated tissue movements during organ formation.

(iii) Dorsal lip guidance theory. According to **Grant** (1978), the morphogenetic centre for gastrulation movements seems to reside in cells at the dorsal lip. If this region is removed from an early anuran gastrula, the remaining tissues continue developing but do not invaginate normally. **Cooke** has suggested that the dorsal lip region does not exert any mechanical guiding influence on surrounding tissues during gastrulation. According to him, the tendency to invaginate seems to be an autonomous property of cells in the marginal zone region.

13-6. CHEMICAL CHANGES DURING GASTRULATION

The gastrulation is the most active phase of embryonic development during which extensive or profound morphogenetic activities of different types of blastomeres occur and rate of metabolism of gastrula as a whole becomes rapidly increased. Its metabolism can be studied under following headings :

13-6.1. Catabolism

The morphogenetic movements during gastrulation cause an increased expenditure of energy rich ATP molecules and consequently, an increase in oxidation. It has been found that the oxygen consumption during the gastrulation of frog and sea urchin shows a further increase as compared with the blastula stage. During gastrulation the food reserves (*e.g.*, glycogen and yolk) are oxidized for the manufacturing of ATP molecules.

(i) Oxidation of glycogen. One of the substances particularly involved in the supply of energy during gastrulation in amphibians and other animals, is glycogen. It has been shown in frogs that amount of glycogen becomes considerably diminished in the invaginating cells of the dorsal lip of the blastopore. Rapid breakdown of the glycogen in the dorsal lip suggests a particularly active respiration in this area.

(ii) Breakdown of yolk. Besides glycogen, the yolk of the blastomeres is broken down for energy metabolism or assimilation. The breakdown of yolk granules has been investigated in amphibian embryos both electron microscopically and biochemically. With the electron microscopy, it can be seen that in amphibian embryos, the first change in the yolk platelets consists of the disappearance of the amorphous or granular peripheral layer which contains, besides proteins, considerable quantities of ribonucleic acids. The disappearing material goes into solution in the cytoplasm and becomes available for synthetic processes. The solubilization of the peripheral layer occurs in the invaginating chordamesoderm during gastrulation in the neural plate during late gastrulation and early neurulation and still later in the epidermis. The solubilization of the crystalline core (''main body'') of the yolk platelets occurs considerebly later, and in endoderm, it is delayed till just preceding the stage when the larvae start feeding.

13-6.2. Anabolism

Throughout gastrulation, the volume of the embryo does not change appreciably. Every expansion in one direction occurs at the expense of a contraction in another direction. Anabolism mainly includes great amount of **transcription** (DNA-dependent RNA synthesis), small amount of DNA **replication** and biosynthesis of certain new types of proteins.

13-7. NUCLEAR ACTIVATION AT GASTRULATION

The gastrulation stage witnesses the onset of the following nuclear events : 1. The rate of nuclear replication decreases sharply in gastrula stage of many embryos. For example, length of the cell cycle increases from 15 to 30 minutes in *Xenopus* cleavage to 3.5 hours in the gastrula stage. More than 60 per cent of toad blastula cells are in some mitotic phase, but only 7 per cent of gastrula cells are in mitosis. As the cell cycle lengthens, the rate of DNA synthesis diminishes. For example, in the frog embryo (gastrula), there is fivefold decrease in the rate of DNA synthesis ; a decrease due to exhaustion of nucleotide reserves at the end of the blastula stage.

Thus, all phases of the cell cycle are lengthened, including mitosis. The G_1 and G_2 periods exhibit the greatest change, so that the proportion of the cycle devoted to S period or DNA-synthesis is very much reduced. This implies that in gastrula more time is available for trasncription of DNA and the embryonic genome (*i.e.*, paternal as well as maternal genes) becomes active during gastrulation.

Nucleoli first make their appearance during late blastula or early gastrula stages (Table 13-1). This marks the onset of stable ribosomal RNA accumulation, a visible sign that embryonic genome is actively transcribing. Still another sign of an activated embryonic genome is the appearance of **paternal proteins** at gastrulation. For example, in the hybrid cross between sea urchins — *Paracentrotus lividus* females and *Psammechinus microtuberculatus* males, no **paternal antigens** are found during cleavage, but 24 hours after fertilization, at the mesenchyme blastula stage, paternal antigens are detected in hybrid embryos. In this cross, messages from the paternal genome are probably synthesized and translated after cleavage is over (see **Grant**, 1978).

13-7.1. Gene Activities during Gastrulation

The nuclear genetic factors or genes, remaining dormant throughout the period of cleavage, begin to manifest themselves during gastrulation, and in ever-increasing measure control the processes of development from this stage onward. This does not mean that the processes taking place during cleavage are not affected by genes but the effect of the genes is then indirect — it is an action through the mRNA molecules and egg organization which are synthesized during oogenesis by maternal genes.

Table 13-1. The developmental stage of various animals during which the nucleolus appears (source **Grant**, 1978).

Category	Genus	Stages of appearance of the nucleolus
1. Mollusc	*Limnaea*	24-cell stage
	Helix	Blastula
	Chiton	Prototroch, apical tuft
2. Tunicate	*Ciona*	Neurula
	Ascidia	Metamorphosis
3. Echinoderms	Sea urchin	Mesenchyme blastula
4. Insect	*Chironomus*	Blastoderm
	Drosophila	Blastoderm
5. Amphibian	*Triturus*	Gastrula
	Rana	Gastrula
	Xenopus	Gastrula
6. Fish	*Fundulus*	Gastrula
	Coregonus (white fish)	Gastrula
	Misgurnus (loach)	Mid-blastula
7. Bird	Chicken	Late blastula
8. Mammal	Mouse	2-cell stage
	Rat	8-cell stage

Various indirect evidences (such as observation of cleavage without the participation of nuclei and hybridization experiments) have suggested that (i) a nuclues containing maternal and paternal genes, if, removed from the egg, the non-nucleated egg or its fragment continues to divide up to morula stage, but does not reach to blastula or gastrula stage. This shows that nucleus is necessary for development to go beyond cleavage.

(ii) The paternal genes cause the production of some entirely new kinds of proteins during gastrulation which make the blastomeres biochemically different from each other (chemo-differentiation).

(iii) During gastrulation many lethal mutations make their phenotypic expression into death of the gastrula.

(iv) Hybridization in amphibians often leads to developmental arrest, and actinomycin arrest development in frogs, sea urchins and ascidians.

The principal genes acting during gastrulation may involve those coding for cell surface components, either protein constituents of the membrane (adhesive sites, receptors), extracellular matrices adherent to the surface, or contractile elements in the surface that contribute to cell and tissue movement.

13-7.2. Cell Properties Changed by Gastrulation

Gastrulation rearranges all organ-forming areas of the blastula into three sheets, each with different **inside-outside relationships** and **recognition properties**. For example, isolated amphibian blastula cells are similar morphologically and are not very adhesive. They do not attach readily to a substratum and are amoeboid, forming broad **lobopodia.** After gastrulation, new cell types appear in each germ layer with different surface properties. Cells become more adhesive ; gastrula ectoderm cells now attach firmly to a glass substratum and flatten out. Still other cells appear (*e.g.*, bottle cells) with fine **filopodia** forming ruffled membranes as they glide along the substratum.

The surfaces of gastrula cells undergo a change ; their charge densities are greater than those of blastula cells, which means gastrula cells migrate more rapidly in electric fields. Invaginating chordamesoderm cells exhibit the greatest change in charge density, which may correlate with the fact that negatively charged sulphated mucoproteins, such as **chondroitin sulphate**, are actively synthesized in these cells during invagination.

Lovtrup (1965) has suggested that the primary cell transformations at the gastrula stage involve changes at the cell surface ; and amoebocyte-like blastula cell becomes a more motile bottle-cell with filopodia and ruffled membranes. These latter cells organise as a **morphogenetic centre**. The amoebocyte also transforms into another cell type, an adhesive cell or **colonocyte**, which assembles into cohesive epithelial sheets, possibly through some type of junctional specialization such as tight junctions.

13-8. ANTAGONISM BETWEEN CYTOKINESIS AND CELL MOVEMENT

Why gastrulation does not begin until cleavage has ceased ? In the very beginning of development, during cleavage and early blastula, cells do not appear to be able to move under any circumstances. Not only do they not move *in situ*, but they are never observed to move when transplanted to a new site or placed in culture. Thus, for them, differentiation of motile capacity is an essential aspect of starting to move. Possibly during cleavage and early blastula, cells are dividing so rapidly that their contractile microfilaments and microtubules, a major part of the machinery of movement, are so tied up in the business of cytokinesis that they are simply not available for other activities such as translocation. Therefore, the beginning of gastrular cell locomotion during early development of sea urchins and other embryos such as those of amphibians and teleosts, coincides with a decline of rapid cell division. The sea urchin provides an impressive example. The first blastomeres to cease cleavage are the micromeres and their descendants, the primary mesenchyme cells, are the first to begin movement. After their first appearance at 16-cell stage, the micromeres (four small blastomeres at the vegetal pole of the egg) undergo three more cleavages, to yield a total of 32-cells. In contrast, the mesomeres and macromeres of the 16-cell stage continue to divide for upward of six more cleavages, ceasing cleavage much later than the micromeres and most significantly their descendants do not start movements until considerably later than the micromeres. Thus, *those blastomeres that cease cleavage first are the first to move and those that cease cleavage last are the last to move.* Moreover, during gastrulation, when there is hardly any cell division at all, cell movements are well marked. Indeed there is so little cell division during gastrulation that even if it is blocked, as by colchicine, certain cell movements continue anyway.

With current interest in the importance of the ECM (extra-cellular material) in development there have been frequent suggestions that changes in it could stimulate cell to move. **Hyaluronic acid** is a case in point. A large local increase in hyaluronic acid concentration often coincides with the onset of cell locomotion. For example, at the precise time that neural crest cells of the prospective mesencephalon begin to move into the adjacent cell-free space, this space becomes filled with glycosaminoglycans (GAGS), of which hyaluronic acid is a major constituent. Likewise, in the developing eye, migration of neural crest-derived fibroblasts into the corneal stroma is preceded and accompanied by a swelling of stroma, with its orthogonally (or right-angled) arranged collagen fibrils, and an increase in hyaluronic acid (which is synthesized by the endothelial cells whose migration preceded that of fibroblasts). Further, reduction of GAG synthesis with the drug **6-diazo-5-oxonorleucine** (**DON**) leads to cleft palate, apparently because proliferation and migration of palatal fibroblasts are prevented. And finally, migrating fibroblasts of the endocardial cushion of the heart normally traverse a cell-free space filled with GAGS, especially hyaluronic acid (the so-called "carida jelly"), during later heart development. If one interferes with synthesis of the

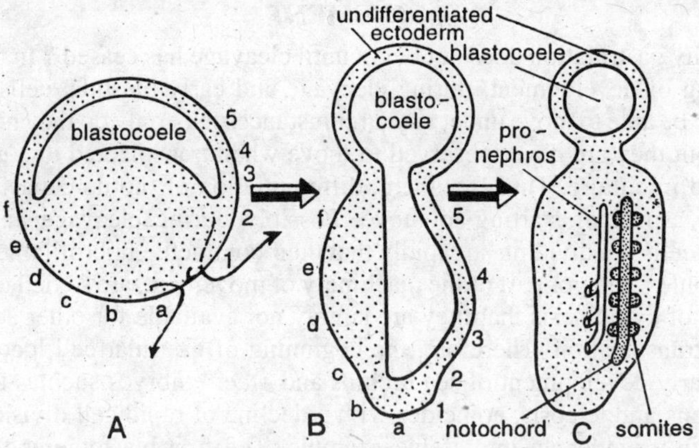

Fig. 13.14. Results of exogastrulation. A—Section of the early gastrula with the regions above and below the dorsal lip designated by numbers and letters. The dotted lines and arrows indicate the movement of regions 1 to 5 and a to f; B—The presumptive notochord, somites and endoderm fail to move interior of gastrula and ectoderm fail to form nervous system; C—Pronephros, notochord and somites form in remaining tissue (after Bodmer, 1968).

hyaluronate, this migration is prevented. The apparent dependence of the migration of primary mesenchyme cells of sea urchins on the presence of sulphate GAGS provides another such example (**Manasek**, 1975 ; **Trinkaus**, 1984).

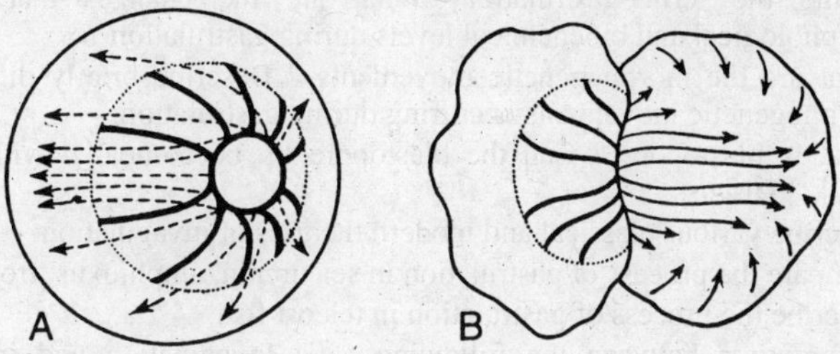

Fig. 13.15. Diagrammatic illustration of the morphogenetic movements. The heavy lines indicate movement of cells on the surface; the dotted lines show the movement of cells inside the gastrula; A—Normal gastrulation; B—Exogastrulation (after Berrill, 1971).

13-9. SIGNIFICANCE OF GASTRULATION

The significance of gastrulation is twofold. First it brings the presumptive structures from a position on the external surface of the blastula into their normal positions in the embryo. A second important aspect of gastrulation is that it is necessary for formation of the nervous system. This can be demonstrated by producing abnormal gastrulae, called **exogastrulae**, of frog or other amphibian. In some amphibian eggs, if the outer protective membranes are removed, the eggs form exogastrulae. In an exogastrula, the endodermal and mesodermal blastomeres evaginate (*viz.*, grow outward) rather than invaginate and the mesoderm and endoderm are not covered with ectoderm, the ectoderm remains isolated from the rest of the tissues.

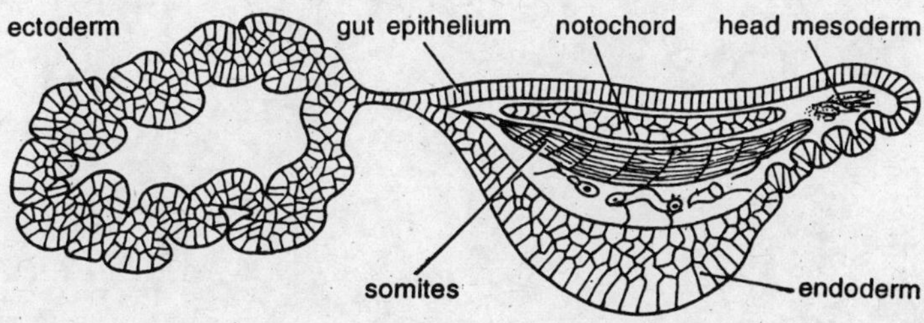

Fig. 13.16. Main differentiations in an exogastrulated embryo (Holtfreter, 1933).

The important aspect of exogastrulation is that endodermal and mesodermal structures develop to a great extent, but no nervous structures form. Neither the neural plate nor any of the sensory structure develop from the ectoderm if chorda-mesodermal area does not move inward and makes contact with the neural ectoderm.

REVISION QUESTIONS

1. Define the term gastrulation. What are the changes that occur at the morphological and biochemical levels during gastrulation ?
2. What are the morphogenetic movements ? Describe briefly different types of morphogenetic movements occurring during gastrulation.
3. What is blastopore ? Can the blastopore lip be compared with the primitive streak ? Discuss.
4. Describe various classical and modern theories of invagination.
5. Compare the process of gastrulation in sea urchin, amphioxus, frog and chick.
6. Describe the process of gastrulation in teleost fish.
7. Differentiate between the following : (i) Invagination and involution ; (ii) Epiboly and emboly ; (iii) Ingression and immigration.
8. Write short notes on the following : (i) Dorsal lip ; (ii) Development in annual fish ; (iii) Role of ECM in gastrulation ; (iv) Nuclear activation at gastrulation ; and (v) Exogastrulation.

14

NEURULATION, MORPHOGENESIS AND GROWTH

At the end of gastrulation, epidermal ectoderm covers the cylindrically shaped embryo. The neural plate lies in the mid-dorsal region. The mesoderm remains closely associated with the notochord, forming the so-called chordamesodermal mantle. The latter exists in the roof of archenteron (the embryonic gut) in association with the endoderm. The gut of the embryo, composed of endodermal cells, extends from the head to tail region. Once the blastopore fuses in the amphibians (with round gastrula), the gut is a closed cavity. In meroblastic animals (with flattened gastrula) the ventral area of the gut is open and rests on the yolk. Regardless of the type of gastrula formed (*i.e.*, round or flattened), the gut is properly oriented ; the region that gives rise to the foregut is located in the anterior region of the embryo, and the part that forms the cloaca is positioned posteriorly. Thus, **dorsoventral** and **anteroposterior axes** are established by this time in the body of the embryo.

In summary, all vertebrate embryos continue to follow generalized pathways at this stage of development, and as a result establish similar **basic embryonic forms**. This general arrangement exists in all vertebrate embryos by the end of gastrulation, regardless of their phylogenetic status and regardless of the amount of yolk present in the original egg or of the types of morphogenetic movements utilized to bring about the architectural structure. From this basic generalized plan, **organ primordia** or **primary organ rudiments** develop.

14-1. TUBULATION AND NEURULA

The basic embryonic form of all vertebrates consists of five **tubes** that eventually differentiate and give rise to the adult structures. They are present in both round and flattened gastrulae. These tubes are oriented around the solid, rod-like notochord which forms the axis of the body (Fig. 14.1). The **neural tube** forms dorsally to the notochord and **endoderm tube** ventrally to it. Two **mesodermal tubes** are located one on each side of the notochord and endodermal tube. The **epidermal tube** encases the entire embryo.

The process of formation of primary organ rudiments or tubes is called **tubulation**. The process of tubulation of the major organ forming areas occurs synchronously and it includes following main processes : (a) Neurogenesis, (b) Notogenesis, (c) Mesogenesis.

14-1.1. Neurogenesis

The formation of neural tube (**neuralization**) and its further differentiation into brain, spinal cord and different sensory organs such as olfactory (nose), optic (eye) and auditory (ear) organs, are collectively included in the process of **neurogenesis**.

The neural tube is derived from the ectoderm along the mid-dorsal axis of the vertebrate gastrulae. During neuralization, the ectoderm of mid-dorsal axis becomes thicker than the remaining ectoderm and called **neural plate**. In different vertebrates, the neural plate may differentiate into neural tube by following two methods :

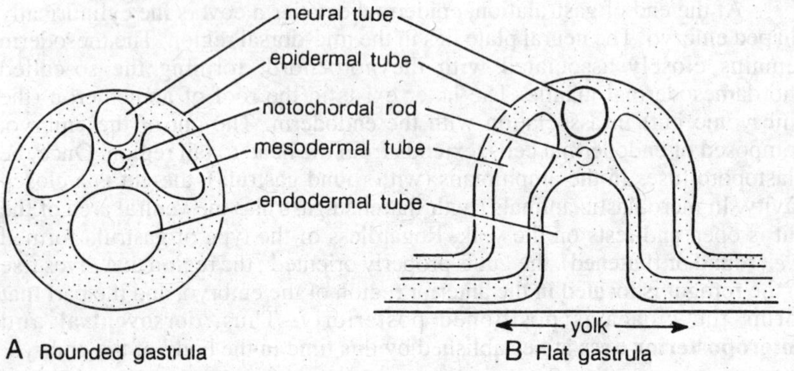

Fig. 14.1. Comparison of basic embryonic forms of animals developing from two types of gastrulae. A—Round gastrula; B—Flat gastrula. Both types of gastrulae are composed of five tubes oriented around the solid notochord by undergoing the process of tubulation or neurulation (after Phillips, 1975).

(i) Thickened keel method. In teleost, ganoid and cyclostome fishes, the process of neuralization occurs by thickened keel method. In them the neural plate material becomes aggregated in the form of a **thickened elongated ridge** or **keel** along the mid-dorsal axis of the embryo. The keel separates from rest of the ectoderm and sinks below the overlying skin ectoderm. Eventually the keel of neural cells transformed into hollow and elongated neural tube.

(ii) Neural fold method. In the majority of vertebrates (*e.g.*, amphibians, reptiles, birds and mammals), the margins of thicker neural plate elevate and its centre is depressed, forming a shallow or **neural groove** along the length of the body. Later the margins of the neual groove meet dorsally in the median line forming a tube called **neural tube**, which subsequently separates from the parent ectoderm. The anterior part of neural tube later on differentiates into the brain, while its posterior part differentiates into spinal cord.

Neural crest cells. After the fusion of epidermis and neural tube, there remain certain number of cells, which are neither included in the neural tube nor in the epidermis. These cells can be traced back to strips running along the crest of the neural folds. After the neural tube has become separated from the epidermis these cells are found as an irregular flattened mass between the neural tube and the overlying epidermis. These cells are known as **neural crest cells**. The neural crest cells give origin to ganglionic cells, pigment cells, mesoderm of the head and branchial area in the adult of most vertebrates (See Chapter 18 for more details).

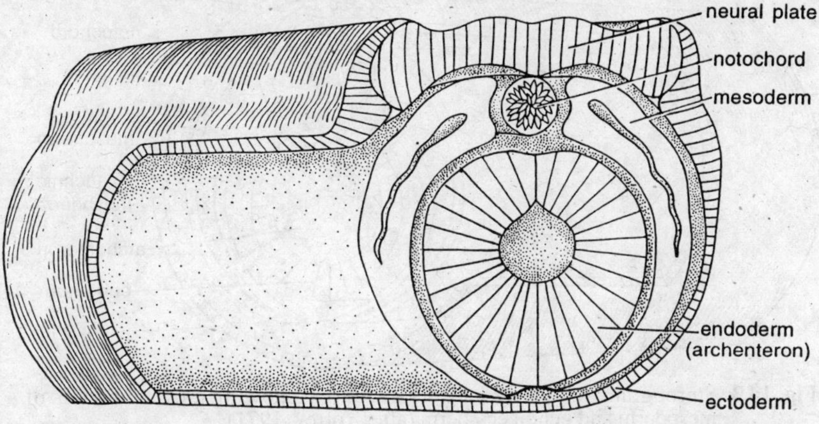

Fig. 14.2. Stereogram of early embryonic organization (after Torrey, 1971).

Neural induction. Spemann and **Mangold** (1924) performed certain heteroplastic transplantation experiments between two species of Urodela, namely *Triturus cristatus* and *Triturus taeniatus* and from their classical experiments, they concluded that in these animals the formation of neural tube was depended on some stimuli from the underlying roof of the archenteron containing the cells of chorda-mesoderm. They called the neural plate as **inducing organ** and the chorda-mesoderm cells as **primary organizers** or **neural inductors**. Later on, it has been found that — (i) Chorda-mesoderm cells act as the neural inductors in many animal groups. (ii) For inducing the neural plate to become neural type, it is necessary that chorda-mesoderm cells must remain in physical contact with the neural plate cells. (iii) The nature of neural induction is chemical. So, for the neural induction, some chemical substances possibily RNA, some protein or ribosomes pass or diffuse from the chorda-mesodermal cells to the neural cells. The recent evidences have favoured proteins as possible chemical neural inductors or as they are normally called, **evocators**. (iv) The evocator proteins are supposed to act only as stimulators, they only ''programme'' the genomes of the neural-plate cells for their self-differentiation into the neural tube. (v) The neural inductor, *i.e.*, the chorda-mesoderm, imposes a regional specificity on the inducing organ, the neural plate. Its anterior part induces head organs and is called a **head inductor** and its posterior part induces trunk organs and tail buds, so is called the **trunk** or **spinocaudal inductor**. The head inductor is further differentiated into the **archencephalic inductor,** which induces fore brain, eyes and nose rudiments, and the **deuterencephalic inductor,** which induces hindbrain and ear vesicles.

14-1.2. Notogenesis and Mesogenesis

The developments of notochord and mesoderm are called **notogenesis** and **mesogenesis** respectively. The chorda-mesoderm lies wedge-like between

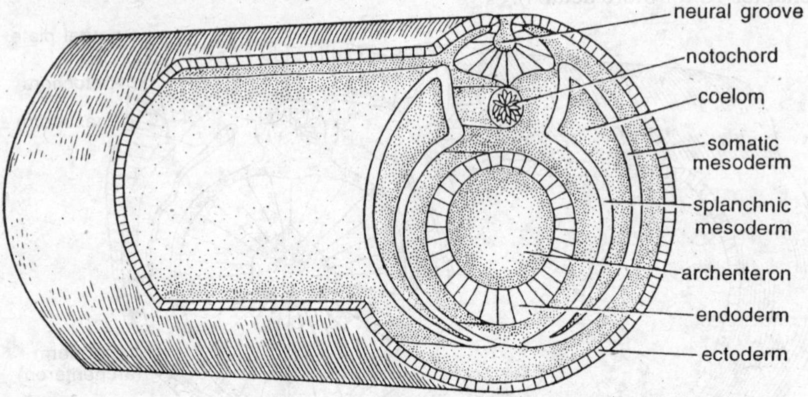

Fig. 14.3. Stereogram of embryonic organization, showing differentiation of mesoderm and neural ectoderm (after Torrey, 1971).

the roof of the gut, and the overlying ectoderm. Progressively, then it spreads ventro-laterad. In the mean time, the median dorsal region of the chorda-mesoderm segregates from that which is more lateral (Fig. 14.2). It is this median strip that rounds up and forms the **notochord**, lying between gut roof and the developing neural tube above.

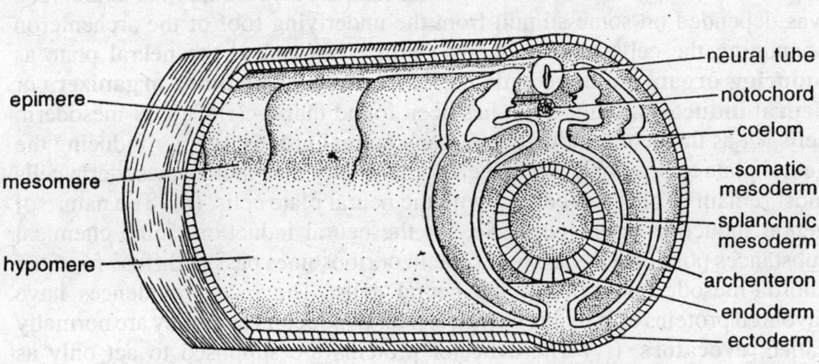

Fig. 14.4. Stereogram of embryonic organization, showing later differentiation of mesoderm and neural tube (after Torrey, 1971).

The lateral mesoderm continues to spread and eventually the spreading sheet on either side meet in the ventral mid-line. At the same time, these mesodermal sheets split into two layers with a cavity in between the two. One layer called **splanchnic mesoderm** lies next to the gut endoderm, the other layer called **somatic mesoderm,** lies next to the ectoderm, the space between the two layers is the **coelom**.

Not only does the mesoderm split up into two layers, but on each side of the embryonic body, the mesoderm becomes sub-divided horizontally into three distinct regions. Most dorsal and flanking the neural tube and notochord is the **epimere;** most ventral and flanking the gut is the **hypomere;** and, intermediate to epimere and hypomere is a narrow **mesomere**. Almost hand in hand with this dorsoventral sub-division of the mesoderm comes the first indication of the basic segmental character of the vertebrate. This takes the form of a series of transverse divisions of the epimere and mesomere. The hypomere does not become segmented. The divisions of epimere are of some significance. The lateral wall of epimere is called **dermatome,** its dorsal median wall comprises the **sclerotome**, and its ventro-median wall is the forerunner of the **myotome** (Fig. 14.4 and Fig. 14.5).

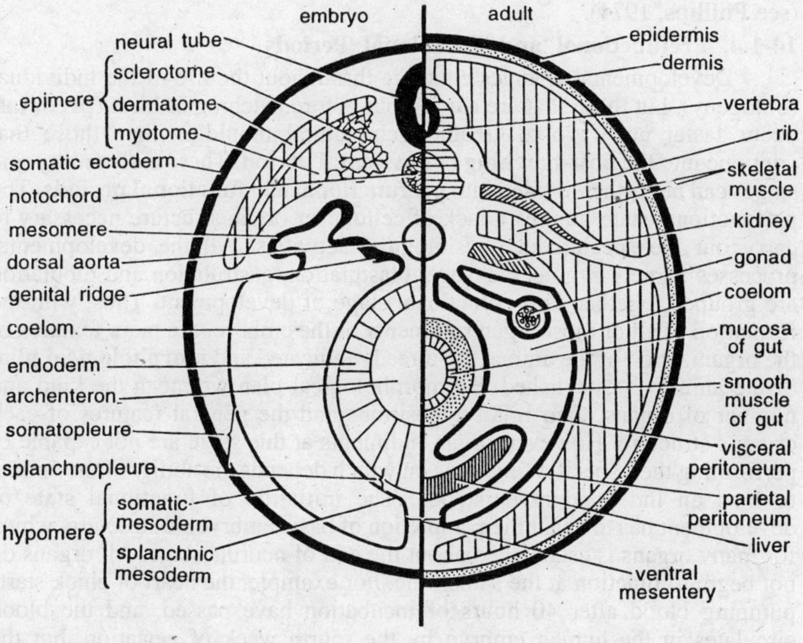

Fig. 14.5. Cross sectional diagram serving to correlate embryonic and adult structures (after Torrey, 1971).

As the late gastrula develops into the **neurula**, the body of the embryo flattens laterally and extends in an anteroposterior direction. The dorsal region of body arches, pushing the neural tube, notochord and somites more dorsally in the body. The anterior extension of the embryo is known as the **cephalic** or **head region** ; a similar projection in the posterior region becomes the **caudal** or **tail outgrowth**. The remainder of the embryo is designated as the **trunk area**.

14-1.3. Flexion of Embryo

Associated with differentiation of general body regions is the overall positional change of the body. Except in the fish, the anterior region of the neural tube becomes bent or **flexed** toward the ventral position (This is called **flexion of embryo**). This flexure involves the mid-brain and results in its growth perpendicular to the axis of the notochord. The midbrain bends towards the cardiac region in what is known as the **cephalic** or **cranial flexure**, readily observable in the 48-hour chick embryo. The **cervical flexure** appears in the region of the hindbrain at a later stage of development. Associated with the flexures is a general bending or twisting of the body region of most vertebrates, beginning at the head end of the embryo and progressing posteriorly until the anterior and posterior regions become curved ventrally toward one another, producing a comma-or C-shaped embryo. The embryos of higher vertebrates, those enclosed within extraembryonic membranes, also undergo **rotation** or **torsion**, as a result of which, the embryo comes to lie on its side usually the left one. This rotation occurs in chickens, as well as in pigs and other mammals (see **Phillips**, 1974).

14-1.4. Prefunctional and Functional Periods

Developmental changes continue throughout the life of the individual (ontogeny), but those that are carried out before hatching or birth (**prenatal**) occur faster and involve greater architectural moulding than those that continue into the **post-hatching** or **post-natal period**. These early embryonic stages can be further divided into **prefunctional** and **functional periods**. The prefunctional state implies a lack of cellular or organ structure necessary to carry out the specialized physiological activities. All the developmental processes namely zygote, cleavage, blastulation, gastrulation and tubulation are grouped together as prefunctional stage of development. Thus, with the formation of all or most organ rudiments of the embryo, the main features of the organization of the animal are already indicated and **morphological plan** of the animal is established. By morphological plan we mean the kind and number of organs, their relative positions and the general features of each organ's structure. However, organ rudiments at this stage are not capable of performing their specific functions on which depends the ability of the animal to lead an independent existence. The initiation of functional state of development has to await the completion of basic embryonic structure, which for many organs is established about the end of neurula stage. All organs do not begin to function at the same time. For example, the heart of chick starts pumping blood after 40 hours of incubation have passed, and the blood circulates in the human embryo by the fourth week of gestation, but the development and differentiation of skeletal muscles and initiation of its contraction begin later.

14-2. MORPHOGENESIS

Morphogenesis (Gr., *morph* = form ; *genesis* = generation) is the developmental process which leads to progressive attainment of changes in the body form (shape) and in the organization of various body parts and organs. For example, during cleavage and up to the blastula stage the embryo retains roughly the shape of the egg from which it started developing. It is spherical in most cases, sometimes oval if the egg is oval, or elongated, as in the case of many insects and cephalopods among the molluscs, which have elongated eggs. The internal structure of the blastula is also simple, consisting of one layer of cells around a cavity.

With the onset of gastrulation the embryo's structure changes. Through the development of the archenteron and then the separation of the mesoderm, the embryo acquires a more complicated internal structure. The structure becomes further complicated with the formation of the primary organ rudiments, which closely follow gastrulation. Gradually the external shape of the embryo starts to change. As primary organ rudiments give rise to secondary and tertiary organ rudiments, the internal structure and the external shape of the embryo approach those of the adult animal. Even during organogenesis, while the various organs are being formed, the shape of the embryo as a whole undergoes far reaching changes. During organogenesis of vertebrates, the following main changes occur in the embryo : 1. Elongation of body (due to stretching and elongation of notochordal rod). 2. Formation of tail. 3. Subdivision of body into head and trunk. 4. Development of appendages. 5. Separation of the embryo proper from the extraembryonic part (*e.g.*, fish embryo).

Since the individual acquires the characteristic form over the entire period of development, some developmental biologists equate morphogenesis with development itself. However, development includes also differentiation and growth. Therefore, it seems more appropriate that the term morphogenesis should be restricted only to the generation of form.

14-2.1. Process of Morphogenesis

The form of an organism depends on two main factors : the form of the individual cells (essentially differentiation) and the relative position of the different cells. So, the process of morphogenesis is operative at various levels : the entire organism (such as a frog body) ; an organ in the body (such as hand, heart, liver or kidney) ; a tissue in an organ (such as bone or muscle in hand), or individual cells (such as osteocytes, myocytes and lens fibre cells). Therefore, one should consider not only the generation of the form of individual organism but also the generation of the form of each individual cell (differentiation). It is the cells that collectively generate the form of an organism.

Thus, in a broader sense morphogenetic processes involve changes both at the level of the cells and at the supracellular level, both of which can be studied in the following activities :

1. Cell shape. The shape of the cell is adapted to the function the cell performs. Thus, nerve cells possess long axons (neurites) and dendrites (dendrons) ; erythrocytes are spherical or disc-shaped ; muscle cells are elongated ; epithelial cells are cubical flattened and leucocytes may be amoeboid. The cell shape also depends on the environment of the cell. For example, leucocytes are spherical when in circulating blood but become amoeboid in an extravascular location. Metazoan cells in tissue culture tend to acquire a spherical shape in suspension but become fibroblastic when attached to substratum.

The shape of a cell mainly depends on its epigenotype or its chemical constitution, however, there are some other unresolved regulating factors which affect the cell shape. Thus, in some cases the physical environment such as pressure from neighbouring cells determines the cell shape. In other cases, the inherent factors seem to play a role.

2. Cell death. In many situations the process of cell death serves as important morphogenetic role. For example, it can help in the separation of parts such as the separation of digits in an amniote limb. Cell death is also useful in the integration of tissues into organs, *e.g.*, the formation of heart from two endocardial tubes. Likewise, during metamorphosis, cell death helps to alter the shape by removing unwanted tissues.

3. Morphogenetic movements. Being supracellular events, the morphogenetic movements involve the combined activity of many cells, normally at the level of the tissues. The following types of morphogenetic movements have a role in morphogenesis :

(1) Morphogenetic movements in epithelia. During the early stages of cleavage, the blastomeres are rather loosely connected to one another, but before gastrulation and as the morphogenetic movements set in, the blastomeres in most animals join to form more closely knit structures in the form of layers of cells, the **epithelia**. Large or small parts of epithelia show following types of coordinated movements during morphogenesis : (i) **Thinning out of epithelium**. This occurs during **epiboly** or **overgrowth** of presumptive ectoderm during gastrulation. For example, the presumptive ectoderm of an amphibian blastula consists of several layers of cells ; the number of cell layers becomes reduced during gastrulation and still further reduced during neurulation. In the process, some of the cells originally lying in the interior become inserted or intercalated between the superficially lying cells. The whole change is, thus, brought about by a rearrangement of the cells of the epithelium. (ii) **Local thickening of the epithelium**. The formation of a neural plate takes place by local thickening of the epithelium. The thickening is brought about the epithelial layer flowing toward the region in which the organ rudiment is to appear, *i.e.*, the mid-dorsal region of embryo. As the thickening becomes distinct, the epithelial cells may become elongated in direction perpendicular to the surface of the epithelial layer. For example, in urodele amphibians, cells of neural plate become columnar and single cells stretch from the external to the internal surface of the neural plate, in spite of increased thickness. Thus, here also a rearrangement of cells of epithelium takes place. (iii) **Separation of epithelial layers**. Epithelial or masses of cells may be subdivided into parts by the appearance of crevices between groups of cells, the cells on the opposite sides of a crevice losing connection with one another. Normally the crevices occur either parallel to the surface of the epithelium or perpendicular to it (Fig. 14.6). In the first case, the epithelial layer is split into two layers lying one on top of the other. The original crevice may be increased by the secretion of fluid into it and may become a more or less spacious cavity. This is the case with the formation of parietal and visceral layers of the lateral plate mesoderm and coelomic cavity between them in chick and other vertebrates. An instance of crevices appearing perpendicular to the surface of the epithelium is found in the development of mesodermal somites.

(iv) **Folding of the epithelial layer**. The invagination of endoderm to form

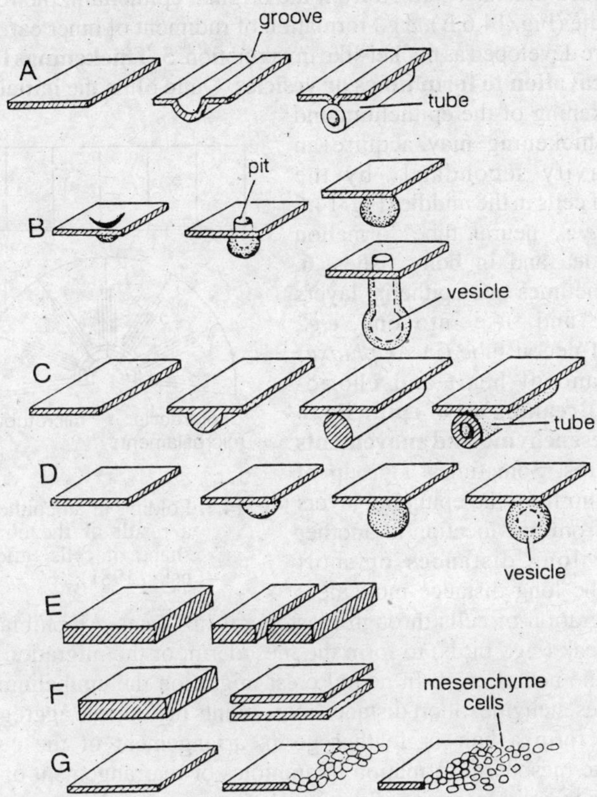

Fig. 14.6. Different types of transformations occurring in embryonic epithelium in the course of development (morphogenesis). A—Groove giving rise to a tube; B—Pit giving rise to a vesicle or to a tubule open to the exterior; C—Development of a tube from a solid elongated thickening of the epithelium; D—Development of a vesicle from a solid circumscribed thickening of the epithelium; E—Splitting of epithelium perpendicular to surface into sections; F—Splitting of epithelium tangentially into two layers with a space in between; G—Breaking up of epithelium to form separated mesenchyme-like cells (after Balinsky, 1981).

archenteron as it occurs in animals such as sea urchin or *Amphioxus* and also in the early stages of gastrulation in the amphibians is obviously a case of infolding of an epithelium. Infolding involves a change in shape of cells (Fig. 14.7). Direction of folding always remains towards the originally proximal surface of the epithelial layer. Thus, in the ectoderm and later in epidermis, the folding is directed inward. While in the invaginated parts of the embryo, as in the whole of the endoderm, the folding is directed outward, away from the lumen of the alimentary canal. Further, if a linear fold becomes closed in and separated from the original epithelium, the resulting structure is a hollow tube (Fig. 14.6 A), *e.g.*, formation of neural tube. If a pocket-like depression becomes closed in and separated from the original epithelium, the result is a hollow vesicle (Fig, 14.6 B), *e.g.*, formation of rudiment of inner ear. Most of the glands are developed as pocket-like invagination. **5. Thickenings followed by their excavation to form tubes or vesicles**. Quite often the initial stage is a solid thickening of the epithelium and this solid thickening may acquire an internal cavity secondarily by the separation of cells in the middle (Fig. 14.6 C and D), *e.g.*, neural tube formation in Myxinoidea and in bony fishes. **6. Fusion.** Sometimes two epithelial layers come close and fuse into one, *e.g.*, formation of neural tube (in *Amphioxus* and frog) and of heart and chorio-allantois. **7. Breaking up of epithelia to produce mesenchyme and movements of cell groups**. Sometimes a group of cells breaks up from the epithelial layers and moves from one location to another either over long distances or short distances. The long distance movement includes migration of cells through blastoporal lips (*e.g.*, Amphibia) or the primitive streak (*e.g.*, birds) to form the mesoderm, or the migration of cells away from the neural crest. In neural crest migration the epithelium breaks down into mesenchyme. Short distance movements include the aggregation of the cells to form a hair or feather germ, arrangement of the cells into epithelial-like mesoderm formation in amniotes or rearrangement of cells to form cavities in solid mass of cells as in the formation of the neural canal in bony fishes or amniotic cavity in many animals.

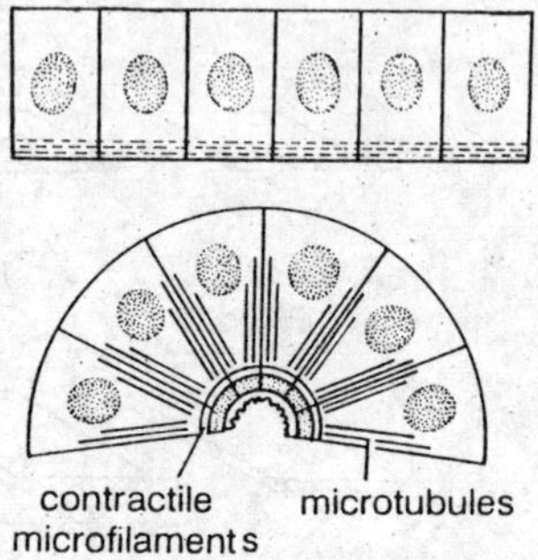

Fig. 14.7. Folding in an epithelium as a result of the change of shape of cells (after Balinsky, 1981).

Embryonic mesenchyme cells. Cells that leave the mesoderm layers of the embryo and migrate singly or in groups into the spaces of body are called **mesenchyme cells**. They are predominantly mesodermal in origin, although neural crest, ectoderm and even to a limited extent, endoderm also contribute to them. These loosely aggregated cells are usually star-shaped (stellate) with extending processes that contact one another. Unlike other cells of the body (*e.g.*, muscle, nerve, etc.), mesenchyme cells secrete a semifluid **intercellular substance** that becomes caught in the meshes of the network formed by their processes. Especially in the early stages of development, mesenchyme cells undergo amoeboid movement and wander over the body, inserting themselves between developing organs. They are **packing tissue** of the early embryo and fill in the spaces between the various body tubes (*i.e.*, neural, notochordal, gut, etc.) in the primitive embryonic form.

The mesenchyme cells not only act as packing tissue, but they also undergo differentiation and develop into various types of specialized cells. Most of the cells that differentiate from mesenchyme have names that end in "blast" (which means germ in Greek), indicating a formative stage of a specific cell type which in the adult functions in different way (Table 14.1).

Table 14-1. Contribution of mesenchyme to the body's structure (source **Phillips,** 1975).

Embryonic source of mesenchyme cells	**Adult tissue**		
	Type of cell derived from mesenchyme	**Type of tissue formed**	**Examples**
1. Myotome	Myoblast	Muscle	Cardiac, smooth, skeletal.
2. Dermatome	Hemocytoblast	Stem blood cells	Erythrocytes, leucocytes, blood platelets.
3. Sclerotome	Mesenchyme	May remain in adult body as reserve cells	Important in body repair.
4. Splanchnic mesoderm	Mast cells	Connective tissue	Synthesizes and stores heparin.
5. Somatic mesoderm	Lipoblast (Fat cells)		Adipose tissue.
	Macrophages (Histiocytes)		Phagocytic cells important in body's defence.
6. Neural crest cells	Fibroblasts	Connective tissue proper	
		Loose	Areolar packing tissue, serous membrane, endothelium, fascia.
		Dense irregular	Dermis, capsules of organs, sheaths of tendons, periosteum, perichondrium.
		Dense regular	Ligaments, tendons.
	Chondroblasts	Cartilage	Embryonic skeleton : hyaline, elastic and fibrous cartilage.
	Osteoblasts	Bone	Skeletal support of vertebrates.
	Osteoblasts		Remodelling of bone.

(2) Cell sorting (or self-assembly and selective affinity). The ability of biological materials to undergo self-assembly is evident at all levels, from the molecular (the folding of proteins) through the subcellular (the assembly of organelles and membranes) to the cellular and even to organismic level (the organization of a regenerating limb from a blastema). We will concentrate here on self-assembly of cells into populations that organize themselves into tissues, organs and embryos.

In a classic tissue-culture experiment of **Townes** and **Holtfreter** (1955) when cells from two tissues are dissociated and mixed and left to develop, they sort themselves out tissuewise (Fig. 14.8). For example, they separated presumptive layers of amphibian gastrulae and early neurulae into single-cell suspensions and recombined the cells in various combinations. Much to their surprise, the cells sorted out so that like cells associated together : mesodermal cells associated only with mesodermal cells, endodermal cells with endodermal, etc. Even more astonishing was the observation that the cells also

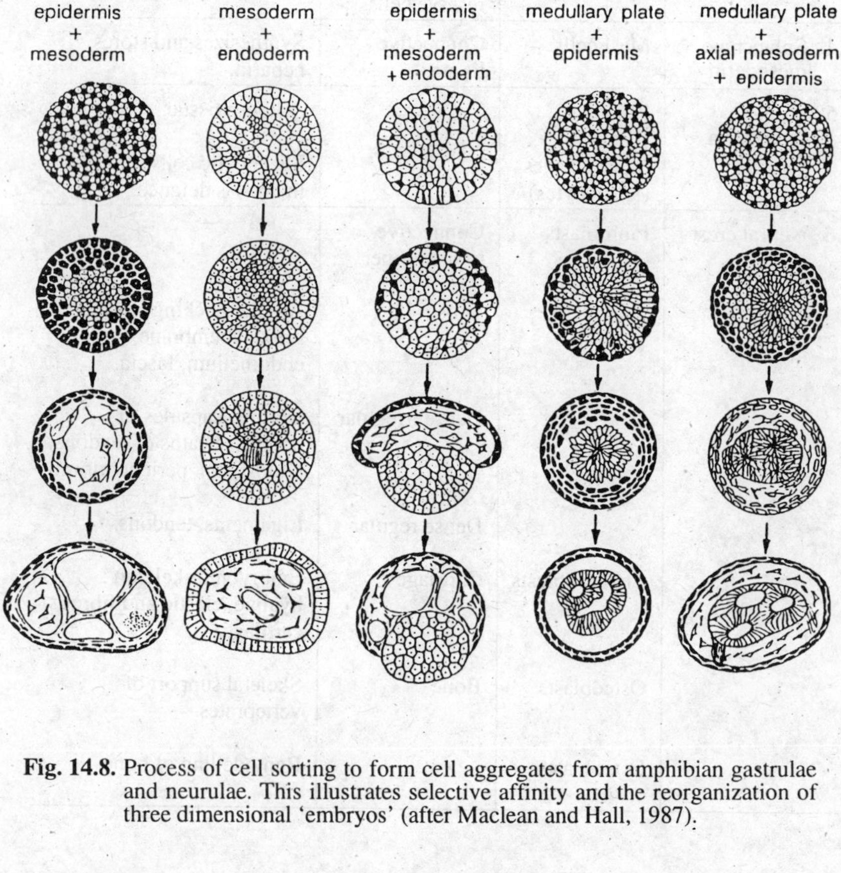

Fig. 14.8. Process of cell sorting to form cell aggregates from amphibian gastrulae and neurulae. This illustrates selective affinity and the reorganization of three dimensional 'embryos' (after Maclean and Hall, 1987).

sorted out to specific regions of the aggregates. Epidermal cells moved to the outside, endodermal cells moved to inside and mesodermal cells moved between the epidermal and endodermal cells. Furthermore, these cell layers formed distinctive structures, the endodermal cells forming a hollow internal tube, the epidermal cells forming a thin sheet surrounding the aggregate, and the mesodermal cells forming blocks of mesenchyme in between. Such structural aggregates bear a striking resemblance to the organization of normal embryos. This phenomenon of cell-type specific aggregation is known as **selective affinity**.

If similar types of dissociation-reassociations are performed with blastulae (*i.e.*, with younger stages of embryos), no selective affinity is seen. Any blastula cell will associate with any other cell to form a random mix of cells. Selective affinity is, therefore, a property of differentiating cells and coincides with when the embryo's own genome is activated selectively in different regions of post-blastula embryo. Therefore, selective affinity is the first overt sign of cytodifferentiation.

As selective affinity is a differentiative property, cells should become increasingly 'choosy' about their association as development proceeds. This can be tested by dissociating tissues of older embryos into single-cell suspensions using enzymes such as trypsin, pancreatin or collagenase, or chelatings agents such as EDTA (Ethylenediaminetetra acetic acid) and culturing mixtures of cells under conditions of gentle rotation to maximize contacts between them. Under such conditions, cells from the entire presumptive neural ectoderm of the gastrula will aggregate with one another. By the early neurula stage, future brain cell will no longer aggregate with future spinal cord cells. Further, by the late neurula, forebrain cells will no longer associate with mid-or hindbrain cells and so on. Thus, selective affinity is progressive.

The surface of each cell seems to be very specific due to certain binding sites. The cells apparently recognize each other due to these specific binding sites. Tissue culture data support such an interpretation. Further evidence for this comes from the loss of contact inhibition of locomotion in cancer cells. Cell surface charge also appears very important in cell sorting in tissue culture. Recently, a class of Ca^{2+}–dependent cell-cell adhesion molecules, termed **cadherins**, have been identified and shown to be tissue-specific in both embryonic chick and the foetal mouse. One of them (N-cadherin) is expressed during morphogenetic events such as gastrulation, neurulation and infolding of the epithelium during the formation of organs as the lens (**Hatta** and **Takeichi**, 1986). Morphogenetic movements are thought to basically dependent on the selective affinity of cells and the cell movements.

14-3. GROWTH

In normal usage growth refers to a gradual increase in size or number. Biological growth is usually considered as an increase in that part of the living system which is in its own turn capable of growth. The measurement of growth is done by determining changes in wet or dry weight (mass) of the organism.

In a way, **growth** is a permanent increase in size. The foetal calf grows, the pregnant cow does not. The cow's increase in size is not permanent but is a physiological fluctuation or **modulation**, albeit a relatively long-term one.

Growth is best measured by increase in protoplasmic content, usually as dry weight, although many other measures are possible. So, growth has been defined by **Hall** (1978) as a permanent increase in size of any parameter that is measurable. Thus, growth of an organism can be viewed and measured at three different levels : the organs and tissues, the cells, and the subcellular structures or the protoplasm.

14-3.1. Protoplasmic or Sub-cellular Growth

This type of growth involves the following processes : 1. Endocytosis by cells of the substances from their surroundings. This may be of small molecules (*e.g.*, water, salts, sugars, etc.), large molecules (*e.g.*, peptides and oligopeptides, fatty acids, etc.) or macromolecules including the entire nuclei (in insect oocytes). Endocytosis leads to an increase in the weight and the molecular population of the cells. 2. Synthesis of molecules other than DNA for the internal use of a cell or for extracellular secretion. These molecules may be small, large or macromolecules such as proteins (enzymes, collagen fibres, etc.), fats or lipids, carbohydrates (glycogen) and ribonucleic acids. 3. Synthesis of genetic material or DNA replication.

14-3.2. Cell Growth

The cell is usually believed to be the unit of life. It, therefore, appears logical to consider growth at the cellular level, *i.e.*, change in cell size. One can distinguish the following three basic ways in which growth of a multicellular animal may be related to the growth of its cells :

1. Auxetic growth. The volume of the animal's body may increase owing to the growth of the individual cells, without an increase in their number (owing to cessation of mitosis in somatic cells of the young after its hatching). This rather rare form of growth is found in nematodes, rotifers and Larvacea (Ascidia). For example, in the development of nematodes, cell divisions stop in the early stages of organogenesis. The number of cells in the fully grown nematode is the same as in a young one which has just emerged from the egg. The number of cells in each organ rudiment may be definitely fixed in this type of development. For instance, the whole excretory system of a nematode consists of only three cells.

2. Multiplicative growth. The animal's growth may be the reuslt of an increase in the number of its constituent cells. The increase in number is brought about by mitotic division of all the cells, while the average size of the cell remains the same or nearly so. This type of growth is quite common in embryos, and it is especially characteristic of the prenatal growth of higher vertebrates.

In fact, cleavage is usually not considered as growth, since, the mass of the fertilized egg is merely subdivided into smaller and smaller units (see **Phillips**, 1975). Though, at the end of cleavage the total mass and size of the organism remains more or less the same, when cell multiplication, morphogenesis and differentiation may have occurred.

3. Accretionary growth. In this case, the growth of the animal is based to a greater or lesser degree on the activity of special cells, retaining their ability to divide mitotically, while other cells have lost this ability more or less completely and, thus, cannot proliferate any more. These latter cells are the differentiating cells of body, performing various physiological functions necessary for the maintenance of the animal's life. The former cells may be called the **reserve** or **stem cells**, for they are undifferentiated cells which provide a supply of new cells, capable of supplementing and replacing, in case of necessity, the functioning differentiated cells.

The epidermis of terrestrial vertebrates may serve as an example of this type of growth. In the outer layers of the epidermis, the cells do not divide and do not grow. Their cytoplasm becomes keratinized, thus, forming a protective layer on the surface of the skin. The fully keratinized cells are no longer vital and are perpetually being peeled off. The **Malpighian** or **generative layer** of the epidermis consists of cells which are not keratinized and which possess the capacity to proliferate. Later, the outermost cells of the layer replace the keratinized cells, undergoing keratinization in their own turn.

Likewise, the cells of bone tisue (the osteocytes) do not grow and do not divide. The growth of bone depends on the activity of the cells of the periosteum, which are capable of dividing and can become osteocytes, while they secrete additional quantities of intercellualr bone matrix.

There are many organs, in which centres are found as regions or layers which alone contain proliferating cells. For example, in the vertebrate eye, the proliferating cells are found in a ring in the region of the ciliary body. From there new cells are added on to both the retina and the iris. In the intestine of vertebrates too, proliferating cells are found at the bottom of the intestinal glands.

Finally, the blood corpuscles, in particular the erythrocytes, are differentiated non-dividing cells, and their supply is continuously replenished or renewed at the expense of the dividing non-differentiated cells in the erythropoietic tissues (*i.e.*, red bone marrow).

Various trends of cell growth. In organ rudiments it very often happens that the initial group of cells laid down for an organ or part of the embryo undergoes a period of proliferation by mitotic division and after a while the division ceases, and the cells of the organ rudiment begin to differentiate. This trend of growth can be observed in the development of the nervous system of vertebrates, in the development of the pancreas and in the development of many other organ rudiments. However, generally when cell death equals cell growth, no further increase in over all growth occurs, stabilizing the growth (weight) of the organism. The transition from the undifferentiated and proliferating cells to non-proliferating and differentiated cells may be quite abrupt, occurring in a matter of hours, but such a transition has the following trend :

1. In some type of tissues, the cells, after entering the phase of differentiation, continue to divide by mitosis, at least for some time. For example, **chondrocytes** of hyaline cartilage may divide several times after being enclosed in cartilage's matrix which they produce as a visible sign of differentiation. Similarly, **fibroblasts** of the connective tissue are supposed to possess the ability to proliferate even when they are enclosed by their own secretions —the fibres of connective tissue.

2. Some differentiated cells, such as nerve cells, although having lost the ability to divide by mitosis, retain the capacity for growth (increase in size).

Thus, the nerve cells never divide but, after division stops, they increase in size to a considerable degree, becoming the largest cells in the vertebrate body, except for the oocytes.

3. A special case of growth, after mitosis has ceased, is presented by cells becoming polyploid in certain tissues ; the genome becomes duplicated once or several times without the cells dividing. For example, among insects salivary gland cells of the *Drosophila* and other Diptera are giant cells with polytene chromosomes. In vertebrates, some liver cells contain a double amount of DNA and, thus, are **tetraploid** (polyploid) and likewise, the megakaryocytes of the bone marrow, producing blood platelets, are also large-sized polyploid cells.

14-3.3. Cellular Control of Growth

There are three aspects of growth : when growth starts, rate of growth and when growth stops. To understand how tissues and organs in a multicellular organism begin to grow, one needs to study them before they begin to grow. This is because growth is the result of coordinated activity of cells, based on the following parameters : number of cells, how many cells divide and how often they divide, the amount of cell death, matrix secretion, etc.

1. A fundamental parameter for growth is the number of cells in the primordium (in case of animals having indeterminate or regulative development) or in the organism (in case of individuals with determinate development). Even when cell number is determinate, as it is in *Caenorhabditis elegans*, many more cells are produced than actually contribute to the organs, the excess being removed by programmed cell death. Thus, out of 671 cells about one-sixth of the cells undergo programmed cell death and are removed as the embryos transform into adults (**Sulston** *et al.*, 1983).

2. Each organ rudiment has its own intrinsic rate of growth regulation and that rate too, is based upon factors, such as initial size of the rudiment ; what proportion of its cells divides ; the length of cell cycle and the minimum duration between its subsequent cell divisions. Cell kinetics are critical at these early stages of growth. Stem cells retain such proliferative ability throughout life, while neoplasia often sets proliferation onto an uncontrollable course.

3. The number of cells in a primordium is often a result of cell migration. Neural crest cells, primordial germ cells, and somatic cells are three examples from vertebrate development, but migration is also seen in invertebrates, as in the ingrowth of primary mesenchyme during gastrulation of sea urchin and in the migration of neoblasts to amputation surfaces during regeneration in planaria. Adhesive and locomotory properties of the migrating cells, and the nature of the substrate through which they migrate tend to influence how many cells reach the organ rudiment.

4. Inductive tissue interactions also play a role since many cells do not display specific growth properties until they have been induced to differentiate. In fact, the initial consequence of many inductive tissue interactions is stimulations of proliferation within the induced cells (*e.g.*, localized stimulation of mesenchymal cell proliferation provided by the scleral papillae in the eyes of embryonic chicks ; **Fyfe** and **Hall**, 1983).

The **cessation of growth** can be controlled in a number of ways. An organ may run out of stem cells, as happens in the growth plates of mammalian long bones. Once the last stem cells have differentiated, all potential for subsequent growth is gone. Note that retaining a stem cell population necessarily means unlimited or **indeterminate growth**. Intrinsic, cell-specific **inhibitors** stop growth by inhibiting proliferation of stem cells. A whole class of tissue-specific growth inhibitors, the **chalones** which act via negative feedback to suppress mitosis, have been postulated as regulators of final size. Some, such as the **epidermal chalone** (**Houck**, 1976) and **erythropoietic chalone** are evidently known but none have been isolated and chemically known, so far (**Maclean** and **Hall**, 1987). Chalones accumulate as the tissue grows, until they reach a threshold concentration, when they switch off further division of the particular cell type. For growth to restart, as it can do following injury or during regeneration of some tissues, chalone concentration has to fall.

14-3.4. Growth Factor

Hormones are one obvious class of growth factors, exerting their stimulation through promotion of cell division. Many hormones also stimulate cell differentiation. For example, **somatomedin** stimulates both DNA synthesis and synthesis of luxury or cell-specific molecules by chondroblasts. Hormones, therefore, provide one mechanism whereby growth is coupled to differentiation.

A whole class of hormone-like factors that funciton as **growth regulators** have now been identified. These are also called **growth factors** and most of them are small polypeptides. Some of them act on a single cell type, some on several cell types. **Nerve growth factors** stimulate the proliferation of sensory and sympathetic ganglia but also enhance outgrowths of neurites from the nerve cell body. **Erythropoietin** stimulates the proliferation of precursors of erythroblasts. **Thrombopoietin** stimulates the proliferation of blood platelets. **Fibroblast** and **epidermal growth factors** act as stimulators of mesenchymal and epithelial cell growth (see **Maclean** and **Hall**, 1987).

14-3.5. Limitation of Size and Form and Allometric Growth

Certainly, there are great variations in the sizes of animals, particularly between different species, but often even within them. These depend on many factors such as genetic constitution, nutrient supply and toxicity, the ratio of surface to volume and structural limitations (**Sussman**, 1963).

Further, the growth of different organs and parts of the same animal (embryo) very seldom goes on at the same rate. As a rule some grow faster and some grow slower. The result is that the proportions of the animal change with growth. For example, in vertebrates the central nervous system and the sense organs are distinguished by their particularly low rate of growth, and the size of these organs diminishes relatively throughout the whole embryonic and post-embryonic development. In the early embryo of a chick, for instance, the head is at first quite as large as the rest of the body, and a large portion of the head is made up by the relatively enormous eyes. However, at the time of hatching, the head becomes much smaller than the body, and in the adult fowl the head is relatively small.

J. Huxley (1932) suggested that if two parts of an animal grow at different rates their size at any given moment are in a simple relationship to each other. The relationship is expressed by the following fomula, known as the **formula of allometric growth** :

$$y = b\,x^k$$

where y is the size of one of the organ, x is the size of the other organ, b is a constant and k is known as **growth ratio** or **allometry constant**, because it shows the relations of the rates of growth of the two parts which are being compared. If the growth ratio equals 1 the two organs grow proportionally or **isometrically**. The constant b in this case shows the proportion of one organ to the other : $b = y/x$. If the growth ratio does not equal 1, the growth is disproportional or **allometric**, the relative size of the two organs changing as growth proceeds. If the growth ratio is greater than 1, organ y grows at a quicker pace than organ x. In this case, organ y is called **positively allometric** to organ x. If the growth ratio is less than 1, organ y grows at a slower pace than organ x ; it is **negatively allometric** to x.

It has been found that organs attaining especially large proportions in certain animals often owe their increased size to positive allometric growth. In the fiddler crab, *Uca pugnax*, one of the chelae in the males attains an extraordinarily large size, up to 38 per cent of the weight of the body as a whole. The large chela is positive allometric with respect to the rest of the body, with a growth ratio of 1 : 62. This means that in a young male crab the chelae are very nearly equal in size and that the relative size of the larger chela increases as the crab grows (Fig., 14.9).

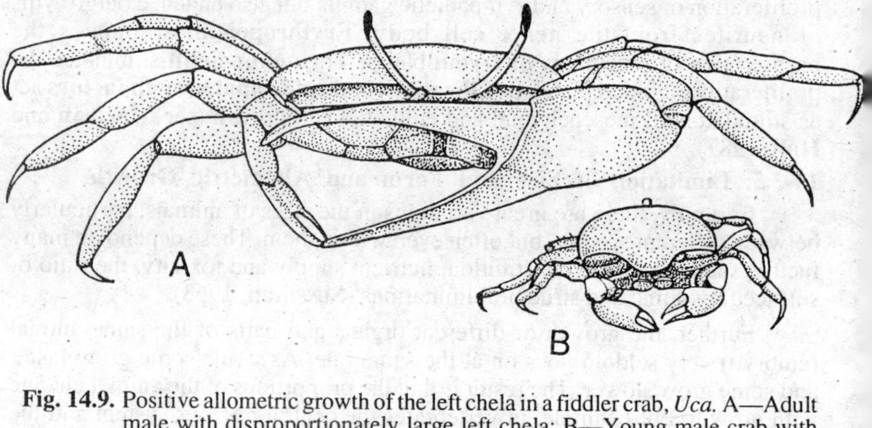

Fig. 14.9. Positive allometric growth of the left chela in a fiddler crab, *Uca*. A—Adult male with disproportionately large left chela; B—Young male crab with both chela of equal size (after Balinsky, 1981).

Allometric growth is also intimately related with the phenomenon of **polymorphism**, *i.e.*, differences in form between adults. **Huxley** (1932) has provided a simplest example of this relationship in the earwig, *Forficula*, in which two types of male adult occur, one with short, the other with long, abdominal forceps. The forceps follow an allometric growth. There is sometimes an extra-ecdysis in males, producing adults which are not only larger overall, but in which the forceps are disproportionately large. Likewise, the soldier caste of polymorphic social insects such as the ant *Camponotus castaneus* is produced by allometric growth. Such soldiers have relatively enormous heads and jaws in comparison to bodies (**Wilson**, 1971).

14-4. DIFFERENTIATION

The establishment of the germ layers by the end of gastrulation gives rise to layers of cells that are differentiated but remain prefunctional. For example, the cells that make up the neural plate at this time are different from the surrounding epidermal ectoderm and from their previous state of presumptive neural ectoderm. They are still unable to function, however, they cannot carry out physiological activities until they undergo a second more special type of differentiation, called **histological differentiation** or **cytodifferentiation**. As a result of this second level of differentiation, new proteins appear in the cell and a specialized cell results. These special proteins allow certain cells to perform one function and not another, *i.e.*, muscle cells contract, nerve cells conduct and gland cells secrete.

Cytodifferentiation is achieved as follows : 1. Under the influence of maternal genes, the egg cytoplasm is organized. 2. The egg cytoplasm is selectively distributed among blastomeres, so that the daughter cleavage nuclei come to lie in different cytoplasm. 3. The daughter nuclei undergo changes in their morphological and/or functional state under the influence of different cytoplasmic stimuli and produce a new set of RNAs. These RNAs ultimately govern the differentiation of egg and sperm cell. Certain cells such as neurons and erythrocytes undergo extreme specialization and become end cells and can no longer indulge in mitotic divisions. Such cells are called **terminally differentiated cells** (For more information about cytodifferentiation, see Chapter 24).

14-5. HISTOGENESIS AND ORGANOGENESIS

After gastrulation and neurulation the embryo enters the next phase of its development, called **organogenesis**. Histogenesis is the name given to the process by which the undifferentiated or pre-functional cells of the germ layers differentiate into specialized functional cells of body. In other words, histogenesis is the coordinated differentiation of groups of cells into a tissue, organ or appendage. It is the first step in organogenesis.

Each tissue consists of a group of similar cells that carry out comparable functions united by intercellular substances. The cells of the tissues are specialized cells, but epithelial and connective tissues are considered less specialized than nervous and muscle tissues. Their general form and internal structure are modified during development in such a way that each tissue can carry out a particular function. The appearance of myosin, for example, responsible for a cell's ability to contract, is the first apparent difference in the cells of future muscle tissue.

The formation of an intercellular substance, usually secreted by the cell and often referred to as the **matrix** of the tissue separates each cell from a neighbouring cell. Epithelial cells, which are closely associated with one another in lining external and internal surfaces, have little intercellular substances. In some cases, however, it is the matrix that is important in giving form and function to the body. The intercellular substance of bone and cartilage are examples of this type of tissue. Blood is a unique tissue in which the cells, the formed elements of blood, are suspended in fluid matrix (the plasma). The derivation (*i.e.*, specialized cells or **cell types**) of three germinal layers can be summarized as follows :

A. Ectoderm

1. Epidermal ectoderm. Epidermis of skin, skin derivatives such as hair, feathers, nails, scale, claws, sebaceous glands, sweat or sudoriferous glands and mammary glands ; epithelium of oral cavity (stomodaeum), nasal cavity, para-nasal sinuses, salivary glands and anal canal (proctodaeum) ; olfactory organs, lens of eye, part of cornea, internal and external ear of mammal (i.e., ear vesicles or labyrinth) ; anterior pituitary (adenohypophysis).

2. Neural ectoderm (Neural tube). Brain, spinal cord ; peripheral nerves (cranial and spinal nerves), ganglia ; rods and cones of retina of eye, optic nerve ; various receptors of skin ; auditory and taste receptors ; posterior pituitary (neurohypophysis).

3. Neural crest cells. Sensory neurons and satellite cells of the spinal ganglia ; cholinergic (*i.e.*, acetylcholine-producing) neurons and satellite cells of the ganglia of the parasympathetic nervous system ; andrenergic (*i.e.*, noradrenaline-secreting) neurons and satellite cells of the ganglia of the sympathetic nervous system ; Schwann and glial cells of the peripheral nervous system ; cells of adrenal medulla that produce adrenalin and noradrenalin ; parafollicular or C-cells of the thyroid gland (*i.e.*, cells that produce the calcium-regulating hormone, **calcitonin**) ; all of the pigment cells (chromatophores) of body (*e.g.*, of hair, feathers and skin) except those of the pigmented retinal epithelium ; cartilage of visceral and facial skeletons ; bone of the cranofacial skeleton ; dermal connective tissue of the face and ventral regions of the neck ; ciliary muscles of the eye ; portion of the striated muscles in the face and neck ; dental papillae and odontoblasts of the teeth ; fibroblasts of the corneal stroma ; fibroblasts of endothelia ; myoblasts and fibroblasts of arteries arising from the aortic arches ; fibroblasts of glands such as pituitary, salivary, thyroid, parathyroid and thymus ; mesenchyme of the dorsal fins of amphibians.

B. Endoderm

Epithelium of digestive tract (*e.g.*, pharynx and tonsils) and its derivatives, the liver, pancreas and gall bladder ; epithelium lining respiratory tract (*i.e.*, larynx, trachea, bronchi, and alveoli of lungs) ; epithelium of urinary bladder and some endocrine glands such as thyroid, parathyroid, pancreas and thymus ; epithelium of auditory tube, tympanum and eustachian tube ; primordial germ cells of some animals ; epithelium of prostate and bulbourethral gland, vagina, urethera and associated glands.

C. Mesoderm

1. Notochord. Becomes well developed in *Amphioxus* ; in other chordates it is either obliterated (Ascidians) or replaced by vertebral column (vertebrates).

2. Epimere. (1) Dermatome—dermis of skin. (2) Sclerotome—vertebral column. (3) Myotome—skeletal muscle and appendicular skeleton (in fish).

3. Mesomere. Excretory organs (*e.g.*, kidneys), excretory ducts, urethera, urinary bladder (in fish) ; epithelium of gonads (ovaries and testes) and genital ducts ; adrenal cortex (*i.e.*, steroidogenic cells secreting steroid hormones such as progesterone, estrogen and androgen).

4. Hypomere. (1) Somatopleure (Somatic mesoderm) : Parietal peritoneum, *i.e.*, lining of pleura, pericardium and peritoneum ; limb muscles (higher animals) and trunk muscles (in chick). (2) **Splanchnopleure (Visceral mesoderm)** : Visceral peritoneum ; mesenteries ; heart, cardiac muscles ; bone marrow, blood cells, blood vessels, smooth muscles of viscera. (3) **Coelom** : Body cavity.

Formation of organ systems. Groups of specialized cells (tissues) can function independently, but usually the four fundamental tissues combine to form the **organs** of the animal. An organ (*viz.*, kidney, heart or stomach) is a large functional unit whose combination of tissues enables it to carry out some general function. Certain organs, in turn, cooperate with other related organs to constitute **organ system**. The organ systems carry out the main functions of the body, such as respiration, excretion and reproduction. The organ systems of adult vertebrate body may be categorized by the following scheme, reflecting their general germ layer derivation :

A. Ectodermal layer

1. Integumentary system;
2. Nervous system

B. Endodermal layer

3. Digestive system;
4. Respiratory system

C. Mesodermal layer

5. Muscular system;
6. Skeletal system;
7. Circulatory system;
8. Excretory system;
9. Reproductive system.

REVISION QUESTIONS

1. Why is it difficult to define the biological growth ? Describe growth at the molecular, cellular and tissue level.
2. Why is biological growth studied at the level of cell ? Explain the three types of cell growth, *i.e.*, auxetic, multiplicative and accretionary growth.
3. How the growth of cell is maintained ?
4. Write an essay on morphogenesis ?
5. What is tubulation ? Describe the process of formation of primary organ rudiments.
6. Give the summary of the cell types which are derived from the three fundamental germinal layers.
7. Write short notes on the following :

   (i) Allometric growth ; (ii) Cell sorting ; (iii) Flexion of embryo ; (iv) Histogenesis ; (v) Morphogenetic movements ; (vi) Embryonic mesenchyme; (vii) Neural crest cells ; (viii) Differentiation.

15

EMBRYOGENESIS OF A SIMPLE ASCIDIAN

The ascidians are those marine prevertebrate chordates which belong to class **Ascidiacea** of subphylum **Urochordata**. The knowledge of the development of ascidians has thrown unexpected light on their true taxonomic position. The adult ascidian is a sessile animal with no organs of locomotion and no notochord but with a nervous system of a very primitive nature. The adult animal had been classed as a near relative of molluscs until **Kowalevsky** (1866) discovered that the larvae of the ascidians possess a well-developed dorsal brain and spinal cord, a definite notochord and lateral bands of muscles in the tail (*i.e.*, organs that are typical for the vertebrates). The ascidians, therefore, considered as belonging to the same phylum as the vertebrates, the phylum Chordata. The developmental process that transforms an advanced larva of ascidians into the primitive adult is called **retrogressive metamorphosis**.

In this chapter, we are mainly concerned with the embryogenesis of the members of order **Ascidiae simplices**, *i.e.*, *Herdmania, Ascidia, Phallusia, Ciona, Styela (Cynthia)*, etc. There exists no detailed account of the embryogeny of the Indian ascidian species, *Herdmania,* but the embryogeny of many simple ascidians, like *Styela, Ciona, Phallusia* is well worked out. The pioneer work of **Conklin** (1905, 1911), who first of all followed the development of *Styela partita, Ciona intestinalis* and *Phallusia*, cell for cell, until the attainment of the larval form, is found to apply well to all the Ascidiae simplices. Therefore, we will follow a general account of embryogeneis of Ascidiae simplices.

15-1. REPRODUCTION

Most simple ascidians are hermaphrodite and the gonads or **ovotestes** are paired structures. In *Herdmania*, two large, lobulated gonads lie at the left

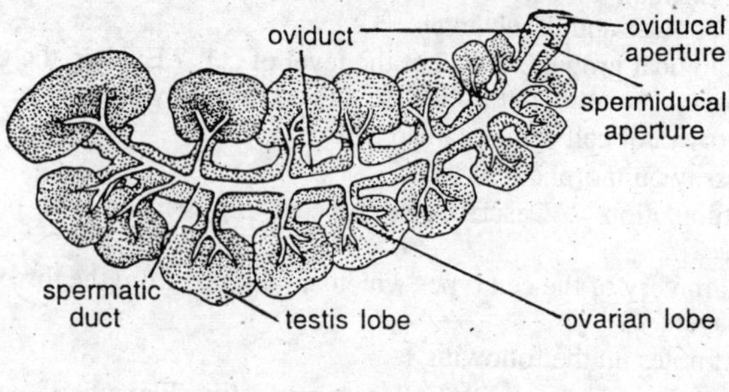

Fig. 15.1. Left gonad of *Herdmania* seen from inner side (after Jordan and Verma, 1981).

and right sides of animal body and their genital ducts open into the atrial cavity, at the base of the atrial siphon. Each lobe of gonad has testicular tissues at the outer side and ovarian tissues towards inner side (Fig. 15.1.). The subsequent spermatogenesis and oogenesis in the testicular and ovarian tissue, produce **sperm** and **ova**, respectively. Most simple ascidians, including *Herdmania*, are **oviparous** and cross fertilization occurs in sea water, external to the animal body. Cross fertilization is ensured by **protogyny** of *Herdmania* and other ascidians, *i.e.*, the ova are matured and released from the ovary earlier (*i.e.*, in the young animal), while the spermatozoa are matured in older animal and thus, self-fertilization is avoided.

15-1.1. Gametes

A mature **spermatozoon** or sperm of *Herdmania* is flagellated, microscopic and about 4 μm in length. Each has an anterior broad and nucleated **head** capped by a beak-like **acrosome** in front, a middle piece with **neck** and a very long straight **tail**. According to **Das** (1936), the sperms of *Herdmania* are **polymorphic** with at least three types of sperms having acrosome either shorter, equal or longer than head.

The **ova** or **unfertilized eggs** of simple ascidians are **microlecithal** and **bipolar** having distinct animal and vegetal poles. They measure between 0.1 to 0.3 mm in diameter. The ripe ovum of *Styela partita* is laid in the germinal vesicle stage and is full of ooplasm that is well differentiated into variously coloured distinct regions. This is called **egg cytoplasmic localization** (see **W.R. Jeffery**, 1988). For example, ooplasm rich in slaty gray yolk granules is located in the centre and is called **endoplasm** since it contains materials for the development of prospective **endoderm** and **notochord**. Surrounding the

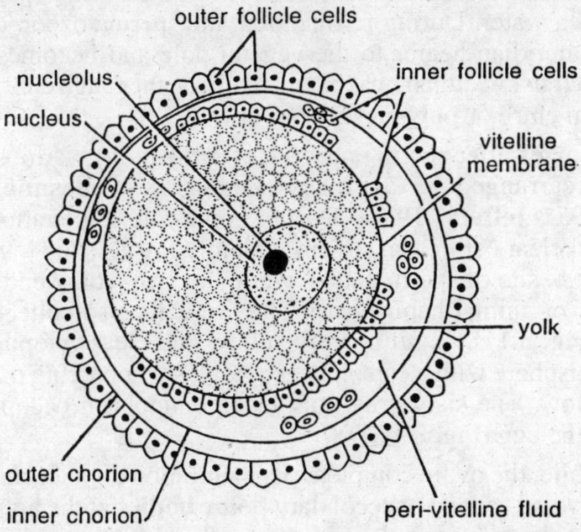

Fig. 15.2. Structure of a mature ovum of *Herdmania* (after Jordan and Verma, 1981).

yolky endoplasm is a peripheral or cortical layer of protoplasm in which are embedded yellow pigment granules and many minute mitrochondria. This yellow cortical layer of ooplasm is called **myoplasm** since it is found to contain materials for prospective **mesenchyme** and **muscle cells**. A region of clear cytoplasm surrounds the germinal vesicle (or nucleus of the ovum) at the animal pole and is called **ectoplasm**, since it contains materials for the prospective **epidermal** and **neural cells**. The position of egg nucleus close to the animal pole stresses the polar organization of the egg, but there is no trace of a bilateral organization : all meridians of the ovum are exactly alike. Mature ovum of *Herdmania* (Fig. 15.2) is about 0.3 mm in diameter and contains an eccentric nucleus with a prominent nucleolus. It is surrounded by the following three types of **egg membranes** : 1. **Vitelline membrane** ; 2. **Inner chorion** ; and 3. **Outer chorion**. The thin vitelline membrane is secreted by the ovum itself and closely applied to it. The follicular cells of the ovary secrete the two chorion membranes which are **primary egg membranes**. The ovum lies eccentrically in the **perivitelline fluid** enclosed by the space between the vitelline membrane and the inner chorion. The perivitelline space also contains a large number of small-sized, red-granules-containing cells, called **inner follicle cells** or **test cell** , some of which float in the fluid but most of them attached to the vitelline membrane. These cells nourish the ovum and secrete an enzyme which digests the chorions to facilitate hatching. Attached to external surface of the outer chorion are many large-sized vacuolated **outer follicle cells**. These cells serve to keep the ovum floating in water.

As soon as the ovum enters the sea water, it begins to mature ; the germinal vesicle (GV) breaks down and the first metaphase spindle appears at the animal pole. The ovum remains in this state until fertilized.

15-1.2. Fertilization

In ascidians, **cross-fertilization** and **external-fertilization** take place, when mature sperms of older individuals and ripe ova of younger individuals are liberated in their cloaca and pass out through their atrial apertures into the surrounding sea water. During fertilization, the spermatozoon penetrates the ovum at any meridian nearer to the vegetal pole and becomes the **male** or **sperm pronucleus**. Ascidian ova (or eggs) rely on the tough chorion to prevent multiple sperm entry or **polyspermy (Grant**, 1978).

Five minutes after sperm entry (*i.e.*, fertilization), various cytoplasmic regions are rearranged by a process known as **ooplasmic segregation** (**Conklin**, 1905 ; **Jeffery**, 1984). During ooplasmic segregation, the yellow myoplasm and clear ectoplasm stream violently towards the lower or vegetal hemisphere (*i.e.*, site of sperm entry), while the gray yolky endoplasm flows into the upper or animal hemisphere. The test cells also collect at the lower hemisphere (Fig. 8.11 C). Ultimately yellow and clear cytoplasm collect in the lower hemisphere with a cap of clear cytoplasm appearing over the yellow cytoplasm below. The slaty gray yolk-laden endoplasm occupies the upper two-thirds of the egg (Fig. 8.11 D).

Meanwhile, the ovum completes its maturation division and releases the first or primary and second or secondary **polar bodies** at the animal pole. The resultant haploid nucleus of the ovum is called **female** or **egg pronucleus** which descends to meet the sperm pronucleus just below the equator. Further, as the male pronucleus penetrates deeper in the cytoplasm on its course toward the female pronucleus, the yellow myoplasm reverses its movement and stream upward, but this time on one side of the egg only, namely, on the side of egg where the spermatozoon has entered in the egg. Here, this myoplasm extends into a **yellow crescent**, called **mesodermal crescent** just below the egg's equator. The yellow crescent marks the future ventral-posterior region of zygote and thereby establishes the **dorsal-ventral plane** of the egg to give it a distinct **bilateral symmetry**. At the same time, another crescent but of light gray cytoplasm appears subequatorially on the opposite side of egg. Since this crescent gives rise later to notochord, so is also called **notochordal crescent**. Subsequently, the ectoplasm moves back into the animal hemisphere and endoplasm returns to its original position in the vegetal hemisphere.

Thus, due to ooplasmic segregation, fertilized egg of *Styela* becomes **bilaterally symmetrical** and **mosaic** or **determinate type** and it shows the following four presumptive areas with predictable future : 1. an upper hemisphere of clear cytoplasm containing little yolk granules, few mitochondria and materials released by germinal vesicle (GV) ; 2. a mitochondria rich yellow crescent postero-ventrally ; 3. a gray crescent antero-dorsally, and 4. a vegetative area of slaty gray yolk cytoplasm (Fig. 8.11).

15-2. FATE MAPS AND CELL LINEAGE MAPS OF ASCIDIANS

Because of the natural differences in pigmentation, the developmental fates of above-mentioned four ooplasmic regions have been worked out with

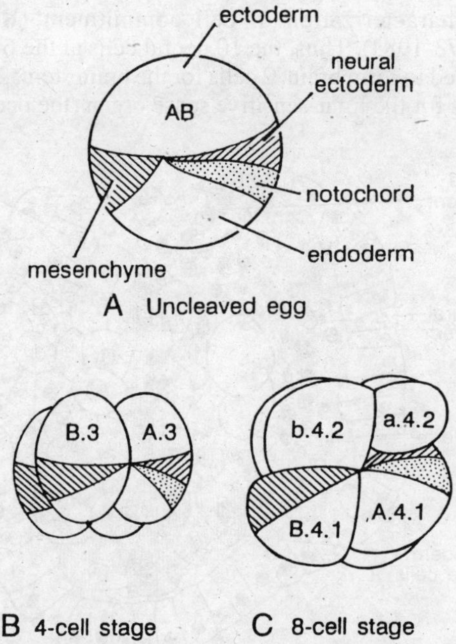

Fig. 15.3. Mode of distribution of the cytoplasm in cleaving egg of *Styela*. A—Uncleaved fertilized egg; B—4-cell stage; C—8-cell stage (after Maclean and Hall, 1987).

considerable precision. In a fate map of *Styela* (Fig. 15.3) the upper clear plasm represents prospective epidermal ectoderm, so is called **ectoplasm,** the yellow crescent is prospective mesoderm, so is called **mesoplasm** and the dark gray yolky plasm is prospective endoderm, so is called **endoplasm** (**Conklin**, 1929). The pale gray crescent cytoplasm actually encompasses two developmental prospects an upper neural ectoderm (called **neuroplasm**) and a lower notochord (called **chordaplasm**).

The evidence that these prospective organ regions actually possess biochemical differences foreshadowing the parts they are to produce, is demonstrated by centrifuging eggs before they begin cleaving. Such treatment serves to shuffle the regions so that they come to occupy abnormal positions. An embryo will still develop from a centrifuged egg, but its tissues and rudiments of organs have a chaotic arrangement. It is an orderly chaos, however, in the sense that each part develops in the position where its cytoplasmic precursor came to lie as the result of centrifugation (Fig. 15.4).

Conklin could construct a fate map of embryo of *Styela partita* up to 64-cell stage which shows that 10 neural and 4 notochordal cells with pale gray cytoplasm ; 10 endodermal cells with dark gray cytoplasm ; 14 mesodermal cells with yellow cytoplasm ; and 26 ectodermal cells with clear cytoplasm. These early fate maps have been extended by **Reverberi's** experiments involving isolation of individual cells, and by **Whittaker's** construction of **cell lineage maps** using cell isolation coupled with biochemical characterization of cell commitment (**Reverberi**, 1977 ; **Whittakcr**, 1973-1983). Thus, thc 10 neural cells at the 64-cell stage consist of 4 cells destined to form brain, 2 cells for the brain stem, 2 cells for the spinal cord and 2 cells for the light-sensitive sense organ (the ocellus).

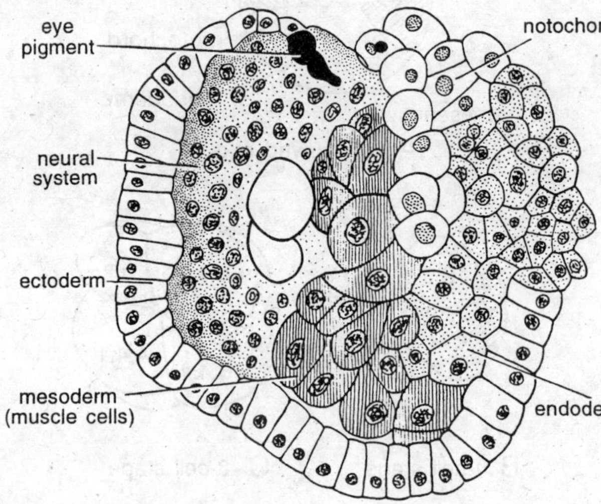

Fig. 15.4. An embryo develops from a centrifuged fertilized egg of *Styela partita* (after Balinsky, 1970).

Whittaker stained the enzymes specific for particular cell types and used these histospecific markers for cell determination : **acetylcholinesterase** (**AChE**) for larval muscle cells, **tyrosinase** for pigment cells (*i.e.*, brain melanocytes) and **alkaline phosphatase** for endodermal cells (*i.e.*, gut cells). Either cell of a 2-cell embryo can produce AChE, but by the 8 - cell stage only the 2 cells closest to the vegetal pole of the embryo (*i.e.*, cells B.4.1 and A. 4.1 in Fig. 15.3) are capable of synthesizing AChE. That these are the only cells determined for muscle cell differentiation at the 8-cell stage was confirmed by ablating them. The resulting embryos were unable to form muscle cells (**Maclean** and **Hall**, 1987).

15-3. DEVELOPMENT OF EGG FROM CLEAVAGE TO LARVA FORMATION (HATCHING)

The **cleavage** of ascidian egg is **holoblastic** (complete), somewhat **unequal** and **determinate**, *i.e.*, fate of the blastomeres is predetermined. The first cleavage plane of *Styela* is meridional and it passes through the polar axis and bisects the mesodermal (*i.e.*, yellow crescent) and chordaneural (*i.e.*, gray crescent) crescents. The second set of cleavage divisions occurs at the right angles to the first, but is vertical cutting off two slightly smaller blastomeres at the posterior side and two larger ones in front. Division of each of the four blastomeres is in the horizontal plane, producing an eight cell stage that consists of an upper (apical) tier of slightly smaller blastomeres and four larger lower (basal) blastomeres. The eight-celled embryo undergoes three more sets of cleavages and consequently a 64-celled compact **blastula** is produced.

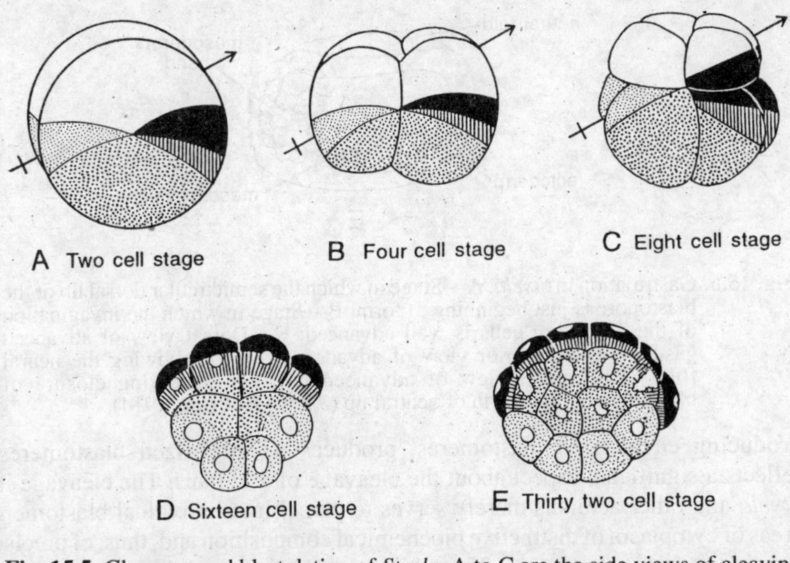

Fig. 15.5. Cleavage and blastulation of *Styela,* A to C are the side views of cleaving eggs, while, D to E are the views of the cleaving eggs from vegetal pole (After Torrey, 1971).

Peculiar feature about ascidian cleavage. The fact that an egg of *Styela* has very small amount of yolk, but its holoblastic cleavage, instead of

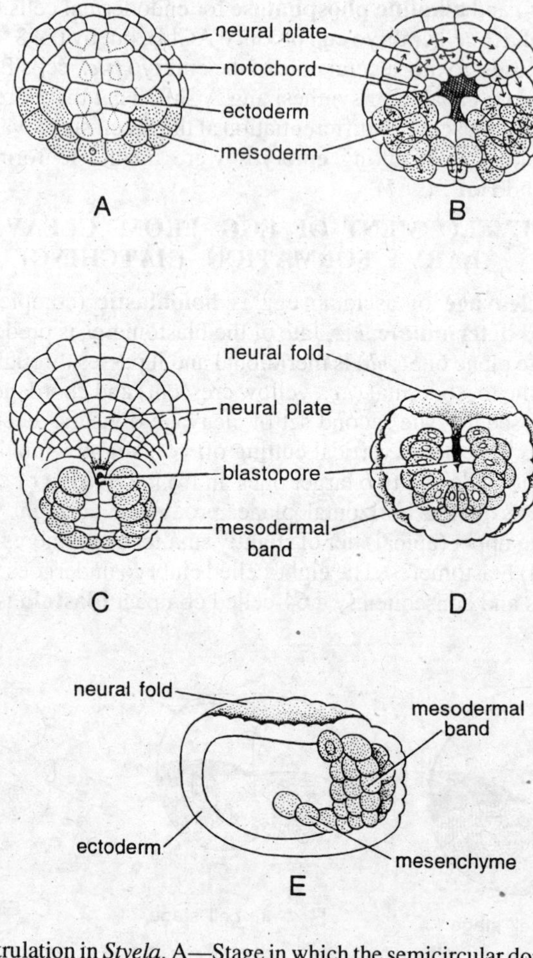

Fig. 15.6. Gastrulation in *Styela*. A—Stage in which the semicircular dorsal lip of the blastopore is just beginning to form; B—Stage in which the invagination of the endoderm cells is well advanced; C—Dorsal view of advanced gastrula; D—Posterior view of advanced gastrula, showing the neural folds; E—Lateral view of advanced gastrula, showing closure of blastopore by upgrowth of ventral lip (After MacBride, 1914).

producing equal-sized blastomeres, produces unequal-sized blastomeres, reflects a significant aspect about the cleavage of ascidians. The cleavage in *Styela* and other acidians merely serves to parcel into individual blastomere areas of cytoplasm of distinctive biochemical composition and, thus, of precise developmental potentiality. That this parcelling is real indeed has been demonstrated by experiments in which certain blastomeres are injured so as to prohibit them from further developmental participation. If, for instance, the four right-hand blastomeres of an 8-cell stage are injured, the left-hand blastomeres produce only a left half embryo, or if the four rest blastomeres are immobilized, the four in the front produce only those parts they are normally destined to produce, namely, epidermis, neural tube, notochord, and endoderm.

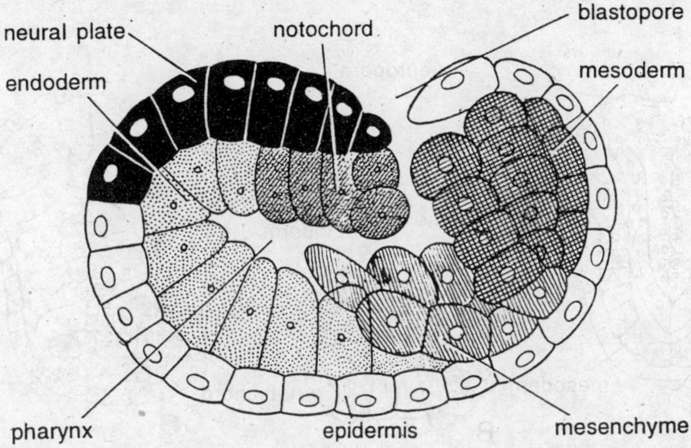

Fig. 15.7. Optical sagittal section of gastrula of *Styela partita* (After MacBride, 1914).

Gastrulation by **invagination** occurs following the sixth cleavage and is completed by the time of the seventh cleavage, *i.e.*, between the 64-cell and 128-cell stage (at the mid-gastrula stage, there are about 76 cells). On the completion of gastrulation, the large ventral opening, the **blastopore** of the gastrula closes, and the embryo begins to lengthen along the embryonic axis indicated by the plane of the first cleavage. The upper surface or **neural plate** of the elongating gastrula then invaginates along the mid-line to form a groove which subsequently becomes a tube, the **neural tube**, and the stage during which neural tube formation occurs, is called **neurula**. The anterior portion of neural tube remains wider and so, develops into **brain-vesicle** ; the remainder portion of neural tube is narrow and constitutes the spinal cord. For a short time, the front part of the brain-vesicle communicates with the exterior by a pore termed the **neuropore** (Fig. 15.8). The hinder end of the neurula becomes extended into a **tail**, into which the neural tube, notochord, mesoderm and mesenchyme are extended. With further embryonic development, the neurula develops into an ascidian **tadpole larva** ; which bursts the egg membranes and becomes free swimming.

Neural induction. G. Reverberi (1960, 1961) has investigated that if pairs of blastomeres are isolated at the eight-cell stage, the anterior vegetal pair, posterior animal pair, and posterior vegetal pair differentiate in keeping with their prospective fate. However, the anterior animal pair do not differentiate into neural tissue as expected and require the presence of adjacent anterior vegetal blastomeres (having the chordaplasm) in order to differentiate and to become nervous tissue (see **Berrill** and **Karp**, 1978). According to him, the destiny of the presumptive "neural ectoderm" is not fixed at the eight-cell stage, at which stage all the other territories are already fixed or "determined". The presumptive "neural ectoderm" is converted into the neural system only under the inductive action of the immediately underlying chordal (and endodermal) cells.

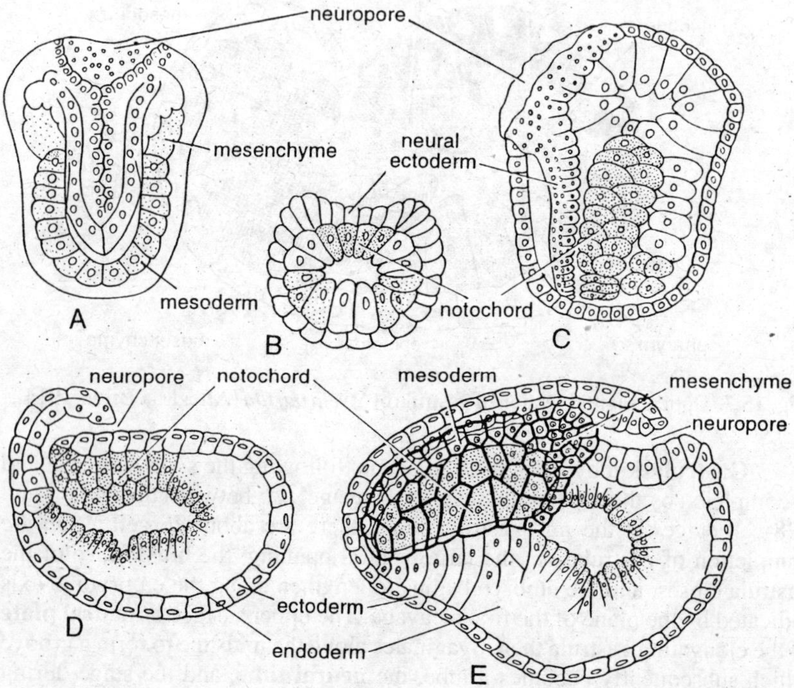

Fig. 15.8. Neurula stage of development of ascidian. A—Dorsal view and B—Side view of neurula of *Styela*, and C—Cross section of open neural-plate stage or younger embryo; D—Early and E—Late neurula of *Clavelina* from side, showing neuropore, neural tube, notochord, and superimposed tail mesoderm and mesenchyme (after Berrill, 1971).

15-4. TADPOLE LARVA AND RETROGRESSIVE METAMORPHOSIS

15-4.1. Structure of Tadpole Larva

The fully developed tadpole larva of ascidians is transparent, minute, about 1.5 mm long and highly motile (Fig. 15.9). Its entire body is covered by a thin **tunic** or **test** secreted by the ectoderm. The larval body shows two distinct divisions : an anterior short oval **head** or **trunk** and a posterior long **tail**.

1. Head or trunk. The cylindrical trunk of larva has at its anterior end three stalked **adhesive papillae**, two being dorso-lateral and one ventro-lateral. These are made of secretory ectodermal cells and serve for attachment

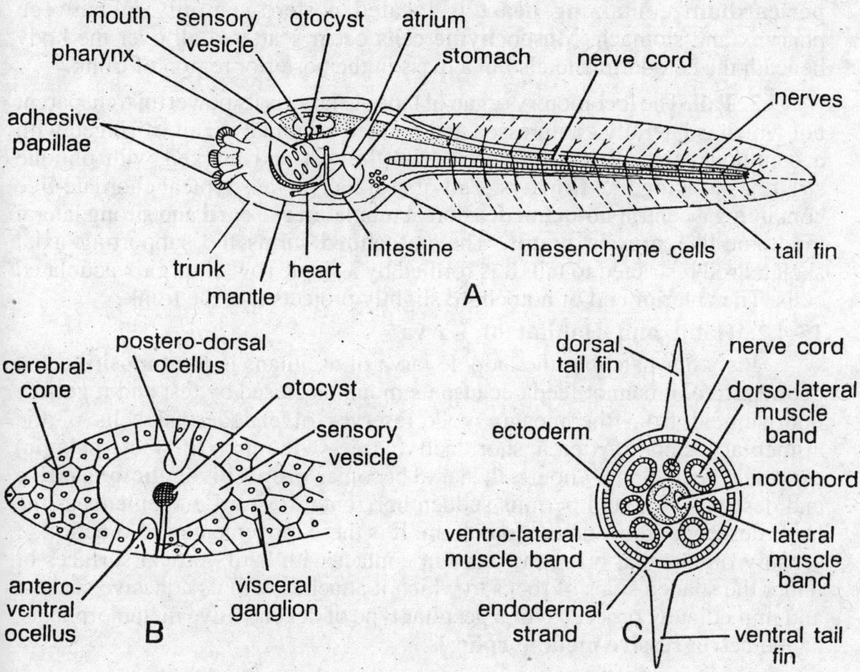

Fig. 15.9. A—Tadpole larva of *Herdmania* in left lateral view; B—Sensory vesicle of larva enlarged; C—Tail of larva in T.S.

during metamorphosis. Situated in the dorsal part of trunk is an oval, hollow **sensory vesicle**, **brain vesicle** or **brain** which is an anterior extension of nerve cord or spinal cord of the tail. Its wall is made of a single layer of cells. The sensory vesicle encloses two types of larval sensory organs : 1. One **ocellus** or simple eye which is a photoreceptor having a cornea, a lens, and pigmented retina to distinguish light from dark surfaces. Tadpole larva of *Herdmania* contains two ocelli : one postero-dorsal and other antero-ventral (Fig. 15.9 B). 2. One gravity-receptor, the **otocyst** or **otolith** which serves to maintain equilibrium. These organs make up the sensory ganglion that controls larval behaviour (see **Grant**, 1978). In front of sensory vesicle is a conical solid mass of nerve cells, called **cerebral cone**. Behind the vesicle is a thickened solid mass of nerve cells called **visceral ganglion**, which extends posteriorly up to the tip of the tail in the form of nerve cord. **Mouth** is situated antero-dorsally leading into a rudimentary **alimentary canal** comprising a narrow branchial siphon, a large sac-like pharynx, a short narrow oesophagus, a swollen stomach, a slender intestine and a small rectum. The **anus** opens into the left side of atrium. The **pharynx** has a well developed ventral **endostyle** and initially one pair of large **gill-slits** on its lateral sides. Each gill-slit ultimately splits to form six **stigmata** on either side. The **atrium** develops as an ectodermal pouch to surround the pharynx laterally and dorsally and leads outside by a posterior mid-dorsal **atrial aperture** or **atriopore**. The **pericardium** containing **heart** is located postero-ventrally in between pharynx and stomach. **Mesenchyme cells** occur scattered all over the body beneath the ectoderm and also in a mass in the posterior region of trunk.

2. Tail. The locomotory organ of tadpole larva is its powerful transparent tail which is laterally compressed and pointed terminally. Tail is fringed with a continous vertical **tail fin** formed by the test and marked with oblique striations looking like **fin-rays.** Tail of ascidian larva is typical chordate-like containing a central **notochord,** a dorsal tubular **nerve cord** and strong lateral myotome-like **muscle bands.** The **notochord** forms the supporting axial skeletal rod restricted to tail. It is formed by a single row of turgid vacuolated cells. The anterior end of notochord slightly projects into the trunk.

15-4.2. Habit and Habitat of Larva

Just after hatching the tadpole larva of ascidians is **photopositive** and **geonegative**. It cannot feed because its mouth is closed by test and it gets its nourishment from the meagre yolk reserves of endodermal cells of the alimentary canal. After a short active free-swimming (in fact, **pelagic**) existence lasting 3 to 4 hours, the larva becomes **geopositive, photonegative** and descends. The tail permits sudden horizontal forward movements as the larva descends from surface to bottom. It is these movements and function of sensory organs that bring the larva in contact with hard vertical surfaces or under the shaded sides of rocks to which it attaches with its adhesive suckers and immediately proceeds for a peculiar type of degenerative metamorphosis, called **retrogressive metamorphosis**.

Larva as a site selector. The pelagic larvae of ascidians are very selective when it comes to invading a new site. The site must be of the right consistency (*i.e.*, mud or sand for molguloid ascidians and rock for other ascidians) ; otherwise they cannot attach and undergo metamorphosis. For many ascidian larvae (*e.g.*, *Herdmania, Ascidia* ; etc.) vertical surfaces such as pilings are favoured sites. These larvae are adapted to select a preferred site, a role often superseding all other roles. Such larvae are locomotory but do not feed. They are dispersed by currents, but their pelagic lives are too short for extensive dissemination. Because it cannot feed, a larva must find a site and metamorphose within a few hours or days ; otherwise it dies.

Thus, the ascidian larva has evolved as a most efficient site selectors. This tadpole-like larva with neural tube, myotomes, pharyngeal gill-slits, eye and otolith resembles a chordate body plan as it probably is the ancestral type for all vertebrates (**Grant**, 1978).

Further, the success of the ascidian species which are ecologically adapted for hard vertical surfaces, depends on the facility with which these larvae find and colonize a new site. However, some ascidians such as the molguloids, colonize sand or mud bottoms. These exposed surfaces are so widely distributed that the organisms do not need to be precise site selectors.

Their larvae are structurally simple, without eyes and tails, and descend to the bottom in response to gravity, attach indiscriminately and metamorphose. In fact, the existence of tailless larvae among ascidians that do not require preferred sites is additional confirmation that site selection is the principal adaptation for the typical tadpole larva (see **Berrill** 1955 ; **Grant**, 1978).

15-4.3. Retrogressive Metamorphosis

Metamorphosis (Gr., *meta* = after + *morphe* = form + - *osis* = state) is the shape change in form during postembryonic development and in many cases, signals a dramatic change in habitat of the animal such as a change from a pelagic to benthic existence. Metamorphosis of the ascidian larva is unique and begins almost explosively. It involves transformation of an active, non-feeding, pelagic, **lecithotrophic** (*i.e.*, that feed on its own yolk reserves) and tailed larva having many advanced features such as axial notochord, dorsal neural tube and special sense organs, into an inert, sedentary or sessile, simple (primitive) and **plankotrophic** filter-feeding adult with only a phyrynx with stigmata and endostyle, both indicating the chordate features of adult ascidian. This type of metamorphosis which shows degenerative or retrogressive changes from larva to adult is called **retrogressive metamorphosis**. It involves the following two types of changes :

1. Retrogressive metamorphic changes. They involve destruction and disappearance of some of the larval structures such as follows :

(i) Long tail of larva with caudal fin shortens and finally disappears.

(ii) Caudal muscles, nerve cord and notochord disappear as they break down and are consumed by phagocytes. Thus, with the resorption of tail go the most obvious traces of the tunicate's chordate affinity. **Grant** (1978) has suggested that a sudden contraction of longitudinally oriented micro-filaments in the epidermal cells of the tail results in complete resorption of all tail structures (Fig. 15.10).

(iii) Larval sense organs (the ocellus and the otolith) are lost and sensory vesicle is transformed into an adult cerebral ganglion.

(iv) Adhesive papillae disappear completely.

The biological phenomenon in which the structures which are relevant only for the embryo or larva and have no significance for the adult animal, are lost at hatching, birth, or metamorphosis, is called **caenogenesis** (**Cohen**, 1967). It allows larvae to exploit quite different environments from the adults of some species.

2. Progressive growth. Cells in the larval head give rise to the adult. Some larval features are retained and developed into specialized adult structures :

(i) Due to loss of tail, the trunk becomes pear-shaped and four **larger ectodermal ampullae** grow out of its four corners. These ampullae firmly anchor the metamorphosing tadpole to the substratum. Soon two more smaller **ectodermal ampullae** appear dorsolaterally.

(ii) Anterior region between point of attachment (adhesive papillae) and mouth exhibits rapid growth, while original dorsal side with atriopore stops growth. This causes shifting of mouth through 90°. The body too rotates so that the general form of the adult sessile organism is assumed.

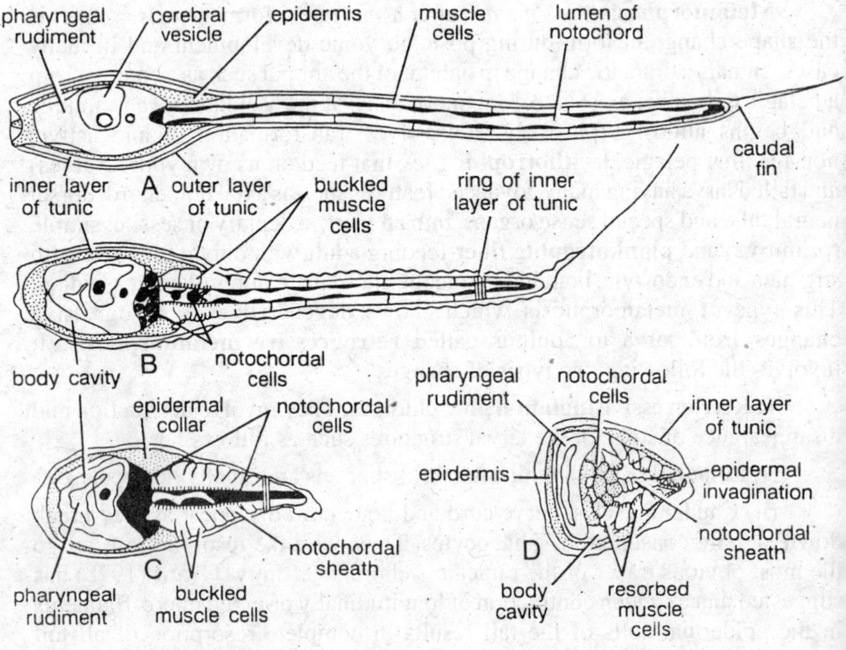

Fig. 15.10. Diagram showing various steps of retrogressive metamorphosis in the ascidian *Boltenia villosa.* A—A mature larva; B—Two to three minutes after the beginning of metamorphosis. Note the contraction of tail due to contraction of microfilaments in epidermal cells in tail causing compression and buckling of muscle cells; C—Young ascidian after ten to twelve minutes of the beginning of metamorphosis; D—Twenty-nine to thirty minutes after beginning of metamorphosis. Almost complete adult-like form; muscle and notochordal cells are resorbed (after Grant, 1978).

(iii) Adult **neural glands** and **nerve** or **cerebral** ganglion are formed by the neural tube.

(iv) With the absorption of its test covering, the **mouth** becomes functional and filter-mode of feeding begins by incoming ciliary water currents.

(v) **Pharynx** greatly enlarges, develops blood vessels and **stigmata** multiply rapidly, forming the **branchial sac**.

(vi) **Stomach** enlarges, **intestine** elongates and gets curved and **liver** develops.

(vii) **Atrial cavity** becomes more extensive.

(viii) **Circulatory system** with heart and pericardium develops.

(ix) **Gonads** and gonoducts develop from the mesodermal cells.

(x) **Test** or **tunic** spreads to cover entire animal, becomes thick, tough and vascular and attaches the animal to the substratum by forming a foot if necessary.

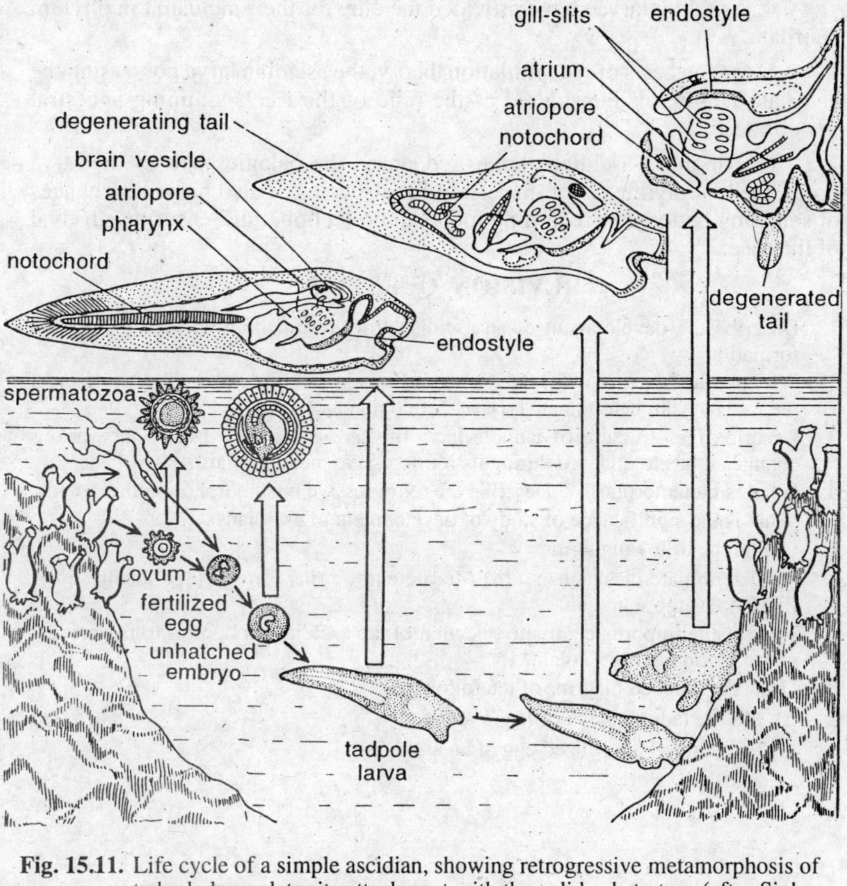

Fig. 15.11. Life cycle of a simple ascidian, showing retrogressive metamorphosis of tadpole larva, later its attachment with the solid substratum (after Sinha, Adhikari and Ganguly, 1973).

Thus, foregoing metamorphic changes mark the beginning of a sedentary, actively feeding, sexual adult form.

15-5. EMBRYOLOGICAL SIGNIFICANCE

The embryogeny of simple ascidians bears following significant points :

1. Ascidians provide best examples of mosaic eggs with well organized, pre-patterned and well differentiated ooplasm and highly determinative type development.

2. They are the only chordates which include retrogressive metamorphosis in their embryonic development.

3. Their development establishes the fact that the egg cortex is the site of morphogenetic patterning, relating to polar, bilateral and general organization of the developing egg. Further, cleavage tends to segregate cytoplasmic territories having different (cortical) histogenetic properties.

4. Ascidian larvae have provided the clue for their inclusion in phylum chordata.

5. On the basis of recapitulation theory, the ascidian larva possessing the chordate features is considered as the relic of the free-swimming ancestral vertebrates.

6. The adult ascidians being sedentary, the pelagic habit of the larva provides the only means of dispersal of the species. It also provides chances of selecting better sites regarding food and protection, thus, ensuring survival of the race.

REVISION QUESTIONS

1. Describe the development of an ascidian from fertilization up to tadpole larva formation.
2. Why are the fate maps of ascidians so well worked out ? Explain the relevance of the fate maps in determinate type of development of ascidians.
3. Describe the structure of tadpole larva of *Herdmania* and discuss the various changes it has to undergo during its retrogressive metamorphosis.
4. What is metamorphosis ? Describe the retrogressive metamorphosis of ascidians.
5. What is the significance of study of development of ascidians ?
6. Define the following terms.

   (i) Determinate cleavage ; (ii) Mosaic egg ; (iii) Cytoplasmic localization ; (iv) Ooplasmic segregation.
7. What is the importance of site selection of the ascidian larva prior to the process of metamorphosis ? Explain.
8. Draw the labelled diagrms of the following :

   (i) Mature tadpole larva of *Herdmania* ;

   (ii) Fate map of a fertilized egg of an ascidian.

16

EMBRYOGENESIS OF *AMPHIOXUS*

The tropically and temperately distributed, sand-inhabiting, marine and fish-like creatures, which belong to the subphylum **Cephalochordata** of phylum Chordata, are collectively called "*Amphioxus*" (= *Branchiostoma*). An adult *Amphioxus* has typical chordate characters — (i) a notochord which extends into the head, (ii) a dorsal tabular neural tube, and (iii) a well-developed gill-apparatus in the pharynx. Its muscles are arranged as in fish, in series of blocks called **myotomoes**.

Amphioxus is an **oviparous** animal (*i.e.*, producing eggs that develop and hatch externally) and has a **indirect development** producing a free-living, ciliated and self-nourishing larva which has to undergo the process of metamorphosis (*i.e.*, progressive metamorphosis like the insects and frog) to become an adult-like creature. Its embryonic development (such as cleavage, blastulation, gastrulation, tubulation and organogenesis) has general resemblance with that of vertebrates on one hand and with the echinoderms on the other hand. Likewise, the processes of fertilization and ooplasmic segregation of *Amphioxus* are similar to the Ascidians; however, in contrast to the latter, its development is indeterminate or regulative type, *i.e.*, separation of two blastomeres at 2-cell stage of *Amphioxus* produces twins.

16-1. REPRODUCTION

16-1.1. Gonads

In *Amphioxus*, the sexes are separate, but, there exists no well marked sexual dimorphism between male and female sexes. In the adult animals, the testes or ovaries remain located on both sides of the body between the body wall bordering the atrium and myotomes (Fig. 16.1). There are approximately 28 gonads on each side and they remain extended from 12 to 41th myotomes.

The primordial germ cells, which are contained later in the mature gonads as eggs and sperms, arise in the embryo as small clumps on the lower anterior region of the myotomes. At first they are contained within the cavity of the myotome (myocoel), from which they are later separated to become enclosed in a separate space—the **gonocoel**. Neither the testes nor the ovaries ever develop ducts.

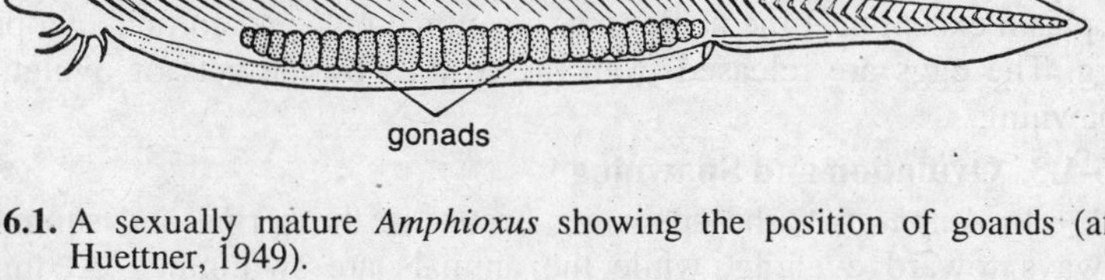

Fig. 16.1. A sexually mature *Amphioxus* showing the position of goands (after Huettner, 1949).

16-1.2. Gametes

The primordial germ cells of testes or ovaries undergo spermatogenesis or oogenesis, as the case may be, and produce spermatozoa or ova, respectively. The **spermatozoa** of *Amphioxus*, have typical structure of flagellate spermatozoa (*i.e.*, contain head, mid-piece and tail), except that, its head being spherical and mid-piece being very short (Fig. 16.2B). The oogenesis of *Amphioxus* is peculiar in the respect that the maturation divisions occur in the egg (oocyte) only after their release from the ovaries (*i.e.*, after ovulation). First maturation division occurs just after the ovulation, while second maturation division, though started before fertilization, completes only after the entry of the sperm in the ovum.

The unfertilized ovum or secondary oocyte of *Amphioxus* is 0.10 mm to 0.12 mm in diameter (Fig. 16.2 A). It is **microlecithal** and **isolecithal**. Its ooplasmic contents remain bounded by a plasma membrane, which in its turn remains surrounded by a thin membrane of mucopolysaccharides, called **vitelline membrane**.

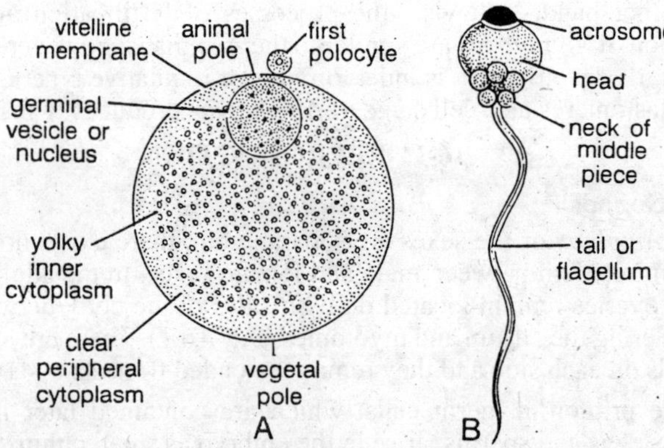

Fig. 16.2. *Amphioxus.* A—Unfertilized ovum; B—Sperm.

The ovum of *Amphioxus* has well determined polarity—the large-sized egg nucleus or germinal vesicle lies at one pole, the **animal pole** near the egg plasma membrane. The opposite side of the egg from the animal pole is called **vegetal pole**. The yolk granules remain uniformaly distributed throughout the ooplasm except near the nucleus at the animal pole and cortical cytoplasm of egg. The eggs are released from the ovaries by the acts of ovulation and spawning.

16-1.3. Ovulation and Spawning

In *Amphioxus*, the spawning occurs in the spring and summer, and always toward evening, while the animals are swimming. At this time, muscular contractions occur in the walls lying above gonocoel cavities and, thus, cause the eggs to burst through the walls of gonads at certain points which are termed as the **cicatrices**. The cutis wall of the atrium is also ruptured in these regions and the eggs, thus, reach the atrial cavity and from thence the exterior.

As soon as the egg comes in contact with the sea water, the vitelline membrane separates from the plasma membrane of the egg and a space called **perivitelline space** is developed in between the vitelline membrane and plasma membrane of the eggs. Now, the eggs are ready for fertilization.

16-2. FERTILIZATION

In *Amphioxus*, the fertilization is external and occurs in the sea water. It includes three simultaneous events—(i) cortical activation and elevation of fertilization membrane ; (ii) completion of second maturation division of the egg and (iii) fusion of male and female pronuclei of amphimixis.

As soon as the eggs are released, they are attracted by numerous sperms which have been shed into the water near the spawning female. One or more sperms make their entry through the contents of the egg at the vegetative pole. The mere entry of sperm in ooplasm is sufficient to cause activation of egg for the cortical reaction. During cortical reaction the cortical granules are burst open and fertilization membrane is elevated outside the egg plasma-membrane (**Sobotta**, 1897). The sperm entry into the ooplasm also stimulates the second maturation division, which is standstill at metaphase to become completed and eventually the second polar body (polocyte) is released. The second polar body, however, is retained in the **secondary perivitelline space** (a space formed in between egg plasma membrane and fertilization membrane) which helps to mark the animal pole and to orient the egg.

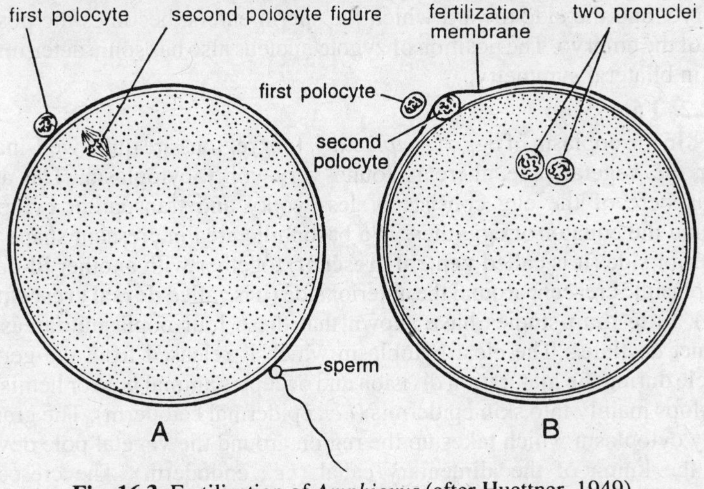

Fig. 16.3. Fertilization of *Amphioxus* (after Huettner, 1949).

Meanwhile, sperm nucleus enlarges to become sperm pronucleus. The egg nucleus also becomes pronucleus. The two pronuclei (*i.e.*, nuclei of sperm and egg) migrate towards the equator of the egg and fuse to form **zygote**

nucleus. The point of this fusion is generally little above the equator of the egg, and slightly toward the side which will eventually be the posterior end of the embryo. The fertilization of *Amphioxus* resembles very closely with that of *Styela*, except that it lacks ooplasmic pigmentation. The zygote nucleus of *Amphioxus* is also surrounded by yolk-free transparent cytoplasm (nucleoplasm) which is supposed to be released from the breakdown of germinal vesicle during maturation division (**Conklin**, 1932). The fertilized egg of *Amphioxus* becomes **telolecithal**.

16-2.1. Organization of Fertilized Egg

The fertilized egg of *Amphioxus* has well-marked polarity, bilateral symmetry, and well established cytoplasmic organization. The **polarity**, (*i.e.*, differentiation of animal and vegetal poles of the egg) is traceable to point of attachment of the oocyte in the ovary, *i.e.*, the vegetal pole remains on the unattached side. This is a matter of considerable interest because, as **Conklin** has pointed out in many invertebrates that it is the vegetal pole which is attached with the wall of ovary. He further suggests that this reversal may mark the initiation of the later reversal in dorso-ventral symmetry which places the nerve cord in chordates on the dorsal side instead of the ventral. This seems reasonable, since such a profound and early appearing difference as this, must certainly have its origin very far back in the ontological process.

Besides polarity, the fertilized egg of *Amphioxus* has well-marked **bilateral symmetry**. The entrance of the sperm slightly to one side of the vegetal pole establishes a third point on the egg with reference to the two poles already present, and so determines a median plane. Later development shows that this median plane of the egg becomes the median plane of the future embryo, and the side toward which the sperm enters, becomes the posterior side of the embryo. The position of zygote nucleus also has some determinative role in bilateral symmetry.

16-2.2. Fate Map

In the ooplasm of egg of *Amphioxus*, four regions are quite well-marked. Near the vegetal pole, there is found very yolky cytoplasm. The animal hemisphere of the egg consists of less yolky and transparent cytoplasm. Around the equator of egg, there is a basophilic area in crescent shape at the posterior side (it is called **ventral crescent**) together with another ill-defined cytoplasmic crescent around the anterior side (it is called **dorsal crescent**, Fig. 16.4). **Fate map** studies have shown that these four ooplasmic areas have distinct destinies. The clear cytoplasm which is released from the germinal vesicle during the maturation division and occupies animal pole or hemisphere develops mainly into skin epidermis (*i.e.*, epidermal ectoderm). The granular, yolky cytoplasm which takes up the region around the vegetal pole develops into the lining of the alimentary canal (*i.e.*, endoderm). The crescent of basophilic cytoplasm of posterior side develops into muscles and the lining of the body cavity and, thus, represents a mesodermal area. On the opposite side of mesodermal crescent occurs a notochord crescent and above the notochord crescent occurs a crescent area which develops into the nervous system.

Thus, dorsal crescent includes both notochordal and neural ectodermal materials. By applying local vital staining technique **Tung Wu** and **Tung** (1962) have found that the presumptive mesodermal area (the ventral crescent) is not restricted to one-half of the egg only but reaches farther around the equator. On the opposite side, a zone giving rise to the notochord could be detected and above that a crescentic area which develops into the nervous system (Fig. 16.4). The fate map of *Amphioxus*, thus, displays close similarity with that of ascidians and amphibians.

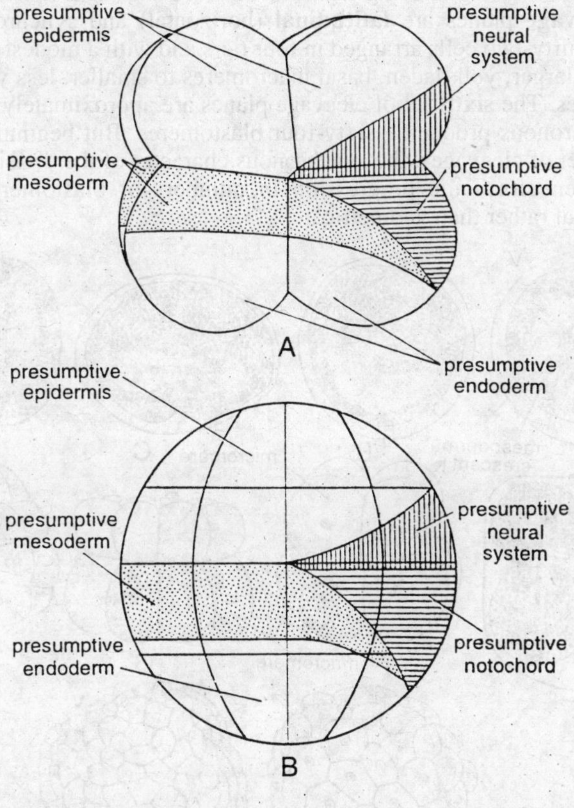

Fig. 16.4. Fate maps of *Amphioxus belcheri* at the eight-cell stage (A) and at the 32-cell stage (B) (after Balinsky, 1970).

16-3. CLEAVAGE AND BLASTULATION

The cleavage of the eggs of *Amphioxus* is of holoblastic and of almost equal type. It has been worked out by **Hatschek** (1811), **Kowaleweski** (1867), **Cerfontaine** (1906), and **Conklin** (1932). Further, the embryogenic development of *Amphioxus*, is of **indeterminate** type (**Tung**, *et al.*, 1958, 1960).

The **first** set of cleavage is **meridional** which cuts through the egg along its median axis, starting at the postero-ventral side of egg. Both resultant blastomeres have equal size and establish the bilateral symmetry of the adult animal, one producing the right side and the other the left. The **second** set of cleavage is again **meridional**, but at the right angle to the first and divides the first two blastomeres into four equal-sized blastomeres. The **third** set of cleavage planes involve four blastomeres and they are **latitudinal, horizontal** or **transverse**, dividing each of the four blastomeres into a smaller **micromere** at the animal pole and a larger **macromere** at the vegetal pole. Eight blastomeres are thereby produced. The **fourth cleavage** planes are **meridional**, and all the eight blastomeres divide synchronously. They result in a sixteen-cell stage having eight micromeres and eight macromeres. The **fifth cleavage** planes are **latitudinal** (horizontal) and synchronous. They result in thirty-two cells arranged in four tiers and with a modest gradation in size from larger, yolk laden, basal macromeres to smaller, less yolky, apical micromeres. The **sixth** set of cleavage planes are approximately **meridional** and synchronous producing sixty-four blastomeres. But beginning with the **seventh** set of cleavages, the synchronous character of the division declines, so that henceforth the progression in numbers of blastomeres becomes arithmetical rather than geometrical.

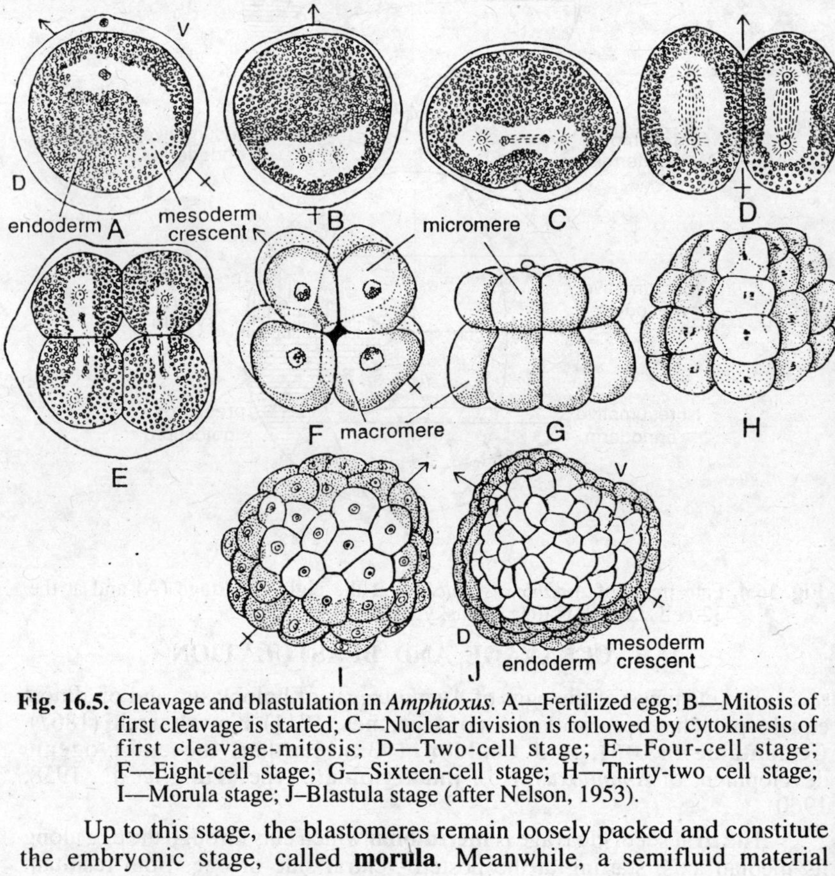

Fig. 16.5. Cleavage and blastulation in *Amphioxus*. A—Fertilized egg; B—Mitosis of first cleavage is started; C—Nuclear division is followed by cytokinesis of first cleavage-mitosis; D—Two-cell stage; E—Four-cell stage; F—Eight-cell stage; G—Sixteen-cell stage; H—Thirty-two cell stage; I—Morula stage; J–Blastula stage (after Nelson, 1953).

Up to this stage, the blastomeres remain loosely packed and constitute the embryonic stage, called **morula**. Meanwhile, a semifluid material accumulates in the centre of the mass of cells (blastomeres) of morula. This serves to push all the blastomeres outward, so that they become arranged in single layer, called **blastoderm**, enclosing a central liquid-filled cavity, the **blastocoele**. To form the blastoderm, each blastomere of morula performs some morphogenetic movement and each assumes a columnar shape. The resultant, hollow, spherical or pyriform (pear-shaped) embryonic stage in which a single-cell thick blastoderm encloses a fluid-filled blastocoele, is the **blastula** of *Amphioxus*.

A fully-developed blastula of *Amphioxus* is called **coeloblastula** and it contains a large blastocoele and a blastoderm consisting of a single layer of columnar cells. The cells at the vegetal pole are somewhat larger than at the animal pole and the blastoderm is, therefore, thicker at the vegetal pole. Further, the cells that are formed by ventral or mesodermal crescent are not columnar but small-sized, spherical amoebocytes (Fig. 16.6). Various types of intercellular junctions such as tight junctions and gap junctions are established between the columnar cells of the blastoderm of the blastula of ventral side (see Chapter 10). The joint tension of a supracellular membrane and difference in cell adhesivity of different blastomeres of blastoderm are found responsible to give a pear shape to the blastula of *Amphioxus* (see **Lovtrup**, 1974).

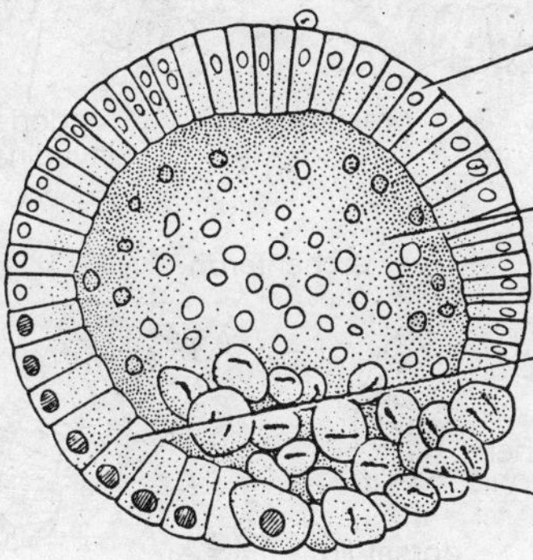

Fig. 16.6. Blastula of *Amphioxus* (after Balinsky, 1970).

16-4. GASTRULATION

The process of gastrulation in *Amphioxus* is started when there are about 800 cells in the embryo (*i.e.*, between 9th and 10th cleavage). It involves three basic types of morphogenetic movements of the embryonic cells : 1. **invagination** (emboly) of endodermal cells (which change their shape to become first columnar and then wedge-shaped, *i.e.*, elongated and wide at their inner surfaces) ; 2. **epibolic** expansion of the ectodermal cells ; and, 3. **involution** (emboly) of notochordal and mesodermal cells. Gastrulation also involves the obliteration of blastocoele and formation of a new cavity called **gastrocoel** or **archenteron** which remains communicated to the exterior by the **blastopore** (see Chapter 13, Section 13-4.4 and Fig. 13.6).

In *Amphioxus*, the process of gastrulation is initiated when the blastoderm at the vegetal pole becomes flat and subsequently bends inwards, so that whole embryo, instead of being spherical becomes converted into a cup-shaped structure, having a large cavity, the **archenteron**, in open communication with the exterior by the **blastopore**, on the side that was originally the vegetal pole of the embryo. The cup has double walls, an external and internal epithelial layers, both of which remain continuous with each other over the rim of the cup-shaped embryo, the **gastrula**. The circular rim of the blastopore is termed the **'lip'**, thus, with reference to the future orientation of the embryo, the prospective mesoderm lies in the ventral lip of the blastopore and the prospective notochord lies in the dorsal lip of the blastopore.

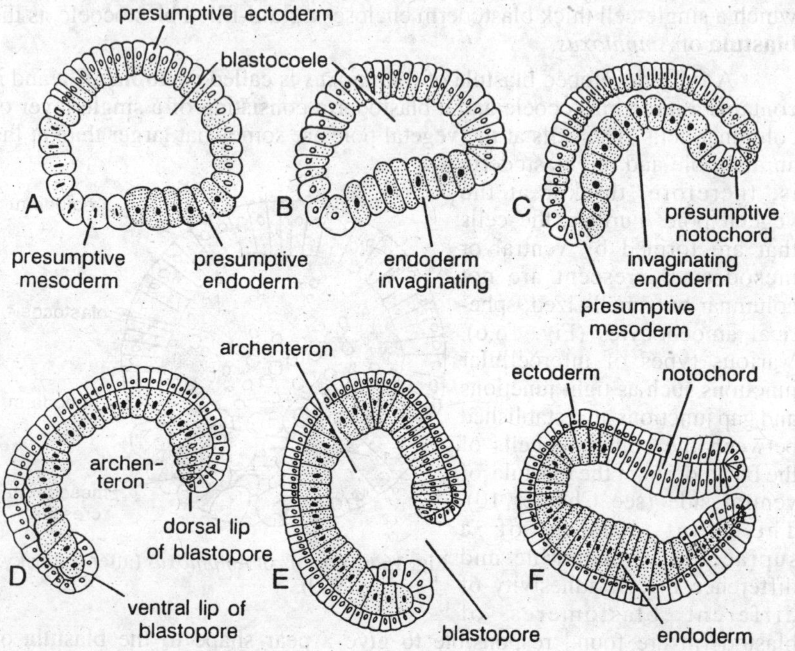

Fig. 16.7. Gastrulation of *Amphioxus*. A series of consecutive stages (after Balinsky, 1976).

As the endodermal plate continues to move inward, notochordal cells in the mid-dorsal region of the blastopore **involute**, move inward along with the endoderm. This serves to bring the prospective notochordal material to the interior and beneath the prospective neural ectoderm. Due to continued invagination of endoderm, the prospective mesoderm in the ventral lip of the blastopore is also involuted and brought into the interior.

As the presumptive notochord and the mesodermal crescent shift into the interior of the gastrula, they also change their position relative to each other (Fig. 16.8). In the blastula, these two areas lie on opposite sides of the embryo. Now the lateral horns of the mesodermal crescent converge toward the dorsal side of the embryo and come to lie on both sides of the presumptive notochord. In the next stage that follows the contraction of the rim of blastopore, the embryo becomes elongated in the antero-posterior direction, all the various presumptive areas participating in this elongation.

One effect of the reorientation of the notochord from a transverse to a longitudinal band is the steady pulling of the mesoderm from the other side in a dorso-anterior direction. With still further elongation of the gastrula, there is created an internal primitive gut tube containing a notochordal strand dorsally, a strip of mesoderm on eihter side of that, and a major component of endoderm. Externally, neural ectoderm lies in the dorsal mid-line above the notochord, and epidermal ectoderm occupies the remainder. In other words, the gastrula, at this stage, consists of two layers—an outer **epiblast**, consisting of neural and epidermal ectoderm, and an inner **hypoblast** encompassing prospective notochord, mesoderm and endoderm. The endodermal cells remain united through the hyaline membrane, with the presumptive myotomal (mesodermal) and notochordal cells (see **Lovtrup**, 1974).

The blastopore is very broad in the initial stage of gastrulation, but soon the lips of the blastopore begin to contract, so that the opening which leads into the archenteron becomes smaller and is eventually reduced to an insignificant fraction of the original orifice.

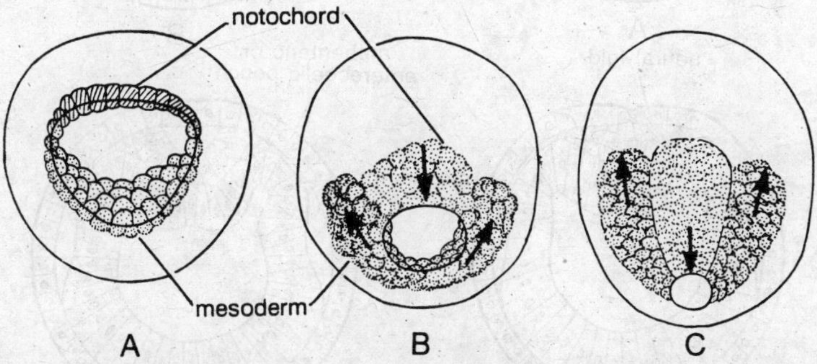

Fig. 16.8. Change in the relative position of the presumptive mesoderm and the presumptive notochord during closure of the blastopore in *Amphioxus* (After Balinsky, 1981).

16-5. TUBULATION AND ORGANOGENESIS

The germinal layers formed as a result of gastrulation are the source of material for the development of all organs of adult animal. The organs of adult *Amphioxus* are not formed directly from the germinal layers, but they develop in a synchronous and gradual manner. The processes of tubulation and organogenesis of different organ systems of *Amphioxus* can be studied as follows (Fig. 16.9) :

16-5.1. Formation of Neural Tube

The beginning of development of the central nervous system is indicated by the flattening and thickening of the prospective neural ectodermal cells of dorsal region of the elongated gastrula. Thus, the prospective neural ectodermal cells which occur just above the notochord become transformed into elongated filocytes and this process begins in the posterior end, spreading gradually forwards. Such cell transformation is initiated by neural induction (**Tung**, *et al.*, 1960, 1962) by underlying notochordal cells. Later on, these neural ectodermal cells become separated from the cell of epidermal ectoderm in the form of an elongated plate, the **neural plate.** The neural plate sinks below the level of the remainder of the ectoderm and is then covered by the free edges of the epidermal epithelium. The neural plate forthwith rolls itself

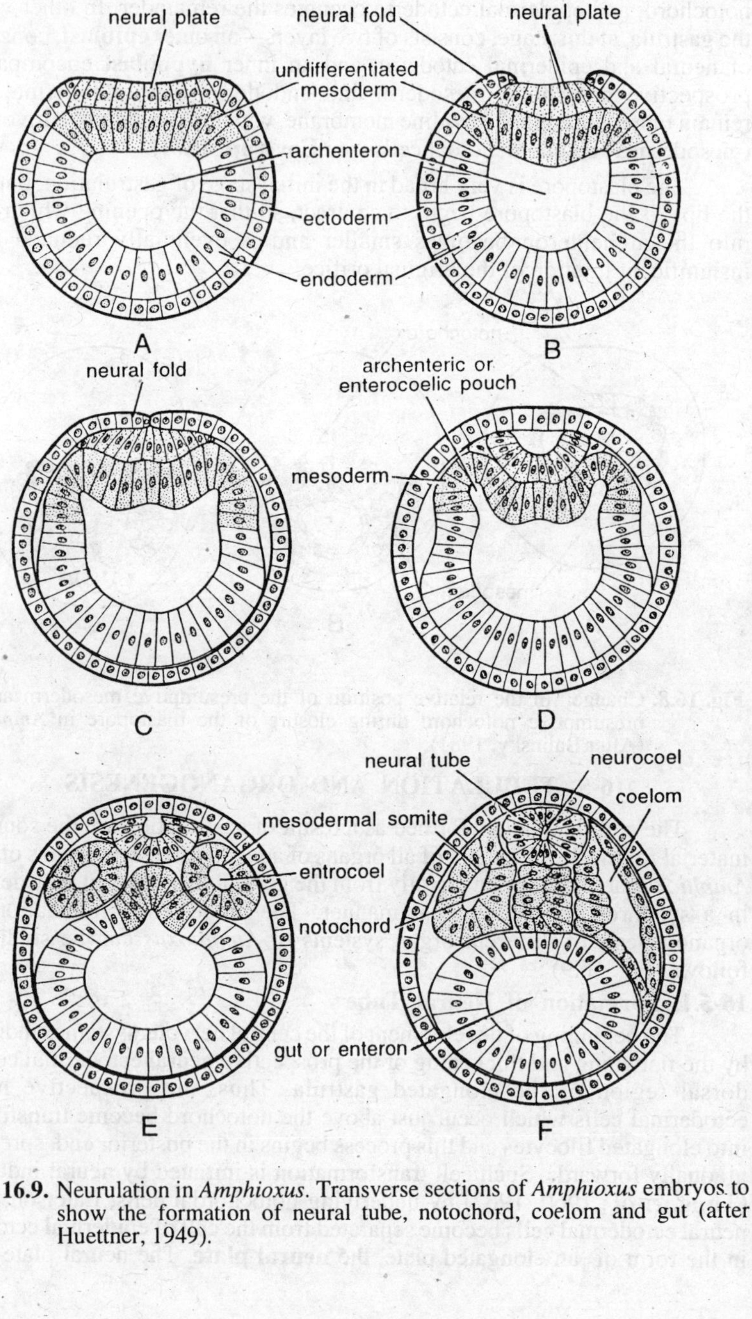

Fig. 16.9. Neurulation in *Amphioxus*. Transverse sections of *Amphioxus* embryos to show the formation of neural tube, notochord, coelom and gut (after Huettner, 1949).

into a tube, the lateral edges folding themselves upward fusing along the mid-line. The neural plate is then transformed into the **neural tube**, enclosing a longitudinal canal, the **neural canal** or **neurocoel** (Fig. 16.10). The neural tube does not close completely at the anterior end but leaves an opening, the **neuropore**, which remains open until the later stage of development.

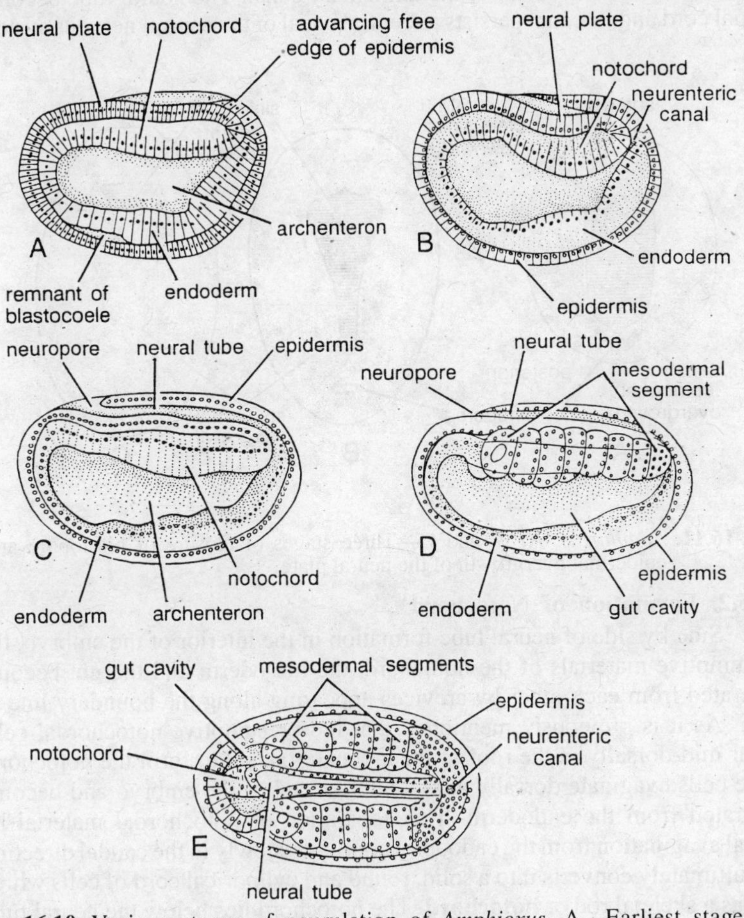

Fig. 16.10. Various stages of neurulation of *Amphioxus*. A—Earliest stage of neurulation; B—Intermediate stage of neurulation; C—Almost completed neurula; D—A later stage of C showing division of dorsal mesoderm; E—Complete neurula seen from the dorsal side (after Balinsky, 1981).

The epidermal epithelium, which spreads over the neural plate cells, derived from the sides and from the area posterior to the neural plate. Because the hind end of the neural plate borders on the blastopore, the posterior portion of epidermis covering the neural plate is derived from the region below (ventral to) the blastopore (Fig. 16.11). As the epidermis shifts forward over the surface of the neural plate, it passes over the blastopore. The blastopore becomes cut off from the exterior and opens into a space lined by the walls of the neural tube. The canal, thus formed, connects the blastopore and the archenteron with the cavity of the neural tube and, therefore, is called the **neurenteric canal**. The neurenteric canal persists only for a short time until the cavity of the neural tube becomes completely separated from the cavity of the archenteron, which in turn becomes the cavity of the alimentary canal. The neural tube becomes spinal cord and its canal persists as central canal of the central nervous system.

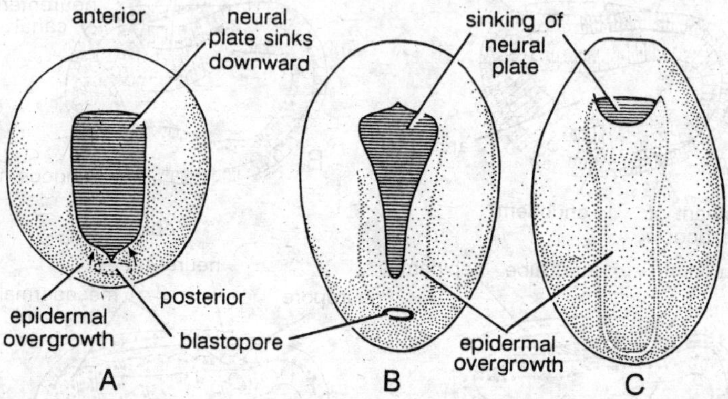

Fig. 16.11. *Amphioxus.* A, B, and C—Three stages of closure of blastopore and epidermal overgrowth of the neural plate.

16-5.2. Formation of Notochord

Side by side of neural tube formation in the interior of the embryo, the presumptive materials of the notochord, the mesoderm and the gut become separated from each other by crevices appearing along the boundary line of each. As it is previously mentioned that the presumptive notochordal cells occur mid-dorsally in the roof of the archenteron. so, to form the notochord, these cells evaginate dorsally at the anterior end of the embryo and become separated from the endoderm. This separation of notochordal material by dorsal evagination from the endoderm continues slowly in the caudal direction and ultimately converts into a solid, round and cylindrical cord of cells which acts as a skeletal rod or **notochord**. The notochord lies below the neural tube and between the mesodermal somites and resting in a slight depression along the dorsal aspect of the endodermal tube. It remains extended from the brain region to the caudal end of the body.

In later stages, the notochordal cells develop vacuoles which seem to contain a jelly-like supporting substance. Soon, this rod is surrounded by the fibrous connective tissue, called **notochordal sheath**.

16-5.3. Development of Mesoderm and Coelom

The presumptive mesodermal material also occurs at the dorsolateral side of the archenteron roof and it flanks the notochord from both sides. Both mesodermal bands also get separated from the endoderm by dorsal evagination into the residual blastocoelic space.

Anyhow, two longitudinal **mesodermal grooves** are formed which lie at both sides of notochord and at dorsal side of endoderm tube. Beginning at the anterior end these longitudinal mesodermal grooves soon become transversely divided into distinct **segments** or **somites**, which are roughly cuboidal masses of cells lying on each side of the notochord, one behind the other along the length of the animal's body. Some of these somites (only anterior most two somites) retain within themselves a portion of the archenteron or enterocoel ; the rest are solid blocks of mesoderm. They ultimately become entirely cut off from the rest of the endoderm.

Soon, each somite grows ventrally and becomes differentiated into following parts — (i) Some portion of each somite continues to grow around notochord and neural tube. It becomes thickened to become the **myotome**. (ii) The portion of each somite, which remains near the epidermal ectoderm, is called the **somatic parietal mesoderm** (iii) The portion of each somite, which remains closely associated with the endoderm, is called the **visceral** or **splanchnic mesoderm**. (iv) The space occurring in between the somatic and visceral layer of mesoderm is called **enterocoelic coelom**, because it is originally derived from the archenteron.

A. Further differentiation of myotome. Each myotome consists of a median muscular portion and the laterally placed, thin-walled, parietal part which surrounds the coelomic space or **myocoel**. The muscular portion of each myotome enlarges rapidly and forms > shaped **muscle plate** or **myotome** of the adult *Amphioxus*. The myocoelic portion because contributes important **connective** or **skeletal tissues** to the framework of the body, so, in each segment it gives rise following three diverticula :

(i) A lower sclerotomic diverticulum extends up between the myotome and the medially placed notochord and nerve cord. Its walls differentiate into following two parts :

(a) An inner layer which together with a similar contribution from the somite on the opposite side, wraps around the notochord and nerve cord and subsequently gives origin to a **skeletogenous sheath** of connective tissues which ensheath these structures.

(b) An outer layer which covers inner aspect of the myotome with a **fascia** or **connective tissue covering.**

(ii) A ventral diverticulum extends between the lateral wall of the splanchnocoel (coelom) and the epidermal layer of the body wall, and separates the somatic mesoderm of the splanchnocoel from the epidermal wall. This ventral diverticulum or **dermatomic fold**, together with the external or parietal wall of the myocoel above, forms the **dermatome**. The inner and outer layers of the ventral diverticulum gradually fuse to form the cutis or **dermal layer** of the integument or skin in the ventro-lateral portion of the body, whereas the parietal wall of the myocoel above gives origin to the same dermal layer in the dorsal body region lateral to the myotome.

(iii) The **dorsal sclerotomic diverticulum** forms the **fin-ray cavities** in the dorsal fin. These cavities become entirely isolated from the rest of the myocoelic spaces. Several fin-ray cavities occupy the breadth of a single myotome.

B. Further development of coelom. We have already mentioned that the spaces enclosed by somites are the forerunners of the coelom of adult *Amphioxus*. When each somite differentiates into myotome, somatic and splanchnic mesoderms, the portion of coelom occurring in between somatic and splanchnic mesodermal layers is called **ventro-lateral coelom** or **splanchnocoel**. As the myotomes enlarge, the splanchnocoelic space becomes more and more displaced ventrally and most of it comes to lie on either side of the enteron. Eventually, the splanchnocoels of each pair of somites push ventrally to the lower portion of the enteron, where they ultimately fuse. Gradually, the splanchnocoels of each segment fuse antero-posteriorly and in this way a continuous, antero-posterior, splanchnocoelic space below and around gut tube is formed. A horizontal **septum**, the **interocoelomic membrane** also appears, separating the myocoels above from the splanchnocoelic cavity below.

Thus, in *Amphioxus* the coelom has dual origin : the anteriormost part of the coelom is **enterocoelic** in origin like the echinoderms, since, in this case, coelom is originally derived from the enterocoel or archenteron that exists in the anteriormost somites. The rest of the coelom is **schizocoelic** in origin like the other chordates, *i.e.*, the mesoderm is initially solid and becomes split into two layers with an intervening coelom or **schizocoel**.

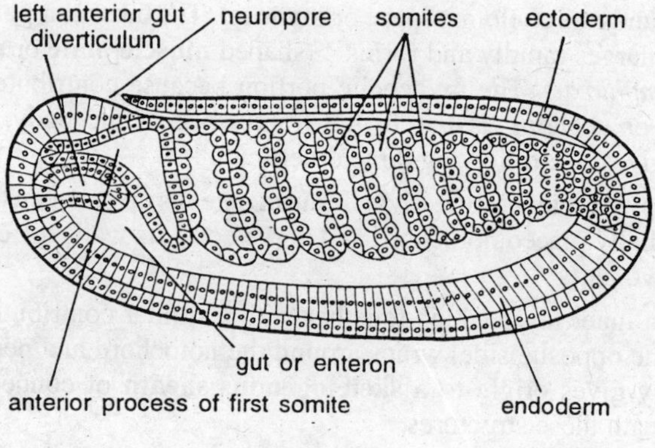

Fig. 16.12. A young larva of *Amphioxus,* showing formation of head cavities, and somites (after Huettner, 1949).

16-5.4. Development of Endoderm or Alimentary Canal

When the notochord and the mesodermal segments dissociate themselves from the endodermal material, the free edges of the endoderm approximate each other and fuse along the dorsal mid-line. The endoderm, thus, becomes converted into a closed sac. The cavity of this endodermal sac becomes the cavity of the alimentary canal. The cavity of the alimentary canal later acquires communication with the exterior by means of the oral and anal openings which break through the body wall, but this does take place until a very much later stage of development.

Further, after the formation of eight mesodermal somites, the dorsal portion of the anterior end of the gut cavity produces a new diverticulum, called **dorsal** or **anterior gut diverticulum**. This endodermal diverticulum soon splits to form two equal-sized diverticula — a left and a right. Shortly after their separation, the **right diverticulum** becomes very thin and extends into the anterior part of the embryo in between the ectoderm and notochord, in front of the first pair of myotomes. Subsequently, it gives origin to the **head cavity** or the cavity of the **oral hood**. The **left diverticulum** remains small and thick-walled and later fuses with an ectodermal invagination to form an adult sense organ called the **pre-oral pit** (Fig. 16.12 and 16.13).

After the separation of anterior diverticulum, the cells of the floor of anterior part of gut become columnar and form an elongated pit, the **hypopharyngeal groove** or **endostyle**. Much later, the embryo develops the liver in the form of hollow diverticulum from the alimentary canal. To form the **phyrynx**, the anterior portion of gut becomes enlarged.

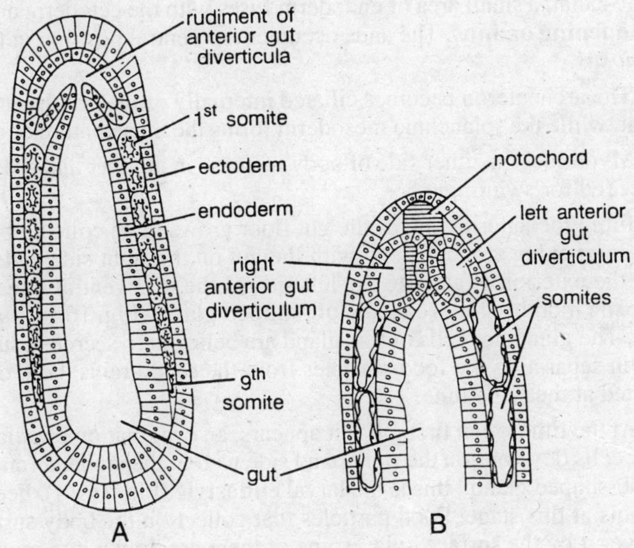

Fig. 16.13. Origin of the anterior gut diverticula in *Amphioxus*. A—Frontal section of young larva with 9 pairs of somites; B—Sagital section through anterior end of larva having 13 pairs of somites (ventral view).

16-6. HATCHING AND POST-HATCHING EMBRYOGENY

In *Amphioxus*, the developing embryo remains inside the vitelline or fertilization membrane up to the formation of two pairs of somites. When the two pairs of somites become completed, the elongated embryo comes out from the egg membranes and this process is called **hatching**. Soon other somites are developed and at nine somite stage, the hatched embryo is called **larva**. The larva swims actively by means of cilia which clothe the entire epidermis.

This ciliated larva of *Amphioxus* lacks mouth, anus and is incapable of feeding. It gradually elongates, becomes laterally compressed, the number of somites increases, and during later stages, cilia disappear. To become adult like *Amphioxus*, each larva undergoes following developmental changes :

1. The **mouth** opening develops at a time when the larva acquires about 16 to 18 pairs of mesodermal somites. It appears at the anterior end at about the level of the second coelomic pouch (*i.e.*, mesodermal somite). On the left hand side of the ventral surface of the head, the ectoderm fuses with the endoderm. At the meeting point of two, a perforation occurs, forming the mouth opening. The mouth opening has ciliated borders. During later stages, mouth shifts from the left side to assume the antero-ventral position, where its aperture is reduced to form the **enterostome**, while its margins grow to form **velum**. The **oral hood** along with **oral cirri** are appeared behind the pre-oral pit. The pre-oral pit is differentiated into **Hatschek's pit** and the left diverticulum itself develops into **wheel organ**.

2. At the caudal end of the larva, following the degeneration of the neurentric canal, a small area of endoderm fuses with the ectoderm and forms the **anal opening** or **anus**. The anus occupies the ventral position in the larva of *Amphioxus*.

3. The archenteron becomes ciliated internally ; **mid-gut diverticulum** grows out, while the splanchnic mesoderm forms the musculature of gut wall.

4. Myotomes on either side of body increase in number and provides all that is needed for swimming.

5. From the anterior end of the gut floor grows out a coiled tube which acquires an opening at its one end with the gut on the right side and at other end with the exterior, slightly to the left side of the mid-ventral line and just ventral to the mouth. This is called **club-shaped gland** (Fig. 16.14 B and Fig. 16.15 A). The glandular cells of this gland are believed to secrete a substance that aids in separating the food particles from the sand grains, both of which are ingested at the same time.

6. At the time when first gill-slit appears, an evagination of ciliated and glandular cells develops on the right hand side wall of the archenteron in front of the club-shaped gland ; this is the larval **endostyle** (Fig. 16.14). Feeding of larva begins at this stage. Food particles that collect on the body surface are carried over it by the surface cilia ; some of them are drawn into mouth. The first gill-slit is meant for throwing out water. Later on, by the time of 8 gill-slits of the first set are formed, the club-shaped gland begins to disappear and the endostyle becomes V-shaped with its apex pointed backward. The endostyle remains in this position on the wall of gut (pharynx) of right side. Eventually, the endostyle extends backwards in the mid-ventral line between the ventral ends of the gill slits.

7. At the time of mouth formation, the endoderm opposite to the first pair of somites (or the endoderm of pharynx) pushes ventrally and fuses with the ectoderm. This area of fusion finally perforates and forms the **first gill-slit**. The gill-slit, once formed, moves up on the right side of the body. The endodermal area, from which the first and later gill-slits make their appearances is called the **branchial rudiment**.

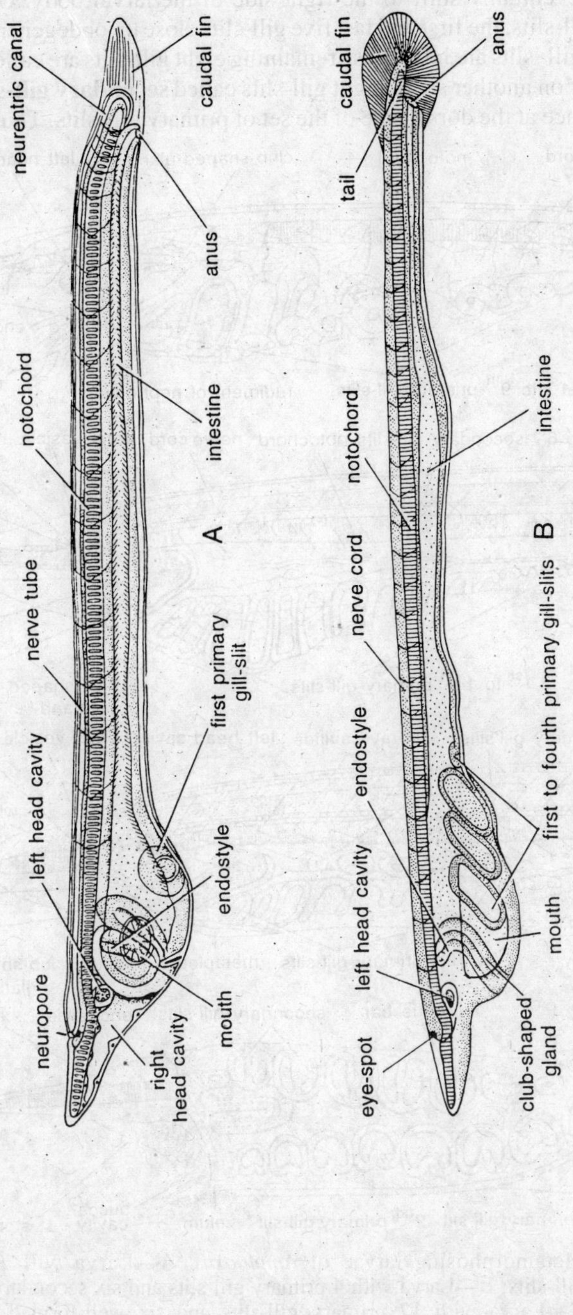

Fig. 16.14. Young larvae of *Amphioxus*. A—Larval stage after opening of mouth and first gill-slit; B—Larval stage in which four primary gill-slits are present (after MacBride, 1914).

Thus, fourteen gill-slits perforate the branchial rudiment one after the other and eventually shift to the right side of the larval body. Out of these fourteen gill-slits, the first and last five gill-slits close up or degenerate, so that, only eight gill- slits are left. These remaining eight gill-slits are called **primary gill-slits**. Soon another set of eight gill-slits called **secondary gill-slits**, makes its appearance at the dorsal side of the set of primary gill-slits. Thus, two rows

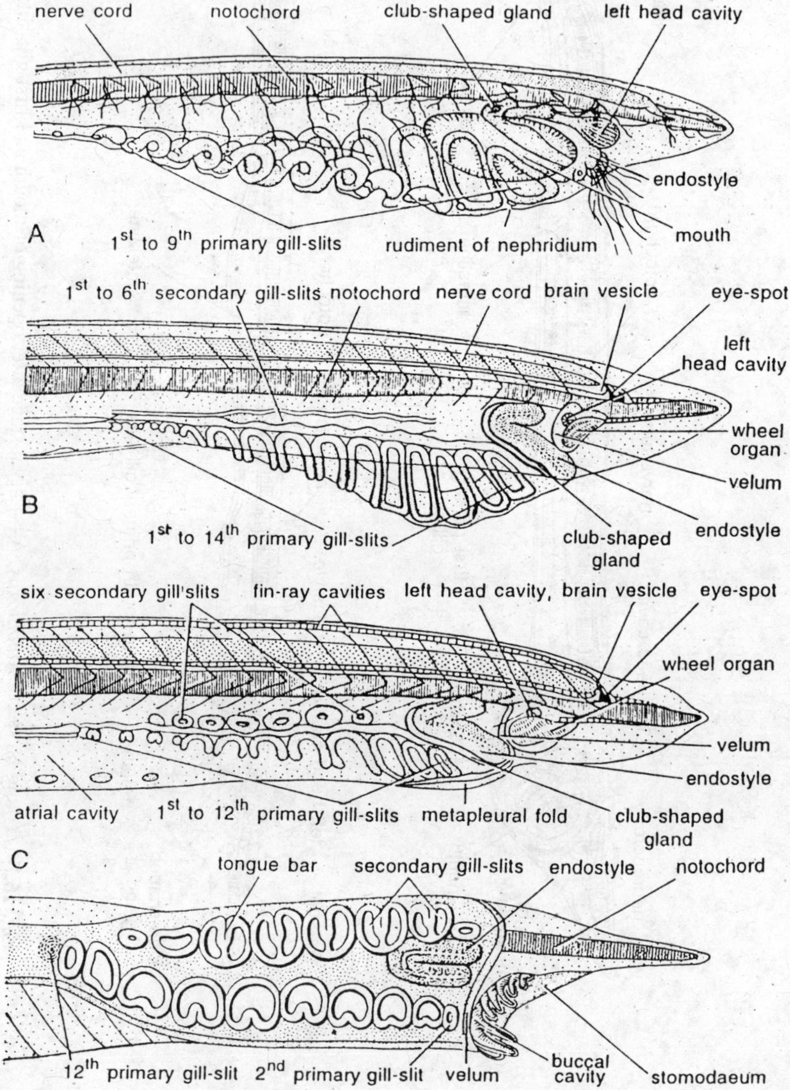

Fig. 16.15. Metamorphosing larvae of *Amphioxus.* A—Larva with 9 primary gill-slits; B—Larva with 4 primary gill-slits and six secondary gill-slits; C—Larva with 12 primary gill-slits and six well-formed secondary gill-slits; D—Larva with fully developed gill-slits (after MacBride, 1914).

of gill-slits are formed on the right side of the body. Eventually, the set of eight primary gill-slits is shifted to the left side of the pharynx to become the gill-slits of that side. The set of eight secondary gill-slits is retained on the right side of the pharynx. These gill-slits, later on, are subdivided by tongue bars and new gill-slits are gradually added at the posterior ends of both sets of gill-slits (Fig. 16.15).

8. During the larval life the excretory organs, the **nephridia** are formed. They consist of tubes ending internally in tufts of **solenocytes** or **flame cells**. Their origin has been described by **Goodrich** (1909). When they are first visible they consist of small pear-shaped masses of cells, which are ectodermal in origin and remain situated at the lower borders of the primitive gill-slits. They rapidly become hollow and long solenocytes grow out from their inner ends. These solenocytes extend up to the left side of the body to the level of dorsal aorta. In the adult the solenocytes become much shorter than in larva. The **Hatschek's nephridium** develops as a small tube behind the pre-oral pit.

9. The **atrium** is derived from the ectoderm. The atrial folds develop in the larva short before the appearance of the secondary gill-slits as two longitudinal thickenings of the body wall on the ventral side. The atrial folds start appearing from posterior end where they are quite close to each other. Anteriorly, as they grow, they enclose the gill-slits in between them. Finally, these folds enlarge and become well-marked. From the inner faces of each fold a solid ridge, called the **ridge** or **epipleura**, grows towards the middle line. They meet and fuse just below the ventral body wall to form a tube, the **atrium**, which is lined by ectoderm. The atrium first remains open on both ends, but, later on the anterior opening closes and posterior opening is retained as the **atriopore**. The atriopore lies at posterio-ventral end of the body.

The atrium gradually develops, obliterates the coelom completely and surrounds the pharynx from all sides except the dorsal side. The **metapleural folds**, *i.e.*, the lower edges of the atrial folds, are first solid, but, soon become hollow by the appearance of lymph spaces in them (Fig. 16.16).

16-6.1. Asymmetry of Larva

The larva of *Amphioxus*, despite having bilateral symmetry like the adult, has a asymmetry in the arrangement of various body organs. For example, (i) The mouth makes its first appearance on the left side, not on the mid-ventral axis of the body. (ii) Two diverticula which are originated from the anterior part of the gut, also show asymmetry, because, the left diverticulum remains smaller, while, right one becomes large-sized. (iii) The gill-slits arose asymmetrically. Both sets of gill-slits first appeared on the right side and they acquire bilaterally symmetrical arrangement only afterward.

16-7. METAMORPHOSIS

The asymmetrical larva of *Amphioxus* swims and feeds on planktonic food material for about three months and then it sinks down. It lies on the bottom and undergoes slow metamorphosis to become a bilaterally symmetrical adult. Most of the developmental changes which take place during metamorphosis have already been discussed along with post-hatching changes. Some very significant metamorphic changes are the following :

1. Larva of *Amphioxus* develops the habit of filter feeding.

2. The tongue-bars subdivide the gill-slits.

3. The development of atrium results in the restriction of coelomic cavity only in the pharyngeal region, in the muscle bands (myocoel) and in gonads (gonocoel).

4. Mouth shift from left to ventral side of body. Oral hood and oral cirri are formed.

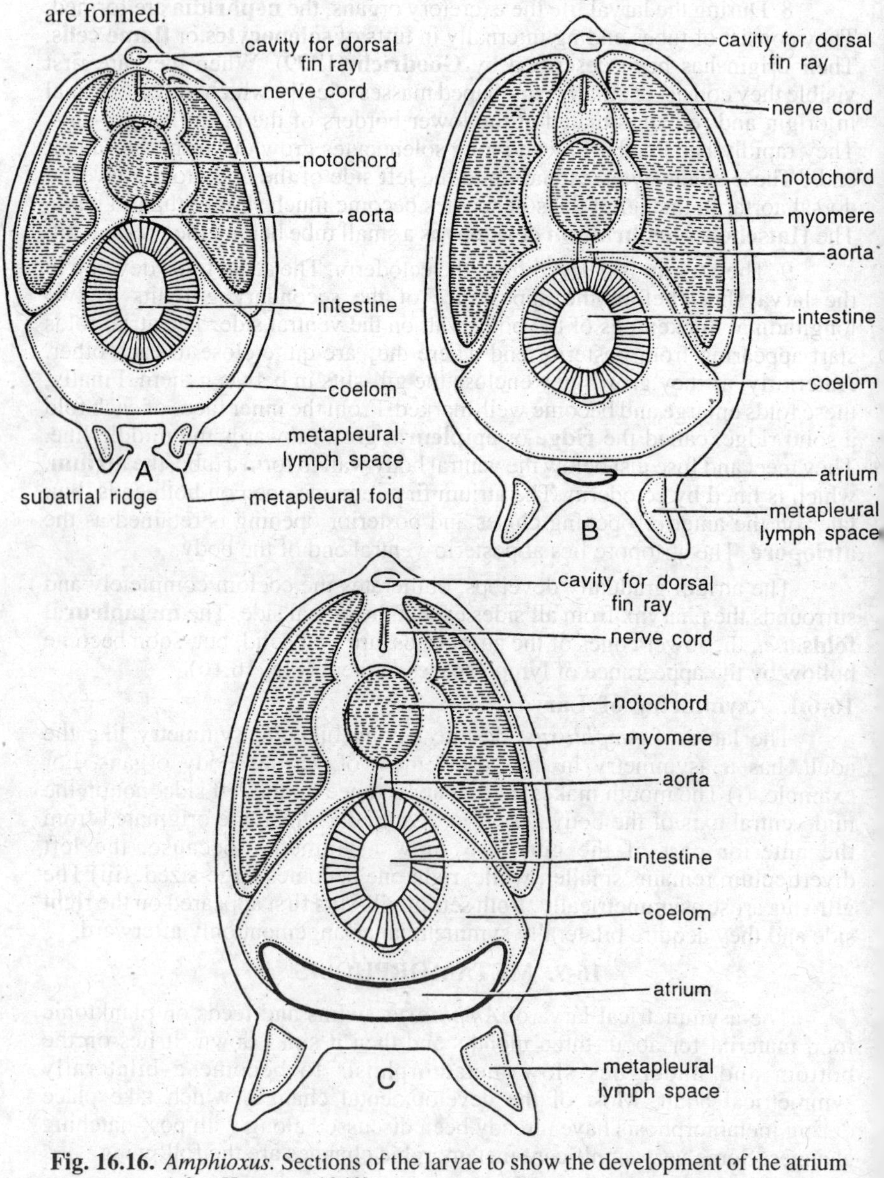

Fig. 16.16. *Amphioxus.* Sections of the larvae to show the development of the atrium (after Huettner, 1949).

5. Gill-slits increase in number and also enlarge in vertical direction.

6. The gonads arise as small buds of cells, which sprout from the anterior lower angles of the myotomes and which project into myocoelic spaces, which are now called **gonocoels**. The gonocoels, soon cut off from the myocoels by a septa and further development of gonads takes place in them.

16-8. EMBRYOLOGICAL SIGNIFICANCE OF *AMPHIOXUS*

The importance of embryology of *Amphioxus* is due to following facts—Its early embryological stages closely resemble to those of such invertebrates as echinoderms and at the same time are representative of vertebrate development in simplest terms. Thus, its embryology helps in understanding the possible links in between the non-chordates and chordates.

REVISION QUESTIONS

1. Describe the structure of ovum of *Amphioxus*. Describe the changes that are brought about fertilization.
2. Describe the development of *Amphioxus* up to gastrulation.
3. Describe the process of embryogenesis of *Amphioxus* up to the formation of three germinal layers.
4. What is larva ? Describe how a larva is formed in *Amphioxus* from three basic germinal layers.
5. Why the larva of *Amphioxus* is considered asymmetrical ? Describe various metamorphic changes of the larva of *Amphioxus*.
6. Describe the process of coelom formation during embryogenesis of *Amphioxus*.

17

EMBRYOGENESIS OF FROG

The frog is an anuran amphibian of subphylum Vertebrata. It lives in freshwater ponds, swamps and has easy availability for embryological studies. The embryology of frog has been most extensively worked out and its development has fetched the attention of many embryologists because of following reasons : (i) An understanding of frog and other amphibian development is fundamental to the interpretation of chick and mammal development. (ii) Its early development furnishes an excellent transition between the corresponding stages in *Amphioxus* and in those animals which are more highly evolved. (iii) Its later development (*viz.*, metamorphosis of gill-breathing, aquatic fish-like tadpole larva into lung-breathing, amphibious, tetrapod-like adult animal) suggests a possible evolutionary transition from gill-breathing vertebrate type (*i.e.*, fish) to lung-breathing vertebrate type (*i.e.*, tetrapod). (iv) The eggs of frog and other amphibians are easily available and they may be studied in the laboratory from the moment of fertilization onward. (v) The amphibian embryos are relatively straight, not coiled like other vertebrates. This makes their description easier. (vi) In most experimental embryological studies, the frog eggs are generally used, having been either collected in abundance from freshwater ponds during early spring or obtained by means of induced ovulation, (*i.e.*, by injecting pituitary contents of female frog in a sexually mature female during non-breeding seasons) in all the seasons of the year.

17-1. REPRODUCTION

In frog, the sexes are separate. The sexual dimorphism is well marked, *i.e.*, the male frog has **copulatory** or **nuptial pads** on thumbs of the forelimbs and also possesses **vocal cords** for attracting female frogs for copulation during the breeding seasons. The gonads and gametogenesis of both sexes are of following types :

17-1.1. Testes and Spermatogenesis

The male gonads are testes which are paired, small-sized, white, and ovoid bodies. They remain located in the coelomic cavity near the upper end of the kidneys, where they are held in place on the dorsal body wall by a double sheath of peritoneal mesentery, the **mesorchium**. Histologically, each testis is composed of numerous fine convoluted tubules, the **seminiferous tubules**, which are lined by germinal epithelium which includes spermatogonial cells and Sertoli cells. The spermatogonial cells undergo spermatogenesis and produce spermatozoa, which have a typical structure of a flagellate spermatozoon of Chordata. The matured spermatozoa are conveyed from the seminiferous tubules to **vasa efferentia**. By ciliary movement of the lining of vasa efferentia, the mature sperms come in **vas deferens**. Just prior to mating they are stored in the **seminal vesicles** (Fig. 6.1).

17-1.2. Ovaries and Oogenesis

The female gonads of frog are two ovaries, which remain attached to the dorsal side of the body cavity in its anterior portion by a double peritoneal sheath called **mesovarium**. Unlike the testes, however, the ovaries always vary greatly in size and appearance, depending upon the time of year. During breeding season, *i.e.*, summer, they assume large size due to storage of numerous, mature eggs. After ovulation in spring, they become small-sized.

Histologically, each ovary consists of many compartments, whose outer walls are formed of connective tissue or **stroma**, and interior space is filled by **oocytes**. Each of these oocytes is surrounded by a single layer of flattened cells, the **follicle cells** which make the **ovarian follicle**. The oocytes undergo oogenesis and mature ova are released from the ovaries (*i.e.*, ovulation) into the coelomic space from where they are conveyed by the ciliary action on the peritoneum to the exterior by **oviducts**. Oviducts are long convoluted tubes whose size and convolutions are somewhat increased during breeding season. They open anteriorly into the coelom by a ciliated funnel, the **infundibulum**. Posteriorly, they open into the cloaca. The lumen of the duct is lined by tubular glands (which form gelatinous coats around the passing ova) and the ciliated epithelium for the extrusion of ova. At the posterior end, each duct widens and its walls become thinner and very elastic. These dilated regions, known as the **uteri**, serve for storing the ova just prior to extrusion (Fig. 6.2).

The pre-embryonic development of the egg or **oogenesis** of a frog or any other amphibian is a slow process.It requires three seasons during which time it passes through three phases : **multiplication, growth** and **retention**. Multiplication phase started from the tadpole stage and exists up to older adult stage. During this phase, the oogonial cells undergo mitosis and form **primary oocytes**. The primary oocytes enter the prophase of meiosis-I and before the completion of meiosis-I, they enter the most important phase of oogenesis, the growth phase. During this phase primary oocyte increases tremendously in its size. The chromosomes which are occurring in synapsis, become large-sized lampbrush chromosomes and large amount of rRNA, mRNA and tRNA are transcribed around the DNA of chromosomal loops. Certain proteins are also accumulated inside the nucleus. As a result, the volume of nuclear sap increases and nucleus of oocyte becomes inflated and is called **germinal vesicle**. The increased transcription of rRNA is related with number and size of nucleoli per nucleus, both of which are usually increased during this phase.

The volume of cytoplasm of the oocyte of frog also increases qualitatively as well as quantitatively. The amount of yolk platelets and glycogen, etc., increase with tremendous rate and a definite polarity is determined in terms of yolk distribution — the yolk tends to accumulate at one pole, the **vegetative pole** and occupies two-third (*viz.*, 2/3) space of egg, pushing the germinal vesicle eccentrically, which lies in a comparatively yolk-free cytoplasm at another pole of egg, the **animal pole**. Besides yolk, amount of glycogen, nucleic acids (DNA, mRNA, rRNA and tRNA), mitochondria, etc., gradually increase in the ooplasm. When, growing oocytes attain the size 1500 to 2,000 μm, follicle cells form a non-cellular **vitelline membrane** around the egg's plasma membrane.

The primary oocytes reach their full growth toward the end of the second season, but, they are not immediately laid. The third phase is, therefore, one of quiescent retention of these primary oocytes in the ovary until the next breeding season. Then, sensitized by hormones of pituitary and stimulated by the clasp of the male, the ripe ova erupt through the epithelium covering of ovary as the ovarian follicles break and enter the body cavity. This is called **ovulation**. The ovulation is accompanied by the dissolution of the germinal vesicle and the spreading of its liquid contents (nuclear sap) across the animal hemisphere. The **spindle** of meiosis-I forms and the first polar body is given off. As the egg enters the oviduct, a **spindle** of mieosis-II forms and second meiotic division progresses as far as the metaphase. At this stage, it pauses, and no further progress is made until and unless the ovum is fertilized.

Spawning. Each frog egg, as it comes from the cloaca of female, is spherical, unicellular, 1.75 mm in diameter, **mesolecithal** and **telolecithal**. It is surrounded by the vitelline membrane (primary egg membrane) and many layers called tertiary egg membranes of adhesive jelly which were added as it passed through the oviduct. The outermost jelly layer is absorptive and tends to hold the eggs together. A cluster of frog eggs which remain stick together is called **spawn**. A spawn of *Rana tigrina*, may have 3000 to 4000 ova. The spawn is laid, when during pseudocopulation or **amplexus**, male clasps the female and sheds a suspension of sperm over the eggs as they leave the cloaca.

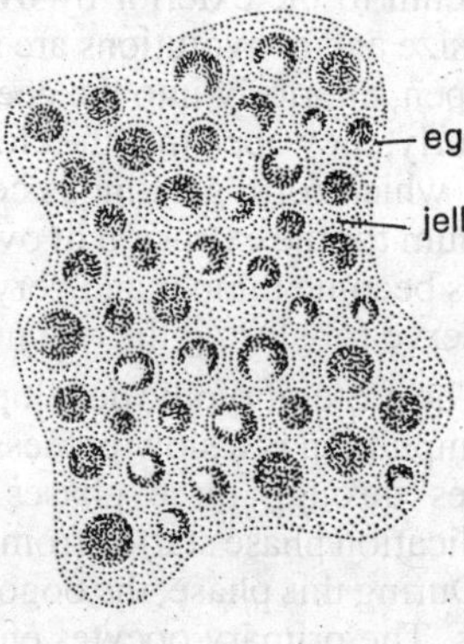

Fig. 17.1. Spawn of frog.

On reaching water, the jelly layers of fertilized eggs absorb water, begin to swell and lose much of their stickiness. These jelly layers of eggs serve various important functions — 1. They serve to attach the eggs to each other and to debris, so that they are not readily washed out. 2. They protect the eggs from the mechanical injuries and make the eggs unedible or distasteful to water snails and other aquatic creatures. 3. The jelly layers serve as insulators to protect the egg from the drastic effects of radiant energy and heat of the sun rays. 4. Being light in weight, jelly helps the ova to float on the surface of pond water. 5. Jelly contains entrapped air bubbles which may help in the respiration of ova.

17-2. ORGANIZATION OF THE UNFERTILIZED EGG

An unfertilized egg of frog is spherical and has a well-marked **polarity** and **radial symmetry**. Its cytoplasm, *i.e.*, **ooplasm** remains differentiated into following two parts :

1. Cortex; 2. Endoplasm.

1. Cortex of the egg. The egg cortex includes the plasma membrane of the egg and adjacent highly viscous or gel-like peripheral layer of cytoplasm called **ectoplasm** of 2 to 3μm thickness. Its material basis is below the limits of resolution of the electron microscope, but chemically, it has been found to be a colloidal complex of mucopolysaccharides, and TP polymers like microfilaments, which requires calcium ions for its stability and physiological integrity. It remains thicker around the animal pole of the egg and becomes gradually thinner towards the vegetal pole of the egg (Fig. 17.3). The egg cortex of frog contains some membrane bounded, spherical bodies of 2μm diameter containing acid mucopolysaccharides and are called **cortical granules**. These cortical granules remain arranged in a layer close to the plasma membrane. They are manufactured by the Golgi apparatus and their mucopolysaccharide contents are fairly homogeneous, finely granular and floccular (*i.e.*, like a tuft of wooly hair). In the egg cortex of frog, just beneath the layer of cortical granules, there occurs a layer of dark brown pigment granules, due to which an unfertilized egg outwardly consists of a dark brown animal region which extends below the equator and a whitish vegetal region which has little pigment granules in the cortex.

The cortical layer of ooplasm remains fairly stable and is not readily displaced by cytoplasmic streaming (cyclosis) or in moderately strong centrifugation. Due to its stability, it plays the role of a fixture in the egg cell and has been found to remain in the same position for continuous period of growth and development. The larger cytoplasmic inclusions are either embedded or attached to the cortex by the help of cortical microfilaments (see **Lovtrup**, 1974) and so they remain in a fixed position. Recent work has found the cortex as a site of morphogenetic patterning relating to polar, bilateral (which arises during fertilization) and general organization of the developing egg.

2. Endoplasm of the egg. The internal portion of ooplasm is called **endoplasm**. It remains in the physical state of a suspension, with a continuous liquid phase (containing in solution various large and small particles, ranging in size from simple inorganic ions to protein molecules) and larger inclusions such as mitochondria and ribosomes dispersed more or less at random in the continuous phase. These inclusions are freely movable in natural conditions owing to cytoplasmic streaming or cyclosis. It consists of a cup-shaped mass of white yolk platelets called **vitelline cupola** (**Berrill**, 1971) at the vegetal pole. The vitelline cupola occupies most of the vegetal pole and its central concavity remains opened toward the animal pole. Its cavity is occupied by a somewhat pigmented mass of cytoplasm, while the endoplasmic region of the animal hemisphere is occupied by less pigmented cytoplasm. Yolk platelets are largest near the vegetal pole, smaller and less densely packed towards the animal pole. There is accordingly a polar axis, the **animal-vegetal axis** with egg constituents arranged radially around it and exhibiting a concentration gradient from pole to pole.

The polarity and concentration gradients of different chemical substances from pole to pole of the unfertilized egg of frog are revealed by following facts — 1. The germinal vesicle (egg nucleus) is located near the animal pole of the egg and it is here that the polar bodies are given off. 2. The yolk granules are few and small-sized in animal hemisphere, while they are larger and densely packed in the vegetal hemisphere. 3. The cortical layer along with pigment granules remains thick around the animal pole and remains thin around the vegetal pole. 4. Besides the concentration gradients of yolk, cortical gel and pigment granules, there exists a gradient in the distribution of the ribosomes.

17-3. FERTILIZATION

In frog, the fertilization is **external** and occurs in water outside the body of the oviparous female. It involves the penetration of one complete sperm into an ovum anywhere around the animal hemisphere. To recall, an egg or ovum is laid at the secondary oocyte stage at a time when the second maturation division has already reached halfway. Spermatic fluid is shed by the male, involved in amplexus (**insemination**) almost simultaneously with the **oviposition** (*i.e.*, release of ova, about 500 to 600 at a time, from the cloacal aperture of female) and no time is lost in bringing about the fertilization. In the case of delay, the jelly coats around the ovum become too thick for the sperm to pass through them and the ovum also starts to show degenerative changes with the loss of power of development.

Monospermy is the rule in mesolecithal ova of the frog. In the act of fertilization, the sperm becomes attached to the jelly coats of ovum due to fertilizin-antifertilizin reaction (see Chapter 8) and locally dissolve them by acrosomal lysins (enzymes) of sperm to create a passage for sperm entry up to the vitelline membrane of ovum. Recently it has been found that the vitelline membrane is made permeable for the sperm in the body of female, while '**coelomic eggs**' (*i.e.*, ovulated eggs present in the coelom) enter the uppermost section of the oviduct. In the middle part of oviduct, the eggs are enveloped by the jelly coats. Some substance contained in the jelly coats provides the necessary condition for spermatozoa to go to the egg (**Balinsky**, 1981). In fact, jelly layers of the frog's egg depend on the presence of disulphide bonds for maintaining their structure. It has been suggested that the lysins of broken acrosome of frog's sperm may utilize a disulphide bond as a substrate. According to **Gusseck** and **Hedrick** (1971), the fusion of plasma membranes of the sperm and the egg occur by sulphydryl-disulphide bond interchange (**J.B. Phillips**, 1975).

Further, as soon as the tip of the acrosome filament of sperm touches the oolemma of ovum, the latter forms a small crator-like projection called the **cone of reception** or **cone of fertilization** which engulf the sperm by a complex process resembling to phagocytosis. Sperm entry is followed by the following sequence of events (often a complex series of cortical and cytoplasmic rearrangements) :

1. Lifting of vitelline membrane and rotation of egg. The entry of sperm in the ovum has caused some cortical reactions in the latter due to which the vitelline membrane is get separated from the egg plasma membrane and a perivitelline space is established in between the two membranes. This perivitelline space is filled with the liquid substance that is released by exocytosis of cortical granules. This frees the egg to orient itself with respect to gravity, so that the darker animal pole faces upwards and the pale vegetal pole downwards. Such a rotation of fertilized egg is called "**rotation of orientation**". This is first visible manifestation of fertilization.

2. Formation of fertilization membrane. Contents of burst cortical granules adhere to the vitelline membrane making it impermeable for late coming sperms (*i.e.*, it checks the polyspermy) (**Wolf** *et al.*, 1976). Such a transformed vitelline membrane is called the **fertilization membrane.**

3. Completion of maturation division of egg. The stimulus of the entrance of sperm inside the egg incites the germinal vesicle or egg nucleus which had paused in the metaphase, to complete its second maturation division. It gives off second polar body at the animal pole and becomes vesicular to become **egg pronucleus.**

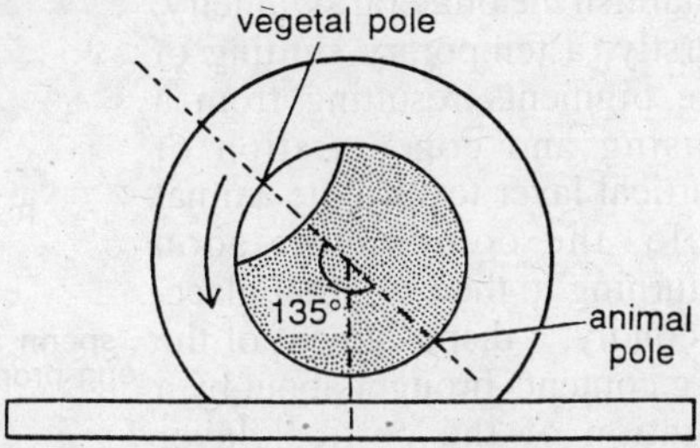

Fig. 17.2. Rotation experiment to determine the median plane of the frog's egg. An unfertilized egg in its jelly membrane if mounted on a glass plate; the egg axis forms at an angle of 135^0 with the vertical (after Berrill, 1971).

4. Penetration and copulation path. The whole spermatozoon usually takes a minute or two to get entirely inside the ovum. The sperm tail disintegrates, sperm nucleus becomes vesicular by absorbing water from the ooplasm and is called **sperm pronucleus**. It rotates so that proximal pair of centrioles come to lie at its forwarding end and develop asters. As the sperm pronucleus and centrioles travel (glide) centripetally to meet the egg pronucleus, they carry a trail of cortical pigment along with them. This is called **sperm track**. The straight sperm track is called the **penetration** or **entrance path**. However, in the case of an eccentrically placed egg pronucleus, the path followed by male pronucleus is first straight, then inclined at an angle. The inclined pigmented trail is called the **copulation path.**

5. Amphimixis. The sperm pronucleus along with its centrioles moves towards the egg pronucleus and ultimately both pronuclei fuse together to form **zygote nucleus**.

6. Bilateral symmetrization. During fertilization, another most important process called **bilateral symmetrization** is also occurred. During this process the radially symmetrical unfertilized egg becomes bilaterally symmetrical and it has been found that the bilateral symmetry of uncleaved fertilized egg remains intact during embryogenic development. Thus, the bilateral symmetry of the adult animal originates during fertilization.

First sign of the origin of bilateral symmetry in the fertilized frog egg is the appearance of heavily pigmented **sperm entry point** (**SEP**) on one side of the animal hemisphere, while the side opposite to SEP becomes lighter in pigmentation. The sperm-entry half of the egg eventually forms the **posterior** and **ventral structures** of the larva, with other half forming **head** (anterior) and **dorsal structures** (**Smith**, 1988).

Following fertilization, certain very significant changes occur in the egg cortex to establish the bilateral symmetry. Firstly, a temporary shifting of the pigments resulting from a raising and concentration of cortical layer toward the animal pole, the components soon returning to their original place, secondly, a displacement of the egg contents brought about by a rotation of the cortical layer around the underlying layers, with relative concentration of the pigmented layer towards a particular meridian. This is accompanied by a corresponding thinning of the cortex on the opposite side of the egg, thereby producing a crescent-shaped surface known as the **gray crescent**, which resembles the yellow crescent of the *Styela* egg. The plane passing through the centre of the gray crescent and the animal pole defines the median plane of bilateral symmetry in each case, *i.e.*, it coincides with the embryonic axis and is the only plane which separates the egg into two equivalent parts, each containing half the crescent material. In *Styela* the gray crescent is made visible by its contained yellow, mitochondrial granules. In the frog, the gray crescent represents material no longer covered by the pigmented cortical layer, and it is gray because it lacks reflective yolk platelets.

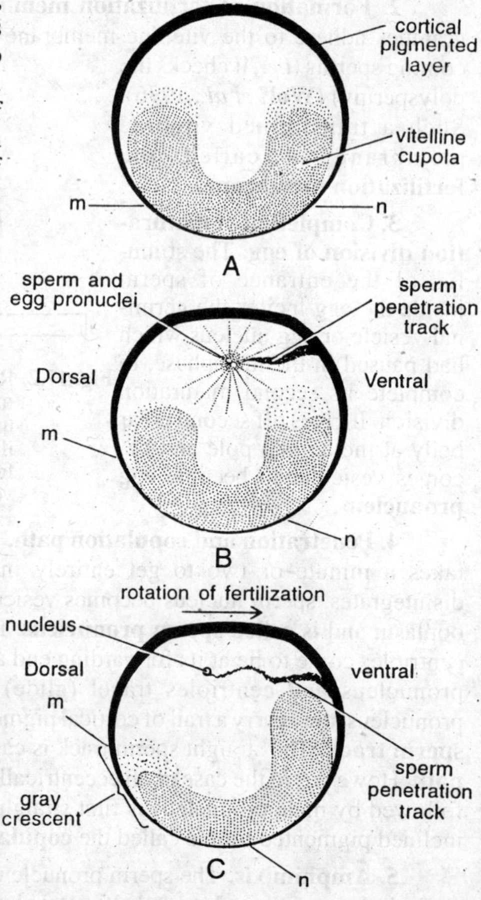

Fig. 17.3. A—Diagrammatic axial section of an unfertilized egg of *Rana fusca,* m-n representing the interior limit of the cortical pigment layer; B—The reaction of the egg to the spermatozoon when it enters. Note in particular the 100 inclination of m-n, and that the vitelline horn has moved closer to the future dorsal side; C—The rotation of symmetrization of cortex and formation of gray crescent (after Berrill, 1971).

According to **Pasteels** (1964), who has worked intensively in this field, two periods should be recognized between fertilization and the first cleavage : (1) The labile period between fusion of the pronuclei and the appearance of the gray crescent and (2) the time when bilateral symmetry is irreversibly fixed.

Thus, due to slight rotation of pigmented cortex around the endoplasmic mass, the median axis of the bilateral symmetry is established and becomes irreversibly fixed. Such a cortical movement which fixes the median axis of the bilateral symmetry is called the **"rotation of symmetrization."**

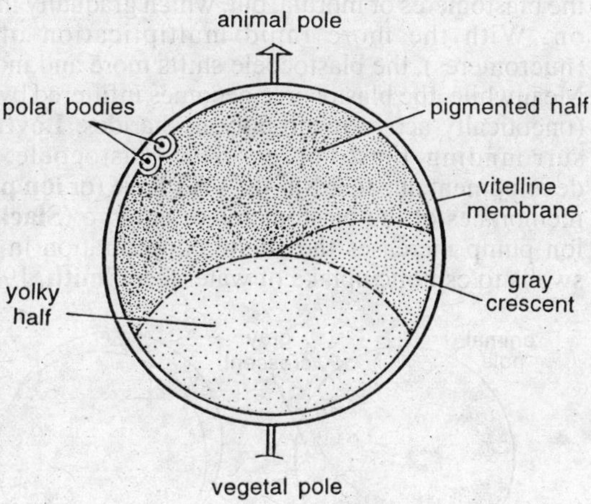

Fig. 17.4. A fertilized egg of frog (after Huettner, 1949).

17-4. CLEAVAGE AND BLASTULATION

The cleavage of a fertilized and telolecithal frog egg is of **holoblastic** and **unequal** type. The **first cleavage plane** is **meridional**. This cleavage furrow first appears near the animal pole and progressively extends towards the vegetal pole of the egg. It cuts the egg through its median animal-vegetal polar axis and results in two equal-sized blastomeres. The **second set** of cleavage divisions is likewise holoblastic and equal, and the cleavage planes are **meridional**, but at the right angles to the first plane. But with the **third set** of divisions, the unequally distributed yolk makes its influence felt. The mitotic spindles orient parallel to the polar axis, and displaced as they are toward the animal pole. The four blastomeres not only cleave **latitudinally**, but unequally. The eight-cell stage, therefore, consists of four small-sized, yolk-poor, animal cells, the **micromeres** and four large-sized, yolk-rich, vegetal cells, the **macromeres**. By so doing, thus, third set of cleavage is set up a differential. The **fourth set** of cleavage planes are **meridional, holoblastic**, but, **unequal**. They divide yolk-poor micromeres more rapidly, than yolk-rich macromeres. The micromeres divide rapidly, due to the fact that they are free from the burden of cleaving through a yolk-dense region. The macromeres divide slowly, since their cytoplasm is now loaded with yolk throughout. Yolk platelets have the double effect of conferring physical inertia on the dividing mass and diluting the active cytoplasm with metabolically inactive inclusions. The fourth set of cleavages, thus, producing a 16-cell stage.

The blastomeres in the early cleavage stages tend to assume a spherical shape like that of the egg before cleavage. Their mutual pressure flatterns the surfaces of the blastomeres in contact with each other but free surfaces of each blastomere remain spherical, unless these outer surfaces are also compressed by the vitelline membrane. At this stage the whole embryo acquires a characteristic appearance reminiscent of a mulberry and so it is called **morula**.

It is about the fourth and fifth cleavage stages that a small space, the **blastocoele**, appears between the blastomeres of morula. Actually, this blastocoelic space first may be represented just by narrow crevices between the blastomeres of morula, but, which gradually increases as the cleavage goes on. With the more rapid multiplication of the animal blastomeres (micromeres), the blastocoele shifts more and more towards the animal pole. Meanwhile, the blastocoele becomes infiltered by water and albuminous fluid (oncotically active muco-polysaccharides, **Lovtrup**, 1974), secreted by the surrounding blastomeres. The blastocoele enlarges througout early development due to a **Na^+-K^+- ATPase** (or **ion pump**) present in the internal membranes of the blastomeres of embryo (**Slack** and **Warner**, 1973). This ion pump increases the solute concentration in the blastocoele which then swells to osmotic uptake of water (see **Smith**, 1988).

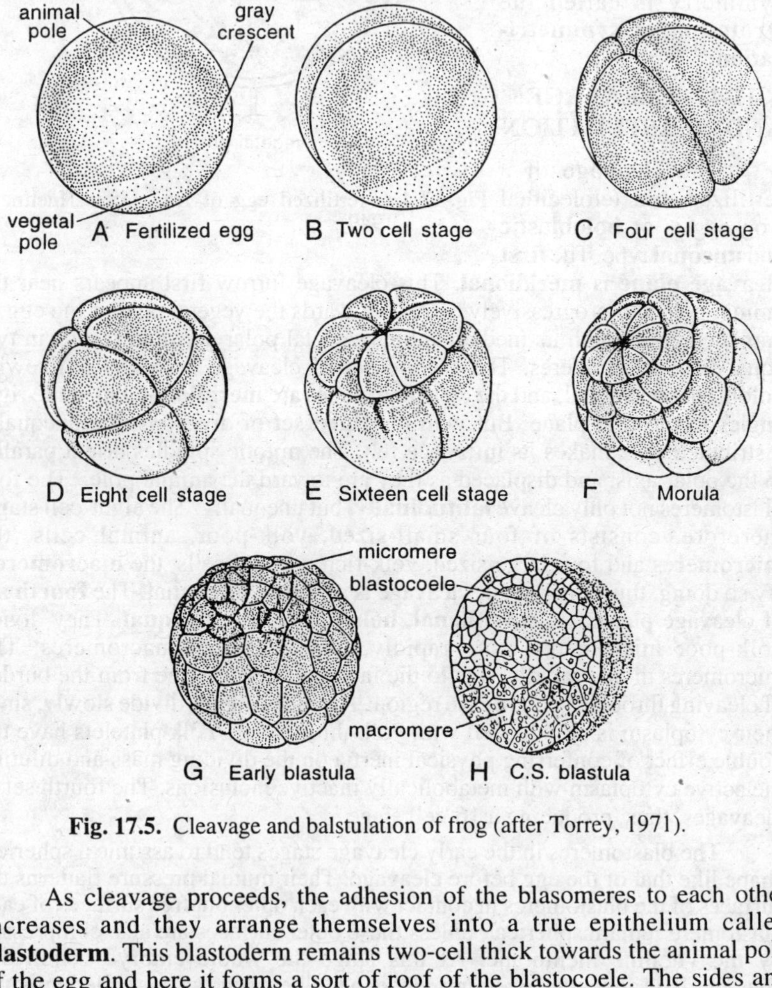

Fig. 17.5. Cleavage and balstulation of frog (after Torrey, 1971).

As cleavage proceeds, the adhesion of the blasomeres to each other increases and they arrange themselves into a true epithelium called **blastoderm**. This blastoderm remains two-cell thick towards the animal pole of the egg and here it forms a sort of roof of the blastocoele. The sides and floor of the blastocoele are occupied by multilayered blastoderm of large yolky blastomeres. The resultant embryo having fluid-filled blastocoele and blastoderm is the **blastula** of frog.

In frog (*e.g., Xenopus*), it has been observed that 11 subsequent cleavages take place **synchronously** every 35 minutes. However, after the 12th cleavage or when embryo (blastula) containing about 4096 cells, the cell cycle becomes longer and **asynchronus,** the cells become motile and RNA-synthesis (*i.e.,* transcription) starts (**Newport** and **Kirschner**, 1982). This turning point in development is called **mid-blastula transition** (**Gerhart**, 1980). The process of gastrulation starts when there are about 20,000 cells (**Smith**, 1988).

17-5. FATE MAPS OF BLASTULA OF FROG

In the blastula, the blastomeres which have to form different germinal layers and different organs of the adult frog, have their representation at the external surface of the blastula. The fate of each type of blastomeres has been observed by artificial-vital staining methods of **Vogt** (1925) and prospective organ region maps or fate maps have been prepared for the blastula of frog. According to the fate map studies, the whole surface of the blastula of frog can be divided into followed three areas :

1. Prospective ectoderm area. The entire pigmented area on and around the animal pole is the prospective **ectoderm**. The **neural ectoderm**, which develops into the central nervous system, occurs largely on the future dorsal side of the blastula, while the **epidermal ectoderm**, which develops into the skin epidermis of the embryo, occupies the antero-ventral side of the blastula. The material for the sense organs is also contained in these two areas. Inside the neural ectoderm area occurs a small sub-area which develops into the eye of the embryo. Likewise, inside the epidermal ectoderm area, there occur the sub-areas for nose, sucker, ears and ectodermal part of mouth.

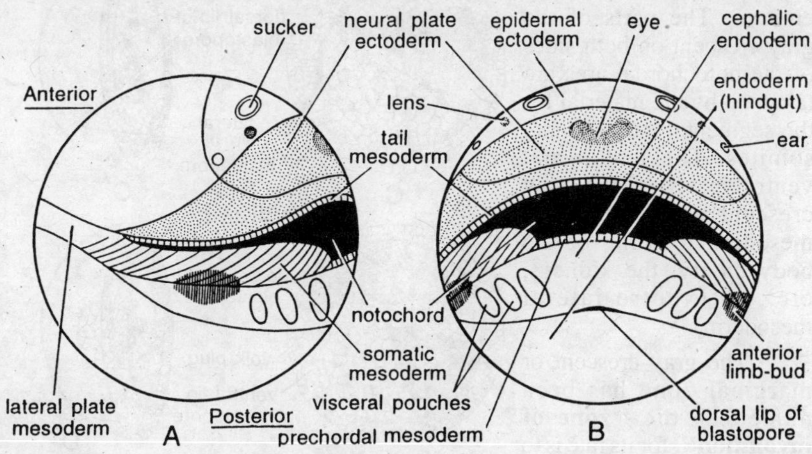

Fig. 17.6. Fate maps of balstula of frog. A—Viewed from the lateral side; B—Viewed from the dorsal side (after Huettner, 1949).

Gilchrist (1968) has called the zone of prospective ectoderm as **"zone of expansion"**, because, during gastrulation this zone expands downward and then converges towards the vegetal pole. At the end of gastrulation, the ectodermal zone gives rise to the entire outer layer of gastrula.

2. Prospective notochord and mesoderm area. The pigmented animal hemisphere (the prospective ectodermal area) is followed by a crescentic area of gray colour, the **marginal zone**, which goes all along the equator of the blastula, and has blastomeres for the formation of **notochord** and **mesoderm** of the frog embryo. The notochordal cells occupy the large area of dorsal side of the gray cresecent. Beneath the notochordal area and toward the vegetal pole there lies a narrow strip of cells which form the pre-chordal connective tissues of the embryo and, hence, this strip is called **pre-chordal plate**. At the upper side of notochordal area and toward the animal hemisphere, lies another narrow strip of cells which develops into the tail mesoderm of the frog embryo. The parts of the gray crescent on both sides of the notochordal area are taken up by the material for the segmental muscles, *i.e.*, somites. The lateral and ventral parts of the gray crescent give rise to the mesodermal lining of the body cavity, the kidneys, etc., (ventro-lateral mesoderm).

The gray crescent or marginal zone has been named as the **"zone of involution"**, for its fate is to roll around and beneath the lips of the blastopore (**Gilchrist**, 1968).

3. Prospective endodermal area. The entire

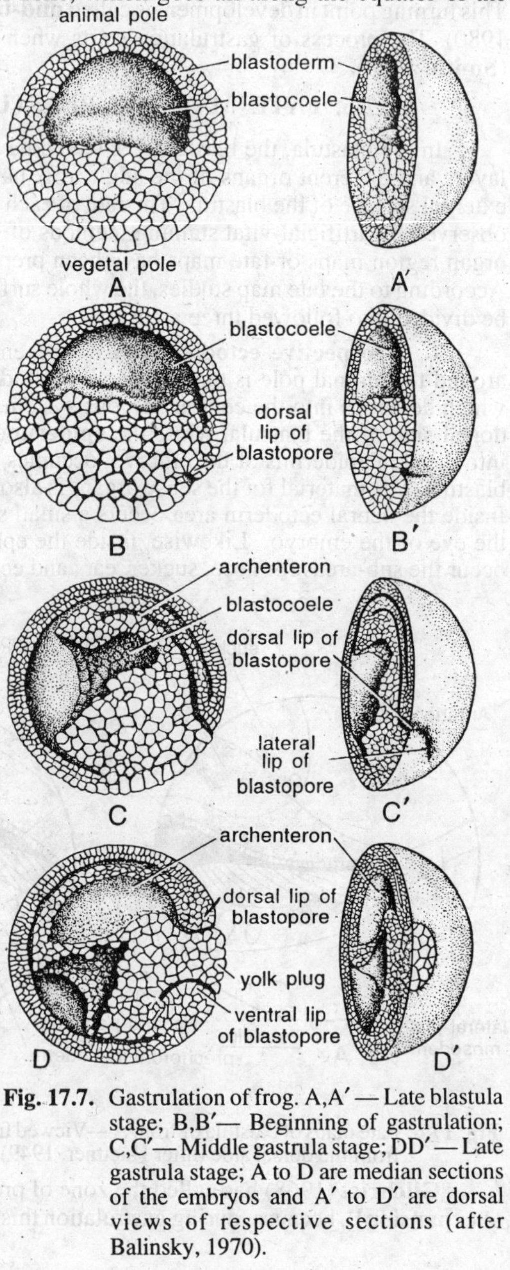

Fig. 17.7. Gastrulation of frog. A,A′ — Late blastula stage; B,B′— Beginning of gastrulation; C,C′— Middle gastrula stage; DD′ — Late gastrula stage; A to D are median sections of the embryos and A′ to D′ are dorsal views of respective sections (after Balinsky, 1970).

non-pigmented area on and around the vegetal pole of frog blastula is the prospective **endoerm**. Just beneath the pre-chordal plate area of mid-dorsal part of gray crescent, occurs the endodermal cells for the anterior parts of the alimentary canal—the endodermal lining of the mouth, gill region and pharynx. Rest of the endodermal area includes the materials for mid-gut, hind-gut and other organs such as liver, pancreas, urinary bladder and certain endocrine glands.

Gilchrist (1968) has called the prospective endodermal zone, the **"zone of invagination"**, because, during gastrulation, it sinks and glides into the blastula of frog.

Pregastrulation. In the late blastula, shortly before the appearance of the blastopore, certain pregastrulatory activities have been observed by **Vogt** (1929). Thus, pre-gastrulation includes three processes—displacement of the marginal zone, 'contraction' of the vegetal area and inclination of the animal-vegetal axis.

17-6. GASTRULATION

The process of gastrulation of frog is somewhat more complicated than *Styela* and *Amphioxus*. It includes three kinds of morphogenetic movements, namely **invagination of endoderm, involution of chordamesoderm** and **epiboly of ectoderm**.

17-6.1. Invagination of Endoderm and Formation of Blastopore

In frog, the gastrulation is inaugurated by the deformation of certain prospective endodermal cells in a circumscribed area below the equator or in

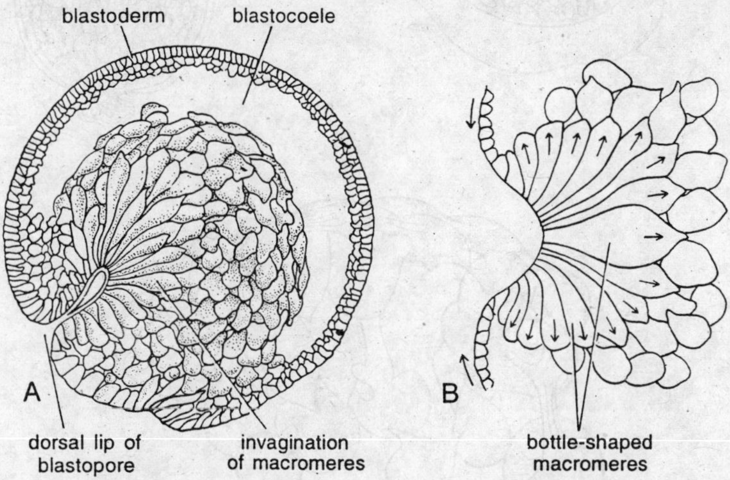

Fig. 17.8. Invagination of endodermal cells in an amphibian egg. A—Slightly schematized section through an advanced gastrula; B—The mechanics of gastrulation, showing active streaming of endodermal cells by bottling process (after Berrill, 1971).

the borderline of the vegetal region and exactly below the mid-dorsal point of gray crescent of blastula. These cells assume the elongate shape of a bottle and move toward the interior of the blastula. Their steadily elongating necks remain attached to the surface of the blastula with the outermost cementing layer which is discovered by **Holtfreter** (1943) and is found to lack in any electron-dense material, hence, remains unrevealed by electron microscopic studies of **Balinsky** (1960), but, recently confirmed to occur (see **Lovtrup**, 1974). Therefore, as the bulky-cell bodies move inward, a pull exerted along their attenuated necks creates an indentation at the surface. With continued multiplication and attenuation of bottle cells, the indentation or **invagination** deepens. Because, by internal movements to be described shortly the invagination will be expanded to form the cavernous **archenteron** or **gastrocoel**, the original indentation is called the **blastopore**, because, it forms a external opening for the gatrocoel. We have already seen that the region of the animal pole is the future anterior end of the embryo, so the blastopore marks

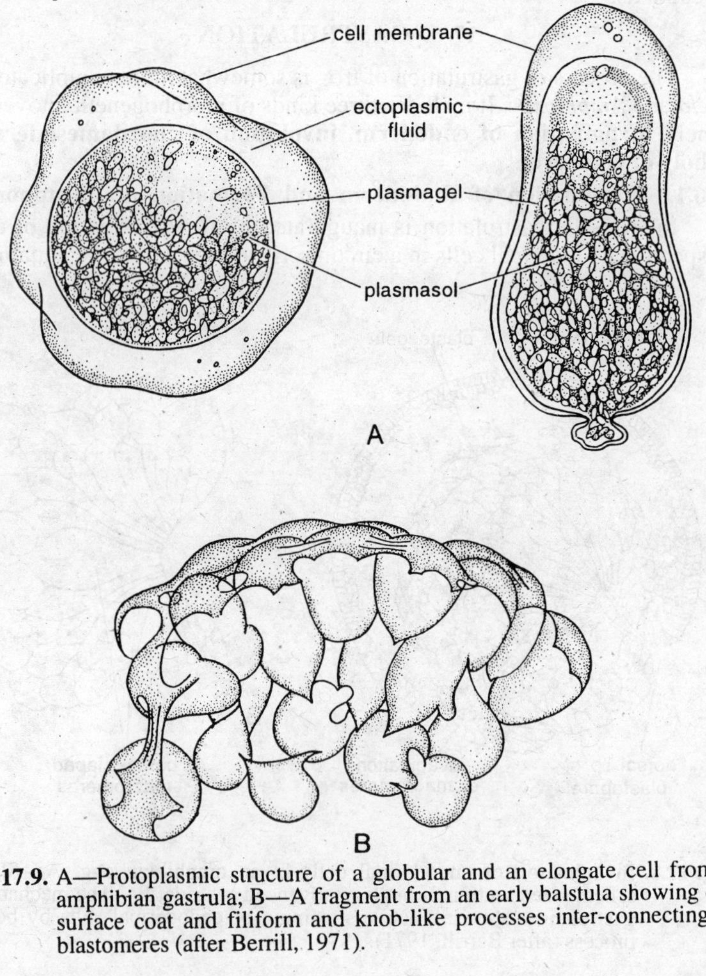

Fig. 17.9. A—Protoplasmic structure of a globular and an elongate cell from an amphibian gastrula; B—A fragment from an early balstula showing dark surface coat and filiform and knob-like processes inter-connecting the blastomeres (after Berrill, 1971).

the posterior end. And because the prospective notochord and neural ectoderm identify the dorsal side of the embryo, the area immediately above the blastopore may be termed as the **dorsal lip** of the blastopore.

Gradually, the blastoporal invagination extends circulolaterally on each side, so that the blastopore becomes crescentic, then horse-shoe-shaped and finally circular. Accordingly, the materials bounding the blastopore on the lateral sides represent its "**lateral lips**" and the material on the ventral side, the "**ventral lip**". Obviously, however, the terms dorsal, lateral and ventral refer only to the topography of a continuous circular lip.

It should be noted here that different parts of endoderm perform different kinds of morphogenetic movements to migrate inside the gastrula. Thus, the part of the endoderm lying in the marginal zone (the prospective fore-gut area) is mainly absorbed into the original pit-like invagination of the blastopore, but, some of endoderm of fore-gut involute over the dorsal lip along the chorda-mesoderm. The rest of the prospective endoderm of vegetal region passes into the interior of the embryo more or less passively and there it comes to lie in the floor of the gastrocoel. Actually, it is gradually covered up by the epibolic growth of ectoderm.

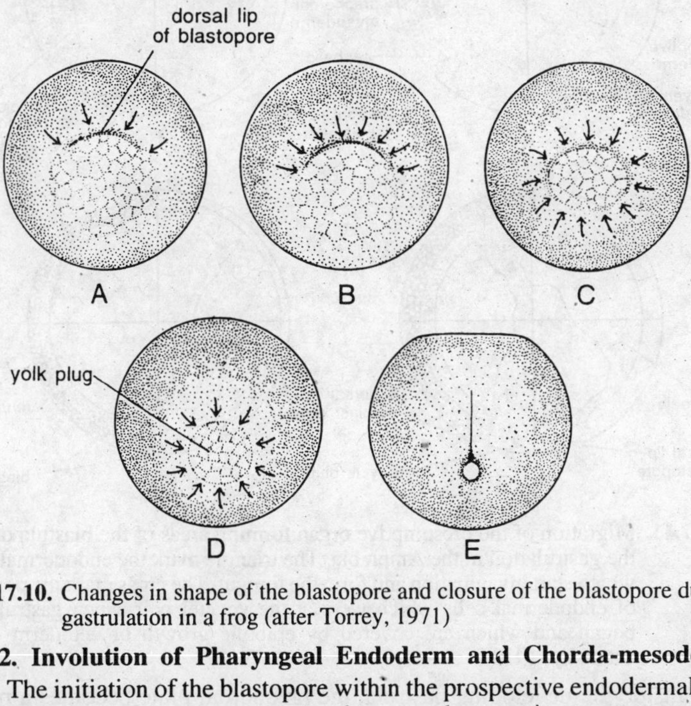

Fig. 17.10. Changes in shape of the blastopore and closure of the blastopore during gastrulation in a frog (after Torrey, 1971)

17-6.2. Involution of Pharyngeal Endoderm and Chorda-mesoderm

The initiation of the blastopore within the prospective endodermal zone sets in the most important gastrular morphogenetic movements, the **convergence** or flow of surface material towards the lips of blastopore and the **involution** of that converged material inside the blastoporal lip. To recall, the endodermal cells abuting the dorsal lip of the blastopore form the prospective pharyngeal endoderm, which is followed by pre-chordal plate, notochord and tail mesoderm. When the dorsal lip is formed, the pharyngeal endodermal cells are the first cells which "turn the corner" or undergo involution over the dorsal blastoporal lip. These cells move to the interior and are followed by prechordal material which converges in to replace it. Soon, pre-chordal material also involutes and is trailed by the anterior part of the prospective notochord and tail mesoderm. As these materials flow inward around the dorsal lip, they become considerably narrowed and in the process of forward flow, greatly elongated. The prospective pharyngeal region later forms the most anterior part of the advancing gastrocoel. In the later stages of gastrulation, the oral and the pharyngeal endoderm expands so as to form the spacious **foregut** whose lateral, ventral and anterior walls consist of a thin layer of endoderm. Only part of the dorsal wall of the foregut is taken up by the prechordal plate and the anterior tip of the notochord.

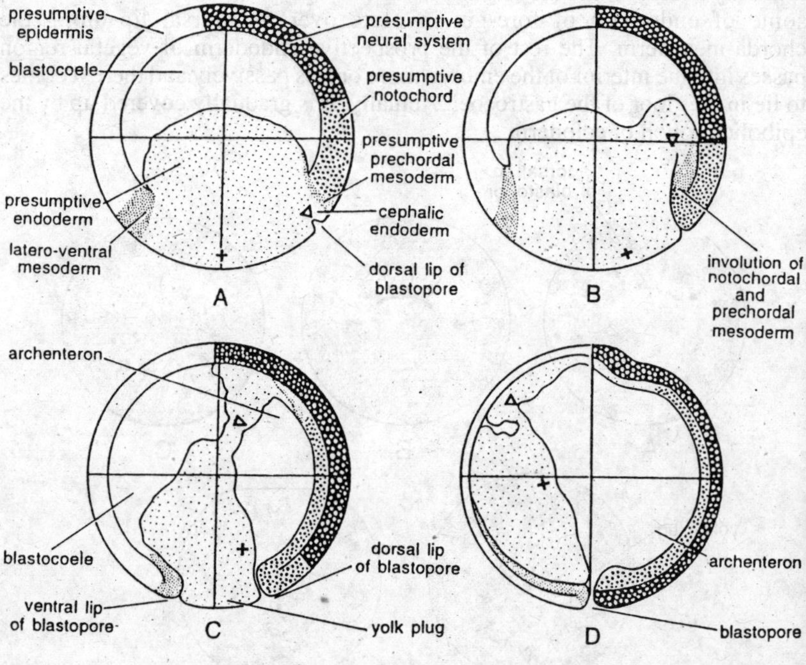

Fig. 17.11. Migration of the presumptive organ forming areas of the blastula during the gastrulation in the Amphibia. The triangle mark the endodermal cells which start invagination and form the foregut. The cross mark the position of endodermal cells which were at the vegetal pole when gastrulation began and which are covered by epibolic growth of ectoderm (after Balinsky, 1970).

Inside the developing gastrula, the prechordal plate becomes a part of the archenteron roof in front of the anterior end of the notochordal material. The anterior part of notochord stretches along the dorsal side of the archenteron, forming the mid-dorsal strip of the archenteron roof. Later it is joined by the cells of posterior part of the notochord which involutes or rolls over the dorso-lateral lips of the blastopore. Eventually, the presumptive notochord undergoes a very considerable elongation in the transverse direction. It becomes concentrated on the dorsal side of the embryo. The tail mesoderm remains near the blastopore, which marks the posterior end of the embryo.

Most of the mesoderm (*i.e.*, trunk somites and ventrolateral mesoderm) invaginates into the interior by rolling over the lateral and ventral lips of the blastopore. Once inside the embryo, the whole kinds of mesoderm namely, notochord, prechordal mesoderm, somites and ventrolateral mesoderm move from posterior side (blastoporal end) towards the anterior side as a single unit called **chorda-mesodermal mantle**, penetrating between the ectoderm on the outside and the endoderm on the inside. Eventually, the mesoderm occupies whole place between the endoderm and ectoderm except a small portion at the anterior end of the embryo which remains free from the mesoderm and it represents the place at which the mouth is formed, during later developmental stage.

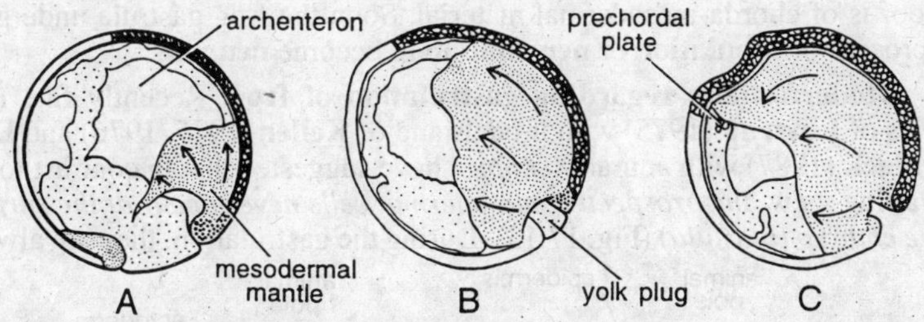

Fig. 17.12. Internal streaming of the mesodermal mantle during the later phases of gastrulation of an amphibian (after Torrey, 1971).

17-6.3. Epiboly of Ectoderm

Throughout gastrulation the embryo retains its spherical shape and a uniform size. This means that as soon as the prospective pharyngeal endoderm and chorda-mesodermal mantle move out of the surface, their places are taken by ectoderm. To accomplish this both neural and epidermal ectoderm greatly increase their surfaces. The expansion of ectoderm is an active process, and the increase of surface area goes on at the expense of a thinning out of the epithelial layer. The presumptive epidermis expands in all directions, but in the case of the presumptive nervous system the expansion is mainly in the longitudinal direction, *i.e.*, toward the blastopore. In the transverse direction on the other hand, the presumptive nervous system area contracts. As a result, the whole nervous system area changes its shape and becomes oval, elongated in an antero-posterior direction.

17-6.4. Closing of the Blastopore

It is in consequence of the removal of endodermal and mesodermal materials from the surface and the compensating spreading of the ectoderm that ectoderm finally arrives at the circular lip of the blastopore, that is, the **yolk-plug** (white coloured endodermal cells as seen from the blastopore) comes to be bounded by ectoderm. Meanwhile, due to forward movement of early invaginated pharyngeal endoderm, and its extension in the antero-posterior direction following changes occur in the gastrula : (1) The gastrocoel gradually deepens and it eliminates completely the blastocoele. (2) As the endodermal mass accumulates on the future ventral side, the centre of gravity is shifted and the embryo rotates so as to bring its dorsal side uppermost. (3) The protruding yolk plug gradually withdraws to the interior, and as it does so, the diameter of the blastopore steadily contracts until at the end of gastrulation, the blastopores recognizable only as a narrow slit.

17-6.5. Gastrula

Thus, gastrulation converts the radially symmetrical monoblastic blastula into a spherical, bilaterally symmetrical triploblastic gastrula having a head-to-tail axis. It is clothed externally by ectoderm and containing endoderm and mesodermal components internally. The part of the ectoderm that lies lenghtwise as a broad, mid-dorsal band is the fore-runner of the nervous system ; the remainder is the potential epidermis. The gastrocoel (or archenteron) represents the lumen of the gut to come. For the present, the walls and floor of the primitive gut or archenteron are composed of endoderm, but, its roof is of chorda-mesodermal material. Soon the frog gastrula undergoes the process of **tubulation** or **neurulation** to become **neurula**.

Modern views regarding gastrulation of frog. Recently fate map studies of **Lovtrup** (1975) with urodels and of **Keller** (1975, 1976) and **Dale** and **Slack** (1987) with anuran *Xenopus*, have suggested that in contrast to the prevailing view, *the prospective mesodermal cells never occur at the surface of the embryo (blastula)* (Fig. 17.13). During the gastrulation, they are always

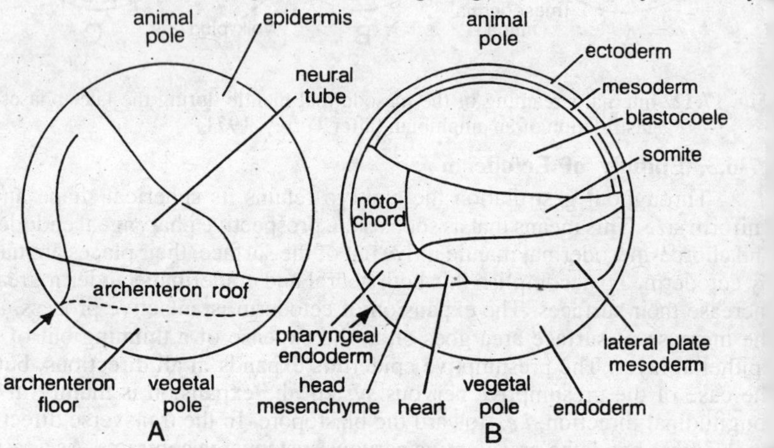

Fig. 17.13. Fate map of the *Xenopus* late blastula–early gastrula. A—Fate map of superficial layer; B—Fate map of deep layer of cell (after Smith, 1988).

beneath the prospective endodermal sheet which is at the surface prior to its invagination and involution, and between the involuted endoderm and superficial ectoderm after involution. Moreover, the prospective mesodermal cells begin moving into the interior before the appearance of the blastoporal depression in a kind of **ingression** (*i.e.*, mesodermal cells may sink inside as individuals). During ingression, mesoderm acts as a flowing cell stream and it initiates the process of **invagination** and **involution**. This advance (*i.e.*, ingression) toward the newly formed blastopore (convergence) and as the mesodermal cells enter the zone of involution (*i.e.*, lip of blastopore), they lose their regular arrangement, become motile and move actively into the interior, forming the mesodermal mantle of the wall of newly forming archenteron. Since the cells of mesodermal mantle are also motile, they continue this movement forward. Since both the overlying prospective ectoderm and endoderm are attached to the moving underlying mesoderm, both of them are moved passively downward toward the blastopore, where the endoderm undergoes involution because of this continuing attachment (**Trinkaus**, 1984). Such a working hypothesis, therefore, accounts both for the *epiboly of the prospective ectoderm* and *convergence and involution of prospective endoderm*.

17-7. NEURULATION

The process of neurualtion, though, specifically is the process of neural tube formation, is the phase of development during which mesoderm and endoderm also undergo primary differentiation into particular structural entities. In frog it includes following processes, all of which occur in a simultaneous manner :

(1) Neural tube formation ; (2) Tubulation of chorda-mesoderm ; (3) Tubulation of endoderm.

17-7.1. Neural Tube Formation

At the close of gastrulation, when the yolk plug finally disappears and the blastopore closes to a dorsoventral slit, the amphibian embryo is still almost spherical and evenly pigmented. Shortly afterward, the presumptive area of the nervous system becomes differentiated from the rest of ectoderm (epidermal ectoderm) in the form of a pear-shaped **medullary** or **neural plate**. The ectodermal epithelium moves toward the dorsal side of the embryo. At the same time, the cells of the neural plate change in shape and become elongated and arranged themselves into a columnar epithelium. Thus, they become different from the cells of the epidermis, which remain more or less flat and arranged as a stratified epithelium usually two cells thick.

During this process the embryo lengthens along its anteroposterior axis, and at the same time the edges of the neural plate become thickened and raised above the general level as ridges called **neural folds**. Continuing the process, the nueral plate narrows in a transverse direction, especially in its posterior parts and the neural folds become higher, so that the neural area contained between them forms a wide **neural groove**. Subsequently, the neural folds meet each other in dorsal middle line and fuse. In this way the neural tube is formed, finally the lateral epidermal ectoderm of each side also meets and fuses at the mid-dorsal line, while the neural tube segregates and comes to lie below the completed epidermal layer. The **neurula** is the stage from the beginning to the end of the above process.

The process of neural tube formation, includes thickening of neural ectoderm to form neural plate infolding or invagination of neural plate, splitting of neural folds from epidermis, and segregation of neural crest cells from neural ectoderm and epidermis. Finally, some regionalization becomes visible in the neural tube with sub-division of the brain into **prosencephalon** (fore-brain), **mesencephalon** (mid-brain) and **rhombencephalon** (hind-brain).

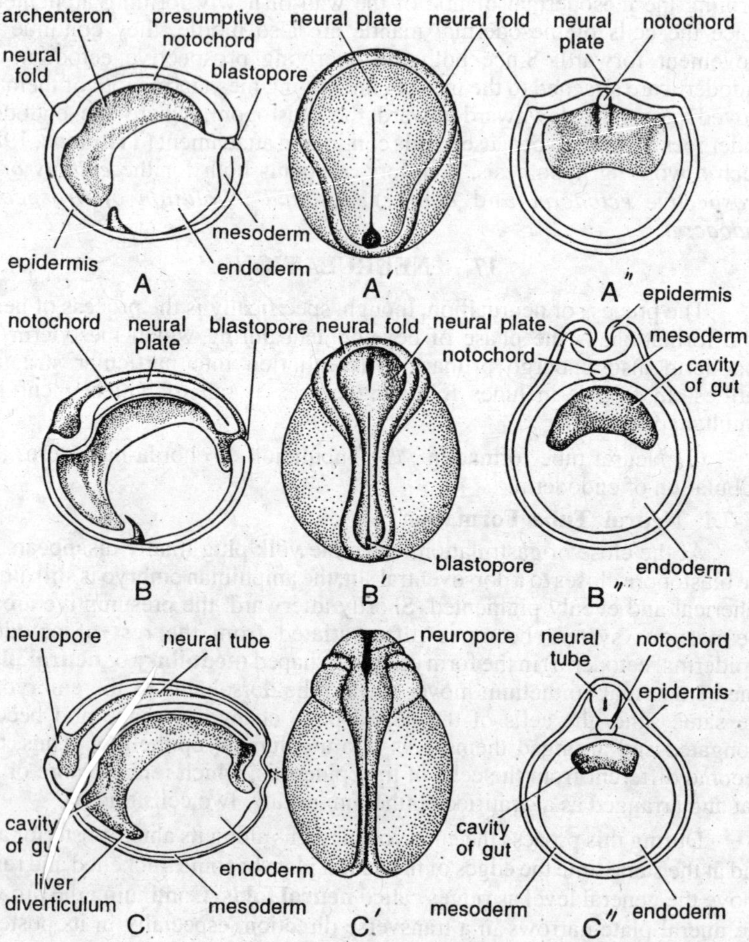

Fig. 17.14. Three stages of neurulation in a frog embryo. The drawings in the middle show whole embryos in dorsal view. The drawings on the left show the right halves of embryos cut in the median plane. The drawings on the right show the anterior halves of embryos cut transversely. A, A′, A″ — Very early neurula; B, B′, B″ — Middle neurula; C, C′, C″ — Late neurula with neural tube almost completely closed. The asterisk in C indicates the point at which the anal opening will break through (after Balinsky, 1970).

Neural crest cells. Tissue from the folds of the neural plate that comes to lie between the dorsal part of the neural tube and the dorsal epidermis is the neural crest. The neural crest cells undergo extensive migration and form the autonomic nervous system, melanocytes and parts of the skull (see **Smith**, 1988). Investigations of **Raven** (1936) have shown that the neural crest cells end up in the spinal and sympathetic ganglia (neurocytes) ; in **pia mater** ; in epidermal basement lamellae (collagenocytes) ; in myxoblasts of dorsal fin and head and melanocytes.

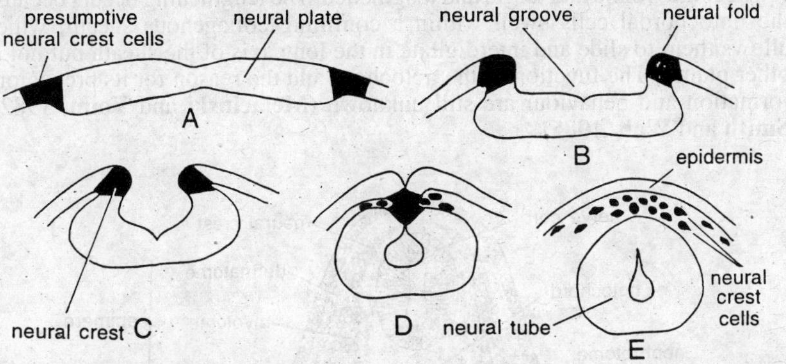

Fig. 17.15. Stages in the formation of the neural plate and neural tube in amphibians (after Balinsky, 1970).

Neural induction. The process of neurulation (*viz.*, formation of neural plate and its conversion into neural tube) is initiated or induced by chorda-mesoderm which lies in the roof of archenteron of the late gastrula of frog and it is called **neural inductor**. Because, the primordial chorda-mesoderm has the capacity of self-differentiation combined with complex inductions (neural and others), therefore, it is called **primary organizer** (**Spemann,** 1938).

17-7.2. Tubulation of Chorda-mesoderm

At the time of the closure of the blastopore, the chorda-mesodermal mantle separates itself from the endoderm and the mesoderm attains its definitive position between the endoderm and the ectoderm. To make the separation complete, the notochord splits at its anterior end from the prechordal plate and the prechordal plate itself separates from the adjoining endoderm which forthwith closes under the prechordal plates. During neurulation of frog, following major morphogenetic changes occur in chorda-mesoderm :

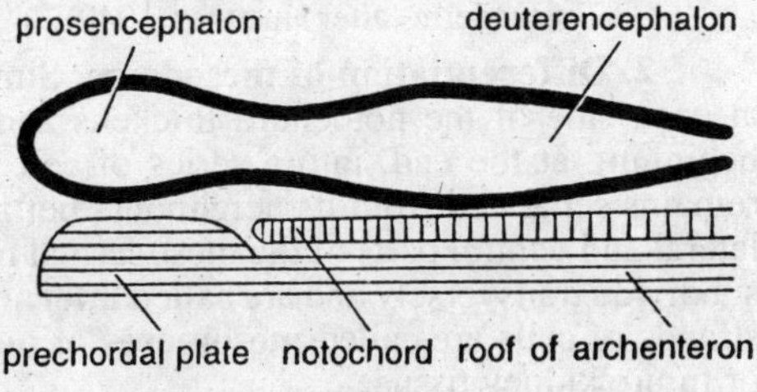

Fig. 17.16. Schematic diagram demonstrating the subdivision of the brain archencephalic and deuterencephalic parts and the relation of these to the neural inductor system (after Torrey, 1971).

1. Formation of notochord. After its separation from the prechordal mesoderm as a narrow rod of cells, the rudiment of notochord separates from rest of the chorda-mesodermal mantle and notochordal cells transform into colligocytes and a collagenous sheath is deposited around them (see **Lovtrup**, 1974). Shortly, after, fluid-containing vacuoles appear in the notochordal cells which press the nucleus and cytoplasm against the cell borders, as a result the whole rod becomes round and turgid and lengthened. The lengthening occurs because the notochordal cells swell within a confining collagenous sheath, which allows them to slide and interdigitate in the long axis of the sheath but not in other planes. The function of the notochord and the reason for its precocious formation and behaviour are still unknown (**Melacinski** and **Youn**, 1982 ; **Smith** and **Walt**, 1985).

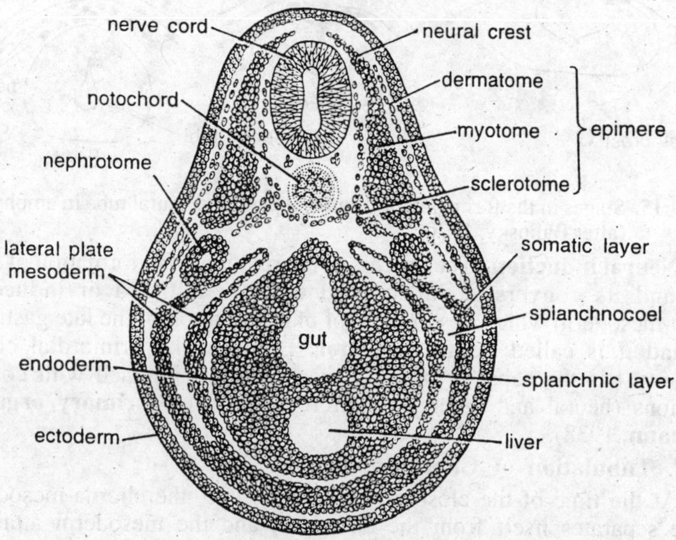

Fig. 17.17. T.S. young tadpole through trunk region showing differentiation of mesoderm (after Huettner, 1949).

2. Differentiation of mesoderm. Simultaneously the tip of mesoderm at each side of the notochord thickens and subdivides in transverse plane, beginning at the end, into a series of cell masses or **somites**. Each somite remains separated from its neighbours but remains joined to the thinned out lateral and ventral parts of the mesodermal mantle on each side (which do not subdivide transversely and are called **lateral plates**) by strands of cells. These strands of cells are called **mesomeres** or **nephrotomes**. They are the source of future kidney tissue.

Subsequently, the part of each somite (**epimere**) next to the notochord separates and becomes the **sclerotome** or skeleton-forming tissue around the notochord, while the remaining major part of the somite becomes the **myotome**, the cells of which differentiate into the striated muscle fibres of the somatic muscles. The outermost, narrow strip of somite cells, which lie just beneath the epidermis, **dermatome**, later on forms the dermis of the skin.

Lastly, the lateral mesodermal plate of each side splits into two layers, an outer or **parietal** layer next to the ectoderm and an inner, **visceral** layer next to the endoderm. The slit-like cavity between the two layers is the **coelomic cavity**. The inner layer gives rise to the smooth muscles of the intestine and to the blood and blood vessels.

17-7.3. Tubulation of Endoderm

In contrast to ectoderm and mesoderm, the endoderm is slower to differentiate and eventually forms the pharynx, lungs, stomach, liver and intestine. At the end of gastrulation it is still an open trough. As neurulation begins, the free margins of the endoderm unite in the dorsal middle line beneath the notochord to complete the formation of the definitive gut, or **enteron**. The floor of the enteron persists as a thick layer of large yolk- filled cells. Anteriorly the enteron makes contact with the anterior ectoderm, where later on, the mouth invagination forms and fuses with the blind end of the enteron to form a continuous gut cavity. Later, the lungs, liver and pancreas develop from evaginations from the gut. The pharyngeal pouches are formed as lateral outpushings of the foregut and in later development these give rise to the **gill-clefts.**

During neurulation, the embryo already begins to stretch in length. At the same time it is flattened from side to side and also becomes smaller in a dorso-ventral direction. The volume of the embryo does not change appreciably. The stretching is greatest in the posterior part of the embryo. As the neural tube takes part in this general stretching, it becomes still more thinned out in its posterior parts, so that the difference between the presumptive brain and the presumptive spinal cord is increased. The part of the embryo above the blastopore becomes elongated beyond the blastopore and this elongation becomes the rudiment of the tail, the **tail bud**. Now the neurula is called a **tail-bud embryo**.

17-8. POST-NEURULAR DEVELOPMENT OF FROG

The development of neurula into the tadpole includes the following two events :

17-8.1. Formation of Tail-bud Stage and Hatching

The neurula of *Rana tigrina* is 3.2 mm long and develops into tail-bud embryo, which consists of bulges of gill plates ; two optic bulges, one on each side ; stomodael groove at antero-ventral margin of head and a pair of **adhesive organs**, **cement glands** or **oral suckers** in the form of pigmented conical protuberances, each developing from the ectoderm, ventral to the mandibular arch. Later on a single V-or U-shaped adhesive sucker is formed due to meeting of two pits of suckers. At this stage, there are only a pair of myotomes in the trunk of the embryo.

Soon (within 1 hour and 15 minutes) tail of embryo elongates and is covered dorsally, ventrally and terminally by epidermal **tail fins** ; dorsal profile of embryonic body becomes concave ; trunk contains about 12 pairs of myotomes and each gill plate has rudiments of three gill clefts (**Agarwal** and **Niazi**, 1977). Above the gills appear invaginations forming **auditory pits** (or vesicles of internal ears). At the front of head are present a pairs of **olfactory**

pits. There also forms a simple S-shaped **heart** tube having no chambers. Ventral portion of embryo's trunk bulges out due to yolk-filled endodermal cells on the floor of midgut. Gut is a straight tube leading outside through a slit-like **cloacal aperture** (proctodaeum or cloacal tail-piece) ventrally at the base of tail (Fig. 17.18).

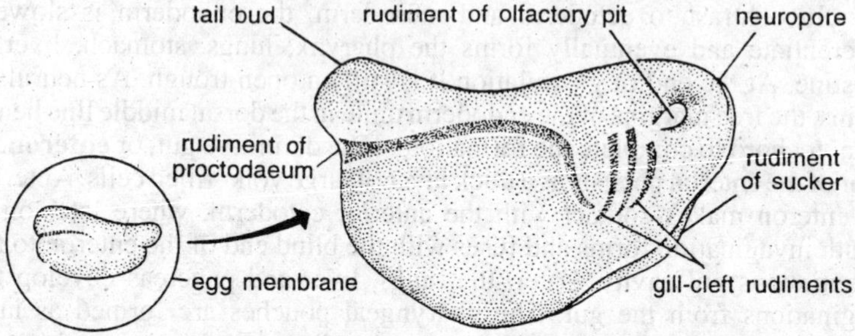

Fig. 17.18. An unhatched tadpole larva of frog (after Huettner, 1949).

A 4.5 mm long embryo has a fully developed tail with its tail fins and myotomes which extend up to half length of the tail. A pair of buds of external gills are erupted from each gill plate. Stomodaeum becomes cup-shaped and V-shaped oral sucker becomes enlarged. This embryo comes out from the egg's vitelline membrane in the process of **hatching**. The hatched embryo or young one of frog is the **larva** which is called **tadpole** (Old English, *tade* = toad ; *pole* = head). Hatched tadpoles of frog lack mouth and are unable to feed ; they get attached with the leaves of aquatic plants or other objects by the help of their adhesive suckers and nourish themselves from the yolk still present in the endodermal cells of midgut.

The process of hatching of tadpole is probably aided by the hatching enzyme secreted by its skin which dissolves the egg membranes. **Marshall** (1928) has described the method of hatching of frog as follows : "The tadpole possesses a frontal gland immediately in front of the horse-shoe-shaped cement organ on the underside of its head. Fixing this cement organ to one point by the egg-membrane, it performs movements around this fixed point by the aid of the cilia which cover the greater part of its body and rubs the frontal gland over this part of the vitelline membrane. Ultimately the membrane is softened and gives way, allowing the tadpole to creep out and adhere to weeds." According to **Agarwal** and **Niazi** (1977), at the temperature ranges between 28°C to 32°C, hatching in Indian bullfrog (*Rana tigrina*) takes place within 17.30 hours of development of the fertilized egg.

17-8.2. Post-Hatching Development of Tadpole

The fish-like young tadpole undergoes speedy development (*i.e.*, growth, morphogenesis and cytodifferentiation) which includes the following two steps :

1. External gill stage of tadpole. Due to development, in a 5.5. mm long tadpole the following structures are visible :

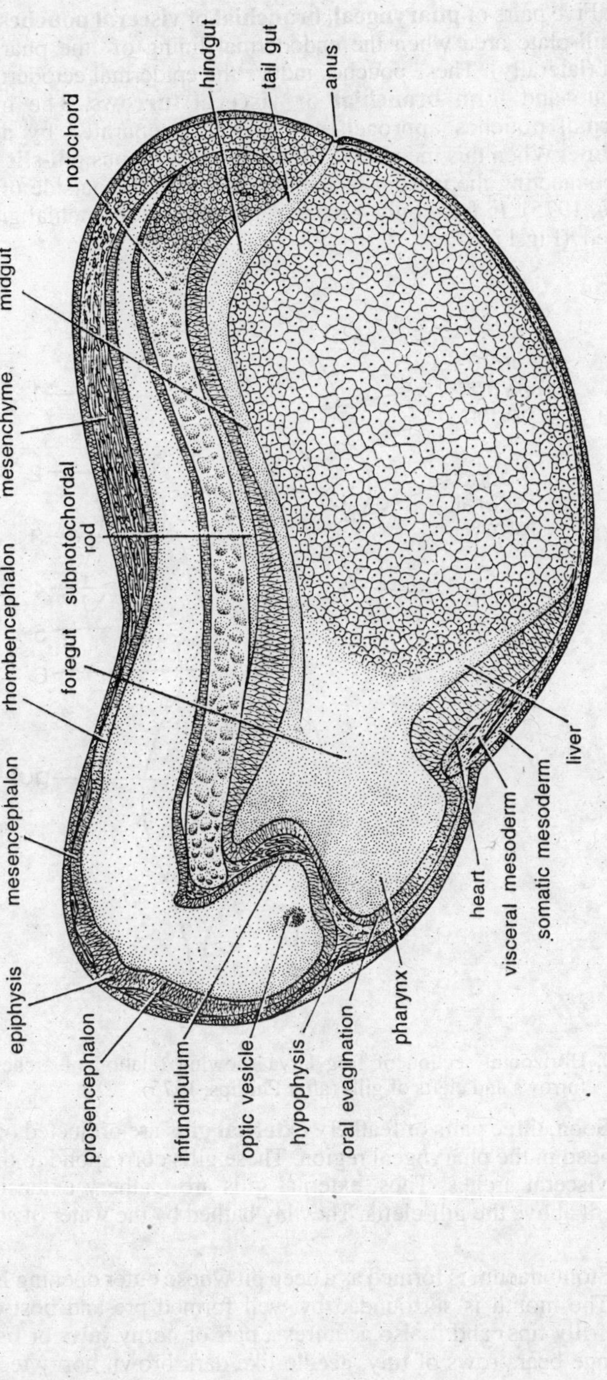

Fig. 17.19. Median longitudinal section of an early tadpole larva (8 mm) (after Huettner, 1949).

1. Five pairs of **pharyngeal**, **branchial** or **visceral pouches** are formed in the gill-plate area when the endodermal lining of the pharynx pushes outward (laterally). These pouches induce the epidermal ectoderm of skin to invaginate and form **branchial** or **visceral furrows**. The invaginating endodermal pouches approach one another, separated by a **branchial membrane**. When this membrane ruptures, a continuous **gill-slit** or **gill-cleft** forms, connecting the interior of the pharynx with the outside of the animal (**Phillips**, 1975). In frog, only 4 pairs of gill-clefts or branchial gill-clefts are perforated (Fig.17.20).

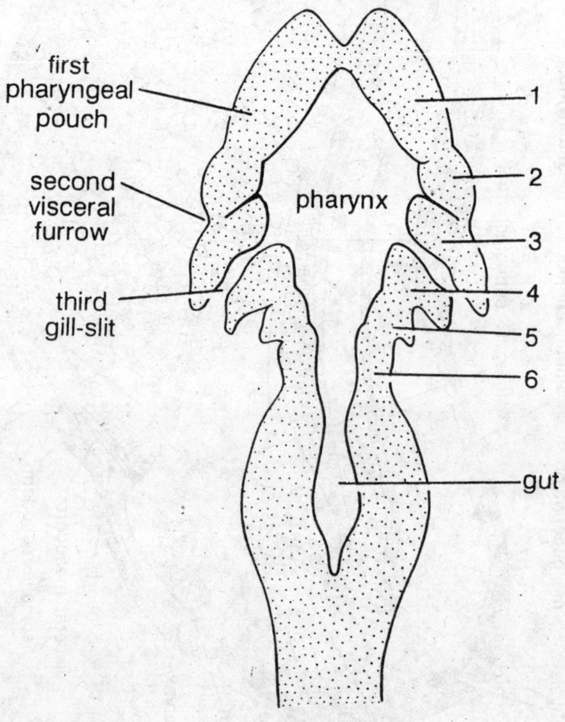

Fig. 17.20. Horizontal section of frog larva showing relation of arches, pouches, furrows and clefts of gills (after Phillips, 1975).

2. Soon, three pairs of feathery **external gills** are projected out from the sides of head in the pharyngeal region. These gills correspond to the 3rd, 4th and 5th visceral arches. Thus, external gills are feathery extensions of the integument above the gill-clefts. They lay bathed by the water of surrounding medium.

3. Stomodaeum is formed as a deep pit whose outer opening is the larval **mouth**. The mouth is surrounded by well formed pre-and post-oral **labial fringes** (frilly lips) and it also acquires a pair of horny **jaws** or **beaks**. Each labial fringe bears rows of tiny, needle-like dark brown horny teeth. A 10.5

mm long tadpole contains three rows of pre-oral labial teeth and three rows of post-oral labial teeth.

4. The adhesive sucker gradually disappears.

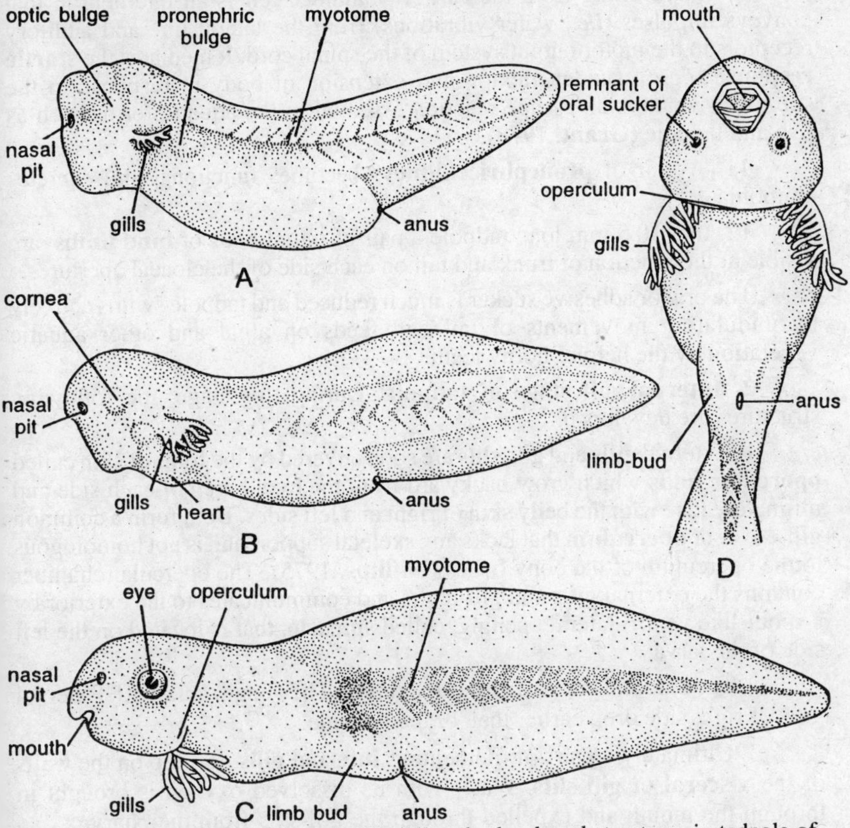

Fig. 17.21. Formation of external gills, tail and other larval structures in tadpole of frog. A—Larva at hatching; B—8 mm long tadpole; C—11 mm long tadpole; D—Ventral view of 11 mm long larva (after Huettner, 1949).

5. The larval gut is differentiated into pharynx, oesophagus, stomach, liver, gall-bladder and intestine. As an adaptation for the herbivorous mode of feeding, the intestine of tadpole becomes very long and lies coiled like a watch spring. This is to increase the absorptive surface of the intestine.

6. Myotomes increase in number and extend up to the tip of the tail.

7. **Melanophores** (*i.e.*, melanin pigment secreting pigment cells or chromatophores) appear in the skin of the dorso-lateral surface of head, trunk and tail of tadpole.

8. The cornea becomes transparent and eye lens becomes visible as a white spot.

9. A lateral line sensory system becomes visible on either side of tail.

10. A pair of giant neurons called **Mauthner cells** are developed in the hind-brain of tadpole. These cells receive many afferent inputs from the so-called lateral line sensory system and the vestibular branch of auditory cranial nerve coming from the ear. A Mauthner cell is an interneuron that conveys impulses (*i.e.*, water vibrations) from the lateral line and auditory receptors to the motor output system of the spinal cord. It mediates the **startle response** (*i.e.*, a sudden flexion and extension of body in response to the stimulus) which is an escape reaction essential to the aquatic forms such as fish and tadpole (**Grant**, 1978).

11. A pair of **pronephric kidneys** becomes functional and excretes ammonia.

12. In a 10.5 mm long tadpole a pair of initial buds of **hind limbs** are visible at the junction of trunk and tail on each side of the cloacal aperture.

The oral or adhesive sucker is much reduced and tadpole swims actively by undulating movements of tail and feeds on algal and other aquatic vegetation by the help of horny teeth.

2. Internal gill stage of tadpole. During this stage, the following structures are developed :

1. External gills and gill-slits are soon covered by the folds of skin called **opercular folds** which grow backward from the hyoid arch of each side and ultimately fuse with the belly skin of right and left sides. Both form a common gill-cover or **operculum** that lacks any skeletal support and is not homologous to the operculum of the bony fishes (**Phillips**, 1975). The opercular chamber contains the external gills and gill-clefts and communicates to the exterior by a spout-like ventro-lateral opening, called **spiracle**, that is located on the left side of the larva.

2. The external gills are of short duration. They begin to fall off. External gills of right side drop earlier than of the left side.

3. Ultimately 4 pairs of filamentous **internal gills** develop on the walls of the **visceral** or **gill slits**. Water with its dissolved oxygen is brought in through the mouth and expelled through the gill-slits from the pharynx ; so that, there exist a constant flow of water over the surface of the vascular internal gills.

4. In the growing tadpole (32 mm to 40 mm long) the intestine contains six and half coils which largely contribute to the bloating appearance of the body of tadpole.

5. Due to morphogenesis different parts of the hind-limbs such as thigh, shank, ankle, foot and five toes become well formed in the tadpole of 40 mm long (within 33 days of egg development).

6. Soon a pair of buds of fore-limbs appear behind the head but remain concealed by the operculum. In the mature tadpole (40 mm long), a pair of fully formed **fore-limbs** become visible through the transparent skin in ventral view. As development proceeds further, the left fore-limb emerges through a window formed by enlargement of the spiracular opening. The right fore-limb emerges through a window formed by dissolution of skin below its elbow.

7. In the mature tadpole a pair of **lungs** also develops from the pharynx so that the larva uses both gills and lungs for breathing. The tadpole frequently comes to water surface to gulp air for buccal and pulmonary respiration.

Thus, embryogenesis converts a fertilized frog's egg into an aquatic fish-like, tailed tadpole larva, which has no resemblance with the terrestrial, lunged frog. To become adult-like animal, therefore, it has to pass through another process of embryogenesis, called **metamorphosis**.

17-9. METAMORPHOSIS

Frog has an **indirect development** in which egg develops into a strikingly different organism (the larva or tadpole) from the adult. Tadpole leads an independent free life and has a different habitat and a different way of life than the adult frog. The developmental process of transformation of a larva into an adult is called **metamorphosis**. In fact, metamorphosis is the post-embryonic activation of morphogenetic processes associated with a dramatic change in habitat and consequent way of life. Such changes in environment and activities demand equally rapid transformations of the structures and functions of the living machinery during a brief discrete period.

The frog like other amphibians and insects has a **progressive metamorphosis** (in distinct contrast to the retrogressive metamorphosis of the Ascidians). In progressive metamorphosis of frog the metamorphosis is associated with a transition from an aquatic to a terrestrial mode of life and from a herbivorous to carnivorous mode of feeding. It has a task to transform an aquatic, herbivorous, tailed tadpole larva into a terrestrial, carnivorous and tailless frog. This transformation or metamorphosis involves numerous structural, biochemical and physiological changes. These changes include the destruction of existing structures, construction of new structures and modification of larval structures. At cellular level, metamorphosis is accomplished by cell death, cell proliferation and cell differentiation. Many genes active in larva become switch off ; whereas many genes inactive in larva, start their activity. Since metamorphosis requires a harmonious expression of many genes in different tissues, it is essentially controlled by hormones such as TH or thyroid hormones (*e.g.*, thyroxine).

In *Rana tigrina* the entire process of metamorphosis is completed within a week (**Agarwal** and **Niazi**, 1977). It involves essentially the following two types of changes :

17-9.1. Anatomical Metamorphic Changes

Such changes fall into the following three categories :

A. Destructive or regressive changes. The organs or structures that are necessary during larval life but no longer needed in the adults, are altered or may disappear completely :

1. The long tail of tadpole with its dorsal and ventral fin-folds is resorbed and disappears without a trace.

2. The internal gills are resorbed, gill-clefts are closed and the peribranchial cavities disappear. The opercular fold too falls off. Some aortic arches are reduced.

3. The horny teeth, labial fringes and horny lining of jaws are shed with the larval skin.

4. Intestine reduced from 9× body length to 2× body length.

5. Lateral line sensory system disappears ; the Mauthner cells of brain also degenerate.

6. Some blood vessels disappear (see Chapter 18).

Resorption of different organs takes place due to **autolysis**, *i.e.*, synthesis of the lysosomal enzymes and hydrolases (cathepsin, collagenase, nucleases, etc.). Amoebid macrophages of local mesodermal origin engulf (phagocytose) the debris of the disintegrating cells.

B. Constructive or progressive changes. These involve growth and morphogenesis of the following structures :

(i) The limbs increase in size and differentiate further. (ii) The middle ear develops in connection with the pharyngeal pouch (the pouch situated between the mandibular and the hyoid arch). (iii) The tympanic membrane develops, supported by the circular tympanic cartilage. (iv) The eyes protrude on the dorsal surface of the head and develop eyelids and nictitating membranes. (v) The tongue develops from the floor of the mouth. (vi) The quadrate cartilage is rotated backwards so that there is considerably increased gap enabling the frog to prey upon large insects. (vii) With the changes in the organs of respiration, corresponding changes take place in the vascular system. The afferent and efferent vessels develop their direct connections and more and more blood passes into lungs. The heart becomes three-chambered. Aortic arches take on the pattern of the adult frog.

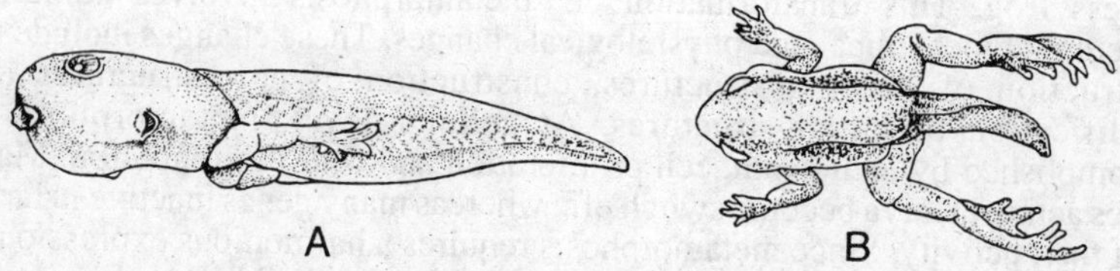

Fig. 17.22. Metamorphosis in frog. A—A mature tadpole larva; B—A young froglet (after Marshall, 1928).

C. Modification of larval structures (Adaptive changes). Following organs of the frog function both in the larva and the adult, but change their

Fig. 17.23. Metamorphosis of a tailed frog into a tailless frog (after Huettner, 1949).

differentiation during metamorphosis : 1. Change in the hardening, pigmentation and adaptation for dehydration takes place in the skin. The **skin** of the tadpole is covered with a double-layered epidermis. During metamorphosis, the number of cells in the epidermis increases and the surface layers become cornified. Multicellular mucous and serous glands develop as pockets sinking from the surface into the subcutaneous connective tissue layer, the dermis. The pigmentation of the skin also changes and new patterns of colours appear.

2. The intestine, which is very long in tadpole, as in most herbivorous animals, becomes greatly foreshortened, and most of the coils which it forms in the tadpole become straightened out.

3. In the head, widening of mouth, development of adult type jaw, and repositioning of eyes take place. The snout assumes the adult shape.

4. Trunk becomes narrower than the head.

5. The larval pronephros (kidneys) change into **mesonephric kidneys** of adult.

6. Erythropoeitic site changes from kidney to spleen.

7. The brain becomes more highly differentiated.

8. At the cellular level, cell modifications are evident in eye lids, limbs, lungs, tongue, eardrum, operculum, skin, liver, pancreas and intestine. Probably no cell or tissue or organ remain entirely unaffected.

9. There is no change in the lungs during metamorphosis.

17-9.2. Biochemical or Physiological Metamorphic Changes

Hand in hand with morphological changes during metamorphosis go the following physiological changes : 1. In the frog, the endocrine function of the pancreas starts at metamorphosis and this is connected with the increased role of the liver in the turnover of carbohydrates (glycogen).

2. A profound change takes place in the excretory mechanism. In the tadpole larva, the end product of nitrogen metabolism is ammonia, which is easily disposed of by diffusion in an aquatic medium but in a terrestrial animal it might accumulate and becomes dangerous because of its high toxicity. Metamorphosed frogs, however, excrete most of their nitrogen in the form of urea and only a small amount as ammonia. The change over occurs in the late stages of metamorphosis and is, of course, due to changed function of the liver which performs the synthesis of urea (**Munro**, 1939).

3. The liver of the froglet starts the secretion of a copper-binding serum protein called **ceruloplasmin** (see **Goel**, 1984).

4. In skin begins the synthesis of melanin and serotonin (a local vasoconstrictor).

5. Peptic activity starts in the stomach of froglet for the digestion of animal tissue.

6. In eyes, modification takes place in the visual pigments — the larval **porphyropsin** is replaced by adult **rhodopsin**.

7. In red blood cells, the respiratory pigment haemoglobin also changes. Thus, **larval haemoglobin** that has higher affinity for oxygen and insensitivity to acid (*i.e.*, independent from the pH) is replaced by **adult haemoglobin**

which has lower affinity for oxygen and shows **Bohr effect** with pH changes (*i.e.*, highly sensitive to acid).

If we consider the normal habitat of tadpoles, we can readily see the advantage of this change. Tadpoles often live in stagnant pools with low oxygen content and at times, also high carbon dioxide content in water. A high oxygen affinity, therefore, is an advantage. Because low oxygen and high carbon dioxide are likely to occur at the same time, a pronounced Bohr effect would be undesirable, for it would decrease the oxygen affinity of the blood when a high affinity is most needed (**Schmidt-Nielsen**, 1983).

17-9.3. Hormonal Control of Metamorphosis

The process of metamorphosis is essentially under hormonal control. Hormones from the **hypothalamus** (*e.g.*, neurosecretions like **TRF** or thyrotropin-releasing factor and **PIH** or prolactin-release-inhibiting factor or hormone), the hypophyis (*e.g.*, **TSH** or thyroid-stimulating hormone and the **PH** or prolactin hormone) and the **thyroid** (*e.g.*, **TH** or thyroid hormones such as thyroxine) regulate the process of metamorphosis. Thyroxine hormone affects the tissues directly, causing the degeneration and necrosis of some cells and stimulating the growth and differentiation of others. Iodine is not only essential for the working of thyroxine hormone, it has been found that iodine alone can cause metamorphosis in frog. If the tadpoles are kept in water containing the element iodine or if a weak solution of iodine is injected or an iodine crystal is implanted in a tadpole, it undergoes metamorphosis.

Certain environmental factors also affect the process of metamorphosis in frog. These factors include starvation, temperature, crowding, illumination, chemicals and dietary supplements. For an instance, illumination probably affects the ''biological clock'' of hormone through the hypothalamus.

REVISION QUESTIONS

1. What are tertiary membranes around frog's egg and how they are formed ?
2. Give an account of fertilization is frog. Explain significance of fertilization.
3. Describe the development of frog up to the formation of three germinal layers.
4. What do you mean by gastrulation ? What is its significance in the process of embryogenesis ? Describe the process of gastrulation of frog.
5. What are the formative movements ? Describe them in context with the development of frog.
6. What are the fate maps ? With the help of well-illustrated diagrams describe the fate of different germinal layers of frog.
7. Tabulate the different tissues and organs that develop from each of the three primary germinal layers of frog.
8. What is grey crescent ? What is its significance ?
9. Describe the process of neurulation or tubulation in frog.
10. What is the tail bud stage ? Describe the development of neurula up to the stage of hatching.
11. Describe the structure of tadpole of frog at the time of hatching and the changes it undergoes before the process of metamorphosis.
12. How is tadpole adapted to aquatic life ? Discuss structural similarities between frog's tadpole and fish.

13. How is tadpole adapted to herbivorous diet ?
14. Why is the intestine of tadpole very long and highly coiled ?
15. What is tadpole ? Give an account of metamorphosis in frog.
16. Enumerate various structural and physiological changes which occur during metamophosis of frog's tadpole. Add a note regarding the hormonal control of metamorphosis in frog.
17. How will you differentiate between the following :
    (i) Embryo and larva ; (ii) Ovulation and spawning ; (iii) Male frog and female frog ; (iv) Blastula and gastrula ; and (v) Invagination and involution.
18. Give one specific term for the following statements :
    (i) A group of eggs enclosed in jelly ;
    (ii) Eggs with moderate amount of yolk ;
    (iii) Eggs in which yolk is unequally distributed ;
    (iv) Union of egg and sperm ;
    (v) The division of zygote ;
    (vi) Embryo with a small white mass of macromeres projecting out of a small aperture.
19. What will happen :
    (i) if frog's tadpole is kept in freshwater with no plants ?
    (ii) if jelly of spawn of frog is damaged ?
    (iii) if vocal sacs of male frog are damaged ?
20. Write short notes on the following :
    (i) Grey crescent ; (ii) Fate map; (iii) Gastrulatiron ; (iv) Presumptive organ forming areas ; (v) Yolk plug stage; (vi) Blastopore ; (vii) Tadpole.
21. Draw a well-labelled diagram of V.S. of fully formed gastrula of frog.
22. Draw a well-labelled diagram of L.S. of a tail bud stage of frog.
23. Draw a labelled diagram of tadpole larva at internal gill stage.

18

DETAILED ACCOUNT OF ORGANOGENESIS OF FROG

In the last chapter, it has been mentioned that during neurulation (tubulation) of frog, the germinal layers, namely ectoderm, mesoderm and endoderm become tubulated and differentiated into primary organ rudiments of their respective organ systems. These primary organ rudiments, further, subdivide into **secondary organ rudiments** which gradually develop and differentiate into organ systems of adult frog. These processes of organogeny occur during the stage of conversion of neurula into tadpole larva and metamorphosis of tadpole larva into adult frog. The organogenesis of different organs of frog can be studied separately for each germinal layer as follows :

18-1. DEVELOPMENT OF ECTODERMAL ORGANS

The neurula of frog has three kinds of ectodermal tissues namely, epidermal ectoderm, neural ectoderm and neural crest cells. During organogenesis, these ectodermal tissues form a variety of body organs in the following manner :

18-1.1. Development of Epidermal Ectoderm

When the epidermis is segregated from the neural plate and neural crest cells during the process of neurulation, it is still a very complex rudiment of many organs. Most of it becomes the epidermis of skin, but, a number of other structures, such as sensory organs (olfactory, auditory, lateral line, etc.), lens and cornea of eye, median fins, external gills, adhesive sucker, etc., are also derived from it. Epidermal ectoderm also gives rise to the lining of the orifices of body such as the mouth and anus. The development of some important epidermal organs takes place as follows :

(a) Development of skin. The epidermal epithelium of young tadpole is two-cell thick : the outer cellular layer is called **periderm** and it is a ciliated, protective and transitory layer. The inner layer of cuboidal cell is called **stratum germinativum** or **stratum Malpighi** that retains the ability to proliferate (*i.e.*, to divide mitotically).

The integument or skin of the adult frog is composed of two main layers : superficial stratified epithelium, the **epidermis** and an underlying dense fibrous **dermis.** The dermis is vascular and supports, cushions and nourishes the avascular epidermis. Both of these layers of skin are derived from different germinal layers : epidermis is derived from the ectoderm and dermis is formed by the mesoderm (which is derived either from head mesenchyme, dermatome, neural crest cells or mesenchyme originating from lateral and ventral somatic mesoderm). The formatin or morphogenesis of skin provides an excellent example of tissue interaction during development. For example, the ectoderm depends on the mesoderm for the direction of its differentiation (**Rawles,** 1955). Experiments have shown that it is dermis which determines whether the epidermis clears and forms the cornea of the eye; becomes keratinized and gives rise to skin or so on.

The epidermis of frog contains only one type of differentiated cells which produce both mucus and keratin filaments (**Flaxman**, 1972). After cell division, as the epidermal cells migrate from the basal layer (the stratum germinativum) to the surface of skin, they secrete and release mucus into the intercellular spaces. Along the way to the surface of animal, however, the mucus-secreting cells transform into horny cells filled with keratin. As a result of keratinization or cornification, the cuboidal epidermal cells flatten, their nuclei degenerate, the cells lose their capacity to divide and fibrous keratin replaces the cytoplasm. Ultimately a layer of dried, dead, horny cells, called **stratum corneum**, covers the surface of the skin.

In dermis two well-defined layers become clear. The outer layer called **stratum spongiosum** consists of loosely arranged collagen fibres and contains blood vessels, nerves, pigment cells (chromatophores), lymph spaces and glands but little fat. The inner layer called **stratum compactum** is more compact and ties the skin to underlying skeletal muscles. In fact, the innermost layer of epidermis, the stratum germinativum becomes adhere firmly with a non-cellular layer called **basement membrane** which comprises two distinct layers : 1. **Basal lamina** which is the layer in contact with the epithelial basal plasma membrane. It consists of a fine feltwork of fibrils of **collagen Type IV** which are embedded in an amorphous glycoprotein ground substance. 2. **Reticular layer.** This layer is inner to basal lamina and consists of fine **reticulin** fibres which are also embedded in a ground substance. This layer merges with underlying connective tissue of dermis. Both of these layers have dual origin : the basal lamina is synthesized by ectodermal epidermal cells and reticular layer is formed by mesodermal connective tissue (**Wheater** *et al.*, 1979). Basement membrane provides structural support for epithelia and may constitute an important barrier to the passage of matrials between two types of tissues *(i.e.*, epidermis and dermis). It is never penetrated by blood vessels, so, epithelium is dependent on the diffusion of oxygen and metabolites from underlying connective tissue. It also acts as a substratum on which move the fibroblasts of connective tissues and the epithelial cells during wound closure (**Trinkaus**, 1984).

Some type of neural crest cells, called **melanoblasts,** invade epidermis and dermis to give rise to certain pigment-containing cells, called **chromatophores.** Chromatophores are stellate cells having long slender processes, called **dendrites.** They have following distribution in the skin of adult frog : 1. Epidermis contains only **melanophores** (secreting black **melanin** pigment). 2. Dermis contains three types of chromatophores : (i) **Iridophores** or **guanophores** are reflecting cells containing guanine, hypoxanthine or adenine or combination thereof. (ii) **Xanthophores** or **lipophores** which produce yellow pigment, the **xanthophyll.** (iii) **Erythrophores** containing red pigment. Like any other vertebrate, in frog colour-change results when these different chromatophores interact. In fact, there exist units of chromatophores called **dermal chromatophore units** (**Taylor** and **Bagnara**, 1972), in which the xanthophores form a basal layer and iridophores are located in one or more rows in the middle. The characteristic olive green colour of the dorsal skin of frog is a combined effect of melanin, xanthophyll and guanine.

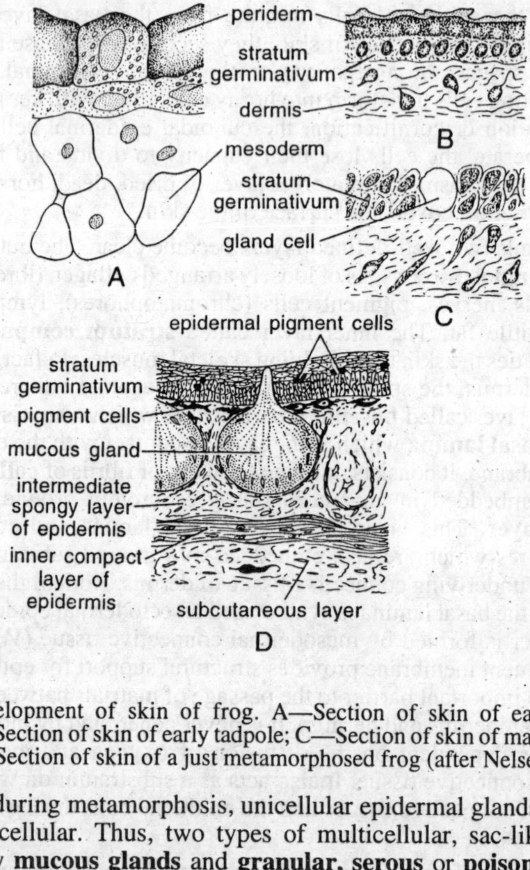

Fig. 18.1. Development of skin of frog. A—Section of skin of early neurula; B—Section of skin of early tadpole; C—Section of skin of mature tadpole; D—Section of skin of a just metamorphosed frog (after Nelsen, 1953).

Lastly, during metamorphosis, unicellular epidermal glands of tadpole become multicellular. Thus, two types of multicellular, sac-like, exocrine glands namely **mucous glands** and **granular, serous** or **poison glands** are derived from the epidermis and are located in the loose dermal layer of skin. Mucous glands of frog are small and composed of low cuboidal epithelium. They secrete a viscous, slimy material. On the other hand, the serous glands are larger, irregulary spaced, and composed of cuboidal cells having granular cytoplasm. They secrete a more watery secretion which is poisonous, ranging from mildly irritating to toxic.

(b) Development of epidermally derived sensory organs. Most of the sensory organs of frog, such as the olfactory organ (nose), the auditory organ (ear), lateral line organs, and lens and cornea of eye, etc., are derived from the epidermis. Most of these structures are formed from the plate-shaped, epidermal thickening called **placodes.** The placodes are the anlages or primordia of the organs or their some parts. In the early neurula stage the placodes are arranged in an almost continuous horse-shoe-shaped swelling around the open cranial neural folds. In most cases, the initiation and further development of the placodes depend on induction by the tissues beneath it.

By the time the neural fold closes, the generalized swelling becomes identified as contributing to specific sense organ. The **nasal placode** is located anterior to the neural plate and is stimulated to form olfactory sac. The optic vesicle contacts the epidermis and induces it to thicken as **lens placode** which develops into the lens of the eye. An **auditory placode** located laterally to the hind brain gives rise to the ear. A **lateral-line placode** associated with the auditory placode becomes the lateral line sense organs of aquatic animals. Several placodes give rise to cranial ganglia (*e.g.*, these contribute to the fifth and seventh to tenth cranial nerves; see **Phillips**, 1975).

(i) Development of olfactory sensory organ or nose. The nose rudiments appear just after the closure of the neural tube as two epidermal thickenings, the **olfactory placodes,** at the antero-lateral sides of the presumptive telencephalon. The central part of each olfactory placode invaginates and forms a **olfactory** or **nasal sac,** which remains open to the exterior by the **external naris** or **choana.** Some of the cells in the wall of the olfactory sac differentiate into sensory cells. These cells take the form of rod-like projections ending in sensitive brushes extending to the surface. Processes develop on the proximal end of each cell (end closest to the brain) and grow into that part of the forebrain which is destined to become the **olfactory bulb/tract.** In the tract the processes (axons) synapse with the mitral cells (*i.e.*, large nerve cells) which carry the impulses to the appropriate centre in the brain. These processes of the nasal epithelial cells leading from the sac to the telencephalon make up the first cranial nerve, the **olfactory nerve.**

From the ventral wall of each olfactory sac, arises an elongated tube called **nasal canal.** Each nasal canal stretches backward and downward towards the oral cavity and opens into the latter by **internal** or **primitive choana.** The epithelial lining of the nasal canal becomes thin and sharply segregated from the thicker epithelium giving rise to the sensory parts. The **Jacobson's organ** develops from a medioventral part of olfactory sac and has sensory function.

(ii) Development of ear. Each ear of frog consists of an internal ear, called **ear labyrinth** and a **middle ear.** The middle ear includes the eustachian tube, the ear ossicle (columella) and the tympanic membrane. The ear labyrinth is derived from the **auditory placodes,** while the middle ear is derived from the branchial apparatus.

Development of internal ear. The auditory placodes are formed by thickenig of sensory layer of head epidermis of tadpole, against the sides of hind brain. Each auditory placode separates from rest of the epidermis and invaginates to form a closed vesicle, the **otic** or **ear vesicle.** The ear vesicle is the rudiment of the internal ear which composed of the organ of equilibrium and organ of hearing. When first formed, it is pear-shaped and its pointed end remains directed upward and later on develops into the **endolymphatic duct;** the expanded portion of the otic vesicle gives rise to **membranous labyrinth.** Soon each ear vesicle starts expanding, pushing away the surrounding loose mesenchyme. Parts of the wall of ear vesicle become very thin; the epithelial cells become flat and form the membranous area of the labyrinth. The medio-ventral wall of each labyrinth becomes thicker, the cells of this area become columnar and give rise to pathches of sensory epithelium which form the **maculae** of the internal ear. The expansion of an ear vesicle, however, remains unequal, so that, it becomes constricted in some places and bulges out in others. As a result, the shape of the organ becomes increasingly complicated. The lower part of ear vesicle becomes divided into **sacculus** and **utriculus.** The utriculus becomes drawn into three mutually perpendicular folds, the

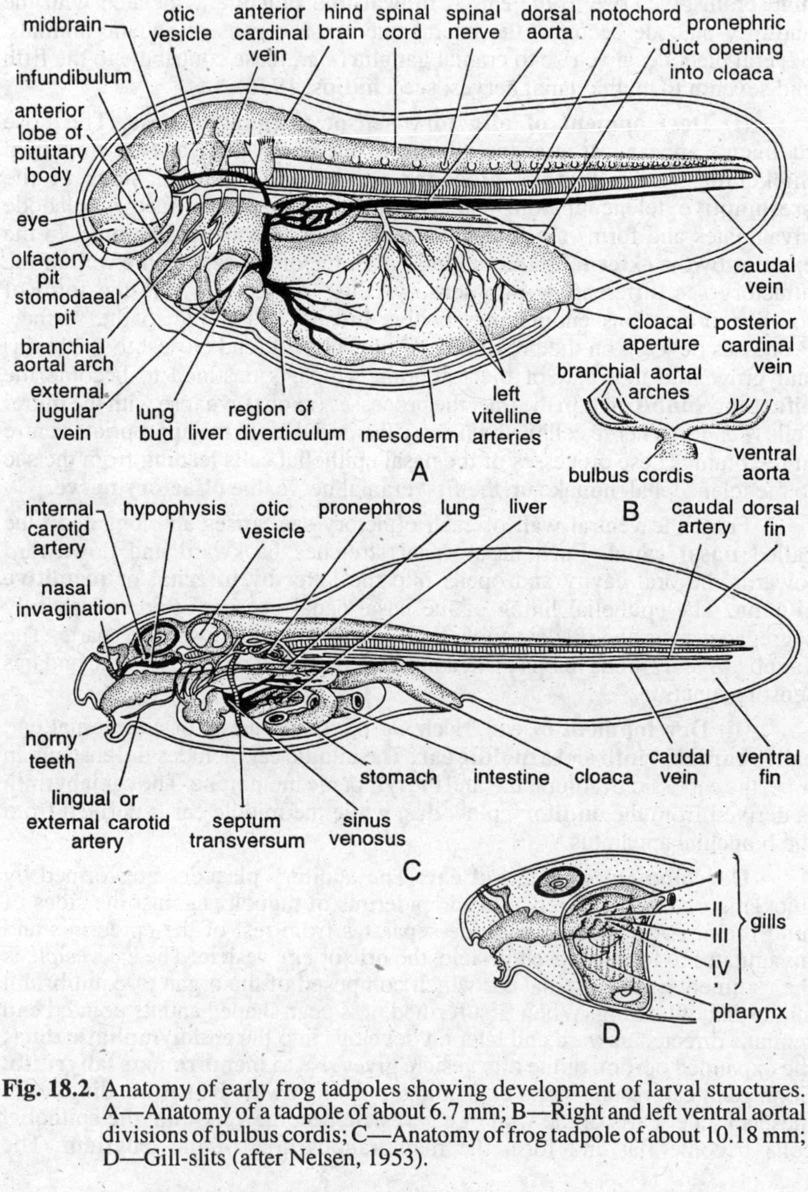

Fig. 18.2. Anatomy of early frog tadpoles showing development of larval structures. A—Anatomy of a tadpole of about 6.7 mm; B—Right and left ventral aortal divisions of bulbus cordis; C—Anatomy of frog tadpole of about 10.18 mm; D—Gill-slits (after Nelsen, 1953).

rudiments of the **semi-circular canals.** A hollow outgrowth of the sacculus forms the rudiment of **lagena.**

Neuroblast cells differentiate from the otic vesicles and form a club-shaped mass on the median wall. These cells are destined to become the **acoustic** or **auditory ganglion.** Processes develop on the cells of the ganglion and penetrate the hind brain (rhombencephalon) giving rise to the root of the eighth cranial nerve (**auditory nerve**).

As the ear vesicles expand to produce the internal ears, they become surrounded by mesenchyme cells which later on give rise to cartilage and produce the **cartilaginous labyrinths** or ear capsules which surround and protect the internal ears. However, a space exists between the membranous labyrinth (derived from epidermal ectoderm) and cartilaginous labyrinth (derived from sclerotome). This space is filled with a fluid, called **perilymph**, a fluid very similar to the endolymph of membranous labyrinth.

Development of middle ear. In frog, like other terrestrial chordates, development of the middle ear takes place from the **first pharyngeal gill pouch** which lies immediately ventral to the developing labyrinth and indeed its dorsal wall is pressed against the ventral side of the sacculus/lagena. This pouch does not break through to the outside to form a typical gill-slit or cleft; instead it exists as an expanded lateral cavity, the primordium of the **middle ear** or **tympanic cavity.** The addition of mesoderm to the thin membrane left between the pouch and the anterior ectoderm converts it into the **tympanic membrane, ear drum** or **tympanum.** Since a gill-pouch is a out pocketing of the pharynx, the internal lining of tympanic membrane is the endoderm. Its outer surface, however, is derived from superficial epidermal ectoderm. The resulting tympanic membrane thins out and responds sensitively to air waves. The remnant of the attachment of the gill-pouch to the pharynx becomes narrowed and remains as the **eustachian tube.**

Mesenchyme condensations give rise to the primordium of the transmitting apparatus (ear ossicle) of the middle ear. In frog, a piston-like slender bone, called **columella auris** crosses the tympanic cavity from the tympanum to a membrane-covered opening in the ear capsule, called the **oval window** (**fenestra ovalis**).

In frog, there is no cochlea like the amniotes and the auditory function of labyrinth is performed by two sound-receptors located in the sacculus : 1. **amphibian papilla** and 2. **basilar papilla**. Both of these receptors contain a **tectorial membrane** and **hair cells** (bearing kinocilia) which are innervated by sensory fibres of VIII cranial nerve (see **Kent**, 1987).

Auditory inductions. The development of the ear of frog is the outcome of a chain of embryonic inductions. The **primary inductor** of the chain is the roof of archenteron which causes the development of hind brain from the neural ectoderm. The hind brain acts as a **secondary inductor** and stimulates the development of the ear vesicles from the epidermal placodes and the ear vesicles are found to function as **tertiary inductors** in conjunction with head mesoderm to cause the formation of cartilaginous ear capsules around the internal ears. The middle ear requires no induction for its development.

(iii) Development of lateral line organs. The lateral line sense organs of the tadpole develop from the auditory placodes. Part of the epidermis adjoining the auditory placodes becomes thickened and from these epidermal thickenings develop the **lateral line sense organs,** which remain distributed over the head in several rows and a row of these lateral line sense organs stretches backward along the side of the entire body and tail.

(c) Development of adhesive sucker. The oral or adhesive sucker of an early tadpole is a horse-shoe-shaped or V-shaped glandular structure which has a deep furrow with thick-edges. Histologically, it contains very elongated columnar cells which produce a copious secretion of mucoprotein. The sucker is derived from the epidermis lying at the antero-vental margin of an early tadpole.

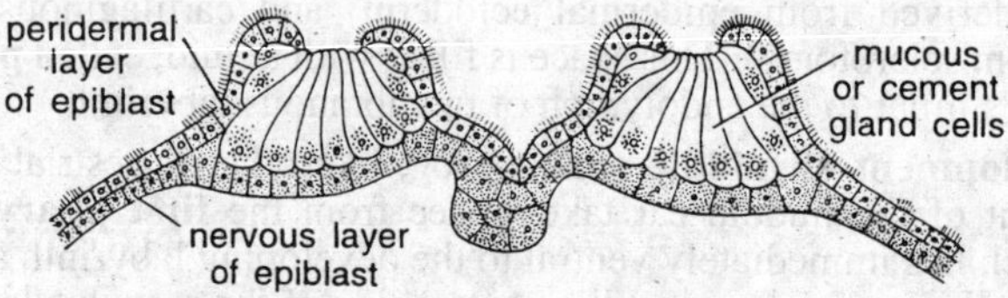

Fig. 18.3. Cross-section of an adhesive sucker of tadpole (after Nelsen, 1953).

(d) Development of external gills. The external gills of tadpole are protrusions of epidermis with connective tissues, blood vessels and muscles inside. They develop as outwardly directed pockets above the gill-clefts. The original outpushings of epidermis form the shaft of the external gill, and on this shaft develop secondary branches which are formed at first as solid, outwardly directed thickenings of the epidermis of the shaft. These secondary branches become hollowed out and are penetrated by mesenchymal tissues. The shaft of frog tadpole is short, therefore, its external gills appear as a bunch of filaments. The external gills become disappeared in the mature tadpole.

(e) Development of median fins. The median fin folds (dorsal, ventral and caudal fins) of tadpole have epidermal epithelium and mesenchyme. A fin fold first appears as an epidermal ridge-like, longitudinal thickening, which increases by shifting upward and toward the mid-line of the adjacent strips of the epidermis. The cells moving in from the right and left flanks remain separated as two layers of epithelium except at the crest of the ridge. There is, however, no hollow in between the two layers. The fold is hollowed out a bit later by the two epithelial layers separating in the middle, and then connective tissues of neural crest origin penetrate into the fold. These fin folds lack fin-rays and disappear during metamorphosis.

18-1.2. Development of Neural Ectoderm

It is convenient to subdivide the mature nervous system of frog into a central nervous system (CNS) and a peripheral nervous system (PNS). CNS comprises the spinal cord, brain and organ containing special sensory receptors (photoreceptors in retina, hair cells in the papilla and so on). PNS comprises sensory ganglia, autonomic ganglia and the network of peripheral nerves which link cutaneous, connective tissue and muscle sense organs to CNS. Brain and spinal cord are continuous structures and have a common embryonic origin.

The neural ectoderm of neural tube of neurula is the complex rudiment of central nervous system, peripheral nervous system and eye. These organ systems develop by following methods :

A. Development of central nervous system. By the end of neurulation the neural ectoderm is in the form of a dorsal, hollow neural tube that extends under the surface epithelium of the embryo from its anterior to its posterior end. The tube when formed is of unequal diameter throughout; its anterior, expanded thick-walled part develops into brain, while, its posterior, thin-walled narrow part develops into spinal cord.

The neural tube undergoes a general organization. In fact, at early stages one can distinguish certain concentric layers and longitudinal bands in the walls of both the presumptive brain and spinal cord. Three concentric layers are (1) an inner **germinal layer** of actively mitotic cells, lining the central canal (neurocoel) of the tube; (2) a middle cellular region, the **mantle layer** of cells proliferating from the germinal layer, and (3) and outer **marginal layer** consisting of fibres from the cells located in first two layers.

Some of the mantle layer cells are **neuroblasts**, which sprout **neurites**, the axon and dendrites, to become **neurons**. The rest are **spongioblasts** that develop into **neuroglial cells.** As neuroblasts differentiate, their axons grow into and add to the marginal layer, which, therefore, consists chiefly of nerve fibres. Because many axons become surrounded by a fatty myelin sheath, the marginal layer looks white when fresh and is called **white matter.** The cytoplasm of the cell bodies of the mantle layer causes this zone to look grey, hence, the name **grey matter.** Some of the nerve fibres or axons that grow into the marginal layer turn upward or downward in the spinal cord or brain to synapse with neurons elsewhere in CNS. The earliest of these fibres lie close to the neruocoel and extend only one segment, a primitive condition. The lateral and larger fibres become aggregated in long ascending and descending **fibre tracts,** each composed of functionally related fibres.

The primitive neural tube can also be divided into six longitudinal bands. A relatively thin roof (**roof plate**) and floor (**floor plate**) form the dorsal and ventral areas of the tube. Each lateral wall consists of a dorsal **alar plate** and a more ventral and thicker **basal plate,** both plates being separated by **sulcus limitans.** The alar plate receives incoming (sensory) impulses (by second order sensory neurons), whereas the basal plate gives rise to motor neurons. When all nerve cells have been formed that will ever be formd in the cord or brain, the cells of the germinal layer cease to divide. The undifferentiated cells that remain adjacent to the central canal (neurocoel) become **ependymal cells.** The ependyma is the supportive lining of the neurocoel (Fig. 18.4).

(a) Development of brain. Soon after the formation of neural tube, the brain rudiment becomes constricted into an anterior brain vesicle, the **prosencephalon** and a posterior brain vesicle, the **deuteroencephalon.** The deuteroencephalon is further sub-divided into **mesencephalon** and **rhombencephalon.** These three brain vesicles, namely prosencephalon, mesencephalon and rhombencephalon are the rudiments of fore-brain, mid-brain and hind-brain, respectively.

(i) Development of fore-brain. The prosencephalon is the rudiment of **telencephalon** (cerebral hemispheres) and **diencephalon.** It divides into an anterior telencephalon and more posterior diencephalon. The telencephalon gives origin to two antero-lateral pouches, the **telencephalic vesicles** which later develop into **cerebral hemispheres.** Each cerebral hemisphere contains a pocket-like cavity, the **lateral ventricle** which is the remnant of original neurocoel. The lateral ventricles of both sides remain freely communicated with the cavity of diencephalon by a broad opening called **inter-ventricular foramina** or **foramina of Monro.** This opening becomes narrow during later stages. Subsequently, the cerebral hemispheres grow forward, their walls get thickened and **olfactory nerves** make their anterior ends which now become **olfactory lobes.**

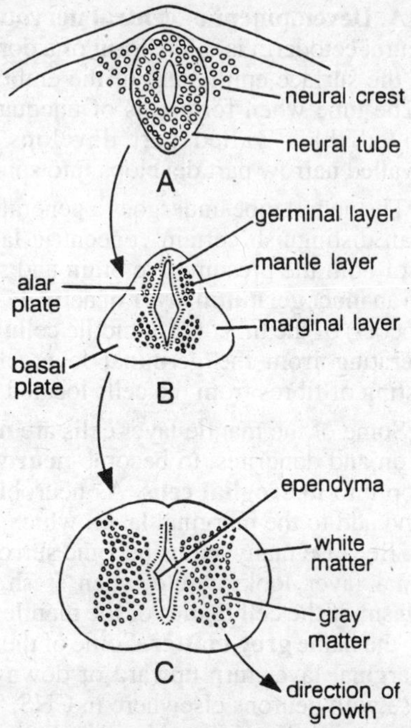

Fig. 18.4. Embryogenesis of spinal cord. The alar plate contains association neuroblasts with which incoming (sensory) nerve fibres will synapse. The basal plate contains motor neuroblasts, some of the axons of which are growing in the direction of the arrow in C to become part of a nerve (after Kent, 1987).

The diencephalon possesses a large cavity, called **third ventricle.** Most of its antero-dorsal wall becomes non-nervous, membranous and vascular **choroid plexus** which bulges down in its third ventricle. Only the posterior dorsal section of diencephalon wall remains nervous. The mid-dorsal part of diencephalon wall gives out dorsally a long, finger-like, outgrowth or evagination, the **epiphysis.** The end section of epiphysis becomes a rounded mass, the **pineal body,** while, its stalk becomes constricted and obliterated later on. The ventral wall or floor of diencephalon gives out laterally two outpushings, the **optic vesicles,** which later develop into retina of eye and optic nerve. At its mid-ventral line, the floor of diencephalon gives out a funnel-like outpushing, called **infundibulum** which fuses with an outgrowth from the stomodaeal invagination (**Rathke's pouch**) to form the **hypophysis** or **pituitary.** The walls of the diencephalon on the sides and posterior to the infundibulum become **hypothalamus.**

(ii) Development of mid-brain. The mesencephalon does not subdivide further like the fore-brain and hind-brain rudiments. Its walls remain thick and its roof gives rise to an important nerve centre, the **tectum,** anterior part of which is the primary centre of visual organ, the eye. But, the posterior part of tectum is found to receive nerve fibres from ear. The tectum gives out two swellings the **corpora bigemina** or **optic lobes.** The cavity of mid-brain becomes narrow and is called **aqueduct of Sylvius.**

(iii) Development of hind-brain. The rhombencephalon divides and gives rise anteriorly the **metencephalon** and posteriorly the **mylencephalon.** From the roof of metencephalon develop two **cerebral outpushings,** which later form the **cerebellum.** The mylencephalon develops into **medulla oblongata.** The cavity of rhombencephalon expands especially anteriorly just behind the mid-brain and becomes the **fourth ventricle.** The roof of medulla becomes non-nervous, thin and vascular **choroid plexus.**

(b) Development of spinal cord. Posterior to the brain, the neural tube gradually develops into spinal cord. When the neural tube is first formed its walls contain a single layer of pseudostratified neural epithelium. This layer develops into three layers : germinal, mantle and marginal, as already described.

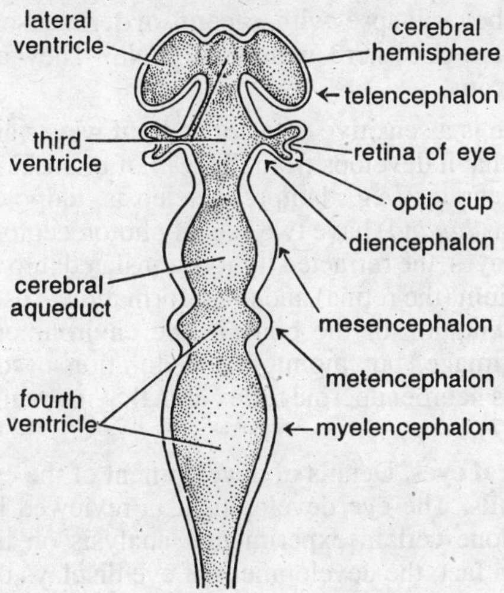

Fig. 18.5. Formation of five basic parts of frog's brain (after Torrey, 1971).

B. Development of peripheral nervous system. The cranial and spinal nerves and their ganglia have double origin. Their **motor fibres** are derived from the cells of neural tube origin, while the **sensory fibres** of all cranial and spinal nerves are derived from neural crest cells. Except olfactory and optic cranial nerves, the ganglia of all peripheral nerves remain outside the central nervous system and have their derivation in neural crest cells.

1. Development of motor components. Many of the axons that sprout from neuroblasts in the basal plate grow out of and away from the neural tube to make contact with striated muscles. These become motor fibres of the cranial and spinal nerves. Because these fibres sprout from neuroblasts within CNS, their cell bodies are within the adult brain or spinal cord.

Some of the fibres that sprout from neuroblasts in the basal plate grow out of the neural tube to make contact with neuroblasts in autonomic ganglia. These are **pre-ganglionic fibres** of the autonomic nervous system. The neuroblasts in autonomic ganglia sprout **post-ganglionic fibres** that grow toward, and innervate smooth muscles and glands. Thus, all motor neurons have their cell bodies in the cord or brain with one exception—post-ganglionic neurons of the autonomic system, have their cell bodies in autonomic ganglia.

Most neuroblasts of autonomic ganglia are migrants from the neural crests. A few, in frog, migrate outward from the basal plate of the neural tube.

2. Development of sensory component. At the time the neural tube is closing to form a tube, a longitudinal ribbon of neurectoderm separates from the developing tube dorsolaterally; on each side and soon segments to form a metameric series of **neural crests,** one pair in each body segment. Some neural crest cells become neuroblasts that give rise to sensory neurons of spinal and cranial nerves. In doing so they pass through a bipolar stage in which one process grows into the alar plate of the cord or brain and the other grows into a sense organ. Thus, there is established a neuronal connection between sense organ and central nervous system. Since each neural crest gives rise to a large number of sensory neurons, the result is a sensory ganglion on the nerve close to CNS. Neurons whose cell bodies are in sensory ganglia are **first-order sensory neurons**, since, they conduct an impulse from a sense organ to the CNS. Inside the cord or brain they synapse with **second-order sensory neurons** that conduct the impulse elsewhere within the CNS. They are sometimes called **association neurons.**

C. Development of eye. Eye is a sensitive detector of light waves and differs from other sense organs in that it develops from the wall of brain. Frog and many other cold-blooded verebrates (*e.g.*, lampreys, teleosts, tadpoles, *Sphenodon* and some lizards such as *Iguana*) have two sets of photoreceptors, lateral and median eyes. In **lateral eyes**, the refracted light is translated into an image on a photo-sensitive epithelium (the retina) and the information is used in maintaining an appropriate orientation of the body in the environment. **Median** or **third eye** forms no image, but monitors the duration of the photoperiods. Here, light stimulates neuroendocrine reflex arch that maintains biological rhythms (see **Kent**, 1987).

1. Development of the lateral eyes. Details of development of the eye were first studied by **George Walls.** The eye development is reviewed by **Coulombre** (1965) and he has done certain experimental analysis on the development of lens and cornea. In fact, the development of eye displays the full morphogenetic repertoire of embryonic sheets, including inductive interactions, sheet movement, changes in cell shape and finally, differentiation of specific cell phenotypes of lens and retina (see **Grant**, 1978). Thus, the morphogenesis of the whole eye can be subdivided into several phases : In the **first phase**, inductive effects separate the specific material of a particular rudiment such as the retinal rudiment within the neural plate and lens and cornea within the surface ectoderm. In the **second phase,** cell differentiation occurs within each rudiment, *e.g.,* the differentiation of optic vesicle cells into pigment and retinal epithelia and the differentiation of the lens. The **third phase** is of supracellular character and is concerned with the formation of the ciliary body and corneal curvature, the differentiation and growth of sclera and cornea and the general growth of eye (**Berrill** and **Karp**, 1976).

The eye of frog develops from three primordia. An outgrowth from the brain forms the **retina.** An ingrowth of surface (skin) ectoderm forms the **lens**. The local mesoderm (mesenchyme) gives rise to the fibrous and vascular coats (sclera and choroid). The different components of frog eye develop in the following manner :

(i) Development of retina of eye. The eyes of frog's embryo are determined at the neural plate stage. Vital stain mapping of frog's neural plate shows that the presumptive eye-forming region is located in a median position in the extreme anterior part of the neural plate. The same procedure has also shown that the area from which the future lens develops lies outside the presumptive neural plate, lateral to and in front of prospective eye rudiment. Soon the presumptive eye area is divided into right and left components, the **definitive optic-vesicle areas,** both remaining attached to each other by a short stalk. As neural folds rise, the eye primordia are displaced laterally, retaining their connections by the small median strip of tissue which develops into the **optic chiasma,** the region of crossing of optic nerves.

If developing eggs are chemically treated with lithium chloride, alcohol, chloretone, etc., or morphologically altered so that the two lateral parts of the prospective optic vesicle areas fail to become separated by an intruding wedge of median neural plate and mesoderm, the prospective eye area develops into a single large median eye. This phenomenon is called **cyclopia.**

Further, shortly before the neural folds fuse in the midline, the future lateral eyes emerge as two lateral evaginations or outpocketings of the diencephalon of prosencephalon or fore-brain. These two outpocketings are called **optic vesicles,** each of which containing a cavity, the **optocoel** that is continuous with the cavity of brain (**prosocoel**). A constriction soon separates

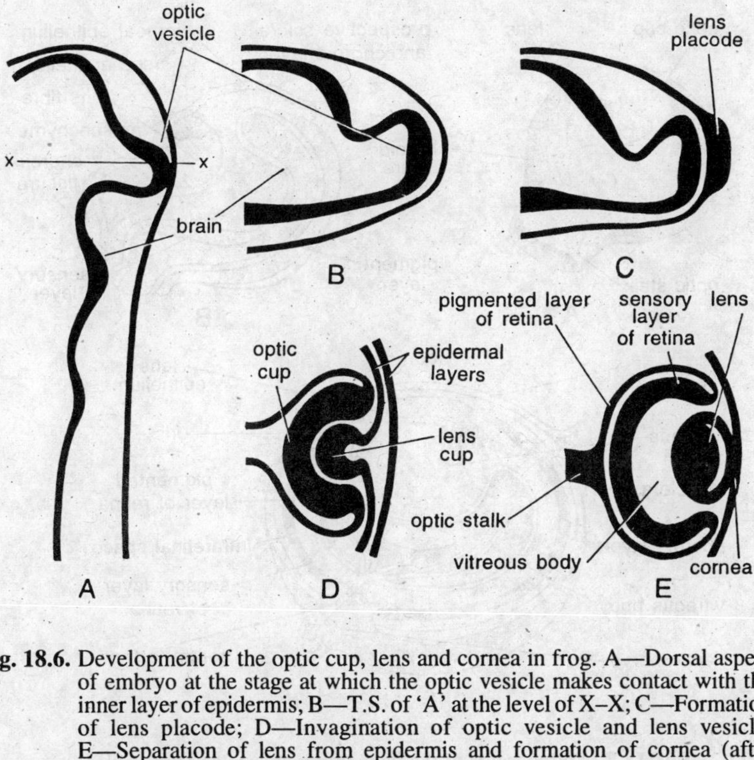

Fig. 18.6. Development of the optic cup, lens and cornea in frog. A—Dorsal aspect of embryo at the stage at which the optic vesicle makes contact with the inner layer of epidermis; B—T.S. of 'A' at the level of X–X; C—Formation of lens placode; D—Invagination of optic vesicle and lens vesicle; E—Separation of lens from epidermis and formation of cornea (after Dehaan and Ursprung, 1965).

the proximal end of optic vesicle from brain and it becomes the **optic stalk.** Once the anterior neuropore has closed, internal accumulation of fluid plays a part in expanding the optic vesicle. Each optic vesicle balloons out toward the head ectoderm, the most distal portion of it thickens into a columnar layer, while the more proximal region of it thins out. When the optic vesicle contacts the head ectoderm, its distal wall flattens and invaginates, so that, the thickened layer contacts the inner surface of the thin layer. The optocoel is obliterated and a double-layered **optic cup** is formed. The cells of indented wall of optic cup undergo mitosis and form the sensory layer of the retina. Mesenchymal pigmented cells invade the thin outer layer of the optic cup to form the **pigmented layer** which prevents scattering of light and blurring of image. Ultimately sensory retina is fused with pigmented layer.

The rim of optic cup later develops into the edge of **pupil**. The cavity of the optic cup is the future **posterior chamber** of the eye, which is, later on, filled up by **vitreous body.** The opening of eye cup is very large at first, but later on, the rim of the cup bends inward and converges, so that the opening of pupil is constricted and reduced to its final relative dimension. The rim of the optic cup surrounding the pupil is considerably thinned out to become iris.

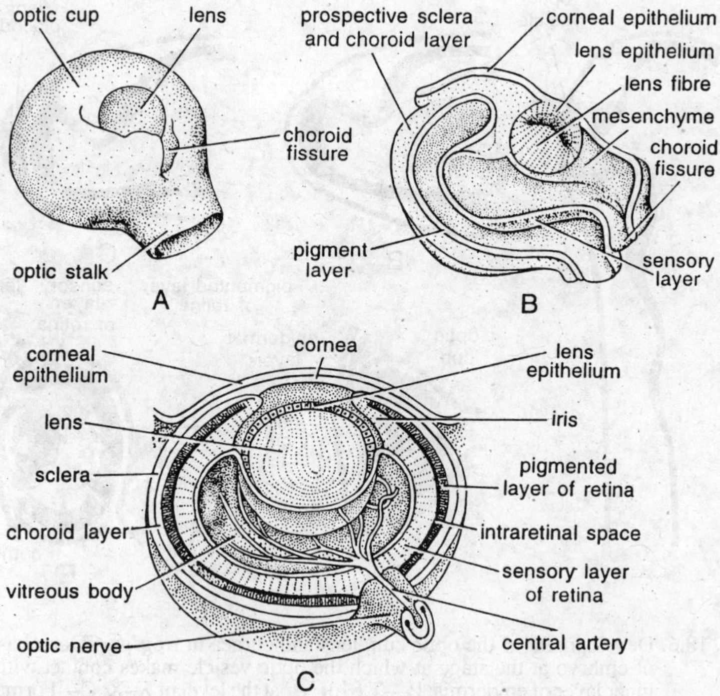

Fig. 18.7. Stereograms of the developing eye (after Torrey, 1971).

The cavity of the optic cup remains open ventrally, forming a groove extending to optic stalk. This groove is developed into **choroid fissure.** This fissure serves for the entry into the posterior chamber of the eye of a blood vessel and of mesenchymal cells, which are found later in the vitreous body. During later stages, these mesodermal cells disappear from the vitreous chamber. In the mature eye, the edges of choroid fissure fuse and a small tube forms within the already tubular optic stalk. The central or hyaloid artery and vein remain in this canal and become main blood vessels of the retina. Sometimes, choroid fissure persists in the adult eye, due to which a gap is left on the ventral edge of the pupil. This deformity is called **coloboma** of the iris and it does not affect the vision. In humans, it is resulted by hereditary factors (see **Balinsky,** 1981).

Development of retinal architecture. The primary function of retina is image formation. It lies only in the posterior portion of the eye and consists of an inner nervous tissue layer and an outer pigment layer. The presumptive pigment layer in the retinal zone of the optic cup remains one cell thick. Each epithelial cell of this layer (*i.e.*, cuboidal cell) depletes its moderate glycogen reserves and acquires the machinery to produce the melanin pigment granules (*i.e.*, tyrosinase activity in the Golgi area). It also sends long processes on the vitreous surface to interdigitate with the outer segments of the rods and the cones. The cell processes secrete non-sulphated mucopolysaccharides. From the scleral side inwards the differentiated pigment layer cells have a basal zone, a nuclear zone, an apical zone and a zone of cell processes.

The neural retina or nervous layer of retina is one of the most complex tissues in the vertebrate body. It comprises the following three distinct zones of neurons : These three zones, named in the order in which they conduct impulses, are the **photoreceptor neurons** (rods and cones), the **bipolar neurons** (*i.e.,* horizontal ganglion cells, bipolar ganglion cells, amacrine cells) and the **ganglion neurons** (Fig. 18.8). The dendrites of the photoreceptor neurons are called rods and cones because of their shapes. They are visual receptors highly specialized for stimulation by light rays. When information has passed through the photoreceptor neurons, it is conducted across synapses (**outer plexiform layer**) to the bipolar neurons in the intermediate zone of the nervous layer of the retina. From here it is passd to the ganglion neurons via synapsis of **inner plexiform layer.** Finally, the ganglion neurons transmit their signals through optic nerve fibres to the brain in the form of nerve impulses.

The presumptive neural retina becomes thicker because of the increased mitotic rate. Mitoses are more frequent on the outer (scleral) side of the retina, and near the margin of the eye cup. Autoradiographic experiments with tritiated thymidine show that these cells migrate to the scleral side to divide and to return to the vitreous side to take up their final positions. As a result, the differentiation of the neural retina proceeds from the vitreous to the scleral side. The first cell types to differentiate are the neuroglial cells of Muller. The next to differentiate is the ganglion cell layer. These cells send out the centrally directed axons. These axons run in parallel arrays under the inner limiting membrane (formed by neuroglia) and course through the optic stalk to form the optic nerve. The receptor cells, the rods and cones, are the last to differentiate, while the retina differentiates the waves of cells death occur. The significance of programmed cell death is not fully understood but it may concern the orderly establishment of the definitive neuronal connections.

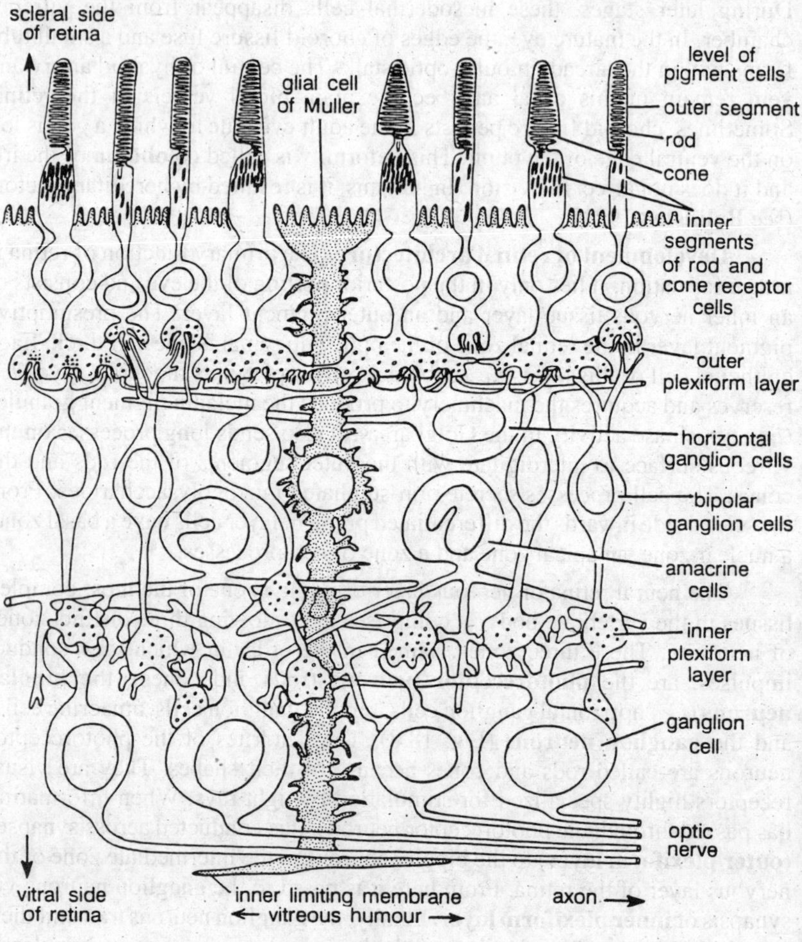

Fig. 18.8. Summary diagram of synaptic contacts between the principal types of retinal cells in the human retina (after Berrill and Karp, 1976).

Cytogenesis of rods and cones. The cytogenesis or terminal differentiation of receptor cell begins with the protrusion of a cytoplasmic bud, the prospective **inner segment** or the receptor cell, through a pore in the external limiting membrane (which is formed by neuroglial cells). Here again, further differentiation is sequential and centrifugal. The inner segments of the rods and cones are first formed, and from them the outer segments later arise. Several specialized cell organelles such as myoid, ellipsoid, paraboloid and oil droplets develop in the inner segment. The **myoid** is an acetylcholinesterase rich (**Shen,** *et al.,* 1956; **Yoshido,** 1958) contractile region in the basal portion of the inner segment which in adult positions the photosensitive elements relative to the pigmented retinal epithelium during photochemical adjustments to different light intensities (**Detwiler,** 1943). The **ellipsoid** or **lentiform body** is a tightly packed aggregation of mitochondria. The **paraboloid** is a dense collection of glycogen granules and is derived from a granular endoplasmic

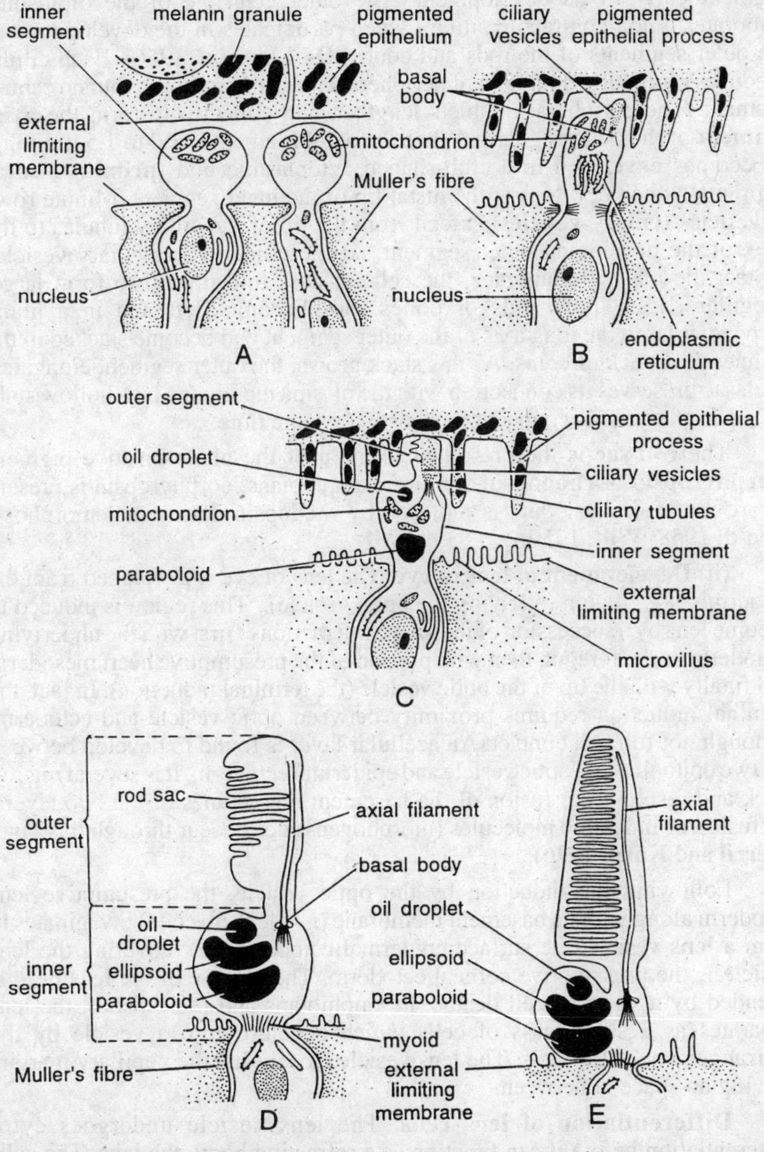

Fig. 18.9. Rod and cone cytogenesis. A–E—Illustrating the successive stages in the maturation of the cone (after Dehaan and Ursprung, 1965).

reticulum (**Carasso,** 1960). The **oil droplets,** which are usually confined to the cones, are spherical globules of **carotenoids,** such as astacene, xanthophyll and galloxanthin (**Wald** and **Zussman**, 1937, 1938); they appear late in development between the mitochondria of the ellipsoid and the tip of the inner segment. The oil droplets serve as focusing lenses and as optical filters.

Another organellae, the **diplosome** (basal body) migrates into the inner segment early in its development. The outer centriole of the diplosome elaborates a non-contractile cilium which participates in the development of the outer segments of the rods and cones (**De Robertis,** 1956). Each cilium produces nine pairs (doublets) of peripheral ciliary microtubules but no central filament. The peripheral doublets lengthen and collectively form the **axial filament** of the outer segment. When the axial filament is 1.0 to 1.5 μm long, it becomes enveloped in a club-shaped cytoplasmic bud (**primitive outer segment**) which develops at the distal tip of the inner segment. Minute rows of vesicles (**ciliary vesicles**) extend from the tip of the ciliary tubules to the apex of the developing outer segment. At the points where ciliary vesicles contact the plasma membrane, the cell surfaces fold inward to form large, irregular sacs. These rods or cones sacs become flattened in a plane perpendicular to the long axis of the outer segment and become stacked in the manner of a stack of coins. As this stack grows, the outer segment elongates. Each sac preserves its connection with the plasma membrane by a hollow stalk at its periphery. This connection persists for some time.

The **rod-sac** is the presumed location of the photosensitive pigment (**porphyropsin** or **rhodopsin**). In some amphibians, porphyropsin is present before metamorphosis and is replaced by rhodopsin during metamorphosis (**Wald,** 1958; **Wilt,** 1959).

(ii) Development of lens of eye. The lens of eye is developed from the epidermis (*i.e.*, region overlying the optic vesicle)., This region is induced to become lens by a succession of inductive interactions : first with the underlying endoderm of the foregut, then with portion of the presumptive heart mesoderm and finally with the tip of the optic vesicle (the terminal induction). In fact, the terminal induction requires proximity between optic vesicle and ectoderm, although not full cell contact. An acellular layer is found to develop between the two epithelia, *i.e.*, optic vesicle and epidermal ectoderm. It is several micron thick and results from fusion of the basement membranes of the two layers. Diffusion of inductive molecules (morphogens) does occur through this layer (**Berril** and **Karp**, 1976).

Following the induction by the optic vesicle, the presumptive lens ectoderm along with its basement membrane (*i.e.*, lens placode) invaginates to form a **lens vesicle**; the surface epidermal ectoderm now covering the lens vesicle is the presumptive corneal ectoderm. The closure of the lens pore is attended by a wave of cell deaths. In amphibians and bony fishes, the lens separates as a solid mass of cells which develops into a vesicle by the rearrangement of the cells. The lens vesicle closely fits the pupillary margin, leaving no space in between.

Differentiation of lens cells. The lens vesicle undergoes cyto-differentiation before it can function as a refracting body, the **lens.** The cells on the corneal side of the lens vesicle remain more or less unchanged and form the lens epithelium. However, the cells on the vitreous side start elongation, become columnar and form the lens fibres. During such a terminal differentiation (*i.e.*, lens fibre formation) the nuclei degenerate and the cytoplasm becomes hard and transparent. This occurs due to gradual accumulation of secretory product of lens cells, the proteins, called **crystallins** of which there are several types such as alpha, beta, delta or gamma. During differentiation of a lens fibre, possibly the last portion of the nuclear DNA to break down is one that codes for the crystallin mRNAs. The lens fibres become arranged in a very orderly concentric fashion, somewhat like the skins of an onion.

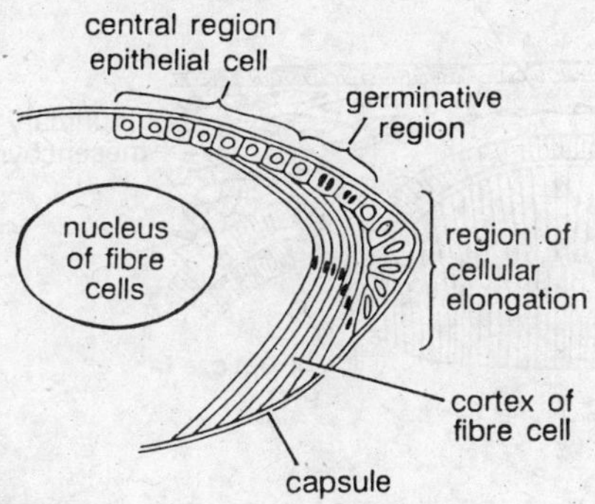

Fig. 18.10. Different zones of the lens of the adult vertebrate (after Berrill and Karp, 1976).

Morphologically, a fully formed eye lens lacks blood vessels and is composed of an outer mitotically active single layer of **epithelial cells**; a zone of cellular elongation or **equatorial region,** composed of cells in the process of developing into fibre cells; and **inner** or **cortex fibre cells** (Fig. 18. 10). After the embryonic lens has been formed, fibre cells are continuously laid down throughout life in the zone of cellular elongation, where cuboidal epithelial cells continuously transform to the elongated fibre cells characterized by gamma-crystallin protein. Finally, since the fibre cells lose their replicative activity, they essentially enter a permanent stationary phase and can only proceed to death. The epithelial cells of central region retain their ability to replicate.

Thus, in the lens, new fibre cells continually form from the peripheral equatorial zone of elongating epithelial cells; these in turn are derived from the adjoining germinative region of lens epithelium. In this way, fibre cells are systematically laid down, layer upon layer, throughout life. The oldest fibres located in the centre of the lens form the **lens nucleus,** while the fibres differentiated later on form the **lens cortex**. The lens bow is due to the central positioning of the nuclei in the fibres. The basement membrane surrounding the lens vesicle forms the lens capsule.

(iii) Development of the cornea of eye. The cornea of eye, like the skin, is derived from an outer epidermal layer. In contrast to normal skin, cornea becomes transparent because pigment and gland cells do not develop and mesenchyme below remains unvascularized. In amphibians and chick embryos, the cornea becomes transparent after lens detaches from the epidermis. Presence of lens is necessary for corneal development. If the lens and eye cups are removed at early stages, the head ectoderm undergoes normal skin development and becomes pervaded with glands and pigment (**Grant**, 1978).

In tadpole, the epidermis forms an **external cornea** which remains continuous with the skin, while mesenchyme forms an **internal cornea** which remains in continuation with the sclera. Both corneal epithelia remain separated by a narrow space. During metamorphosis these layers fuse under the influence of thyroxine (**Kaltenbach,** 1953), and they become transparent, so that rays of **light** may enter the eye.

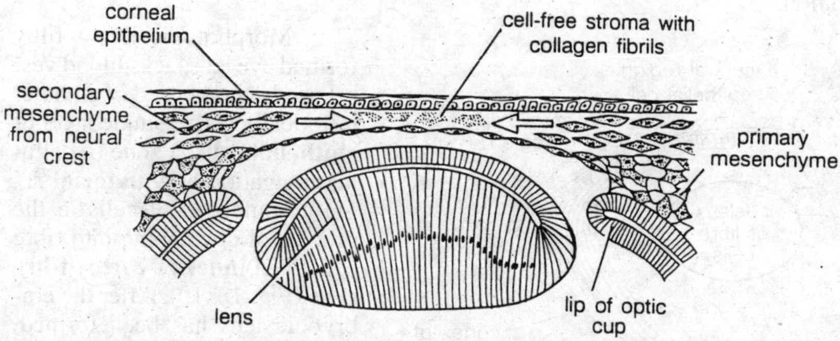

Fig. 18.11. Section through the lens and neighbouring tissues showing migration of neuro-ectodermal fibroblasts into collagenous stroma under the cornea (after Ede, 1978).

The fully developed cornea consists of three kinds of cells, together with nerve fibres and a very small number of extracellular components produced by fibroblasts, mainly collagen fibrils and mucopolysaccharides which become deposited in the interfibrillar space. According to **Coulombre** (1965) these components of cornea assembled in time and space in such a manner that the cornea develops a large number of strikingly different functional characteristics such as vascularity, tensile strength, deturgescence, a tissue specific population of ions, regenerative capacity, a characteristic interference pattern in polarized light, an appropriate refractive index, a precisely controlled curvature that contributes to its refractory power and transparency.

(iv) Development of choroid coat and sclera. The choroid coat and sclera of frog eye develop from the mesenchyme cells accumulating around the eye ball. The interior layer of mesenchyme cells gives rise to a network of blood vessels surrounding the pigmented retinal epithelium and is called **choroid coat.** The outer layer of mesenchyme forms a fibrous capsule, the **sclerotic coat** or **sclera,** around the eye. The sclera provides protection to eye and the eye muscles are inserted on it.

Chain of optic inductions. During the development of frog eye, following chain of inductions are found to be operative :

(i) The **primary inductor,** the archenteron roof induces the neural plate and, therefore, also the eye-cup-rudiment which is the part of the neural plate.

(ii) The eye-cup-rudiment develops into the optic vesicle which acts as **secondary inductor,** and induces the epidermis to form **lens placode.**

(iii) The lens placode acts as **tertiary inductor,** it induces the neurosensory layer of optic vesicle to invaginate and when develops into lens, the latter induces the epidermis and mesenchyme for the formation of cornea, choroid coat and sclera. Though, the lens is principally responsible for the induction and maintenance of the cornea; however, corneal epithelium becomes progressively more independent of its inductor.

2. Development of the median eye. Median eye is an evagination of the roof of diencephalon. Primitively, it was one of two evaginations constituting an **epiphyseal complex.** The more anterior evagination is the **parapineal**, the posterior one is the **pineal.** It is usually the parapineal that is photosensitive.

In larval frog, the third eye or median eye is called the **frontal organ** or **stirnorgan**. Authorities are uncertain whether this is the pineal or parapineal part of the epiphyseal complex. At metamorphosis, the photoreceptive part of this organ regresses, leaving only the glandular component that produces the hormone **melatonin** in the larva (**Kent**, 1987).

18-1. 3. Development of Neural Crest Cells

The neural crest is the quintessential vertebrate embryonic tissue, for it embodies the fundamental problems of totipotency, determination, differentiation, cell migration, morphogenesis and epigenetic control of gene expression by extracellular matrix products and epithelial-mesenchyme interactions. A tissue that is restricted to the vertebrates, the neural crest is the region of the embryo where future neural and epidermal ectoderm meet in the neural folds; the neural crest cells or NCCs are the cells in the neural crest.

Experimental evidence from amphibian embryos shows that it is the exposure to the combined actions of neuralizing and epidermalizing influences that elicits the formation of the neural crest cells. Thus, chordamesoderm (the future notochord) induces ectoderm to become neural ectoderm; lateral mesoderm induces ectoderm to become epidermal ectoderm; and the combined action of the two induces ectoderm to become the neural crest.

One of the most remarkable aspects of the behaviour of the neural crest cells is their extensive migration away from the neural tube, a migration that begins before or just as the neural folds close into a neural tube. Various recent studies have shown that neural crest cells essentially migrate through some type of extracellular matrices and on the basement membrane of the epithelia. Migration is initiated by neural crest cells losing their connections to the basement membrane of the neural tube and decreasing their adhesiveness to one another (**Newgreen** and **Gibbins,** 1982). Once free of the neural tube, they preferentially migrate along fibronectin-rich extracellular matrices. In addition to using fibronectin, migrating neural crest cells utilize a cell-surface galactosyl-transferase (which they possess in large amounts) to bind to terminal N-acetyl glucosamine residues in the extracellular matrices and on the basement membranes. This binding is used as a further mechanism for directing cell migration (**Runyan** *et al.*, 1986).

It is still not clear what the signal is that terminates migration, but once at their final site, the neural crest-derived cells differentiate into one of a very considerable variety of cell types (Table 18-1), including neurons, Schwann cells (Neuroglial cells), hormone synthesizing cells, pigment cells, chondroblasts, osteoblasts, myoblasts, connective tissue fibroblasts and odontoblasts (see Chapter 14). These differentiative properties are clearly regionalized. For example, the cranial but not the trunk neural crest is capable of forming chondroblasts, osteoblasts and odontoblasts and of depositing cartilage, bone and dentine respectively. Likewise, within the trunk neural crest, those cells that migrate beneath the endoderm differentiate into pigment cells; those that migrate between somites and neural tube differentiate into dorsal root ganglia; those that migrate over the somites differentiate into sympathetic ganglia; and those that migrate between the somites differentiate into sympathetic ganglia and cells of the adrenal medulla. Cervical neural crest cells differentiate into cholinergic neurons of the parasympathetic nervous system and possess acetylcholine as their neurotransmitter, while trunk neural crest cells differentiate into adrenergic neurons of the sympathetic nervous system and have noradrenalin as their neurotransmitter.

Table 18-1. Cell types that arise from the neural crest cells (after **Maclean** and **Hall**, 1987).

1. Sensory neurons and satellite cells (= one of the neurilemmal cells surrounding nerve cells in PNS) of the spinal ganglia.
2. Cholinergic neurons (producing the neurotransmitter acetylcholine) and satellite cells of the ganglia of the parasympathetic nervous system.
3. Adrenergic neurons (producing the neurotransmitter noradrenaline) and satellite cells of the ganglia of sympathetic nervous system.
4. Schwann and glial cells of the peripheral nervous system.
5. Cells of the adrenal medulla that produce adrenalin and noradrenaline.
6. Parafollicular cells of the thyroid gland (*i.e.*, para-follicular or C cells of the thyroid which produce the calcium regulating hormone, calcitonin.
7. All of the pigment cells of the body except those of the pigmented retinal epithelium.
8. Cartilage of the visceral and facial skeletons.
9. Bone of the craniofacial skeleton.
10. Dermal connective tissue of the face and ventral region of the neck.
11. Ciliary muscles of eye.
12. Portion of striated muscle in the face and neck.
13. Dental papillae and odontoblasts of the teeth.
14. Fibroblasts of the corneal stroma.
15. Fibroblasts of endothelia.
16. Myoblasts and fibroblasts of arteries arising from the aortic arches.
17. Fibroblasts of glands such as pituitary, salivary, thyroid, parathyroid and thymus.
18. Mesenchyme of the dorsal fins of amphibians.

The extracellular products with which neural crest cells interact during and after their migration are often those of epithelia. There is now a considerable literature on the requirement for epithelial-mesenchymal interactions before the neural crest derived cells can differentiate into tissues such as cartilage, bone and dentine (**B.K. Hall**, 1984, 1987). Thus, for cartilage to differentiate into the mandibular and facial skeletons of different species, neural crest-derived cells have to interact with products of pharyngeal endoderm in amphibians, products of cranial ectoderm or pigmented retinal epithelium in birds and products of mandibular epithelia in mouse (**Hall,** 1987). Likewise, differentiation of bone-forming osteoblasts in the mandibular skeleton of both the embryonic chick and mouse requires that neural crest-derived mesenchymal cells interact with the mandibular epithelium. Such interactions require a very close association between the mesenchymal cells and the basement membrane of the epithelia, suggesting a mechanism of exchange of components or interaction with components of the basement membrane (**Smith** and **Thorogood**, 1983).

Most types of neural crest cells are **bipotential**, for example, neural crest cells of neural tube are bipotential for both neurogenic and melanogenic differentiation, but during migration the cell's fate becomes restricted to just one option. Such a restriction occurs due to interaction between neural crest cells and extracellular products of somites (see **Maclean** and **Hall,** 1987). Lastly, we can consider the following examples to understand how the organs are formed by the neural crest cells in the frog :

(a) Formation of pigment cells. To give characteristic pigmentation to the frog, the neural crest cells which have no yolk but pigment derived from egg, transform into **melanophores** and **guanophores.** Both of these pigment cells remain confined to connective tissues of skin, peritoneum and in the walls of blood vessels and give characteristic colouration to the body.

(b) Formation of visceral skeleton. The visceral skeleton of frog (lower jaw, ear ossicles and branchial skeleton) is derived from neural crest cells. The mass of the neural crest cells behind the eye, that is, in between the eye and the first branchial pouch, becomes the **mandibular arch**, the upper part of it

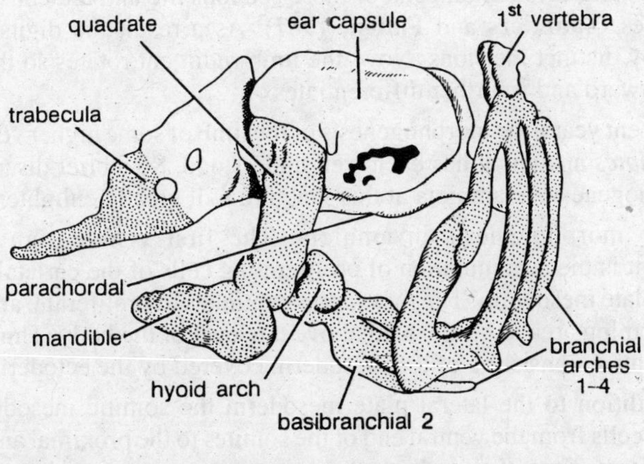

Fig. 18.12. Skull of a salamander larva, indicating parts developing from neural crest cells (stippled) (after Balinsky, 1976).

becomes **quadrate,** while its lower part develops into **Meckel's cartilage** or **mandible** proper. The mass of neural crest cells lies in between the first and second branchial pouches, becomes the **hyoid arch,** and so forth. The mass of neural crest cells moving downward in front of the eye contributes to the formation of the anterior half of the trabeculae of the skull (Fig. 18.12). The second branchial of frog is not derived from neural crest cells but from mesoderm.

18-2. DEVELOPMENT OF MESODERMAL ORGANS

The mesoderm is a complex rudiment of various important body organs such as limbs, endoskeleton, heart, blood vessels, kidney, reproductive organs, all kinds of connective tissue, coelom, etc. The development of some important mesodermal organs takes place by following methods :

18-2.1. Development of Appendages or Limbs

The paired fore-limbs and hind-limbs of frog make their appearance in tadpole larval stage and derived from the epidermal ectoderm and the mesoderm of somites and lateral plate. The development of a limb is started by thickening of lateral plate mesoderm just beneath the epidermis of a presumptive limb area which may lie either behind the branchial region (for fore-limb) or just in front of anus (for hind-limb). Soon, the epidermis lying over the mesenchyme mass becomes slightly thickened and bulges outward to form a **limb-bud.** Due to active multiplication of mesenchymal and epidermal cells, each limb-bud grows and gets flattened. After a limb-bud has grown so far that its length exceeds its breadth, the differentiation of its subordinate parts sets in. Its distal portion flattens and broadens to become a circular **hand** or **foot plate,** which soon becomes pentagonal, the projecting points indicating the rudiments of the digits. While the tips of the digit rudiments keep growing out further, extensive necrosis (cell death) occurs in the intervening sections. Mesodermal and ectodermal cells of these sections die and are consumed by macrophages (**Saunders** and **Fallen,** 1967). As a result the digits become separated by distinct incisions. Now, the limb rudiment rotates so that its tip bents downward and it further differentiates.

In recent years, the morphogenesis of the limb of some higher vertebrates such as *Calotes* and chick has been well understood. So a brief discussion of limb morphogenesis in amniota at this juncture will greatly enlighten us.

Limb morphogenesis in amniota. The first visible sign of limb development is the accumulation of proliferating cells of the parietal layer of the lateral plate mesoderm. The mesoderm continues to proliferate, and causes the limb forming area to be elevated above the level of the body. Thus, a **limb ridge** is formed, consisting of the mesoderm covered by the ectoderm.

In addition to the lateral plate mesoderm the somitic mesoderm also contributes cells from the ventral end of the somites to the proximal and central limb mesoderm. Thus, the forelimbs tend to receive more mesoderm from the somites than the hindlimbs. The somitic mesodermal cells are found to contribute the musculature (myotubes), while the lateral plate mesodermal cells form cartilage, bone, soft tissue and muscle connective tissue of limb.

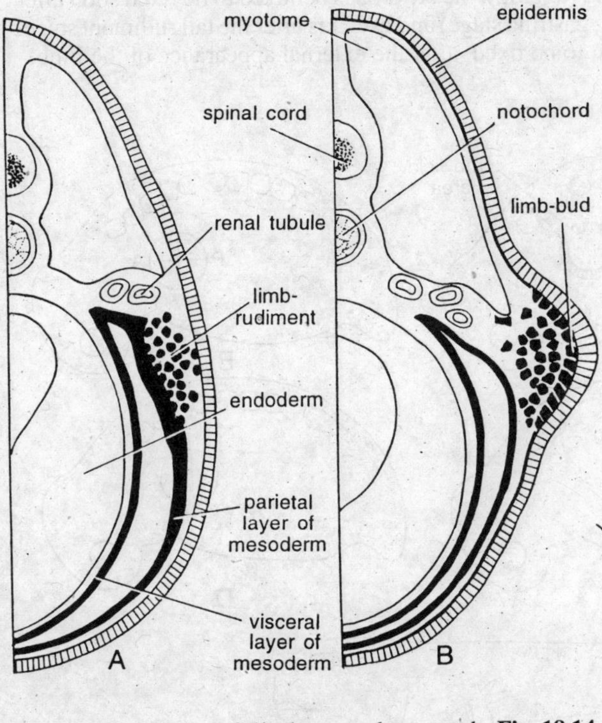

Fig. 18.13. Origin of limb-mesoderm and formation of a limb-bud of an amphibian embryo (after Balinsky, 1970).

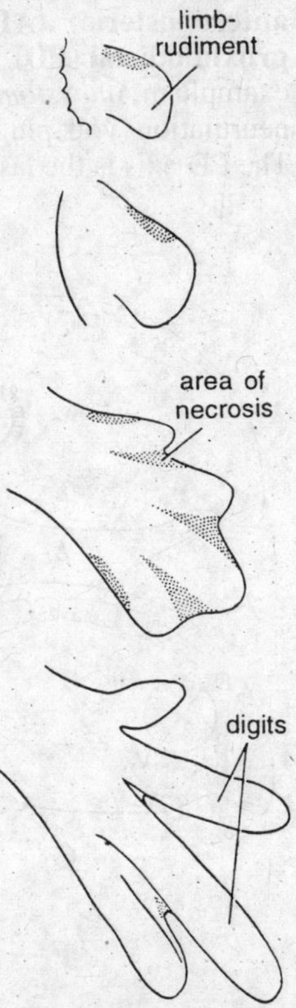

Fig. 18.14. Differentiation of a limb-bud into hind-limb of a vertebrate embryo (after Balinsky, 1970).

Further, in the embryos of some tetrapods (*Calotes* and chick) limbs develop from the anterior and posterior ends of the **Wolffian crests** or **ridges** in the flank region. However, in amphibians the potency to develop limbs is present in the entire flank region.

1. Limb field. During the development of certain organs of the embryo such as a limb or an eye, an area of embryo becomes committed to follow a certain path of development. It is called a **morphogenetic field.** If this area is transplanted, it develops along a specific predetermined path. Such ''determined areas'' or ''fields'' have certain interesting properties : (1) They include more than 'presumptive areas' that ultimately develops in a particular organ (*i.e.*, limb). (2) These areas have considerable regulative ability. Thus, it is possible to combine two such areas or remove half or more of a given area and still get a well formed organ. (3) Moreover, these areas, if necessary, can incorporate competent cells from neighbouring areas to form a particular organ. A limb field is clearly demonstrated in the amphibian *Ambystoma* (Fig. 18.15).

2. Limb polarity. A developed limb has three polarities : **anteroposterior (AP**; also called craniocaudal), **dorsoventral (DV)** and **proximodistal (PD)**. These polarities are not determined simultaneously. For example in *Ambystoma* (axolotl), the AP axis is believed to be fixed soon after neurulation (yolk plug gastrula stage) and DV axis after the tail rudiment stage. The PD axis is the last to be fixed, after the external appearance of the limb.

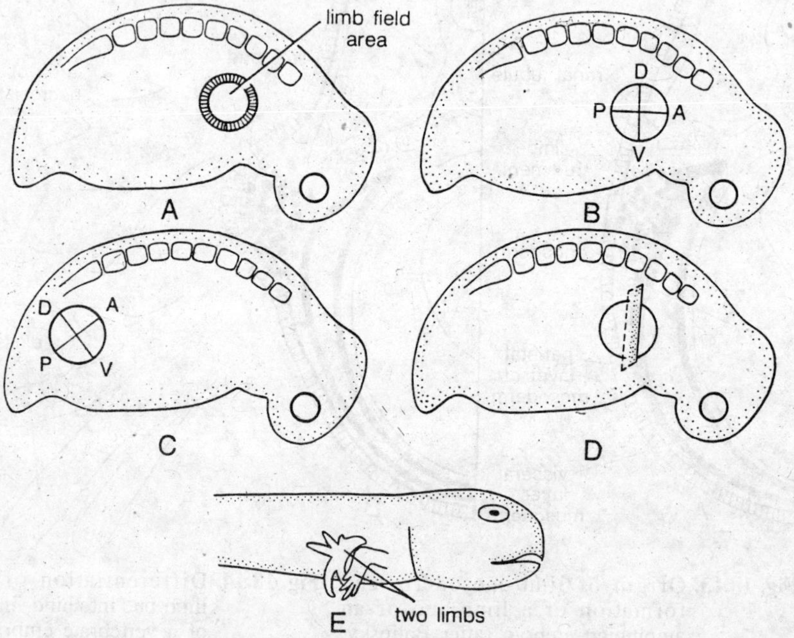

Fig. 18.15. Limb field in *Ambystoma.* A—Limb field area (hatched + stippled) is larger than presumptive limb area (LA : stippled); B–E—Experiments with limb field at tail bud stage; B—Limb disc transplanted at a normal limb site (orthotopic) in another embryo or C—at a different site (heterotopic) such as flank, produces a normal limb. At the time of transplantation dorsoventral (DV) or anteroposterior (AP) or both axial polarities of the limb disc can be reversed to test if these were determined; D—Limb field is split into two by a membrane or extraneous tissue. E—It (*i.e.,* D) produces two limbs.

3. Growth of limb. The growth of the limb bud is asymmetrical from the very beginning. The limbs grow more rapidly along the PD axis than along the AP or DV axis. The dorsal surface of the early limb bud appears convex, while the ventral surface is rather flat. Slightly later the limbs undergo DV flattening at the distal end, and come to acquire a free margin. A distinct ectodermal ridge called **apical ectodermal ridge** or **AER** develops around the margin of limb bud. The AER is located somewhat ventrally and becomes nipple-shaped at

the distal end. As development proceeds, the distal end of the limb becomes paddle shaped. In the blade of the paddle the rudiments of the digits are formed, each digit pointing to a corner in a paddle.

4. Growth of mesoderm. The growth of different parts of the limb mesoderm is reflected in their characteristic cell proliferation rates. There is no migration of the mesoderm from one area to another, but only a relative expansion of each area.

5. Growth of AER. The apical ectodermal ridge (AER) is formed probably due to the action of the underlying mesoderm. The AER is two cell layer thick, but is pseudostratified. The cells of AER of amniotes (*e.g.*, rat) are metabolically active. This is reflected in the fact that these cells contain more RNA, more glycogen and high contents of enzyme alkaline phosphatase than other epidermal cells (**Milarire,** 1956).

The mesoderm has a dominant role in limb development. In early stages it has the potency to form limb in cooperation with non-limb ectoderm; however, the limb ectoderm does not cooperate with non-limb mesoderm in the similar way. Mesoderm tends to induce, maintain and organize the AER in the competent ectoderm through the action of the **apical-ectodermal-ridge-maintenance-factor (AEMF).** The AER tends to play a crucial role in limb development as an ''inductor'' of limb outgrowth, and by maintaining the subjacent mesoderm cells in a state of developmental plasticity.

Cell death has an important role in limb development. The interdigital necrotic zone influences the separation of the digits and the extent of the webbing. The cause of cell death is genetic and not a physiological stress. The cells become determined for death in the same manner as for differentiation.

18-2.2. Development of Appendicular Endoskeleton

Side by side of external changes in limb rudiment, internally a skeleton of **procartilage** is formed by the mesenchyme of differentiating limb bud in the form of separate skeletal units. The procartilaginous limb skeleton soon becomes cartilagionus and ultimately becomes bony. In fact, chondrogenesis (cartilage synthesis) proceeds in a proximodistal sequence but most of the mesopodial (wrist) cartilages appear well after the metapodial ones (*i.e.* metatarsus and metacarpus).

The formation of limb skeleton generally proceeds in a proximodistal direction. In frog, the first skeletal part to become differentiate is **stylopodium (humerus** or **femur**). Part of the **zeugopodium (radius** and **ulna** of fore- limb; and **tibia** and **fibula** of hind-limb) are laid down next. The **autopodium** (hand or foot) differentiates considerably later. In autopodium, the **metacarpals** or **metatarsals** develop more rapidly than **carpals** or **tarsals.** In the digits, however, the proximal phalanges are laid down earlier than the distal ones.

The **pectoral** and **pelvic girdles** develop in intimate connection with the limbs. The mesodermal material for each kind of girdle is derived from the peipheral parts of that mesenchyme mass, the central part of which develops into the limb-buds.

18-2.3. Development of Axial Skeleton

The axial endoskeleton (skull and vertebral column) of frog passes through following three phases during its development : **(i) notochordal phase, (ii) cartilaginous phase,** and **(iii) bony phase.** It is entirely mesodermal in origin. The cartilaginous axial endoskeleton of early tadpole is derived from **sclerotomes** which become converted into mesenchyme. The mesenchyme spreads out as a continuous sheath along the notochord, enveloping the spinal cord above and caudal vein below the notochord in caudal region. The mesenchyme gives rise to number of cell types—such as **fibroblasts, chondroblasts** and **osteoblasts.** The fibroblasts are primarily responsible for the formation of reticular tissue, tendons, ligaments and areolar tissue; the chondroblasts and their products form cartilage; and the osteoblasts form the bone. All the three cell types secrete products that form a **matrix** outside the cell, so that the bulk of tendon of cartilage or bone is non-cellular. Living cells occupy the cavities within the matrix in all these cases.

(a) Development of skull. The visceral part of frog skull is derived from branchial skeleton, while the neural portion of cranium is formed by head mesoderm. The head mesoderm forms a cartilaginous **chondrocranium** (Fig. 18.16) which is made of following independent rudiments—(i) a pair of **trabeculae** or **prechordals** (ii) a pair of **parachordals;** (iii) the **capsules** of **sensory organs** (olfactory capsule, optic capsule and auditory capsule). All of these chondrocranial rudiments are derived from sclerotomes, and they develop around the anterior portion of notochord. Later on, they unite to form a **neuro-chondrocranium** of tadpole. To the neuro-chondrocranium, later are added the visceral components and a cartilaginous skull is formed (Fig. 18.17). In adult frog some cartilage plates of larval skull become ossified and some new dermal bones are formed and added to it.

(b) Development of vertebral column. The vertebrae are formed by sclerotome which forms a continuous mesenchymal sheath around the notochord. In this mesenchymal sheath appear the nodules of cartilage, called **arcualia,** in close apposition to the external surface of the notochord. Typically the arcualia appear in double pairs, one pair is formed dorso-laterally to the notochord and another pair ventro-laterally. The dorso-lateral arcualia grow out dorsally along the side of spinal cord and unite above the spinal cord to form the **neural arch.** In the caudal region, the ventro-lateral arcualia grow downward and unite underneath the caudal vein to form the **haemal arch.** In the cervical and thoracic region, the ventrolateral arcualia give rise to two lateral outgrowths—the rudiments of **ribs**. At the same time the proximal parts of all four arcualia of each vertebra spread out around the surface of the notochord and forms the body (**centrum,** etc.,) of a vertebra.

18-2.4. Development of Circulatory System

The cardio-vascular or circulatory system of frog passes through two phases in its development—a **prefunctional phase** and a **functional phase.** During the prefunctional phase, the heart and major blood vessels differentiate independently of each other. During functional phase, heart and blood vessels unite and circulation of blood is started. Different parts of circulatory system of frog develop in following manner :

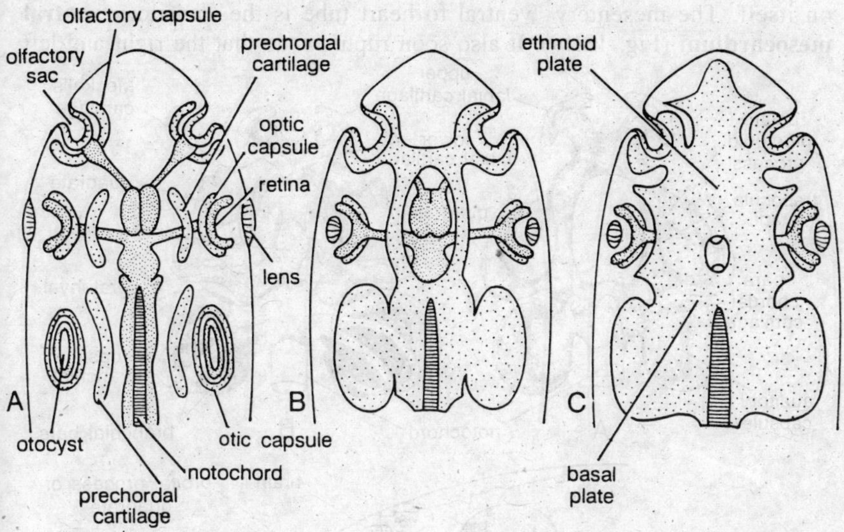

Fig. 18.16. Early stages in development of a cartilaginous neurocranium. A—Prechordals and parachordal plates and nasal, otic and orbital (optic) capsules first appear; B—Fusion of these plates and capsules form a cartilaginous floor consisting of basilar and ethmoid plates that incorporate the notochord; C—Chondrocranium (after Kent, 1987).

(a) Development of heart. The heart of frog develops at the ventral side of pharynx. It is formed by mesenchyme of lateral plate origin. At the end of neurulation, the lateral plate mesoderm separates into **somatic (parietal)** and **splanchnic (visceral) mesoderm,** both enclosing the coelom. The splanchnic mesoderm moves toward the midline where it thickens under the gut tube. Loose mesenchyme cells derived from the splanchnic mesoderm congregate in the midline (*i.e.*, middle of the mesoderm-free zone lying just below the pharynx) between the converging right and left sides of ventro-lateral mesoderm (Fig. 18.18). These mesenchyme cells form a longitudinal tube, called **endocardium** (*i.e.,* the endothelial lining of the cavity of heart). At its anterior end the endocardial tube divides to form two **ventral aortae**; its posterior region (*i.e.,*prospective sinus venosus) becomes continuous with the vitelline veins that have grown into the area.

While the endocardial tube is being formed, the edges of the lateral plate mesoderm continue to push ventrally under the gut (also the endocardium) until they reach the midline. The mesoderm from two sides first fuses ventrally; to the endocardial tube. Later, the splanchnic mesoderm encloses the endocardial tube and fuses in the dorsal region as well. This double-layered membrane (which is formed from the right and left sides of splanchnic mesoderm) is the **dorsal mesocardium** (a part of ventral mesentery). It remains attached to the anterior and posterior regions of the developing heart, but soon it ruptures in the middle area leaving the heart tube free to bend on itself. The mesentery ventral to heart tube is the transitory **ventral mesocardium** (Fig. 18.19). It also soon ruptures, so that the right and left

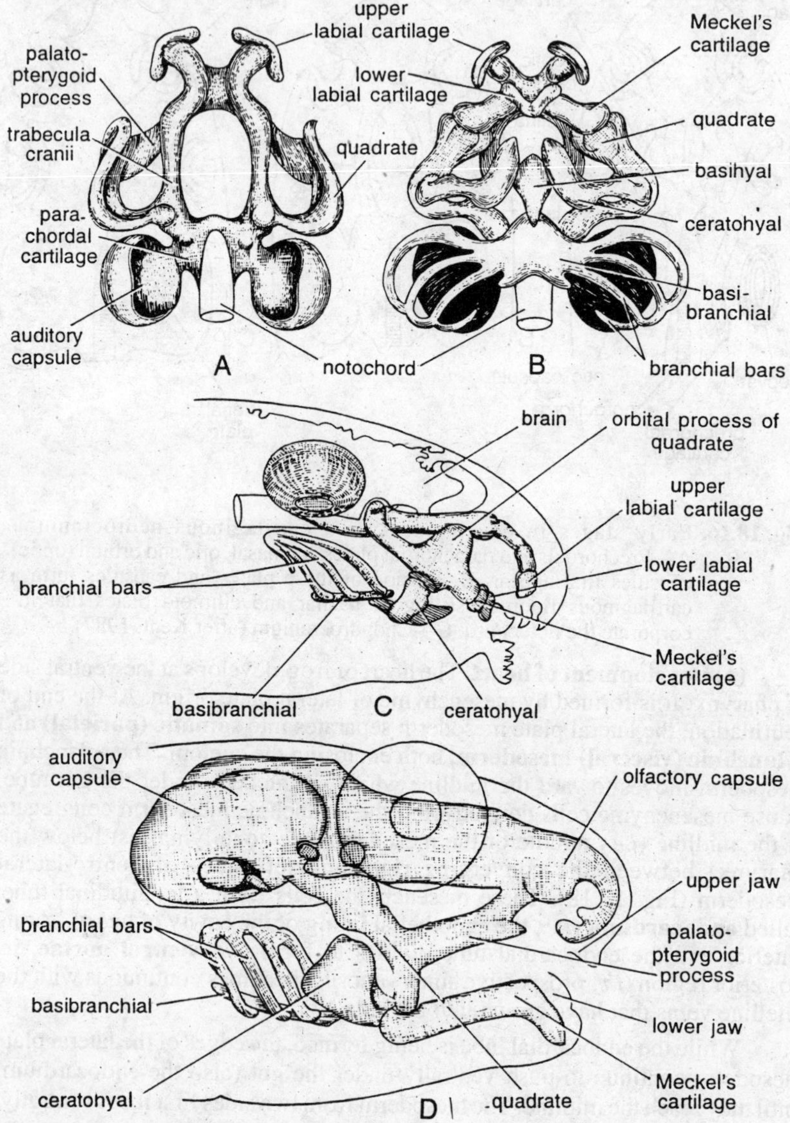

Fig. 18.17. Development of skull of frog. A—A skull of 12 mm tadpole in dorsal view; B—Skull in ventral view; C—Right side view of skull; D—Skull of a tailed frog toward the end of metamorphosis (after Marshall, 1928).

pericardial cavities become continuous, except in the region of the remaining dorsal mesocardium. The splanchnic mesoderm adhereing to the endocardial tube gives rise to **epimyocardium.** By the time the frog reaches 5 mm, this layer, in turn, separates into the **myocardium,** which will give rise to the striated involuntary muscles of the heart, and the **epicardium** or visceral peritoneum. The anterior end of pericardial coelom remains blind, while the

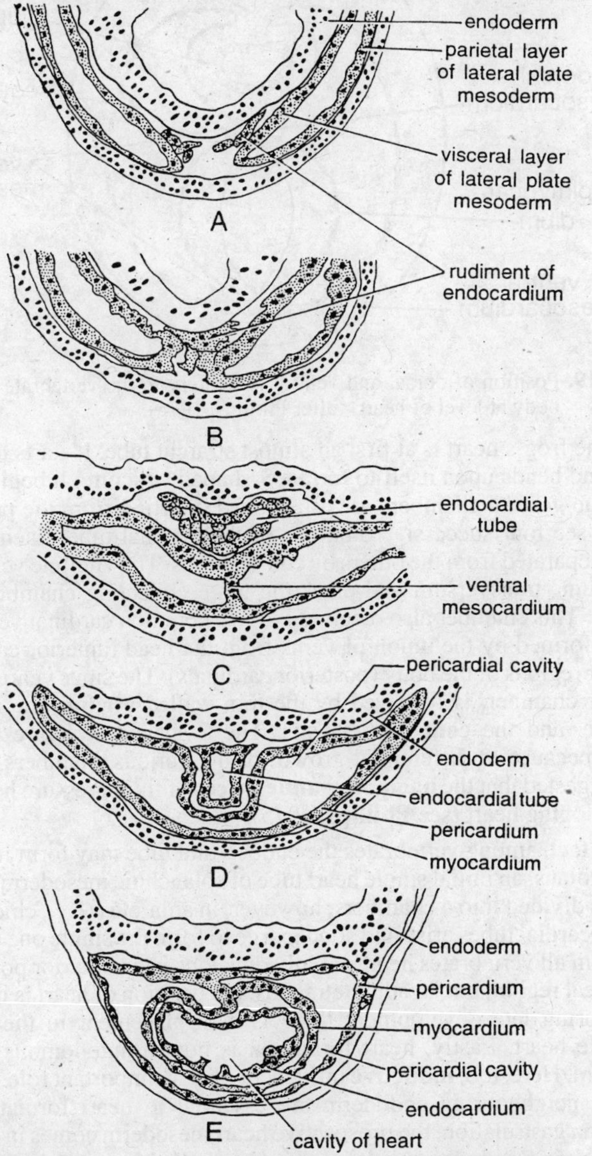

Fig. 18.18. Development of the heart in amphibian embryo (after Balinsky, 1970).

posterior end gets separated with rest of the body coelom by a connective tissue wall called **septum transeversum.**

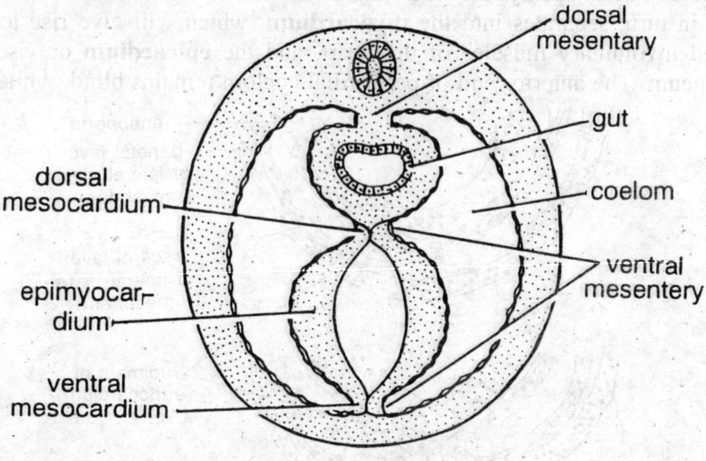

Fig. 18.19. Position of dorsal and ventral mesentery in the vertebrate body (T.S. of body at level of heart) (after Phillips, 1975).

The frog's heart is at first an almost straight tube. It starts to increase in length and bends upon itself to form a S-shaped structure. It begins to pulsate before the general chambers are established. Starting from the posterior end, one can see four successive dilations of the original tube. Each chamber of heart is separated from the others by constrictions. The vitelline veins penetrate the septum transversum and enter the large collecting chamber, the **sinus venosus.** This chamber also receives the two common cardinal veins (ducts of Cuvier) formed by the union of veins from the head (anterior cardinals) and posterior regions of the body (posterior cardinals). The sinus venosus, the most posterior chamber, is followed by the thin-walled **atrium,** the thick-walled **ventricle**, and the **conus arteriosus.** The S-shaped structure of the heart appears because of differential growth of the various chambers. It has also been suggested that the **haemodynamic forces** of fluid pressure helps to shape the developing heart (see **Phillips,** 1975).

In fact, among vertebrates the endocardial tube may form in two ways : in amphibians, an initial single heart tube of splanchnic mesoderm appears and becomes divided into 4 chambers; however, in amniotes (*e.g.*, chick) right and left endocardial tubes arise and soon merge to form the single one (Fig. 18.21). Further, in all vertebrates heart initially develops in an anterior position, in the pharyngeal region, but, in adult tetrapods, the position of heart is in the thorax. Thus, during late development there is a displacement in the position of vertebrate heart. Lastly, heart formation is not an autonomous property of mesoderm. Here, too, **inductive** influences play an important role. Interactions with the neighbouring endoderm are essential to heart formation. During amphibian gastrulation, the prospective heart mesoderm comes in contact with the underlying endoderm as it involutes into the blastoporal lip. The paired heart-forming areas can be identified in an early gastrula from the fate maps; and their morphogenesis depends on continuing inductive influences from endoderm. For example, if the entire endoderm is excised from a neurula, heart formation is prevented. The nature of the inductive influences is not clear. **De-Haan** (1972) suggested that endoderm in chick furnishes a substratum on which precardiac mesoderm migrates (see **Grant**, 1978).

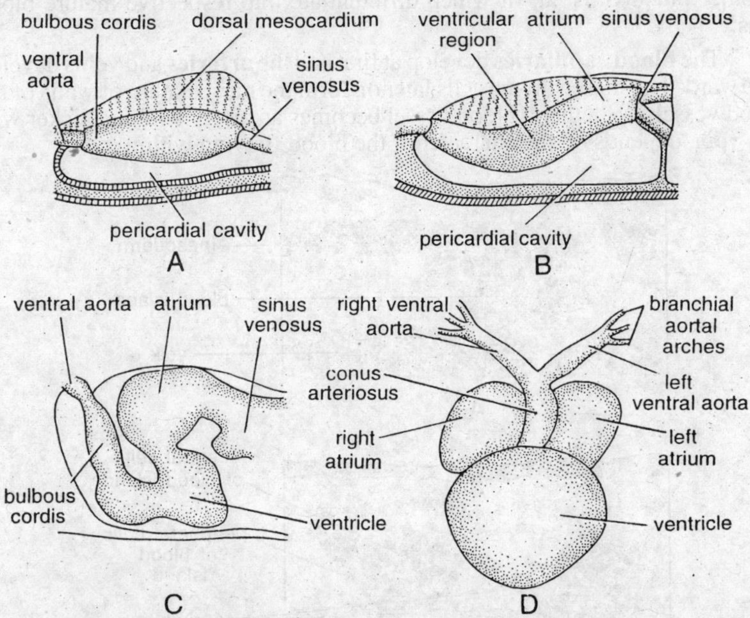

Fig. 18.20. Heart development of frog (after Nelsen, 1953).

(b) Development of blood vessels. The blood vessels and blood of frog are originated early in development by so called **blood islands** of mesenchyme

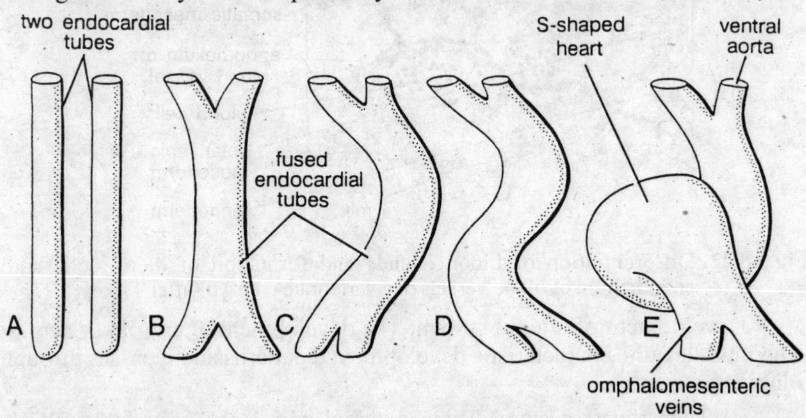

Fig. 18.21. Development of the heart from a paired rudiment in the chick embryo.

cells called **angioblasts** or **vasoformative cells** of splanchnopleuric origin (Fig. 18.22). The outer cells of these blood islands flatten and become the endothelial lining of vitelline blood vessels. The inner cells of the blood islands round up and become free, basic staining cells, called **haemocytoblasts.** These haemocytoblasts subsequently develop in haemogloblin containing primitive red cells (**erythroblasts**); primitive white blood cells (**leucoblasts**) and variety of other blood cells, all of which differentiate into respective mature blood cells.

The **blood capillaries** develop at first and the **arteries** and **veins** develop afterward. The mesenchyme cells later on form the musculature of walls of the blood vessels. Whether a blood vessel becomes an artery with its thicker wall or a vein, depends on the character of the blood flowing within it.

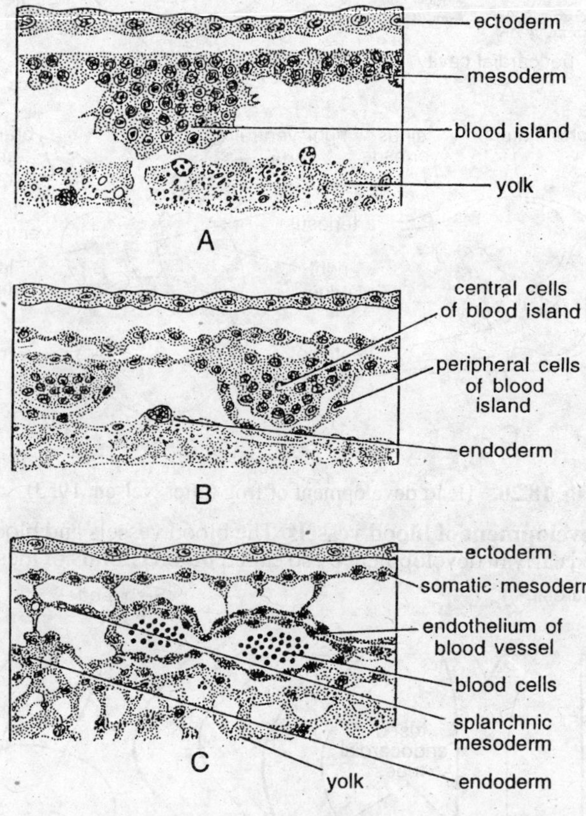

Fig. 18.22. Differentiation of blood islands and formation of blood cells and endothelium of blood vessels of a vertebrate embryo (after Torrey, 1971).

Development of arterial system. The development of arterial system of frog presents quite characteristic developmental patterns and it passes through following stages :

(i) The arterial system during external-gills stage of tadpole. While tadpole is breathing by means of gills, its circulation resembles in all respects with that of fish (Fig. 18.23). The **venous blood** returned from the body enters the posterior end of the heart or sinus venosus; from this it passes into the second or auricular chamber, thence to the ventricle, and from there to the truncus arteriosus. The truncus arteriosus on reaching the anterior end of the pericardial cavity, bifurcates into right and left branches. Each of these again divides into **first, second** and **third afferent branchial arteries,** which carry venous blood into the external gills and their branches. The **fourth afferent branchial artery** remains rudimentary.

In the external gills, the oxygenated blood passes through short wide capillary loops into four pairs of **efferent branchial arteries,** which thus form along with afferent branchial arteries, three pairs of **complete branchial arches** and one pair of **incomplete** (*i.e.* **fourth) branchial arches.** All efferent branchial arteries open into a pair of **dorsal aorta** in the roof of pharynx. The dorsal aortae of the two sides run forward as the **carotid arteries** to supply the head and brain, and also run backward in the roof of the pharynx. The aortae of two sides meet and unite about the junction of head and body to form the single **systemic arch** which supplies arterial blood to all body organs.

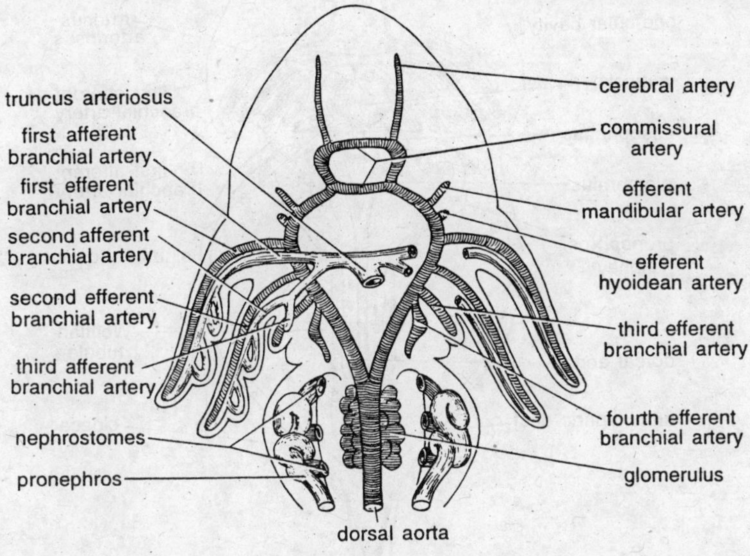

Fig. 18.23. The arrangement of afferent and efferent branchial arteries in a tadpole larva in its external gill stage (after Marshall, 1928).

2. The arterial system during internal-gills stage of tadpole. After the external gills are shriveled up, and the opercular or internal gills become well established in the tadpole, following modifications occur in the blood circulation of tadpole—The truncus arteriosus divides at once into right and left branches, which run straight outward in the floor of the pharynx. Each of these branches divides after a short course into three vessels and the hinder most vessel again into two. In this way the four afferent branchial vessels of the first, second, third and fourth branchial arches are formed respectively (Fig 18.24).

Each afferent vessel runs outwards and upwards in its own arch. The efferent branchial vessels lie immediately in front of the corresponding afferent vessels with which they are connected by very numerous capillary loops in the substance of the internal gills. At their upper ends the efferent vessels open, as before into the dorsal aorta.

As the lungs become considerably developed during late tadpole stage, they receive blood by the **pulmonary arteries** which are formed by branches of fourth efferent branchial vessels. The blood from the lungs returns directly to heart by a pair of **pulmonary veins** which open into left auricle.

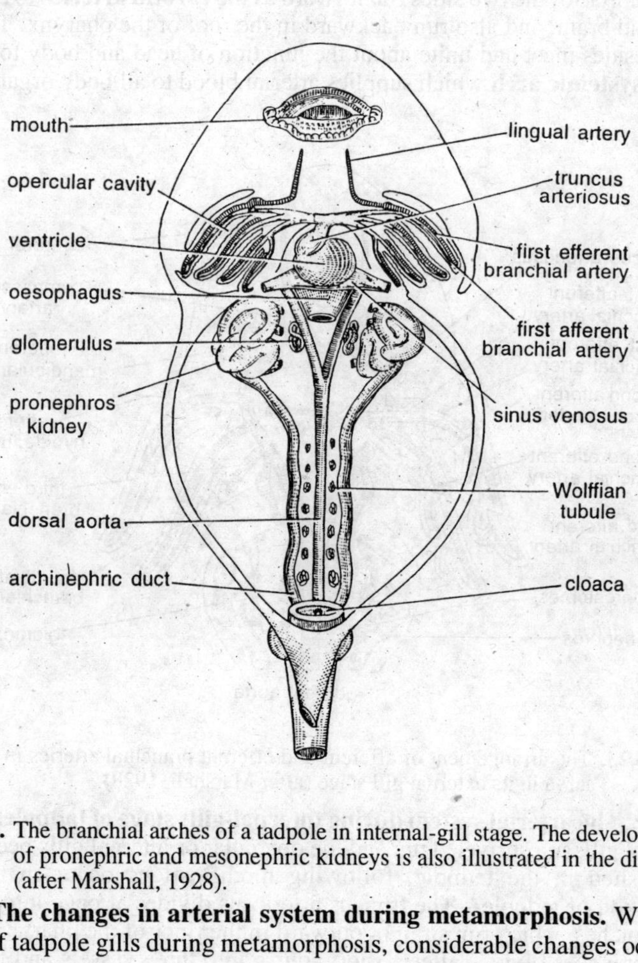

Fig. 18.24. The branchial arches of a tadpole in internal-gill stage. The development of pronephric and mesonephric kidneys is also illustrated in the diagram (after Marshall, 1928).

3. The changes in arterial system during metamorphosis. With the atrophy of tadpole gills during metamorphosis, considerable changes occur in arterial system (Fig. 18.25). Thus, the aortic arches from I to VI in the tadpole are represented as I or mandibular (rudimentary), II or hyoidean (rudimentary), and first, second, third and fourth branchial arches. During metamorphosis, of these, mandibular and hyoidean arches and V aortic arch (= third branchial arch) disappear; while III aortic arch (or first branchial arch) is represented by **carotid arch**; IV aortic arch (or second branchial arch) by **systemic arch** and VI aortic arch (fourth branchial arch) by **pulmocutaneous arch** in adult frog (Table 18-2). Further due to the development of a spiral valve in the conus arteriosus or truncus arteriosus, the latter divides into a **pulmonary trunk** and a **systemic trunk** (see **Phillips**, 1975).

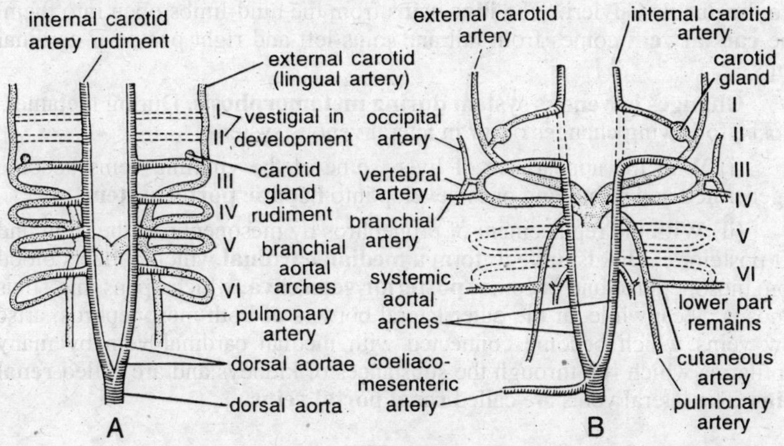

Fig. 18.25. Modifications of aortal (branchial) arches in the frog. Pattern of branchial blood vessel of tadpole during metamorphosis (A) and after metamorphosis (B) (after Nelsen, 1953).

Table 18-2. Fate of aortic arches in frog.

Aortic arch	Embryonic (Tadpole) (Branchial arches)	Adult (Frog)
I	Mandibular (rudimentary)	Disappears
II	Hyoidean (rundimentary)	Disappears
III	First branchial	Carotid arch
IV	Second branchial	Systemic arch
V	Third branchial	Disappears
VI	Fourth branchial	Pulmocutaneous

Lastly, the development of aortic arches in higher animals provides an excellent example of embryonic structure that can best be explained in the light of the evolutionary past. Their development is quite accordant with De Beer's modification of the old adage to read "*ontogeny represents fundamental steps in the ontogenesis of ancestral forms, especially when these steps are of structural and functional importance to the individual*" (**De Beer**, 1954).

Development of venous system. The main larval veins of frog are anterior and posterior cardinals, Cuvierian ducts, iliacs and the median caudal. The **anterior cardinal vein** comes from the head and is formed by the union of external and internal **jugular veins,** coming from the dorsal and ventral parts of the head. The **posterior cardinal veins** develop in relation to the pronephros and collect the blood from the posterior body organs. The anterior and posterior cardinal of each side unite to form the **common cardinal vein,** called **ductus Cuvieri.** The **lateral abdominal veins** extend from the posterior end to open into the ductus Cuvieri. The **iliac veins** from the hind-limbs open into them. The **caudal vein** comes from tail and joins left and right posterior cardinal veins.

Changes in venous system during metamorphosis. During metamorphosis, following changes occur in lateral venous system :

(i) With the appearance of liver, some of the vitelline veins become degenerated and remaining ones develop into **hepatic portal system.**

(ii) With the replacement of pronephros by mesonephros, the right and left posteior cardinals unite to form a **median cardinal** which collects blood from mesonephros and becomes **posterior vena cava,** which opens into sinus venosus. Meanwhile, at the outer lateral borders of both mesonephros, arise new veins which become connected with median cardinal vein by many capillaries which run through the substances of kidneys and are called **renal veins.** The lateral veins are called **renal portal veins.**

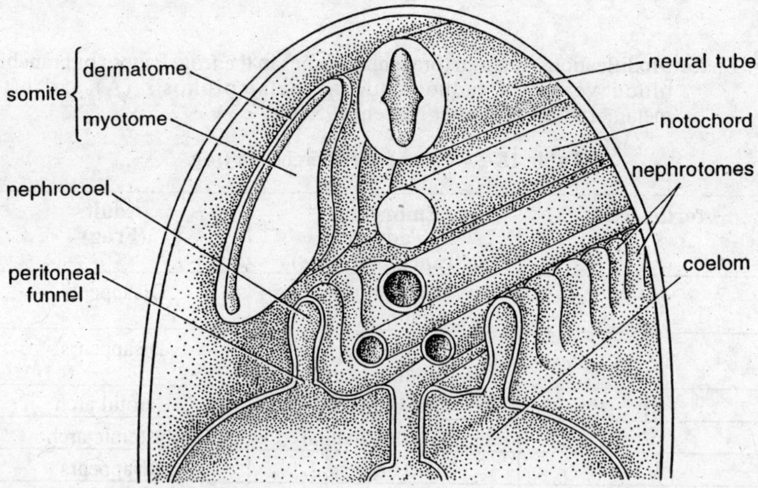

Fig. 18.26. Stereogram of vertebrate embryo showing the differentiation of nephrotomes (after Torrey, 1971).

(iii) With the appearance of legs, each renal portal vein is joined by **iliac vein,** which divides at its distal end into the **femoral** and **sciatic veins.**

(iv) When tail is lost, the part of the posterior vena cava posterior to the kidneys (mesonephroi) vanishes, so that most of the blood from posteior body region has to pass through the renal portal vessels and the abodominal veins.

18-2.5. Development of Kidneys

The excretory organs or kidneys of frog develop into two subsequent phases. The tadpole has **head kidney** or **pronephros,** and adult frog has **Wolffian bodies** or **mesonephros.** The **pronephroi** are globular organs embedded in the dorsal wall of the body at its anterior end. Each pronephros is a convoluted tube with glandular walls, opening into the body cavity by three ciliated mouths or **nephrostomes.** Each pronephros posteriorly sends a **archinephric** or **pronephric** or **segmental duct** which opens into the cloaca.

In tadpoles of about 12 mm length, the adult kidneys or mesonephroi begin to form in the hinder part of the body as a series of paired tubules, which grow towards and open into the archinephric duct, which is now called **mesonephric duct.** These **mesonephric tubules** or **uriniferous tubules** increase in number, size and complexity and all become bounded together by connective tissue to form compact kidneys of the fully formed tadpole. At the same time the pronephroi diminish in size and undergo degenerative changes, and by the time of the metamorphosis have almost completely disappeared. The mesonephroi become **opisthonephroi** (a kidney formed by uriniferous tubules of mesonephros and metanephros). By a series of further changes the ureters and generative duct of the adult become established.

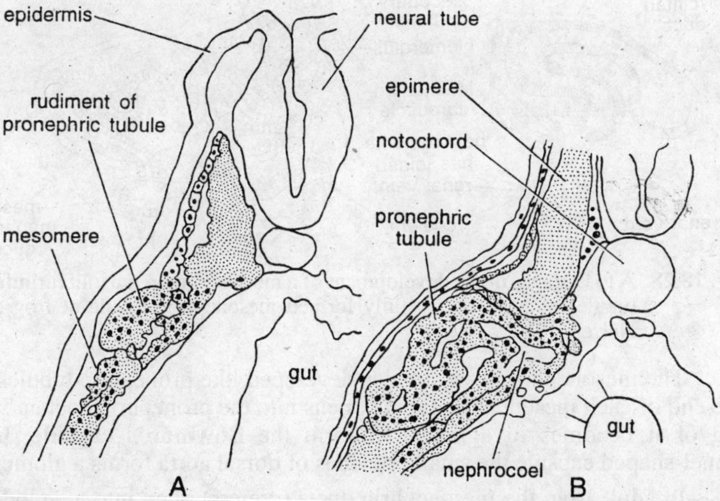

Fig. 18.27. Early development of a pronephric tubule in the frog (after Balinsky, 1970).

The pronephros and opisthonephros kidneys of frog are developed from **mesomere** or **nephrotome.** The pronephroi are formed by mesomere of anteior region, *i.e.,* 2nd to 4th somites, while meso-and opisthonephroi are formed by mesomere of posterior body region. To form **pronephric renal** or

uriniferous tubules, the dorsolateral wall of nephrotome becomes drawn out into hollow and elongated tubules, the cavities of which are the extensions of the nephrocoel. The distal (outward) ends of these tubules soon turn backward and then fuse with each other to form a common excretory duct, the **pronephric duct.** One end of these uriniferous tubules opens into coelom as nephrotomes and opposite to the pronephroi an irregular sacculated outgrowth, the **glomerulus,** arises from the aorta on each side.

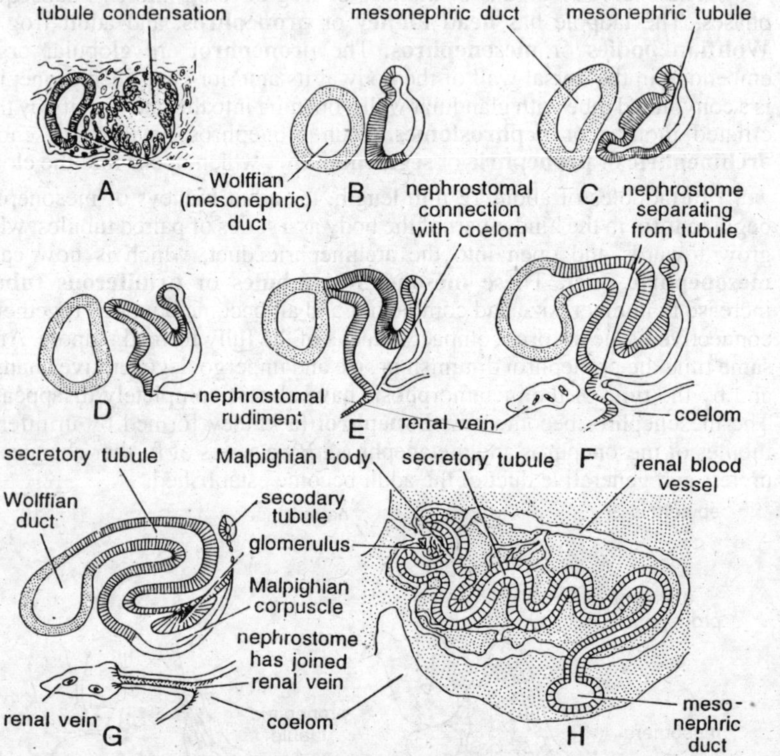

Fig. 18.28. A to G stages of the development of a mesonephric renal unit (uriniferous tubule) of frog. H is a fully formed mesonephric tubule of frog (after Nelsen, 1953).

The mesonephric renal tubules developed like pronephric tubules. The one end of each mesonephric tubule opens into the pronephric duct and other end of it becomes invaginated to form the **Bowman's capsule.** In the funnel-shaped capsule the small branches of dorsal aorta forms a glomerulus.

In adult frog, the mesonephric ducts (ureters) open into a endodermal sac-like protrusion of cloaca called **urinary bladder.**

18-2.6. Development of Gonads

Like other vertebrates, the gonads of frog develop from the upper edge of visceral layer of lateral plate mesoderm in the posterior half of the body. The gonad-rudiment appears as a thickened longitudinal strip of columnar shaped mesodermal cells from the mesodermal epithelium of body cavity and is called **germinal ridge.** It occurs in between dorsal mesentery and rudiment of mesonephros. Soon the cells of germinal ridge get arranged into a compact mass several cells thick and it protrudes into the coelomic cavity. In a gradual manner, the connection between it and the peritoneal wall becomes constricted laterally and the gonad remains suspended from the peritoneal wall only by a double layer of peritoneum—the **mesorchium** and **mesovarium.**

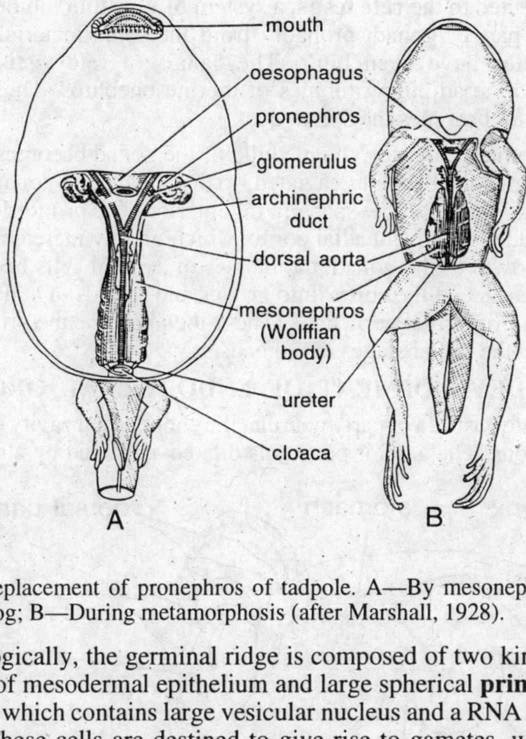

Fig. 18.29. Replacement of pronephros of tadpole. A—By mesonephros of tailed frog; B—During metamorphosis (after Marshall, 1928).

Histologically, the germinal ridge is composed of two kinds of cells—typcial cells of mesodermal epithelium and large spherical **primordial germ cells,** each of which contains large vesicular nucleus and a RNA and yolk-rich cytoplasm. These cells are destined to give rise to gametes, while ordinary mesodermal cells become either Sertoli cells of adult testes or the superficial epithelium and follicle cells of the ovaries. These primordial germ cells have an extragonadial origin primarily occurring in dorsal mesentery, but later migrating to the germinal ridge.

Differentiation of gonads. After arriving in the germinal ridges, the primordial germ cells become embedded in the epithelium of the germinal ridge, which, simultaneously becomes convex toward the coelomic cavity and, on its dorsal side, forms a hollow which is filled initially with loose undifferentiated mesenchyme. The latter is soon replaced by compact strands called **primitive sex cords** of mesenchyme cells. These sex cords form the **medulla** of fully developed gonad. The epithelium of germinal ridge forms the surface of the gonad and is called **cortex.** In gonadial cortex numerous primordial germ cells remain embedded. Up to this stage, the development of male and female gonads remains without any difference, so, such gonads are called **indifferent gonads.** It is the onset of histological differentiation which makes testes and ovaries increasingly different.

In the male embryos, the primordial germ cells migrate from the gonadial cortex into the primitive sex cords of the medulla, which forthwith become hollowed out and are ultimately converted into seminiferous tubules. The primordial germ cells undergo spermatogenesis to manufacture spermatozoa, while, the Sertoli cells are derivd from the sex cords. The seminiferous tubules become connected to the **rete testis,** a system of very thin tubules developing in the dorsal part of gonad, probably from the same material from which seminal tubules have been built. The canals of rete testis, soon form connections to the adjoining tubules of the mesonephros. The mesonephric duct of the latter becomes the sperm duct.

In the female embryos, the medulla of the gonad becomes reduced and degenerating medulla becomes caviated, so that the mature gonad acquires the form of a sack with a hollow space in the interior. The primordial germ cells remain embedded in the gonadial cortex which greatly increase in thickness. During the growth of the gonad, the masses of cortical cells become split up on the inner surface of the cortex into groups and strands of cells surrounding one or several primordial germ cells. These then become the primary oocytes which enter in the initial stages of oogenesis.

18-3. DEVELOPMENT OF ENDODERMAL ORGANS

In the late gastrula or early neurula, the endodermal cavity contains three unequal portions. The anteior portion is dilated and lined by a relatively thin

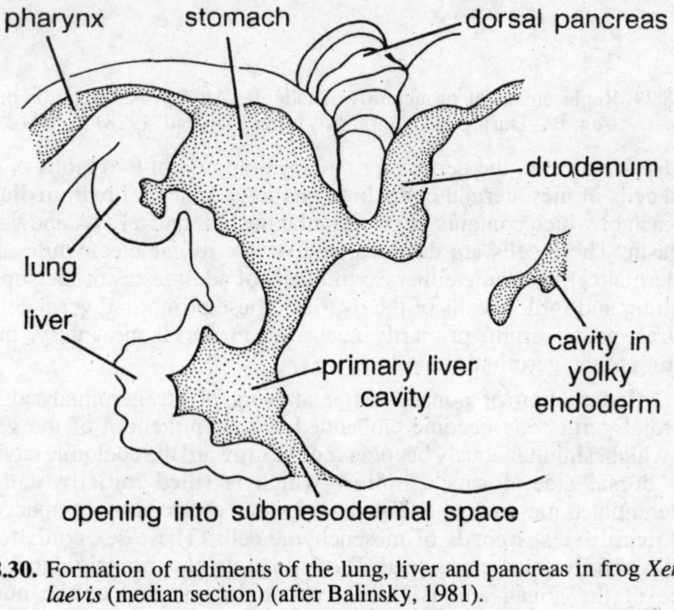

Fig. 18.30. Formation of rudiments of the lung, liver and pancreas in frog *Xenopus laevis* (median section) (after Balinsky, 1981).

endodermal epithelium. This part is called **foregut.** The following portion is **midgut,** the cavity of it is narrow and remains lined by a thin epithelium dorsally. But, its ventral wall contains many large-sized yolk-laden cells. The most posterior part of cavity which adjoins the blastopore forms the **hindgut.**

In the late neurula stage, the ventral wall of foregut flattens out, folds slightly upward and originates two pocket-like processes. The large anterior one, lying immediately underneath the brain, gives rise to the **buccal cavity** and the **branchial region** (pharynx). The posterior pocket develops into the rudiments of **liver, pancreas,** part of **stomach** and **duodenum.** The liver diverticulum extends as a funnel-shaped invagination in postero-ventral direction, drawing into it the rudiments of anterior part of oesophagus, stomach, and duodenum, which now become tubular.

18-3.1. Development of Pharyngeal Pouches

The part of the gut in front of the stomach rudiment also elongates, but instead of becoming tubular it is flattened dorsoventrally and expanded side ways. Most of this cavity becomes the **pharynx.** A series of vertical ridge-like thickenings appear on the sides of pharynx. These are **visceral arches** and six in number on each side.

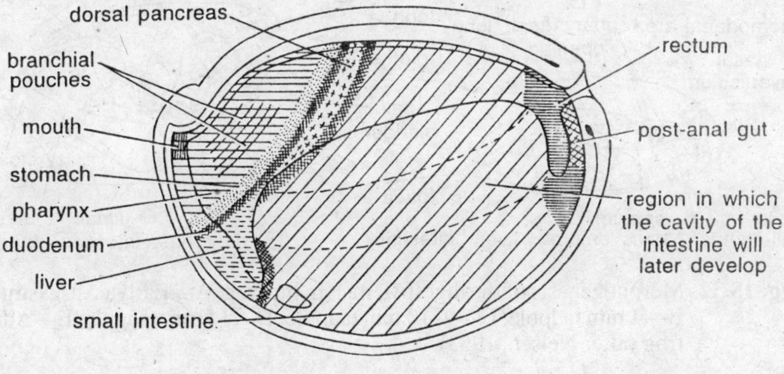

Fig. 18.31. Fate maps of the endodermal organs in the neurula of a frog. The embryo is represented as being cut in half in the median plane and viewed from side to cut surface (after Balinsky, 1970).

The most anterior is the **mandibular arch** and gives rise later on to the lower jaw; the second is the **hyoid arch** and the succeeding four are the **first, second, third** and **fourth branchial arches.** About the time of hatching, the external gills grow out from the outer surfaces of the first and second branchial arches and a little later from the third branchial arch as well.

At the same time lateral walls of pharynx thrown out paired endodermal outpushings (five in number) lying between the visceral arches. The outer wall of the last four endodermal pouches fuse with epidermis and form **branchial membranes** which later become perforated to communicate the pharyngeal cavity with the exterior. These perforations of endodermal pouches, thus, become the **gill-clefts,** which are four in number.

During metamorphosis, these gill-clefts obliterate and their parts are used up in the development of other structures. Thus, the first pair of pharyngeal pouches become **eustachian tube,** the third and fourth pair of pharyngeal pouches become **thymus** and **parathyroid** endocrine glands. The **thyroid** develops from the pharynx as a ventral pocket in the floor of the latter.

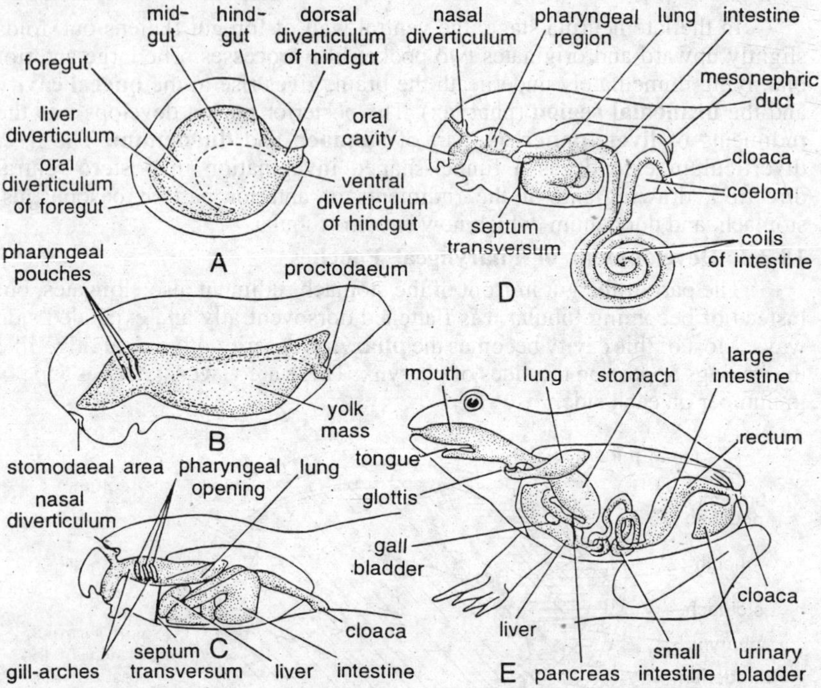

Fig. 18.32. Morphogenesis of the digestive tract in the frog. A—Early post-gastrula; B—4 mm tadpole; C—9–10 mm tadpole; D—15 mm tadpole; E—Adult frog (after Nelsen, 1953).

18-3.2. Development of Midgut

As the main portions of the alimentary canal begin to take shape, the canal as a whole becomes twisted in a characteristic way. At an early stage, the stomach has already assumed a slanting position. Subsequently, the posterior end of the stomach is shifted to the left, while the adjoining part of the duodenum comes to lie transversely, going from left to right, The distal part of the duodenum is then bent in such a way that it leads to the anterior end of the intestine, which is more or less in a dorsal position. The alimentary canal, thus, performs a complete spiral revolution, which may by referred to as the **gastrointestinal loop.** The posterior part of the intestine also becomes twisted into folds and loops as the result of the elongation of the alimentary canal which exceeds the length of the body.

18-3.3. Development of Mouth and Anus

The development of mouth and anus occurs quite late. The **mouth** opening breaks through, where the anterior end of the foregut touches the epidermis just beneath the front end of neural tube. The epidermis here sinks to form a pocket-like depression, the ectodermal mouth invagination or the **stomodaeum.** The ectodermal and endodermal epthelia fuse here and become the **oropharyngeal membrane,** which later on becomes thin and disappears to form the **mouth opening.** The stomodael invagination also produces a small pocket called **Rathke's pocket** at its dorsal side. The pocket, later migrates to the lower side of diencephalon to fuse with **hypophysis.**

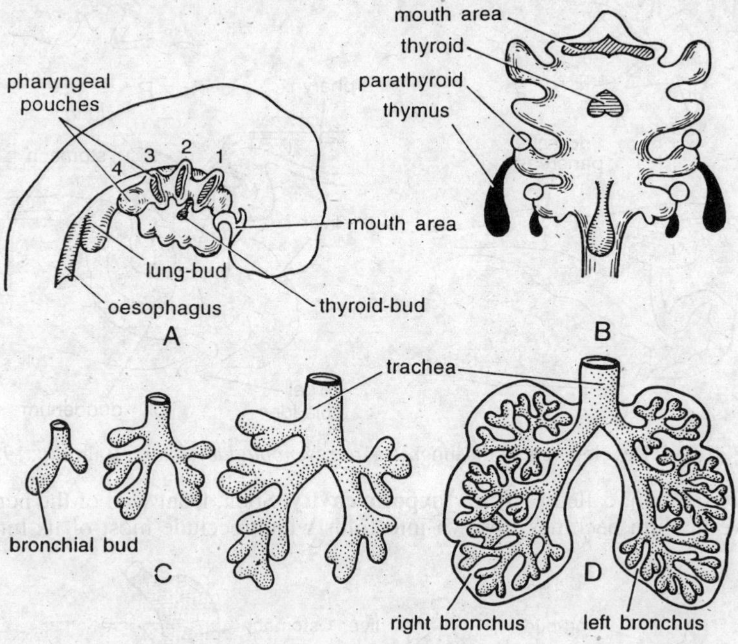

Fig. 18.33. Embryonic frog pharynx showing the places of origin of various structures in the adult. A—Lateral view; B—Ventral view; C—Stages in the development of lungs (after Nelsen, 1953).

The **anus** or **proctodaeum** is formed due to invagination of epidermis near the blastopore. The inner end of this invagination opens through the posterior end of hindgut and, thus, forming the **cloaca aperture.**

18-3.4. Development of Accessory Organs of Alimentary Canal

1. Development of lungs. The lungs are terrestrial respiratory organs of adult frog. They develop from a rudiment which is a pocket-like evagination of the endodermal epithelium on the ventral side of the alimentary canal just posterior to the branchial region. The lung rudiment at first projects straight downward and then bifurcates at its tip. The two branches grow out to the sides and backward. The unpaired medial part of the rudiment becomes the **trachea,** while the two branches develop into two **bronchi.** The lungs develop as sac-like expansions at the ends of the bronchi, the wall of which becomes greatly folded. Later on these folds are traversed by blood vessels and nerves.

2. Development of liver. The liver develops from the anterior walls of the liver diverticulum, which bulges forward to enlarge the cavity of liver

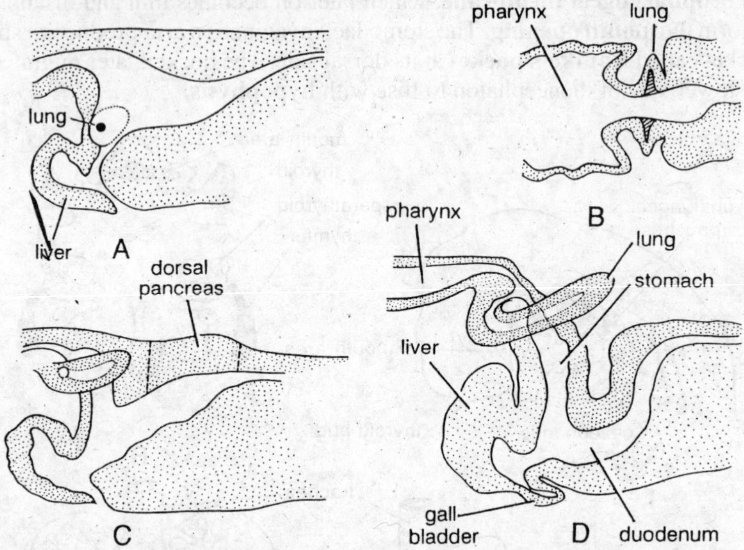

Fig . 18.34. Development of the lungs in a frog, *Xenopus laevis* (after Balinsky, 1981).

diverticulum, called **primary hepatic cavity.** Soon, front wall of the hepatic diverticulum becomes thrown into folds which occlude most of its lumen,

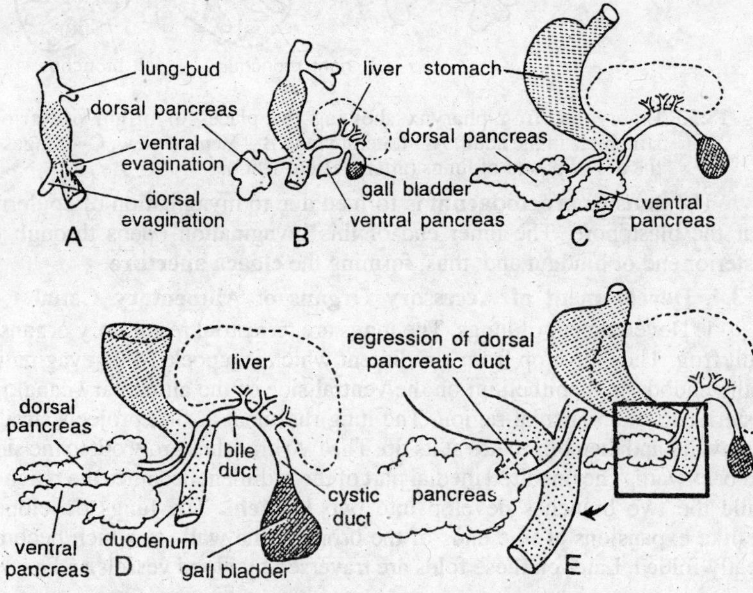

Fig. 18.35. Development of liver and pancreas of frog.

leaving only its posterior part open. At the same time, the original communication between the primary hepatic cavity and the duodenum is constricted and is transformed into **bile duct.** The posterior remnant of the primary hepatic cavity becomes the cavity of **gall-bladder.** The anterior portion of primary hepatic cavity develops into liver.

3. Development of pancreas. The pancreas develops from two rudiments—a ventral rudiment and a dorsal one. The ventral rudiment is derived from duodenum, while the dorsal rudiment is produced by a section of the roof of the midgut, posterior to the rudiment of the stomach. The ventral rudiment is pocket-like and it distally becomes subdivided into a system of proliferating epithelial tubules. Its lumen becomes the duct of the **ventral pancreas** which later on becomes united with the **dorsal pancreas,** forming a system of alveoli and ducts of pancreas.

4. Development of urinary bladder. The urinary bladder of adult frog arises shortly before the metamorphosis as a ventral outgrowth from the hindgut. This outgrowth soon bifurcates at its distal blind end and becomes the urinary bladder.

Thus, the digestive and respiratory systems have a common embryonic origin from the endoderm although they function in entirely different ways. Certain glands (exocrine as well as endocrine) also originate from the primitive gut. Digestive glands such as liver and pancreas form as outgrowths of the original endodermal tube. Endodermal cells compose the endocrine glands, thyroid and parathyroid, as well as the two important lymphoid structures, the thymus and bursa of Fabricius (of birds) (see **Phillips,** 1975).

REVISION QUESTIONS

1. Describe the development of skin of frog.
2. Describe the development of central nervous system of frog (or of any other vetebrate).
3. Give an account of the development of brain in frog.
4. With the help of suitable diagrams describe the development of eye in frog.
5. Describe the process of differentiation of neurons and retinal cells.
6. What changes occur in the cytoplasm and nucleus during the differentiation of the lens fibres ?
7. Give an illustrated account of development of heart in frog.
8. Describe the metamorphic changes in the aortic arches of frog.
9. Describe the development of the limb in a vertebrate.
10. Describe the development of different digestive organs from the endoderm.
11. Write short notes on the following :
    (i) Lens differentiation; and
    (ii) Cytogeneis of photoreceptors.

19

EMBRYOGENESIS OF CHICK

The chick or fowl *(Gallus ferrugineus)* belongs to order **Galliformes** of class **Aves** of subphylum **Vertebrata**. Its embryology has been worked out more extensively only because of following reasons : its eggs are of convenient size ; its different developmental stages can be easily got in laboratory conditions for experimental purposes ; its embryology bears many resemblances with those of reptiles and mammals ; and lastly its eggs are easy to procure and remain available throughout the year. Despite these reasons, the embryogeny of chick is important due to its following phylogenetic significance :

19-1. PHYLOGENETIC SIGNIFICANCE OF CHICK EMBRYOLOGY

Uptill now, we have studied the embryogeny of those animals which lay their eggs in aquatic media (*i.e.*, Ascidians, *Amphioxus*, Frog, etc.), but now onward we will study the embryogeny of those animals which are terrestrial and lay their eggs on the land (*e.g.*, reptiles, birds and monotreme mammals).

The evolution of land laid eggs marked a great advance in the history of life on earth. It is likely that the first land-laid vertebrate eggs were those of amphibious ancestors of the reptiles, which lived in water and feed on fish. Thus, there were advantages in placing the eggs on land for the embryos, unlike water larvae they were not subjected to the fouling of the water and a lack of oxygen. Neither they were in danger of being consumed by the voracious predators which inhabited the ponds.

But there were also great hazards to be met and problems to be solved in laying eggs on land. The newly hatched young had to be well enough developed to crawl on land and fend for themselves ; and this required a large store of nutrient in the egg in the form of yolk and well extended embryonic period. The eggs and embryos also had to endure dryness and the harshness of rugged surroundings ; and for this purpose they were supplied with a watery envelope of albumen, tough outer membranes and often a hard shell. Furthermore, the embryos needed special adaptations to protect their delicate outer surfaces and to carry on respiratory exchange of gases with the surrounding air. They needed receptacles to retain the nitrogenous excretory wastes which if they accumulated within the embryo proper would poison them. To serve these several functions the embryo developed four **foetal membranes**, namely the **chorion, amnion, yolk sac** and **allantois**. Amnion is the most important foetal membrane since it serves to keep the embryo moist and to protect it from desiccation. On the basis of amnion, the vertebrates are classified into following two groups — **(i) Anamniota.** The vertebrates without amnion, *e.g.*, Cyclostomata, fishes and Amphibia. **(ii) Amniota.** The vertebrates with amnion, *e.g.*, reptiles, birds and mammals.

Further, a reorganization of the circulatory system was also needed for gill no longer be used for respiration. Finally, there was evolved a method of excretion whereby wastes could be eliminated with the least possible loss of water and useful substances. This was accomplished by substituting a relatively insoluble waste, the **uric acid**, for the highly soluble wastes, **ammonia** and **urea**. Eggs which meet these specifications are known as **cleidoic eggs** (**Needham J.**, 1931). Cleidoic means box-like. Thus, a cleidoic egg becomes self-sufficient except for oxygen intake and CO_2 outflow. Such cleidoic eggs are the characteristics of reptiles, birds and monotreme mammals. The embryogeny of cleidoic eggs of chick, thus, presents a representative case, which applies well to all the animals belonging to these orders.

19-2. REPRODUCTION

The sexes of fowl are separate, and sexual dimorphism is well marked in between male fowl (called **rooster** or **cock**) and female fowl (called **hen**).

19-2.1. Testes and Spermatogenesis

The gonads of cock are testes which are paired and ovoid. They remain suspended from the dorsal body wall by the mesorchia like that of frog. Each testis of cock has typical vertebrate morphology, *i.e.*, having seminiferous tubules in which the spermatozoa are manufactured and which remain connected with sperm duct (vas deferens) by vasa efferentia. The vas deferens ends in the genital papilla located in the cloaca. The spermatozoa of chick have typical morphology of a flagellate sperm.

Copulation. During copulation, the cock or rooster crouches on the back of the hen and while the latter elevates her tail, and, thus, turning the cloaca upward, the cock or rooster turns his tail downward, so that the two cloaca become apposed. The spermatozoa are then transferred and injected into the vagina of the hen. They travel upward and remain in the upper portion of oviduct where they will fertilize the ova as the latter enter the oviduct. In the oviduct of hen, the spermatozoa may remain alive for two weeks.

19-2.2. Ovaries and Oogenesis

The gonads of hen are ovaries. An adult hen has only one ovary and one oviduct and that is also of left side of body. The ovary of right side remains degenerated. The retention of one ovary in the adult birds, is an important factor of several flying adaptations of them, because, two ovaries might have many ripe large-sized eggs which may increase the weight of the body.

Anatomically, the ovary of hen contains a well-differentiated connective tissue (**stroma**) which is supplied with blood vessels and interspersed with many smaller and larger aggregates of cells, the **follicles**. Each follicle has a large central cell (**oocyte**) surrounded by a peripheral layer of smaller cells (**follicular septum** or **granulosa**). During **oogenesis**, the oocyte grows in size due to accumulation of yolk. At first a sphere of **white yolk** is accumulated in the centre of oocyte. Then gradually layer upon layer of **yellow yolk** alternating with white yolk are added in a daily cycle until the cell — the "yolk" of the egg, as we commonly know it — has become a sphere of an inch and a half or so (about 4 cm) in diameter. In six days it increases its mass as much as two hundred folds. This increase in mass of oocyte is at the expense of follicular cells. Biochemical analysis has revealed that the proteins of the egg yolk are synthesized in the mother's liver (as a common precursor **vitellogenin**), circulated to the ovary and deposited as yolk-fluid and **yolk** granules in the oocyte.

The nuclei of the youngest oocytes are at the centre of each cell but as each oocyte grows in size its nucleus moves from the centre keeping close to one side namely the **animal pole.** The cytoplasm is at first distributed throughout the oocyte ; but as growth takes place it becomes more and more limited to a **"germinal disc"** or **nucleus**. For a time a thin layer of cytoplasm remains also at the periphery of the cell, but ultimately it disappears except for the disc. Beneath the disc and extending to the centre of the oocyte is a column of white yolk known as **latebra**. When a oocyte reaches its full size a membrane, the **zona radiata**, forms between the oocyte and surrounding follicle cells. The term "immature ovum" is generally used for the primary oocyte which is ready to leave the ovary.

Normally the ovary of a hen contains follicles with their oocytes at various stages of development. The youngest follicles are microscopic, occurring in the centre of the ovary ; the oldest are full-sized yolks, occurring at the periphery and ready to ovulate.

1. Ovulation, maturation and fertilization. After its complete development, the ovarian follicle breaks through the ovarian wall thereby, thus, rupturing the follicular envelops and liberating the immature ovum into the coelomic cavity near the infundibulum or ostium of the oviduct. The break in the enveloping layers of the oocyte occurs due to contraction of muscles surrounding it and that also occurs at a definite elongated spot containing practically no blood vessels. In hen, the ovulation is, therefore, seldom accompanied by the rupture of a capillary vessel and the shedding of a droplet of blood.

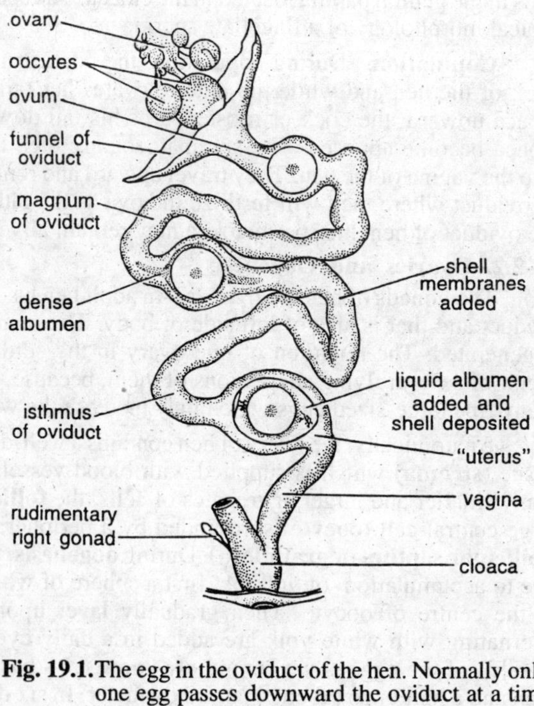

Fig. 19.1. The egg in the oviduct of the hen. Normally only one egg passes downward the oviduct at a time (after Gilchrist, 1968).

Immediately prior to this, the germinal vesicle (female nucleus) disappears and first maturation division gives rise to a small-sized first **polocyte** and a large-sized **secondary oocyte**. The **second maturation division** occurs after ovulation and when an immature ovum or secondary oocyte enters in the oviduct. The entry of the oocyte in the oviduct occurs by active movement of ostium which virtually grasps and swallows the egg (secondary oocyte).

During second maturation division which occurs in the anterior end of oviduct, the secondary oocyte is surrounded by many sperms and several sperms enter in it (**polyspermy**). Simultaneously second maturation division is completed and second polocyte is extruded from the contents of the **ovum**. Though polyspermy occurs as a rule in chick and other birds, but the nucleus of only one spermatozoon fuses with the female pronucleus and, thus, **amphimixis** occurs. The nuclei of other sperms become degenerated subsequently.

As a fertilized egg passes downward inside the oviduct two events occur simultaneously, firstly, the fertilized egg undergoes the process of cleavage, blastulation and even gastrulation ; and secondly various accessory coverings are added to the developing egg by the oviducal wall.

2. Addition of accessory layers around the fertilized ovum. As the fertilized egg (developing egg or so called 'yolk') passes down the first or convoluted and glandular part of the oviduct (**magnum**), a thick layer of dense albumen is deposited around it. This stringy albuminous layer adheres to the vitelline membrane and extends as cords before and behind it, along the longitudinal axis of oviduct. As the egg continues its descent, the spiral arrangement of the folds in the oviduct wall of magnum causes the fertilized egg to rotate, twisting the adherent albumen in spiral strands, called **chalazae**, projecting at either ends of the yolk. But according to **Gilchrist** (1968), such rotations of egg occur in the posterior portion of oviduct, the **uterus**, due to the contraction of muscles of uterine walls. He further states that during rotation of egg only the albumen envelope revolves around the ovum which remains freely suspended in the former and incapable of rotation due to the fact that the cytoplasm of its blastodisc is lighter in weight than rest of the yolk.

The rotation of egg envelope, thus ,determines the orientation of the embryo. A rule dating back to **von Baer** (1828) states that if the egg is held with its pointed end to the right, the future antero-posterior axis of the embryo will be at right angles to the long axis of the egg, and the head will point away from the observer. Accordingly, the sharp end of egg of hen is on the embryo's right, the blunt end is on its left. The chalazae become coiled clockwise on the future embryo's left, counter clockwise on its right. The chalazae also remain instrumental in keeping the ovum well placed in the centre of the albuminous mass after it completes its formation.

Additional albumen is deposited in concentric layers about the ovum moving along the length of the oviduct. The **shell membranes,** which are the sheets of matted organic fibres, are added to the egg as it progresses through the **isthumus** part of oviduct. The porous calcareous shell is added while the egg passes through the shell gland portion of the oviduct (the **uterus**). The calcareous shell gives the typical "ovoid" contour to the egg, one end of it being slightly blunt than the other. Within the egg, there forms a air-space in between two shell membranes at the blunt end of the egg.

The oviduct leads ultimately to the cloaca, a common chamber into which the excretory ducts also open. The developing egg remains in the body approximately for 22 hours before it is laid. The development of eggs of hen occurs at the body temperature, *i.e.,* about 38°C.

19-3. STRUCTURE AND CHEMISTRY OF THE FERTILIZED EGG

The eggs of hen are **polylecithal** (macrolecithal) and **telolecithal.** The polarity of an egg is well-marked, the animal pole having very small amount of active cytoplasm in the form of disc and a zygote nucleus and is called **germinal disc** or **blastodic.** Most of the space of animal pole and vegetal pole of the egg is occupied by the yolk. Most of the yolk is liquid, but about 23 per cent of it, is in the form of solid ''yolk-spheres''. The yolk as a whole contains 48.7% water, 32.6% phospholipids and fats, 16.6% proteins, 1% carbohydrates and 1.1% other chemical molecules. The proteins of chick yolk are chemically closely related to those of the amphibian's yolk, *i.e.*, they occur in the form of **phosvitin** (phosphoprotein) and **lipovitelline** or **livetin** (lipid protein). The fatty portion of yolk is predominantly neutral fat (50%), the rest being the **phosphatids, cerebrosides** and **cholesterol**. The lipids are the greater part of the solid constituents of the yolk and serve as the major source of food for the embryo. The major carbohydrates of the yolk are free **glucose** and **polysaccharides**, combined with the phospholipids, phosphoproteins and cerebrosides.

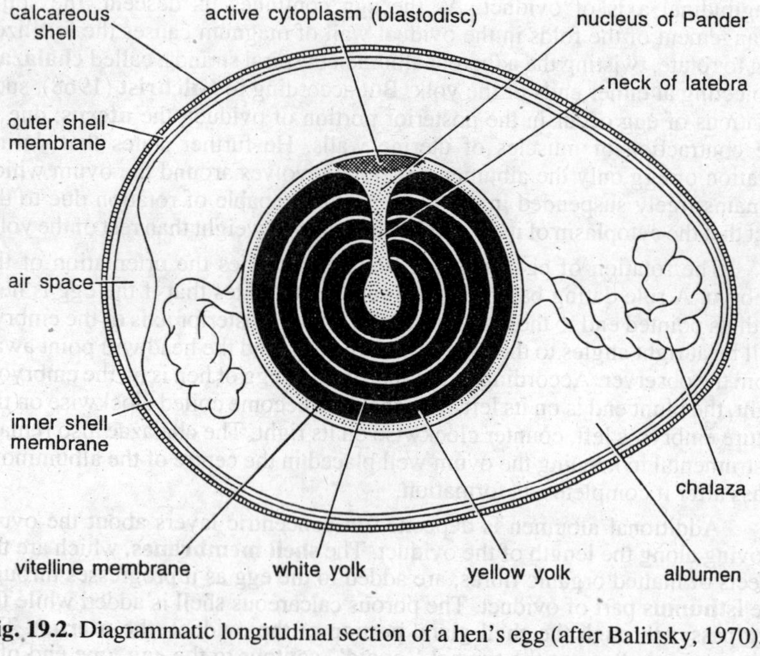

Fig. 19.2. Diagrammatic longitudinal section of a hen's egg (after Balinsky, 1970).

The yolk and blastodisc are ultimately bounded by a plasma membrane. Besides the plasma membrane, the egg is bounded by no less than five **accessory membranes**. The plasma membrane is followed by a **vitelline membrane** which is a converted form of **zona radiata**. According to **Balinsky** (1970) the vitelline membrane of hen's egg is of double origin. The inner layer of vitelline membrane is produced in the ovary, in the space between the oocyte and the follicle cells. This layer is composed of very rough fibres. An outer more finely fibrous layer is then formed on top of the first when the egg enters the upper portion of the Fallopian tube (**Bellairs**, 1963).

The next egg membrane is the white or **albumen** of the egg. This layer is differentiated into two distinct layers — a dense inner layer and a thin outer layer. The 85% of the egg-white is water ; the rest (15%) is a mixture of several proteins, mostly **albumins** which make up 94% of the dry weight. A denser part of the egg-white forms the spiral strands, the chalazae. The albumen plays a role in nutrition of the embryo, it serves as a water store and it also provides envelope protecting the embryos from mechanical and chemical injuries. Albumen also has bactericidal qualities (**Phillips**, 1975). Further, the yolk and albumen of egg contain a variety of **enzymes, vitamins, pigments** and **phosphorus**.

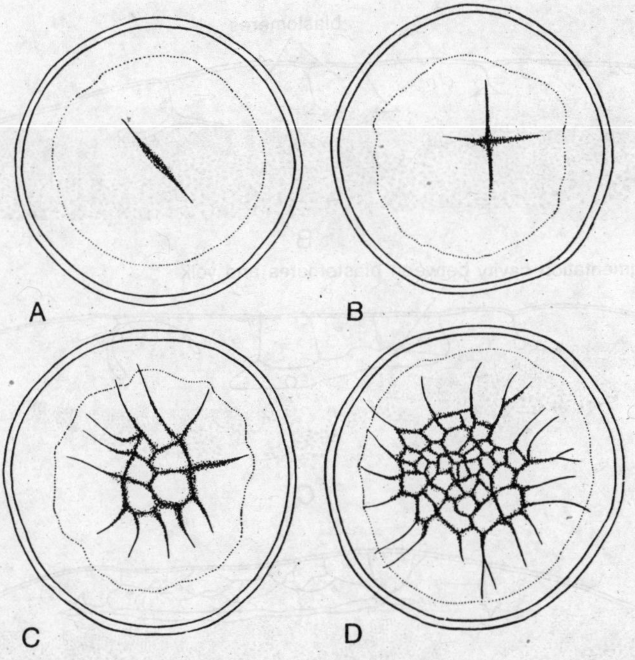

Fig. 19.3. Surface view of germinal disc of hen's egg showing early cleavage. A—Two-cell stage; B—Four-cell stage; C—Twenty-cell stage; D—A late cleavage stage (after Balinsky, 1970).

Next to the egg-white come two layers of **shell membranes** which contain fibres of **keratin** matted together. Both shell membranes enclose a large **air space** towards the blunt end of the egg.

The outermost layer is **calcareous shell** which is pierced by a large number of fine **pores** which are filled by an organic (protein) substance related to collagen. In an average hen's egg the pores have a diameter of 0.04 to 0.05 mm and their total number is estimated at about 7000.

19-4. CLEAVAGE AND BLASTULATION

In chick, the **discoidal meroblastic cleavage** remains restricted to the germinal disc, the yolk remains in uncleaved state and is encompassed eventually by the growing tissues of the embryo. The **first cleavage furrow**

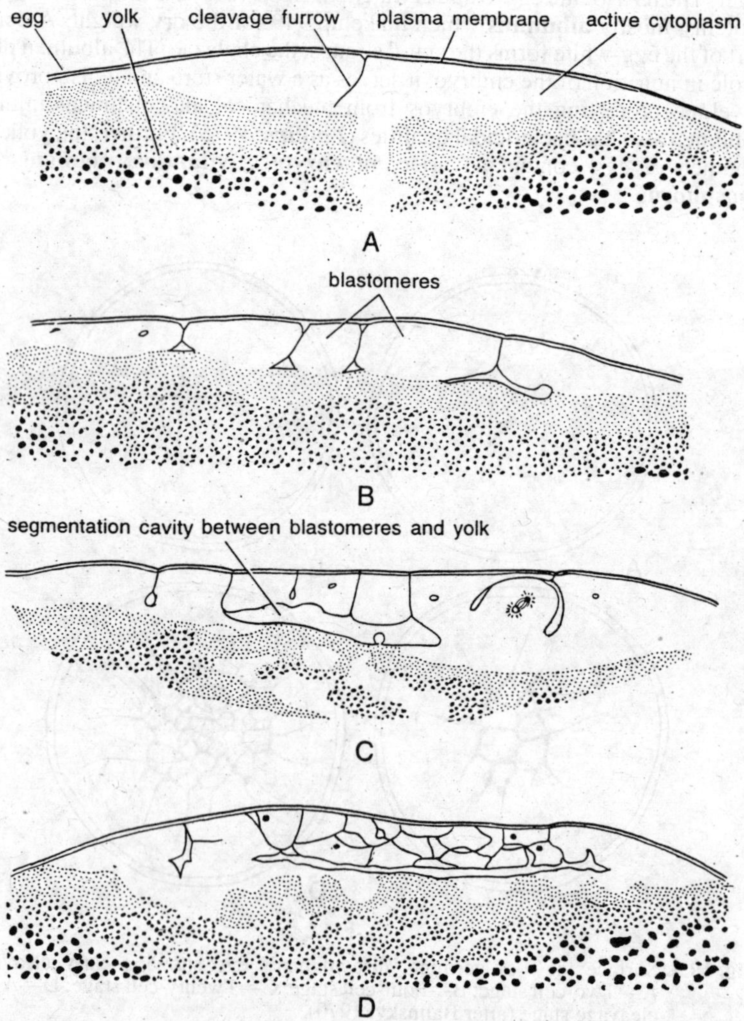

Fig. 19.4. T.S. of germinal disc of hen's egg showing early cleavage (after Balinsky, 1970).

consists of slight **meridional** incision near the centre of the germinal disc. This furrow passes yolkward but does not reach the lower portion of the disc where the cytoplasm is filled with coarse yolk granules. The **second cleavage plane** is again **meridional** but at the right angles to the first. The **third set** of cleavage furrows is **vertical**, cutting across the second set of meridional furrows, and consequently tends to parallel the first cleavage furrow. The **fourth set** of cleavage furrows is also **vertical** and although not synchronous, but it proceeds gradually to form eight **central blastomeres** which are surrounded by twelve **marginal blastomeres**. Thus, the meridional and vertical cleavage separate the daughter blastomeres from each other, but not from the yolk, so that the central blastomeres are continuous with the underlying yolk at their lower ends, and the blastomeres lying on the circumference are, in addition, continuous with the uncleaved cytoplasm at their outer edges. As the nuclei at the edge divide, more and more cells become cut off to join the ones lying in the centre, but the new blastomeres remain in continuity with the uncleaved yolk underneath.

In a latter stage of cleavage, the blastomeres of the central area become separated from the underlying yolk in one of the two ways—either slits may appear beneath the nucleated parts of the cells or else the cell division may occur with **horizontal (tangential)** planes of fission. Thus, in later stages of cleavage the central cells of the blastodisc become cut off on all sides but for some time, at the edge (the **periblast**) there is syncytial region where cells still remain in open communication. Further cleavage produces a **blastoderm** — a round disc, 5-6 cells deep in the centre (the **epiblast**) but only 1 to 2 cells

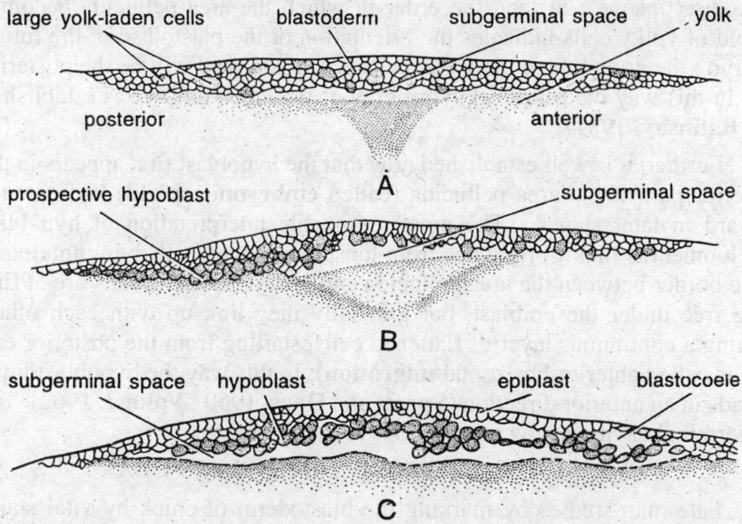

Fig. 19.5. Longitudinal sections through blastoderms of successive stages showing origin of hypoblast and blastocoele (after Torrey, 1971).

deep at the periphery. The blastoderm contains **formative** or **embryonic cells** which later on form the embryo proper. Very soon, a **subgerminal space** or **cavity** (also called **segmentation cavity**) is formed only under the central blastomeres or epiblast and is thought to be originated due to accumulation of fluid from the albumen through the vitelline membrane (see **Goel**, 1984). This cavity separates the yolk from the overlying blastomeres of epiblast.

During later part of cleavage two very important events occur simultaneously — differentiation of blastoderm into two cellular sheets namely epiblast and hypoblast and superficial demarcation of two areas, namely area pellucida and area opaca in the chick blastoderm. The process of differentiation of the blastoderm into the area pellucida and area opaca has been best investigated in the embryos of hen by **Eyal-Giladi** and **Kochav** (1976). Such a transformation occurs while the egg is still in the uterus of the laying hen. About 12–14 hours after the egg reaches the uterus (*i.e.*, 6 to 8 hours before the egg is laid), some cells on the inner (under) side of the blastoderm become detached from the rest of the blastoderm and fall on the floor of the subgerminal cavity (This process is called **delamination**). The cells to do so are the ones which contain the greater amounts of yolk ; the cells remaining in the intact blastoderm are the ones with less yolk. This shedding of the yolky cells from the blastoderm begins at what will be posterior edge of the area pellucida and gradually spreads forward, until the whole of the area pellucida becomes free of yolky cells. In consequence, the epithelial layer (*i.e.*, epiblast) in the region of the area pellucida becomes thinner (having fewer layers of yolk-poor cells) and transparent. And since, in the area opaca the yolky cells are not shed, they exist in the epithelium and render this part of the blastodisc opaque. In fact, the order in which the area pellucida becomes devoid of yolky cells indicates the orientation of the blastodisc of the future embryo : the point where the yolky cells start to be shed will be the posterior end. In this way the **bilateral symmetry** of the chick embryo is established (see **Balinsky**, 1981).

Further, it is well established now that the hypoblast first appears in the posterior part of the area pellucida (called **embryonic shield**) and extends forward in later stages. The most reasonable interpretation of hypoblast development is that it originates from loosely arranged cells lying internally at the border between the area pellucida and area opaca. The cells are at first lying free under the epiblast, but gradually they link up with each other, forming a continuous layer of flattened cells, starting from the posterior end (this is called anterior horizontal **migration**). In this way the hypoblast layer spreads in an anterior direction (**Spratt** and **Haas**, 1960 ; **Vakaet**, 1962). The subgerminal cavity is now homologous to the blastocoele.

19-5. FATE MAPS OF DISCOBLASTULA

Fate map studies by marking the blastoderm of chick by vital stains (**Wetzel**, 1929 ; **Pasteels**, 1940), fine carmine or carbon (charcoal) particles (**Spratt**, 1946) or radioactive thymidine have revealed that blastomeres of area opaca (also called **area vitellina**) have nothing to do with the formation of any part of the embryo proper. Instead of it, they destine to form **non**-or **extra-embryonic** ectoderm which during later developmental stages forms extra-embryonic membranes and blood vessels **(area vasculosa)** to involve in the digestion of the yolk.

On the other hand, the epiblast and hypoblast of area pellucida have different fates in the course of embryonic development. The epiblast has mixed blastomeres for **ectoderm, mesoderm** and also for the **endoderm**, while the hypoblast is composed of exclusively **endoderm** forming blastomeres.

Recently, by the use of tritiated thymidine marking, fate maps of chick/bird embryos have been obtained which are far superior in detail and reliability to any which have been based on vital staining or carbon particle marking. The fate map in Figure 19.6 is based on this work. This map refers to the epiblast only. In the fate map of epiblast of chick, roughly in the centre of area pellucida lies a small area which will produce the **notochord**. Posterior to this, in the median plane exists an elongated oval area which is presumptive **endoderm**. This endoderm will form the gut with its subordinate parts. Further toward the posterior edge of the area pellucida lies the **extra-embryonic endoderm** which will form part of the lining of the yolk sac. To the right and left of the presumptive notochord and endoderm and also posterior to the presumptive yolk sac lie the various subdivisions of presumptive mesoderm. Thus, nearer to the notochordal area is situated a ill-defined section producing the **prechordal** or **head mesoderm**. Posterior and lateral to that section are the presumptive **somite** areas, and posterior to these is the presumptive **lateral plate mesoderm**. The right and left lateral mesoderm areas are in continuity with each other posterior to the endodermal area. Toward the posterior edge of the area pellucida a fairly large crescentic area is situated which will give rise to **extra-embryonic mesoderm**, particularly to the extra-embryonic blood vessel system (*i.e.*, area vasculosa). The presumptive **ectoderm** occupies the anterior and lateral parts of the area pellucida. Thus, a roughly semicircular area anterior to the notochordal material is the presumptive **neural plate** area. In a band surrounding this is the presumptive embryonic **epidermis**. Still further outward, forming almost a complete ring at the outer edge of the area pellucida and flanking the area opaca lies the **extra-embryonic ectoderm**.

19-6. GASTRULATION

The process of gastrulation in chick is in a highly modified form than that of *Amphioxus* and frog. This is mainly due to the presence of a massive amount of yolk in the chick's egg. As already described in the Chapter 13, the chick's gastrulation is very complicated and prolonged one, proceeding in several steps. However, the pace of avian gastrulation is so fast that it is well completed by the second day of incubation.

In fact, there still exists a great confusion regarding the purview of the process of gastrulation in chick. For example, **Berrill** and **Karp** (1976) believed quite logically that it includes two main events —the primitive streak formation and the epibolic growth of cells around the yolk mass. However, **Ede** (1978) has regarded the beginning of gastrulation in the division of blastoderm into area pellucida and area opaca. Some authors such as **Grant** (1978) and **Trinkaus** (1984) have regarded the initiation of the process of gastrulation from the stage of formation of hypoblast. **Balinsky** (1981) has taken a safe middle course by branding the formation of hypoblast as merely a pregastrular morphogenetic movement. **Phillips** (1975) and **Goel** (1984) have quite correctly emphasized that the differentiation of blastoderm into epiblast and hypoblast layers both enclosing a **blastocoele** is an event of late blastulation stage. Anyhow, it would be quite safe to conclude that gastrulation of chick includes the following two types of morphogenetic movements :

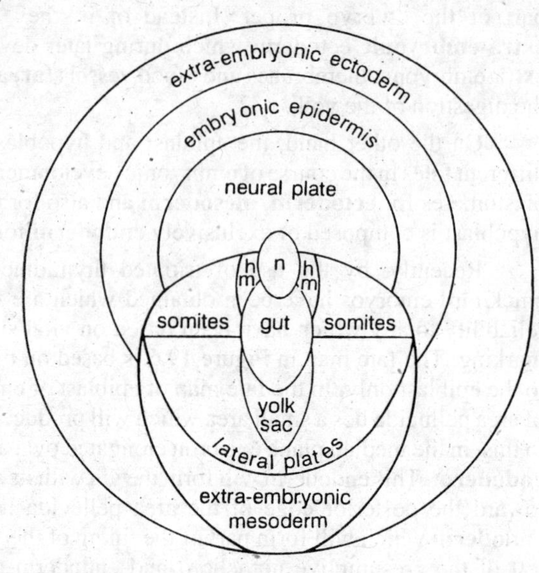

Fig. 19.6. Fate map of the epiblast of a bird which has been prepared by radioactive thymidine marking; hm, presumptive head mesoderm; n,

1. Emboly which involves only the epiblast and includes convergence, invagination and involution (ingression) through bottling process by the mesentoblast cells (*i.e.*, progenitors of both endoderm and mesoderm). Thus, emboly is exhibited by the formation, elongation and ultimate recession of the primitive streak and also the formation of the head process. It results in the formation of embryonic endoderm, mesoderm and notochord.

2. Epiboly which includes the overgrowth of ectoderm (*i.e.*, residual epiblast) and also of hypoblast (or extraembryonic endoderm) over the yolk mass.

19-6.1. Formation of Primitive Streak

Various prospective mesodermal and endodermal cells of the epiblast converge toward the posterior border of the pellucid area and form a conical thickening in the midline as cells pile up in a **morphogenetic centre** called the **initial primitive streak**. This region initiates the definitive embolic movements (such as invagination and involution) of the gastrulation. The primitive streak appears after 6 to 7 hours of incubation. The formation of primitive streak is a progressive affair. The cells composing it are constantly changed. The primitive streak grows because of proliferation of cells as well as of the addition of cells that migrate to it from various parts of area pellucida due to convergence. The elongated axis of the primitive streak foreshadows the antero-posterior axis of the future embryo.

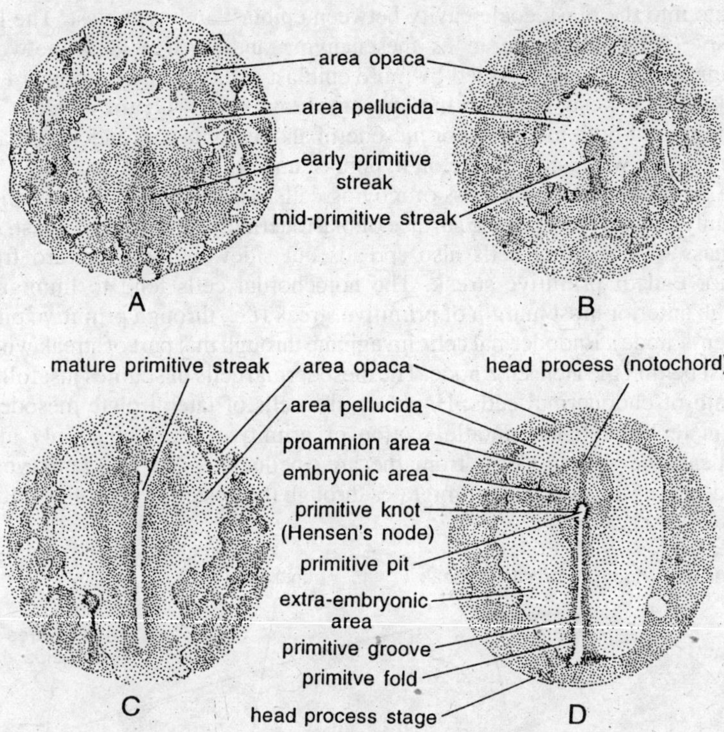

Fig. 19.7. Surface view of chick blastoderm showing development of primitive streak (gastrulation) and head process (start of neurulation). A—Initial streak (stage 2); B—Intermediate streak (stage 3); C—Definitive streak (stage 4); D—Head process stage (stage 5, embryo of 19 to 22 hours of incubation (after Patten, 1958).

After 12 to 13 hours of incubation, the primitive streak has attained an intermediate stage, extending from the posterior margin to the centre of pellucid area. At this time a longitudinal groove called **primitive groove** flanked on both sides by the **primitive folds**, appears in the middle of the streak. Meanwhile, the most anterior portion of the groove thickens into a depressed knot of cells, called **primitive knot** or **Hensen's node**. In this region, the primitive groove terminates abruptly in a funnel-shaped **primitive pit**. The streak now sharply delineated in the midline of the pear-shaped area-pellucida, extending to three-fourth the length with primitive groove and primitive pit. This stage of embryo is called **definitive primitive streak** and it is usually completed after 18 to 19 hours of incubation.

The primitive groove appears as the epiblast cells become flask-shaped and invaginate into the blastocoele. Their basal ends are highly motile, whereas their apical ends remain fixed in the surface by the junctional contacts. In this way, they pull neighbouring cells into groove, migrating as a cohesive sheet. Ingressing cells detach from the thickened inner layers in the groove and migrate into the blastocoele cavity between epiblast and hypoblast. The less adhesive cells immigrate in as mesenchyme and form a loose network. Immigrating cells are replaced by more epiblast cells converging toward the streak area. These immigrating **mesentoblast cells** (**Grant**, 1978), the progenitors of both mesoblast or mesoderm and hypoblast or endoderm, also make contact with and migrate on hypoblast and epiblast surfaces. Thus, the entire primitive streak is a mass of moving cells. The direction of movement is mainly downward from the surface of blastoderm toward the hypoblast, but the mass of migrating cells also spreads out sideways and forward from anterior end of primitive streak. The notochordal cells tend to immigrate through anterior most margin of primitive streak (*i.e.*, through primitive pit of Hensen's node). Endodermal cells invaginate through that part of streak which lies just behind the Hensen's node. The mesodermal cells of somites just follow the path of endodermal cells. However, the cells of lateral plate mesoderm invaginate through the middle section of primitive streak, but only after disappearance of endoderm from the area pellucida. The extraembryonic mesoderm (of the yolk sac) immigrates through the posterior part of primitive streak.

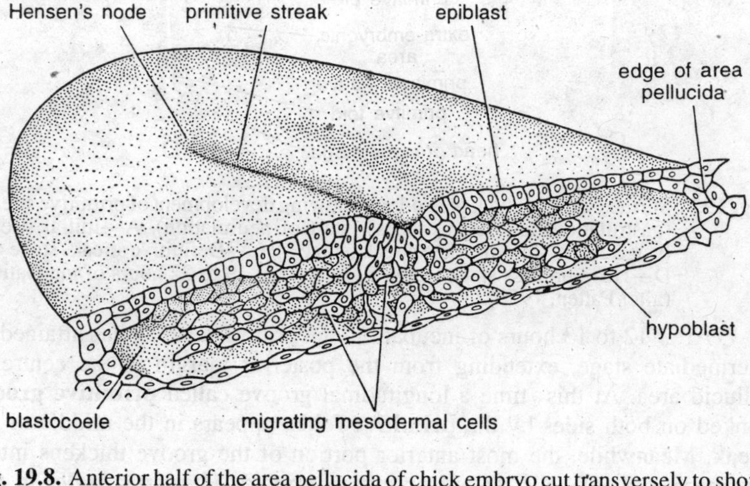

Fig. 19.8. Anterior half of the area pellucida of chick embryo cut transversely to show the migration of mesodermal and endodermal cells from the primitive streak (after Balinsky, 1970).

Meanwhile, some hypoblast cells expand into the area opaca to become extraembryonic endoderm (the lining of yolk sac), while other hypoblast cells attach to the mesentoblast cells and are carried along by the latter's migration. The combined effect is a rolling in of epiblast cells (*i.e.*, involu'ion) over the lips of primitive groove. The expansion pressures in the epiblast force the groove to become narrow and extended anteriorly.

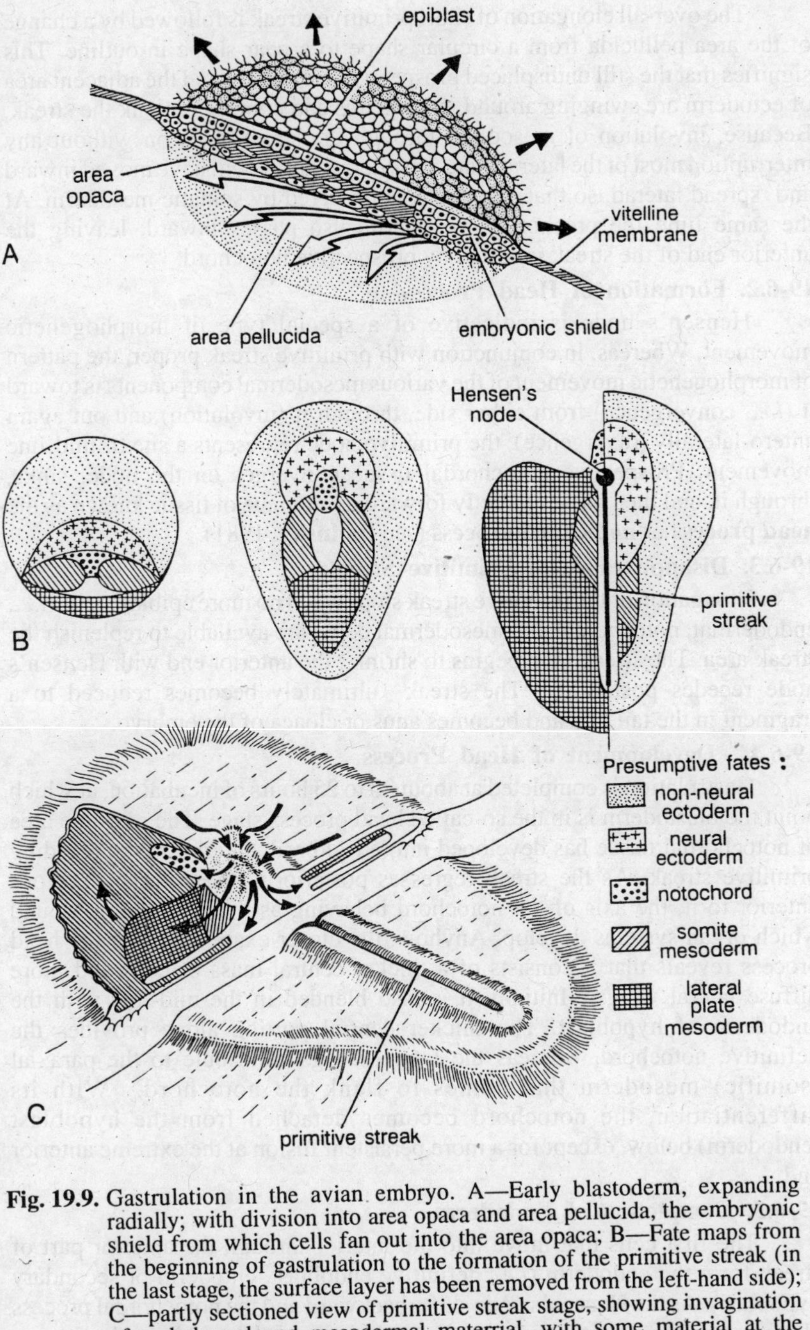

Fig. 19.9. Gastrulation in the avian embryo. A—Early blastoderm, expanding radially; with division into area opaca and area pellucida, the embryonic shield from which cells fan out into the area opaca; B—Fate maps from the beginning of gastrulation to the formation of the primitive streak (in the last stage, the surface layer has been removed from the left-hand side); C—partly sectioned view of primitive streak stage, showing invagination of endodermal and mesodermal materrial, with some material at the anterior end of the streak (Hensen's node) moving anteriorly to form the notochord (after Ede, 1978).

The over-all elongation of short primitive streak is followed by a change of the area pellucida from a circular shape to a pear-shape in outline. This signifies that the still undisplaced prospective mesoderm and the adjacent area of ectoderm are swinging around in long arcs and coming to flank the streak. Because, involution of mesoderm through the streak goes on without any interruption most of the lateral plate mesoderm shortly will have moved inward and spread laterad, so that, the streak is bordered by somatic mesoderm. At the same time the prechordal mesoderm also passes inward, leaving the anterior end of the streak invested by prospective notochord.

19-6.2. Formation of Head Process

Hensen's node is indicative of a special type of morphogenetic movement. Whereas, in conjunction with primitive streak proper, the pattern of morphogenetic movement of the various mesodermal components is toward it (*i.e.*, convergence) from either side, through it (involution) and out again antero-laterally (divergence), the primitive node represents a site of mid-line movement. Prospective notochordal tissue converges on the node, sinks, through it, and then passes directly forward as a tongue of tissue known as the **head process** or **notochord process** (see **Balinsky**, 1981).

19-6.3. Disappearance of Primitive Streak

Elongation of the primitive streak stops when no more epiblast cells (*i.e.*, endodermal, notochordal and mesodermal cells) are available to replenish the streak area. The streak then begins to shrink ; the anterior end with Hensen's node recedes posteriorly. The streak ultimately becomes reduced to a fragment in the tail bud and becomes anus or cloaca of the embryo.

19-6.4. Development of Head Process

Gastrulation is completed at about 20 to 25 hours of incubation, at which point the blastoderm is in the so-called head process stage. The mid-line area of notochordal tissue has developed into a rigid rod, anterior to the receding primitive streak. As the streak regresses posteriorly, the embryo develops anterior to it, the axis of the notochord behaving as the scaffolding around which organ systems develop. Anyhow, the closer examination of the head process reveals that it consists of a thicker central mass of cells and more diffuse lateral wings. Initially it is also blended in the mid-line with the endoderm of hypoblast. The thicker central portion alone provides the definitive notochord, whereas the lateral wings contribute to the paraxial (somitic) mesoderm that comes to flank the notochord. With its differentiation, the notochord becomes detached from the hypoblast (endoderm) below, except for a more persistent fusion at the extreme anterior end.

19-6.5. Completion of Endoderm

The first cells that move into the interior through the anterior part of streak form the endoderm (*i.e.*, definitive endoblast, entoderm or secondary hypoblast). As the Hensen's node shifts backward and the notochordal process elongates in its wake, the presumptive endoderm of the middle and posterior part of the gut, located just behind the node move inside as an endodermal strip underneath the notochord. Thus, endoderm has dual origin. The original hypoblast at the floor of the blastocoelic cavity makes a microcontribution to the gut ; most endodermal tissue of gut is derived from cells that have immigrated through the primitive streak. In chick no **archenteron** is formed during the process of gastrulation.

In fact, in chick, the endoderm formation rapidly outstrip that of the mesoderm, so that an endodermal layer is complete when the mesoderm is still only partly invaginated. This premature formation of the endoderm determines the orientation of the primitive streak (**Berrill** and **Karp**, 1976).

19-6.6. Structure of the Fully Formed Gastrula

The stage of disappearance of the definitive primitive streak may be said to mark the end of gastrulation and beginning of the neurulation. The fully formed chick gastrula consists of three germ layers—**ectoderm, chorda-mesoderm** and **endoderm**. The ectoderm and chorda-mesoderm remain in continuity along the axis of the primitive streak, just as they are at the lateral lips of an amphibian blastopore. The endoderm is also united with the mesoderm and ectoderm at the anterior end of the streak and at its posterior end.

19-6.7. Mechanics of Different Avian Morphogenetic Movements

1. Mode of epiboly. Epiboly of area opaca and yolk sac of an avian egg is certainly the most extensive of all spreading movements that take place during animal morphogenesis. For in it, the originally tiny blastoderm expands and spreads entirely over the truly immense yolk of these large cleidoic eggs. Imagine the extent of spreading required to encompass the yolk of an ostrich egg. The task is far too enormous to be accomplished during the brief period of early development that ends with the completion of gastrulation ; in fact, it (*i.e.,* epiboly) continues long afterward, even during advanced stages of embryogenesis and organogenesis. For example, in the chick embryo, four days of development are required for the blastoderm to cover a yolk about 3cm in diameter, involving an increase in area over 200 fold.

In recent years, minute details of mechanism of epiboly of chick have been worked out by various workers. **New** (1959) has confirmed the fact that the upper surface of the marginal region of chick blastoderm adheres to the under surface of the vitelline membrane and spreads on it during epiboly. The blastoderm normally spreads as a result of the activity of its locomotory margin (*i.e.,* **edge cells** or **marginal cells**) over a passive inanimate substratum, the inner surface of the vitelline membrane. The attached margin of a spreading epiblast is about 3-cell wide and these edge cells are separated from their substratum, the vitelline membrane, by 6–7 nm, as suggested by transmission electron microscopy (TEM). In addition, both **Downie** and **Pegrum** (1971) and **Bellairs** (1982) have noted the projections of the plasma membrane into the vitelline membrane, which they called **attachment plaques**. These plaques may provide the edge of epiblast the grip it requires for locomotion ; also they may prevent detachment and retraction of the sheet as it is placed under more and more tension during epiboly. It is well established now that non-marginal cells of epiblast are put under much tension by the spreading margin (*i.e.*, by edge cells) and they are passively pulled by the margin. In fact, ectoderm cells, which form a cohesive sheet throughout epiboly, are joined to each other only by focal tight and close junctions after an 18-hour period of incubation, when the ectoderm is not greatly stretched. However, by three days, when the stretching is more intensed these junctions are replaced by zonal tight junctions, terminal bars or belt desmosomes and spot desmosomes. **Downie** (1976) has reported that all cells except the edge cells of the epiblast of the spreading area opaca and later, the yolk sac, proliferate mitotically at the high rate. This results in a great increase in the volume of the epiblast. Since no mitosis occurs in the edge cells, so the same edge-cells present at the beginning are responsible for all of epiboly (see **Trinkaus**, 1984).

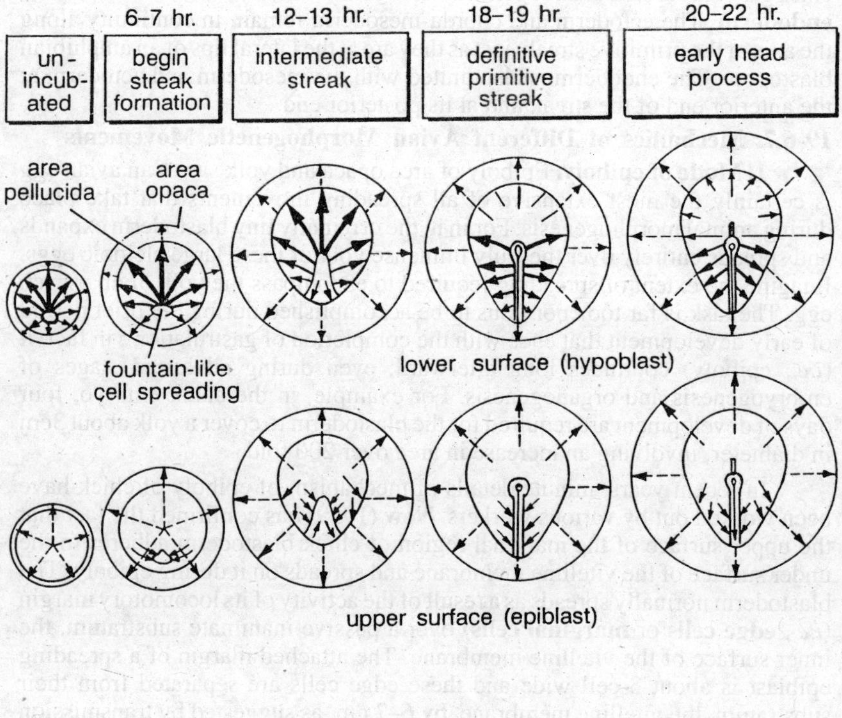

Fig. 19.10. Dye-marking experiments of **Spratt** and **Hass** (1960) to illustrate morphogenetic movements of lower surface of blastoderm (hypoblast) in relation to upper surface (epiblast) before and during primitive streak formation. The movements of these layers are shown in relation to general growth expansion (dotted arrows) of the entire blastoderm. The thickened arrows from the anterior portion of the primitive streak mark the ingressing cells of presumptive endoderm and mesoderm as well as the movement, of the hypoblast. These ingressing cells are derived from the epiblast as they converge toward the primitive streak and descend into the tissues below (after Grant, 1978).

2. Mode of spreading of hypoblast. The blastoderm of an unincubated hen's egg is composed of a cohesive epiblast underlain by a ring of loosely packed hypoblast cells. The hypoblast is the precursor of part of the endoderm. **Spratt** and **Hass** (1960), using carbon marking, have shown that during the first few hours of incubation the hypoblast moves forward and radially on the undersurface of the epiblast from a thickened posterior region. But the epiblast does not orient or give direction to hypoblast movement. **Spratt** and **Hass** believed that intensive proliferation in the thickened posterior part of hypoblast provides cell direction and cell migration. Cells are thought to start movement passively as a result of a push due to increased population density of the rapidly dividing growth centre and to continue movement anteriorward because of a continuation of this *vis a tergo* (= compulsion from behind). There is still no evidence to prove that the posterior part of chick hypoblast is in fact a proliferation centre. One worker, **Vakaet** (1970) found no evidence for such proliferation as he proposed that the hypoblast forms primarily by "polyinvagination" or ingression, with cells slipping from the upper to the lower layer at many points. According to **Trinkaus** (1984), the accumulation of migrating hypoblast cells in the posterior region may merely provide a source of cells for the ensuing cell movements. This movement may start as a result of changes in cell surface activity, such as motility and adhesiveness.

The cessation of movement of hypoblast cells takes place according to hypothesis put forward by **Abercrombie** and **Gitlin**. They have shown that if a given cell moves off so that it becomes isolated from the outgrowth, its movement soon ceases, probably because contact inhibition of movement serves to stimulate movement away from the contact. But as soon as the contact is lost the stimulus is gone. Isolation of hypoblast cells from the main mass of the hypoblast in the posterior half of the blastoderm would deprive them of this contact and perhaps, therefore, of their main stimulus to move.

3. Mode of ingression and involution. It has been established for some time now by detailed studies with vital dyes and carbon that the primitive streak of the avian embryo forms as a result of convergence of epiblast cells to the midline of the blastoderm. There they sink in (*i.e.*, undergo ingression) and migrate out laterally between the epiblast and the hypoblast to form the mesoblast. **Rosenquist** (1966) has reinvestigated this problem by labelling epiblast cells with tritiated thymidine and grafting them to unlabelled blastoderms. These experiments have confirmed the previous studies and have also shown that some endoderm also originates from the epiblast at the primitive streak. At this juncture, the epiblast is a simple pseudostratified columnar epithelium, whereas the hypoblast is a simple squamous epithelium. The mesoderm (or mesoblast) is a mesenchyme in the sense that the cells are more loosely organized. Mesoblast cells are derived from the basal surface of the epiblast in the region of the primitive streak.

Balinsky and **Walther** (1961) opined that mesoblast cells of the epiblast are invaginated by becoming bottle or flask-shaped. Such a change in cell shape is brought about by contraction of an electron-dense layer found in the cytoplasm of the neck region of flask-shaped cells. This would force the bulk of the cytoplasm downward, causing at the same time adoption of the bottle shape and ingression of the cells into the interior of the blastoderm. This proposal could also explain the small elongate invagination of the epiblast known as the primitive groove.

Microtubules are present in the cells of the primitive streak, as well as in other cells of the epiblast, and are arranged parallel to the apico-proximal axis of the cell (**Granholm** and **Baker**, 1970). **Granholm** has shown that the flask cells, down in the primitive groove itself, are packed with longitudinally oriented microtubules, particularly in their elongate, narrow apical portions, and has suggested that they are important in the maintenance of the elongate shape of these cells.

How do prospective mesoblast cells break out of the highly cohesive sheet of the epiblast and slip inside ? All streak cells whose apical surfaces are still at the surface of the streak possess an apical tight – junctional – complex (= a typical zonula occludens plus a more extended proximal region of close apposition and gap junctions as well). Predictably, the apical surface of these tightly joined cells become highly convoluted as they constrict. In order for lateral cells of the highly cohesive epiblast sheet to reach the streak, to undergo the ingression that results in their involution and in the formation of the outwardly migrating mesoblast, they must undergo a truly striking movement of convergence. The mode of movement of convergence is still not understood (see **Trinkaus**, 1984).

In fact, **Trelstad**, **Hay** and **Revel** (1967) showed that the stellate mesoblast cells that leave the basal layer of the epiblast at the primitive streak are distinctly polarized and at first are flask-shaped. The leading end of these cells develop **filopodia** (filamentous pseudopodia). These filopodia do not contain microtubules but contain filaments about 5 nm in diameter. Thus, all mesoblast cells are connected to each other and to cells of the primitive streak by focal tight and close junctions, especially along their trailing surfaces. These observations clearly dismiss the notion that mesoblast cells break away from the primitive streak and migrate as individuals. They form a loosely connected network.

Comparison of Blastopore and Primitive Streak

Many embryologists consider the primitive streak of chick and the blastopore of frog to be **homologous** structures with comparable functions. For example, the movement of cells toward the blastopore in the frog resembles the migration of chick cells toward the primitive streak. The cells of the frog involute through the lips of the blastopore ; the cells of the chick move into the interior through the primitive groove. In both cases, surface cells undergo convergent streaming, disappearance at the streak or blastopore and then take up a position inside the embryo between the epiblast and hypoblast, establishing a third germ layer.

The large amount of yolk in the chick egg is responsible for the main difference between the two structures, that is, the blastopore and the primitive streak. The avian blastopore (*i.e.*, primitive streak) is elongated, while amphibian blastopore is circular. The streak is never open as in frog but is comparable to fused lips of the blastopore. Unlike the frog blastopore, it never communicates with the archenteron.

In fact, the dorsal lip of the amphibian gastrula and Hensen's node appear to correspond to one another, since, it is in these respective areas that the notochordal material makes its way inside the embryo. The primitive pit represents the dorsal opening of the blastopore (neurentric canal). The primitive groove and folds of streak are comparable to the apposed lateral lips of the closed blastopore. The presumptive mesoderm of frog involutes over the lateral lips of the blastopore ; the presumptive mesoderm of chick converges to the primitive streak and involutes from there. The posterior end of primitive streak too may be compared with the ventral region of the blastopore, where later on the anus or cloaca forms.

Even the regression of the primitive streak can be compared with the closure of the amphibian blastopore. When the amphibian blastopore closes, the less active regions of morphogenetic activity, such as lateral and ventral lips, close first, whereas the dorsal lip, the most active site of cellular activity, close last. A comparative situation occurs in chick as the primitive streak regresses. Regression occurs faster in the posterior portion of the streak. Hensen's node remains functional until the end (**Phillips**, 1975).

Lastly, in the both cases (*i.e.*, frog and chick) the power of inducing neural tissue is greater in the anterior end of the neural inductor, *i.e.*, the dorsal lip of the blastopore in amphibians and the Hensen's node of the primitive streak of the chick. In fact, implantation of Hensen's node into a competent area pellucida (*i.e.*, ectoderm) induces a secondary embryo (see **Grant**, 1978).

19-7. NEURULATION AND TUBULATION

The neurulatiorn and tubulation processes of chick include the same events as in amphibians and during it following processes are set in — (i) The **ectoderm** differentiates into **neural plate** and **epidermis**. (ii) Neural plate is converted into **neural tube.** (iii) The **chorda-mesoderm** divides into **notochord** and **mesoderm.** The latter further differentiates into its different subkinds. (iv) The endoderm develops into primitive gut. The following processes can be studied in detail :

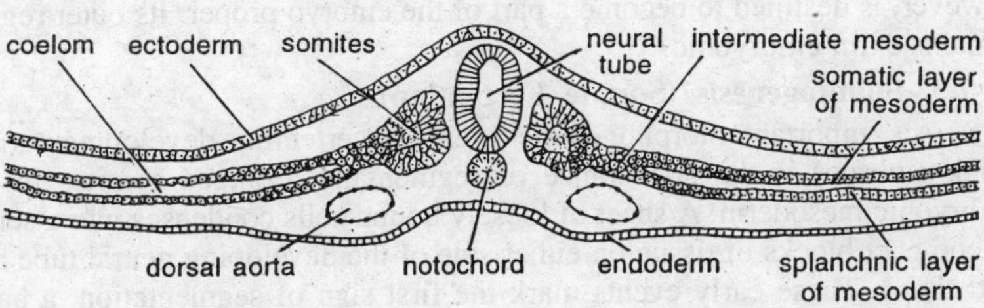

Fig. 19.11. Cross section of an early chick embryo, illustrating the relationships of the germ layers (after Saunders, Jr., 1968).

19-7.1. Neural Tube Formation (Neurulation)

The neural tube formation proceeds in an antero-posterior direction. The neural plate appears in the brain region (*i.e.*, the ectoderm anterior and lateral to the head process), while the gastrulation movements are still in full swing. As Hensen's node recedes farther and farther, parts of the neural plate become differentiated and the anterior parts of the neural plate proceed to close into a tube, the **neural tube**. The neural tube is formed in a very typical way, *i.e.*, the central axis of the neural plate sinks inward as the **neural groove,** while the margins of the neural plate, *i.e.*, **neural folds,** rise up and move toward the mid-line and fuse to form a neural tube.

It was established by **Karfunkel** in 1974 that the mechanism of neural tube formation through elongation of neural plate cells and apical contraction, brought about by microtubules and microfilaments, is the same in birds as in amphibian embryo. More details have been added regarding the fusion of the neural folds in a scanning transmission electron microscope (STEM) study in 1975 by **Bancroft** and **Bellairs**, who showed that as the folds approach each other in the dorsal mid-line, single threads are projected across from the surface of one cell to make contact with similar threads, or the surface, of a cell from the other side. Establishing this degree of communication probably plays some part in guiding the folds into final apposition with each other, and final adjustments are made by tongue-like processes which insinuate themselves into gaps between the cells (**Ede**, 1978).

19-7.2. Notogenesis and Mesogenesis

While the neural plate is folding into the neural tube, the chorda-mesoderm is also differentiating. Its most anterior part, the **prechordal mesoderm** gives rise to much of the mesenchyme of the head. Behind it lies the **notochord**. The notochordal cells become separated from rest of the adjoining sheets of mesoderm and differentiate into notochord.

On each side of the notochord are the three longitudinal strands of mesoderm. (1) The thicker medial strands are axial mesoderms or **epimere,** which merge anteriorly into the mesenchyme of the head. (2) Lateral to the axial mesoderm are strands of intermediate mesoderm or **mesomeres**. In an early stage they are thin and difficult to recognise. (3) Farthest to the side is the lateral mesoderm or **hypomeres.** Only a small part of the lateral mesoderm, however, is destined to become a part of the embryo proper. Its outer region remains extra-embryonic.

19-7.3. Somitogenesis (Somite Formation)

An important morphogenetic event in vertebrate development (*i.e.*, chick embryo) is the appearance of segmentally arranged somites in the embryonic mesoderm. A sheet of loosely bound cells condenses into a series of compact blocks of tissue on either side of the developing neural tube and notochord. These early events mark the first sign of segmentation, a basic feature of the vertebrate body plan. The process has been examined in the chick embryo by **Lipton** and **Jacobson** (1974).

As the Hensen's node regresses, cells detach from the nodal region, anteriorly and condese into the presumptive notochord, segregating the mesoblast into right and left halves. Meanwhile, ectodermal cells in the epiblast above migrate toward the midline and begin to condense into the presumptive neural plate. The migrating mesoblast cells, beneath the epiblast, attach to the basal lamina on the inner surface of condensing neural plate cells and condense into a presumptive somite by forming tight junctions. These somite cells elongate after attaching to the basal lamina, while the more flattened, unattached mesoblast cells below continue to migrate. More cells accumulate in the future somite area, possibly by virtue of their greater adhesive tendencies. Non-somite cells beneath the presumptive epidermis show few signs of condensation or packing, which suggests that the neural plate cells influence the behaviour of somite cells below.

As the neural tube begins to roll up, the adherent somite cells remain for a while but soon detach from the basal lamina and a space, probably filled with matrix materials, appears between the somite and tube. The formation of segmentally arranged clefts in the somite block begins dorsally as mesoderm cells detach from the undersurface of the neural plate at periodic intervals. Clefts also appear from the lateral side and cut inward toward the midline to carve out the tissue block. Presumably in cleft regions, cells become less adhesive with neighbours on one side and join the more compact mass of the somite. Later the surface of the somite is surrounded by a basal lamina, to give it additional integrity, and long fibres extend from the somite toward the notochord developing between a pair of somites, probably stabilizing somite patterns.

The first three pairs of somites are formed in the paraxial mesoderm beneath a broad region of the neural plate in the presumptive hind-brain region, and more pairs of somites are demarcated by intersomitic clefts as the node regresses, forming a series of cuboidal blocks, becoming triangular in cross section, as they conform to their situation adjacent to the notochord and the neural tube on their inner face and to the overlying ectoderm to the outside (**Ede**, 1978).

The subsequent developmental fate of somites is complex. Within the somite mass which thickens dorsoventrally a cavity, called **sclerocoele** is established in the centre of somite block and the cells become arranged as an epithelium around it, which has the appearance of being multilayered (pseudostratified) because the nuclei migrate from one end of the cell to the other in different phases of cell division cycle. Thus, the walls of the cavity differentiate in later development into distinct layers. The outer wall (toward the epidermis) thickens and becomes the **dermatome**. Its cells are the progenitors of all connective tissues of the skin. The inner wall becomes externally thick and enlarged, segregating into two discrete masses, an inner loose mass of mesenchymatous cells, the **sclerotome**, which migrate and surround the notochord and spinal cord to develop into the segmented vertebral column (bone and cartilage). A more condensed mass of the wall, the

myotome, becomes the body musculature. Since the somites are segmentally arranged, the differentiating muscle cells organize into longitudinal parallel bundles in the trunk of the lower aquatic vertebrates. These bundles become the swimming muscles of the tail in fish and larval amphibians. Each set of muscle segments is innervated by segmentally arranged spinal nerves leaving the spinal cord. In these forms, the myotome represents the bulk of the somite, whereas in the higher amniotes, the sclerotome predominates. The myotomes lose their segmental arrangement as amniote embryos adapt for a different life on land with new locomotory patterns.

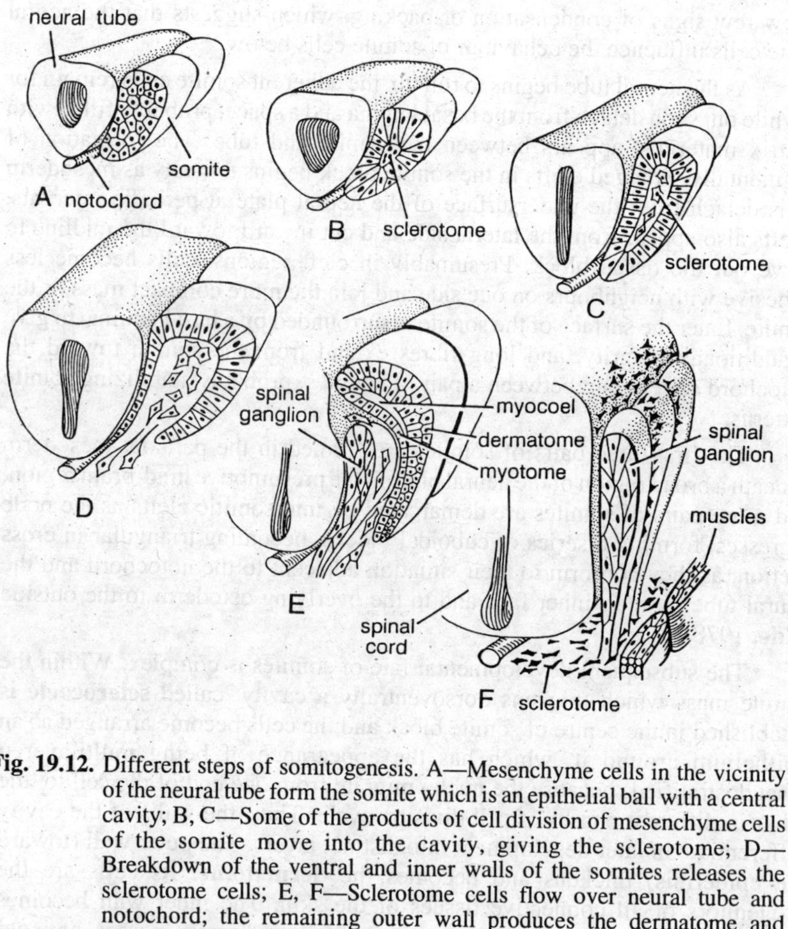

Fig. 19.12. Different steps of somitogenesis. A—Mesenchyme cells in the vicinity of the neural tube form the somite which is an epithelial ball with a central cavity; B, C—Some of the products of cell division of mesenchyme cells of the somite move into the cavity, giving the sclerotome; D—Breakdown of the ventral and inner walls of the somites releases the sclerotome cells; E, F—Sclerotome cells flow over neural tube and notochord; the remaining outer wall produces the dermatome and myotome (after Ede, 1978).

19-7.4. Formation of Primitive Gut

The primitive gut makes its first appearance as **archenteron** which has endodermal layer dorsally and yolk mass ventrally. As the neurulation

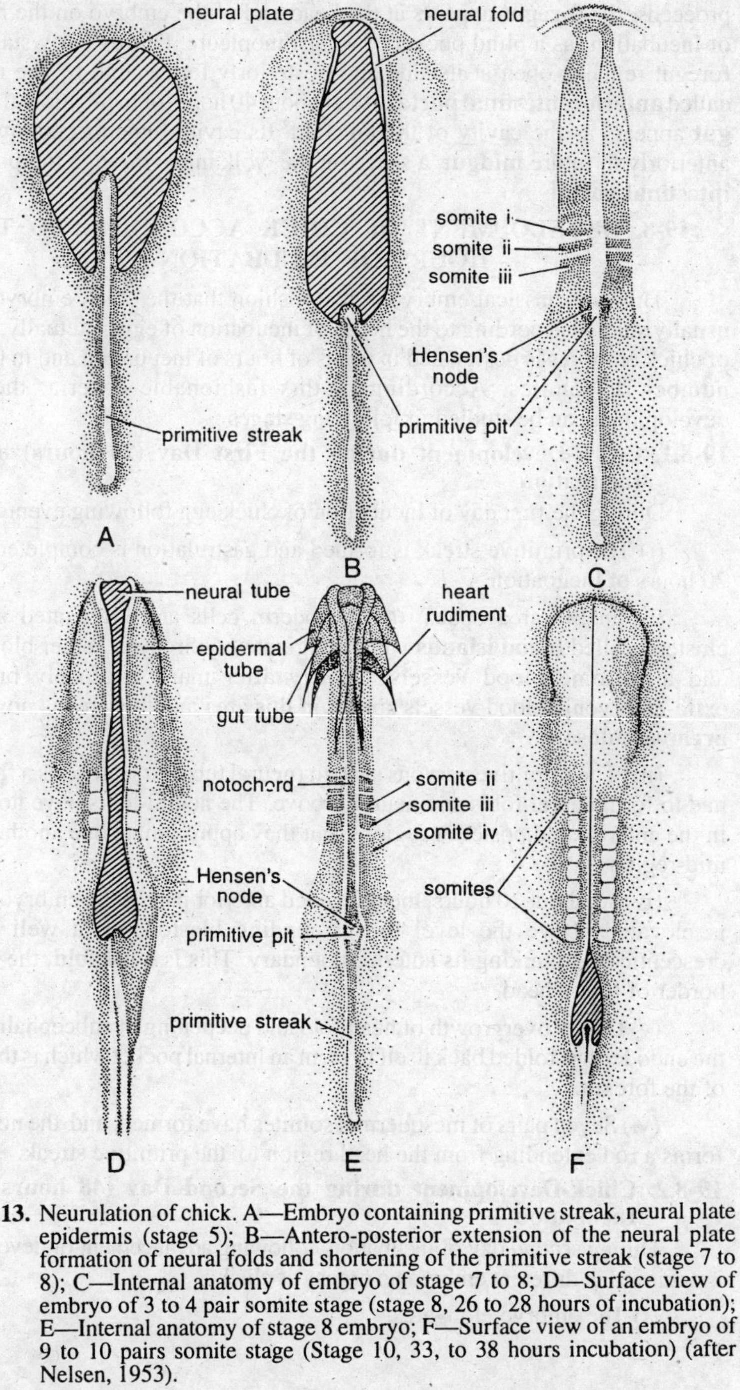

Fig. 19.13. Neurulation of chick. A—Embryo containing primitive streak, neural plate epidermis (stage 5); B—Antero-posterior extension of the neural plate formation of neural folds and shortening of the primitive streak (stage 7 to 8); C—Internal anatomy of embryo of stage 7 to 8; D—Surface view of embryo of 3 to 4 pair somite stage (stage 8, 26 to 28 hours of incubation); E—Internal anatomy of stage 8 embryo; F—Surface view of an embryo of 9 to 10 pairs somite stage (Stage 10, 33, to 38 hours incubation) (after Nelsen, 1953).

proceeds, the **foregut** appears at the head end of the embryo on the first day of incubation as a blind pocket of splanchnopleure. During early stages, the foregut remains open at its side and posteriorly to the yolk by the opening called **anterior intestinal portal**. After about 40 hours of incubation, the **hind-gut** appears as the cavity of the tail fold. Its cavity remains communicated anteriorly to future **midgut**, a space on the yolk mass, through the **posterior intestinal portal**.

19-8. DEVELOPMENT OF CHICK ACCORDING TO THE HOURS OF INCUBATION

This is a classical embryological fashion that the chick embryology is usually studied according to the hours of incubation of eggs. Actually, the age of chick embryos is designated in terms of hours of incubation and in terms of number of somites. According to this fashionable criteria, the chick development can be studied in following stages :

19-8.1. Chick-Development during the First Day (24 hours) of Incubation

During the first day of incubation of chick egg following events occur :

(i) The primitive streak is formed and gastrulation is completed within 20 hours of incubation.

(ii) In the area opaca, the mesoderm cells are aggregated in small clusters, called **blood islands** which originate erythrocytes, other blood cells and also some blood vessels. Shortly after many profusely branched extra-embryonic blood vessels appear in this area and this area is now called **area vasculosa**.

(iii) The primitive nervous system (neural tube) first arose as a flat plate, had to form neural folds and neural groove. The neural folds have not joined in the anterior and posterior regions, but they approximate one another in the mid-region.

(iv) At about 20 hours, the thickened anterior part of the embryonic area is elevated above the level of surrounding blastoderm, a well defined crescentic fold marking its anterior boundary. This is **head fold,** the anterior border of future head.

(v) Due to overgrowth of head fold and deepening of subcephalic space, the endoderm is folded back itself to form an internal pocket which is the cavity of the foregut.

(vi) Seven pairs of mesodermal somites have formed, and the notochord forms a rod extending from the head region to the primitive streak.

19-8.2. Chick-Development during the Second Day (48 hours) of Incubation

During second day of incubation, following advancement of development is achieved in different embryonic organs of chick :

(i) The embryo elongates.

(ii) The neural tube is closed along its most of the length and its anterior end reveals three definite divisions — the **prosencephalon** with **optic vesicles, mesencephalon** and **rhombencephalon**.

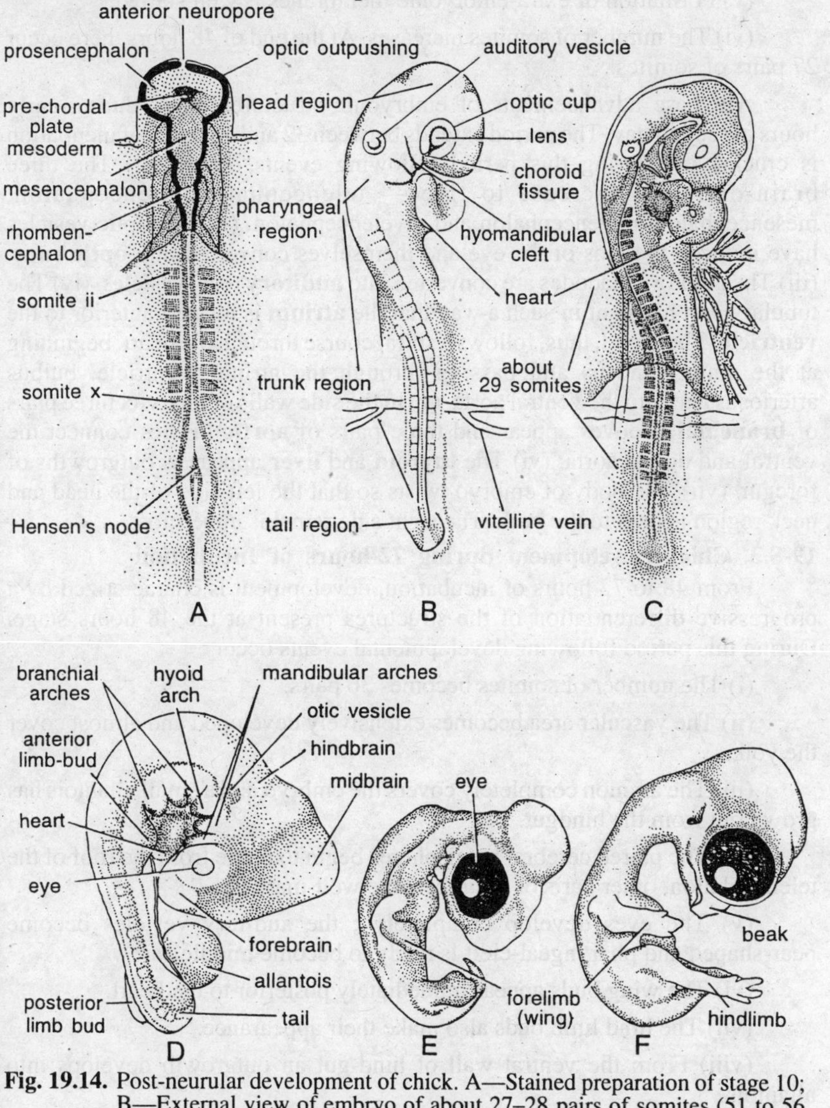

Fig. 19.14. Post-neurular development of chick. A—Stained preparation of stage 10; B—External view of embryo of about 27–28 pairs of somites (51 to 56 hours of incubation, stage 16); C—Internal anatomy of stage 16 (of B); D—Embryo of about 72 to 75 hours of incubation (stage 20); E—Embryo of stage 23; F—Embryo of stage 27 (after Nelsen, 1953 and Balinsky, 1970).

(iii) The auditory placodes become visible in ectoderm adjacent to the rhombencephalon.

(iv) A tubular heart is formed and starts to contract. Large vitelline arteries also extend from heart to area vasculosa.

(v) Formation of extra-embryonic membranes is well started.

(vi) The number of somites increases. At the end of 48 hours there occur 27 pairs of somites.

All these advancements of embryonic structures occur during early hours of second day. The period extends between 32 and 48 hours of incubation is crucial and during this period following events occur : (i) The three brain-divisions increase to five — telencephalon, diencephalon, mesencephalon, metencephalon and myelencephalon. (ii) The optic vesicles have induced the lens of the eye and themselves converted into **optic cups**. (iii) The auditory placodes are converted into **auditory** (ear) **vesicles**. (iv) The tubular heart has bent in such a way that the **atrium** is located anterior to the **ventricle**. The blood, thus, follows a spiral course through the heart, beginning at the sinus venosus and passing through the atrium, ventricle, bulbus arteriosus and into the ventral aorta. (v) In the side walls of pharynx three pairs of **branchial grooves** appear and three pairs of **aortic arches** connect the ventral and dorsal aorta. (vi) The **thyroid** and **liver** appear as outgrowths of foregut. (vii) The body of embryo twists so that the left side of the head and neck region is next to the yolk. The right side remains uppermost.

19-8.3. Chick-Development during 72-hours of Incubation

From 48 to 72 hours of incubation, development is characterized by a progressive differentiation of the structures present at the 48 hours stage. During this period following developmental events occur :

(i) The number of somites becomes 36 pairs.

(ii) The vascular area becomes extensively developed and almost cover the yolk.

(iii) The amnion completely covers the embryo and a small allantois has grown out from the hindgut.

(iv) The paired cerebral hemispheres begin to bulge from the roof of the telencephalon, other parts of brain become well-marked.

(v) The eyes develop completely ; the auditory vesicles become pear-shaped and pharyngeal-cleft is ready to become middle ear.

(vi) The wing buds appear immediately posterior to the heart.

(vii) The hind limb buds also make their appearance.

(viii) From the ventral wall of hind-gut an outgrowth develops into allantois.

19-8.4. Chick-Development from 72-hours of Incubation to Hatching

Further development of chick embryo from 72 hours to the time of hatching is mainly a matter of increased differentiation within different ectodermal, mesodermal and endodermal structures.

19-9. EXTRA-EMBRYONIC MEMBRANES OF CHICK

During the development of chick and other vertebrates, there are produced certain tissues or structures that temporarily or permanently do not enter into the formation of the embryo themselves, but are external to and devoted in one way or another to the care and maintenance of the developing embryo. Collectively these parts are termed as **foetal membranes, extra-embryonic sacs** or **extra-embryonic membranes.**

In fact, the conquest of land by vertebrates became possible only after they were relieved from the need to return to water for reproduction. For this, the vertebrate embryos developed the cellular extra-embryonic membranes to have its own 'pond'. The same extra-embryonic structures later in evolution helped to form a placenta in mammals. In chick and other birds, the extra-embryonic membranes have concerned with nutrition (yolk sac), excretion (allantois), respiration (chorion and allantois) and protection from desiccation and concussions (amnion). The appearance of these new structures (*i.e.*, extra-embryonic membranes), during evolutionary development allowed higher vertebrates to eliminate the vulnerable larval stages from their life cycle thereby ensuring the survival of the adult (**Phillips**, 1975).

19-9.1. Kinds of Extra-Embryonic Membranes

During the chick-development following extra-embryonic membranes are developed :

(i) **Amnion** or **amniotic sac** which surrounds the embryo and provides a kind of private aquarium to protect it from mechanical shocks (such as pressure, abrasion, irritation) and desiccation.

(ii) A **serosa** or **chorion** which forms the outermost membrane surrounding the rest of the embryonic system.

(iii) A **yolk sac** which is filled with the yolk of the egg functions as the first respiratory organ and as a digestive organ for the embryo. It also acts as a haemopoetic organ like the liver and at later stages serves as a place of origin of blood cells.

(iv) An **allantois** which is a precocious urinary bladder by origin, accumulates the embryonic nitrogenous wastes (non-toxic uric acid) and also acts as an embryonic respiratory surface.

All of these extra-embryonic membranous sacs are composite structures in the sense that they involve two germ layers. The amnion and chorion are composed of extra-embryonic ectoderm and somatic layer of mesoderm (both collectively called **somatopleure),** while the yolk sac and allantois, are composed of extra-embryonic endoderm and splanchnic layer of mesoderm (both collectively called **splanchnopleure**).

19-9.2. Development of Extra-Embryonic Membranes

During neurulation of chick, the lateral plate mesoderm splits into an outer somatic layer of mesoderm lying at the inner side of the ectoderm and an inner splanchnic layer of mesoderm lying outer to the endodermal layer.

Both mesodermal layers enclose a **coelomic space** between them. The somatic layer of mesoderm and ectoderm are collectively known as **somatopleure**, while splanchnic layer of mesoderm plus endoderm form the splanchnopleure. During later developmental stages, the somatopleure and splanchnopleure gradually spread peripherally over the yolk mass far beyond the area where the body of the embryo is taking form. As the body of the embryo grows, it becomes separated from the extra-embryonic membranes by the appearance of limiting folds or **sulci** — the cephalic, the caudal and the lateral — limiting folds. Thus, in the anterior region of growing embryo the limiting folds that roll under the head area appear, separating it from the underlying yolk and forming **subcephalic pocket**. Eventually, similar folds develop on either side of body (the **lateral folds**) undercutting it laterally. Posterior limiting folds (the **caudal folds**) form about third day of incubation and lift the tail region from the surrounding yolk area. As a result, the subcephalic, lateral and subcaudal folds define the body of embryo and separate the embryonic from the extra-embryonic region :

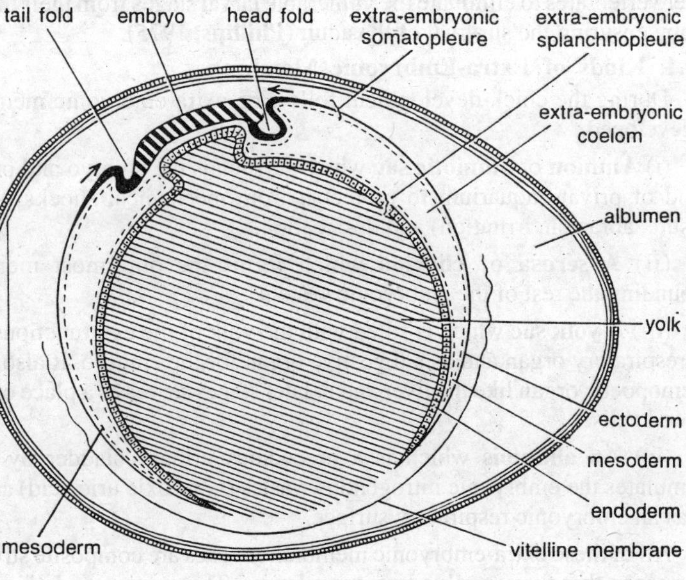

Fig. 19.15. An early chick embryo showing body folds delimiting embryo from extra-embryonic areas (after Torrey, 1971).

Now onward, the somatopleure and splanchnopleure develop into four extra-embryonic membranes of chick by following methods :

1. Development of yolk sac. During neurulation, the gut region of embryo has a dorsal roof and side but no floor. The ventrally open gut rests on the yolk mass. The splanchnopleure which remains closely applied to yolk, extends peripherally and eventually surrounding the entire yolk and form **yolk sac**. Yolk sac is the first extra-embryonic membrane to form. Later on, as the undercutting folds move toward one another, they establish a floor to the gut except for a small region of the midgut. These folds do not completely merge, but a small aperture, the **yolk duct**, remains open into the yolk sac. Once a completely walled gut is formed in the embryo proper, the yolk sac cavity remains in continuity with the gut cavity through the yolk duct which runs in the **yolk stalk** of yolk sac. Despite the fact that the yolk sac is connected to the digestive tract by the yolk stalk, the yolk food reserves are not transmitted to the embryo by this route. Rather the endodermal surface of the yolk sac is thrown into folds (**yolk sac septa**) that penetrate the yolk mass and through the mediation of appropriate digestive enzymes produced by the endodermal cells, the yolk is digested or made soluble and ultimately is absorbed by the endodermal lining of yolk sac and transported to the embryo by **vitelline vein**. As by now, a rich blood circulation develops within the mesodermal component of the yolk sac and the vitelline arteries and veins (the so called **omphalomesentric blood vessels**) traverse the yolk sac. The constriction of the yolk stalk region brings together the vitelline blood vessels (two arteries and two veins) located in the splanchnic mesoderm of both sides of the embryo. Thus, digested yolk is collected by left and right vitelline veins, both of which

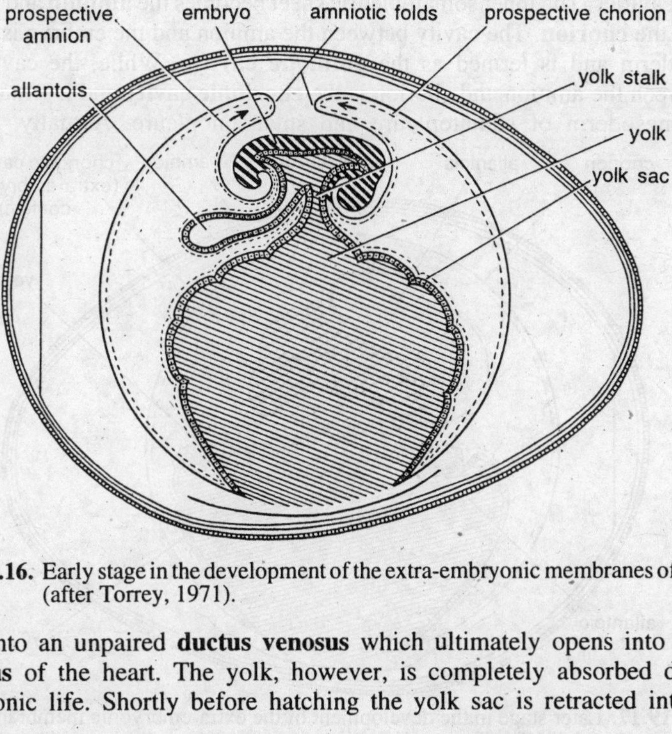

Fig. 19.16. Early stage in the development of the extra-embryonic membranes of chick (after Torrey, 1971).

open into an unpaired **ductus venosus** which ultimately opens into **sinus venosus** of the heart. The yolk, however, is completely absorbed during embryonic life. Shortly before hatching the yolk sac is retracted into the abdominal cavity of the embryo, and the walls of the abdominal cavity close behind it.

2. Development of amnion and chorion. The origin of amnion and chorion is considered together since they are formed simultaneously from the extra-embryonic somatopleure. About the 30 hours of incubation, the extra-embryonic somatopleure is elevated over the embryo by a folding process consisting essentially of a doubling of the somatopleure upon itself. The initial elevation is over the head end of the embryo, producing a double somatopleuric hood, called **cephalic amniotic fold**. As the cephalic amniotic fold gradually extends backward, its caudally extending side limbs called **lateral amniotic folds** arch over the embryo from each side to be joined finally by a similar fold or elevation over the tail, the **caudal amniotic fold**. All these amniotic folds ultimately converge, so as to encase the embryo in two sheets of somatopleure from all sides except the region of yolk stalk. The region where the union of amniotic folds occurs is called **seroamniotic connection**. The point of fusion of the folds leaves a scar, called the **seroamniotic raphe**. According to **Bodemer** (1968), the seroamniotic connection opens before the hatching of embryo and admits albumen into the amniotic cavity, where it is swallowed by the developing chick embryo.

The fusion of the amniotic folds produces two sac-like membranes and two cavities. The inner somatopleuric sheet becomes the **amnion** and the outer one, the **chorion**. The cavity between the amnion and the embryo is lined by ectoderm and is termed as the **amniotic cavity**. While, the cavity lying between the amnion and chorion is the **chorionic cavity** and it remains lined by mesoderm of somatopleure and splanchnopleure. Actually, it is the

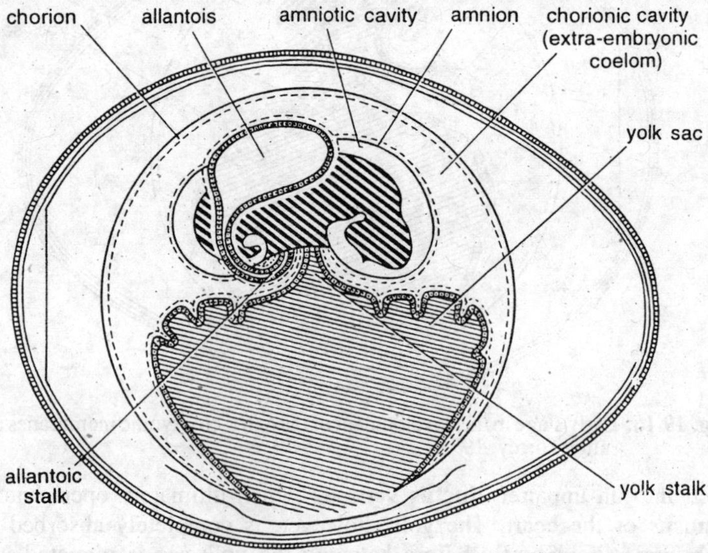

Fig. 19.17. Later stage in the development of the extra-embryonic membranes of the chick (after Torrey, 1971).

extra-embryonic coelom. As the amnion and chorion form, the somatopleure continues to grow peripherally and laterally untill it completely covers the embryo and all the other membranes that are formed at this stage.

Functions of amnion and chorion. The amniotic cavity is at first a very narrow slit between the embryo and the inner wall of the amniotic folds, but soon a fluid is secreted into the cavity which distends it so that the embryo floats freely in the cavity, connecting to the extra-embryonic parts only by the umbilical cord. The advantages that the embryo obtains from the development of amnion, chorion and amniotic cavity are following :

(i) The embryo becomes immersed in a container filled with salty fluid and, thus, can accomplish its development in a fluid medium, although the egg is "on dry land". Thus, the primary function of amnion is protection of embryo from the danger of desiccation.

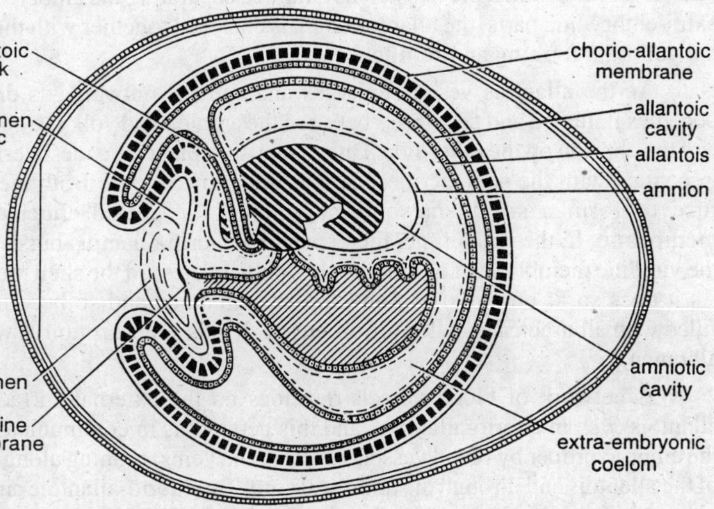

Fig. 19.18. The fully matured extra-embryonic membranes of chick (after Torrey, 1971).

(ii) The fluid of the amniotic cavity is an efficient shock absorber and protects the soft, collapsible and almost skeletonless early chick embryo from the mechanical shocks.

(iii) Amnion isolates the embryo from the shell of the egg, and, thus, protects it from adhesion to the shell or from friction against it.

(iv) During later developmental stages, the mesoderm of amnion forms muscle cells which contract rhythmically, thus, rocking the embryo within the amniotic fluid. The rocking results in the circulation of the amniotic fluid that moves the parts of the embryo about and prevents them from growing together. It is also instrumental in preventing its adhesion to the different embryonic membranes. It may also help to obviate the stagnation of blood in the vessels, a condition which might tend to occur on account of the pressure from the growing organs.

According to **Balinsky** (1970), the formation of the amniotic cavity has a slightly negative effect because it removes the embryo from the surface of the egg and from the source whence it could obtain oxygen. The chorion later on joins the allantois to serve in a capacity of respiratory and nutritional organs.

3. Development of allantois. About the third day of incubation, the region of the future floor of endodermal hindgut begins to bulge as precocious urinary bladder, the **allantois**. Allantois, thus, corresponds exactly in its nature to the urinary bladder of Amphibia. The allantoic outgrowth (or evaingation) consists of endoderm with splanchnic or visceral mesoderm layer covering it from the outside (*i.e.*, splanchnopleure). It grows very rapidly and soon penetrates into the extra-embryonic coelom, the space between the yolk sac, the amnion and the chorion. The distal part of the allantois expands and remains connected with the hindgut of the embryo by means of a narrow **allantoic stalk**. When the body folds contract, separating the embryo from the extra-embryonic parts, the allantoic stalk is enclosed together with the stalk of the yolk sac, forming an **umbilical cord**.

As the allantois vesicle enlarges and spreads outward, its distal part becomes flattened and penetrates between the amnion and yolk sac on one side and the chorion on the other side. Due to this, the splanchnic mesoderm comes in contact with the somatic mesodermal lining of chorion. Both mesoderms fuse to form a single mesodermal membrane called **chorioallantoic membrane**. In the mean time, the expanding chorio-allantois bursts through the vitelline membrane of egg and pushes outward toward the shell membrane. As it does so, it progressively envelops the albumen and so becomes a sac filled with albumen and called **albumen sac**. It aids in absorption of water and albumen.

A network of blood vessels develops on the external surface of the allantois (*i.e.*, in chorio-allantois) and this network is in communication with the embryo proper by blood vessels (arteries and veins) running along the stalk of the allantois and through the umbilical cord. The chorio-allantoic circulation continues until the young chick breaks the egg-shell and begins to breath the surrounding air. Then the umbilical vessels close, the circulation ceases, and the allantois dries up and separates from the body of the young chick. At the time of hatching, the allantoic vesicle with its contents (*i.e.*, the uric acid) is detached from allantoic stalk and left attached to the broken shell.

Functions of allantois. The allantois serves following functions for the embryo :

(i) It acts as a reservoir of embryonic excretory wastes such as uric acid.

(ii) The chorio-allantoic membranes of allantois and chorion act like a respiratory surface for the embryo.

Thus, the four extra-embryonic membranes are devices which provide adequate nutrition, protection, and satisfaction of metabolic requirements to the developing chick embryo.

Table 19-1. Normal stages of chick development (Source : **Hamberger** and **Hamilton,** 1951).

Stage number	Number of somite pair	Incubation time (Hours) (at 37.8°C)	Structure developed
1	—	5–6	Embryonic shield formed.
2	—	6–7	Primitive streak begins to form.
3	—	12–13	Growth of primitive streak; formation of mesoderm and neural plate.
4	—	18–19	Complete formation of primitive streak; formation of primitive groove, Hensen's node and primitive pit; mesoderm become distinct.
5	—	19–22	Formation of head or notochordal process; appearance of neural folds.
6	—	23–25	Notochord fully formed; head fold appears; differentiation of somatic and visceral (splanchinic) mesoderms.
7	1	23–26	One pair of somites formed; coelom appears.
8	4	26–29	Formation of blood islands and heart tube; fusion of neural folds.
9	7	29–33	Fusion of heart tube; onset of heart beats.
10	10	33–38	Pronephros (kidney) differentiates; heart tubes completely fused to form one heart tube; brain and spinal cord become clearly distinguishable.
11	13	40–45	Degeneration of pronephros; amniotic head folds appear; mesonephros forms.
12	16	45–49	Head turns on its left; heart become S-shaped. By the incubation of 50 hours somite pairs become 20.
13	19	48–52	Head fully turned; allantoic bud appears.
14	22	50–53	Yolk stalk forms.
15	24–27	50–55	Lung primordia appear.
16	26–28	51–56	Hind-limb buds appear.
17	29–32	52–64	Fore-limb buds appear.
18	30–36	65–69	Allantois become distinct.
19	37–40	68–72	Aortic arches form.
20	40–43	70–72	Embryo is completely surrounded by amnion.

Table 19-2. Certain important events in the development of chick.

Incubation time	Development events
4th day	Amniotic contractions begin; pronephros disappears; metanephros arises; chorion and allantois fused.
5th day	Four-chambered heart forms.
6th day	Beak forms; three fingers and four toes become distinct.
7th day	Limbs can move independently; sexual differentiation begins; feathers in seven rows in the pelvic region.
8th day	Eye ball moves; three-rows of tail feathers form.
11th day	Metanephros begin to function.
14th day	Body takes a position along the long axis of the egg.
20th day	Chick begins to breathe and makes some sound; the shell cracks.
21st day	Chick hatches and starts feeding.

REVISION QUESTIONS

1. Describe in detail the structure of hen's egg and explain the function of its various parts.
2. Why is the bird's egg is known as a cleidoic egg ? Draw a fully labelled diagram of an avian egg.
3. Describe the development of chick up to the formation of primitive streak.
4. Discuss the process of gastrulation in chick and compare it with that of frog.
5. Explain how the Hensen's node of chick is homologous to the dorsal lip of frog ?
6. Explain the influence of yolk on the course of development of chick.
7. Describe the developmment of chick's egg till the formation of three germ layers.
8. What are the embryonic membranes ? In what way they are different from egg membranes ?
9. How are the extra-embryonic membranes of chick developed ? What role do they play in the growth and development of embryo.
10. Why the embryos have evolved extra-embryonic membranes ? What problems were created as a result of this and how these have been solved ?
11. Describe the arrangement and mode of development of extra-embryonic membranes in chick and explain their functions.

20

Early Embryogenesis Of An Eutherian Mammal — Rabbit

20-1. OVIPARITY, OVOVIVIPARITY AND VIVIPARITY

In preceding chapters, the embryological development of those animals has been described which were egg-laying and in which most of the embryological development of the fertilized egg occurs outside the body of the mother. All these animals are called **oviparous animals** and such condition of reproduction is called **oviparity**. In this chapter the embryological development of an eutherian mammal, the rabbit, which is a **viviparous** animal, will be described. The viviparous animals are those animals in which the eggs are internally fertilized, the fertilized eggs are retained in a specialized part of oviduct called **uterus**, and the embryological development of these eggs takes place inside the uterus. The developing embryo of such a viviparous animal often establishes a direct connection with the uterine wall and draws its nutritional, respiratory, excretory and other metabolic requirements from the maternal body. All mammals, except prototherian mammals which are oviparous (*e.g., Ornithorhynchus*) or **ovoviviparous** (a reproductive condition of egg-laying animals in which the eggs though develop inside the maternal body but are not supplied with maternal nourishment, *e.g., Echidna*), are viviparous. The viviparity is an adaptation to adverse environmental conditions such as cold, dryness and seasonal changes. It permits the mother freedom to move about to feed and to migrate, a freedom which is not enjoyed by a bird parent during its task of incubation. Viviparity is also found in some elasmobranchs, teleosts and reptiles. In most of these animals, however, the egg is supplied with abundant yolk, but develops within organs of its mother's body, until hatching. Birds have not developed viviparity because of the burden of carrying the unborn young when in flight. However, bats which are mammals carry their offspring in flight.

The resumption of viviparity in mammals has certain its own outcomes, such as their eggs become microlecithal or almost yolkless, the egg shells become wanting and a new organ called **placenta** develops. The placenta is composed of maternal and foetal tissues and serves for the transport of nutrients from maternal to foetal tissues.

20-2. EVOLUTION OF VIVIPARITY IN MAMMALS

There are good evidences which indicate that the mammals are derived from ancestors which were very closely related to the early reptiles. According to **Balinsky** (1970), they must have had large, yolk-laden eggs, and their ontogenesis must have been very like that of reptiles. At some stage in their evolution, some of the mammals have become viviparous, the embryo receiving an adequate supply of nutrition from the mother, while it was retained in the uterus, and the yolk supply of the egg became superfluous. The yolk supply consequently became progressively reduced and eventually disappeared altogether. In this respect, there is a distinct gradation in the three sub-divisions of the class Mammalia.

In the subclass **Prototheria** this adaptation (viviparity) does not go further than the state of ovoviviparity. The eggs, therefore, have a large amount of yolk (*i.e.*, polylecithal eggs), they are meroblastic and have a subsequent development on a reptilian pattern (**Flynn** and **Hill**, 1939).

In the subclass **Metatheria** (the marsupials, etc.) the developing embryos receive nourishment from the mother in the uterus, although this kind of adaptation does not go as far forward as in the next group, and their young ones are born poorly developed. The yolk here has already become superfluous and its quantity in the egg is insignificant. Even so, the egg yolk is not used by the developing embryos as in the reptiles and birds, but, the mass of yolk is ejected from the fertilized egg at the beginning of the cleavage as the egg becomes divided into the first two blastomeres (**Hill**, 1918).

In the most advanced mammalian subclass, the **Eutheria** (placental mammals), the egg is devoid of yolk right from the start. With the disappearance of the yolk, the **microlecithal** or **alecithal** mammalian eggs have reverted to **holoblastic** type cleavage, like *Amphioxus* or amphibians, but the subsequent development bears ample evidences of the former presence of yolk in their eggs, because their many late developmental stages (particularly gastrulation) resembles with that of reptilian or avian meroblastic type, dicoidal cleavage. Because, the eutherian eggs lack yolk, therefore, they have been forced to develop inside the maternal uterus and to draw their different requirements from the body of the mother.

The maternal uterus, however, poses some dangers for the developing embryo. **George W. Corner** (1944), one of the fathers of modern reproductive physiology, put the problem of the human embryo this way : *"Accepting the shelter of the uterus, it also takes the risk of maternal disease or malnutrition, and of biochemical, immunological and hormonal maladjustment. Even before it strikes its roots in the living tissues of the endometrium (lining of the uterus) it has a week's journey to make, as long as a submarine takes to pass beneath the polar ice cap. Like the U-boat it has to carry most of its supplies with it in its trip down the oviduct and the uterine lumen and to add to its difficulties it is surrounded by a far more variable and chemically active medium than ice cold sea water"*.

20-3. EARLY EMBRYOGENESIS OF RABBIT

The rabbit is a lagomorphan eutherian mammal in which the sexes are separate, sexual dimorphism is well marked and the reproductive physiology of both sexes has close resemblances with that of other eutherian mammals and even the man.

20-3.1. Male Reprodutive Organs

The adult male of rabbit has a pair of ovoid male gonads, called **testes**, which under the influence of pituitary gonadotropic hormones, produce the male gametes, the spermatozoa and the male sex hormone — the testosterone and allied androgenic steroids—that determine the secondary sexual phenotype of the male sex. Both testes remain suspended in the **scrotums** or

scrotal sacs, both of which remain situated at the lateral sides of the intromittent organ, the **penis**. Each scrotal sac is the extension of the abdominal coelomic cavity and it remains connected with the latter by means of an **inguinal canal**. The testes consist of numerous long convoluted **seminiferous tubules** which are composed of two kinds of cells—primordial germ cells and Sertoli cells. Primordial germ cells contain large-sized nucleus, RNA-rich cytoplasm and have an extra-gonadial origin. These cells undergo spermatogenesis and form haploid spermatozoa. The Sertoli cells are supporting cells, each of which contains a pale nucleus with a distinct eosinophilic nucleolus and a cytoplasm which has abundant smooth endoplasmic reticulum, a feature associated with lipid metabolism. Recently it has been found that the Sertoli cells can make androgens from precursors and so provide an androgenic environment within the lumen of tubule independent of the interstitial cells of Leydig (see **Craigmyle** and **Presley**, 1975).

The seminiferous tubules occupy 80 per cent volume of testes, the rest being connective tissue with a rich vascular bed, and grouped in clusters between the tubules, are the endocrine **interstitial cells** or **cells of Leydig**. The interstitial cells are clusters of polyhedral cells, each of which contained vacuolated cytoplasm with abundant smooth endoplasmic reticulum. These cells secrete male sex hormones (*i.e.*, steroids).

The ends of the coiled parts of the seminiferous tubules of a testis unite posteriorly in a short straight portion which opens into a plexus of wide-epithelial-lined channels, the **rete testis**, from which many fine **vasa efferentia** lead into **epididymis**. The epididymis is an extremely convoluted tube and is a storage place for spermatozoa. It remains attached with the testis and contains a head region called **caput epididymis** in front of testis and a tail region called **cauda epididymis** behind the testis. The secretion of epididymis contains certain biochemical substances such as sialic acid, carnitine, glyceryl phosphocholine, which are normally not found in other biological fluids. These chemicals are found to have certain effects on maturation and changes in the investing membrane of the sperms. Sperms usually enter in epididymis through caput epididymis and are held in the cauda epididymis until ejaculation.

From cauda epididymis of each epididymis arises the **vas deferens**, which is an epithelial-lined duct with a thick smooth muscle coat and a rich autonomic innervation, and which runs from its junction with the cauda epididymis, in the spermatic cord through the inguinal canal and enters in the abdominal cavity. At the base of bladder the vas deferens ultimately unites with the duct of the **seminal vesicle** or **uterus masculinus** which is a median bag—situated above the neck of bladder, to form the **common ejaculatory duct**. This runs through the substance of **prostate gland** and enters the prostatic portion of the urethra. The urethra of male rabbit, thus, serves to convey seminal products (called **ejaculate** or **semen**) as well as urine. Further the vas deferens contains well-developed muscle layers and is largely responsible for the movement of sperm along the tract. In certain eutherian mammals including man, sperm storage may occur in the proximal end of the vas deferens, as well as in the epididymis. Some species (*e.g.*, cat and dog) have no seminal vesicles (**Turner** and **Bagnara**, 1971).

Finally, there is an intromittent organ, the **penis** which is a muscular cylindrical eractile organ with spongy vascular walls. It hangs in front of anus from the ventral wall of the abdomen and has a single duct the urethra to introduce the semen into the genital tract (vagina) of female during the act of copulation.

Male accessory glands. The semen or ejaculate contains fractions added by various accessory sex glands to the sperm-containing component which has been passed up the vas deferens. The accessory sex glands of male rabbit are following. A large **prostate gland** is situated at the base of seminal vesicle and it pours its secretion containing several lytic enzymes, citric acid, spermine and prostaglandins into urethra by numerous ducts. A pair of **Cowper's glands** or **bulbo-urethral glands** are situated just below the prostate glands on urethra. Urethra also contains multiple glands of Littre. The secretion of bulbo-urethral gland and Littre's glands remains sialomucin rich and is released early in coitus. Primarily it has lubricating function but due to its alkaline pH, it neutralizes any acidity both in the urethra after the passage of urine and in vagina of female (For details see **Jordan** and **Verma**, 1977). Lastly, seminal vesicle too produces a secretion which remains rich in fructose, inositol, vitamine C and fluorescent flavine derivatives. The normal secretion of these glands is androgen dependent. The seminal fluid is, thus, a vehicle for the conveyance of germ cells and perhaps provides an environment which activates and nourishes the sperms and in which they can retain their greatest fertilizing capacity (see **Turner** and **Bagnara**, 1971 ; **Craigmyle** and **Presely**, 1975).

20-3.2. Female Reproductive Organs

The adult female rabbit has a pair of oval **ovaries** which occur in the body cavity and remain suspended from its (coelomic) walls by mesovarium ligaments. Each ovary remains covered by a layer of cubical epithelial cells (germinal epithelium) and it has a compact cortical region (**cortex**) and a more loosely arranged **medulla**. In the cortex there are ovarian follicles in various stages of development, but these are not found in the medulla which consists of blood vessels ensheathed in connective tissue. Near each ovary lies funnel-shaped anterior opening of the **oviduct**, the **infundibulum**. In mammals, the oviduct is called **Fallopian tube**. The Fallopian tubes of both sides open into a median broad and muscular pouch called **uterus**. The uterus of rabbit is **duplex** type, because it has two completely separated horns, two cervixes and one uterine body like man. The uterus is lined by mucous and glandular epithelium. The muscular walls of uterus are richly vascular. Beyond the uterus lies a single passage leading to the exterior, the **vagina**. The vagina passes backwards with which joins the neck of the urinary bladder, thereby forming a common passage called **urinogenital duct** or **vestibule**. The vestibule opens to the outside by a self-like aperture, the **vulva**. At the ventral side of the vestibulae is situated a small rod-like **clitoris** which is homologous to the penis of males. A pair of small **Bartholin's glands** is situated on the dorsal wall of vestibule. The secretion of these glands is discharged in the vestibule which lubricates the vaginal passage. The vagina of female receives the penis of male rabbit during coitous.

20-3.3. Oogenesis and Ovulation

Oocytes derived from the germinal epithelium during foetal life are present in the ovarian cortex at birth. They originate as a population of cells derived by proliferation of the germinal epithelium, which invades the stroma of ovarian cortex and loses its connection with the surface epithelium. It becomes subdivided into small clusters of cells which form the **primordial follicle**. Now, the maturation of the ovum goes hand in hand with the development of the **ovarian follicle**. A young oocyte is surrounded by a single layer of cuboidal ovarian cells, which is designated as the **granulosa layer**, constituting what is termed the **primary follicle**. The granulosa cells then rapidly multiply to produce an increasingly thick, multilayered covering around the oocyte (Fig. 5.1 and Fig. 5.14). Coincidentally a transparent non-cellular area known as **zona pellucida** also arises around the oocyte. Electron microscopic observations of ovarian follicles have revealed that the zona pellucida is an area of interdigited projection of oocyte cytoplasm and follicular cells that probably facilitates exchange of materials between the maturing ovum and the surrounding follicle (Fig. 20.1).

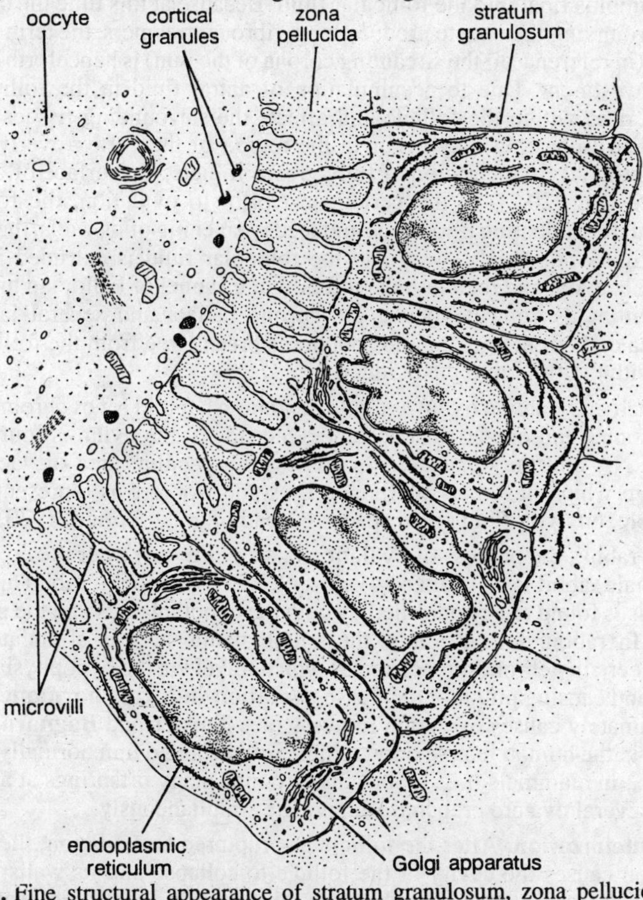

Fig. 20.1. Fine structural appearance of stratum granulosum, zona pellucida and surface of mammalian oocyte (after FitzGerald, 1978).

The appearance of the zona pellucida between the oocyte and the multilayered granulosa marks the secondary stage of follicle development. Clefts now begin to appear within the granulosa. These clefts ultimately coalesce to form a single **follicular cavity** or **antrum**, which from the very start is filled with a fluid. A follicle with an antrum is termed as **tertiary follicle.** Conversion of a tertiary follicle to a mature follicle is largely a matter of increase in size. As the follicle grows, the antrum continues to enlarge in such a way that the oocyte still embedded in a portion of the granulosa layer on one side only. The point at which the oocyte remains attached with the inner wall of the antrum is called **cumulus oophorus**. The granulosa investing the oocyte is now called **cumulus**. While these changes have been going on, the connective tissue framework of the ovary has laid down an additional sheath, the **theca** outside the parietal granulosa.

The oocyte now undergoes first maturation division. As this proceeds, the cumulus gradually becomes undercut so that the oocyte is carried first in a slender stalk of cells and then, still encased in zona pellucida and remnants of the cumulus float into the follicular fluid. Because at this time the cumulus cells surrounding the oocyte produce long, fibrous streamers, the term **corona radiata** (in reference to the streaming corona of the sun) is henceforth applied to the investment. Due to accumulation of antral fluid in the follicle, the pressure rises within the follicle and affects the surrounding tissues so that thining and ischaemia may occur in the region of the cortex overlying the follicle (stigma). In the region of fully grown follicle, the surface of ovary is ballooned (Fig. 20.2). In the period, just prior to ovulation, the follicular pressure falls, reflecting a hormone-induced weakening of its walls, the tissues in the stigma region give way and the follicular fluid is extruded into the peritoneal cavity and then into the infundibular funnel of Fallopian tube. The ovum within its zona pellucida, the corona radiata and several layers of granulosa cells from the cumulus oophorus breaks free from the follicle and passes with the fluid into the oviduct.

Each mammalian ovum as a rule contains a haploid set of chromosomes — autosomes and one X chromosome. One side of the ovum is found to be rich in RNA and the opposite side possesses a peculiar area of vacuolated cytoplasm which has been interpreted by some workers as representing the yolky pole, but with fluid-filled spaces instead of yolk (**Cohen**, 1970).

In rabbit, the process of ovulation takes place only after the copulation (*i.e.*, female rabbit is an **induced ovulator**). The ovulation of rabbit and other mammals is found to be controlled by both the hypothalamus and the pituitary gland. In rabbit during coitus, the hypothalamic area is stimulated, and releases certain neurohormonal agents which travel via the hypophysial portal vessels and causing pituitary to release **ovulation-inducing hormone (OIH)**. OIH ultimately causes ovulation in ovary (see **Turner** and **Bagnara**, 1971). Further, in the human female only a single follicle and ovum normally mature at a time : in mammals (*i.e.*, rabbit) that produce many offsprings at a time or **litters**, several ova are prepared and ovulate simultaneously.

Luteinization. After the follicle has ruptured at ovulation, the release of tension causes the cavity of the follicle to collapse and its walls become contorted. Basement membrane becomes disrupted and capillaries from **theca**

sprout into the residual mass of granulosa cells. Lipid synthesis within the cytoplasm of these cells increases and due to lipid accumulation in its cells the collapsed follicle becomes yellowish coloured body called **corpus luteum**. It secretes a hormone **progesterone**, which prepares an environment for the embryo for implantation in the uterus.

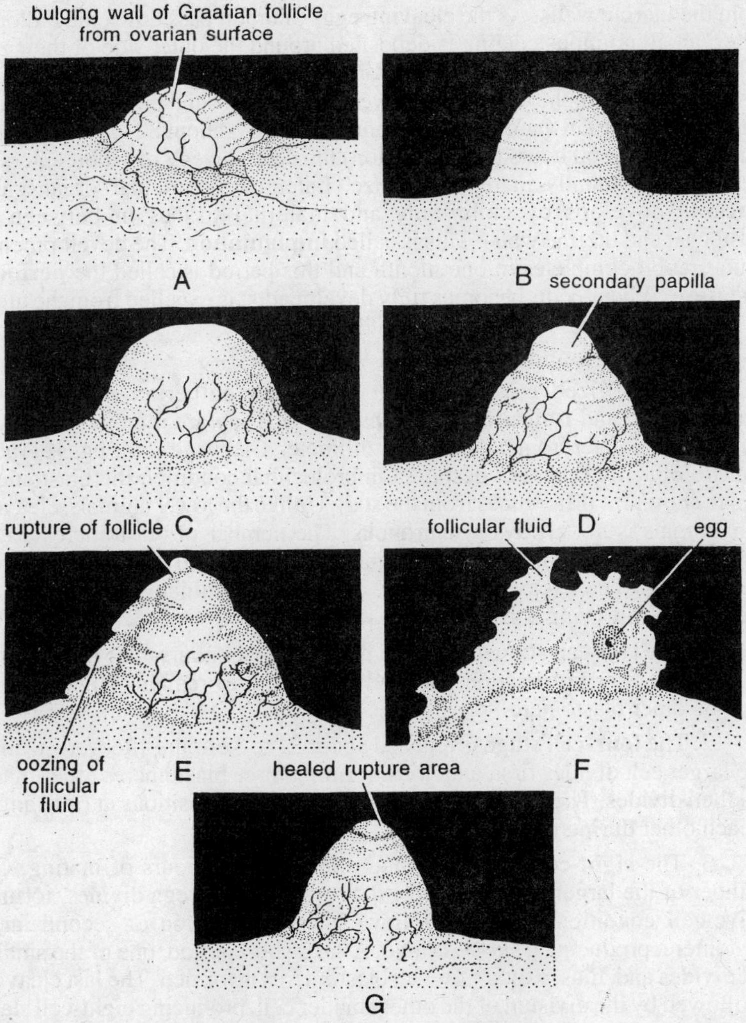

Fig. 20.2. Ovulation in rabbit. Diagram showing external changes in ovarian wall at the time of ovulation (after Nelsen, 1953).

20-3.4. Fertilization

In rabbit, the fertilization is internal and occurs in Fallopian tubes. During copulation, male rabbit sheds its sperms near uterus and from there these sperms move in both Fallopian tubes to fertilize the eggs. In fertilization, complete sperm enters the ovum, the tail of the sperm, however, soon degenerates. The second maturation division of the ovum is then completed and cleavage inaugurated.

The cleaving egg passes down the oviduct to reach the uterus where it would complete its development by drawing its fundamental requirements from the uterine walls. As the cleaving eggs of rabbit pass down the Fallopian tubes, an albuminous coating is deposited around the outer side of their zona pellucida. This albuminuous layer forms an **accessory egg membrane** or covering similar to the albuminous layer deposited around the egg by oviducal cells in prototherian mammals, birds and reptiles. At about 80 to 94 hours after mating (or after 60 hours of fertilization) the cleaving egg of rabbit enters the uterus and gradually increases in size. Later on, a vascular organ called **placenta** develops by embryonic and maternal tissues in between the blastocyst and uterine walls. This is called **implantation**. The development of young rabbit completes in one month and this period is called the **period of gestation**. When foetus becomes fully developed, it is expelled from the uterus by contraction of its muscular walls at **birth**.

20-3.5. Cleavage and Blastulation

The fertilized eggs of rabbit and other eutherian mammals are **microlecithal** and **isolecithal** (**Phillips**, 1975). They are minute in size, *i.e.*, ranging in diameter from 0.08 to 0.15 millimeter. Cleavage of the eggs of rabbit is of **holoblastic** type, but despite a microlecithal condition the blastomeres tend to show size differences from the start. Unlike the lower animals, cleavage in mammals is not always synchronous. The number of blastomeres do not increase by a regular doubling sequence but tend to show arithmetical progression. The cleavage process of rabbit culminates in blastula by undergoing following stages :

1. The **two-cell stage** is reached about 22 to 24 hours after mating or 10 to 12 hours after fertilization. One cell (blastomere) remains slightly larger than the other blastomere.

2. The **four-cell stage** is reached about 24 to 32 hours after fertilization. The larger cell divides first, and, thus, forming three blastomeres ; the smaller cell then divides. The mitotic spindles tend to assume positions at right angles to each other during these cleavages.

3. The **eight-cell stage** is found after 32 to 41 hours of mating. One member of the larger blastomeres of the four-cell stage egg divides, forming a five-cell condition, which is followed by the division of second larger blastomere, producing a six-cell stage. After a short period, one of the smaller cell divides and, thus, a total of seven blastomeres is formed. The last cleavage is followed by the division of the other smaller cell, producing eight-cell stage. The mitotic spindles of each of these cleavage are formed at right angles to one another, thus, showing an independence and asynchrony of cleavage.

Further, in the armadillos each of the 4 or 8 blastomeres produces one of a family of **quadruplets** or **octuplets** ; each free blastomere forms its own little morula, but the embryonic membranes may again fuse and the placenta is often common to all four or eight embryos (see **Cohen**, 1970 ; **Berrill** and

Karp, 1976). It shows that a mammalian egg is **indeterminate** and has **regulative development**.

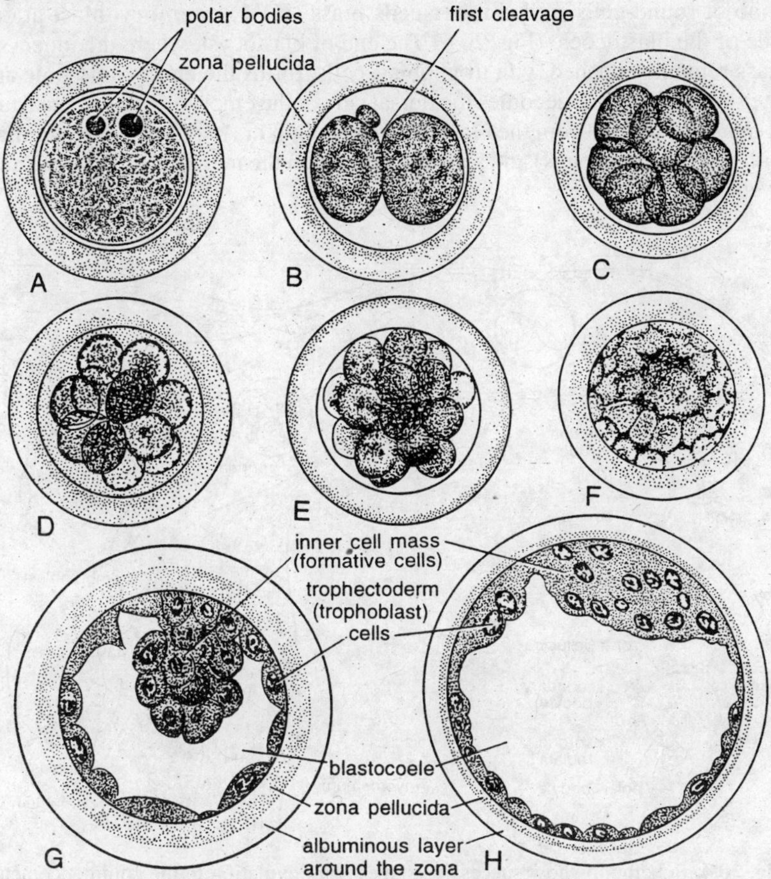

Fig. 20.3. Early cleavages of rabbit (after Nelsen, 1953).

4. The rate of mitotic divisions increases and at about 65 to 70 hours after mating, the 16-cell stage is reached. This 16-cell stage is a solid mass of blastomeres and represents the **morula**. Thus, the zygote, during its journey in the Fallopian tube, cleaves to form the morula by the time it reaches the upper part of the uterus (Fig. 20.4). The period of the morula extends from the 16-cell stage until the blastula formation commences when the cluster of cells numbers 50 to 60 cells.

5. When morula enters the uterus, it lies for about 2 days free in the uterine lumen, bathed by the secretion of the uterine glands. Soon an eccentrically placed fluid-filled cavity, the **blastocoele** appears within the morula which is then known as a **blastocyst** (or **blastula**). Apparently, the blastocoelic cavity is enlarged in its size due to gradual accumulation of the fluid that has passed in the morula from the uterine cavity through the zona pellucida.

The blastocyst consists of a single layer of flat-cells, called **trophectoderm** or **trophoblast**, enclosing the fluid-filled blastocoele and the knob of round-cells, called **inner cells mass** (**ICM**) or **embryoblast** at one side of the blastocoele (Fig. 20.5).The end of blastocyst, where the inner cell mass remains attached with trophoblast cells, forms the **embryonic pole** and later on, develops or becomes the dorsal side of the embryo. Thus, the blastula of mammal differs from the blastula of sea urchin or *Amphioxus* (*i.e.*, animals with microlecithal eggs) and shows kinship with the meroblastic development of chick.

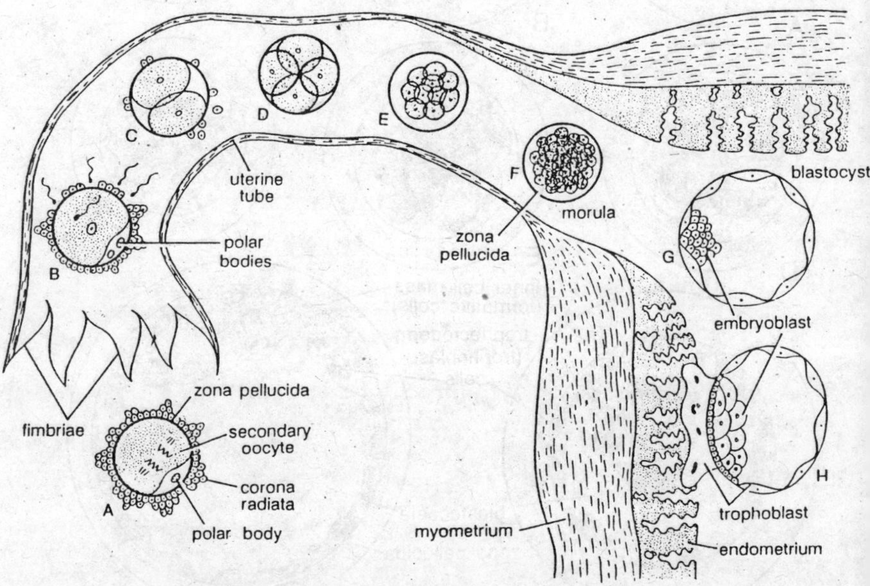

Fig. 20.4. Scheme to show successive stages from ovulation to the commencement of implantation in human beings (after FitzGerald, 1978).

In fact, the blastocyst is an adaptive structure ; it makes possible intrauterine development in the mammal. During the first few days, it is still enclosed by the zona pellucida (a non-cellular secretion produced by follicle cells and oocyte). In mouse, the blastocyst expands, thinning and rupturing the zona pellucida. Also a lytic agent probably emanating from the uterus, lyses the zona pellucida. Loss of zona pellucida allows the mouse blastocyst to "hatch" and exposes the trophoblast cells to the uterine wall (**McLaren**, 1970). At this stage, the trophoblast cells of blastocyst make contact with the endometrium of the uterus and ultimately form the placenta. The attachment of the blastocyst to the endometrium of uterus is called **implantation**. In rabbit, the process of implantation is initiated and controlled by a protein called **blastokinin** which is secreted by the uterus into the uterine cavity (**Berrill** and **Karp**, 1976).

Moreover, the trophoblast cells of mammalian blastocyst have the tendency to stick to any other maternal tissue besides the uterus and to result in further development of embryo. This causes the **ectopic pregnancy** or **ectopia**, *e.g.*, **tubal gestation** in Fallopian tubes, **abdominal gestation** in abdomen and **ovarian gestation** in ovary itself (**FitzGerald**, 1978). To prevent such an abnormality the zona pellucida layer remains around the blastocyst till it reaches the uterine cavity.

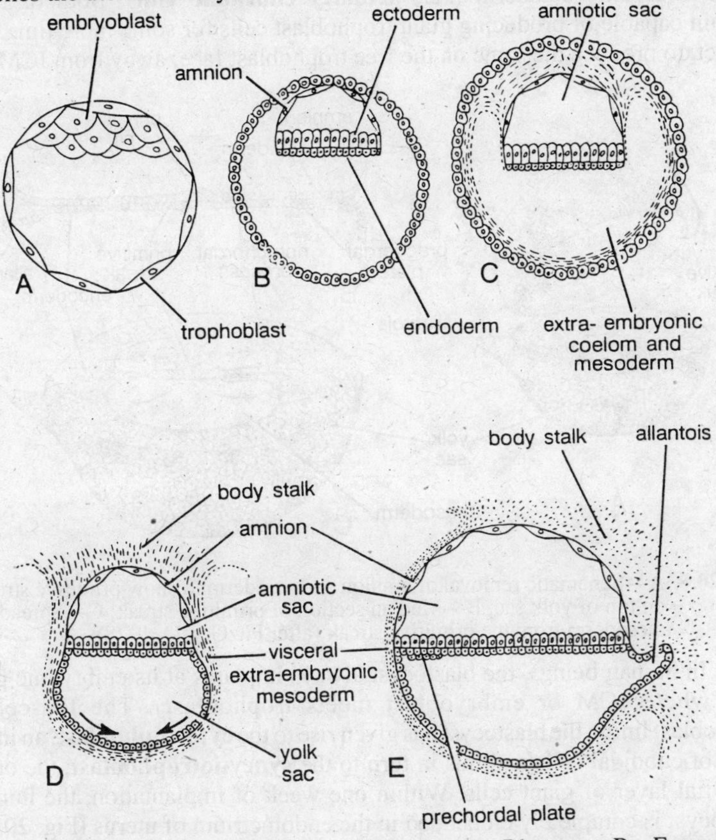

Fig. 20.5. Cleavage and differentiation of blastocyst. A—Blastocyst; B—Formation of ectoderm and endoderm; appearance of amniotic sac; C—Delamination of extra-embryonic mesoderm from trophoblast; D—Formation of yolk sac; E–Appearance of prechordal plate at the cephalic end of the embryo (after FitzGerald, 1978).

Within the rabbit's uterus (which consists of two long, tubular "horns") a solitary cleaving egg or blastocyst usually lodges near the midpoint of the horn containing it, whereas two or more eggs become evenly spaced along the horn.

6. The development of mouse is well worked out ; the time-table of its early developmental events has been shown in Table 20-1. In a 4-day

postcoitus mouse embryo, the mural trophectoderm (*i.e.*, the trophectoderm not in contact with ICM) cease dividing and start endoreduplication or endomitosis of their DNA (polyteny) to form syncytial primary trophoblast giant cells. These cells grow to become bigger than the egg and to invade the uterine tissue. The polar trophectoderm (also called **Rauber's layer**) overlying the ICM, continues dividing and goes on to form the **ectoplacental cone** and extra-embryonic ectoderm (*i.e.*, **primary chorionic villi**). Both of these remain capable of producing giant trophoblast cells for some more time, and in fact do produce the same on the free trophoblast face, away from ICM.

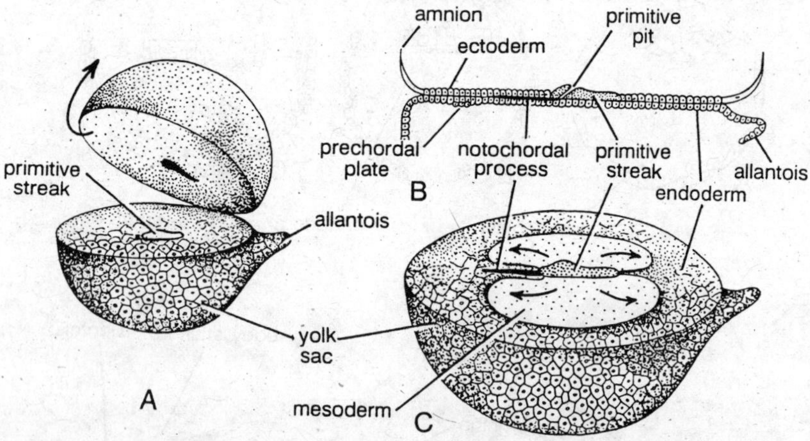

Fig. 20.6. A—Schematic removal of amnion and ectoderm to show primitive streak on top of yolk sac; B—Median section of primitive streak; C—Spread of mesoderm from the primitive streak (after FitzGerald, 1978).

In human beings, the blastocyst always implants at its embryonic pole (*i.e.*, where ICM or embryoblast meets trophoblast). The flat-celled trophoblast lining the blastocyst has given rise to the **cytotrophoblast**, an inner layer of cuboidal cells and this in turn to the **syncytiotrophoblast**, the outer syncytial layer of giant cells. Within one week of implantation, the human blastocyst is completely embedded in the endometrium of uterus (Fig. 20.7).

7. Meanwhile, further development of ICM of blastocyst takes place. The ICM cells facing the blastocoele become flattened and rich in rough endoplasmic reticulum. These are the **primary endodermal cells** which separate from the ICM and start migrating on the inner surface of the trophectoderm. The primary endoderm, thus, forms the lining of the blastocoele, now called the **yolk sac** ; it does not seem to contribute in any major way to the definite endoderm of the foetus.

In mammals (*i.e.*, mouse and man), a cavity, called **amniotic cavity**, appears between the ICM and the overlying polar trophectoderm. It is bounded by a layer of cells, the **amnion** (extra-embryonic ectoderm) on the outer, that is, trophoblast surface. The ICM below the amnion is called the **epiblast** (or

primitive ectoderm) ; it is composed of columnar cells (Fig. 20.5). Below it lies the **hypoblast** which has cuboidal cells of primitive or primary endoderm. Both epiblast and hypoblast form the bilaminar structure, called the **embryonic disc**. The epiblast, as is the case in the chick, gives rise to the entire embryo proper, *i.e.*, it gives rise to the three fundamental germinal layers such as secondary endoderm, mesoderm and ectoderm.

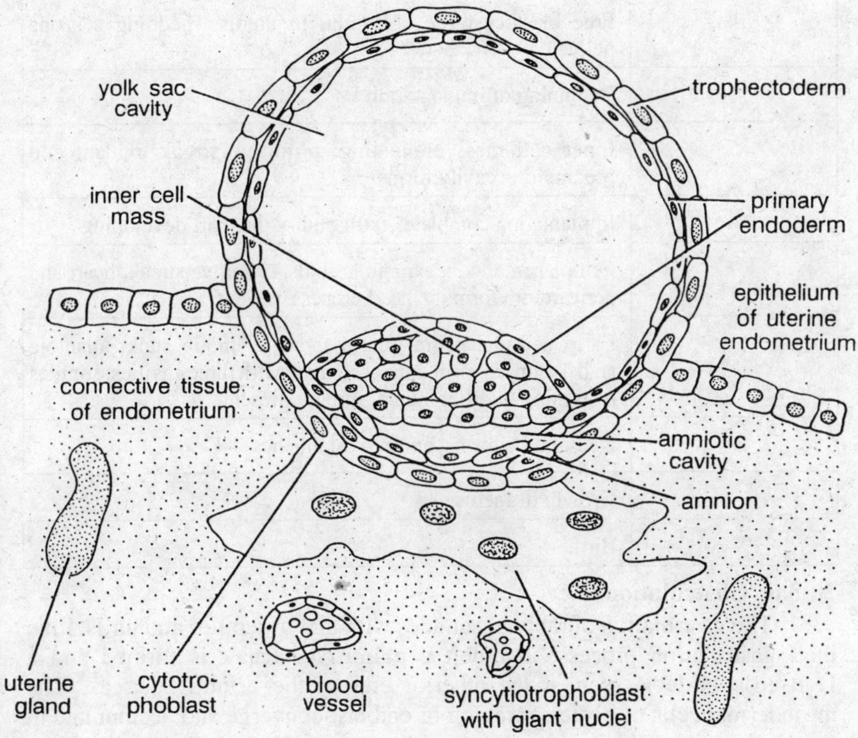

Fig. 20.7. Diagrammatic representation of an implanting blastocyst of man (after Goel, 1984).

8. Very soon, a loose network of cells delaminates, *i.e.*, splits off, from the inner surface of the trophoblast. This is the **extra-embryonic mesoderm;** it comes to line the inner surface of the trophoblast (forming **secondary chorionic villi)** and the outer surface of two vesicles—the amniotic sac and yolk sac. Ultimately fluid-filled loculi develop in the extra-embryonic mesoderm which coalesces to form a new cavity, the **extra-embryonic coelom**, which cleaves the mesoderm into **visceral** and **parietal** layers. The visceral layer of mesoderm surrounds the amniotic sac and yolk sac but it does not pass between them (Fig. 20.10). The parietal layer of mesoderm lines the interior of the trophectoderm, forms connective tissue and blood vessels (*i.e.*, forms the **tertiary chorionic villi**).

Table 20-1. Developmental time table of mouse (source, **Berrill** and **Karp,** 1976)

Days of gestation	Embryonic stage
1	One- to two-cell, in uppermost part of oviduct
2	Two- to sixteen-cell, in transit to uterus
3	Morula, in upper part of uterus
4	Free blastocyst (*i.e.,* blastula) in uterus, shedding of zona pellucida
$4\frac{1}{2}$	Beginning of implantation
5	Inner cell mass elongating, primitive streak evident and proamniotic cavity forms
6	Implantation complete, extra-embryonic part developing
7	Ectoplacental cone, amniotic folds, primitive streak, heart and pericardium forming, head process evident
$7\frac{1}{4}$	Early neurula, neural plate, chorio-amniotic stalk, allantoic stalk begins to form; inner cell mass with three cavities; somites beginning to differentiate; foregut present
$8\frac{1}{2}$	Neural tube forms, embryo establishes
9–11	Growth of foetus
19–20	Birth

20-3.6. Gastrulation

The gastrulation of rabbit resembles with that of reptiles and birds. Like their gastrulation process, in rabbit a **primitive streak** is formed and a **Hensen's node** is seen at the anterior end of the primitive streak. The mesodermal cells of posterior region of epiblast converge and accumulate in the mid-posterior thickening, the primitive streak. These cells soon involute through primitive streak and diverge laterally and anteriorly and occupy position between epiblast and hypoblast. Epiblast also contributes some cells to endoderm (hypoblast) with the convergence and involution of more mesodermal and endodermal cells. The size of primitive streak increases, however, overall size of it does not surpass half the length of blastodisc, as was the case in reptiles and birds. The cells migrating forward from **Hensen's node** remain packed more closely and give rise to the **"head process"**—the **notochord** rudiment (Fig. 20.6).

20-3.7. Neurulation

The primitive streak in rabbit and mammals as in birds, is a transient structure. Having given rise to the mesodermal layer and the notochord rudiment, the primitive streak starts shrinking its anterior end with Hensen's node receding farther and farther back, while in the anterior part of the blastodisc the germinal layers enter the next phase of development, the **neurulation** (Fig. 20.8 and Fig. 20.9).

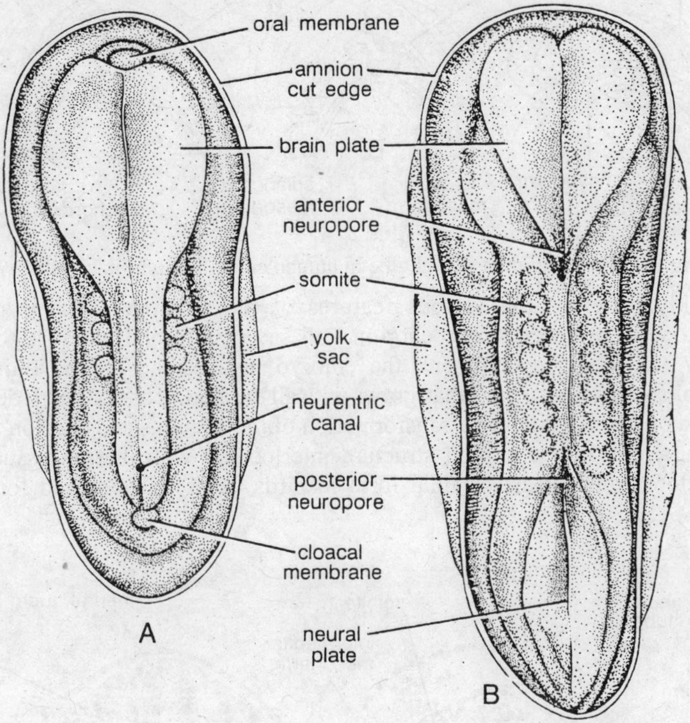

Fig. 20.8. Human embryo with opened amniotic sac from above to show neurulation. A—The neural plate is seen in the floor of the sac, flanked by three pairs of somites; B—The neural folds have united at the middle of the embryo; embryo showing seven pairs of somites (after FitzGerald, 1978).

During neurulation, the **neural ectoderm** gets separated from the **epidermal ectoderm** and forms **neural plate** which invaginates to form the **neural tube**. Side by side, the **mesoderm** differentiates into **epimere, mesomere** and **hypomere**. The hypomere or lateral plate mesoderm splits into an outer **somatic mesodermal layer** and an inner **splanchnic mesodermal layer**, both layers enclose an **extra-embryonic coelom** and themselves forming different **extra-embryonic membranes**, and also intra-embryonic parts. The **endoderm** forms a primitive gut, in a similar manner as has been described in chick.

20-4. PECULIAR FEATURES OF MAMMALIAN DEVELOPMENT

The mammalian egg and its development are noted for several

outstanding characteristics. The egg is very small (about 0.1 mm in diameter) and apparently yolkless, and it develops within the maternal uterus. Yet it

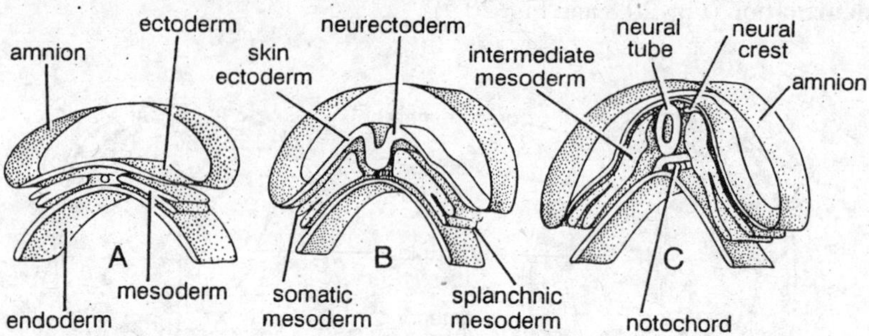

Fig. 20.9. Formation of neural tube in human embryo (after FitzGerald, 1978).

quickly develops the various features of the amniote embryo, such as blastoderm, primitive streak, chorion, yolk sac, amnion and allantois. At a very early stage in development, the embryo invades the uterine lining and establishes itself as a graft that grows with all the avidity of tumour. It develops into a placenta, which is a transformation of the allantois and chorion as seen in the chick into a unified structure interlocking with uterine tissue. This structure which in its original form in birds and reptiles served for waste

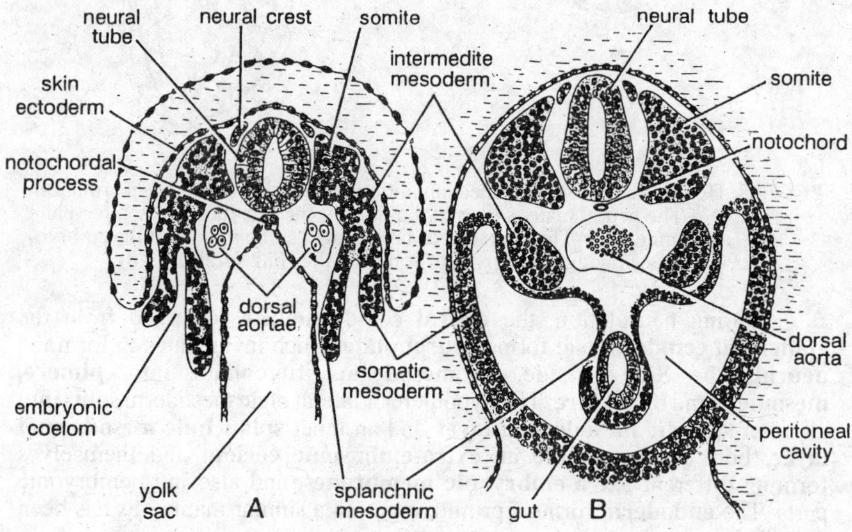

Fig. 20.10. T.S. through the middle of the human embryo to show the formation of head and tail folds and differentiation of mesoderm (after FitzGerald, 1978).

storage and for respiratory exchange, now functions as a selective organ for all metabolic exchange between embryo and mother. The stalk of allantois, containing the artery and the vein that supply the placenta, persists as the umbilical cord.

Further, since there is a lack of yolk in the eggs of viviparous mammals, food must be supplied to the embryo. It depends on the mother for food, oxygen and the elimination of wastes. An early differentiation of tissues in the developing ovum of mammals gives rise to the extra-embryonic membranes. These will function in intimate contact with the maternal tissue and carry out the vital functions of the embryo. By the time implantation occurs two of these foetal membranes, the amnion and the yolk sac, have started to develop. But *gastrulation lags behind.* There is a precocious appearance of extra-embryonic mesoderm in the mammalian embryo, before intra-embryonic mesoderm forms ; in fact, even before the appearance of the primitive streak. Essentially, the developing mammalian egg first establishes the mechanics for maintaining the embryo before it constructs its body form (**Phillips**, 1975).

20-5. DEVELOPMENT OF EXTRA-EMBRYONIC MEMBRANES OF RABBIT

The developing embryos of rabbit and other eutherian mammals are provided with four extra-embryonic membranes namely, **amnion, chorion, yolk-sac** and **allantois** similar to the chick. However, they differ markedly with that of latter in the method of their formation and also in their morphology. Different extra-embryonic membranes of rabbit develop by following methods (Fig. 20.11).

20-5.1. Development of Amnion and Chorion

The development of amnion and chorion occurs after gastrulation and neurulation. The amnion and chorion are formed by the folds of **somatopleure** (somatic mesoderm plus trophoblast cells or ectoderm), called **amniotic folds**. The amniotic fold occurring towards the anterior or cephalic end of the embryo is called the **head fold** and which occurs towards the caudal or tail end, is called **tail fold**. Both of these folds grow over the embryo and fuse at mid-dorsal point of the embryo. After the fusion of amniotic folds, the body of the embryo becomes enclosed in the **amniotic cavity**. The innermost somatopleure having an inner ectoderm facing the amniotic cavity and an outer mesoderm, becomes **amnion**, while the outermost somatopleure of amniotic folds, having ectoderm at its outermost side and mesoderm at its inner side, forms the **chorion**.

This method of amnion formation is found only in primitive eutherian mammals such as marsupials, Artiodactyla, ungulates, pigs, Lagomorpha (rabbit), most Insectivora and lemurs. In other more advanced mammals, the amnion is formed by different method. In them amniotic cavity makes its appearance as a cleft in the inner cell mass. This method of amnion formation is called **cavitation**. The cells which are above the cavity correspond to the amniotic folds (chorion and amnion). The cells which are below the cleft, *i.e.*, between the cleft and the yolk sac, form the embryo proper.

Functions of amnion and chorion. The amnion may be thought of as a membranous tent around the embryo. It contains a fluid, the amniotic fluid, which bathes the outside of the embryo. Its function is to free the embryo from pressures to permit freedom of motion, and to protect the outer surface of the embryo from abrasion.

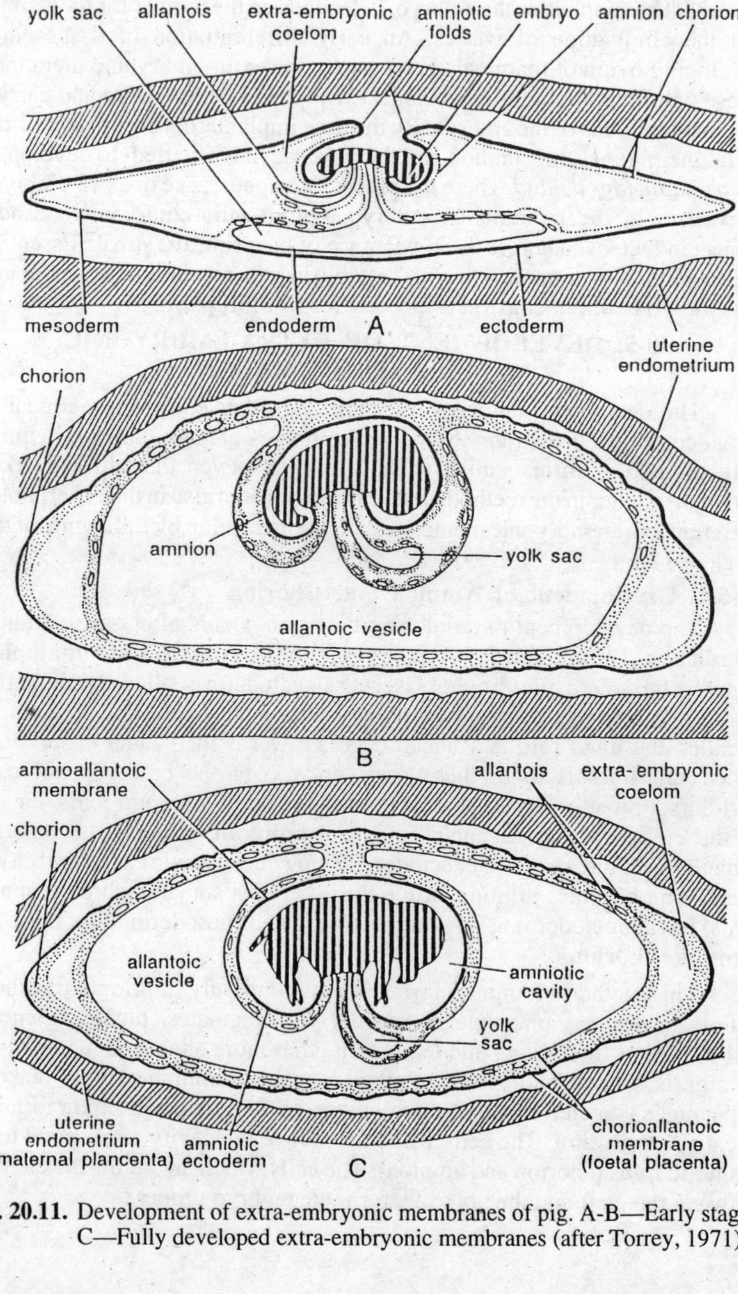

Fig. 20.11. Development of extra-embryonic membranes of pig. A-B—Early stages; C—Fully developed extra-embryonic membranes (after Torrey, 1971).

The chorion, since it is the outer layer, is the place where exchange of substances occurs between embryonic tissue and the maternal environment. In most mammals, the chorion develops more or less complex outgrowth known as **villi**. These villi penetrate into the tissue of the uterus wall and greatly increase the absorptive area of chorion.

20-5.2. Development of Yolk Sac

In rabbit, the endoderm spreads over and encloses the cavity of the yolk sac and the extra-embryonic splanchnopleure of yolk sac has no digestive role like that of chick, but it has a respiratory, haemopoetic (blood cell forming), and storage function.

In most marsupials the wall of yolk sac establishes intimate contact with the uterine wall, forming thereby the **yolk sac placenta**. However, in most placental mammals, the yolk sac is small and is speared from the chorion by the extra-embryonic coelom. The embryo proper remains separated from the yolk mass by body folds and remains connected with the latter by the **yolk stalk**. During later stages, with the gradual development of allantois, the yolk sac starts reducing and finally shrivelled out. This shrivelled yolk sac, yolk stalk and the allantoic stalk constitute the **belly stalk** or **umbilicus**.

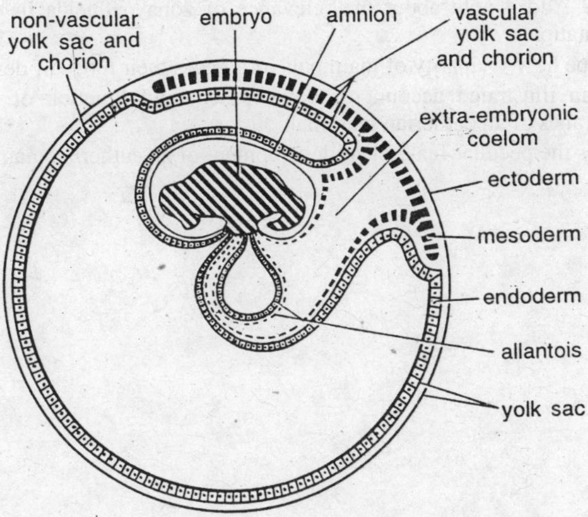

Fig. 20.12. The extra-embryonic membranes of *Opossum* (after Torrey, 1971).

20-5.3. Development of Allantois

The allantois of rabbit and other mammals is composed of splanchnopleure (endoderm + splanchnic mesoderm). It rises from the hind-gut as a small sac which gradually increases in the extra-embryonic coelom, *i.e.*, space between the amnion and chorion. Soon, it grows over the embryo outside the amnion where it enlarges. Later on, mesodermal components of allantoic wall give rise to the blood vessel system of the allantois and fuse with mesoderm of chorion to form the **allantoic chorion.**

Functions of allantois. With the acquisition of viviparity in mammals, the original function of the allantois as a urinary bladder becomes altogether lost. As the maternal organism supplies the embryo with all the necessary nutrition, so it also takes over the removal of the waste products (CO_2, urea, etc.) of metabolism from the embryo. However, allantois of mammals supplies oxygen and nutrients to the embryo, which are taken from maternal blood circulation.

REVISION QUESTIONS

1. The mammalian egg resembles an oligolecithal invertebrate egg, yet its cleavage does not. What is the significance of this cleavage pattern, with particular reference to the evolution of mammal ?
2. Describe the early embryonic development of an eutherian mammal.
3. How do the meroblastic egg of the chick and the holoblastic egg of a mammal acquire a comparable blastula ?
4. Comment on the observation that mammalian eggs do not have large quantities of stored yolk but the embryos have a well developed yolk sac : what functions does the yolk sac perform ?
5. Trace the fate of trophectoderm in a mammalian embryo.
6. What is implantation ? How does the fertilized ovum of a mammal implant itself ? Add a note about the relevance of zona pellucida in the process of implantation.
7. Describe how viviparity of mammals influences their mode of development.
8. Give an illustrated account of development and function of various foetal membranes of an eutherian mammal.
9. Discuss the peculiar features of development of an eutherian mammal.

21

PLACENTA AND PLACENTATION

The **placenta** is a Greek word and it means a "*flat cake*". This name is received from the human placenta which is flat, rounded mass, shaped more or less like a pancake. The term placenta in its broadest sense refers to any region in a viviparous organism where maternal and embryonic tissues of any kind are closely apposed, and which serves as a site for physiological exchanges between parent and embryo (**Kent**, 1987). More aptly the placenta can be defined as a temporary organ which is formed jointly by the extra-embryonic membranes of the foetus and maternal tissues and by which the developing embryo or foetus of viviparous mammals obtains its nourishment from the maternal uterine tissue.

Indeed, the true placenta of the eutherian mammals is a marvel of biologic engineering. Serving as a provisional lung, intestine, kidney and endocrine gland for the embryo, it allows the maternal and foetal blood to come in proximity for the mutual exchange of substances. The placenta permits the entry of foodstuffs, vitamins, oxygen, hormones and antibodies and the exit of carbon dioxide and nitrogenous metabolic wastes. Its actions are highly **selective**, for it allows some materials to enter that are chemically quite similar to others that are rejected.

The mode of formation and fusion of the placenta to the uterine wall is called **placentation**. The phenomenon of placentation is invariably related with the viviparity of the organism. In all viviparous animals, the development of the young one takes place inside the uterus of the mother because the amount of stored yolk of egg does not remain sufficient for the development of an embryo which can lead an independent mode of existence, therefore, the developing embryo has to depend on the mother in lesser or greater degree for nourishment, oxygen supply and other physiological assignments.

The placentae are not found exclusively in mammals but also appear in animals belonging to various groups of the animal kingdom such as in *Peripatus* (Onychophora), *Salpa* (Tunicata), *Mustelus laevis* (Elasmobranchia) and certain lizards (Reptilia). The nature of the tissues entering into the formation of the placenta is not the same in all cases. For example, in placentae of fishes and reptiles, the yolk sac comes in close relation with maternal blood stream and vitelline circulation carries the materials to and from the embryo.

21-1. IMPLANTATION

Before substantial growth can occur, the blastocyst must attach to the uterus and establish the nutritional supply of the embryo. This process of attachment is called **implantation**. To start the process of implantation, the blastocyst is held closely against the uterine luminal (mucosal) epithelium. The uterine capillaries and uterine wall in the immediate vicinity of the embryo become more permeable and cause a local stromal oedema. Soon the mucosal lining of the uterus (*i.e.*, endometrium) around the embryo shows the first sign of a decidual cell reaction (DCR) : the epithelium becomes disrupted and the loosely packed fibroblast-like cells of the stroma are transformed into large rounded glycogen-filled cells tightly bound to each other by gap junctions. The number of cells increases and there is normally an increase in the vasuclarization of the area. The decidual cells, thus, form an "**implantation chamber**" around the embryo and probably have a nutritive role before the establishment of a functional placenta. However, DCR is not a necessary precondition for implantation in all species.

The trophoblast is considered a primary foetal membrane of the mammals. Its origin traces back to the superficial layer of the morula. Later when mesoderm lines its cavity, it becomes the chorion and contributes to the formation of placenta. The trophoblast penetrates the uterine epithelium perhaps by both cytolytic and mechanical activity. The expanding trophoblast in the process of implantation may destroy, pass in between or fuse with the uterine epithelial cells, and destroy and phagocytose the decidual cells. The invasion of the trophoblast through the decidual cells is arrested only when it reaches a layer of healthy decidual tissue and establishes contact with uterine blood vessels. The trophoblast invasion is particularly apparent in the area overlying the inner cell mass or embryoblast and leads to the formation of a distal ectoplacental cone. This is the primordium of the allantoic placenta.

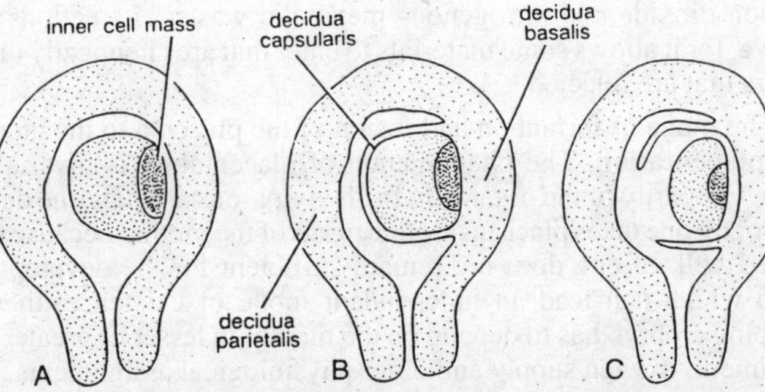

Fig. 21.1. Schematic representation of three types of implantation. A—Centric implantation; B—Interstitial implantation; and C—Eccentric implantation becoming secondarily interstitial (after Goel, 1984).

The pattern of implantation of the blastocyst varies in different species (Fig. 21.1). For example, in most ungulates (pig), carnivores (dog) and monkey, the implantation is **superficial** or **central** (Fig. 21.1A), that is, the blastocyst remains unembedded in the uterine lumen. In some cases such as in hedgehog, guinea pig, some bats, ape and man, the blastocyst burrows into the endometrium and implantation is **interstitial** (Fig. 21.1B). In beaver, rat, squirrel and some other rodents, the implantation is primarily **eccentric**, *i.e.*, the chorionic sac lies for a time in a fold or pocket which looses off from the main cavity ; secondarily it becomes interstitial. Thus, in the human female the maternal component consists of that part of the uterine wall or endometrium that sloughs off at the termination of pregnancy. The name "**decidua**" (meaning "to shed") refers to the pregnant endometrium, with regions of the decidua assigned different names depending on their relationship to the implantation site, as indicated in Figure 21.1C :

1. Decidua basalis refers to that part of the endometrium that is directly underlying the embryo. The part of the embryo which is in contact with decidua basalis, tends to contribute to the placenta.

2. Decidua capsularis refers to thin portion of endometrium on the lumen side of the uterus that covers the implantation site.

3. Decidua parietalis refers to the endometrium lining the uterus other than at the implantation site.

21-2. TWO BASIC TYPES OF PLACENTAE

The placenta from the point of view of its origin, consists of two parts — a **foetal placenta**, furnished by the extra-embryonic membranes and a **maternal placenta**, furnished by the uterine endometrium. Now, it would be obvious that while on the maternal side a single component, the **endometrium** is involved ; on the foetal side, one has to consider the prospective roles of four elements—**amnion, chorion, yolk sac** and **allantois**. The first of these, the amnion, may be ruled out immediately, as it is making no direct contribution to the placenta. This leaves the other three, of which the chorion, because of its most external position, is the membrane making immediate contact with the endometrium. But, we have seen in chick embryo that the chorion plays its role by way of vascular supply, which it acquires from the allantois. In mammals, there are two possible sources of chorionic vascularization—the **vitelline circulation** provided by the yolk sac and **allantoic circulation** provided by the allantois. Thus, it can be said that in mammals, there exist two essentially different main types of placentae—the **chorio-vitelline placenta** and the **chorio-allantoic placenta**. (Fig. 21.2).

21-2.1. Chorio-Vitelline Placenta (or Chorionic Placenta or Yolk-sac Placenta)

This type of placenta is present in most marsupials (*Didelphys, Macropus*) and is the sole means of foetal nutrition in many species. It also occurs as a temporary structure in many eutherians, for example, in insectivores, rodents and horse. Many biologists consider the marsupials, along with the monotremes, the **aplacental mammals**, however, some acknowledged that the inefficient apposion of foetal yolk sac and chorion with maternal tisue functions as a kind of placenta (**Phillips**, 1975).

The chorio-vitelline placenta may begin its development with only one cell layer, the trophectoderm, (which forms **chorion**) enclosing the blastocoele. This is called **chorionic placenta**. Later it becomes bilaminar due to the blastocoele acquiring the endodermal lining. Now it is called **yolk sac placenta**. Still later, in many species, it becomes trilaminar, due to invasion of extra-embryonic mesoderm and its vascular supply between the trophectoderm and the endoderm. At this stage, it is called **chorio-vitelline placenta**.

Thus, in a chorio-vitelline placenta the allantois remains relatively small and never makes contact with the chorion ; whereas the yolk sac becomes very large and gets fused broadly with the chorion. However, the chorion never advances beyond a smooth membrane in close apposition with the vascular uterine lining, the endometrium.

In some cases, the breakdown of the distal (parietal) part of yolk sac brings the endoderm of the proximal (visceral) part of the yolk sac (one underlying the epiblast) in close proximity with the uterine epithelium. Such a placenta is referred to as "**inverted yolk sac placenta**" because the endoderm of yolk sac rather than the chorionic ectoderm faces the uterine endometrium. This type of yolk sac placenta is found in rodents and in the rat it coexists with the chorio-allantoic placenta until term, that is, parturition.

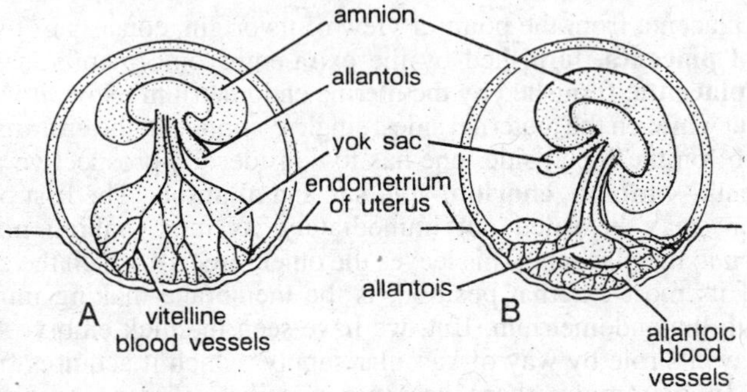

Fig. 21.2. Comparison of two types of mammalian placentae : A—Chorio-vitelline placenta; B—Chorio-allantoic placenta (after Phillips, 1975).

21-2.2. Chorio-Allantoic or Allantoic Placenta

In some marsupials (*e.g., Parameles, Dasyurus*), and all eutherians the yolk sac remains rudimentary and the allantois becomes well developed and vascularized to fuse with chorion and to furnish the latter the blood supply. Such a foetal placenta is called **chorio-allantoic placenta**. In this kind of placenta, the chorion is not smooth but bears root-like vascular processes, the **villi**, which grow out from the chorion into the adjacent maternal tissue.

Thus, in chorio-allantoic placenta, it is that part of chorion which is lined with the allantois that becomes associated with the uterine wall during the gestation period. In it the umbilical (allantoic) vessel transport the materials. As the embryonic body develops, it separates from various membranes until only the **umbilical cord** connects the foetus with the placenta. The mode of formation of chorio-allantoic placenta during development is varied. However, the final foetal part of this placenta consists of chorionic **villi**. Each villus has an outer chorionic ectoderm, an inner allantoic endoderm and a highly vascularized mesodermal connective tissue layer in between. The mesodermal layer has dual origin—somatic or chorionic and splanchnic or allantoic. The ectodermal cells have microvilli to increase the area for absorption of nutrients. The trophoblast that gives rise to the villi often becomes syncytial on its outer border towards endometrium (syncytiotrophoblast) compared to inner cytotrophoblast.

The maternal part of placenta consists of the epithelium of endometrium and the stroma of endometrium with its glands and the uterine blood vessels.

21-2.3. Classification of Chorio-Allantoic Placenta

The chorio-allantoic placenta has been classified into various types ; its classification has been based on its shape or morphology (**T.H. Huxley**, 1864) ; arrangement of villi (**Hubrecht**, 1894) ; histology (**Grosser**, 1927) and electron microscopy.

A. Morphological Classification of Placenta

On the degree of intimacy of foetal and maternal tissues following three types of placentae may be recognised :

1. Non-deciduous placenta or semiplacenta. In most mammals, the implantation is superficial, *i.e.*, the blastocyst lies in the cavity of the uterus in contact with the uterine wall. The contact may be made more intimate by the surface of the blastocyst by forming finger-like outgrowths which penetrate into depressions in the wall of the uterus. Such outgrowths are initially formed by the trophoblast (*i.e.*, the epithelial layer covering the blastocyst), but later on the connective tissue and blood vessels invade the outgrowths. These outgrowths are called **chorionic villi**, the blood vessels of chorionic villi are the branches of allantoic blood vessels in case of chorio-allantoic placenta. (In chorio-vitelline placenta, vitelline blood vessels give their branches to chorionic villi).

At the time of birth, when **parturition** (the separation of the foetus and its membranes from the mother's body) occurs, the chorionic villi are simply drawn out from the depressions in the wall of the uterus, and, thus, maternal and foetal tissue are separated without further damage to the uterine wall and no bleeding occurs. This type of placenta is called **non-deciduate** or **non-deciduous placenta** and is found in pigs, cattle and some other mammals. Further, chorionic villi of a non-deciduous placenta, because lie in apposition with the endometrium, but, do not fuse with it, so such a placenta is also called **semiplacenta**.

2. Deciduous placenta or placenta vera. In other mammals, however, the degree of intimacy between maternal and foetal tissues becomes further increased. The wall of the uterus becomes eroded to various degrees through the action of the trophoblast and the embyronic tissues penetrate into the uterine wall establishing a more intimate contact and facilitating the passage of substances from the mother to the foetus and from the foetus to the mother. Here because the chorionic villi fuse with the eroded uterine mucosa, such placenta is called **placenta vera** (true placenta). At the end of pregnancy the uterine wall is no longer intact and when the foetus with its membranes including the chorion is removed, more or less extensive haemorrhage from the uterine wall ensues (*i.e.*, at birth, when placenta is discharged, the uterine lining also tears away with some bleeding). Such a type of placenta found in higher eutherian mammals is called **deciduate** or **deciduous placenta**.

The maternal tissues which are expelled at birth in the case of deciduate placenta are called **deciduae**. The haemorrhage at parturition is normally stopped by the same mechanism as serves for the expulsion of the newborn, the contraction of the muscular wall of the uterus constricts the blood vessels and, thus, slows down the flow of blood, until clotting of the blood stops the haemorrhage altogether.

3. Contra-deciduate placenta. In *Parameles* and *Talpa* (mole), somewhat modified type of deciduate placenta occurs, which is called **contra-deciduate placenta**. In such case, not only there is a loss of maternal tissue but also of the foetal portion of the placenta, both of which absorbed in situ by maternal leucocytes.

B. Classification of Placentae according to the Distribution of Villi on Chorion

In different mammals the pattern of distribution of villi varies from species to species and accordingly following kinds of placentae have been recognised :

1. Diffuse placenta. In some mammals (*e.g.*, ungulates, pig, sow, mare, horse, lemur, etc.) the villi remain scattered all over the surface of the chorion and their placentae are correspondigly expensive. Such placentae are called diffuse placentae.

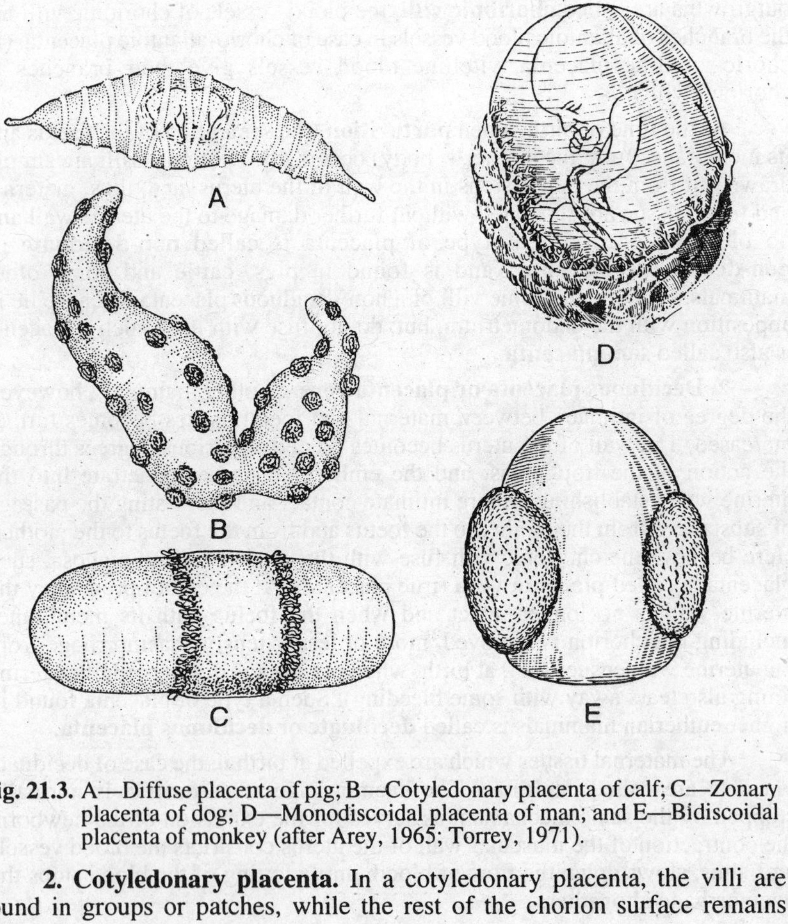

Fig. 21.3. A—Diffuse placenta of pig; B—Cotyledonary placenta of calf; C—Zonary placenta of dog; D—Monodiscoidal placenta of man; and E—Bidiscoidal placenta of monkey (after Arey, 1965; Torrey, 1971).

2. Cotyledonary placenta. In a cotyledonary placenta, the villi are found in groups or patches, while the rest of the chorion surface remains smooth. The rosettes or patches of villi are called **cotyledons**, and the placenta of this type is found in ruminants (cud-chewing) ungulates such as, cattle, sheep and deer.

Camel and giraffe have an intermediate type of placenta in which villi are arranged in cotyledons as well as are scattered.

3. Zonary placenta. In a zonary placenta, the villi are developed in the form of a belt or girdle-like band around the middle of their blastocyst or chorionic sac, which is more or less elliptical in shape. Such a placenta occurs in carnivores (*e.g.*, cats, dogs, etc.). Raccoon has incomplete zonary placenta.

4. Discoidal placenta. In a discoidal placenta, the villi are restricted to a circular disc or plate on the dorsal surface of blastocyst. Such a placenta occurs in insectivores, bats, rodents (rat, mouse), rabbit and bear.

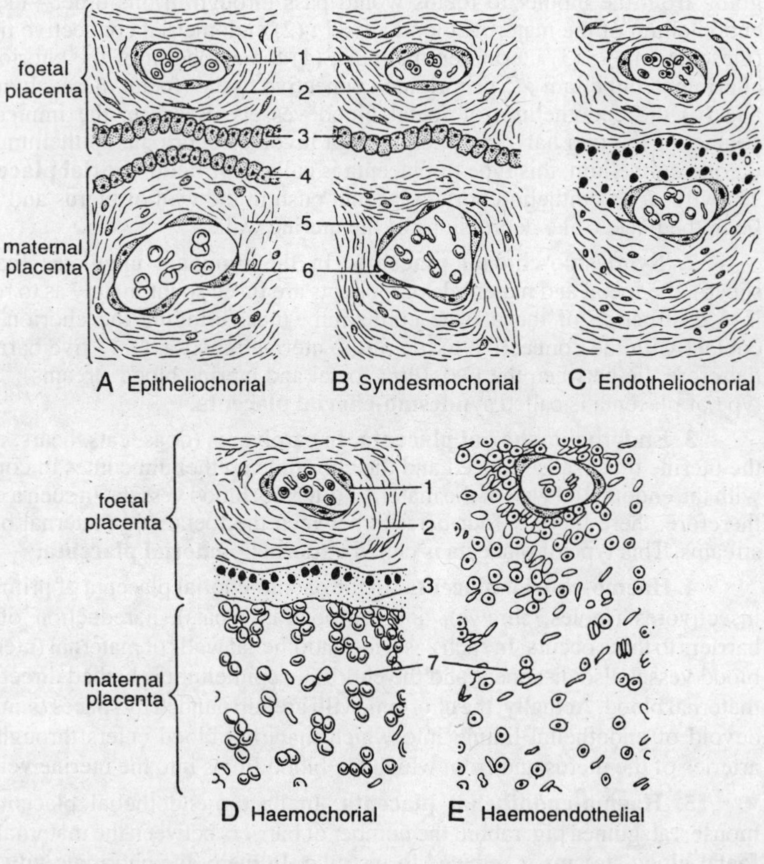

Fig. 21.4. Mammalian placental types arranged in a series to show the progressive elimination of barriers between the maternal and foetal circulations. 1. Endothelium of foetal blood vessel; 2. Chorionic connective tissue; 3. Chorionic epithelium; 4. Uterine epithelium; 5. Endometrial connective tissue (mucosa); 6. Endothelium of maternal blood vessel; 7. Maternal blood pool (after Torrey, 1971).

5. Metadiscoidal placenta. Primates have a special type of discoidal placenta in which villi are at first scattered but later on become restricted to one or two discs. Thus, **monodiscoidal placenta** of man has a single disc-shaped villous area and **bidiscoidal placenta** of monkeys and apes has two disc-shaped villous areas (Fig. 21.3).

C. Histological Types of Placenta

On histological basis, following types of mammalian placentae have been recognised :

1. Epithelio-chorial placenta. The epithelio-chorial type placenta is most primitive type placenta and it is found in marsupials, ungulates (pig, horse, sow, cattle, etc.) and lemurs. In such a case, no fewer than six tissues or membranes (called **barriers**) lie between the foetal and maternal blood streams, therefore, the molecules of nutrients and oxygen, for instance, in going from the mother to foetus would pass through in this order—(1) the endothelium of the maternal blood vessel ; (2) endometrial connective tissue (mesenchyme) ; (3) uterine epithelium ; (4) the ectoderm of the chorion or chorionic epithelium ; (5) chorionic connective tissue (foetal mesenchyme) ; and (6) the endothelium of foetal blood vessels. Because, the immediate contact of the two halves of the placenta involves chorionic epithelium and uterine epithelium, this type of placenta is called **epithelio-chorial placenta**. The villi of an epithelio-chorial placenta push in the wall of uterus and later lie in the pocket-like depressions of the uterine wall.

2. Syndesmo-chorial placenta. In the ruminant ungulates (cattle, sheep), the foetal and maternal components are fused so intimately as to result in a destruction of the uterine epithelium, thus, bringing the chorion into contact with the connective tissue of the uterine mucosa. Only five barriers, therefore, lie between the two, (*viz.*, foetal and uterine blood streams). This type of placenta is called **syndesmo-chorial placenta.**

3. Endothelio-chorial placenta. In carnivores (dogs, cats, bears, etc.), the uterine mucosa is reduced and the chorionic epithelium comes in contact with the endothelial wall of the maternal (uterine) blood vessels. In such a case, therefore, there lie only four barriers between the foetal and maternal blood streams. This type of placenta is called **endothelio-chorial placenta.**

4. Haemo-chorial placenta. In the haemo-chorial placenta of primates, insectivores (moles, shrews), and chiropterans (bats), a reduction of the barriers to three occurs. In such case, the endothelial walls of maternal (uterine) blood vessel also disappear and the chorionic epithelium is bathed directly in maternal blood. Actually, the chorionic villi are surrounded by spaces (sinuses) devoid of endothelial lining, into which maternal blood enters through the arteries of the uterus and from which the blood flows into the uterine vein.

5. Haemo-endothelial placenta. In haemo-endothelial placenta of mouse, rat, guinea pig, rabbit, the number of barriers between the maternal and foetal blood streams is reduced to just two. In them, the chorionic villi lose their epithelial and mesenchymal layers to such a degree that, in most places, the essentially bare endothelial lining of their blood vessels alone separates the foetal blood from the maternal sinuses.

D. Types of Placenta Based on Electron Microscopic Studies

Recent careful electron microscopic observations of already discussed five histological types of eutherian placenta have made the two histological categories, namely syndesmo-chorial and haemo-endothelial, untenable and superfluous. EM has recognized only three histological types of placenta : (1) **Epithelio-chorial placenta** ; (2) **Endothelio-chorial placenta** ; (3) **Haemo-chorial placenta**. Some workers have classified haemo-chorial placenta into three subtypes : **haemomono** -, **haemodi** -, or **heamotri-chorial** placentae on the basis of number of layers of trophoblast covering the foetal endothelium (see **Goel**, 1984).

In summary, mammalian placentae can be classified in several ways, with gradations existing within each category. The classification may be based on the type of blood vessels involved, the shape of the palcenta, whether or not the endometrium is shed at parturition, and the degree of intimacy between chorion and the decidua basalis. For example, the human placenta can be classified as chorio-allantoic, monodiscoidal, deciduate and haemo-chorial.

21-3. HORMONAL CONTROL OF IMPLANTATION

The process of implantation in mammals has been fully investigated only in rats and mice. It is investigated that before occurring of the implantation, the uterine wall is sensitized or prepared by joint activities of two ovarian hormones, namely **progesterone** and **oestrogen**. For optimum uterine sensitization, a high level of oestrogen at the time of oestrus seems to be necessary, followed by a period of several days when progesterone, produced by the developing corpora lutea, is the dominant hormone ; the immediate stimulus for sensitization is given by a very small amount of oestrogen which acts on the uterus on the morning of the 4th day of pregnancy.

Biochemical studies on the uteri of ovariectomized rats treated with oestrogen have shown that oestrogen molecules rapidly attach to cytoplasmic receptor sites in the uterus, and pass from the cytoplasm into the nucleus, where these are associated with the chromatin. **G.P. Talwar** (1969) has reported that oestrogens act to regulate the expression of particular genes within the uterine or other target cells. Estradiol (oestrogen hormone) is suggested to act in nucleus to initiate the transcription of specific sections of DNA (*i.e.*, cistrons) and produce specific kinds of mRNA for the synthesis of specific proteins. Some of these proteins have enzymatic activity and are set free in the uterine lumen, where they dissolve the zona pellucida of mouse blastocyst and activate the latter for implantation. These proteins activate the blastocyst either by increasing its metabolic rate directly or by removing some inhibitor of it. The activated blastocyst in turn has an effect upon the uterus, stimulating the uterine endometrium in its immediate vicinity to undergo decidual cell reaction which is essential if implantation is to occur. Progesterone acts on only those uterine cells which are previously sensitized by oestrogen hormone. Thus, progesterone and oestrogen hormones are of special importance in preparing the uterus for the implantation of blastocysts, in maintaining pregnancy, and in regulating the accessory organs during the reproductive cycle (see **Turner** and **Bagnara**, 1971 ; **Austin** and **Short**, 1972).

Further, it has also been studied that if the uterus has not been sensitized, either because the pregnant mouse is suckling a previous litter, or because the ovaries have been surgically removed earlier in pregnancy, the blastocysts enter a state of dormancy or delay, and implantation is postponed. Delay of implantation during lactation occurs in mice and rats and is also widespread among marsupials where the period of gestation is very short. It ensures that the next litter will not be born until the earlier one has been weaned. A more extended delay of implantation (called **embryonic diapause**) occurs in other groups, *e.g.*, badgers, bears, seals, roe deer and long-tailed weasels, as a device to extend the gestation period so that the young will be born at a time of year optimal for survival (**Austin** and **Short** 1972).

21-4. EMBRYONIC NUTRITION IN MAMMALS

Since the mammalian egg lacks yolk, the embryo has no immediate source of food. During the first few days the embryo metabolizes its endogenous reserves of glycogen and glucose for its energy needs. For this reason the weight of the embryo diminishes, for example, by 40 per cent in the cows.

The **preimplantation** embryo already begins to receive its nutrition from the mother. For example, the contents of yolk sac come to have the same composition as the maternal blood plasma, emphasizing the placental role of early yolk sac. Further, the uterine milk serves as the embryonic food until implantation, which occurs after about 4 days in mouse, 4 to 6 days in man, and 50 to 60 days in horse. The **uterine milk** is a mixture of secretion from endometrial glands and tissue debris. It is rich in glycogen, mucopolysaccharides and lipids. The uptake of the uterine milk is done by specialized cells of the trophoblast.

After implantation, in many species the decidual cells may have a nutritive role before the establishment of a true placenta. These are referred to as **embryotroph**. After the development of a true placenta, the nutrients are taken up by diffusion, active transport and endocytosis (pinocytosis, phagocytosis) by the numerous microvilli present on the absorptive surface of the trophoblast epithelium covering the chorionic villi. Glycogen and glucose appear to be principal source of energy for the developing embryo. The yolk sac placenta of rodents is also characterized by its high glycogen content, which is depleted towards the end of the term.

21-5. PHYSIOLOGY OF PLACENTA (Functions of Placenta)

Placentation is the mechanism by which the foetal and maternal blood circulations are brought very close together for conducting various metabolic functions such as respiration, excretion, and nutrition of foetus. However, there is no mixture of two bloods or fusion of two blood streams. Mother's blood does not circulate in foetus and vice versa. Exchange of substances occurs from one circulation to other through the various **placental barriers** which serve as an **ultrafilter** or **semipermeable membrane**. Such exchange takes place by a variety of transfer mechanisms such as (1) simple diffusion ; (2) facilitated diffusion (accelerated) ; (3) active transport (against concentration gradient) ; (4) pinocytosis (droplet transfer, virtually intact, across the syncytiotrophoblast) ; and (5) leakage (breaking the placental membrane). Various functions of placenta are performed in the following manner :

21-5.1. Nutritive Function of Placenta

The foetus obtains its nutrients from the maternal blood and when the diet is inadequate, then only, depletion of maternal tissue storage occurs. Thus, to avoid maternal tissue storage depletion, a diet rich in essential foods is needed during pregnancy by the mother. Different nutrients are obtained by the following methods :

1. **Glucose** which is the principal source of energy of foetus, is transferred to the foetus by facilitated diffusion.

2. **Lipids** for foetal growth and development are transferred across the foetal membrane or built up in placenta or synthesized in the foetus. **Triglycerides** and **fatty acids** are directly transported from the mother to the foetus in early pregnancy but probably are synthesized in the foetus later in pregnancy. **Cholesterol** is capable of direct transfer through the placenta.

3. **Amino acids** are transferred by active transport. Amino acid concentration is higher in the foetal blood than in the maternal blood. Foetal proteins are synthesized from the transferred amino acids and their level is lower than in mother.

4. **Water** and **electrolytes** such as **sodium**, **potassium** and **chloride** cross through the foetal membrane by simple diffusion. **Calcium**, **phosphorus** (phosphate), **iron** and **iodine** (iodide) cross over placenta by active transport.

5. **Water soluble vitamins** (*e.g.*, ascorbic acid) are transferred by active transport but the **fat soluble vitamins** are transferred slowly by simple diffusion.

6. **Hormones** such as insulin ; steroids from the adrenals; thyroid hormones ; chorionic gonadotrophin or placental lactogen cross the placenta at a very slow rate.

Most macromolecules (antibodies) and protein-bound nutrients such as **cholesterol**, iron-transporting protein **transferrin**, **vitamin B_{12}**, etc., are taken in by placenta by **receptor-mediated endocytosis**. The same cellular mechanism is involved in the penetration of **toxins** (*e.g.*, diphtheria toxin) and certain **viruses** into the foetus (**De Robertis** and **De Robertis, Jr.** 1987).

21-5.2. Respiratory Function of Placenta

Although, the foetal respiratory movements take place *in utero* (L., *within the womb)* there is no gaseous exchange. Intake of oxygen and output of carbon dioxide takes place by **simple diffusion** across the foetal membrane. Oxygen from maternal blood diffuses into the foetal blood, where oxygen concentration is comparatively less than in maternal blood. Similarly, diffusion of carbon dioxide from the foetal blood into the maternal blood stream is also facilitated by placenta.

21-5.3. Excretory Function of Placenta

Waste products from the foetus such as **urea**, **uric acid** and **creatinine** are excreted to the maternal blood via the placenta by simple diffusion. The kidneys of the mother eliminate these waste products of foetal metabolism along with her own waste products.

21-5.4. Enzymatic Function of Placenta

Numerous enzymes which protect the foetus are elaborated in the placenta, *e.g.*, **diamine oxidase**, **oxytocinase** and **phospholipase-A_2.**

21-5.5. Endocrine Function of Placenta

Placenta of mammals acts temporarily as a new endocrine organ. It secretes a variety of hormones. For example, in the rat on twelfth day of the pregnancy, placenta secretes the hormone, **rat chorionic mammoluteotrophin** (which is responsible for the maintenance of the corpora lutea) and later **progesterone** (which is responsible for pregnancy to continue to term). Likewise, horse placenta secretes the **pregnant mare serum gonadotrophin (PMSG)**, a luteotrophic hormone, in addition to **progesterones** and **oestrogens**. In some animals, such as rabbit, the placenta does not seem to secrete any hormone.

Human placenta produces a variety of protein and steroid hormones :

A. Protein hormones : 1. Human chorionic gonadotrophin (HCG); 2. Human placental lactogen (HPL) ; 3. Human chorionic thyrotrophin (HCT) ; 4. Human chorionic corticotrophin (HCC) ; 5. Pregnancy specific β-1 glycoprotein (PS βG) ; 6. Relaxin.

B. Steroid hormones : 1. Oestrogens ; 2. Progesterones.

HCG is secreted by the syncytiotrophoblast of the placenta about 10 days after ovulation. This hormone prolongs the life of corpus luteum (*i.e.*, it stimulates the corpus luteum to sustain the secretion of progesterone). HCG is also found to have a **immuno-suppressive** activity which may inhibit the maternal processes of immunorejection of the foetus (see **Dutta**, 1987). Traces of HCG hormone can be detected in the maternal serum or urine after 7 to 10 days of fertilization ; HCG contents greatly enhanced between 60 to 70 days of pregnancy (This is called **pregnancy test**).

Progesterone helps in the maintenance of pregnancy and prevents premature parturition. In fact, progesterone stops the menstruation and the release of ovum from the ovary until the pregnancy. **Relaxin** hormone is produced by the placenta in very small amounts (it is mainly produced by the ovary) and it causes the relaxation of the symphysis and sacroiliac joints during pregnancy and also reduces the tension of cervix.

21-5.6. Storage of Substances by Placenta

Besides serving the vital functions for the developing embryo/foetus, placenta has to furnish certain its own metabolic needs. For example, in the sheep the oxygen consumption by placenta (25–35 ml/kg/min) is more than that by the foetus (6–8 ml/kg/min). Apparently, this oxygen is being utilized by the foetus in its energy metabolism. Placenta is also found to participate in the metabolism of glycogen, lipids and proteins. It stores a variety of materials such as fat, glycogen and iron (**Williams** and **Warwick**, 1980).

21-5.7. Barrier Functions of Placenta and Mode of Transport of Antibodies, Pathogens and Drugs

Foetal membranes have long been considered as a protective barrier to the foetus against noxious agents circulating in the maternal blood. In general, substances of high molecular weight of more than 500 daltons are held but there are exceptions. For example, **leucocytes** are more numerous in the blood of the umbilical vein than in that of umbilical artery, suggesting that they may migrate from the maternal blood through the placental barrier into the foetal capillaries. Likewise, **antibodies** and **antigens** in immunological quantities are found to traverse across the placental barrier in both directions. Thus, antibodies developed in the blood of mother against certain diseases such as **diphtheria**, **measles**, **small pox**, and **scarlet fever**, are passed from mother into the foetus and cause passive immunization of the foetus. Similarly, anti-Rh factor or **Rh-antibody** of Rh-negative mother enters in the blood circulation of Rh-positive foetus and results in anaemia by destroying red blood cells of the foetus. The transfer of larger molecules (*i.e*, maternal immunoglobulins or IgG) takes place by receptor-mediated endocytosis or by coated-vesicles (**De Robertis** and **De Robertis, Jr.,** 1987).

Maternal infection during pregnancy by **viruses** (*i.e.*, rubella or German measles, chicken pox, measles, mumps, poliomyelitis) ; **bacteria** (*Treponema pallidum* (syphilis), Tubercle bacillus) ; or protozoa (*Toxoplasma gondii*, malarial parasite) may be transmitted to the foetus across the so called placental barriers and affect the foetus *in utero*.

Similarly, almost any drug used in pregnancy can cross the placenta barrier and may have deleterious effect on the foetus. For example, the sleeping drug, **thalidomide**, which was used by ladies as a sedative and to avoid nausea and morning sickness during early pregnancy (25 to 44 days) was found to be a teratogen, *i.e.*, it caused deformities in limb development of the foetus, perforation of anus and development of a defective heart. The children born to such mothers had flipper-like limbs (**phocomelia**) and were called **thalidomide babies**. Drugs like **quinine** and **aspirin** ; ionizing radiations (exposure to **X-ray** and **radium**) ; **alcohol** and **cigarette smoking**, etc., are found to be potent teratogens (**Tortora** and **Anagnostakos**, 1984). If drug happens to be **heroin**, the child is born already addicted to the drug (**Phillips**, 1975).

In fact, placenta during first half of pregnancy is impermeable to bacteria and macromolecules. It, therefore, reduces the chances of appearance of many influences of maternal ill health on the foetus. However, in the later half of pregnancy, the placenta becomes more permeable and there are chances of passing the germs from mother to foetus. For example, during the second half of pregnancy syphilitic spirochetes enter the foetus and causes congenital syphilis in the foetus (**Phillips**, 1975).

21-5.8. Immunological Function of Placenta

The foetus and the placenta contain paternally determined antigens, foreign to the mother but fail to inherit all the mother's antigens. In spite of this, there is no evidence of graft rejection. Chorionic ectoderm or trophoblast of placenta probably offers immunological protection against rejection (see **Dutta**, 1987).

REVISION QUESTIONS

1. What is implantation? How does the fertilized ovum implant itself ?
2. Trace the fate of trophectoderm in a mammalian embryo, and indicate the changing relationship between yolk sac, chorion, amnion, allantois and placenta in a mammal.
3. What is placenta ? Describe various kinds of placenta found in mammals.
4. Explain the importance of the placenta and umbilical cord to foetal growth.
5. Describe various functions of placenta.
6. What type of placenta is found in man and ape ? Describe it.
7. Distinguish between the following :
   (i) Chorion and yolk sac ;
   (ii) Chorio-vitelline and chorio-allantoic placenta ;
   (iii) Syncytiotrophoblast and cytotrophoblast ;
   (iv) Yolk sac and inverted yolk sac placenta.
8. Discuss the different nutritive relationships between mammalian embryo and its mother.
9. How do you justify the statement that mother acts as alimentary canal, liver, kidney and lungs of foetus ?
10. Differentiate between the following :
    (1) Deciduate and non-deciduate placenta ;
    (2) Implantation and placentation ;
    (3) Endothelio-chorial and haemo-endothelial types of placenta ;
    (4) Yolk sac placenta and allanto-chorial placenta.
11. What is the influence of drugs on the foetus, consumed by the pregnant mother ?
12. Write short notes on the following :
    (1) Placenta as barrier ; (2) Pregnancy test ; (3) Uterine milk ; (4) Phocomelia.

22

GRADIENT THEORY

Experiments of **Runnstrom** (1928) and **Horstadius** (1928) demonstrated the existence, in sea urchin eggs, of opposite **morphogenetic gradients**. Certain substances which are required for development, decrease progressively in concentration from the animal to the vegetal pole. Other substances have the opposite distribution, with a maximum concentration at the vegetal pole. Realization of this fact led to the **double-gradient theory** (**Brachet**, 1974).

22-1. DOUBLE-GRADIENT THEORY

The double-gradient theory states that there exist in the sea-urchin's egg, two factors or principles which are mutually antagonistic yet interacting with each other at the same time, and that the normal development of the egg is depended on a certain equilibrium between the two principles. Each principle has its centre of activity at one of the poles of the egg. The activity of each principle gradually diminishes away from the centre, producing a gradient of activity. The two gradients of activity decline in opposite directions : the one from the animal pole, the other from the vegetal pole. Taking into account this type of distribution of activities, the factors or principles have themselves been called **gradients**. (L. *gradi*—to step; meaning a proportional rise and fall). The two gradients are, therefore, the **animal gradient** with a centre of activity at the animal pole, and the **vegetal gradient** with a centre of activity at the vegetal pole. This double-gradient system was found to control morphogenesis and differentiation of sea-urchin's egg only when it includes its both components (*i.e.*, animal and vegetal gradients) which maintain an equilibrium. For example, if the animal gradient is weakened or suppressed by certain mechanical methods (*i.e.*,

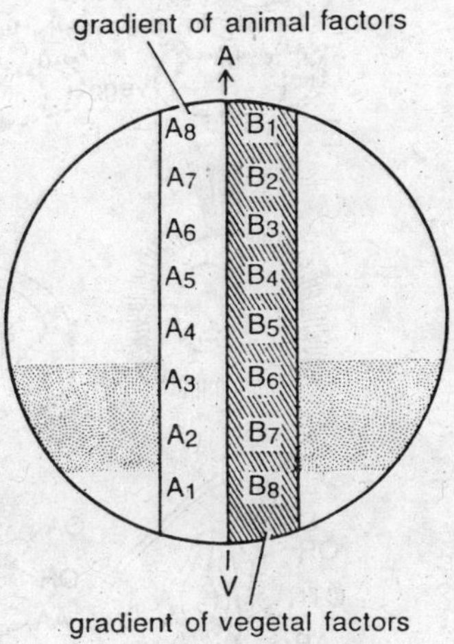

Fig. 22.1. A two-factor hypothesis of the organization of egg. Two factors, A, an animal factor, and B, a vegetal factor, are present in the egg. Factor A converts regions of the egg into ectoderm, B into mesoderm and endoderm. A is highest at animal pole and grades off toward vegetal pole A_8, A_7....A_1; B is highest at the vegetal pole, and grades off toward the animal pole, B_8, B_7....B_1, (after Bodemer, 1968).

removal of blastomeres from animal pole) or by some chemical means (*i.e.*, exposition of fertilized egg of sea-urchin into sea water containing some lithium salts), the vegetal gradient becomes preponderant and the embryo is

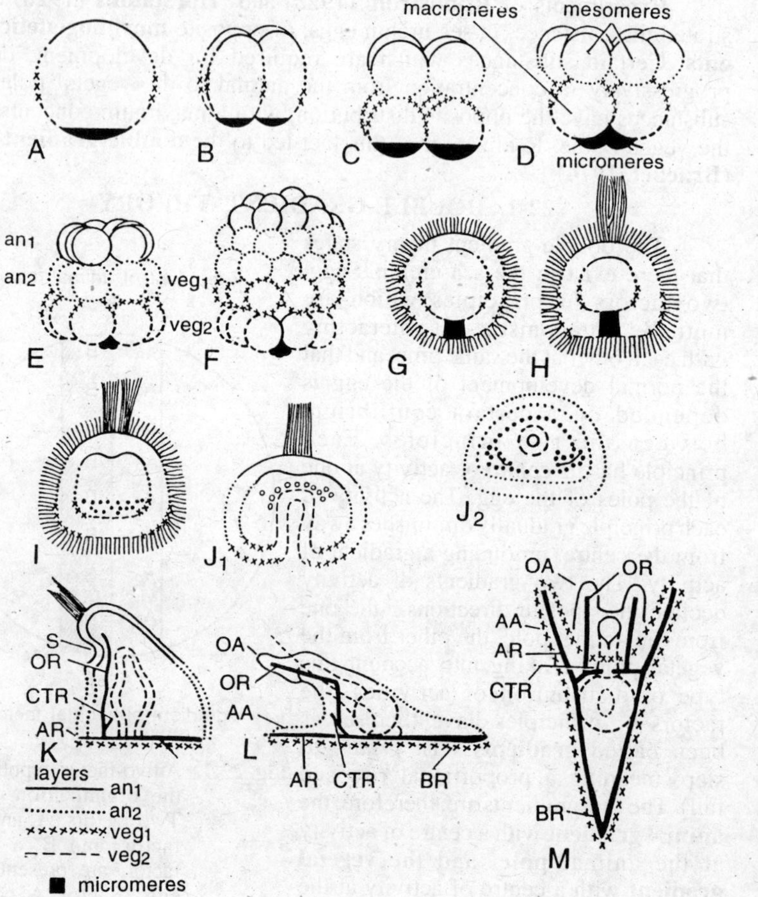

Fig. 22.2. Normal development of sea-urchin *Paracentrotus lividus.* A to F showing differentiation of layers, G and M showing later development; A–Uncleaved egg; B–4-cell stage; C–8-cell stage; D–16-cell stage; E–32-cell stage; F–64-cell stage; G–Young blastula; H–Later blastula with apical organ, before formation of the primary mesenchyme; I–Blastula after formation of the primary mesenchyme; J_1–Gastrula : Secondary mesenchyme after two triradiate spicules have formed; J_2 –T.S. of same gastrula, bilateral symmetry established; K–The so called prism stage; L–Pluteus larva from the left side; M–Pluteus from the anal side; AA–Anal arm; AR–Anal rod; BR–Body rod; CTR–Central transverse rod; OA–Oral arm; OR–Oral rod; S–Stomodaeum (after Berrill, 1971).

vegetalized, *i.e.*, endodermal parts (archenteron, etc.) become greatly developed at the expense of ectoderm. Further, if the vegetal gradient is weakened or suppressed by certain mechanical or chemical means (*i.e.*, exposition of unfertilized egg of sea-urchin into artificial sea water lacking calcium ions but containing sodium thiocyanate (NaSCN), the animal gradient becomes preponderant and the embryo is **animalized**, *i.e.*, it develops to excess parts pertaining to animal gradient, such as the tuft of stiff cilia at the animal pole. Other structures of the embryo, such as ciliary bands and the skeleton of calcareous spicules, can only develop if both gradients are active, and the development is more nearly normal the more the two gradients approach to the correct equilibrium.

22-2. EXPERIMENTAL PROOFS FOR DOUBLE-GRADIENT THEORY

The double gradient concept of **Runnstrom** and **Horstadius** has been supported by following micromanipulative experiments of many experimental embryologists :

1. When the blastomeres of the sea-urchin egg are separated in the two-cell statge or in the four-cell stage, each of them develops into a complete **pluteus larva** of diminished size (Fig. 22.3). This may be related with the fact that the first two cleavage planes are **meridional**, passing through the main animal-vegetal axis of the egg : All of the first two or four blastomeres, therefore, get equal portions of the three cytoplasmic regions of the egg (*i.e.*, ectodermal, endodermal, and mesodermal) or each blastomere has received both gradients in equal amount.

A different result is observed if the egg is separated into the animal and vegetal halves—after the third cleavage, the third cleavage plane being equatorial. In this case both halves produce, as a rule, defective embryos. The four blastomeres of the animal half develop to form an ectodermal vesicle, it does not produce a gut. Even purely ectodermal structures are abnormally differentiated (*i.e.*, the ciliary band is not developed ; the tuft of long cilia on the animal pole, on the other hand, develops excessively, growing over a much greater surface than it does normally, (Fig. 22.4B). The four vegetal blastomeres together develop into a blastula which gastrulates and forms a ovoid embryo with a disproportionately large endodermal gut but without a mouth. There may be a few irregular skeletal spicules but no arms and no ciliary band. In a highly vegetalized condition (Fig. 22.4B) ; the vegetal part develops still more abnormally—it produces a large endodermal gut and a small ectodermal vesicle. The gut does not lie inside the ectodermal vesicle but is evaginated to the exterior (*viz.*, turned inside out) due to an interference with the normal process of gastrulation. This phenomenon is called **exogastrulation**. Further, in a vegetalized embryo, though mesenchyme cells have migrated into the interior, they usually produce no skeletal spicules or only very small and abnormal ones (**Horstadius**, 1928, 1935).

These results have indicated that each of the two halves lacks some parts (of animal or vegetal gradient) contained in the other half and defective larvae are produced due to disturbance of equilibrium of animal and vegetal gradients because of mechanical separation of blastomeres of animal and vegetal poles. Further, cutting the fertilized but uncleaved egg equatorially into animal and vegetal halves, however, showed that this difference in potential already existed shortly before cleavage but was not established closer to the time of fertilization. The implication is that something in the newly fertilized egg becomes concentrated in the vegetal half within a short period.

2. Micromanipulative recombination experiments of Horstadius. Horstadius (1935 and 1939) made his most of the operative experiments with the developing sea-urchin eggs at the 32-and 64-cell stages. The 16-cell stage of sea-urchin egg consists of a ring of eight animal blastomeres (**mesomeres**), four large blastomeres or **macromeres** and four small **micromeres** at the vegetal pole. At the next division the animal half consists of two rings of eight cells each called **an$_1$** and **an$_2$** for identification (Fig. 22.2E). At the 64-cell stage the macromeres have also divided into two rings of eight cells, called **veg$_1$** and **veg$_2$** (Fig. 22.2F). For convenience, therefore, the developing sea-urchin egg may be said to consist of five layers : an$_1$, an$_2$, veg$_1$, veg$_2$ and the micromeres. The development of these five layers representing strata along the animal-vegetal axis, has been studied separately and in various combinations as follows :

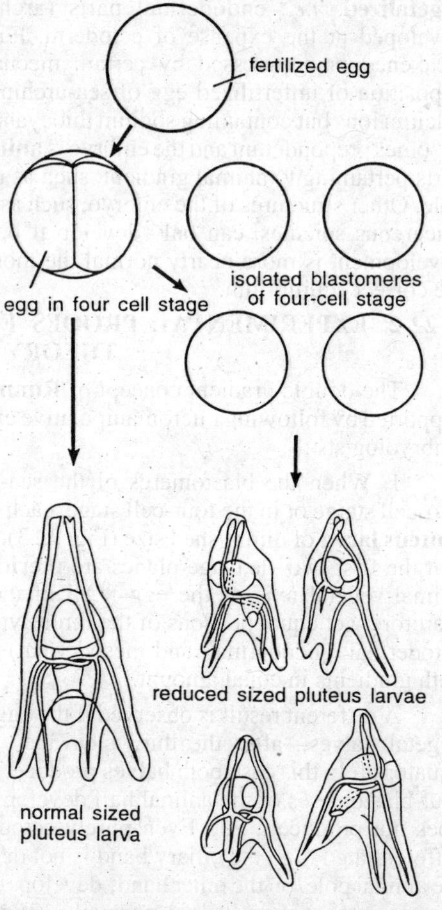

Fig. 22.3. Development of isolated blastomeres of four-cell stage of sea urchin egg. Each develops into a pluteus larva of normal proportion but reduced size (after Berrill, 1971).

(a) The isolation results of Horstadius are following :

an$_1$ — develops to form a blastula entirely covered with long stiff cilia, but develops no further, *i.e.*, it is unable to invaginate to form the gastrula.

an$_2$ — develops like an$_1$ except that long cilia do not cover all the blastula surface.

veg$_1$ — may become a ciliated blastula only or may develop a small archenteron by invagination.

veg_2 — gives rise to ovoid larva with an archenteron and one or two spicules, *i.e.*, veg_2 which normally gives rise only to endoderm and mesoderm now produces ectoderm as well.

micromeres — continue to divide to form a specific number of small isolated cells representing the primary mesenchyme cells they would normally have become.

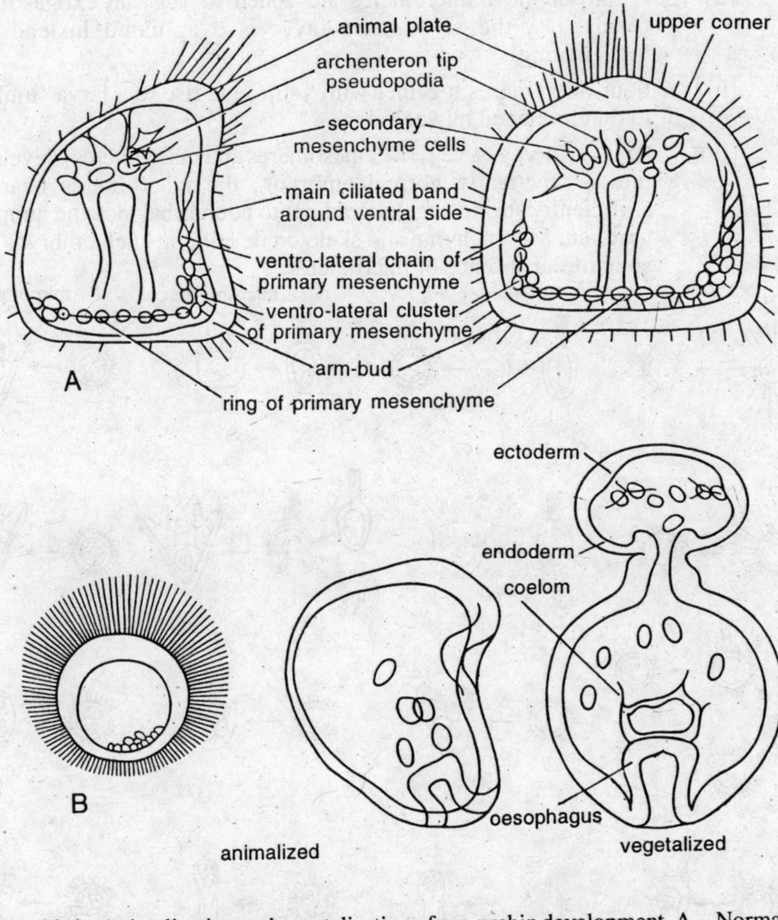

Fig. 22.4. Animalization and vegetalization of sea-urchin development. A—Normal advanced gastrula showing distribution of primary mesenchyme cells; B—Abnormal larvae induced by various chemical agents. Animalized larvae (left two) with long cilia and mesenchyme cells only or with very reduced endoderm which forms a small invagination; vegetalized exogastrula larva (extreme right) with only ectoderm and a corresponding increase in size of the endoderm (after Berrill, 1971).

Thus, the development of an_1 and an_2 (mesomeres) shows the preponderance of the animal gradient and animalization. While, the development of veg_1 and veg_2 (macromeres) shows the preponderance of vegetal gradient in these blastomeres.

(b) The combination results of Horstadius are following :

(i) If four micromeres are added to an animal half (mesomeres) or to an entire presumptive ectoderm, *i.e.*, $an_1 + an_2 + veg_1$, gastrulation occurs and a normal larva develops.

(ii) If one or more micromeres are added to veg_2 an exogastrula forms, *i.e.*, the archenteric layer is evaginated instead of invaginating.

(iii) Four micromeres together with veg_1, give rise to a larva similar to that produced by veg_2 alone.

(iv) $an_1 + an_2 + veg_1 + veg_2$ (*i.e.*, mesomeres and macromeres) develop into a practically normal embryo ; the macromeres bear a sufficiently strong vegetal gradient to counterbalance the animal gradient. Mesenchyme and skeleton develop in such embryos in spite of the absence of micromeres.

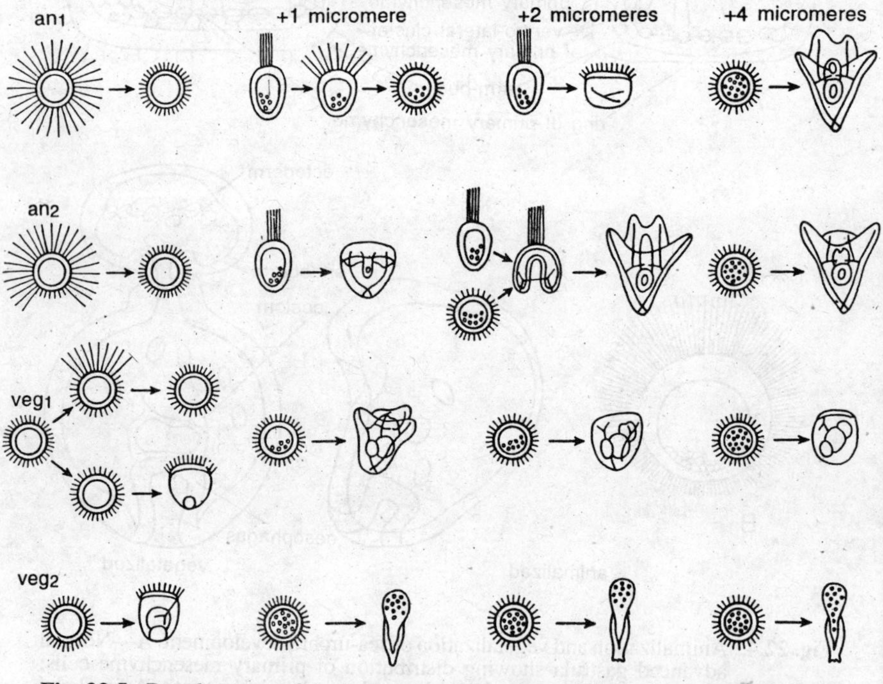

Fig. 22.5. Development of the layers an_1, an_2, veg_1, veg_2, isolated (left column) and with one, two, and four micromeres implanted (after Berrill, 1971).

Accordingly, the most normal development, culminating in a pluteus larva, results from veg_2 alone, veg_1 plus one micromere, an_2 plus two micromeres, and an_1 plus four micromeres. The most illuminated combination remains of mesomeres ($an_1 + an_2$ = ectoderm) plus micromeres (mesenchyme), since it develops into a complete and more or less normal pluteus, and even the gut of such embryos develops in spite of the absence of the macromeres which should normally have supplied the material for the gut. However, the embryo possesses the two gradients ; the animal gradient borne by the mesomeres and vegetal gradient borne by the micromeres ; and the possession of the two gradients seems to create the necessary conditions for normal development.

22-3. EVOLUTION OF GRADIENT CONCEPT

The idea of occurrence of the gradient system in the ooplasm of certain echinoderm eggs has developed in a very gradual manner. It was introduced in 1910 by **Boveri** who considered that the vegetal region of the sea-urchin egg exerts a dominant control over more animal regions, the influence decreasing toward the animal pole. However, he could not give any plausible reason for his concept and it was in 1928 when **Horstadius** and **Runnstrom** individually developed the idea of **double-gradient-system**, which has been discussed in detail in the preceding sections. **C.M. Child** (1940) observed a single physiological gradient called **metabolic axial gradient** in sea-urchin's egg to control morphogenesis in it. By using the rate of reduction of vital dyes, particularly Janus green, as an indicator of the rate of oxidative metabolism, he showed that such a **reduction gradient** (*viz.*, metabolic axial gradient) is present in the oocytes, blastomeres and blastula of sea urchins and starfishes. In recent time, many modern experimental embryologists, **Lallier** (1964), **Gustafson** (1965), **Runnstrom** (1967), **Horstadius** (1969), **Josefsson** (1969), etc., have added much to our understanding of gradient system.

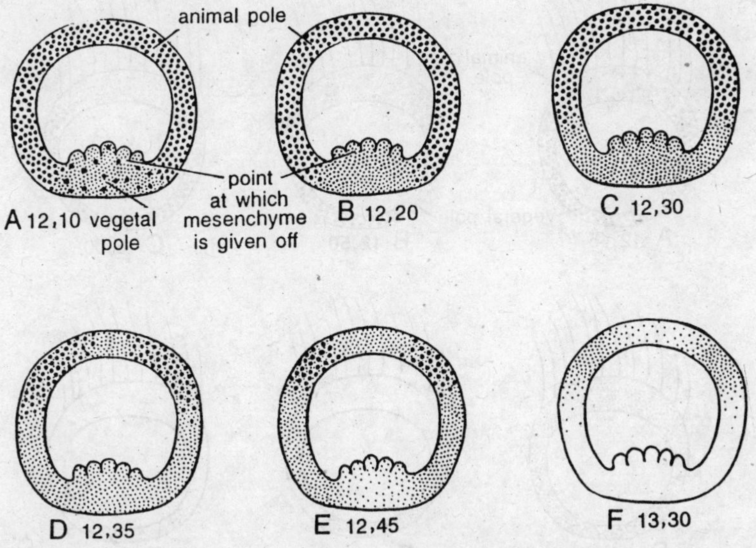

Fig. 22.6. Reduction of Janus green in a late blastula of a sea urchin. Large dots represent blue colour of Janus green, small dots show a complete fading of the stain. Figures indicate time of observation (after Balinsky, 1970).

22-4. THEORIES ABOUT PHYSICO-CHEMICAL NATURE OF THE ANIMAL-VEGETAL GRADIENT SYSTEM IN SEA-URCHIN EGGS

The existence of animal-vegetal gradients in the eggs of sea-urchin and other echinoderms has been demonstrated by various direct methods than the indirect methods of micromanipulation, isolation and recombination experiments of Horstadius and others. By employing certain direct methods for demonstrating animal-vegetal gradients, some knowledge has been gained about the chemical and physical nature of gradient system, as follows :

1. One suggestion was that the animal gradient might be cortical and vegetal gradient might be in the interior. But this view was held invalid, for when more than fifty per cent of the cytoplasm was withdrawn from the sea urchin eggs by micropipette during an experiment and such eggs subsequently have been fertilized, no deviation from normal development (*i.e.*, normal larvae) could be detected. The typical development may, thus, be a consequence of unaltered conditions in the cortex.

2. One of the methods of direct demonstration of the existence of gradient system is to study the reduction of dyes by the embryo under the conditions of an aerobiosis, according to the experiments of **C.M. Child** (1936, 1940) and **Horstadius** (1952, 1955). Sea urchin embryos slightly stained with the vital dye **Janus green (diethyl safranin azodimethyl aniline)** are placed in a small chamber sealed off with petroleum jelly. After some time all the free oxygen contained in the chamber is used up as a result of the respiration of the

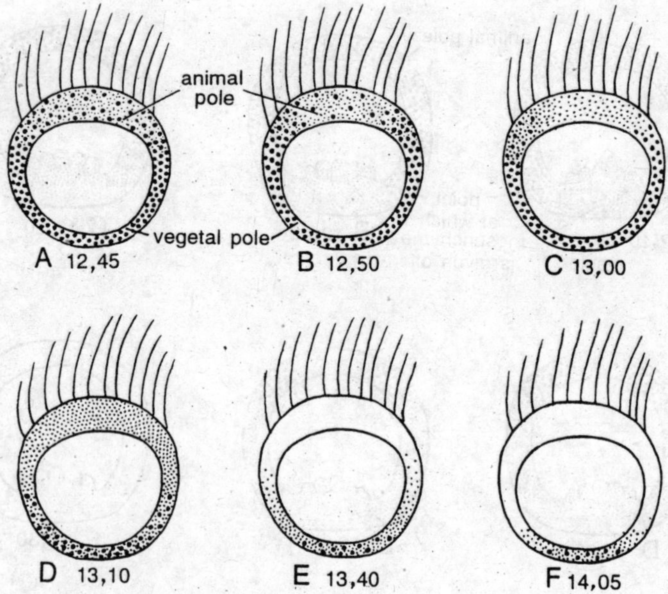

Fig. 22.7. Reduction of Janus green in animalized sea-urchin embryo (after Balinsky, 1970).

embryos. After that the embryos respire using Janus green as an acceptor of hydrogen. The Janus green, thus, is reduced to a red dye–diethyl safranin and further to a colourless substance, **leucosafranin**. Consequently, the light grayish blue colour of Janus green first changes into red, and then, as a second step in the reduction of the dye, the colour disappears completely.

The reduction of Janus green, shown by the colour change, does not occur simultaneously in the whole embryo but follows a very characteristic sequence. For example, in the late blastula and early gastrula of the sea urchin, the change in colour is first noticeable at the vegetal pole, at the point where the primary mesenchyme is given off (Fig. 22.6). Then the colour change spreads out, involving the whole vegetal hemisphere, and reaches the equator. At this stage a spot of changed colour appears at the animal pole and also gradually increases, so that the bluish colour remains only in the form of a ring lying well above the equator in the animal hemisphere, eventually, this ring also disappears. The red colour begins to fade in the same sequence, starting first from the vegetal pole and then from the animal pole (**Child,** 1936 ; **Horstadius,** 1952). The order in which the dye is reduced in the embryo shows a correspondence to the postulated animal and vegetal gradients.

In further experiments of **Horstadius** (1955) of similar nature, when either the animal or vegetal gradient is suppressed, the corresponding gradient of reduction also disappears. In isolated animal halves of sea urchin's eggs the animal tendencies of development are predominant, and it is found that such animalized embryos start reducing Janus green at the animal pole only ; there is no centre of reduction at the vegetal pole (Fig. 22.7). In isolated vegetal halves the animal tendencies of development are suppressed and the vegetal tendencies are supreme. Correspondingly, the only centre of reduction is at the vegetal pole ; no centre of reduction appears at the animal pole (Fig. 22.8). Embryos animalized or vegetalized chemically show reduction gradients similar to isolated animal and vegetal halves of the egg.

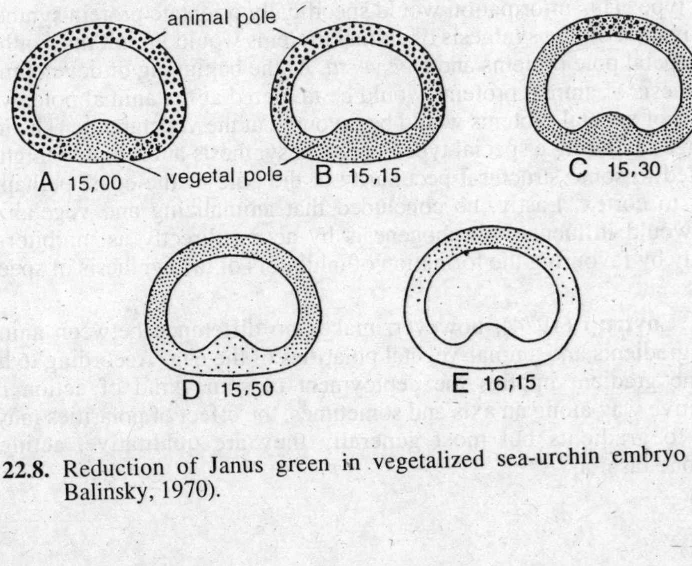

Fig. 22.8. Reduction of Janus green in vegetalized sea-urchin embryo (after Balinsky, 1970).

As evident from these experiments, the micromeres are carriers of the highest point of the vegetal gradient and they preserve this property after transplantation. Accordingly, the micromeres can also serve as a centre of a reduction gradient. If a group of four micromeres is implanted laterally and the embryo is tested for reduction of Janus green, it can be seen that in addition to the two normal gradients, one from the vegetal pole and one from the animal pole, a third gradient appears, having the implanted micromeres as its centre but spreading out from these to the adjacent cells. **Horstadius** (1952), thus, concluded that every modification of the gradient system of the embryo which was postulated from the morphogenetic behaviour of the embryo, was reflected in the gradients of reduction of Janus green.

The animal and vegetal gradients are, thus, considered to be definite metabolic processes or systems of metabolic reactions which involve oxidation and which have their highest point of intensity at the animal and vegetal poles respectively.

3. Certain amino acids, such as glutamine and lysine, have an animalized effect, while others have a vegetalized effect suggesting that both difference in type and differential rates of protein synthesis may underlie the phenomenon of metabolic (oxidation-reduction) axial gradient. This is supported by the discovery by **Horstadius** and **Josefsson** (1969) that specific animalizing and vegetalizing substances, namely, one containing a tryptophan and the other a nucleotide, respectively, are present in mature unfertilized eggs.

4. According to **Berrill** (1971), the animal factor promotes synthesis of ribonucleic acid and proteins, while the vegetal factor seems to inhibit the onset of these processes for a while, so that the time pattern of early protein synthesis will be different in the two regions. In a recent review, **Lallier** (1964) pointed out that the introduction of the idea of **inhibition** in the analysis of the development of the sea-urchin egg suggests an interpretation of differentiation and the mode of action of the animalizing and vegetalizing agents. At the very beginning of development the nuclei would emit information of animal or vegetal type. This information would specifically orientate protein synthesis. The activation of the synthesis of animal proteins would inhibit the synthesis of the vegetal pole proteins and *vice versa*. At the beginning of development, the synthesis of animal proteins would be favoured at the animal pole, while synthesis of vegetal proteins would be favoured at the vegetal pole. The local advantage favouring a special type of protein synthesis at each pole might be connected to some structural peculiarity at the pole of the eggs, perhaps in relation to cortex. Lastly, he concluded that animalizing and vegetalizing agents would influence morphogenesis by acting directly as inhibitors or indirectly by favouring the formation of inhibitors of the synthesis of specific proteins.

5. **Lovtrup** (1974), however, makes no difference between animal-vegetal gradients and animal-vegetal polarities of the egg. According to him, the name gradient implies the deployment of some kind of action in a quantitative way along an axis and sometimes, the effect of polarities may be similar to gradients but most generally they are qualitative, acting in all-or-none fashion.

22-5. VEGETALIZING AND ANIMALIZING CHEMICAL AGENTS

There are certain chemical agents which are found to cause vegetalization or animalization of sea-urchin's egg. For example, **Herbst** (1892) discovered that lithium ions added to sea water caused an increase in the amount of endoderm and led to exogastrulation (vegetalization). Likewise, vegetalization of the egg may be caused by sodium azide (**Child**, 1948) and dinitrophenol. These chemicals are called **vegetalizing agents.** Certain other chemical agents such as sodium thiocyanate, increase ectoderm at the expense of endoderm (*i.e.*, cause animaliztion of egg) and are called **animalizing agents**. Certain other animalizaing agents are — 1. Some metals (zinc, and mercury, etc.) ; 2. Some proteolytic enzymes (trypsin, chymotrypsin, etc.,) ; 3. Many acidic dyes (sulfonic group (HSO_3) containing dyes such as Evans blue, chlorazol sky blue, trypan blue, congo red, etc. and carboxyl group (COOH) containing dyes such as uranin and rose bengal) ; 4. Some sulfonated organic compounds (germanin, etc) ; and 5. Some anionic (acid) detergents. Animalization can also be brought about by treatment with SCN before fertilization or by treatment with O-iodosobenzoic or 2, 5-thiomethylcytosine during cleavage stage. Animalization may be achieved by extreme dilutions of a chemical ; for instance, 1 : 50,000 in the case of evans blue.

22-5.1. Mode of Action of Vegetalizing Agents

Sodium azide and dinitrophenol belong to a group of enzyme poisons. The azide is known to inactivate the cytochrome oxidase system, while the dinitrophenol disturbs respiration by preventing oxidative phosphorylation, that is, the formation of energy rich bonds between phosphoric acid and adenosine diphosphate (*viz.*, with the formation of ATP molecules). This might mean that vegetalization is based on the disturbance of oxidative processes in the embryo (**Lindahl**, 1936) or more specifically on the disturbance of the process of phosphorylation (**Horstadius**, 1953).

Further, lithium ions found to suppress the consumption of O_2 by early gastrula (**Lindahl**, 1936) and accumulation of inorganic phosphate in lithium-ion treated embryos shows that these embryos are incapable of utilizing the energy of oxidation for phosphorylation and synthesis of ATP. **Gustafson** and **Hjelte** (1951) reported that the relative amounts of various amino acids change as a result of vegetalization in lithium ion-treated embryos.

22-5.2. Mode of Action of Animalizing Agents

Most animalizing agents are found to attack proteins or to form compounds with proteins, especially with basic proteins. They are supposed to block the active sites of proteins and, thus, preventing them from interacting normally with other cell constituents. Even the steric configuration of protein molecules may, thus, be altered, since these are based on lateral bonding between molecules and their parts (**Lallier**, 1956, 1957).

22-6. EXAMPLES OF GRADIENTS IN OTHER ANIMALS

Various types of gradients have been discovered in insects, amphibians and other vertebrates :

22-6.1. Double Gradient Model in Insects

Some promising progress in the direction of double gradient, has been made by **Sander** (1975) and **Kalthoff** (1975), working on the embryogenesis of insects. In insect eggs, the yolk is distributed uniformly, so that there is no animal-vegetal axis, but there is a clear antero-posterior axis along which the embryo develops. Even at an early stage the embryo, which forms along the ventral side of the cellular blastoderm, is clearly divided in this axis into easily distinguishable regions which will give rise to the head, jaw, thorax and abdominal segments, so that, any change in their lineal order produced by experiments can be easily seen. It has been shown that the nuclei are **equivalent** during multiplication phase in the interior of the egg, so that regional differences in later development must arise through differences established in the cortical layer of cytoplasm which becomes populated by the nuclei and which then by cytoplasmic cleavage gives the cellular blastoderm.

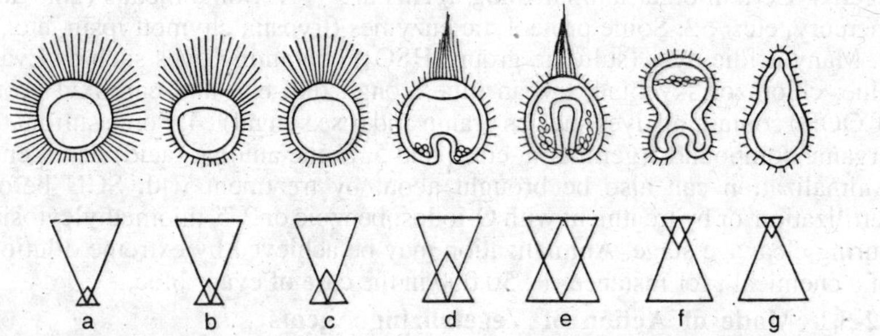

Fig. 22.9. Sea-urchin larvae of about the same age but varying in the animal and the vegetal tendencies. e—Normal gastrula; a-d—Larvae animalized to varying extent; f–g—Larvae vegetalized to a different extent. In (g) the endoderm reduced to vegetalized small knob. The hypothetical double gradient system corresponding to each larva is indicated (after Berrill, 1971).

Sander (1975) showed that in the leafhopper, *Euscelis*, even more clearly than in the sea urchin, regulation following experimental interference by no means always leads to normal development but may produce monsters, notably one in which the anterior structures are replaced by a mirror image of those in the posterior half, giving a "**double abdomen**" embryo. As in the sea urchin, a **double-gradient model** explains the facts best, with the highest level of one gradient at the posterior end of the egg and the highest level of the other at the anterior end, with specification of the various parts according to the relative concentration of two factors along the antero-posterior axis.

The *Euscelis* egg is elongated and may be constricted by a ligature at any point along its length (Fig. 22.10). Furthermore, the posterior tip cytoplasm is marked by the presence of a ball of symbiotic bacteria which may be pushed anteriorly into any other position. In Sander's experiments the regions were scored as **X** (extra-embryonic membrane anterior to the embryo proper), **A** (head), **B** (jaws), **C** (head), **D** (anterior abdomen) and **E** (posterior abdomen). Constriction at the late blastoderm stage produces two partial embryos which together make up all the structures of a normal whole embryo, but done at the late nuclear division stage, the result is that the parts produce less than they would normally, the middle structures being absent altogether.

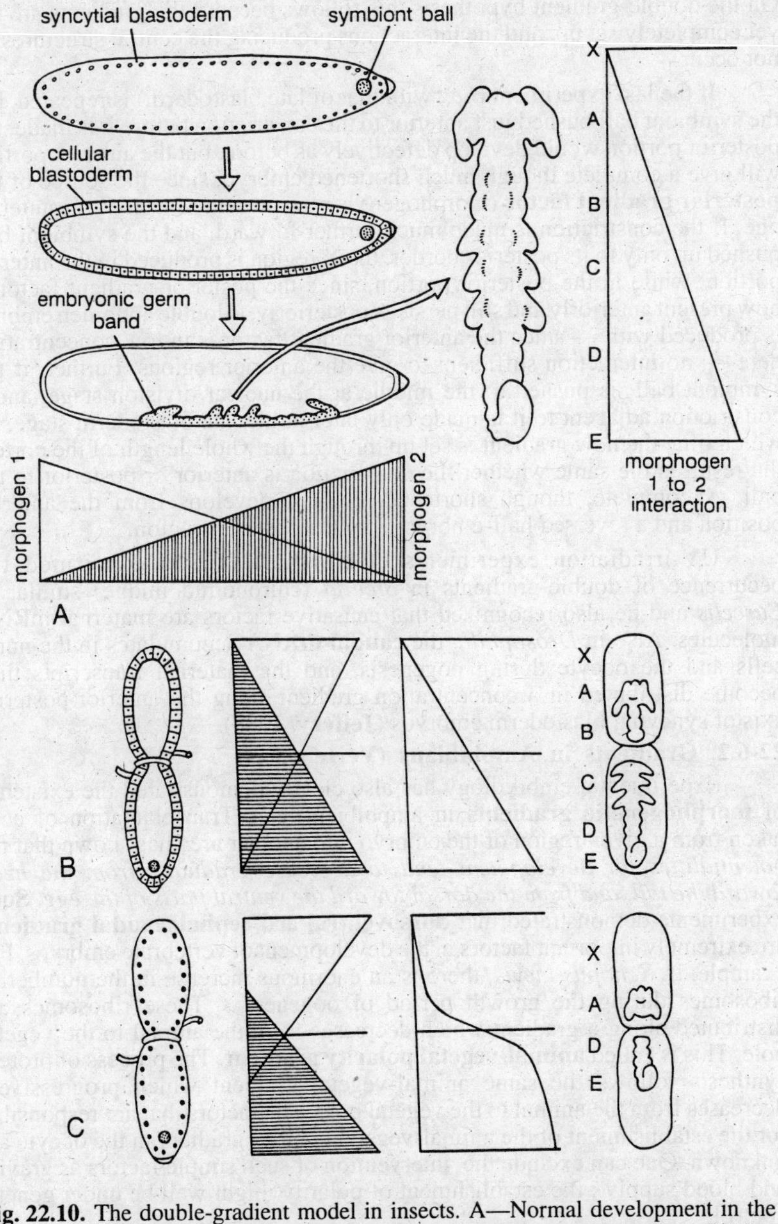

Fig. 22.10. The double-gradient model in insects. A—Normal development in the leafhopper *Euscelis* showing syncytial blastoderm (late nuclear division stage), cellular blastoderm (late blastoderm) and embryonic germ band stages, with the ball of symbiont bacteria situated posteriorly. The two hypothesized morphogen gradients are indicated below and both are combined in a graph indicating the product of their interaction (right); B and C — Development following constriction at the late blastoderm stage (B) and the late nuclear division stage (C) (after Ede, 1975).

On the double-gradient hypothesis this follows because the gradients are not yet completely set up, and the interactions producing the central structures do not occur.

If the last experiment (*i.e.*, with that of late blastoderm) is repeated, but the symbiont ball pushed just anterior to the constriction before it is made, the posterior portion would develop defectively as before, but the anterior portion will give a complete though much shortened embryo, since the source of the **posterior gradient factor** (morphogen) is now present as well as the anterior one. If the constriction is made much further forward, and the symbiont ball pushed up only to its posterior border, the X region is produced in the anterior portion, while in the posterior portion, since the posterior gradient factor is now present anteriorly and still persisits posteriorly, a double abdomen embryo is produced with — since the anterior gradient factor is in low concentration here — no interaction sufficient to give the anterior regions. Further, if the symbiont ball is pushed to the middle at the nuclear division stage, and a constriction adjacent to it is made only later at the late blastoderm stage, by which time the new gradient is set up through the whole length of the cortex, the result is the same whether the constriction is anterior or posterior to the ball. A complete, though shortened, embryo develops from the anterior position and a reversed half-embryo from the posterior region.

UV-irradiation experiments of **Kalthoff** (1975) confirmed the occurrence of double-gradients in *Smittia* (chironomid midge) similar to *Euscelis* and he also recognized that causative factors are maternal mRNA molecules. Also in *Drosophila,* the **caudal mRNA** accumulates in the nurse cells and the oocyte during oogenesis, and the maternal transcripts then become distributed in a concentration gradient along the anterior-posterior axis of syncytial blastoderm embryos (**Jeffery**, 1988).

22-6.2. Gradients in Amphibians (Vertebrates)

Experimental embryology has also clearly demonstrated the existence of **morphogenetic gradients** in amphibian eggs. Transplantation of cells taken from a given region of the embryo into another area has shown that the *potentialities for development tend to decrease gradually from the head toward the tail, and from the dorsal toward the ventral parts of the egg*. Such experiments demonstrated that **dorsoventral** and **cephalocaudal gradients** are extremely important factors in the development of vertebrate embryos. For example, in *Xenopus laevis*, there is an enormous increase in the number of ribosomes during the growth period of oogenesis. These ribosomes are distributed along a gradient which decreases from the animal to the vegetal pole. This is called **animal-vegetal polarity gradient**. The process of protein synthesis follows the same animal-vegetal gradient which progressively decreases from the animal to the vegetal pole. The factors that are responsible for the establishment of the animal-vegetal polarity gradient in the oocyte are unknown. One can exclude the intervention of such simple factors as gravity and blood supply ; the establishment of polarity might well be under genetic control.

This preformed polarity gradient is important for future development, since the head of the tadpole and the adult will form at the animal pole and the tail at the vegetal pole. It coresponds to the antero-posterior, cephalo-caudal gradient of the embryo (**Brachet**, 1974).

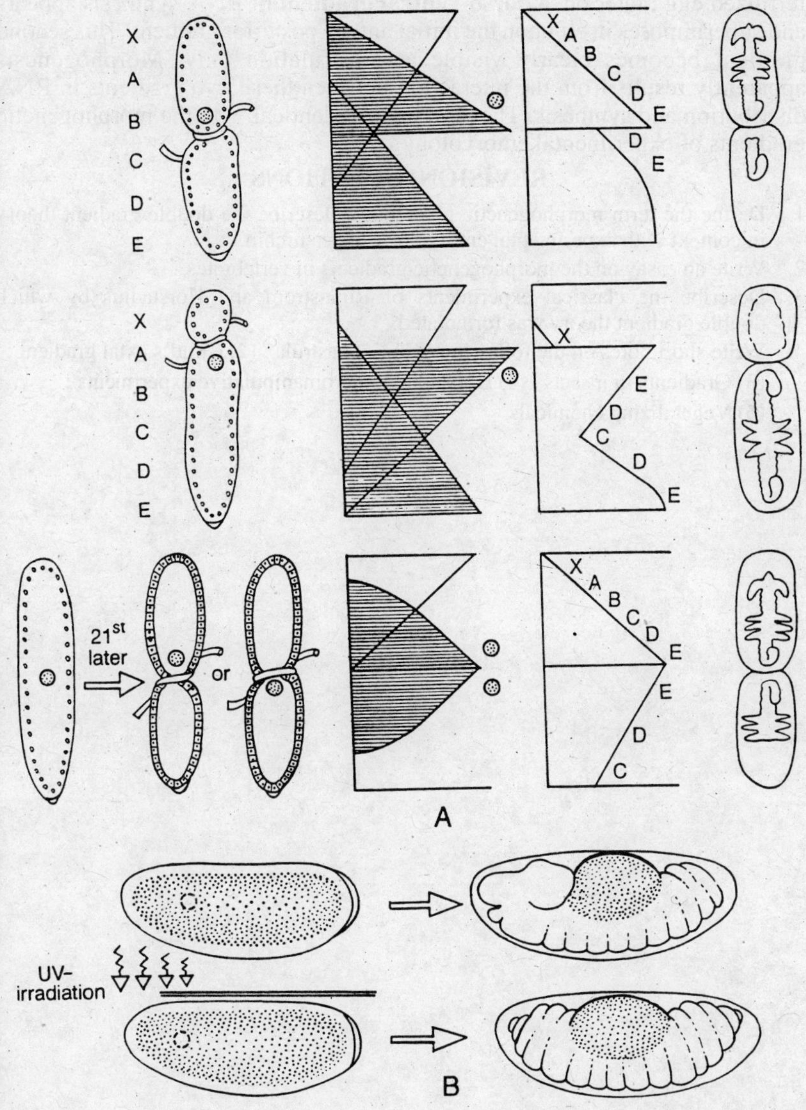

Fig. 22.11. Double gradients and double monsters in *Euscelis*. A—Development following constriction after the cytoplasm around the symbiont ball has been moved anteriorly; B—Normal development in the chironomid midge *Smittia* and production of a double-abdomen embryo by UV-irradiation of the anterior tip of the egg (after Ede, 1978).

In most vertebrate embryos, an antero-posterior or cephalo-caudal gradient in RNA distribution corresponds to the initial polarity gradient of the fertilized egg ; later on, a **dorso-ventral gradient** of RNA-synthesis appears and superimposes itself upon the initial antero-posterior gradient. This second gradient becomes clearly visible at grastrulation only. Morphogenesis apparently results from the interaction between these two gradients in RNA distribution and synthesis. They are indeed identical with the morphogenetic gradients of experimental embryology.

REVISION QUESTIONS

1. Define the term morphogenetic gradients ? Describe the double gradient theory in context with experimental embryology of sea urchin.
2. Write an essay on the morphogenetic gradients in vertebrates.
3. Describe the classical experiments of Runnstrom and Horstadius by which double gradient theory was formulated.
4. Write short notes on the following : (1) Exogastrula ; (2) Child's axial gradient ; (3) Gradients in insects ; (4) Horstadius' micromanipulative experiments ; (5) Vegetalizing chemicals.

23

EMBRYONIC INDUCTIONS AND COMPETENCE

During the embryogenesis of the multicellular organisms, different kinds of interactions between the different types of embryonic tissues at one or other embryonic stages have been observed by many embryologists. One kind of such a tissue-interaction called **inductive interaction** or **embryonic induction** has been reported to be a widespread developmental phenomenon in many invertebrates, prevertebrate chordates and vertebrate chordates. The embryonic induction is a morphogenetic effect by which one embryonic tissue transmits a chemical stimulus that influences another embryonic part to produce a structure that otherwise would not come into existence. The embryonic tissue which exerts such an inductive influence (evocative action) is called an **inductor** and the chemical substance that is emitted by an inductor is called **morphogen** (**Maclean** and **Hall**, 1987). The tissue on which a morphogen of inductor acts is known as the **responsive tissue**. Further, the embryonic induction is a mechanism of determination through which a control of the pattern of the protein synthesis is imposed on a responsive tissue and new kinds of cells with new behaviours are created. It occurs especially in organs assembled as a composite from different sources and is useful in bringing about orderly development and the correct timing and fitting together of parts.

In a way, the induction is an interaction between two cell types in which one (the inductor or evocator) influences the fate of the other (the competent or responsive cell). In regulative embryos (*e.g.*, *Xenopus* and other amphibians), wherever two distinct groups of cells with different developmental histories come together in development, *e.g.*, by rearrangement of the germ layers through the morphogenetic movements, especially when both are organized as sheets (epithelia) or one is an epithelium and the other adjacent mesenchyme, one is likely to affect the development of the other ; the effect may be reciprocal. However, in a completely mosaic embryo, in which each region develops autonomously, this would not be so. But **Raven** (1958) has showed that the formation of the shell glands in the ectoderm of snail *Lymnaea* is induced by contact with foregut endoderm (see **Ede**, 1978).

23-1. HISTORICAL BACKGROUND

One kind of embryonic induction called **neural induction**, first of all reported in an amphibian by a German embryologist, **Hans Spemann** and his student, **Hilde Mangold** in 1924. For the discovery of neural induction, **Spemann** received Nobel Prize in 1935. **Spemann** and **Mangold** performed certain heteroplastic transplantations between two species of newt, namely *Triturus cristatus* and *Triturus taeniatus* and reported in the dorsal lip of their early gastrula the capacity of induction and organization of presumptive neural ectoderm to form a neural tube and also the capacity of evocation and organization of ectoderm, mesoderm and endoderm to form a complete secondary embryo. They called the dorsal lip (containing cells of chordamesoderm and endoderm) of an early gastrula of newts, the **primary organizer**, as it was first in the sequence of inductions and as it was having the capacity to organize the development of a second or twin embryo. Later on, the primary organizer was reported to exist in many animals, *e.g.*, in frogs (**Schotte**, 1930 ; **Dalcq** and **Pasteels**, 1937) ; in lampreys or cyclostomes (**Bytinki-Saltz**, 1937 ; **Yamada**. 1938) ; in bony fishes such as *Fundulus, Salmo* and *Perca* (**Luther**, 1935 ; **Oppenheimer**, 1936) ; in sturgeon (**Ginsburg** and **Dettlaff**, 1944) ; in birds (**Waddington**, 1933, 1952) and in rabbit (**Waddington**, 1934). More recently, primary organizer and neural induction have been investigated in certain prevertebrate chordates, such as ascidians (**Reverberi, Ortolani** and **Fairinella-Ferruzza**, 1960) and *Amphioxus* (**Tung, Wu**, and **Tung**, 1932). In 1960 and 1963 **A.C.G. Curtis** investigated that the organizer of gastrula of *Xenopus laevis* can be distinguished in the cortex of gray crescent of a fertilized egg. **Holtfreter** (1934, 1947, 1968, 1988) ; **Nieuwkoop** (1966) and **L.G. Barth** and **L.J. Barth**, (1968, 1969, 1974), have provided great amount of informations about the chemical nature of embryonic inductions. Recent investigations of various workers such as **Slack** (1983, 1985, 1987), **Sargent** (1986), **Smith** (1988), etc., have added much to our understanding of the phenomenon of induction.

Spemann (1918) working with *Triturus taeniatus* performed an experiment that was almost identical with the one that, under his supervision, was repeated by **Hilde Mangold**. However, his transplantations were performed homoplastically and not, as those of **Mangold**, in heteroplastic combinations (**Mangold** used *Triturus cristatus* as donor which has pigmented embryonic cells). Like **Mangold**, he found that, at the site of transplantation, there appeared a complex of notochord, somites and neural tissues. However, the crucial question : where did the neural tissue come from ? — from the transplant by way of self- differentiation, or from the host, by way of induction — could not be answered , because in this homoplastic combination the tissues derived from the transplant could not be distinguished from those of the host. However, **Spemann** (1918) had a notion that in the embryo, some influence emanates from the chorda-mesoderm, which perhaps travels from cell to cell and which has a determining effect on the presumptive neural plate.

Thus, it was neither **Lewis** nor **Spemann**, but **Hilde Mangold** was the first to produce clear evidence of the inductive capacity of the chorda-mesoderm (see **Holtfreter**, 1988).

23-2. TYPES OF EMBRYONIC INDUCTIONS

Very recently **Lovtrup** (1974) has classified different kinds of embryonic inductions into following two basic types—endogenous and exogenous inductions.

1. Endogenous induction. Certain embryonic cells gradually assume new diversification pattern through the inductors that are produced by them endogenously. Due to these endogenous inductors, these cells undergo either self-transformation or self-differentiation. Such endogenous induction has been reported in mesenchymal cells of ventral pole of Echinoidea and in small sized, yolk-laden cells of dorsal lip of amphibian blastopore. It is found to depend on several factors such as unequal divisions, small cells, yolk-laden cells and sublethal cytolysis, etc.

2. Exogenous induction. When some external influence or a cell or tissue is impressed its diversification pattern upon neighbouring cells through the process of contact induction, then this phenomenon is called **exogenous induction**. An exogenous induction may be **homotypic** or **heterotypic** depending on the fact that whether the inductor provokes the formation of its own kind of tissue or it provokes the formation of different kind of tissues, respectively (**Grobstein,** 1964). In case of homotypic exogenous induction, a diversified (differentiated) cell produces the inductor which serves to maintain the state in the cell proper and which differentiates other adjacent cells according to it after crossing the cell boundaries (*e.g.*, neural ectoderm). The best example of the heterotypic exogenous induction is the formation of a secondary embryonic axis by an implanted presumptive notochord (*i.e.*, primary inductor) in amphibians.

23-3. EXAMPLES OF EMBRYONIC INDUCTIONS

In different animals, the embryonic inductions have been found to be caused either by blastomeres (*e.g.*, sea urchin, *Ascidia*, etc.), or by undifferentiated germinal layers (*e.g.*, neural induction by chordamesoderm to ectoderm in most prevertebrate and vertebrate chordates), or by mesenchyme (*e.g.*, epithelio-mesenchymal interactions during the organogenesis of kidney, pancreas, salivary gland, skin, thyroid and thymus) or by spinal cord (*e.g.*, induction of vertebral cartilage by the spinal cord). Certain important types of embryonic inductive interactions of some animals are following :

23-3.1. Embryonic Induction in Non-chordates

Among non-chordates, embryonic inductions have been studied well only in certain echinoderms :

Induction by micromeres in sea-urchin. In the sea-urchin, when the micromeres were transplanted to the animal half of a developing egg, they caused prospective ectoderm to develop proportionately into ectoderm, endoderm and marginal mesoderm. Further, in the developing egg of sea-urchin the blind end of the advancing archenteron is found to induce the mouth evagination.

23-3.2. Embryonic Induction in Prevertebrate Chordates

Among prevertebrate chordates the embryonic inductions have been investigated in ascidians and *Amphioxus*.

(i) Induction by blastomeres in ascidians. An ascidian egg has strongly developed polar axis, or animal-vegetal-yolk gradient (*i.e.*, yolk-poor animal pole and yolk-rich vegetal pole) and it also has a posterior "mesodermal" ooplasmic crescent and anterior "neural" and "chordal" crescents, one above the other, all more or less at the equatorial level of the egg. At the eight-cell stage there are four animal blastomeres and four vegetal blastomeres of approximately equal size ; the mesodermal crescent material lies mainly in the two posterior vegetal blastomeres, the chordal and neural crescent material is mainly in the anterior vegetal blastomeres, and there is no special plasm in either anterior or posterior animal blastomeres.

During normal embryogenesis of the ascidian, different blastomeres of eight-cell stage embryo have following presumptive fates — (i) the two animal anterior blastomeres produce head epidermis, palps, and the brain with its two pigmented sensory structure ; (ii) the two animal posterior blastomeres produce only epidermis ; (iii) the two vegetal anterior blastomeres produce only notochord, spinal cord, and part of the intestine ; (iv) the two vegetal posterior cells produce mesenchyme, muscles, and part of the intestine. Blastomeres of the eight-cell stage when isolated and allowed to develop as pairs, however, present a more complex situation than first appeared. For instance, it was observed that while all territories (intestine, notochord, muscles, mesenchyme) have their destiny already fixed and consequently differentiated according to their presumptive fate, the destiny of the neural system is not fixed ; the two animal anterior blastomeres, if isolated, do not give rise to brain or palps or sensory spots but on the contrary, give rise only to epidermis. Moreover, one obtains from an isolated animal quartet an ectodermal vesicle, always lacking a brain. Consequently, it is concluded that the formation and differentiation of brain by two animal anterior blastomeres is dependent on the induction of two vegetal anterior blastomeres, which act in ascidians as **neural inductors**.

Nature of neural inductor of ascidians. The two vegetal anterior blastomeres give rise to diverse tissues, namely endoderm, notochord, and spinal cord. At the 64-cell stage the prospective material for these three tissues has been segregated. When the four prospective spinal cord cells of this stage are destroyed, the embryo still forms a normal brain and sensory organs. Accordingly, the inductor should lie in the notochord or in the endodermal cells. But since destruction of the six prospective endodermal cells of the 64-cell stage fails to inhibit the normal development of the brain, only the chordal cells are left, as far as experimental proof is concerned. The prospective notochordal blastomeres or cells are, thus, considered as **neural-inductors** of ascidians. Further, the destiny of the presumptive "neural" ectoderm is converted into the neural system only under the inductive action of the immediately underlying chordal (and endodermal) cells (**Reverberi**, 1960). The induction is consequent on a contact between correlated territories.

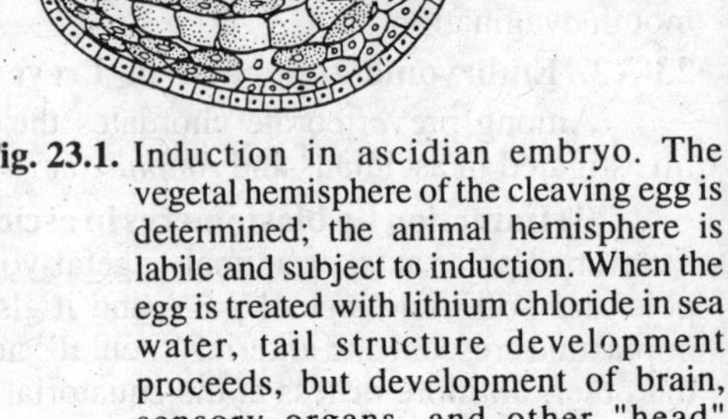

Fig. 23.1. Induction in ascidian embryo. The vegetal hemisphere of the cleaving egg is determined; the animal hemisphere is labile and subject to induction. When the egg is treated with lithium chloride in sea water, tail structure development proceeds, but development of brain, sensory organs, and other "head" structure is suppressed (after Berrill, 1971).

(ii) Embryonic induction in *Amphioxus*. Like ascidians, in *Amphioxus* the developmental potencies of the animal hemisphere are influenced to a large extent by the inductive stimulus from the material of the vegetal hemisphere, although the animal half can also exert an influence on the vegetal half of the egg to some extent. In the experiments conducted by **Tung, Wu** and **Tung** (1960), the animal hemisphere was rotated 90° to 180° on the vegetal hemisphere at the eight-cell stage, without changing the animal-vegetal axis. In spite of the shift of the original presumptive neural blastomeres to a posterior position, normal embryos developed, *i.e.*, the blastomeres moved to the anterior position are induced to develop neural structure. In fact, the whole of the prospective *Amphioxus* ectoderm remains undetermined during cleavage stages, since both the nervous system and tail form in normal relationship in spite of shifting the animal hemisphere blastomeres. The vegetal hemisphere or presumptive endoderm-mesoderm, on the other hand, exhibits early self-determination and has an inductive effect on the animal hemisphere.

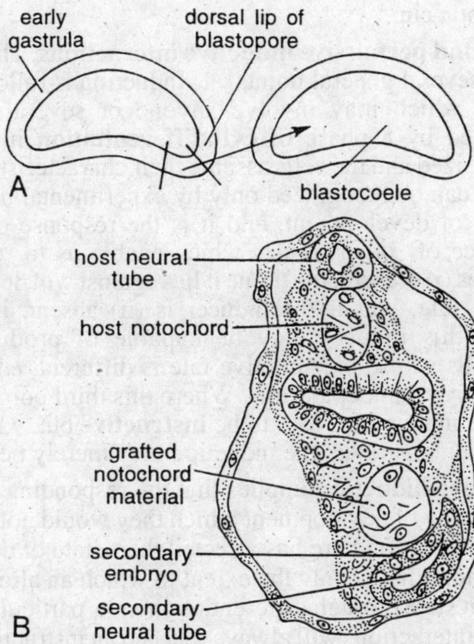

Fig. 23.2. Neural induction in *Amphioxus belcheri.* A—Diagram of operation : cells lying inside dorsal lip of the blastopore transplanted into the blastocoele of an embryo in the same stage of development; B—Transverse section of the host embryo with a secondary neural tube induced by the grafted notochord material (after Balinsky, 1970).

In another kinds of experiments the same investigators (**Tung**, *et al.*, 1962) transplanted pieces of tissue from the inner surface of the dorsal blastopore lip of an early gastrula of *Amphioxus*, into the blatocoele of another embryo in the same stage of development and observed that a secondary embryo developed in the ventral region of the host with a notochord and mesoderm produced by the graft and a neural tube induced from host tissue. Eventually, they concluded that chordal tisssue of *Amphioxus* gastrula possesses the power of neural induction, while mesodermal and endodermal tissues have little, if any, such inductive power.

23-3.3. Embryonic Induction in Vertebrate Chordates

Presently various types of inductive interactions are known to occur in vertebrates. These can be described in the sequence of their discovery such as follows :

1. Induction of Vertebrate Lens

Spemann (1901) showed that in many amphibians, the lens of eye is formed in the ectoderm only as a response to an inductive interaction resulting from contact with the optic cup (see Chapter 18). **Jacobson** (1958) has shown that this classic induction process is in fact only the culmination of a sequence of inductive interactions (leading to the formation of the lens in the correct location at the right time), which begins with a reaction to the anterior archenteron roof in early gastrulation, followed by one to the presumptive heart mesoderm in the late gastrula and concludes only with the response to contact with the optic cup.

Instructive and permissive inductive interactions. The optic cup/lens system illustrates several general points : 1. Induction is followed by a phase of **organogenesis** which may involve anyone or several of the cellular activities, succeeded by a phase of **cytodifferentiation** in which specific proteins are synthesized and the cells assume their characteristic differentiated form. 2. Induction can be recognized only by experimental interference with the normal course of development, and it is the response of tissues to the presence or absence of other tissues which enable us to characterize the process. If inducer is (i) present, the tissue it lies against will develop according to its presumptive fate, but, if the inducer is (ii) absent, it will not ; but furthermore, (iii) this inducer may be capable of producing the same development in cells whose presumptive fate is different, *e.g.*, formation of lens from presumptive flank epidermis. Where this third point can be shown experimentally, the induction is said to be **instructive** but, where only points (i) and (ii) can be demonstrated, the induction may merely be **permissive**.

The **instructive induction** implies that the responding cells have been directed into a pathway of development which they would not otherwise have followed, *i.e.*, that induction here has affected their state of determination. A **permissive induction** affects only the extent to which an already determined cell can express that state through differentiation in its particular pathway, and several permissive interactions will always succeed an instructive one. Besides the optic cup/lens system, many interactions between epidermis and underlying dermis are clearly instructive, *e.g.*, where chick dermis from feather-forming skin region which is implanted beneath the corneal epithelium of the eye induces feather formation in that tissue (see **Ede**, 1978).

Maintenance inductions. If the retina is removed from the eye of an adult frog, the lens regresses and degenerates : thus, persistence of the lens depends upon lifelong interactions with the adjacent retina. Such interactions are sometimes referred to as **maintenance inductions**. They do not initiate the cell type but are required to maintain the population of differentiated cells (**Maclean** and **Hall**, 1987).

2. Neural Induction

(a) Discovery of neural inductor. Among vertebrates, the phenomenon of embryonic induction has been investigated first in amphibians and later on, it was searched in other vertebrates. In amphibians, though induction of a lens by the optic cup rudiment was the first to be reported by **Spemann,** (1901–1912), the classical experiment upon which the concepts of "embryonic induction and organizer" are founded, came later. This was the discovery in 1924 by **Spemann** and his student, **Mangold** that the dorsal blastoporal lip from a salamander (or newt) early gastrula, when implanted into a ventral or lateral position of another gastrula, turns inward, develops into a notochord and somites, and induces the host ectoderm to form a neural tube, *i.e.*, it induces the formation of the axial organs of a more or less complete secondary embryo. This secondary embryo accordingly may have a notochord, somites, kidney tubules, each of which is composed partly of grafted, partly of induced host material. In this experiment, thus, the graft becomes "self-differentiated" (*viz.*, it forms the very structures it would have given rise in its normal setting—notochord, mesoderm, etc.) but it induces the adjoining host tissue to form spinal cord and other structures, including somites and kidney tubules.

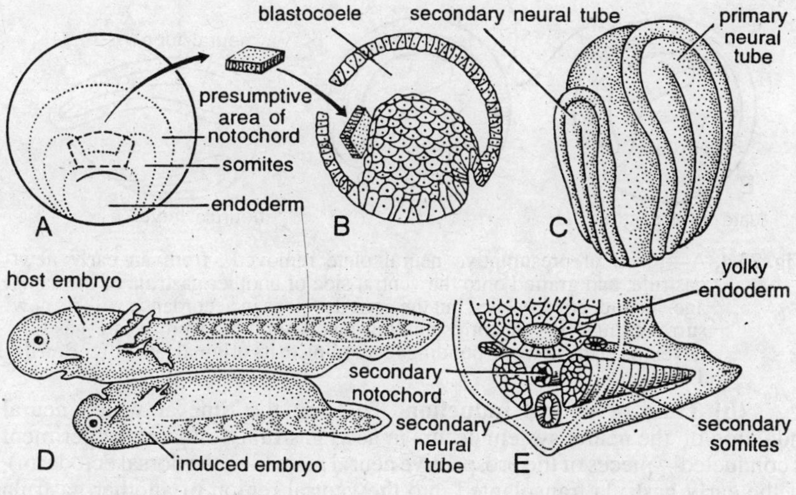

Fig. 23.3. Induction of a secondary set of axial organs by a graft of prospective archenteron roof, A, B—Transplantation of a piece of the upper blastoporal lip into another gastrula; C,D,E—The self differentiation plus inductions of the graft into a secondary embryo (after Torrey, 1971).

Visualizing these results, **Spemann** and **Mangold** spoke of the dorsal lip as the "**organizer**" or "**primary organizer**" because, it caused the development of a complete secondary embryo. It is 'primary' in the sense of being the first in sequence of inductions. However, it is not an "organizer" in the sense attributed to it by its discoverer (see **Ebert** and **Sussex**, 1970).

Actually, the organization of the secondary embryo results from a series of both inductive interactions and self-differentiative changes in the host and donor tissues. Today, therefore, the more general term **embryonic induction** or **inductive interaction** are preferred. Accordingly, since the primary organizer of **Spemann** and **Mangold**, induces the ectoderm to become differentiated into neural tube, it is called **neural inductor** or **primary inductor**. Then neural inductor of newt (*Triturus* sp.) is found to be composed of presumptive material for notochord and dorsal mesoderm (*i.e.*, chordamesoderm).

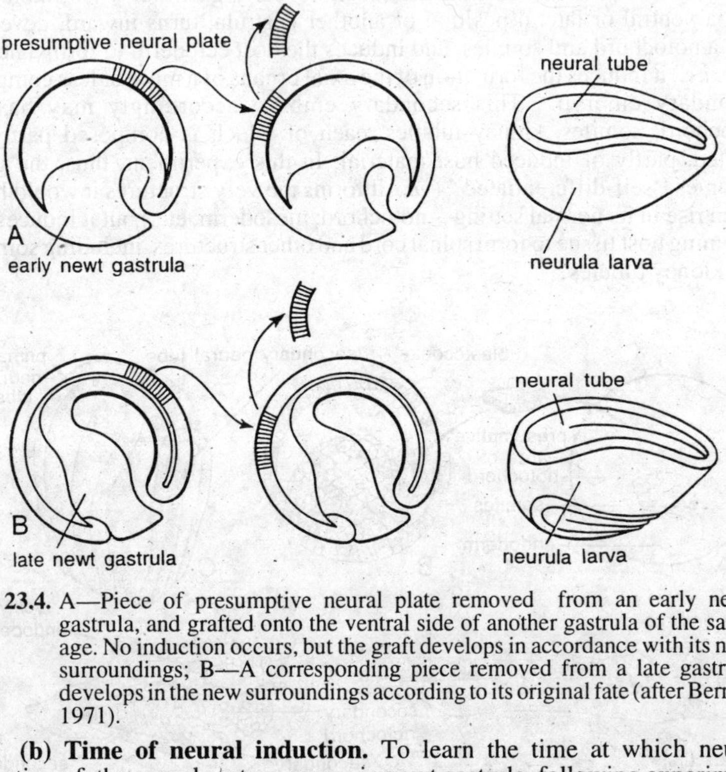

Fig. 23.4. A—Piece of presumptive neural plate removed from an early newt gastrula, and grafted onto the ventral side of another gastrula of the same age. No induction occurs, but the graft develops in accordance with its new surroundings; B—A corresponding piece removed from a late gastrula develops in the new surroundings according to its original fate (after Berrill, 1971).

(b) Time of neural induction. To learn the time at which neural induction of the neural system occurs in newt gastrula, following experiment is conducted—pieces of the prospective neural plate (*i.e.*, the dorsal ectoderm), of the early gastrula transplanted into the ventral region of another gastrula (Fig. 23.4 A) develop according to their new position (*i.e.*, they become belly epidermis). Further, when the pieces of neural ectoderm removed from a late gastrula and transplanted to a ventral position (Fig. 23.4 B) develop according to their original position, *i.e.*, they become neural plate. These experimental results, thus, show that neural induction occurs during the time when the material initially present in the dorsal lip moves inward and forwards to the anterior region of the embryo.

(c) Morphology of neural inductor. Vital-staining experiments of **Vogt** with newt eggs have shown that the material successively forming the dorsal blastoporal lip moves forward as the archenteric roof. Transplants taken from this region are also able to induce a secondary embryo or the belly of a new host, *i.e.*, the archenteron roof acts as a primary inductor in essentially the same way as does the dorsal lip tissue proper.

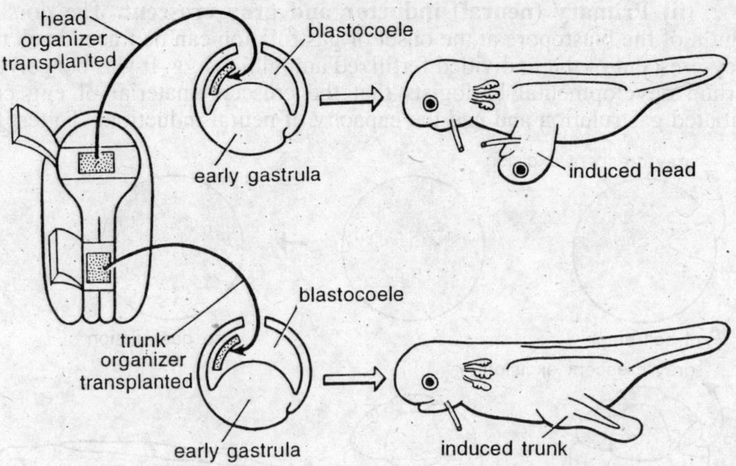

Fig. 23.5. The separation of the neural inductor (primary organizer) into head and trunk organizers (after Bodemer, 1968).

(i) Regional differentiation of neural inductor. The inductions of neural inductor are found to be regionally specific and the regional specificity is imposed on the induced organs by the inductor. In other words, the inductive capacity of the blastoporal lip varies both regionally and temporally. Most of the dorsal and dorsolateral blastoporal material is necessary for a graft to induce a more or less complete secondary embryo. If only prospective endomesodermal material of dorsal lip is grafted into another gastrula, it induces a partial embryo containing ectodermal head structures such as brain, sense organs, mouth structures, etc. Moreover, explants or grafts of the median portion of the dorsal lip tend to form eyes and olfactory pits, lateral parts tend to form posterior cephalic structures and still more lateral parts of the marginal zone tend to induce spinal cord and mesenchyme of a trunk-tail.

As invagination continues, however, and the dorsal lip no longer consists of prospective head endomesoderm but progressively becomes prospective trunk mesoderm, it acts as a trunk-tail inductor. The most caudal-region of the archenteron roof, in fact, specifically induces tail somites and probably other mesodermal tissues. The archenteron roof (containing chordamesoderm and some endodermal cells), thus, induces entirely different classes of tissues : various neural and mesoectodermal tissues by its anterior region and various mesodermal tissues by its most posterior region. From this fact, **Lehmann** (1945) concluded that differences in specific induction capacities exist between head and trunk level of the archenteron roof, and are related to the regional differentiation of the neural tissue into **archencephalic** (including fore-brain, eye, nasal pit), **deuterencephalic** (including hind-brain, ear-vesicle) and **spinocaudal** components. Accordingly, the archenteron roof can be said to consist of an anterior **head inductor** including an **archencephalic inductor** and a **deuterencephalic inductor** and a **trunk** or **spinocaudal inductor**.

(ii) Primary (neural) inductor and gray crescent. The dorsal lip region of the blastopore at the onset of gastrulation can be traced back to the grey-crescent of the undivided fertilized amphibian egg. It was conceived by certain developmental biologists that the crescent material of egg cortex initiated gastrulation and had the capacity of neural induction. To test these

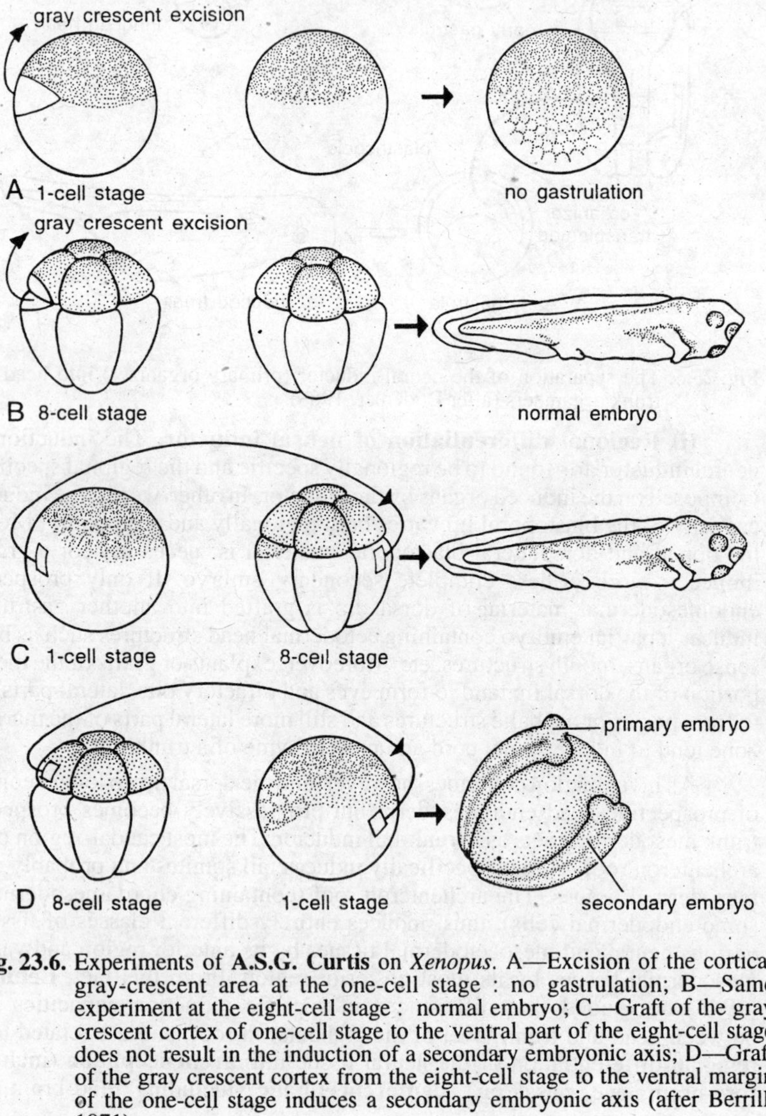

Fig. 23.6. Experiments of **A.S.G. Curtis** on *Xenopus*. A—Excision of the cortical gray-crescent area at the one-cell stage : no gastrulation; B—Same experiment at the eight-cell stage : normal embryo; C—Graft of the gray crescent cortex of one-cell stage to the ventral part of the eight-cell stage does not result in the induction of a secondary embryonic axis; D—Graft of the gray crescent cortex from the eight-cell stage to the ventral margin of the one-cell stage induces a secondary embryonic axis (after Berrill, 1971).

hypotheses, **A.S.G. Curtis** (1960, 1963) performed a series of experiments of transplanting parts of the cortex of the fertilized egg of the clawed-toad, *Xenopus laevis* at the beginning of cleavage. In one of the experiments, the grey-crescent cortex was excised from the fertilized egg, then it was observed that the cell division though proceeded undisturbed, the gastrulation failed to take place. In other experiment, the grey-crescent cortex of a uncleaved fertilized egg was excised and transplanted into a ventral position of a second egg, so that the egg receiving the graft had two grey crescents on opposite sides. As a result, egg cleaved to form a blastula which underwent two separate gastrulation movements to produce two separate primary nervous systems, notochords and associated somites (*i.e.*, a primary and a secondary embryo were formed). Similar experiments conducted on the eight-cell stage, however, showed that something had happened during the short interval represented by the first three cleavages. Grey-crescent cortex of the eight-cell stage still retained its inductive capacity when grafted to younger stages, but the eight-cell stage itself no longer reacted to grafts of the crescent material. Also, excision of the grey crescent at this stage no longer inhibited subsequent gastrulation and normal development, the missing crescent properties being replaced from adjacent cortical regions. Accordingly **Curtis** concluded that a change in cortical organization spreads across the surface of the egg during the second and third cleavages, starting from the grey crescent, when this change is completed, interaction, probably of a biophysical nature, can take place among various parts of the cortex.

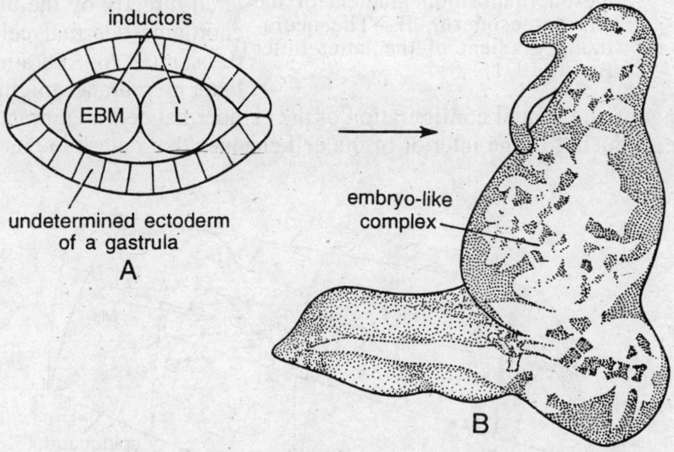

Fig. 23.7. The effect of two heterogeneous inductors on the undetermined ectoderm of an amphibian gastrula. A—Scheme of the experiment : both inductors are wrapped in a sandwich of ectoderm; B—An embryo-like complex has developed. EBM, extract of bone marrow; L, piece of guinea pig liver (after Berill, 1971).

(d) Gene activation in neural inductors. There are sufficient evidences that the component tissues of neural inductor namely presumptive chordamesoderm and endoderm, become differentiated prior to ectodermal cells. It is known that during late blastula stage, in the presumptive chordamesodermal and endodermal cells, the rate of transcription of mRNA and differential activation of genes become increased manifold. While the differentiation of ectodermal cells is set in only after mid-gastrulation and because of this fact the ectodermal cells retain their potencies to differentiate into either epidermal or neural tissues or even mesodermal cells up to early gastrular stage.

(e) Mechanism of neural induction. Since the roof of the archenteron causes the overlying ectoderm to develop into neural tissue, it suggests a direct action upon the ectodermal cells. To understand the nature of this direct action (induction) of the chorda-mesodermal cells of archenteron roof upon ectoderm, following two broad possibilities are considered by different embryologists :

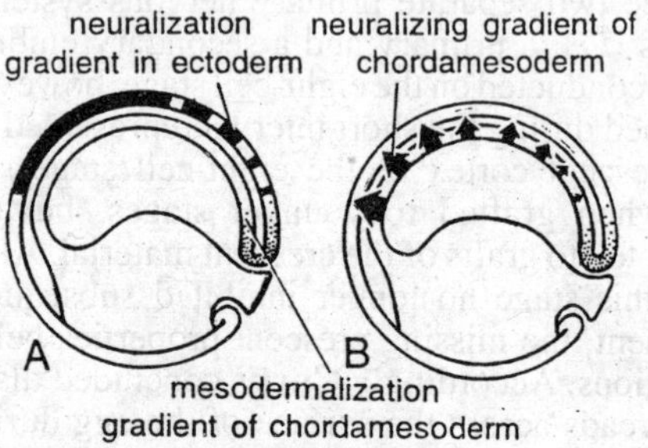

Fig. 23.8. A—The neuralization gradient in the ectoderm overlying the mesodermalization gradient of the chordamesoderm; B—The neuralizing gradient of the latter (after Berrill, 1971).

(i) One of the broad possibility is **surface interaction** of the cell at the inductive interface. The contact of the two cellular layers may provide a device whereby the structural pattern or geometry or behaviour of the ectodermal cell membranes is altered directly by the underlying chordamesodermal cells. Thus, the spatial configuration of the latter membranes might induce a change in the spatial configuration of the ectodermal cell membranes, this in turn producing in the interior of the cell changes that determine its develop-

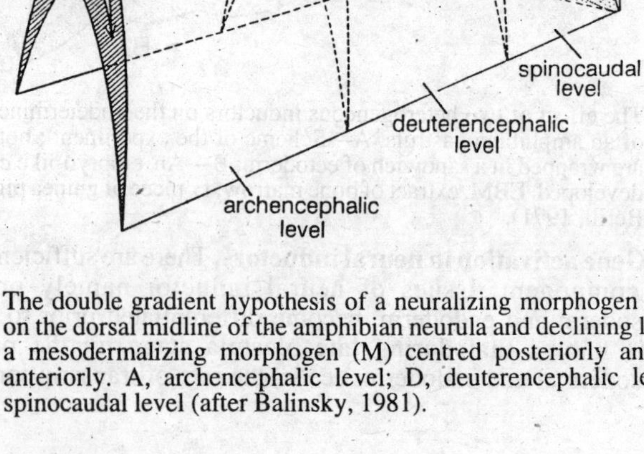

Fig. 23.9. The double gradient hypothesis of a neuralizing morphogen (N) centred on the dorsal midline of the amphibian neurula and declining laterally and a mesodermalizing morphogen (M) centred posteriorly and declining anteriorly. A, archencephalic level; D, deuterencephalic level and S, spinocaudal level (after Balinsky, 1981).

ment into neural plate. A morphological arrangement of this kind could account for quick and effective transmission of the inductive effect.

(ii) The another broad possibility is a **chemical mediation** of the inductive effect. Thus, a chemical substance or substances produced and released by inducing chordamesoderm cells at the archenteron ectoderm interface may act upon or enter the ectodermal cells to initiate cellular activities leading to neural development. A great deal of evidences favour the idea of an exchange of material between cells and also suggest that a diffusible substance may act as an effective inductive stimulus.

Chemical basis of neural induction. Many attempts have been made to elucidate the mechanism of induction and to identify the chemical substance or substances presumed to be involved. The results of numerous studies have been mostly disappointing and often confusing. It was rather disconcerting when it was found that many different tissues, embryonic or adult, from a great variety of different species, were capable of inducing nervous tissue in amphibian embryos. Moreover, some foreign tissues were found to be much more potent inductors after they had been killed by procedure such as heat or alcohol treatment. This fact remains against the concept of a universally present ''masked'' organizer, normally released for action only in the primary inductor region. Finally inorganic agents such as iodine and kaolin, and then even local mechanical injury, and also exposure to saline solutions of excessively high or low pH, also cause neural differentiation in cultures of undifferentiated ectoderm. Thus, according to **Holtfreter** (1945) an enormous variety of entirely unspecific substances—organic acids, steroids, kaolin, methylene blue, sulphydryl compounds, which had nothing in common except the property of being **toxic** to subectodermal cells—produced neurulation in explants. These findings establish the early grand concept of a master-chemical embryonic organizer of Holtfreter's **sublethal cytolysis**, namely, the idea that reversible cell injury liberates the neural inductor, but not the mesodermal inductor. However, most recently **Holtfreter** (1988) has concluded that inquiries into the chemical nature of inducing agents have led us astray. The fundamental principles underlying embryonic induction have remained obscure.

To identify the molecule which causes the neural induction, different chemical molecules of either grey crescent or dorsal lip or chordamesoderm are first separated by different biochemical methods and then afterwards the inductive capacity of each molecule was tested separately (**Niu**, *et al.*, 1953, 1956). Likewise, the chemical principles of abnormal or foreign inductors were analysed by two schools, namely the school of **Yamada** in Japan and the school of **Tiedemann** in Germany. The general conclusion of their experiments remains that evocator or inducing substance is of protein nature (see **Balinsky**, 1970).

When it was found that like the neural inductor, a variety of unnatural or abnormal inductors of inorganic, organic or mechanical nature could cause induction of ectodermal cells to become developed into neural tube, then exhaustive attempts were made by different embryologists to understand the real mechanism of neural induction. They have forwarded many theories to explain the mechanism of neural induction. Some important theories about the mechanism of neural induction are following :

(i) Protein denaturation theory of neural induction. According to **Ranzi** (1962, 1963) notochord formation and neural induction are related to protein denaturation. The amphibian grey crescent, the site of notochord formation, is a centre of high metabolic activity, and, according to **Ranzi**, centres of greater metabolic activity correspond to sites of protein denaturation. Neural induction by the notochord may also involve protein denaturation.

(ii) Double gradient theory of neural induction. Two groups of investigators, represented by **Toivonen** (1955, 1968) and **Yamada** (1961) assume that two chemically distinct factors are involved in the action of the primary inductor, one a "**neuralizing agent**" and the other a "**mesodermalizing agent**". Both groups have worked mainly with denatured bone marrow and liver as the inductors. The general interpretation is as follows — the regional specificity of the embryonic axis arises from the interaction between two gradients : the neuralizing principle has its highest concentration on the dorsal side of the embryo and diminishes laterally, while the mesodermalizing principle is present as an antero-posterior gradient with its peak in the posterior region. Anteriorly ; the neuralizing principle acts alone to induce fore-brain structures, more posteriorly the mesodermalizing principle acts along with the neuralizing one to induce mid-brain and hind-brain structures ; while even more posteriorly the high concentration level of the mesodermal gradient produces spino-caudal structure.

(iii) One factor hypothesis of neural induction. Nieuwkoop (1966) using living notochord as the inductor, postulates only one factor which first evokes ectoderm to form neural tissue and later causes ectoderm to transform into more posterior and mesodermal structures. In one experiment, consisting of combining isolated gastrular ectoderm with a piece of notochord and then removing the notochord tissue after varying lengths of time, it was found that only 5 minutes exposure to the inductor caused a part of the ectoderm to transform into brain and eye structures.

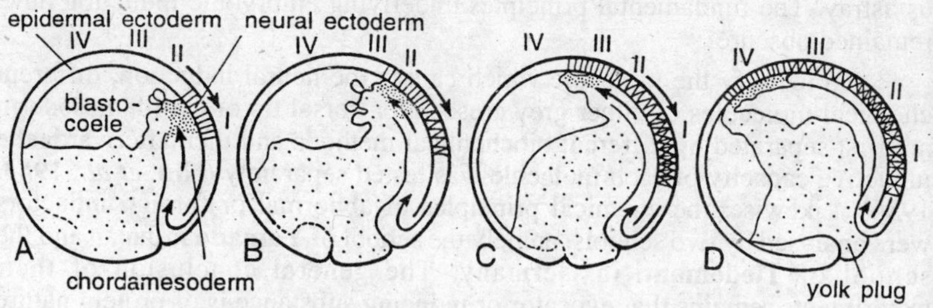

Fig. 23.10. Oppositely directed movements of invaginating archenteron roof and epibolically extending ectoneuroderm at four successive stages of the gastrulation process (A–D). The interaction of both layers leads to a caudo-cranial spreading of an activating inductive principle in the ectoderm when it overlies the prechordal portion of the archenteron roof. This primary action is followed by a caudo-cranial spreading of a superimposed transforming inductive principle emanating with increasing intensity from successively more caudal regions of the archenteron roof (Nieuwkoop, 1966).

(iv) Sexen and **Toivonen** (1962) investigated that (a) most of the inductively active tissues can be separated chemically into fractions yielding different inductive actions ; (b) correspondingly, the fraction which yield archencephalic inductions are chemically different from preparations inducing spino-caudal or mesodermal structures ; and (c) same chemical or physical treatments of "combined" inductors (deuterencephalic or spino-caudal) will always destroy selectively the capacity to induce caudal structures, and following these treatments the tissues yield a pure archencephalic inductive action.

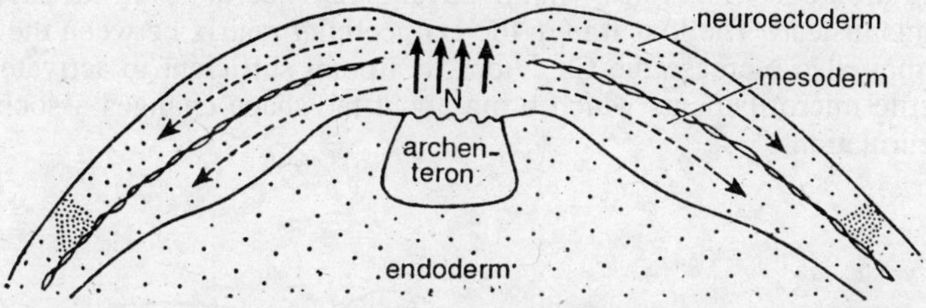

Fig. 23.11. Medio-lateral spreading of an inductive action within the mesoderm and a similar spreading of the neuralizing action in the overlying ectoderm. Parallel vertical arrows indicate the main site of the neuralizing inductive action (N) in the area of intimate adhesion of the two germ layers (Niuewkoop, 1966).

(v) Ionic theory of neural induction. L.G. Barth and **L.T. Barth** (1968, 1969), who remained long associated with the problem of primary-inductor analysis, have advanced the ionic theory of neural induction. Their working hypothesis is that the actual process of induction may be initiated by release of ions from bound form, representing a change in ratio between bound to free ions within the cell of the early gastrula. Experimentally, induction of nerve and pigment cells in small aggregates of prospective epidermis of the frog gastrula has been found to be dependent on the concentration of the sodium ion. Also, normal induction of nerve and pigment cells by mesoderm in small explants from the dorsal-lip and lateral marginal zones of the early gastrula is dependent on the external concentration of sodium. Nerve and pigment cells are induced when the culture medium contains 0.088M NaCl; at 0.044M NaCl the mesoderm differentiates into muscle and mesenchyme but nerve and pigment cells induction does not occur. The interpretation is that **normal embryonic induction depends on an endogenous source of ions and that an intracellular release of such ions occurs during late gastrulation**. A forty-fold increase in intracellular sodium ions and a three-fold increase in calcium ions takes place during gastrulation. The ratio of the two kinds of ions may be more significant than the actual concentration.

In 1974, **Barth** and **Barth** proposed a model of induction during normal gastrulation in *Rana pipiens* (Fig. 23.12). According to this model, in the blastula, ectodermal cells surrounding the blastocoele cavity release ions such as Na^+, K^+, Ca^{++} and Mg^{++} into the blastocoelic fluid from their inner surfaces. During gastrulation, the invaginating chordamesoderm becomes closely applied to the inner surface of the ectoderm overlying the archenteron roof and establishes an extracellular matrix between the two layers. While ventral ectodermal cells still lining the reduced blastocoele cavity continue to release ions into the cavity, the ions released from the inner surface of the presumptive neural ectoderm accumulate between the two layers instead and the extracellular concentration of Na^+ and other ions probably increases sufficiently to insure a higher concentration inside the cells, enough to induce. Ectodermal cells lining the blastocoele cavity, containing a lower concentration of sodium ions, will not develop into neural tissue but become epidermis instead. The ions trap in the extracellular matrix between the two layers may also increase the Ca^{++} ions to a point sufficient to activate the contractile microfilaments which bring about the shape changes associated with neurulation.

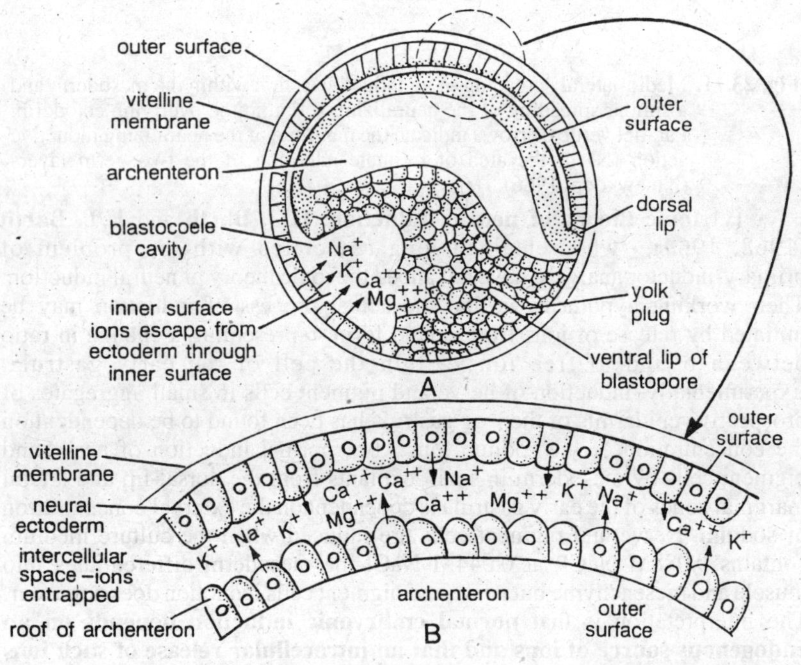

Fig. 23.12. Role of ions in primary induction according to model proposed by **L.G. Barth** and **L.J. Barth** (1974). A—Late gastrula of *Rana pipiens,* when roof of archenteron is inducing neural plate in overlying ectoderm; B—An enlarged view of the region of induction shows the accumulations of ions within the extracellular matrix with a consequent higher ion concentration within the ectoderm cells themselves. This initiates neural induction (after Grant, 1978).

Ions tend to affect **adenylcyclase** enzyme of cell surface which regulates intracellular levels of cyclic AMP. The cAMP tends to affect cell permeability; cation flow is altered by changes in levels of cAMP inside or outside the cell. Such mutual interactions could affect cell division and morphogenesis. For example, lithium ion is known to have many morphogenetic effects on amphibian development such as chordamesoderm and endoderm do not invaginate but spread outside, forming exogastrula (**Grant**, 1978).

(vi) Neurotransmitter theory. **McMohan** (1974) proposed a hypothesis of induction which suggests that molecules which function as neurotransmitters (*e.g.*, serotonin, acetylcholine) in physiological activity, are secreted by the inducer cells and act at the surface of responding cells and control their intracellular concentration of cyclonucleotides such as cyclic AMP and cyclic GMP whose effects and metabolism are intricately interwoven with the ionic conditions within and outside the cell. These cyclonucleotides act upon the gene regulatory mechanism to lead development into particular development pathways (**Ede**, 1978).

(f) Occurrence of neural inductor in different vertebrates. Although the ''primary organizer'' or ''neural inductor'' was first discovered in urodele amphibians, it was soon found that the dorsal lip of the blastopore and the roof of the archenteron of other vertebrates has the same function in development. In particular it was found that the chordamesoderm in all vertebrates induces the nervous system and the sense organs. Neural inductor has been investigated in following vertebrates.

1. In **cyclostomes**, particularly in lampreys, whose cleavage is holoblastic and whose blastula and gastrula resemble those of amphibians, the property of neural induction lies in the presumptive chordamesodermal cells of dorsal lip of the blastopore (**Bytinski Saltz**, 1937 ; **Yamada**, 1938).

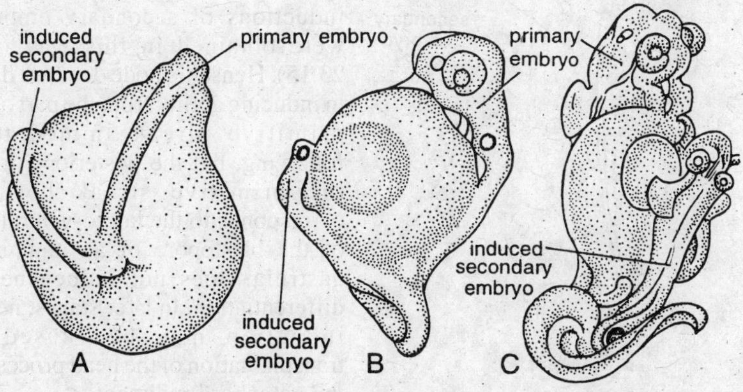

Fig. 23.13. Induction of secondary embryos by grafting of the primary organizer or neural inductor in the lamprey (A) and the perch (*Perca*) (B, C), (after Balinsky, 1970).

2. In **bony fishes**, inductions of secondary well-developed embryos were produced by transplanting the posterior edge of the blastodisc, which corresponds to the dorsal lip of the blastopore, into the blastocoele of another embryo (Fig., 23.13, B,C; **Oppenheimer**, 1936, in *Fundulus* and *Perca*), or by transplanting the chorda-mesoderm and endoderm (**Luther**, 1935, in *Salmo*). Neural inductions were also obtained by transplanting the dorsal lip of the blastopore in the sturgeon (**Ginsburg** and **Dettalff**, 1944).

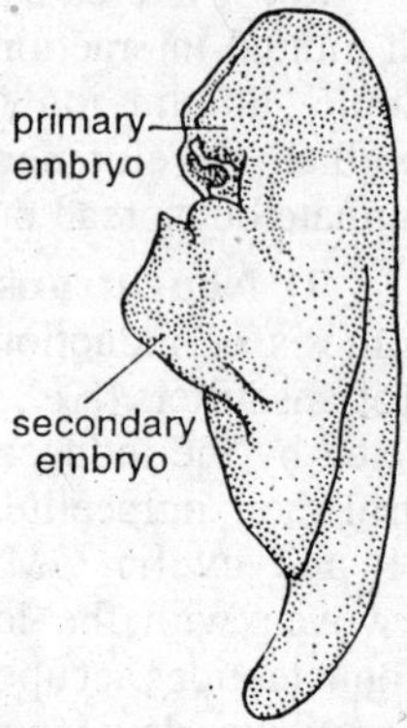

Fig. 23.14. Induction of a secondary embryo by means of a grafted neural inductor in a frog (after Balinsky, 1970).

3. In **frogs**, the induction of a secondary embryo can be performed by the dorsal lip of the blastopore transplanted into the blastocoele of a young gastrula, in very much the same way as in newts and salmanders (**Schotte,** 1930).

4. In **reptiles**, there is no experimental proof of occurrence of neural inductor, but, it is generally expected that in them archenteron has the same inducing activity as in other vertebrates (see **Balinsky,** 1970).

5. In **birds**, anterior half of primitive streak which corresponds to the lips of blastopore in amphibians, is found to act as neural inductor or primary organizer (**Waddington**, 1933, 1952). To test the inducing ability of the primitive streak, whole blastoderms were removed from the egg in early gastrulation stages and cultivated *in vitro* on a blood plasma clot. Parts of the primitive streak from another embryo were then inserted between the epiblast and hypoblast. Very good inductions of secondary embryos were obtained in this way (Fig. 23.15). Hensen's node does not differ in inducing ability from the part of the primitive streak immediately following, but the posterior third of the primitive streak because corresponds to the latero-ventral lips of the blastopore of an amphibian gastrula, does not induce neural differentiation. In later stages, neural induction may be caused by transplantation of the head process — the notochord rudiment.

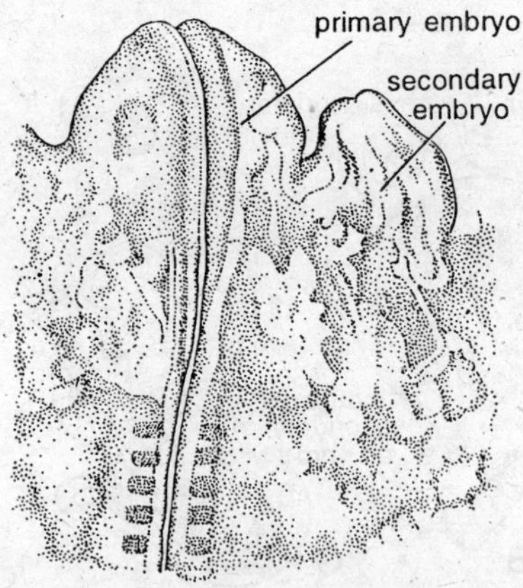

Fig. 23.15. Induction of a secondary embryo by means of a grafted primitive streak in a bird. The secondary embryo is on the right (after Balinsky 1970).

Other experiments of **Waddington** and his collaborators (1933, 1952) clearly established that the primitive streak of birds is dependent on the underlying hypoblast for its formation. This process of induction is something for which no analogue has been found so far in other vertebrates.

6. Because of technical difficulties, only very inadequate information is available on the primary organizer and neural induction in **mammals**. A successful neural induction has been performed in a rabbit embryo by cultivating the early blastodisc on a plasma clot and implanting the primitive streak of the chick as an inductor (**Waddington**, 1934). The experiment proves that the tissues of a mammalian gastrula have the competence for neural induction. It is practically certain that the primitive streak and the cells migrating from the streak (chordamesoderm) are the source of the inducing stimuli. This suggestion may be supoported by the fact that the anterior end of a somewhat later rabbit embryo (with two pairs of somites) induced a neural plate in a chick embryo when placed under a chick blastoderm (**Waddington**, 1936).

3. Mesodermal Induction

In *Xenopus*, mesoderm specification occurs around the 128-cell stage (**Nakamura** *et al.*, 1970). The early *Xenopus* embryo can be considered (by the fate map studies) to consist of two cell types : presumptive ectoderm in the animal hemisphere and presumptive endoderm in the vegetal hemisphere (**Jones** and **Woodland**, 1986). By the early blastula stage, a third cell type — mesoderm — has arisen, and it is possible to show that this is formed as the result of an inductive interaction.

Mesoderm induction was first demonstrated by **Nieuwkoop** (1969, 1973) in urodele embryos. Explants of blastula's animal pole tissue form epidermis in isolation, but when they are placed adjacent to cells from the vegetal hemisphere, large amounts of mesodermal cell types are formed, which according to cell marking experiments, are derived from the animal pole component (Fig. 23.16) (**Nieuwkoop** and **Ubbels**, 1972).

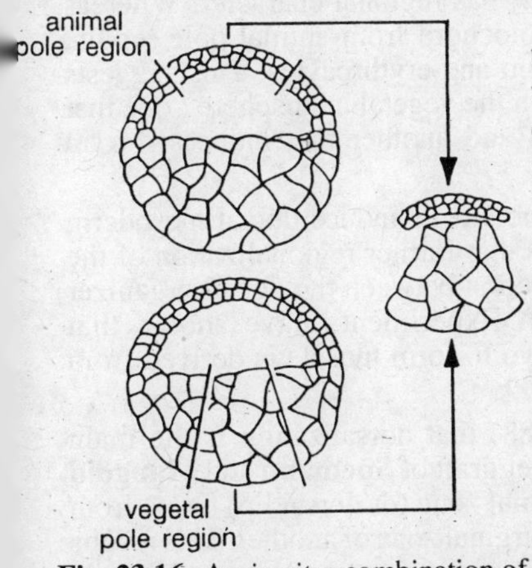

Fig. 23.16. An *in vitro* combination of animal and vegetal pole regions of amphibian blastula (after Smith, 1988).

The determination of germ cells in urodeles requires at least two steps of inductions : First endoderm induces ectoderm to become mesoderm as part of the overall vegetal-animal gradient of induction that characterizes urodele embryos. Then primordial germ cells are induced to form within a portion of mesoderm. Dorsal and ventral endoderm have different inductive roles. Ectoderm that lies beneath dorsal endoderm is induced to form notochord, somatic mesoderm and a portion of kidney (pronephric) tubular mesoderm, while the endoderm associated with ventral endoderm is induced to form the balance of the pronephric tubules, mesenchyme and the mesoderm from which the primordial germ cells will develop. The specificity resides within the inducing ventral endoderm, since germ cells can be induced from any region of the animal half of the blastula when the region is associated with ventral endoderm (**Maclean** and **Hall**, 1987).

A recent study on mesoderm induction in *Xenopus* has made use of cell lineage labels and cell-type specific markers (*e.g.*, antibodies specific for notochord and muscle) to demonstrate clearly that animal pole cells that normally make ectoderm can be converted to mesoderm (**Dale** *et al.*, 1985). This shows that mesoderm induction is a genuine instructive interaction. The results of animal-vegetal combination experiments with *Xenopus* were found to be dependent on the stage of the embryos. For example, in combinations made at the early blastula stage, the animal pole components always formed muscle and about one-half the cases formed the notochord. A few contained erythrocytes, mesothelium and pronephros. However, at the mid-blastula stage, virtually all the mesoderm was present as muscle, but mesoderm induction was poor at the early gastrula stage, with only a few positive cases, each producing small amount of mesenchyme. The lack of mesoderm induction in these latter cases was found due to a loss of the ability of the animal pole component to respond to induction ; early blastula animal pole tissue was induced to form mesoderm by gastrular vegetal cells, but the inverse combination gave very few positive results. The loss of responsiveness of the animal pole cells to mesoderm induction (*i.e.*, loss of **competence**) is followed by the acquisition of competence to respond to a new signal, *i.e.*, neural induction.

It has not proved as yet that whether all the mesoderm in the amphibian embryos arises through induction. In fact, **Gurdon** *et al.*, (1985) have suggested that in *Xenopus*, some mesoderm, perhaps that derived from the four vegetal pole cells at the 8-cell stage, forms as a result of cytoplasmic localization (*i.e.*, due to the presence of muscle determinants) (see **Smith**, 1988).

The mesoderm induction of *Xenopus* has regional character. Whereas dorsal vegetal cells induce muscle and notochord from animal pole region, ventral vegetal regions induce mesothelium and erythrocytes. This suggests that there are at least two distinct regions in the vegetal hemisphere : one that induces **ventral mesodermal structures** and another that induces **dorsal mesodermal structures**.

(1) Dorsalization. Dorsal vegetal pole cells induce dorsal mesoderm from overlying animal pole cells. It seems that further regionalization of the mesoderm rudiment results from an interaction between the dorsal organizer region and adjacent ventral mesoderm. Experiments have shown that instruction to the ventral half of the embryo to form muscle is derived from the dorsal marginal zone or **organizer**.

It has been argued (see **Smith**, 1988) that dorsalization is the main inductive interaction at work in the organizer graft of **Spemann** and **Mangold** (1924). In the organizer graft, dorsal marginal zone (or dorsal lip) tissue from one embryo is transplanted to the ventral marginal zone of another. This results in the formation of a double-dorsal mirror-symmetric embryo, in which only the notochord, and a few cells in the somites, are derived from the graft. It seems likely that the somite tissue of the secondary embryo arises through dorsalization of adjacent ventral marginal zone cells.

(2) Timing and transmission of mesoderm induction. Experiments aiming at establishing the time at which mesoderm induction occurs in normal development have concentrated on removing presumptive mesoderm from the embryo at different stages. By this means **Nakamura** *et al.*, (1970) suggested that specification of the mesoderm occurs around the 128-cell stage.

Sargent *et al.*, (1986) introduced a more improved technique in which they used three germ layer-specific gene probes to monitor their results : one encoding a **cytokeratin** (ectoderm), one encoding **α-actin** (mesoderm), and one encoding an unidentified éndodermal product. When fertilized *Xenopus* eggs were freed from their vitelline envelope and cultured in a calcium-magnesium-free solution ; they continued to divide, but the cells separated from each other and formed a heap. Calcium and magnesium were returned to the disaggregated embryos at the time when controls commenced gastrulation and all three germ-layer specific markers were subsequently expressed (As also shown by **Gurdon** *et al.*, 1984, calcium and magnesium ions are required during gastrula and neurula stages for α-actin gene expression). By contrast, when the cells of dissociated embryos were dispersed from their heap and kept separate throughout the period up to gastrulation, addition of calcium and magnessium did not permit expression of the α-actin gene. Levels of the ectoderm and endoderm-specific genes, however, were enhanced. Thus, expression of a mesoderm-specific gene during gastrulation seems to require the close proximity of cells at the blastula stage, but not firm adhesion, implying perhaps that *mesoderm induction can occur without specialized cell-cell contact.* The timing of mesoderm induction during normal development could now be studied by disaggregating embryos at different stages and asking at what stage α-actin gene expression becomes independent of cell dispersion. The result so far indicated that mesoderm induction occurs later than the 128-cell stage and is independent of strong calcium-mediated adhesion (**Sargent** *et al.*, 1986).

(3) Mesoderm inducing factor. The nature of the signal produced by the vegetal hemisphere of the amphibian embryo is unknown, but extracts of a variety of tissues and cells have been shown to have mesoderm inducing activity. Until recently the best characterized of these substances had been **Tiedemann's vegetalizing factor** which is isolated from the trunks of 9-to 13-days chick embryos (**Tiedemann** and coworkers, 1959, 1987). This factor is a protein of 28-30,000 M.W., which separates into smaller chains of 13,000 to 15,000 M.W., in formic acid. It binds to heparin when the vegetalizing factor is applied, in the form of an insoluble pellet, to amphibian blastula ectoderm, it causes the formation of a range of mesodermal cell types, including notochord, muscle, kidney and blood. Other heterologous sources of mesoderm inducing factors include guinea pig bone marrow, HeLa cells, and carp swim bladder and extracts of *Xenopus* blastulae and gastrulae.

Recently, successes have been obtained in isolating a purified mesoderm-inducing factor which remains active in soluble form. In this regard the following two exciting discoveries have added new dimension to this problem : (1) A *Xenopus* cell line, called **XTC** is found to secrete a mesoderm-inducing factor (**Smith**, 1987). When animal pole ectoderm is cultured in XTC conditioned medium, it forms muscle and notochord, rather than epidermis. XTC factor is a heat-stable protein of 16,000 MW. (2) **Fibroblast growth factor** (**FGF**) and a variety of other heparin-binding growth factors also induce animal pole ectoderm to form mesoderm (**Slack** *et al.,* 1987). Although there is no evidence yet that *Xenopus* embryos contain FGF, it is noteworthy that heparin which binds to FGF-like molecules, can inhibit mesoderm induction in animal-vegetal combinations. FGF tends to induce ventral mesoderm rather than dorsal, so it is possible that this is the ventral mesoderm induction signal, while XTC factor is the dorsal mesoderm induction.

(4) Gene expression in response to mesoderm induction. Mesoderm induction is found to cause the activation of mesoderm-specific genes, particularly the cardiac actin gene, which is strongly expressed in larval somite muscle (**Mohun** *et al.*, 1984). **Kimelman** and **Kirschner** (1987) have confirmed that FGF induces muscle actin gene expression in isolated *Xenopus* animal pole regions and demonstrated that *Xenopus* embryos contain an mRNA encoding a protein homologous to FGF. They also show that **transforming growth factor-β** (**TGF- β**) enhances the ability of FGF to induce muscle actin, although TGF-β alone has no effect.

(5) Mesoderm induction in other vertebrates. Recently by using grafts of [^{3}H] thymidine-labelled tissue or quail cells, three different inductive interactions have been demonstrated in chick embryos. The first of these, the **induction of primitive streak**, is analogous to mesoderm induction in *Xenopus* (**Slack**, 1983). The primitive streak, both in birds and mammals, represents the site through which the cells of the epiblast pass on their way to form mesoderm and endoderm, and there is evidence that the primitive streak is induced from the epiblast by the underlying hypoblast (**Waddington**, 1933). In the second interaction, analogous to **dorsalization**, grafts of Hensen's node (the presumptive notochord) can be shown to induce lateral mesoderm to form somite (**Hornbruch** *et al.*, 1979), and, as in the amphibian organizer grafts, this experiment also demonstrates the third interaction, *i.e.*, **neural induction** (**Gallera**, 1971). Thus, the interactions we study in amphibian embryos have their counterparts in chick, and probably, mouse development as well (**Smith**, 1988).

(6) Induction of metanephric kidney tubules. The kidneys are formed in a strip of intermediate mesoderm (mesomere) on each side of the embryo which lies between the somites and the lateral plate. From this a series of tubules is formed, divided into three groups produced in sequence : the **pronephros, mesonephros** and — in birds and mammals, in which it is the only one which survives into postembryonic life — the **metanephros**. Tubules from the pro-and mesonephros open directly into a longitudinal Wolffian duct, but those of the metanephros open into a ureter which grows forward from the posterior end of the Wolffian duct into the nephrogenic mesenchyme and there induces the mesenchyme to form the tubules (Fig. 23.17). In a number of mouse and chick mutants, *e.g.*, the *wingless* mutant of the fowl, the ureteric bud fails to grow forward, so that it does not reach the mesenchyme, which, therfore, produces no tubules. **Grobstein** and **Saxen** and their colleagues (1961, 1968) have found that in the formation of metanephric tubules the inductive interaction is permissive rather than instructive.

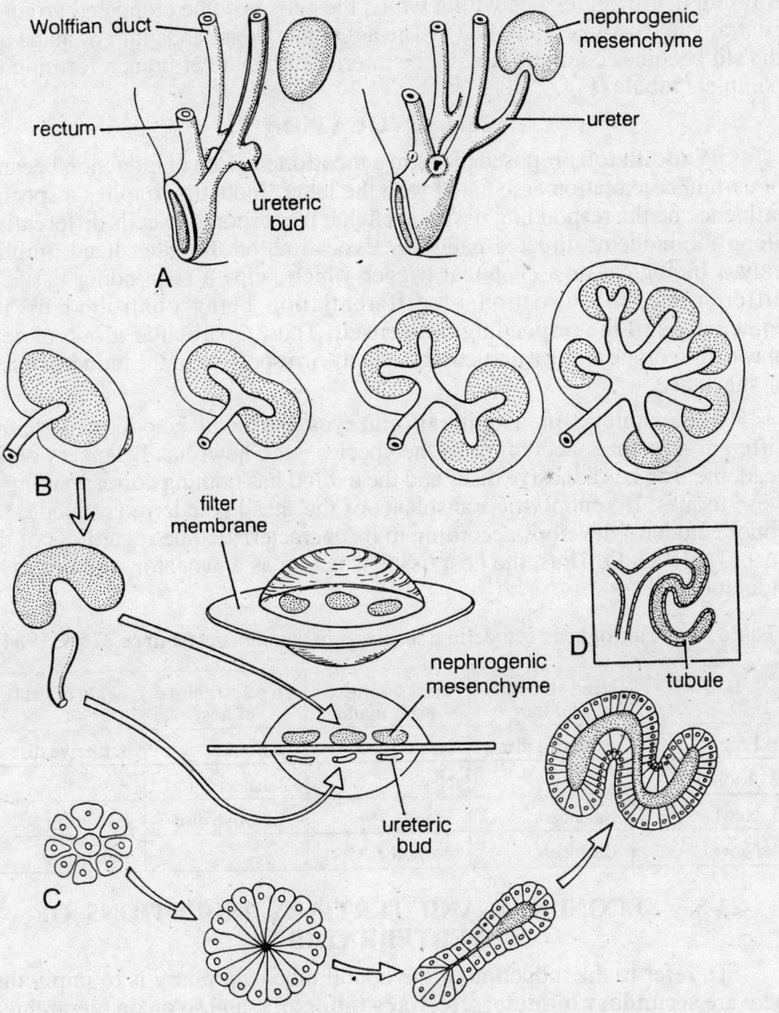

Fig. 23.17. Induction of metanephric kidney in mouse. A—Outgrowth of the ureteric bud anteriorly into the nephrogenic mesenchyme (mesomere); B—(left to right) Induction of tubule formation in the mesenchyme, beginning with condensation of the mesenchyme cells around the branching tips of the ureter rudiment; C—Kidney tubulogenesis observed in nephrogenic mesenchyme cultured on a filter membrane, in proximity to a ureteric bud placed on the other side of the membrane; D—Final stage of tubulogenesis, in which the tubule (stippled) becomes confluent with the ureteric branch which has induced its development (after Ede, 1978).

The inducing tissue, the ureter rudiment, grows into the loosely organized mesenchymal rudiment and forms a branching tree-like structure within it ; the mesenchyme then condenses around the tips of the branches, forming dense aggregates within which the cells become elongated around a central site (**Saxen** *et al.*, 1968). The aggregate then bends into S-shape and the slit becomes continuous with the interior of the ureter branch forming the complete tubule (Fig. 23.17).

23-4. EVOCATION

Waddington proposed the term **evocation** in place of induction because of certain connotation associated with the latter. Induction implies a specific influence on the responding tissue, such that the responding cells differentiate along the inductor-directed pathway. Evocation, on the other hand, implies only a biological or a chemical trigger which helps a responding tissue to differentiate, the direction of differentiation being controlled by the epigenotype of the responding tissue itself. Thus, a particular group of cells may differentiate along a particular pathway, irrespective of the inductor tissue or substance.

For example, in amphibian embryos the head ectoderm develops different structures according to the species : the newt has balancers on its head, the frog has adhesive discs and the axolotl has nothing corresponding to these tissues. If xenoplastic transplants of the head ectoderm are made, the donor ectoderm develops according to its characteristic fate regardless of the host (Table 23-1). Thus, the host tissue is acting as a evocator, and not as an inductor.

Table 23-1. Fate of head ectoderm transplants in amphibians **(source, Goel, 1984)**

Donor	Head structure of donor	Host on which graft made	Head structure of host	Fate of graft
1. Frog	Adhesive disc	Newt	Balancers	Adhesive disc
2. Axolotl	—	Newt	Balancers	—
3. Newt	Balancers	Frog	Adhesive disc	Balancers
4. Newt	Balancers	Axolotl	—	Balancers

23-5. SECONDARY AND TERTIARY INDUCTIONS OF VERTEBRATES

To refer to the induction of the neural tube as primary is to imply that there are **secondary inductors, tertiary inductors**, and so on, in hierarchical array. For example, the brain having been induced by archenteron roof, forms its various regional differentiation under the influence of head endoderm and mesoderm. The hind-brain induces overlying ectoderm to differentiate into the **ear placode**, which then invaginates and, in turn, induces surrounding mesenchyme to form the cartilaginous **auditory capsule**. Another sequence of primary, secondary and tertiary inductions, occurs during the development of eye. The **optic vesicle**, a differentiation of the mid-brain, makes contact with head ectoderm, which becomes induced to form a **lens placode** ; the lens

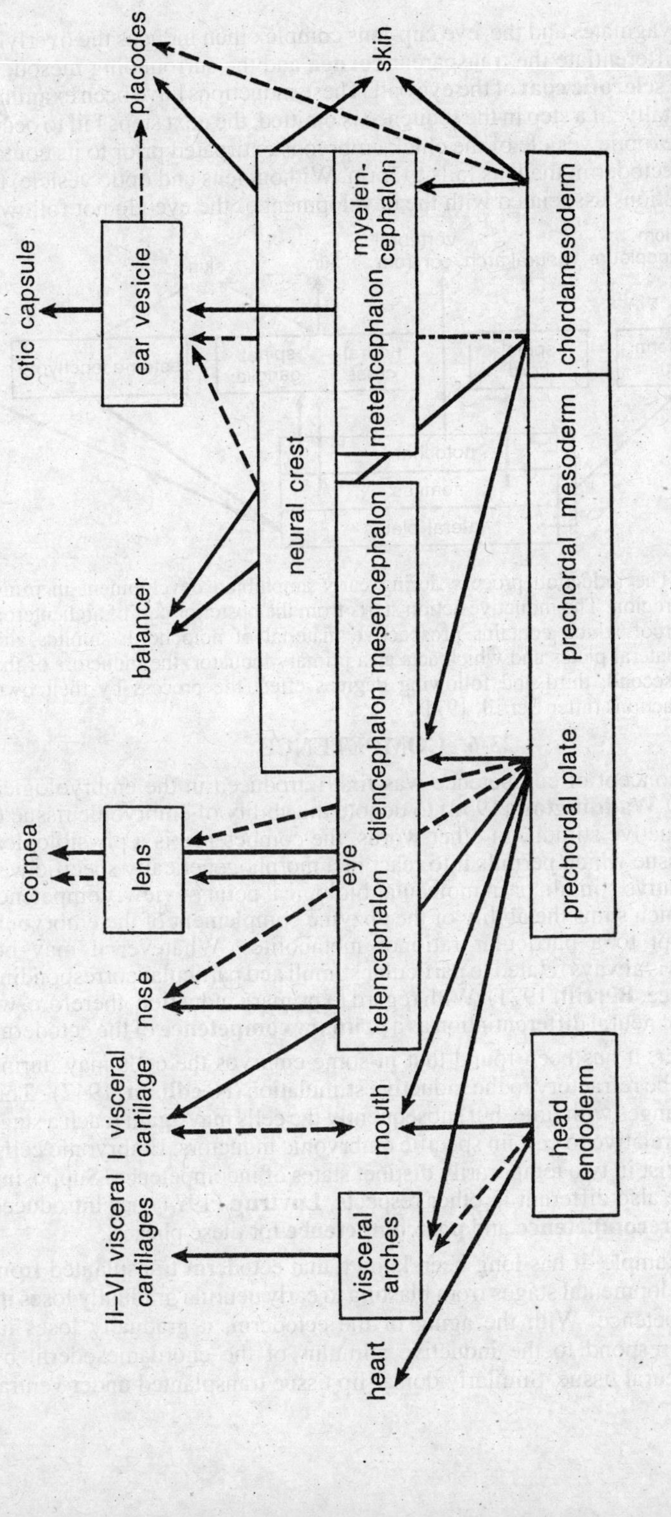

Fig. 23.18. The induction process during early amphibian development in head region. The inductive action starts from the anterior part of archenteron roof (inductor of first degree–primary inductor), the inductors of second, third and following degrees effect the process by their own actions. The different strengths of the inductive actions are indicated by thick and broken arrows (after Berrill, 1971).

placode invaginates and the eye cup-lens complex then induces the overlying tissue to differentiate the transparent **cornea** and the surrounding mesoderm to form the **sclerotic coat** of the eye ball. These inductions have been examined experimentally ; if a step in the sequence is omitted, the next steps fail to occur. Thus, if the optic vesicle of the chick embryo is extirpated prior to its contact with head ectoderm, the lens fails to form. Without lens and optic vesicle, the other inductions associated with the development of the eye do not follow.

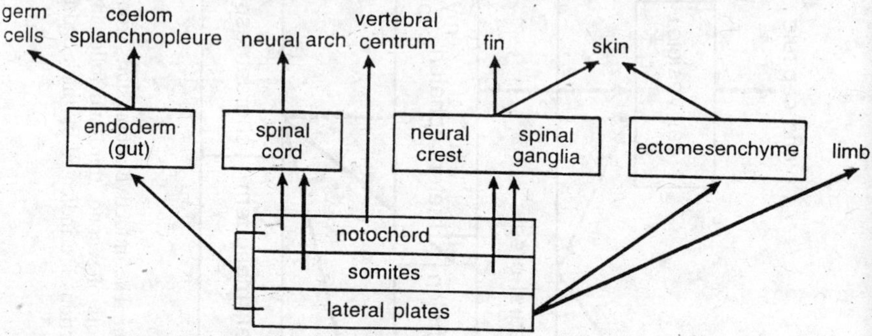

Fig. 23.19. The induction process during early amphibian development in trunk region. The inductive action starts from the posterior part of archenteron roof which contains prospective material of notochord, somites and lateral plates and which acts as a primary inductor; the inductors of the second, third and following degrees effect the process by their own actions (after Berrill, 1971).

23-6. COMPETENCE

The concept of competence was first introduced in the embryological literature by **Waddington** (1932) to denote the ability of embryonic tissue to react to inductive stimuli. In other words, the competence is a physiological state of a tissue which permits it to react in a morphogenetically specific way to determinative stimuli, or in molecular biological point of view, competence is a term which sums the ability of the enzyme complement of the embryonic cell to adapt to a particular ratio of metabolites. Whatever it may be, competence is always related to particular stimuli and particular corresponding responses (see, **Berrill**, 1971). With regard to primary induction, therefore, we may speak of neural differentiation as a **primary competence** of the ectoderm.

Further, it has been found that in some embryos the cells may during early stages be refractory to the inductive stimulation (**Needham**, 1942). This situation changes with time, but subsequently the cells may again reach a stage of inertness relative to certain specific embryonic inductors. Embryonic cells, thus, may exist in two temporarily distinct states of incompetence. Supposing that they are also different in other respects, **Lovtrup** (1974) has introduced the names **precompetence** and **postcompetence** for these phases.

For example, it has long been known that ectoderm transplanted from various developmental stages from blastula to early neurula gradually loses its neural competence. With the aging of the ectoderm, it gradually loses its capacity to respond to the inductive stimulus of the chordamesoderm by becoming neural tissue. Similarly dorsal-lip tissue transplanted under ventral ectoderm at various stages of development, shows that neural competence drops sharply near the end of gastrulation and is lost by the onset of neurulation. Isolated ectoderm unexposed to neural induction and ectoderm transplanted too late differentiate into epidermis.

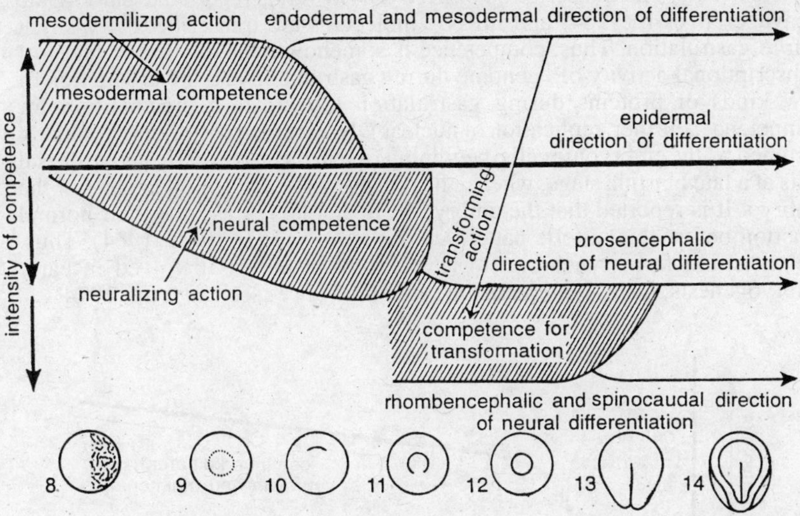

Fig. 23.20. Diagrammatic presentation of the successive phases of competence for respectively, endodermal and mesodermal differentiation, prosencephalic neural differentiation, and rhombencephalic and spinocaudal neural differentiation of the presumptive ectoneuroderm. In the absence of any inductive action, the ectoderm develops exclusively in an epidermal direction (after Berrill, 1971).

The decrease in neural competence with aging of the tissue has been tested by isolated fragments of gastrula ectoderm, maintaining them in isolation for various lengths of time, and transplanting them into different locations in neurula where various strong inductors are present. For example, competence to form brain structures markedly decreases at an age corresponding to the late gastrula stage of controlled embryos. Ectoderm corresponding to the late neurula stage is completely without neural competence ; yet the two neural-crest derivatives, namely, mesenchyme and pigment cells, can still be evoked in ectoderm of the tail-bud stage, older ectoderm, already differentiating as epidermis, is entirely without competence to do anything but proceed towards its intrinsic epidermal fate.

Moreover, the aging of ectoderm does not merely restrict its neural competence but it can bring about new responsiveness not present before. For example, late neurula epidermis, no longer convertible into neural tissue, becomes competent to respond to other inductors : under the influence of eye vesicle, hind-brain, and fore-brain respectively, it differentiates into lens, ear vesicle, and nasal pits, during post-neurula stages of development. Successive states of competence appear synchronously with the succession of primary and secondary inductors.

Molecular Biology of Competence

At molecular biological level it has been investigated that the precondition for competence is the presence of a functional genome. It is also known that up to the stage of amphibian gastrulation, no transcription of rRNAs occurs. The transcription of tRNA, however, is reported to start in late cleavage (**Brown**, 1964) and mRNA molecules are transcribed very slight before gastrulation. Thus, competence is somehow related with the onset of transcriptional activity of genome during gastrulation and the production of new kinds of proteins during gastrulation too create a precondition for competence. Further, replication of nuclear DNA is an activity that is gradually assumed by the embryonic cells, beginning first in the particularly active dorsal cells at a late blastula stage, whence it spreads to the cells of other parts of the embryo. It is reported that the embryonic nuclei do not assume their normal function before this mitotic pattern comes to an end (**Dettlaff**, 1964). Thus, the transition from synchrony to asynchrony in mitosis involved in early embryogenesis, forms a prerequisite for competence (see **Lovtrup**, 1974).

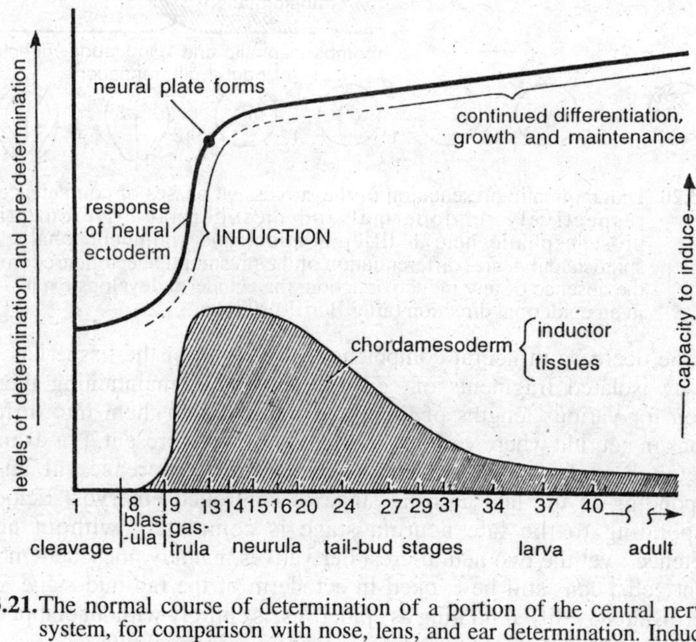

Fig.23.21. The normal course of determination of a portion of the central nervous system, for comparison with nose, lens, and ear determination. Induction begins at a very high level of predetermination. It is intense at early stages, and the course of induction is brief, the organ forming at a very early stage (after Berrill, 1971).

23-7. RECIPROCAL ACTION

In considering a particular induction, one tends to identify one component of the process as the **initiating** or **action system** and the other as the **responding system**. Thus, we speak of induction of lens by optic vesicle, of neural tube by notochord, and so on. Yet when live tissues are in inductive association, it is more accurate to recognize that one is dealing with stimulus-response sequence acting in both directions : that is, there is **reciprocal action** between components of inductive system. The effect of the induced system on the inductor is sometimes rather subtle, however, and often not readily recognized.

The principle of reciprocal action can be illustrated by the interaction of ectoderm and mesoderm in morphogenesis of vertebrate limb. In the chick embryo, when the wing bud arises at sixty hours of incubation, the ectoderm at its apex thickens, forming an antero-posterior ridge. This is called the **apical ectodermal ridge**. It forms in the ectoderm in response to mesodermal induction, for when prospective wing mesoderm is grafted to a wound in the flank, the ectoderm that heals over it, forms a ridge. Flank ectoderm loses its competence to give this response to mesodermal induction at about sixty-five hours of incubation. Mesoderm loses its capacity to induce at about the same time.

The apical ridge itself is now an inductor. If it is removed from the apex of the mesodermal bulge, limb development ceases. If the mesoderm of limb primordium is adorned with two ridges by means of suitable microsurgical techniques, two limbs will grow out from the originally single primordium. The ectodermal ridge, therefore, is said to induce mesodermal outgrowth. Further, the growing mesoderm reacts on the ectodermal ridge to keep it in a thick and inductively active configuration. It supplies a **maintenance factor**, not yet identified, that can pass through filters barring cellular contact; this factor keeps the apical ridge inductively active until the terminal parts of the limb are formed (see, **J.W. Saunders**, 1968). Thus, limb morphogenesis of chick shows that interacting inductive systems maintain, not a one-way communication system, but a dialogue, transmitting developmental clues in both directions and balancing and adjusting them so as to achieve the construction of a harmonious organ.

23-8. MORPHOGENETIC PATTERNS AND FIELDS

A morphogenetic field is the constellation of factors that determines that a part of an embryo shall engage in or have some commitment to follow a specific course of development. For example, the chordamesoderm field can be referred to as the set of properties that resides in a part of the amphibian embryo and determines that this part will invaginate, induce neural tube, and form notochord and associated mesodermal structures ; the limb field would occupy that part of the embryo in which a set of activities has differentiated that confers the potency of limb formation and leads to that developmental end point.

The experimental analysis of many morphogenetic fields reveals their characteristics : they have **strength**, as shown by the ability of organ fields to differentiate their structures in isolation and to incorporate surrounding cells and tissues into their morphogenetic patterns under other circumstances ; they show **heteropolarity**, as reflected in gradients of the field strength along the three dimensions of space within the field ; and they have **regulative ability**, in that the pattern of the whole field may be reconstituted from a part of the fields. The property of regulating ability shows us that the field as a whole may have an irreversible commitment even though the fates of the cells that comprise it are not individually determined.

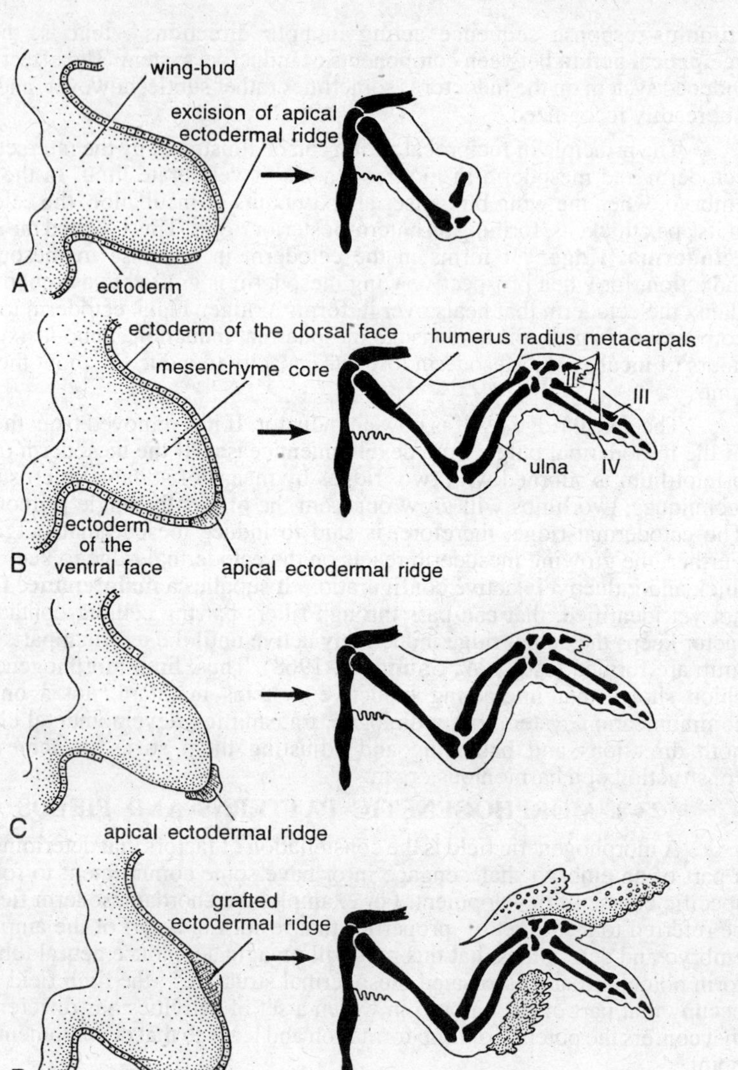

Fig. 23.22. Induction of limb outgrowth by the apical ectodermal ridge in the chick embryo. On the left, diagrammatic cross sections of the wing bud depicting various operations, on the right, the result obtained. A—Excision of the apical ridge results in deficiencies of terminal limb parts; B—In the presence of a normal ectodermal covering, the wing parts appear in their normal proximodistal sequence and pattern ; C— Removal of the dorsal and ventral ectoderm permits outgrowth to continue as long as an apical ridge is present; D—Grafting an extra ridge to the dorsal surface of the wing bud results in the formation of two wings (after Saunders, 1968).

As development proceeds, reactions within the comprehensive fields lead to the establishment of smaller fields of increasing specialization, *i.e.*, concerned with only limited patterns of differentiation. Thus, the vegetal field breaks up into intestinal field, skeletal field, and so on ; the discussion of induction and determination included examples of this. The chordamesoderm field induces the neural field which itself breaks up into smaller fields that may themselves be inductive.

23-9. CONCLUDING REMARK ABOUT INDUCTION AND COMPETENCE

By studying the phenomenon of embryonic induction and competence it becomes evident that inductors play a vital role in embroynic development even though their nature and means of action are poorly understood. It is also equally well-established that the role of responding tissue is greater and even less well understood. Neural induction, to take the primary example, is apparently a triggering process that sets competent ectoderm along the path of neural differentiation. The extent and shape of the inductive material, either as tissue or as inductive substance diffusing from a more restricted region, determine the extent and shape of the affected ectoderm, *i.e.*, both are wide anteriorly and narrow posteriorly, within this territory a patterned structure develops that has no relation to any structural properties of inductor. Yet once established, such an area of neuralized ectoderm undergoes a self-determined patterning, a development typical of so-called embryonic, or morphogenetic fields.

Lastly, certain transplantation experiments with some amphibians, have suggested that (1) transplanted competent tissue responds to a new situation according to its own genetic constitution, *i.e.*, each species has its own specific repertoire of possible cytodifferentiations. (2) The guidelines (morphogenetic fields and general inductors) are essentially the same throughout the entire class of Amphibia, and possibly throughout other classes of vertebrates. Neither fields nor inductors are species-specific. Different species and orders of amphibians are endowed with a set of genetically similar or "**homodynamic**" modes of action and reaction, responsible for general organ pattern of class.

REVISION QUESTIONS

1. What is organizer ? Describe suitable experiments to demonstrate the nature of primary organizers.
2. What is induction ? Give an account of discovery, structure and function of neural induction in vertebrates.
3. Differentiate between the terms induction and evocation.
4. Write an essay on the concept of organizer.
5. Describe various theories to explain the mechanism of neural induction.
6. Explain clearly the gradient theory of neural induction.
7. Write an essay on mesoderm induction in vertebrates.
8. What is competence ? Give an account of competence.
9. Write short notes on the following :

   (i) Hilde Mangold and her experiments ; (ii) Gray crescent and neural induction; (iii) Regional differentiation of primary inductor ; (iv) Lens induction ; (v) Spemann's experiments ; (vi) Induction of metanephric kidney ; and (vii) Competence.

24

DIFFERENTIATION (Cell Commitment, Determination and Cytodifferentiation)

The fully-developed organism consists of a number of different kinds of cells, about 200 in vertebrates, produced in the process known as **cytodifferentiation** or **differentiation** (or **cell diversification**, **Lovtrup**, 1974). The process of differentiation usually involves striking changes in cell shape, the appearance of new subcellular organelles or rearrangement of old ones, synthesis of specific products in the cytoplasm (*e.g.*, haemoglobin in red blood cells), and in some cases their release through the plasma membrane (*e.g.*, extracellular matrix in cartilage). this process depends upon the execution of particular genetic programmes, either by activation of some genes and inactivation of others at the level of transcription, by controlling production of the various messenger RNAs, or by control of their translation into enzyme and structural proteins, *e.g.*, by modification or selection of mRNA, or by controlling ribosomal function in processing mRNA, or in releasing polypeptide chains (**Ede**, 1978).

It is evident that whereas all cells of the mature organism arose from one fertilized ovum by repeated mitotic divisions and probably inherited the same biochemical capability, they become specilized and different during the course of development by the process of differentiation. Differentiation involves the events by which cells and other parts become different from one another and also different from what they were originally (**Balinsky**, 1970). For example, while the neural plate as a whole is being induced by the roof of the archenteron, its cells become different from both the presumptive epidermis and from the parts of the blastoderm which give rise to the neural plate. Thus, during embryogenesis there arise different organs and systems of organs, each made up of tissues in various combinations and arrangements. And each of these tissue comprises certain kinds of cells that are clearly recognizable as different from other kinds : the cells of skeletal muscle are quite unlike those of the liver ; brain cells are distinctive from cells that line the intestine ; there are red blood cells and a variety of white blood cells, and these are all different from fat cells, skin cells, and so on. Moreover, besides its basic functions each kind of cells has its own specialized functions, its biochemical individuality, and its morphogenetic assignment. The specialized functions of cells can be distinguished from the basic functions of life which are common to all living cells. Every cell is capable of performing the processes of metabolism (respiration, synthesis, and so on), and possesses to a certain degree the ability for ameoboid movement, shows irritability and is able to react to external stimuli. Differentiated cells, however, are able to perform special functions, or to perform them in such a manner as other cells cannot. For example, the nerve cells are capable of conducting nervous impulses to great distances and at a high speed. The liver cells secrete bile (besides their other specific functions). The melanophores produce granules or pigment in their cytoplasm. These are the special functions of nerve cells, hepatic cells and melanophores.

24-1. CELL COMMITMENT AND DETERMINATION

The Oxford Engligh Dictionary defines commitment as an "*engagement or involvement that restricts freedom of action*". That many cells have a state of commitment before expressing any sign of differentiation has long been evident from experiments in which cells are removed from their normal location and shown to subsequently develop either along a predictable pathway, or into cells of a predictable fate. So when applied to cells and their destinies, this term is often used synonymously with determination. However, **Maclean** and **Hall** (1987) have used this term to refer both determined and differentiated cells and simply describes their relatively stable dedication to a particular specialized fate, either potential or realized. Although stable, commitment is often acquired gradually in a series of steps, but in other situations the commitment is complete from the time of its inception.

A cell is said to be **determined** when its specialized fate is fixed but the overt demonstration and realization of that fate has not yet become apparent. The state of this specialization is, of course, differentiation itself, so determination is deemed to precede differentiation.

24-1.1. The Epigenetic Landscape : A Model of Determination and Differentiation

In his pioneering analysis of the relation of genetics to embryology in 1940, **Waddington** put forward a useful visual model of the developmental events leading to differentiation which he called the **epigenetic landscape** (Epigenetic denotes the process of increasing complexity that are beyond the gene ; it encompasses the influence of the environment on development and phenotypic expression). It consists of a number of diverging pathways running downhill through a series of valleys, each of which terminates in an isolated location which represents a particular differentiated state. A ball representing a cell's cytoplasmic state, placed at the top of the landscape has the potentiality of ending up in any of the terminal locations, but, as it rolls down and is deflected into one pathway or another, at each point of divergence its potential becomes more and more restricted until it enters a final pathway leading to a unique differentiated state. This process of restriction of possible fates is the phenomenon of determination and the cell in its final pathway is said to be determined. (Fig. 24.1).

The topography of the landscape represents the total genetic programme of the organism ; the slopes which lead the balls from their original positions into the valleys and keep them in the pathways at the bottom of them (**canalization** in Waddington's terminology) represent the various regulating mechanisms which produce normal development, in spite of disturbances which tend to force it off course. Which pathway is taken at any point of divergence depends upon cues supplied by the cell's microenvironment, *e.g.*, inductive interaction with neighbouring cells or extracellular matrix, or response to concentration level of a morphogen or a hormone. In many cases the choice is a binary one, *e.g.*, in early mouse embryo's blastomeres are determined as trophoblast cells or inner cell mass cells, suggesting simple genetic switch mechanisms, but there is no reason to suppose that in as many others the choice is not between multiple alternatives. The landscape will vary greatly, even in its general features, in different organisms, in major embryos entrance into the final pathways of determination will occur very early, through inclusion in the cells of the special plasma (see Chapter 4), whereas in highly regulative embryos it will — except for the special case of the germ cells — be long delayed.

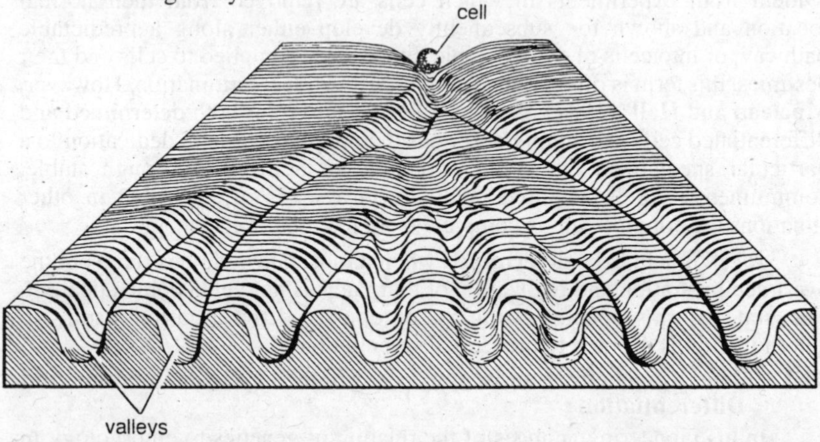

Fig. 24.1. Waddington's epigenetic landscape model of determination and differentiation. The genetic system controlling development provides a sequence of diverging valleys (pathways of determination) down which the cell rolls to its ultimate differentiated state. At first there are many pathways open to it, but the number becomes increasingly restricted as development proceeds (after Ede, 1978).

24-2. CHARACTERISTICS OF DIFFERENTIATION

Further, the ability to perform characteristic or special functions is dependent on the existence of specific mechanisms in the differentiated cells. These mechanisms are sometimes visible in the form of organoids or store products of cells, such as the myofibrils of muscle cells, cilia of epithelial cells of the trachea, long processes of nerve cells, haemoglobin of red blood cells, antibodies of plasma cells and crystalline lens protein of lens cells. These evident morphological properties of cells are also called differentiation. In other tissues, the differentiation becomes visible not so much as a change in the structure of the cell itself, but as the result of the production by the cells of intercellular structures, such as fibres in connective tissue, matrix of cartilage and bone, cuticle on the outer surface of the skin in invertebrate animals. These extracellular parts are called differentiations in the same sense as the organoids of the cells. The special functions of a cell may be the secretion of a substance that does not remain in the tissue as a permanent part but is removed or dissolved in the surrounding medium, such as granules of secretion (a hormone, an enzyme, for example).

24-3. TYPES OF DIFFERENTIATON

From the preceding discussion, it becomes, thus, clear that the emerging differences (due to differentiation during embryogenesis) manifest themselves in the number of ways. First, there is a **morphological differentiation**. As the cells (blastomeres) of embryo multiply, individual cells and groups of cells become structurally different from other cells and group of cells. For example, from a common starting point in generalized ectoderm, nerve cells and epidermal cells acquire distinguishing features of size, shape and internal architecture. Second, there is a **behavioural differentiation**. Although all cells exhibit common basic attributes, such as metabolism and irritability, special functional capabilities are eventually superposed on these general properties. Thus, nerve cells come to transport electrical disturbances, muscles to contract, gland cells to secrete special products, and so on. Third and perhaps most important, is **chemical differentiations** : consider, for example, the egg itself, whatever its class. From the beginning it has a complex chemical composition that is often reflected in a measure of pattern and spatial arrangement ; and as the egg cleaves and progressively becomes converted to blastula, gastrula and definitive embryo, individual cells and areas of cells become biochemically different from one another.

The elabortaion and assortment of chemical differences lie at the root of morphological and behavioural differences that ultimately emerge ; the acquisition of special biochemical properties by the cells precedes the appearance of a recognizable neural tube, heart, or whatever else. Putting it somewhat differently, there comes a time when a given region of an egg or embryo takes on those chemical qualities that dictate the course of development to be followed by that region. The process by which this fixity of developmental prospect is acquired is called **determination**. Differentiation, thus, involves a series of events by which materials become chemically determined and then assume distinctive form and functions.

24-4. CHEMICAL BASIS OF DIFFERENTIATION

Whatever the nature of the chemical differentiation of a cell, it is based on "**differentiation** or **diversification products**" or enzymes that direct the synthesis of the organic compounds, that give the cell its uniqueness. Therefore, it has been said that "*differentiation is the production of unique enzymatic patterns*" (**Spiegelman**, 1948). But because all enzymes are proteins, the foregoing aphorism may properly be altered to read "*differentiation is the production of unique protein patterns*", (**Balinsky**, 1970). Since cell constituents such as carbohydrates, lipids, pigments, and organic acids are synthesized under the direction and control of enzymes, a given pattern of enzymes will dictate the pattern of these organic compounds. Not all the proteins of a cell are enzymes, however, some are so called **structural proteins**, the integral parts of the biochemical architecture of the cell. Further, since the elaboration of the differentiation products in a differentiating cell often remains associated with a suppression of the reproductive capacity of the cell, therefore, they have been called 'luxury', 'unessential', 'obnoxious', or 'lethal' molecules (see **Holtzer** and **Abbott**, 1968 ; **Lovtrup**, 1974). In contrast, the diversification products which may be unessential from cellular point of view are nonetheless indispensable for the formation and maintenance of the metazoan body. The phenomenon of differentiation forms the basis of the ontogenesis and phylogenesis of the

Metazoa, because the diversification products, to some degree in contrast to the essential protein molecules of cells, are of importance for the whole body, being involved in morphogenesis (*e.g.*, extracellular elements), regulation mechanisms (*e.g.*, hormones), food uptake (*e.g.*, digestive enzymes) etc. For this reason the phenomenon of cell diversification is of central significance for the study of embryogenesis (**Lovtrup**, 1974).

Thus, for an analysis of the basic processes of differentiation it is necessary to examine the causes of differentiation in genome, in cytoplasm and also the processes which are involved in the regulation or control of protein synthesis. In other words, we have to examine these processes by which one cellular genotype (*i.e.*, egg) gives rise to many hundreds of different cellular phenotypes.

24-4.1. Equivalence of Cleavage Nuclei

As the mechanism of mitosis, which occurs during cleavage, blastulation, gastrulation and post-gastrulation stages of embryogenesis and also in adult animal, involves a replication of chromosomes and genes, so the cells resulting from a division bear the same hereditary outfit, as the parent cell. That is, the countless cells making up the animal body contain the same genetic information possessed by the original zygote. Proof of this has been provided experimentally in a number of ways (see Section 11-2 Chapter 11). From these experiments, it could only be concluded that all the cells which derive from a zygote, contain identical genotype. Now question arises that how is it possible for cells of identical genetic constitution to become different during the process of differentiation.

A resonable answer to this contradiction is that although all cells have the same complement of genes (genome), all genes are not equally active (*e.g.*, there occur some regulating mechanisms for differential gene action) and differences in the configuration of cytoplasm, *viz.*, regional differentiation of ooplasm of egg somehow directs the course of differentiation.

24-4.2. Role of Cytoplasm of Egg during Differentiation

There are numerous evidences which have suggested that ooplasm has a predominant effect on the process of differentiation. For instance, during nuclei transplanting experiments of **Gurdon** (1966, 1968), the amphibian nuclei transferred to a new cytoplasmic environment, change their function within an hour or two to the new circumstances by assuming the condition and activity characteristic of the nucleus that has been replaced :

(i) This is seen in normal fertilization when, after penetration, the compact sperm nucleus swells and assumes the appearance of the female pronucleus.

(ii) Nuclei from cells of the mid-blastula stage, which normally synthesize DNA but little RNA, when injected into the unfertilized egg (mature oocyte), synthesize RNA but not DNA. When these nuclei are injected into an activated egg, DNA is synthesized.

(iii) Neurula nuclei which synthesize all kinds of RNA stop doing so when first placed in cytoplasm of an uncleaved egg but again synthesize transfer RNA, rRNA and mRNA when transplant embryos are reared through the gastrula and neurula stages.

(iv) Nuclei from frog brain cells, which normally synthesize RNA and never divide, switch from RNA to DNA synthesis when injected into activated eggs. The same nuclei injected into maturing oocyte cease nucleic acid synthesis altogether and are rapidly converted into groups of chromosomes or spindles.

From these experimental results, it becomes evident that the state or nature of the cytoplasm controls the particular state of the nucleus as long as the cytoplasmic condition persists. In all cases, whether an injected nucleus is of the type to support development or not, the immediate visible change is a pronounced swelling. The sperm nucleus increases about fifty fold in volume in half an hour to become 20 μm in diameter, while the egg pronucleus swells to the same size at the same time. Blastula nuclei, still comparatively large, increase in volume about three fold when injected into an unfertilized egg. Nuclei from tadpole intestinal epithelium swell about fortyfold, while brain cells swell about sixtyfold. How such swelling occurs is not known. During enlargement, the chromatin becomes dispersed and some cytoplasmic proteins pass into the swelling nucleus. The primary effect of egg cytoplasm is to promote a single reaction (the polymerization of deoxynucleotide triphosphates into DNA by DNA polymerase enzyme) ; the product of which is easily recognized biochemically or auto-radiographically by incorporation of tritiated thymidine.

Thus, transplantation of nuclei into enucleated oocytes demonstrates that the cytoplasmic environment of the oocyte regulates gene activity in Amphibia, in *Drosophila*, in the mouse, in several species of fish and in the sea squirt, *Ciona*. Nuclei obtained from malignant neoplasms will revert to normal development when exposed to oocyte cytoplasm or when microinjected into, or fused with, cells of early embryos (see **Maclean** and **Hall**, 1987).

Gurdon's hypothesis is that nuclear swelling and chromatin dispersion facilitate the association of cytoplasmic regulatory molecules with chromosomal genes, thereby leading to a change in gene activity of a kind determined by the nature of the molecules that enter the nucleus. Nuclear genes are necessarily lost or permanently inactivated in the course of cell differentiation, and major changes in chromosome function as well as different kinds of gene activity can be experimentally induced by normal constituents of living cell cytoplasm. The identity of such cytoplasmic components and their mode of action are now under investigation. However, it has been found that in normal development, during the telophase of mitosis, chromosomes are reprogrammed with regard to potential gene activity by association with cytoplasmic proteins, for interphase nuclei show no such change except when enlarging in a new cytoplasmic environment.

To what extent chromosomal change takes place in varying cytoplasmic environments, is known only in dipteran insects whose salivary glands have nuclei with giant-banded chromosomes, that exhibit localized areas of enlargement or puffs. The distribution of puffs along the chromosomes exhibits distinctive patterns for cell type and for the stage of its differentiation. The dependence of nuclear activity on the cytoplasmic environment is shown by transplanting nuclei from *Drosophila* salivary gland cells into younger cytoplasm obtained from embryos of different ages. In each case, the puffing pattern changes and is characteristic for each cytoplasm.

Thus, cytoplasm differences in the egg determine the path of differentiation what will take in different cells of the embryo. As the differentiation of cells in the various parts of the embryo involves production of different proteins, and this presupposes a selective depression of the genes responsibile for the manufacturing of the corresponding kinds of mRNA, it follows that the kind of cytoplasm present in the cells determines in the last instance which genes are to come into action. Further, the fact that induction of the neural plate and other embryonic parts by the ''primary organizer or neural inductor'' (*i.e.*, chordamesoderm) is carried out by means of emission of a protein which in first place penetrates into the cytoplasm of the reacting cells, support the idea that cytoplasmic factors are involved in gene regulation during differentiation.

Cytoplasmic Maintenance of Cell Commitment

1. The first and most convincing point is that cellular potency is quite distinct from nuclear potency and that nuclei given new cytoplasmic environments are capable of extensive reprogramming.

2. Gene regulatory proteins are synthesized exclusively in the cytoplasm, and then they return to the nucleus to operate. All of the numerous other proteins involved in transcription and DNA replication have similar origins and destinies. It is, therefore, clear that any of these molecules that have short half-lives are constantly dependent on replenishment from the cytoplasm. The masking and isolation of mRNA in the cytoplasm and the initial production of some proteins in inactive forms provides further scope for the cytoplasm to exert influence on the nucleus in the long term.

3. Whatever signals are received from the cellular exterior are either encountered at the cell surface or accepted into the cytoplasm. In either situation the cytoplasm has the key role of mediating the transmission of further molecular signals to the nucleus from the exterior.

4. In the present context, the cell surface is cytoplasmic and is the interface of the cytoplasm and the external world. Now it is clear that the cell surface is central to such phenomenon as cell sorting out and so mediate the position of a cell in a tissue.

24-5. MOLECULAR BIOLOGY OF DIFFERENTIATION

Cytodifferentiation or cell diversification involves that among the proteins elaborated by individual cells certain of them specifically characterize a particular **diversification pattern**. Thus, it may be inferred that the information passed on from the nucleus to the cytoplasm varies from one cell-type to another. This phenomenon may be explained in two different ways — either the genomes of the cells are different, owing to a mechanism of **differential replication of DNA,** or they are identical, in which case **differential transcription of DNA** must take place (**Lovtrup**, 1974).

Thus, in eukaryotes the functions of genes can be regulated anywhere between gene replication and the release of the completed polypeptide chain from the ribosome. The levels at which regulation may occur may be represented as follows :

DNA → RNA → Protein
DNA < replication transcription translation
DNA → RNA → Protein

Accordingly, the activities of genes can be regulated at the level of genome, at the level of transcription and at the level of translation.

24-5.1. Control of Differentiation at the Level of the Genome

Almost universally the whole genome is retained in the nuclei of differentiated cells, there is real difficulty in determining to what extent it remains totally available for transcription. The changes which can occur at the level of genome have been recently summed up as follows (see **Maclean** and **Hall**, 1987) :

1. Permanent Changes in Genome

(a) The whole nucleus is eliminated. This very unusual process occurs in erythrocytes, keratinocytes, lens fibre cells and blood platelets.

(b) Some part of the genome is eliminated. There being no equivalent coding sequence left within the nucleus. This is also rare and occurs in *Ascaris* and *Parascaris*. Chromatin diminution was first observed in *Parascaris* by **Boveri** (1899), but also occurs in other Nematodes and less commonly, in Diptera (Insecta). The phenomenon described by **Boveri** involves that, in the cell closest to the animal pole, part of the chromosomal material is spilled into the cytoplasm during the second cleavage division. This incident is repeated three more times in one of the cells arising from the other of the first two blastomeres. At the 32-cell stage consequently only two cells have the full gene complement, whereas the remaining ones have undergone chromatin diminution. It has been found that the former are primordial germ cells, the latter are presumptive somatic cells.

Likewise, in certain Diptera it has been found that the cytoplasm located at the vegetal pole, distinguished by a high content of nucleic acids, is instrumental in preventing chromatin diminution in the presumptive germ cells. In this case it is, thus, clearly shown that differential DNA replication involves nucleo-cytoplasmic interaction. Similar nucleo-cytoplasmic interaction in chromatin diminution was observed in *Parascaris* (**Boveri**, 1910).

2. Probably Permanent Changes in Genome

(a) Major changes in genome by **mutation**, **translocation** and **deletion**. For example, O-nu mutant of *Xenopus laevis* is caused due to deletion of the 28S and 18S rRNA genes from one of the chromosomes, as a result this mutant lacks a nucleolus. Hybridization experiments have shown that the lethal homozygote retains less than 1% of the DNA coding for r-RNA. The hybrid, 1- nu, contains about 450 copies or 50% of r-RNA genes, while the wild type, 2-nu, contains all 900 copies of rRNA genes for both 28S and 18S rRNA.

(b) Changes in arrangement of genome induced by activity of **transposable elements**, *e.g.*, in maize, *Drosophila* and retroviral insertional activity in mammals.

3. Relatively Permanent Changes

DNA methylation. DNA may be modified by alkylation or methylation reactions for which the necessary enzymes are known to be present within cells (see **Pollak** and **Sullivan**, 1972). The methylation takes place after the synthesis of DNA is complete and, therefore, the methylation of any particular nucleotide base of the DNA has to be carried out anew in each chromosome once replication has been completed. Since the methylating enzymes have to act at each cell division, DNA methylation seems more a product than an initiating or controlling mechanism of differentiation. Methylation in the promoter gene seems particularly crucial.

4. More or Less Reversible Changes

Chromatin condensation. Many mechanisms are involved, including H_1 mediation of nucleosome association and also large-scale condensation as in heterochromatin of the inactivated X chromosome (*i.e.*, sex chromatin). For example, when H_1 histone undergoes phosphorylation due to activation of the histone kinase enzyme, chromatin becomes much more strongly condensed, as it does in mitotic chromosomes.

The heterochromatin is condensed intercoiled state of chromatin. It contains two or three times more DNA than the euchromatin but remains genetically inactive as it does not direct synthesis of RNA (*viz.*, transcription) and of protein and is often replicated at a different time from the rest of the DNA. In its condensed state, heterochromatin of interphase nucleus has been reported to have some specific role in gene regulation. It is more commonly observed in highly differentiated cells (*i.e.*, cells committed to the synthesis of a very few proteins) than the undifferentiated cells. For example, normal human white blood cells are found to contain large masses of condensed heterochromatin, than the leukemic white blood cells which lack condensed heterochormatin and synthesize many different proteins at uncontrolled rate.

These facts suggest that regulation of differential gene action might involve some mechanism by which specific genes are either ''shunted into'' and ''released from'' condensed physical conformation (heterochromatization) that cannot be transcribed. Recent studies of **J. Frenster** have shown that heterochromatization of chromatin is done by acidic proteins (see **Goodenough** and **Levine**, 1974).

5. Readily Reversible Changes

(A) **Histone acetylation-deacetylation.** It is mediated by deacetylase enzymes and other factors. Acetylation affects **lysine** residue of core histones (H2A, H2B, H3 and H4) of the nucleosomes and tends to loosen the nuclesome structure. It brings about a transition from a silent condensed gene to a transcriptionally active and extended gene (see **Maclean** and **Hall**, 1987).

(B) Association of gene regulatory molecules, polymerases and other factors.

(C) Short term adoption of novel three-dimensional structures, supercoiling, nucleosome phasing and perhaps nucleosome stripping from chromatin.

(D) **Increase or decrease in the quantity of DNA.** Due to increase or decrease in the quantity of DNA molecule, control of differentiation occurs by following methods :

(1) Polytene chromosomes. The giant-banded polytene chromosomes of certain insects are derived from repeated duplication of the chromosome without any separation taking place and consist of up to 10,000 and sometimes even more DNA helices. It is well established that the relative positions of bands of polytene chromosome remain fairly constant and in different kinds of tissues at different stages of embryogenesis or physiological states, certain bands become grossly enlarged to form puffs or Balbiani rings. And those gene loci which exhibit puffs or Balbiani ring formation are actively involved in transcription.

In fact, puffs represent the massive decondensation of chromatin in the proximity of an active tissue-specific (or luxury) gene. The following types of puffing patterns of these chromosomes indicate that cells of the same type have identical programmes of gene activity :

(i) Nuclei from the same salivary gland all have the same puffing pattern, as do nuclei from other larval salivary glands at the same growth stage.

(ii) Nuclei from different tissues of the same larva have different puffing patterns in evidence.

(iii) The puffing pattern of a gland changes with development and these changes take place in concert in all the cells of the gland.

(iv) Agents such as the steroid hormone ecdysone and physical factors such as heat shock are able to alter the puffing pattern of the cells of the salivary gland, and again all of the cells change their pattern in concert.

Thus, **ecdysone** presents an evidence for specific gene regulatory molecule. Ecdysone binds with target cells of a specific receptor protein, which then associates with specific chromatin sites and activates puffing, itself a reflection of localized transcriptional activity. These activated genes are called **primary response genes** and produce various products, some of which shut off the primary response by a negative feedback mechanism, others of which proceed to turn on a larger set of genes, which constitute the **secondary response genes**. So here is a rather simple two tired system of gene regulation by a hormone, a specific hormone receptor protein and a family of gene products that result from the activity of the ecdysone primary response puffs.

The puffing of restricted segments of the *Drosophila* polytene chromosome by heat shock stimulus provides another visible evidence of the activity of genes coding for cell-specific products. Thus, when *Drosophila* larvae are exposed to elevated temperature of about 40°C, certain specific puffs appear, known as **heat-shock puffs**. The products of these puffed genes are assumed to be heat-shock proteins that help to protect the cell against the damaging effects of increased temperature. Heat-shock protein production is widespread in nature, being known in organisms ranging from yeast to the human, but what is so remarkable about the *Drosophila* system is the way in which the heat shock signal activates the heat-shock genes. **Compton** and **Mc-Carthy** (1978) were able to demonstrate that cytoplasm from heat-shocked *Drosophila* tissue culture cells would initiate the production of heat-shock puffs in isolating nuclei from salivary gland, following exposure of the nuclei to such cytoplasm. Capitalizing on this elegant *in vitro* assay, **Craine** and **Kornberg** (1981), have attempted to analyze the nature of the cytoplasmic factors and found them to be both protease sensitive and heat-labile, that is almost certainly of proteins. They further demonstrated that following such exposure to heat-shock cytoplasm, the nuclei are more than 100-fold enriched for heat-shock-gene-specific mRNA. More recently, further work on *Drosophila* heat-shock response has identified a protein called **heat-shock**

transcription factor B that binds to 55-b.p. region upstream of a heat-shock gene TATA box and apparently function by facilitating RNA polymerase II attachment (**Parker** and **Topol**, 1984). This factor seems to have all the characteristics expected of a classical gene regulatory protein.

(2) Gene amplification. In Chapter 5 of "female gonads and oogenesis" it was pointed out that in *Xenopus laevis* each normal cell contains about 900 copies of the genes for rRNA (900 genes, each for 28S rRNA, 18S r RNA and 5.8S rRNA). Such repeated gene sequences are often referred to as **redundant genes** and each cell possesses as part of its genome all of the redundant gene- copies. It was discussed in the Chapter 5 that during oogenesis of amphibians, an amplification of rDNA takes place, so that, the size or number of nucleoli is greatly increased and consequently, large number of ribosomes are manufactured. Further, the chromosomes of amphibian oocytes assume a 'lamp-brush' like configuration to enhance the production of mRNA. These extra copies of rRNA and mRNA molecules are supposed to have some regulatory effect over the course of differentiation involved in embryogenesis (see **Goodenough** and **Levine**, 1974).

24-5.2. Control of Differentiation at the Level of Transcription

At the level of DNA dependent RNA synthesis (transcription) the activities of different genes of genome of developing embryo may be regulated by following methods :

(a) Transcriptional gene regulation by histones. The histones are chromosomal proteins with an unusually high proportion of basic amino acids (arginine, lysine, and histidine). The double- stranded molecules of DNA have free acidic groups of phosphoric acid on their outer surface, and these can establish firm bonds with the NH_2^+ groups of the basic amino acids of histone chains. This intimate association of DNA and histone has led to the suggestion that by associating themselves with the DNA molecule, the histones prevent the DNA from interacting other substances (*i.e.,* RNA polymerase, triphosphoribopurines and triphosphoribopyrimidines, etc.) in the cell and, thus, from serving as templates for the production of mRNA (**Stedman** and **Stedman,** 1950 ; **Block,** 1962). In fact, histones inhibit DNA-primed RNA synthesis in such unrelated forms as bacteria, pea and mammals. It has been suggested that structural genes remain "turned off" before gastrulation because they are associated with histones, and are later "turned on" by changes in the DNA- histone association (see **Berrill,** 1971).

About the possible mechanism of histones in regulating transcription of DNA, the current molecular biological understanding is that local differences in histone distribution may exist along a eukaryotic chromosome, that hormones or other influences (induction, etc.) may effect this distribution, and that transcription may be modified accordingly. They are supposed to serve as **repressors** and are known to be modified by phosphorylation, acetylation, and methylation and some of these modifications might alter DNA-histone interactions and allow certain segments of the genome to be transcribed (**Goodenough** and **Levine**, 1974).

(b) Maclean and **Hall** (1987) have provided the following brief review of the various classes of cell factors that are likely to affect transcription, either by inducing or preventing transcription on specific sequences, or by generally increasing or decreasing transcriptional rates in the cell.

1. RNA polymerases. These enzymes are necessary for transcription and if in short supply, would sharply affect it. They are of three basic types : Type I, Type II and Type III. Most probably there is competition for type II polymerase (for hn RNA and mRNA) by the various promoter sequences that lie upstream of the protein-coding genes or cistrons.

2. Endonucleases. Such enzymes are likely to affect transcription by introducing into DNA nicks that may serve as initiation sites for some polymerases. At low concentration some endonucleases are likely to enhance transcription in cell-free experiments, but at high concentrations the endonucleases will sharply reduce it.

3. Topoisomerases, helicases and other helix-destabilizing proteins. A number of different proteins are known that alter the three-dimensional structure of DNA and render it more available for processing, *i.e.*, replication and transcription.

4. DNA methylase. As mentioned earlier (see Chapter 3), this enzyme is likely to make DNA less available for transcription, and factors that antagonize methylation tend to enhance transcription.

5. Histone acetylases and deacetylases. Such enzymes, by modulating histone acetylation as already discussed, may have profound effects on transcriptional rates.

6. Factors such as ATP. These molecules may show any strict affinity for either nucleic acid or chromatin, but by changing the available energy they may influence transcription rate.

7. Ions and other small molecules. The conformational state of chromatin has been seen to be a major factor in modulating gene activity, and it is known that many ions, especially those of calcium, magnesium, and manganese, directly affect chromatin form or conformation. Many ions also affect the RNA polymerases (**Gilmour**, 1978).

24-5.3. Control of Differentiation at Post-Transcription or Translation Level

Since many reactions are required to transform the information encoded in DNA to the ultimate polypeptide-gene product, many post-transcriptional regulating mechanisms may be considered to exist : (i) Specific protection or degradation of mRNA within the nucleus. (ii) Transport of messenger RNA from nucleus to cytoplasm. (iii) Association of functional messengers bound to polyribosomes with specific regulatory molecules. (iv) Destruction of mRNA on the ribosome. (v) Masking of mRNA. (vi) Activation or inactivation of ribosome. (vii) Alterations in the folding of nascent polypeptide chains either by association with other proteins or with small molecules. (viii) Alterations in the function of specific factors involved in the initiation or termination of protein synthesis.

Accordingly, several factors such as **hormones** might facilitate diffusion of particular mRNAs out of the nucleus, stabilize their life times in the ooplasm, increase the frequency with which they associate with ribosomes and become translated, and so on. For example, some effects of **estrogen hormone** on uterine cells are following : increase in total cell protein, increase in transport of amino acids into cell, increase in protein synthesis activity per unit amount of polyribosomes, increased synthesis of new ribosomes, alteration of amounts of nuclear protein to nucleus, increased amount of polyribosomes per cell, increse in nucleolar mass and number, increase in activity of two RNA polymerases, increase in synthesis of contractile proteins, imbibition of water, increased synthesis of many phospholipids, increased *de novo* synthesis of purines (dependent on new enzyme synthesis), alteration in membrane excitability, alteration in glucose metabolism, increase in synthesis of various mucopolysaccharides (**Britten** and **Davidson**, 1969).

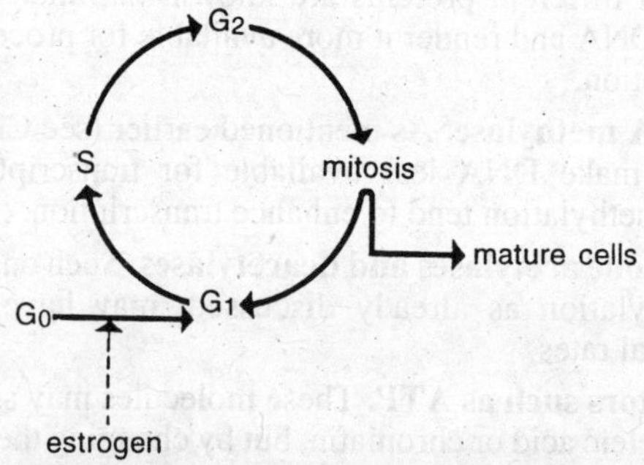

Fig. 24.2. Hormonal stimulation of differentiation : mitosis precedes terminal differentiation process. During diestrus, most of the cells are in what is called a G_0 phases *i.e.,* they can synthesize DNA and subsequently undergo mitosis, but will not do so until the appropriate output of ovarian estrogen is attained. When this level of estrogen is reached, there is a transition to a presynthetic or postmitotic interphase (G_1), that is, a transition to cells that will synthesize DNA (S phase) in the near future : These cells then enter the S phase and subsequently divide (after Berrill, 1971).

Further, it has been suggested that different classes of ribosome translate different mRNAs or that regulator molecules induce the formation of special factors. The late translation of 'Masked' mRNA which transcribed during oogenesis, is most significant example of gene regulation at translation level. Besides this, any factor (availability of different enzymes, tRNA, protein factors, ionic disturbance, etc.) which is involved at any stage of protein synthesis can be influenced by any physiological reason and may regulate gene action at post-transcriptional level.

24-6. LEVELS OF DIFFERENTIATION

Enzymes and specific proteins in differentiating systems fall into three catagories : 1. Fundamental structures and metabolic pathways that exist in all cells. 2. Secondary metabolic pathways found in the cells of a number of tissues. 3. Characteristic cellular function which occurs in single, specific cell types. Since the relative concentration of proteins in all categories changes during differentiation, the overall process must, therefore, involve the activation or modulation of the activity of many genes.

An outstanding feature of cell differentiation in a multicellular animal, whether a *Hydra* or a mammal, is that there is continuous spectrum of types ranging from one type through fine gradations to an alternate type, except in the sense of an undifferentiated or partially differentiated cell becoming fully differentiated. Each cell type falls into a clearly recognizable discrete category, sharply set off from other cell types. This seems to imply, on a genetic level, that as one set of genes is turned on, another set, in some linked way, is turned off. In these terms sets of genes of the first order determine the general cell category, such as the mesenchyme family ; and sets of genes of the second order determine the cell type ; within the general category. It is evident that majority of genes or cistrons of first set of genome (DNA) determine the general-cell-type ; the genes of second set regulate the genes of first set and certain other set of genes have a supracellular or intercellular determinative effect, over and above the first two.

The differentiation process as seen in the pattern of accumulation of specific proteins in the developing mouse and rat pancreas has been intensively investigated by **W.J. Rutter** and **N.K. Wessells** (1969) and others, with regard to both exocrine (digestive enzyme secreting) and endocrine (insulin-secreting) cells. They recognized four levels of differentiation, namely, the **undifferentiated state**, the **protodifferentiated state**, and **differentiated state**, and **modulation**.

(i) The primary regulatory event consists of the conversion of the undifferentiated cell to a cell with essentially pancreatic character. At this stage specific pancreatic proteins are present at detectable levels, even before the onset of histogenesis, suggesting that this event may "uncover" all the genes required for complete pancreatic differentiation.

(ii) The secondary regulatory event involves conversion from the protodifferentiated state to the differentiated state. During this period, which may include 2 days, cells undergo extensive proliferation and then proliferation ceases in those cells which are undergoing differentiation. Specific mRNAs must already be available, since specific proteins were present in the preceding stage. The synthesis of ribosomal RNA, the assembly of functioning ribosomes, or the development of the endoplasmic reticulum could indicate the second regulatory event.

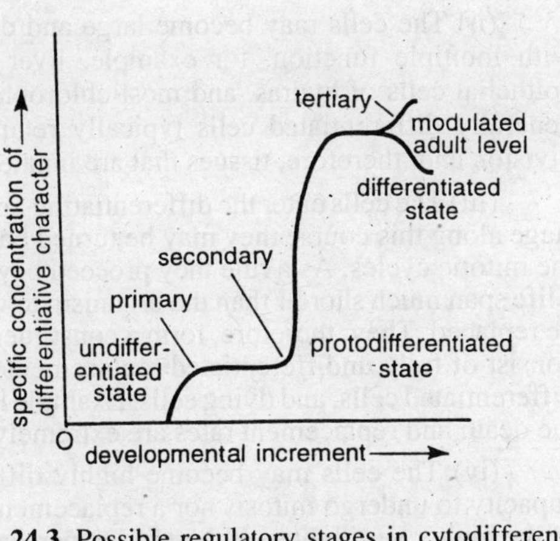

Fig. 24.3. Possible regulatory stages in cytodifferentiation (after Berrili, 1971).

(iii) The tertiary regulatory event is believed to be a mechanism distinct from the others, although it is not clear whether the observed alterations in specific enzymes activity among the differentiated cells (*i.e.*, changes in relative protein concentration) reflect changes in the level of the production of specific mRNA in the cell (transcription) or changes in the efficiency of ribosomal function in protein synthesis (translation). These are changes that occur late in development and may be regarded as modulation of the differentiated state in response to extracellular factors such as metabolites or hormones.

24-7. TISSUE MAINTENANCE AND REPLACEMENT DURING DIFFERENTIATION

The **rate of differentiation** in a particular type of cell is itself under strict regulative control, so that, the differentiated cell persists for a period in keeping with the special circumstances of its existence and the life span of the parental organism as a whole. Except lens and certain other tissues, the cells of many other tissues follow the path of differentiation to inevitable death. In almost all cases, therefore, the survival time of the differentiated or differentiating cell and the mechanism for replacement are vitally important to the maintenance of the tissue and the welfare of the organism. In general, tissues fall into one of several categories:

(i) The cells remain essentially undifferentiated and persist in the stationary or G_0 phase until activated by growth or repair requirements or by hormones. As such they may serve simple functions such as forming membrane ciliated or not.

(ii) The cells may become large and differentiated in a general sense, with multiple function, for example, liver cells of vertebrates, the large epithelial cells of hydras, and most chlorophyll-containing plant cells. Such generally differentiated cells typically retain the capacity for mitotic cell division and, therefore, tissues that are innately self-maintained.

(iii) The cells enter the differentiating and aging pathway. Up to a certain stage along this course they may be turned back by hormones, for example to the mitotic cycles. As a rule they proceed beyond the point of no return, have a life span much shorter than the organism to which they belong and they must be replaced. They, therefore, form a constituent part of tissues which typically consist of fully undifferentiated replacement cells, differentiating cells, fully differentiated cells, and dying cells. Example is red blood cell system, in which the death and replacement rates are extremely high.

(iv) The cells may become highly differentiated but have neither the capacity to undergo mitosis nor a replacement source, such as the nerve cells of the higher vertebrates. If they fail to persist in fully functional condition as long as the rest of the animal, senility ensues.

Accordingly, cell population of the different organism may be **static** (with neither cell loss nor cell displacement, *e.g.*, neurons) ; **expanding** (with new cells accounting for growth, *e.g.*, pancreatic exocrine tissue, kidney tubules and thyroid follicle), or **renewing** (with new cells continually replacing, cell lost, *e.g.*, epithelium of the intestinal villus, pulmonary epithelium, epidermis, blood forming or haemopoietic tissue, reticuloendothelial system, and seminiferous tubules).

24-7.1. Stem Cells

The stem cells are the characteristic cells of renewing cell populations. They are typically "undifferentiated" cells that serve as a source for specialized, non-proliferative tissue throughout the body. There are many kinds of stem cells, and although a stem cell may give rise to multiple derivative types, a particular stem cell, even though undifferentiated, generally belongs to a certain basic category. Blood-forming tissue provides a good example of stem cells. This has been studied mainly in the chick, where the red-blood cells or erythrocytes retain their nuclei throughout cell life, and in the mammal, where the erythrocyte nucleus degenerates and disappears during cell maturation. In these animals, the red bone marrow is the main blood forming or **erythropoietic tissue**.

In the bone marrow an undifferentiated stem cell received a stimulus which directs it towards the erythrocytic series. This is called **determination**, for it is an all-or-none phenomenon. The stem cell is a small, round cell commonly found in several organs. **Erythropoietin** is a proteinous determiner substance which determines the stem cells for erythropoiesis. Once determined, the stem cell becomes a **proerythroblast**, following which a series of steps occur known as **maturation**. In red blood cells, when a cell leaves the compartment of the stem cells, in which it becomes a proerythroblast, it undergoes four cell divisions and six definite stages of morphological maturation. This period corresponds to the appearance of new proteins, of lipoproteins and nucleoproteins, of carbohydrates, and of enzymes relating to multiplication and growth on the one hand and to the synthesis of haemoglobin on the other.

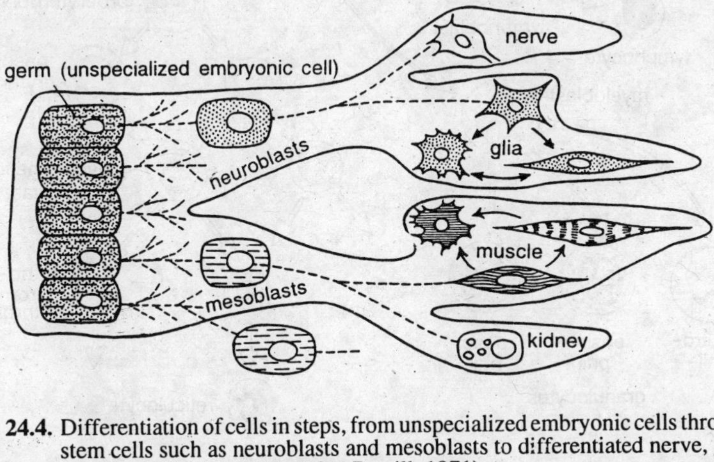

Fig. 24.4. Differentiation of cells in steps, from unspecialized embryonic cells through stem cells such as neuroblasts and mesoblasts to differentiated nerve, glia, muscle, and kidney cells (after Berrill, 1971).

Erythropoietin is a hormone, obtainable from plasma, which serves as a useful experimental tool. It can induce eyrthropoiesis in tissues which, although normally erythropoietic, have been totally suppressed with regard to red blood cell formation. According to **Bessis** (1967), erythropoietin acts on stem cells by initiating transcription of previously repressed genetic loci, *i.e.*, it derepresses. *In vitro*, it causes rapid mRNA synthesis, haemoglobin synthesis, glucosamine incorporation and cellular iron accumulation in a non-haeme form. With regard to the accumulation of iron, there appear to be two possibilities—1. Stimulation of iron accumulation by marrow cells, stimulated by erythropoietin, requires the prior synthesis of a protein which may be needed to ensure and adequate intracellular supply of iron from haeme

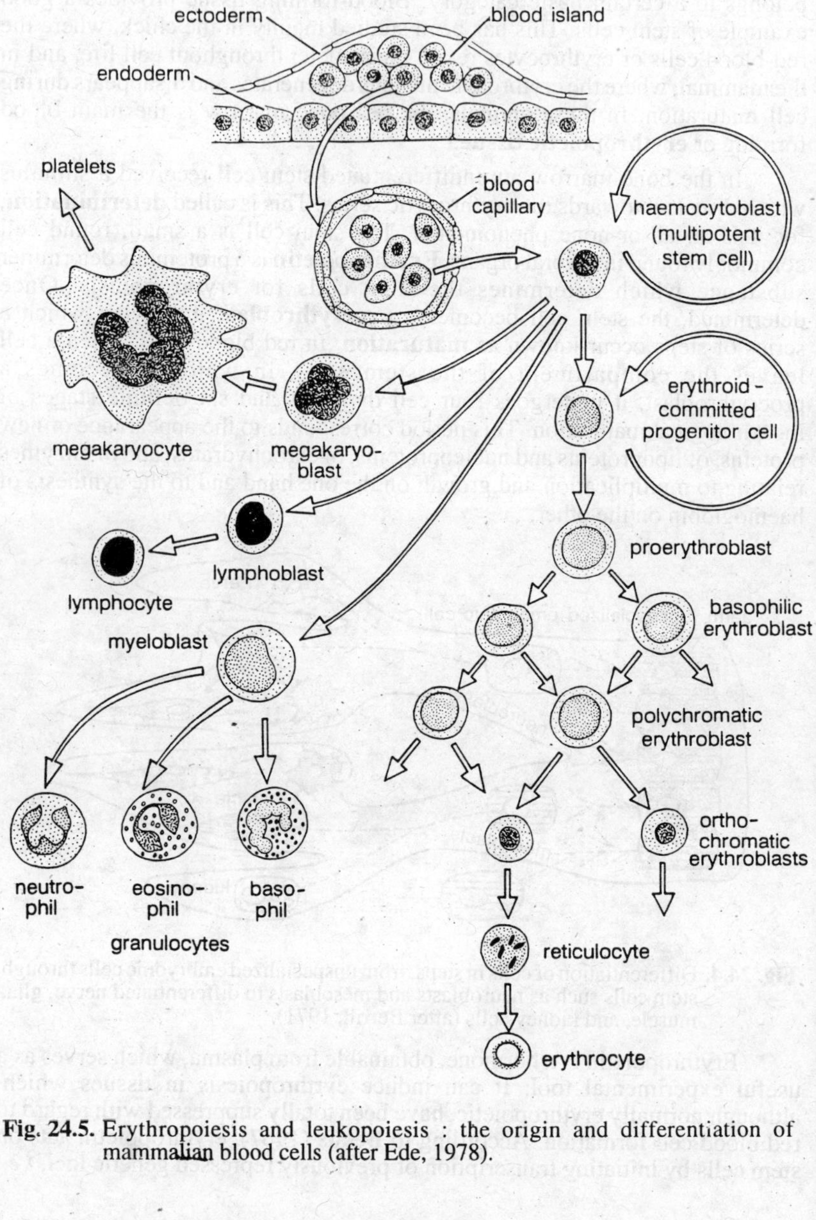

Fig. 24.5. Erythropoiesis and leukopoiesis : the origin and differentiation of mammalian blood cells (after Ede, 1978).

synthesis. 2. Globin may play a key role in the regulation of haemoglobin synthesis. Haeme is a feedback inhibitor of its own synthesis. Globin acts as a scavenger of haeme and channels the whole system toward haemoglobin synthesis.

After 3 days, following the four cell divisions, the cytoplasm of the erythroblast has become filled with haemoglobin. It then expels its nucleus (in mammals) and enters the circulation as a **reticulocyte**. The reticulocyte ripens during the next 2 to 3 days and becomes an adult **erythrocyte**. In man the erythrocyte lives for about 120 days and is then phagocytosed. In both mammals and chick there is a stepwise reduction of transcriptional activity during differentiation, and even in the chick, where the erythrocyte nucleus is retained, there is the total arrest of RNA synthesis in the mature nucleus.

24-8. ROLE OF MICROENVIRONMENTS IN DIFFERENTIATION

The basal layer of cells in the **epidermis** consists of actively proliferating cells which are cytologically and biochemically unspecialized. Individual cells of the basal layer lose their attachement to the basement membrane and appear to be forced up by crowding into the more superficial layers of the epidermis. Only then do they begin to synthesize **keratin.** Here, the basal-layer cells are stem cell. Their differentiation into keratinizing cells is clearly a response to the change in their environmental circumstances, *i.e.,* they become separated by the remaining basal layer from their original contact with the basement membrane and the subepidermal chemical medium. At the same time, their fate is not immediately determined, since exposure of the cells to vitamin A results in differentiation of the keratinizing epidermis into a mucoid-secreting epithelium.

The **cell microenvironments**, therefore, have a great deal to do with the particular direction of differentiation taken by an unspecialized cell. Experiments with isolated, undifferentiated vertebrate pigment cells by **Wilde** (1961) are significant. In no case did a single isolated cell undergoes differentiation in a minute hanging drop of nutrient culture medium. But when two cells are present together in a microdrop, one undergoes differentiation. Apparently, there is an exchange of cues into the microenvironment, extending to the neighbour cell, so that one of the two cells is able to initiate differentiation process.

REVISION QUESTIONS

1. Explain the differences between cell commitment, determination and differentiation.
2. What is differentiation ? Write an essay on differentiation.
3. Describe Waddington's epigenetic landscape model.
4. What are the different aspects of cell differentiation ?
5. Describe the molecular basis of differentiation ?
6. What is a stem cell ? Describe the process of erythropoiesis in mammals.

25

ASEXUAL REPRODUCTION AND BLASTOGENESIS

In preceding chapters, we have discussed different facets of sexual reproduction and subsequent means of embryogenesis in different chordates, but we have not discussed, so far a peculiar mode of ontogenetic developmet —the asexual reproduction and blastogenesis both of which are found to occur, in addition to sexual reproduction and embryogenesis, in certain primitive prevertebrate chordates and many non-chordates of different groups. **Asexual reproduction** is the development of a new individual without the meiosis dependent maturation of male and female gametes and the subsequent specialization and fusion of these gametes. In reference to the multicellular organisms, the asexual reproduction means the development of new individuals at the expense of **somatic cells** — cells of the **soma**, or body as opposed to generative, germ, or sex cells (gametes). In asexual reproduction, some body part (called **blastema**) of the already differentiated organism again embark on an active process of morphogenesis to produce a new individual like the parent organism. Thus, while in sexual reproduction, the new individual develops from one cell — the fertilized or parthenogenetically activated ovum, in asexual reproduction, the new individual develops from a group of cells, the blastema. The development that starts from a blastema has been termed as **blastogenesis** as opposed to embryogenesis, or development from the ovum. Further, the individuals resulting from asexual reproduction are referred to as **blastozooids** ; the individuals developing from the ova are then termed as **oozoids** (see **Balinsky** 1970).

Thus, in asexual reproduction, small groups of somatic cells in the adult retain developmental potential and proliferate into buds that develop into adult organisms. The asexual bud is initiated after a small group of cells (10 to 50) escape parental control. Physiologically released, the cells acquire independent morphogenetic potentialities and develop into a new individual that separates from the parental body. Sometimes, new individuals remain attached to the parent and organize as a colony, a mode of asexual reproduction typical of many coelenterates and tunicates. Such colonies are made up of genetically identical individuals, but, in some, a **phenotypic polymorphism** has evolved ; that is, morphologically distinct individuals (called **zooids**) display different specialized functions such as feeding, reproduction or protection (**Grant**, 1978).

25-1. ASEXUAL VERSUS SEXUAL REPRODUCTION AND THEIR EVOLUTIONARY SIGNIFICANCE

In asexual reproduction, there is only one parent, and there are no special reproductive organs or cells. Each organism, whether male or female is capable of producing identical copies of itself as soon as it becomes an adult. The production of copies is simple, direct and typically rapid. In contrast, sexual reproduction involves two parents as a rule, each of which contributes special sex cells or gametes, that in union develop into a new individnal. The zygote formed from this union receives genetic material from both parents and accordingly is different from both. The combination of genes produces a genetically unique individual, still bearing the characteristics of the species but also bearing traits that make it different from its parents.

Sexual reproduction, by recombining the parental characters, tends to multiply variations and makes possible a richer and more diversified evolution. Asexual reproduction can only produce carbon copies and must await **mutations** to introduce variation into the line. This would seem to explain why asexual reproduction is restricted mostly to unicellular forms, which can multiply rapidly enough to neutralise the disadvantages of non-stop replication of identical products.

In fact, in those asexual organisms such as molds and bacteria that are haploid (bear only one set of genes), mutations are immediately expressed and evolution can proceed quickly. In sexual animals, on the other hand, a gene mutation is seldom expressed immediately since it is masked by its normal partner on the homologous chromosome. There is only a remote chance that both members of a gene pair will mutate in the same way at the same moment.

Asexual reproduction is found only among the simpler forms of life such as bacteria, protozoans, cnidarians, bryozoans, and a few others. It is absent among the higher invertebrates (molluscs and arthropods) and all vertebrates.

Lastly, eggs have no more than two developmental programmes (*i.e.*, larva and adult) ; the developmental repertoire of the asexual bud is richer. For example, a coelenterate egg produces a simple ciliated larva, called **planula** with a few tissue types. An asexual bud of the same species can develop directly into one of many different functional adults specialized for feeding (**gastrozooid**), reproduction (**gonozooid**) or protection (**dactylozooid**).

25-2. DIFFERENT MODES OF ASEXUAL REPRODUCTION IN METAZOA

In different non-chordates and certain prevertebrate chordates a variety of means of asexual reproduction has been observed. A thorough survey of literature has revealed the fact that different mode of asexual reproduction of most animals belong to any one of the following categories : 1. Fragmentation; 2. Fission ; 3. Budding ; 4. Gemmule and statoblast formation.

25-2.1. Fragmentation

Spontaneous fragmentation of the body, followed by regeneration of the missing parts, is a common method of asexual reproduction in sponges, sea anemones, and planarians, and certain polychaete groups such as syllids and sabellids (see **Meglitsch**, 1972). For instance, tips of the branches of some sponges regularly break off and produce masses of cells that become young sponges. Many sponges form reduction bodies during unfavourable periods. The main mass of the body deteriorates, leaving a residual mass that develops into young sponges under favourable conditions.

Certain Anthozoa such as *Metridium* reproduces asexually by fragmentation of the pedal disc, producing young with irregularities in septal and siphonoglyph arrangement. New mouths and whorls of tentacles appear in the coral coenenchyme. They develop septa continuous with the septa of adjacent polyps (see **Meglitsch**, 1972).

In many syllids, the points of breakage are predetermined and can be recognized by peculiarities in the septa between the somites where separation will occur. In such species, fragmentation and regeneration have become highly organized and each fragment can develop a new anterior and posterior end. Further in these annelids the spontaneously formed fragments may be quite numerous ; some species of *Dodececaria* and *Ctenodrilus* produce some single somite fragments. *Chaetopterus* can produce new individuals from any of several of the highly specialized anterior somites (**Meglitsch**, 1972).

25-2.2. Fission

In case of asexual reproduction by fission the new individual is formed from a relatively large portion of the body of the maternal organism, and differentiated organs and tissues or their parts are passed on to the offspring. It occurs most commonly in protozoans, cnidarians, rhabdocoeles (freshwater planarians), annelids, echinoderms and ascidians.

Many protozoans reproduce asexually by fission. In a most common kind of fission called **binary fission**, the parent body divides equally to produce two daughters which regain adult size and form. Where there is a definite body axis, fission may be **transverse**, as in ciliates, or **longitudinal**, as in flagellates. Fission involves karyokinesis, or nuclear division, and cytokinesis, or cytoplasmic division. Certain protozoans such as sporozoan parasites, Foraminifera and Radiolaria reproduce asexually by **multiple fission**, in which repeated nuclear divisions occur without cytokinesis to form a multinucleate organism. The cytoplasm then cleaves to form many uninucleated individuals.

In metazoa, the simplest form of fission includes the separation of an adult individual into two parts of approximately equal size, similar to the binary fission in protozoans. In some corals (Anthozoa) fission occurs in the form of longitudinal fission, with the division plane in the long axis of the body. Fission also occurs in some brittle stars, as for instance in *Ophiactis savignyi*, in which parental individuals break across the disc, and each half then proceeds to regenerate the missing half of the disc and the three arms. In rhabdocoeles (freshwater planarians) and polychaete and oligochaete annelids, the planes of division are transverse to the body axis, so that the worm is divided into anterior and posterior halves (**Berrill**, 1952). Both halves inherit from the parent the skin, a section of the alimentary canal, sections of the nerve cord, and, in annelids, a number of mesodermal segments, with muscles and nephridia, and correspondingly, sections of parenchymatous mesoderm in rhabdocoeles. The posterior individual is at first devoid of a head and, therefore, lacks initially the supraoesophageal ganglion, sense organs (eyes), tentacles and mouth parts, where these occur. The head is produced either subsequent to division or, more often, in preparation for the division, so that the posterior individual is fully developed when two blastozooids separate.

Further, in rhabdocoele turbellarians and some oligochaetes, new fissions may be started by one or both filial individuals even before they are separated from one another (Fig. 25.1). The result is a chain of blastozooids, some of them in different stages of reconstitution (degree of head development). The division may also be very unequal, so that of the two or several blastozooids one may be clearly distinguishable as the parent and the other, as offspring.

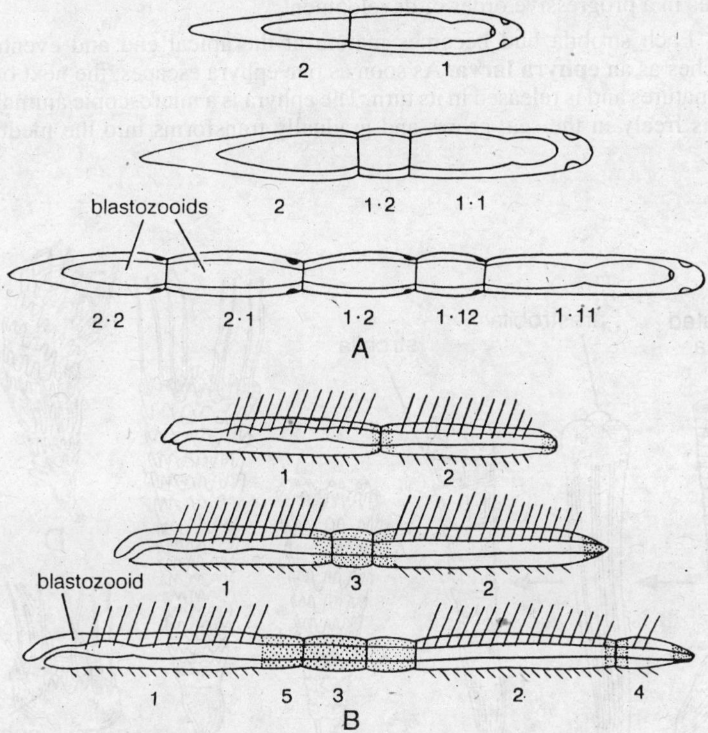

Fig. 25.1. Reproduction by transverse fission in worms. A—Transverse fission in the rhabdocoele turbellarian *Stenostomum*; B—Transverse fission in the oligochaete *Nais*. The numbers show the succession in the formation of the individual blastozooids (after Balinsky, 1981).

Strobilation. Strobilation is a special form of transverse fission in which subsequent transverse divisions occur prior to the earlier divisions being completed and in consequence a number of filial blastozooids are resulted. Strobilation is a classical method of propagation by which, in Schyphozoa (*e.g.*, *Aurelia*), the asexual polypoid generation gives rise to the sexual medusoid generation.

During spring months, as polyps mature, a striking change becomes evident along their slender tubular length (Fig. 25.2). A series of transverse constrictions appears at regular intervals along the cylinder. These constrictions deepen, giving the polyp the appearance of a stack of inverted plates. These gradually transform into radially symmetrical **buds** with marginal lappets, a manubrium, gastral filaments, and rudimentary sense organs. The uppermost bud is most advanced, with other following behind in decreasing order of maturity until those near the base are least differentiated. The polyp is now called a **strobila**, and the process of plate formation is known as **strobilation**. According to **Grant** (1978), strobilation is a modified form of budding in which a series of buds is simultaneously carved out of parental tissues in a progressive order of development.

Each strobila bud becomes mature at the apical end and eventually detaches as an **ephyra larva**. As soon as one ephyra escapes, the next one in line matures and is released in its turn. The ephyra is a microscopic animal that swims freely in the sea, grows and gradually transforms into the medusoid stage.

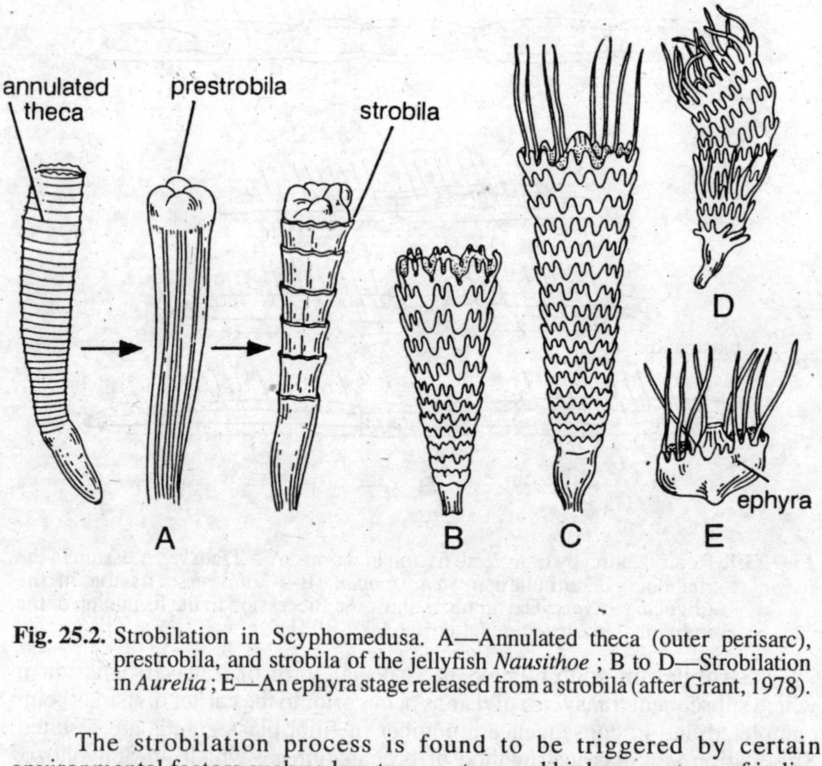

Fig. 25.2. Strobilation in Scyphomedusa. A—Annulated theca (outer perisarc), prestrobila, and strobila of the jellyfish *Nausithoe* ; B to D—Strobilation in *Aurelia* ; E—An ephyra stage released from a strobila (after Grant, 1978).

The strobilation process is found to be triggered by certain environmental factors such as low temperature and high percentage of iodine in sea water (**Spangenberg**, 1967). Iodine is effective as iodide or as thyroxine when added to artificial sea water used as a culture medium for laboratory - maintained *Aurelia* polyps (**Spangenberg**, 1971). Natural sea water normally contains iodine in trace amounts, and since the iodine contents fluctuate with the seasonal growth and decay of iodine-binding sea weeds, iodine may well be the trigger in nature as well (**Berrill** and **Karp**, 1976).

The process of **strobilation** has been reported also in the colonial ascidian *Amaroucium* by **Freeman** (1971). The body of *Amaroucium* has three regions : a **thorax** with branchial basket, an **abdomen** containing stomach-intestine, and the elongate **post-abdominal** region (Fig. 25.3). The ascidian strobilates by a series of ectodermal constrictions which cut the body's abdominal region into small fragments or blastozooids that develop into new zooids.

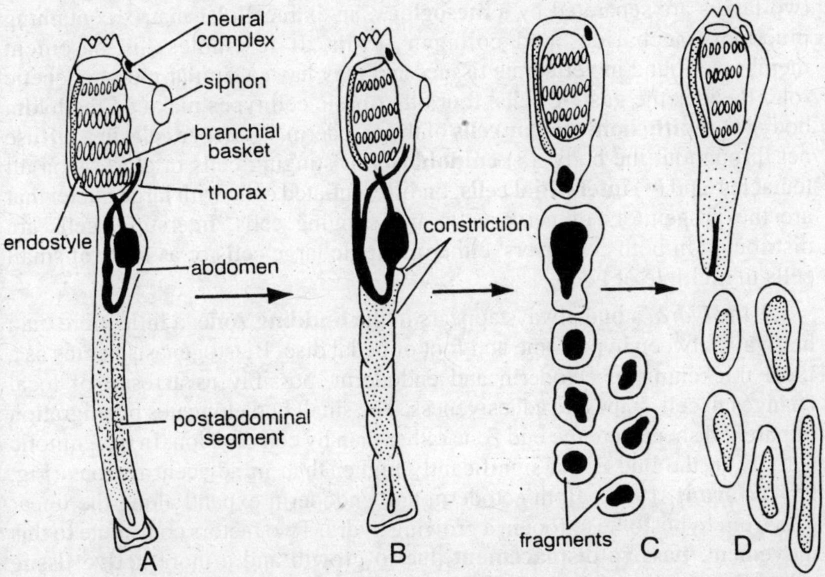

Fig. 25.3. Pattern of strobilation in 14 - to 15- day old (days after metamorphosis) individual. First constrictions appear posterior to stomach mass in abdomen; then within a few hours the entire region behind the thorax is constricted into contracted fragments, each of which soon extends and regenerates into a new individual (after Grant, 1978).

25-2.3. Budding

The budding is a most common method of asexual reproduction of many metazoans such as sponges, cnidarians, cestodes (*e.g.*, larvae of *Echinococcus granulosus*), annelids (*e.g.* Naids and Oeolosomatids), and tunicates (see **Balinsky**, 1970 ; **Meglitsch**, 1972). In budding, the new individual develops from a small outgrowth on the surface of the parent. No organs of the parent are passed on as such to the offspring but the bud is supported by the parental organism at least during the initial stages of its development. **Meglitsch** (1972) has considered budding as an unequal form of fission. Like the fission, it involves preparatory phases before budding and a later period of growth and differentiation.

Budding patterns of the following animals can be taken as typical examples :

1. Budding in *Hydra*. *Hydra* is a simple freshwater cnidarian or coelenterate. It is small, about 10 mm long and 1 mm wide when extended ; yet it is large enough to be readily observed and amenable to operation. It is composed of about 100,000 cells which fall into 17 distinguishable kinds (**Berrill** and **Karp**, 1976). *Hydra* is diploblastic, *i.e.*, a mature animal consists of two layers of cells, an outer **ectoderm** and an inner **endoderm** which is made up of gastroepithelial cells lining the entire gastrovascular cavity. These two layers are separated by a **mesogloea**, an extracellular matrix containing mucopolysaccharides and collagen fibrils. It resembles the basement membrane found in vertebrate tissues and may have a similar morphogenetic role. Besides the **gastric cells**, four other basic cell types make up the hydra body : (1) **epitheliomuscular cells** of the ectoderm, (2) **nerve cells** in a diffuse net throughout the body, (3) **cnidoblasts** or stinging cells in clusters on all tentacles, and (4) **interstitial cells**, undifferentiated cells with large nuclei that are the progenitors of nerve cells and stinging cells. Interstitial cells are distributed in both cell layers, either as single large cells or as nests of small cells in multiple of two.

In *Hydra*, a bud always appears in the **budding zone**, a little more than halfway between hypostome and foot or pedal disc. Blastogenesis begins as a local thickening of ectoderm and endoderm, possibly as a result of local changes in cell shape and adhesiveness. The small bud elongates by migration of parental sheets into the bud zone rather than by cell division. In fact, mitotic activity in the bud is not significantly higher than in adjacent regions (Fig. 25.4) (**Grant**, 1978). Both ectoderm and endoderm expand along the linear axis from hypostome to foot in a growing hydra. Two factors contribute to this movement, **passive displacement** due to growth and a more **active tissue migration**, with each layer moving at different rates. The siting of a bud in the budding region involves diversion of these expanding tissues ; both cell layers change polarity 90 degrees at the junction between parent and bud as the latter grows with its own axial symmetry.

The gastrovascular cavity in the bud endoderm is continuous with the parental cavity. The bud assumes a cone shape with interstitial cells at the tip, under the thinning epithelial muscular layer. This apical region becomes the future **hypostome** of the new individual as cells migrate distally and proximally from it. Some interstitial cells in the prospective hypostome region differentiate into nerve cells which gradually anastomose into an irregular net, sweeping downward from the hypostome into the lengthening column. Nerve cells have a trophic role in growth and nerve net differentiation may correlate with the progressive outgrowth of the bud. **Tentacles** appear at the base of the oral cone or hypostome while a **mouth** breaks through at its centre, and the new individual, now able to feed and live an independent existence, detaches at its base and separates from the parental body.

The entire budding process of hydra takes 50 hours at a temperature of 26°C, and a population of budding hydra will double its number every $3\frac{1}{3}$ days if well fed. As long as environmental conditions of food, space, temperature, etc., remain optimal, budding hydras are virtually immortal : so are their budding progeny.

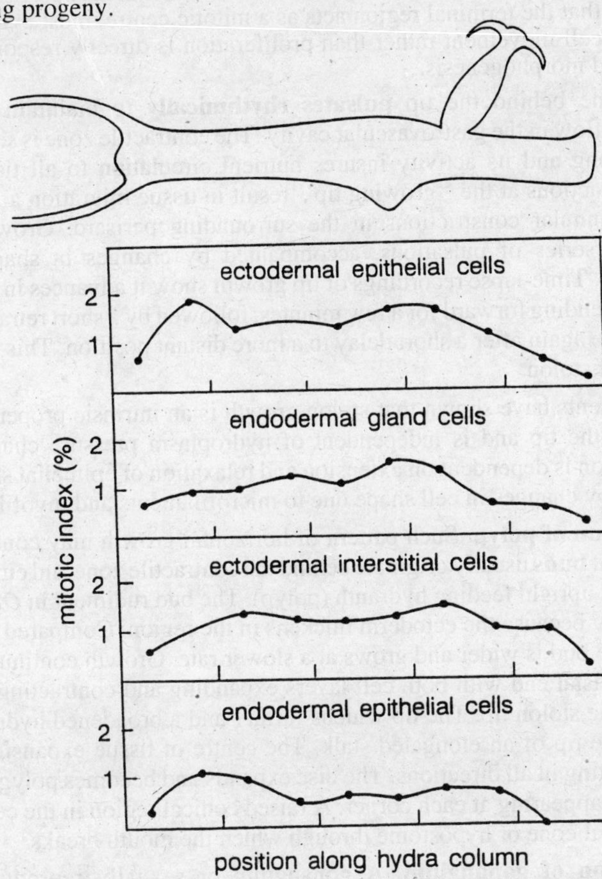

Fig. 25.4. Distribution of cell proliferation along the body column of *Hydra littoralis* in four cell types. No mitotic activity occurs in the tentacles. Note the absence of any gradient (after Grant, 1978).

2. Budding in colonial hydroids. Asexual reproduction by budding also occurs in colonial hydroids such as *Obelia*. The basic morphogenetic structure of *Obelia* is the **stolon**, a two-layered horizontal tube **(coenosarc)** surrounding a gastrovascular cavity. The stolon and all budding individuals derived from the stolon are surrounded by a rigid extracellular secretion or **perisarc**, to which the coenosarc may attach at several places. Stolonic growth is horizontal on a substratum, occasionally sending out branches. As it grows, new perisarc is secreted. Two different functional individuals (polyp and gonangia) extend vertically from **bud discs** along the stolon. The type of bud depends upon the differential contraction tendencies of ectoderm and endoderm in response to environmental signals.

The principal growth zone in a stolon is the terminal and where ectoderm and endoderm cells elongate. As in *Hydra*, proliferation (mitotic cell division) occurs throughout the stolon in both layers and is no higher at the tip. Although some believe that the terminal region acts as a mitotic centre, most evidence suggests that cell movement rather than proliferation is directly responsible for growth and morphogenesis.

The zone behind the tip **pulsates rhythmically** to maintain fluid (hydroplasm) flow in the gastrovascular cavity. The contractile zone is several millimeters long and its activity insures nutrient circulation to all tissues. Periodic contractions at the ''growing tip'' result in tissue migration and are recorded as annular constrictions in the surrounding perisarc. Growth is marked by a series of pulsations, accompanied by changes in shape of ectoderm cells. Time-lapse recordings of tip growth show it advances in short pulsations, extending forward for a few minutes, followed by a short retraction and an advance again after a short delay to a more distant position. This leads to extension of stolon.

Experiments have shown that stolon growth is an intrinsic property of the tissues in the tip and is independent of hydroplasm pressure changes. Growth of stolon is dependent on extension and relaxation of epithelial sheets accompanied by changes in cell shape due to microfilaments and myofibrils.

Formation of polyp. Such pattern of horizontal growth may continue indefinitely, but **buds** usually originate behind the contractile zone and emerge vertically as an upright feeding hydranth (polyp). The bud rudiment in *Obelia* arises vertically because the ectoderm thickens in the region. Compared with a stolon tip, the bud is wider and grows at a slower rate. Growth continues at the widening distal end with both cell layers expanding and contracting just as they do at the stolon tip. The tip widens further and a broadened hydranth develops on the tip of an elongated stalk. The centre of tissue expansion is disc-like, radiating in all directions. The disc expands and becomes polygonal with a tentacle appearing at each corner. A raised conical region in the centre becomes the oral cone or hypostome through which the mouth breaks.

Formation of gonangium. A gonangium or sexually reproducing individual develops at the terminus of stolon. However, from the beginning, its growth differs from that of hydranth. In a gonangial bud, the epidermis always maintains a more rapid rate of expansion than does the underlying endoderm cap. The gonangial rudiment, the last of a series of growth surges, expands transversely more rapidly than it elongates but does not develop tentacles or a manubrium. Instead, the lateral walls detach from the perisarc and evaginate into two-layered folds of tissue which become medusa buds.

Factors regulating budding in *Obelia*. Budding patterns in *Obelia* are controlled in part by temperature, because ectoderm and endoderm respond differentially to temperature changes. At high temperatures, both layers of *Obelia* grow rapidly and synchronously, typical of stolonic growth. At lower temperatures, however, each layer grows at a different rate, producing wide, slow-growing terminal growth from which tentacle rings and polyps develop. At still lower temperatures, linear growth and expansion of the endoderm are suppressed in relation to the ectoderm and a gonangial individual appears.

3. Budding in ascidians. Only adult ascidians bud, a process distinguished by high regulative capacity ; a bud adjusts its develop- mental pattern to the number of cells originating in the initial bud disc. The fewer the number of cells in the disc, the simpler is its final organization and only a few essential organ systems develop.

The ascidian bud consists of two layers. The outer layer is from the adult epidermis. The underlying layer may be derived from any one of three different epithelia : (1) the epithelium lining the atrial cavity on either side of body, (2) the lining of the epicardium or perivisceral cavity adjacent to the heart, or (3) the mesenchymal septum separating afferent blood flow in stolon-like growth. The outer epidermis never participates in bud morphogenesis but acts as a protective envelope surrounding the **totipotent** cell layer below. Its role is limited to gross changes as opposed to the detailed sculpturing of a new individual. Bud morphogenesis is the property of the inner cell sheet which tends to transform into all adult organs by direct development.

Budding patterns in ascidians are almost as diverse as the number of species. We will describe a simple pattern in *Botryllus* which is a colonial form consisting of mature zooids connected via common cloacal cavities into a star-shaped disc pattern (Fig. 25.5). A bud first appears as a local enlargement of 5 to 8 cells of atrial epithelium in the body wall of growing zooids. These cells proliferate rapidly into a small population of

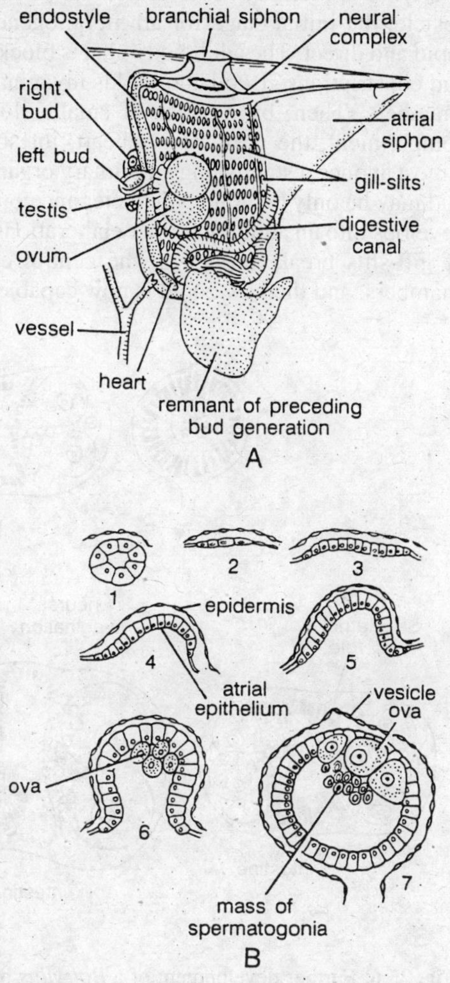

Fig. 25.5. Budding in Botryllus. A—One mature zooid in the colony with mature testis and ovum and two buds, left and right; B—Within the atrial epithelium of the developing left bud another bud disc is appearing to develop into the next generation. (a) surface and side views of initial disc stage, (b) Evagination of disc into a hemisphere, (c) Conversion of hemisphere into closed vesicle surrounded by protective epidermis (after Grant, 1978).

columnar cells surrounded by a squamous epidermis. As soon as the disc of cells achieves a maximum size of 20 to 150 cells, it begins morphogenesis by evaginating into a hemisphere that eventually becomes a hollow sphere. The overlying epidermis, never in contact, is adapted to produce a two-layered vesicle, the initial stage for all morphogenesis. Subsequent development is rapid and direct. The adult body plan is blocked out by a series of simple folds and constrictions. Two lateral folds invaginate to divide the cavity into three chambers. Then, by a series of smaller local invaginations in the central compartment, the presumptive heart, intestine, and neural complex appear almost at once, establishing all primary organ systems. At this stage, the entire bud may be only 0.1 mm in diameter and consists of a few hundred cells. This develops into an adult by further elaborations of the primary rudiments. Rows of gill-slits break through in the central compartment, joining with atrial chambers, and the new zoid is now capable of feeding (Fig. 25.6).

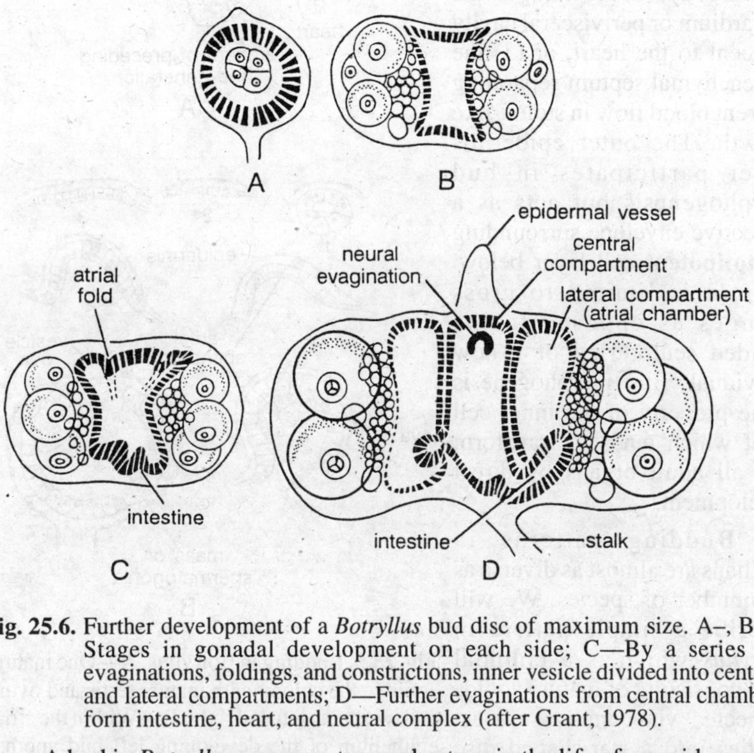

Fig. 25.6. Further development of a *Botryllus* bud disc of maximum size. A–, B—Stages in gonadal development on each side; C—By a series of evaginations, foldings, and constrictions, inner vesicle divided into central and lateral compartments; D—Further evaginations from central chamber form intestine, heart, and neural complex (after Grant, 1978).

A bud in *Botryllus* appears at an early stage in the life of a zooid, usually after the heart rudiment becomes distinct. The new budding generation is initiated even before the parental zooid is completely functional or sexually mature. Budding individuals remain connected to form the typical colony morphology. A parent undergoes dissolution before the next generation of buds appears.

25-2.4. Gemmule and Statoblast Formation

The freshwater sponges and bryozoans produce certain highly specialized asexual reproductive bodies called **gemmules** in sponges (Fig. 25.7) and **statobalsts** in bryozoans, both of which can survive after the maternal individual or colony dies off during an unfavourable season (winter in temperate countries, periods of drought in warmer climates). In such cases, the new individual develops from groups of cells which become completely cut off from the maternal individual and disseminated, so that the development is, from the start, quite independent of the maternal body.

The gemmules and statoblasts are formed in the interior of the parental body from a number of undifferentiated cells which become enclosed in a shell with spicules in freshwater sponges or in a chitinous envelope in bryozoans. On destruction of the parental body, the gemmules or statoblasts are set free, and after environmental conditions have become favourable, the cells contained inside burst out and develop a new individual.

Certain marine Demospongia also produce continuously gemmules which have no hard shell but flagellated epithelium at the flagellated pole and non-flagellated epithelium at the non-flagellated pole of the gemmules. After a free-swimming period, these gemmules settle and attach at the naked end, developing into young sponges (**Meglitsch**, 1972).

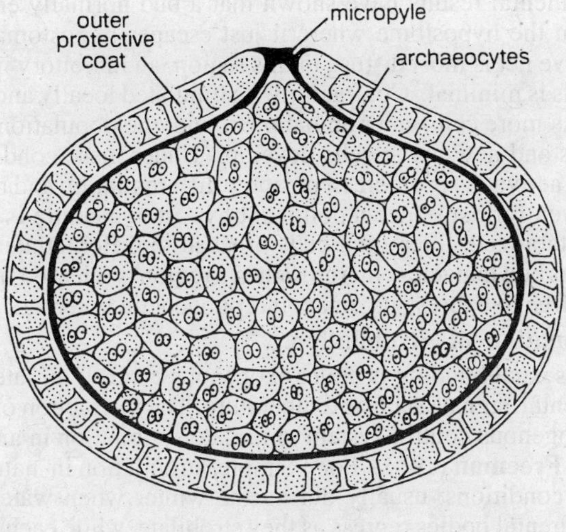

Fig. 25.7. A gemmule of the sponge *Ephydatia fluviatilis* (after Grant, 1978).

25-3. ANALYSIS OF ASEXUAL BUDDING

25-3.1. Regulation of Budding in *Hydra*

Budding is an extension of the normal growth and reproductive cycle of the individual. Most asexually reproducing organisms are characterized by high growth and regenerative potential. Their somatic tissues are constantly undergoing a process of cell renewal ; in a sense, whole animal behaves as a ''**renewal organism**''. Periodic cycles of growth, budding, and dissolution at the organismic level are indicative of a system of massive cell replacement. In most cases, organisms contain mitotically active cells distributed haphazardly (*i.e.*, randomly) throughout the body. As new daughter cells appear, they tend to migrate to more mature regions where they replace tissues depleted by continuous cell death. Most somatic cells are **pluripotent**, *i.e.*, they are able to develop into more than one cell type, whereas some cells (*e.g.*, those in segementation zones) are **totipotent**, *i.e.*, they are able to develop into a complete organism.

Buds always emerge from the *Hydra* body wall in a localized budding zone. The morphology of the budding zone of hydra, however, is no different from adjacent regions prior to the initiation of a bud. The same distribution of cell types is seen, and there is no evidence of local accumulation of interstitial or nerve cells, nor is mitotic activity higher in the bud region.

Grant (1978) has believed that **bud initiation** is based on factors controlling cell shape and tissue migration. A bud first appears as a thickening of the two layers, either as a change in cell shape or as a result of increased cell adhesiveness in the budding zone. Because of enhanced adhesive contacts, the region becomes a ''sink'' into which parental tissue layers preferentially migrate. Cell division is not involved ; bud growth seems to be entirely due to a polarized migration of tissues from the parental body into a secondary axial outgrowth, now oriented perpendicular to the main axis.

Experimental results have shown that a bud normally emerges at that distance from the hypostome where it just escapes hypostomal dominance. Since the nerve net is most diffuse in this region, an inhibitory influence from the nerve cells is minimal. Contractility is stimulated locally and tissues begin to migrate. As more cells enter an incipient bud, the population of interstitial cell increases and new nerve cells develop. Under these conditions, trophic factors (head activator) promote hypostome differentiation and bud outgrowth is induced even further. Once a dominant hypostome centre is established, a new bud is stabilized. Interstitial cells proceed to differentiate stinging cells (cnidoblast) as bud tentacles emerge. When a region is released from inhibitory control, it is free to express its innate morphogenetic potential.

25-3.2. Regulation of Strobilation

Most asexual budding is under exogenous control, initiated in response to environmental signals such as temperature. The interaction of these signals with the endogenous control system to initiate strobilation in an ascidian was examined by **Freeman** (1971). For example, strobilation in nature is initiated under stress conditions, usually during the winter when water temperature drops. The parental bodies regress as they strobilate, while each new fragment resumes the morphogenetic cycle. **Brien** (1968) had suggested that growth is the factor initiating strobilation ; zooids must reach a critical size to trigger the process. However, **Freeman's** experiment indicates that neither growth nor size of the animal is the trigger for strobilation. Strobilation probably reflects a rhythmic contractile process, very much like the pattern of rhythmic growth in coelenterate stolon : it is regulated by an inhibitory nerve centre elsewhere in the parental body.

25-4. SOURCES OF CELLULAR MATERIAL IN ASEXUAL REPRODUCTION

Now an attempt shall be made to trace out the resources that a differentiated organism have in order to start building new parts in a new burst of morphogenetic activity during asexual reproduction are as follows :

1. The development of new parts in asexual reproduction by **fission** can be interpreted more easily, as the reproduction by fission is closely related to regeneration. Both in regeneration and fission, thus, a comparatively large part of the body is isolated and reconstructed into a new complete whole by the regeneration of the missing parts. In both cases, the mechanisms of development of the missing parts remain same. In the Chapter 26 of regeneration, we have discussed that the organs which are produced anew during regeneration in annelids and in turbellarians owe their origin greatly to undifferentiated ''reserve'' cells : **neoblasts** in annelids, **parenchyme** cells in turbellarians. The same cells are available for the development of missing parts in daughter zooids. The neoblasts are actually reported to concentrate in the fission zone of oligochaetes. As in regeneration, the epidermis gives rise to the nervous system and sense organs in the new head of the posterior zooids.

Similarly, during asexual reproduction by transverse fission in the ascidian *Eudistoma* (**Berrill**, 1951), each of the daughter blastozooids receives a section of the parent's body, including epidermis, mesenchyme and a section of the intestinal tube, but lacking the branchial chamber and associated structures such as the atrial cavity, the nerve ganglia and the siphons. These missing parts of each zooid are replaced by regeneration. In the regeneration process of the ascidian a most important role is played by the **epicardium** which is a mesodermal tube lying ventral to the intestine. The mesodermal lining of the epicardium has the properties of an undifferentiated reserve tissue, at the expense of which most of the internal organs in the anterior part of the new zooids develop. Quite interestingly, the branchial chamber does not regenerate at the expense of the remaining part of the endodermal alimentary canal but is also produced by the epicardium and secondarily becomes fused onto the endodermal oesophagus.

2. In **budding**, only very little of the organization of the parent is passed on to the daughter zooid. In a scyphozoan polyp, the fist sign of impending budding is the formation of a pocket-like evagination of the endodermal epithelium which pushes through the mesoglea and ultimately reaches the ectodermal layer. The ectodermal layer now becomes activated, and both ectoderm and endoderm form a conical protrusion on the surface of the body. This protrusion gradually increases and elongates and becomes a cylindrical body with an internal cavity which is in open communication with the gastric cavity of the parent (**Gilchrist**, 1937). At a later stage, a mouth is formed at the tip of the bud and a whorl of tentacles develops around it ; the bud has become a new polyp.

In hydrozoan polyps the method of budding is essentially similar. Both in *Hydra* and *Obelia* cells of the bud come from parental somatic tissues migrating into a new morophogenetic zone with independent axial gradient system. All differentiated ectoderm and endoderm cells, including interstitial cells participate in bud outgrowth. Interstitial cells furnish the nematocysts and nerve cells of the emerging bud.

Each tissue layer in coelenterates retains morphogenetic plasticity. The two layers may be separated, and each can regenerate cell types of the other layer, even to the point of reconstituting a new animal. Cells **dedifferentiate** before they become morphogenetically active as a bud and acquire large nuclei and nucleoli with RNA-rich cytoplasm, a characteristic of actively growing cell (see **Grant**, 1978).

The situation in ascidians is very much like the coelenterates. The adult epithelia supply all bud disc cells including germ cells that differentiate into a mature zooid. In them, asexual budding does not depend on a reserve of undifferentiated embryonic cells ; rather most somatic cells preserve their pluripotent nature in spite of their specialized function and transform into a bud when they escape inhibition of the parental body.

In ascidians, the bud is essentially an outgrowth of the ectodermal epidermis, supported by another tissue which forms a hollow **vesicle** inside the outgrowth. In some cases, parts of the gonads are included as a third element. The epidermis of tunicates is a very specialized tissue which produces the mantle, the cellulose layer covering the surface of the skin in these animals. In budding, the epidermis produces only more epidermis and provides the external covering for the blastozooid. The internal organs are developed from the previously mentioned vesicle formed inside the bud. This inner vesicle is found to be derived from a number of different tissues, namely : (1) an outgrowth of the atrial cavity (*Botryllus*) ; (2) an outgrowth from the pharyngeal wall at the posterior end of the endostyle (*Salpa, Pyrosoma*) ; (3) an outgrowth of the epicardium, which was already mentioned before (*Distaplia*) ; or (4) a group of blood cells (*Botryllus*, **Oka** and **Watanabe**, 1957). In certain ascidians such as *Clavellina* and *Perophora*, the buds develop on long rhizoid-like stolons. These buds consist only of epidermis and a longitudinal mesodermal septum, splitting the cavity of the stolon into two canals which allow for the backward and forward circulation of blood. Buds developing on such a stolon consist of the epidermal covering epithelium and the inner vesicle, which in this case is produced from the cells of the mesodermal septum (**Berrill**, 1951).

The internal vesicle, regardless of its varied origins in different ascidians, produces a variety of organs and tissues : the alimentary canal with the branchial chamber, the atrial cavities, the heart, the nervous system (except in *Salpa*), and sometimes also gonads. It is, thus, evident that several parts in the fully differentiated body of an ascidian retain very wide potentialities for development. These potentialities embrace practically the whole organization of the ascidian except the epidermis.

3. In other groups of organisms such as sponges, undifferentiated reserves of totipotent cells do participate in asexual budding. These cells do not exhibit any specialized function in the adult but appear to be retained for budding and regeneration. An example of these cells is the **archeocytes** in the sponge which form the gemmules in response to adverse conditions (such as changes in temperature or moisture).

The **gemmules** of freshwater sponges are found to contain the **totipotent** cells—cells capable of producing every differentiation found in the animal species in question. The first rudiment of a gemmule in the interior of the sponges appears in the form of an accumulation of **archeocytes**—undifferentiated amoeboid cells which are dispersed in between the differentiated elements of the body of a sponge. Around the central core, consisting of archeocytes, other migrating cells (amoebocytes) are lodged and become arranged in the form of a columnar epithelium which secretes a hard, inner membrane or shell on the surface of the gemmule. The shell is strengthened by the action of special skeleton building cells—**scleroblasts** which derive from the parent mesogloea and deposit in the shell peculiar spicules, the amphidiscs. Both the columnar epithelial cells and the scleroblasts are later withdrawn or they degenerate. A small opening, the micropyle, remains temporarily open in the shell. It has been claimed that feeder or nurse cells called **trophocytes**, enter the interior of the developing gemmule and pass on food reserves to the archeocytes and in their turn disappear. The archeocytes, on the other hand accumulate in their cytoplasm food reserves consisting of glycoproteins in the form of discoidal platelets. When the development of the gemmule is completed, the archeocytes are the only living cells left in it and all the tissues and cell types in the new sponges which develop eventually from the gemmule, including its sex cells, are derived from the archeocytes.

From the above review, it becomes clear that somatic cells of lower invertebrates have high reproductive potential. Except for highly specialized cells such as nerve cells and cnidoblasts, most somatic cells are labile and pluripotent or totipotent.

25-5. COMPARISON OF BLASTOGENESIS AND EMBRYOGENESIS

Most asexual organisms exhibit an alternation of generations—an asexually reproducing stage that buds or strobilates into sexually mature individuals. In them embryogenesis and blastogenesis are found at different stages in the life cycle. The egg and the asexual bud are developmental systems that employ similar morphogenetic processes. They both produce two-dimensional cell sheets that fold into a double-layered vesicle. This double-layered vesicle is the starting point for the construction of multicellular organism in both, with its basic tube-within-a-tube body plan. From this point the similarity within two vanishes.

Comparison of blastogenesis and embryogenesis reveals the fact that blastogenesis is very much simpler than the embryogenesis. This aspect may become more clear from following considerations :

1. The process of producing a new individual is simplest in reproducing by fission, when the blastozooid is derived from half the parental organism and in this way is provided with a large proportion of the organs and parts which are necessary for making the new individual self-sufficient. What has to be done is the regeneration of missing parts. The remnant of the old individual determines the nature, position and orientation of the newly differentiated organs. The polarity and bilaterality of the parent organism prevail in the blastozooids.

2. The task of development is more complicated in the case of budding, since all organs and differentiated parts of the blastozooids have to be produced anew. Nevertheless, the initial system, the bud, always has a higher degree of

complexity than a fertilized egg or even than a blastula as it occurs in embryogenesis. A typical bud always consists of two layers of epithelial cells. The young blastozooid is, thus, already in possession of the concentric stratification of body layers, a condition which in embryogenesis is achieved only after gastrulation.

The two sheets in the bud differ morphogenetically from comparable sheets in the gastrula. The ectoderm of ascidian bud is protective and makes only a modest contribution to the adult body. The bud endoderm, however, folds into most of the organ systems of the adult, including the nervous system.

In one further respect the morphogenetic processes in budding are simpler than in embryogenesis : the new individual inherits its polarity directly from the parent zooid. The point of attachment of the bud to the maternal body or the stolon always becomes the proximal end of the blastozooid.

3. However, in the development of gemmules, the task of producing a new individual becomes most complicated and approaches that of the development of the egg. Thus, the germination of the gemmules occurs after the death and decomposition of the parent animal ; consequently, the polarity of the new individual has to be worked out by itself, and the parent body is no longer there to influence the polarity of the offspring. Further, the complete homogeneity of the contents of the gemmule deprives the new individual of any remnant of morphological organization. The structure of the new sponges has to be established by the interaction of practically independent cells.

4. Most eggs contain the information for two developmental programmes, larval and adult. Embryogenesis, in general produces an efficient larval organization because this stage serves an important adaptive function, such as dispersal or site selection, in the life cycle of the species (see Chapter 27). Where the egg is organized to make a larva, potential adult structures remain undifferentiated, as in insects. The final adult organization is achieved after an elaborate tissue reconstruction during metamorphosis. Most larval tissues are lost, never incoporated into the adult body, while other tissues organize into the adult body plan.

The bud, with a single purpose, develops directly into the adult. It seems to be the more primitive developmental system since it does not include a larval programme (**Grant**, 1978).

REVISION QUESTIONS

1. How is the asexual reproduction different from the sexual reproduction ? Discuss it with respect to evolutionary significance.
2. Describe the asexual reproduction by budding in *Hydra*. Analyse the concepts of dominance, polarity and positional information in *Hydra*.
3. Give an illustrated account of asexual budding in animals.
4. Describe the process of strobilation in *Aurelia*.
5. Critically discuss the issue of source of cells for asexual reproduction.
6. Compare the phenomena of blastogenesis and embryogenesis.
7. Write short notes on the following :

   (i) Positional information theory ; (ii) Head activator ; (iii) Bipolarity in *Hydra* ; (iv) Gemmule ; (v) Budding in *Obelia* ; (vi) Hypostome dominance.

26

REGENERATION

This is one of the evident facts of cell biology that in a healthy, normal adult animal, many kinds of cells need replacement after a species-specific span of time. For example, there are approximately 25×10^{12} red blood cells in the active circulation of the human adult at any one time. About 1 per cent of these cells are replaced daily ; all the red blood cells in the body are replaced every four months. Further, in man a single ejaculate of about 3 cubic centilitre of semen contains approximately 350 million spermatozoa, which must be replaced by cell divison of the spermatogonia. Such replacement process also occurs in skin cells, cells in the lining of the gastrointestinal tract, and cells in the lining of the uterus. Adding all these cell divisions together it is clear that the adult body is losing and producing vast numbers of new cells daily, and, thus, on the cellular level it is a state of **dynamic equilibrium**. This morphological expression of the functional activity of living organs may be considered as **physiological regeneration**. Since the organism must be considered as existing in a state of dynamic equilibrium, it is natural to find extensive development or morphogenetic processes occurring in the adult animal. Some animals in the adult stage retain much of the labile organization of the early embryo and are able to form two or more adults from a single adult. For example, each part of a bisected *Hydra* or planarian worm will develop into an adult *Hydra* or an entire planarian, respectively, just as a divided egg may form two complete embryos. Other animals are able to re-form a part that is removed, just as the embryo is able to replace lost parts. This process of replacing of lost parts, or repairing of damaged body organs is known as **regeneration (Bodemer**, 1968 ; **Balinsky**, 1970 and **Berrill**, 1971), or more specifically as **reparative regeneration**.

Generally the capacity to regenerate is very useful to an animal. It helps the restoration or normalization of important organs after loss or injury ; organs that usually regenerate are those which are definitely advantageous to the animal but whose temporary loss can be tolerated. The power to regenerate is generally more in less evolved animals (those with less complex organization) than in highly evolved (organized animals). It is more in embryonic and larval forms than in adults. (The presence or absence of the regulative capacity in blastomeres should not be regarded as the presence or absence of embryonic regeneration). Closely related species may have widely different regenerative capacities.

26-1. HISTORICAL BACKGROUND OF REGENERATION

The phenomenon of reparative regeneration has fascinated man since antiquity. The earlier historical record of regeneration is of Aristotle, who recorded it in his *De generation animalium* written over two thousand years ago. Recent studies on regeneration are made by **Nicholas ; Locatelli, Berrill, Rose, Weiss, Schmidt, Burnett, Thornton** and many others.

26-2. COMPARISON OF EMBRYOGENESIS AND REGENERATION

The reparative regeneration occurs in various invertebrate and vertebrate animals. Being a specialized kind of blastogenesis, in both animal groups it is characterized by certain processes and events that bear great similarity to the processes and events of embryonic development or embryogenesis. Thus, there is a stimulus to regeneration, there occur cell migration and morphogenetic movements, tissue interactions and cell population dynamics, progressive determination, growth and differentiation like their embryogenesis. The only difference between the two is that regeneration occurs in the tissues of adult animals and in the presence of hormones, while, that does not apply to embryogenesis. Anyway, at a fundamental level both regeneration and embryonic development are epigenetic processes, that is, they involve gene expression and programming.

26-3. PATTERNS OF REGENERATION

Many protozoans, invertebrates and vertebrate animals are found to possess regenerative ability. In all of those animals regenerative process fundamentally is the same, however, **Morgan** (1901) has distinguished following two basic kinds of reparative regeneration (a) **Epimorphosis** or **epimorphic regeneration** occurs when a body part is repaired due to addition of the parts to the remaining piece of that part, *e.g.*, regeneration of amphibian limb and of other vertebrates. (b) **Morphallaxis** or **morphallactic regeneration** is a type of repair involving a reorganization of the remaining part of the body of an animal, *e.g.*, regeneration of new individuals from body pieces of *Hydra*, planaria and other invertebrates.

Some animals such as sponges, sea anemones, corals, turbellarians, annelids and starfishes have such powers of regeneration that voluntary fragmentation (fission) and regeneration serves as a mode of asexual reproduction in natural conditions. Regeneration of autotomized (voluntarily broken away) body parts in certain animals such as starfishes (arm), crabs, insects and spiders (legs) and lizards (tail) is frequent. In vertebrates, the urodele amphibians have an amazing ability to regenerate limbs, tail, eye, lens, jaw, external gills, etc. The regeneraion of internal organs such as liver, and bone occurs in most vertebrates including man. Now let us trace the regeneration in the animal kingdom :

26-3.1. Regeneration in Protozoans and Other Microorganisms

The phenomenon of reorganization has been observed to ooccur in many protozoans. For example, very small segments of many ciliates such as *Stentor, Blepharisma* and *Spirostomum*, are capable of reconstitution. In them, **cell cortex** (ectoplasm) has been found to contain a most significant role in the process of regeneration, because, in experiments when all of the endoplasm of a *Stentor* withdrawn through a small incision in the surface, leaving only the cortex and the nucleus, endoplasm is promptly restored and normal growth and reproduction follow. On the other hand, if all the cortex is stripped off, leaving only endoplasm and nucleus, the cell becomes spherical, survives for a while, but finally dies. However, if a piece of cortex is left on the endoplasm, it gradually spreads around it, reconstituting its lost organization and reproduces normally. In *Stentor* **macronucleus** is also found necessary for regeneration.

In the single-cell marine alga known as Mermaid's cap (*Acetabularia*) has 50 mm length and its body part as small as 1/100 of the total size is capable to regenerate new individuals.

26-3.2. Regeneration in Multicellular Animals

Regeneration in multicellular animals varies from group to group, being most pronounced in the sponges, coelenterates and other lower invertebrates and less developed in higher forms (*i.e.,* higher invertebrates and chordates).

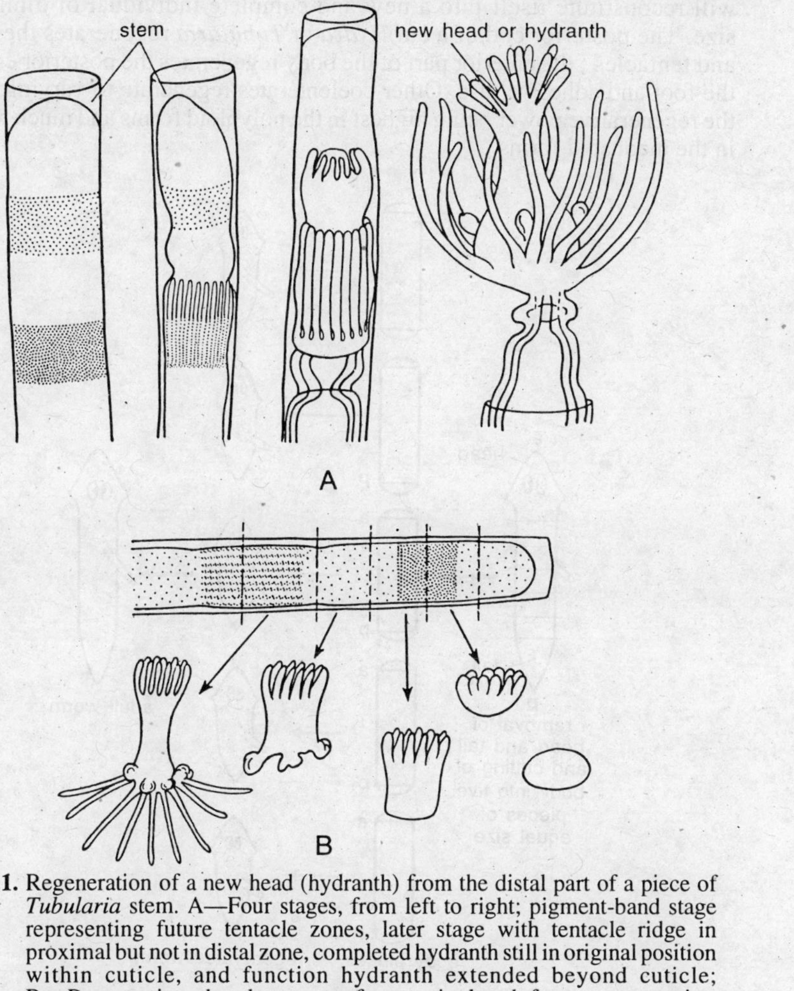

Fig. 26.1. Regeneration of a new head (hydranth) from the distal part of a piece of *Tubularia* stem. A—Four stages, from left to right; pigment-band stage representing future tentacle zones, later stage with tentacle ridge in proximal but not in distal zone, completed hydranth still in original position within cuticle, and function hydranth extended beyond cuticle; B—Prospective development of parts isolated from regenerating *Tubularia* stem (after Berrill, 1971).

Even in those higher animals, regenerative ability is most marked in the embryonic, larval and young stages and least likely to occur in older individuals. Among multicellular animals regeneration occurs in following animal groups :

1. Regeneration in invertebrates. Many invertebrates such as sponges, coelenterates, planarians, nemerteans, annelids, molluscs, nematodes, arthropods and echinoderms are found to possess the ability of regeneration. In the **coelenterates**, the regenerative ability is exceedingly high. It was the regeneration of the freshwater polyp *Hydra*, first discovered by **Trembley** (1744), that attracted the attention of scientists to this phenomenon. A *Hydra* or *Tubularia* may be cut in two or more parts and each body part will reconstitute itself into a new and complete individual of diminished size. The posterior end of a cut *Hydra* or *Tubularia* regenerates the mouth and tentacles ; the anterior part of the body regenerates the posterior end with the foot and adhesive disc. Other coelenterates regenerate to varying extent, the regenerative power being highest in the polyploid forms and much reduced in the medusoid forms.

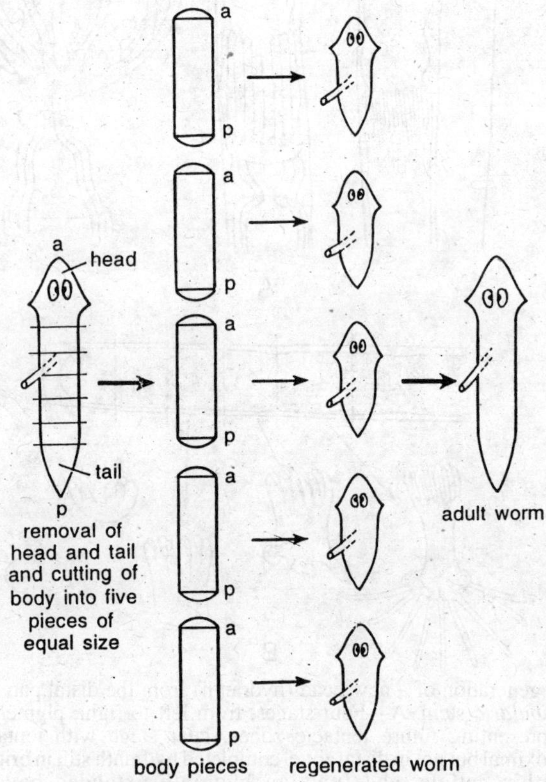

Fig. 26.2. Regeneration in a planarian worm. **Left :** The head and tail are removed and the body cut into five pieces of equal length; **Centre :** Each piece begins to regenerate and forms a small but complete worm. Each worm grows in adult size. Note that each of the five pieces retains the same polarity it had in the whole worm. Anterior (a); posterior (p) (after Bodemer, 1968).

The next invertebrate group remarkable for their high regenerative ability are the **planarians**. Planarian flatworms display a pronounced plasticity in respect to regeneration. Like an egg, this complex organism can be divided into various parts, and each part will differentiate in a series of steps and regenerate a complete organism. Other platyhelminths do not regenerte to any great extent.

The **nemerteans** have a high regenerative ability, and a complete worm may be formed even starting from very small fragments.

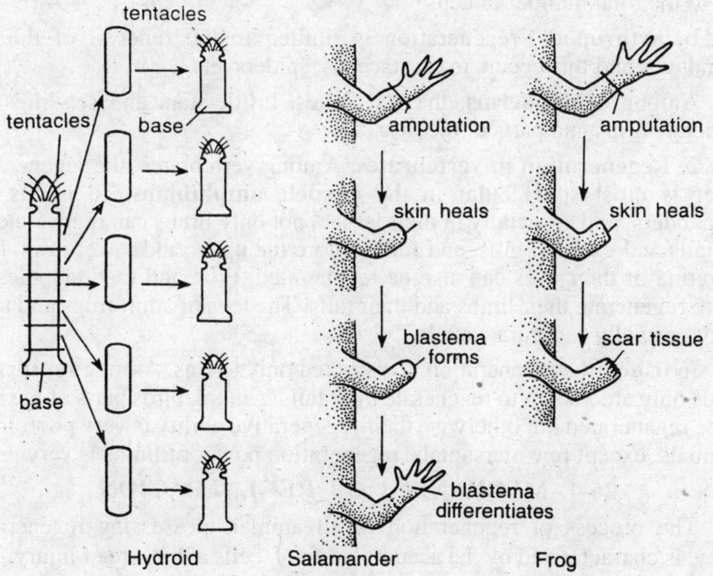

Fig. 26.3. Variation in the regenerative capacities in three different organisms. **Left:** Total regeneration in hydroids. Every part will regenerate into a whole individual; **Centre :** Limited regeneration in salamanders. A limb is amputated and after healing, a blastema forms which differentiates into the missing part; **Right :** The frog-limb stump heals but no blastema forms and no regeneration occurs (after Bodemer, 1968).

Both polychaete and oligochaete **annelids** regenerate anterior and posterior ends after an amputation. If an earthworm or other oligochaete is cut into two halves, the posterior half regenerates the anterior end with the mouth, and the anterior end regenerates a new posterior end. Two new individuals may, thus, be produced from the original one. In the majority of the annelids regeneration is somewhat restricted ; at the anterior cut surface only a species-specific limited number of segements are formed. For example, earthworm *Allolobophora foetida* is found to regenerate its four or five anterior segments. In annelids caudal regeneration is wide spread — nereid polychaetes readily regenerates a new tail, but only rarely a new head ; sabellid and syllid polychaetes, some earthworms and leeches, will regenerate new heads ; and the highly specialized type dwelling *Chaetopterus* will regenerate a complete individual from only one of its first fourteen segments. On the other hand, there is no restriction in the regeneration of posterior segments. *Hirudinea* do not regenerate at all (**Weiss**, 1939).

In **molluscs**, the regeneration is relatively poor. In gastropods, eye stalks with eyes may be regenerated as well as parts of the head or of the foot. The arms of cephalopodes may regenerate but not other parts of their body.

In **nematodes**, the regeneration is very poor. In them only the closure of superficial wounds is possible. Their poor regenerative ability is connected with the high degree of differentiation of the cells of their bodies and the fixed limit to the total number of cells.

In **arthropods**, regeneration is limited to the renewal of the lost appendages and this occurs in crustaceans, spiders, etc.

Among **echinoderms**, the star-fishes, brittle-stars and sea-lilies can regenerate arms and parts of the disc.

2. Regeneration in vertebrates. Among vertebrates, the regenerative power is most spectacular in the **urodele amphibians**. In newts and salamanders, and especially in their larvae, not only limbs can regenerate but also tails and external gills, and furthermore the upper and lower jaws. Lens and retina of their eyes can also be regenerated. Frog and toad tadpoles are able to regenerate their limbs and their tails. The legs of adult frogs and toads do not normally regenerate at all.

In **fishes**, the regeneration is restricted only to fins. Among **reptiles** the lizards only are known to regenerate their tails. Among birds parts of the beak can be regenerated but otherwise their regenerative ability is very poor. In the mammals, except few marsupials, regeneration power of limbs is very poor.

26-4. MECHANISM OF REGENERATION

The process of regeneration in all animals possessing regenerative ability, is characterized by the accumulation of cells at the site of injury. This cellular aggregation is referred to as a **blastema**. It is from blastema that new structures arise. The nature of the blastema and the origin of cells that give its origin are not completely understood in all cases, therefore, only following examples will be considered :

26-4.1. Mode of Regeneration in Hydroid Coelenterates

In hydroid coelenterates the **interstitial cells**, located in the epidermis appear to represent the reserve of embryonic, totiopotent cells, underlying tissue repair and regeneration. In hydroid coelenterates regeneration process includes cell proliferation and migration, blastema formation and differentiation of blastema into lost body part or body. Thus, in *Hydra* regeneration does not occur in the absence of transforming interstitial cells and it is of **morphallactic** type.

26-4.2. Mode of Regeneration in Planarians

In planarians, there is a stock of undifferentiated cells, called **neoblasts**, which have remained in their embryonic state and appears to be omnipotent cells. During the early phases of regeneration the neoblasts divide actively by mitosis and migrate toward the zones of injury to form the regeneration blastema from which the organs are differentiated. The neoblasts are capable of being transformed into somatic as well as germinal cells.

26-4.3. Mechanism of Regeneration in Annelids

In an annelid, when segment is lost, the open wound is sealed by the contraction of the body wall muscles and then a blastema is formed which is composed of mesoderm with overlying ectoderm, together with **neoblast cells** of mesenchymal origin. The blastema produces the ectodermal and mesodermal parts of the regenerating segment, and proliferating endoderm forms the gut. Once a new pygidium has been formed, the blastema takes over the role of the original proliferative region (**Herlent Meewis**, 1964).

In annelid regeneration, the growth is stimulated by both local and systemic influence. It is likely that in annelids, **wound hormone** plays an important role in blastema formation through stimulating the activation and migration of neoblasts. The neurosecretory hormones of brain and nerve cord also have a significant role in annelid regeneration.

26-4.4. Mechanism of Regeneration in Amphibia

Among amphibians, urodeles (salamanders and newts) possess the intense capacity of regeneration of limbs, tail, etc. For example, when a limb of newt is amputated, a process of restoration or regeneration begins immediately in which following periods or phases have been recognized : (a) a phase of wound healing, (b) a phase of blastema formation, and (c) a phase of differentiation and morphogenesis. [Note : Regeneration does not occur if the limb and also the entire girdle is removed. The limb and girdle together seem to constitute a **"regenerative field"** or **"territory".** A part of the field must be left with the organism for the rest of the organs of the field to regenerate].

(a) Phase of wound-healing or preblastema stage. During preblastemic stage, the wound healing process includes clotting of blood and migration of epidermal cells from the basal layer of the adjacent epidermis toward the centre of the wound. The blood clot serves to close off the gaping wound from the external environment. Thus, in less than 24 hours of the amputation, wound is covered with epithelium which is thicker than the adjacent epidermis of the limb.

(b) Phase of blastema formation. During blastema formation phase, cells accumulate beneath this newly formed epithelial covering, and the combined cap of epithelial and adjacent cell mass is known as **blastema**. The blastema has the general properties associated with a limb bud or an egg in so far as it has the potentiality of growing and differentiating into highly organized unit structure.

The production of blastema continues for some time before developmental events become detectable. Epidermal cells continue to accumulate at the apical region, possibly through further migration but certainly in part through mitotic division of cells initially forming the **cap**. Mesenchymatous cells progressively accumulate beneath the cap, so that an epithelium-mesenchyme reacting system is established, resembling that of a limb-bud or feather-bud on a much large scale.

Recent studies have established the fact that the fibroblast-like cells of blastema derive from mature epidermis and from internal tissue that contained fully differentiated muscle, nerve, and cartilage, together with connective-tissue cells (*viz.*, fibroblast or mesenchyme), through an apparently regressive process known as **dedifferentiation**.

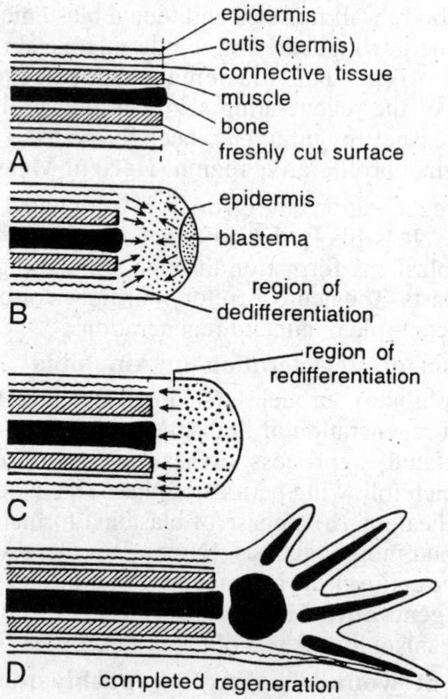

Fig. 26.4. Diagrammatic representation of the cellular changes after cutting the limb of a larval salamander. A—A clean cut traverses epidermis, dermis, connective tissue, muscle, and bone (nerves and blood vessels not shown); B—A blastema forms. The drawing represents the blastema as forming from the dedifferentiated cells of the various tissues of the limb plus cells which migrate in from the epidermis ; C—Stopping of dedifferentiation and redifferentiation of cells of blastema into muscle, bone and so forth; D—Regeneration completed. The blastema has differentiated into the missing part (after Bodemer, 1968).

Dedifferentiation. The source of the mass of mesenchymatous cells comprising the bulk of the blastema has been a problem plaguing the analysis of limb regeneration for many years. Following amputation, the residual differentiated tissue of the **stump**, mostly cartilage and muscle, undergoes extreme disintegration and apparent dedifferentiation. At the same time the mesenchyme cells beneath the epidermal cap increase in number enormously. The problem has been to ascertain the connection between these two events. Following opposing interpretations have been adopted for explaining the dedifferentiation :

(i) Is the apparent dediffrentiation of mature muscle and cartilage cells a true return to an undifferentiated state ? If so, they can obviously give rise to a mass of undifferentiated blastema cells. It has been extraordinarily difficult, however, to establish whether or not true dedifferentiation and subsequent redifferentiation may actually take place.

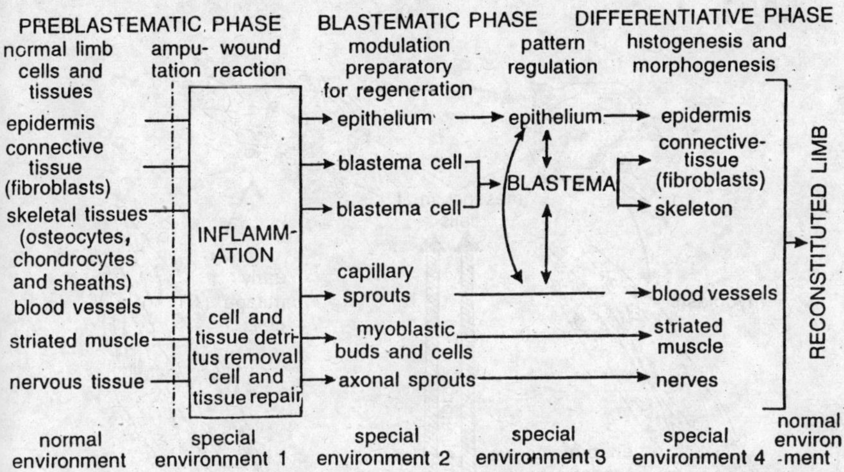

Fig. 26.5. A concept of regeneration process in amphibian amputated appendages with the exception of the final reconstitution of the amputated appendage, this diagram could also serve as a summary of repair process in all vertebrates (after Berrill, 1971).

(ii) Are the disassociated and otherwise changing cartilage and muscle cells degenerating and in fact dying, allowing their products to stimulate and support rapid multiplication of any unspecialized mesenchyme cells present ? A dependent population replacement of this sort would be effective and is known to occur in regenerative processes in some other forms (*e.g.*, ascidians). In this event no transformation of cells is called for, only differential multiplication.

(iii) Do the cartilage and muscle cells merely appear to dedifferentiate, *e.g.*, lose their distinctive histological character but in actuality retain their fundamental character and as such contribute to the blastema ? The phenomenon of **pseudo-differentiation** is commonly seen in tissue culture and has been termed **modulation** by **Weiss**. If this is the case, however, in a blastema, then we are faced with another problem, because the implication is that modulated cartilage, muscle, and other cells, having been mixed together subsequently sort out according to their kind, *i.e.*, like to like, and also become reassembled in the new configuration of the developing limb structure.

The long standing difficulty has been the failure in following the history of individual cells during the crucial period, that is, in tracing individual cartilage or muscle cells from the original stump tissues to whatever their destiny may be, or in determining the source of the mesenchyme cells with any precision. Electron-microscopic studies support the contention that the differentiated cells of the stump do become morphologically undifferentiated cells in the blastema. What is less clear is whether they redifferentiate into new cell type.

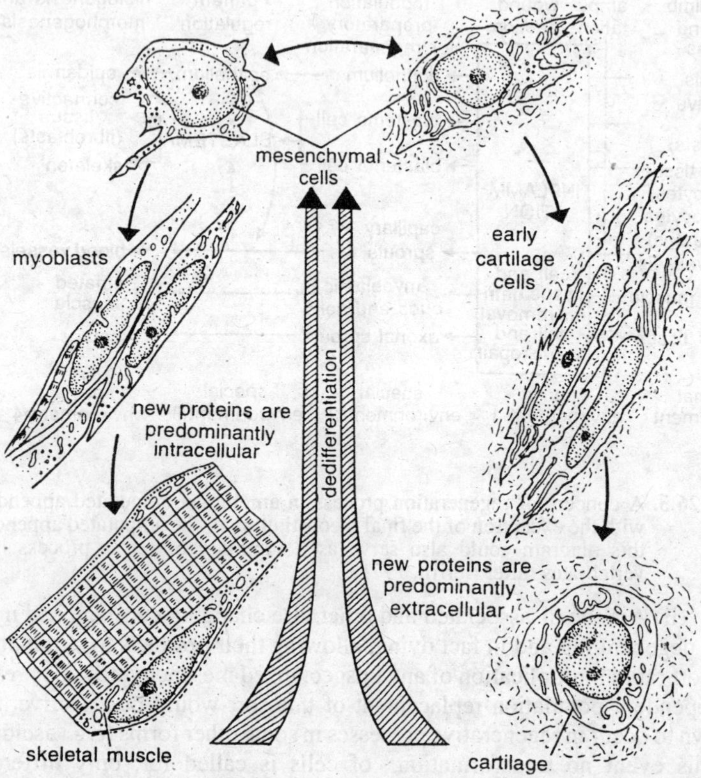

Fig. 26.6. Diagram summarizing the changes in fine structure which occur in the mesenchymal blastemal cells as they differentiate into myoblasts and muscle cells, early cartilage cells and cartilage. The formed cells dedifferentiate into mesenchymal cells when the limb is amputated (after Bodemer, 1968).

C. Phase of redifferentiation and morphogenesis. When the blastema reaches a certain size dedifferentiation ceases, and the established blastema enters into a **growth phase**. Later on, it begins to differentiate and to replace the part of which the limb was previously deprived. Specifically, if the fore-arm is removed, the blastema differentiate directly into the muscle, bone, cartilage and skin of the fore-arm.

Therefore, during amphibian limb regeneration, some types of differentiated cells do 'dedifferentiate' and 'redifferentiate' in other types of cells (This is called **metaplasia**). However, the cells do not, revert to a completely undifferentiated state, for instance, the mesodermal tissue does not produce the epidermal tissue and *vise versa*.

Thus, in the newt limb — blastema passes through several definable stages during the course of regeneration. Morphologically, in the adult newt at 20°C, 15 days after amputation, the blastema is filled with undifferentiated cells ; by 20 days the blastema has become a **cone**, the **palette stage** (a flattend cone) is reached at 25 days ; the **notch stage**, representing the first sign of digits, is reached at 30 to 35 days ; and from then on a digital pattern becomes progressively in evidence, the pre-cartilaginous skeleton condenses, and a complete limb is present by 75 days.

26-5. PHYSIOLOGICAL PROCESSES INVOLVED IN REGENERATION

During the process of regeneration in an animal following two periods with different type of metabolism have been distinguished : first period including the predominantly **destructive metabolism (catabolism)** and second period including the predominantly **constructive metabolism (anabolism)**.

26-5.1. Catabolism of Regeneration

The period of destructive metabolism has great significance in those animals in which dedifferentiation of adult tissues occurs at the beginning of regeneration, as in urodele amphibians. Most prominent is the abrupt increase (*i.e.*, up to four fold during tail regeneration of a tadpole) in the activity of **proteolytic enzymes** (**Bromley** and **Orechowitsch**, 1934). The proteolytic enzymes that are mainly responsible for this increase are **kathepsin** and **dipeptidases**. Through the activity of protolytic enzymes, the cells damaged by the operation are destroyed in the first instance, but the main result is dedifferentiation of part of the healthy tissues adjacent to the wound surface : destruction of the intercellular substances and the highly differentiated parts of cells, such as the muscle fibrils.

As a result of these catabolic processes, the amount of **free amino acids** increases to about double the amount in normal tissues from 16.8 per cent to 35.1 per cent of the total nitrogen (**Orechowitsch** and **Bromley**, 1944). A qualitative change in the structure of the proteins and their products also take place. This is shown by an increase in the amount of reduced sulphydryl compounds, that is, compounds having a reduced-SH group, presumably owing to a reduction of the cystine in the tripeptide glutathione.

Another phenomenon of a physiological nature following amputation is a change in the character of the oxidation in the tissues of the stump. The respiratory quotient, **R.Q.** (ratio of CO_2 produced to amount of O_2 consumed), falls abruptly (**Ryvkina**, 1945). This is a sign of incomplete oxidation in regenerating tissues. In fact, it has been reported that during preblastemic phase of regeneration the amount of **lactic acid** greatly increases in the area of blastema formation because of the absence of adequate means (*viz.*, absence of **cytochrome oxidase**) of oxidizing pyruvic acid in the Krebs cycle (**Okuneff**, 1933). This suggests that an anaerobic environment prevails during preblastemic and blastemic phase of regeneration (**Schmidt** and **Weidman**, 1964). Further, in amphibian regeneration the amount of glycogen and mucopolysaccharides increase in blastema. The glycogen is derived from injured muscle and nerve cells.

The change in the metabolic processes taking place in the earlier phase of regeneration — accumulation of free amino acids, of lactic acid and of free-SH group — all lead to a lowering of the pH in the tissues of the stump and the regeneration blastema.

26-5.2. Anabolism of Regeneration

Catabolic phase of regeneration is followed by anabolic processes of regeneration. Anabolic phase of regeneration is characterized by an increase of oxidation, by a return of the pH to normal, by a higher R. Q., by a decrease in the amount of free-SH groups and of lactic acid. Moreover, the cells of the regeneration blastema become very rich in cytoplasmic ribonucleic acid (RNA), even if the cells from which the blastema is derived do not originally contain large amounts of this substance. The increase in RNA production is mainly due to newly synthesized ribosomal RNA. Accordingly, it is noticed that the size of the nucleoli increases in cells participating in regeneration (**Gurdon** and **Graham,** 1967). When the stage of differentiation of the regenerating part is reached, the amount of ribonucleic acid in the cells diminishes again (**Roskin,** 1950).

Lastly, injury affects the metabolism of carbohydrates and proteins. The electrolyte balance is altered after injury, the tissue releasing large quantities of potassium ions (K^+), while sodium (Na^+) and chloride (Cl^-) ions are conserved by body. These physiological changes are influenced by hormones from adrenal, thyroid, anterior lobe of pituitary and gonadal glands.

26-6. INDUCTIVE INTERACTIONS IN REGENERATION

According to the definition, regeneration is the replacement of lost parts. One could have expected, therefore, that the loss of some part of the body would be the adequate stimulus to set in motion the mechanism which restores the part. But, certain experiments have made it clear that not the absence of an organ but the presence of a wound (open or internal) containing damaged tissues is the immediate cause of onset of regenerative processes. Certain other kinds of experiments have confirmed the fact that the damaging and disintegrating tissues of a wounded body part release some chemical substances which act as **stimulus** for regeneration. Further, recent studies have indicated that during regeneration different kinds of inductive interactions occur in various animals, as follows :

26-6.1. Inductive Interaction in Regeneration of Hydroid Coelenterates

The body of hydroids is characterized by an **axial gradient** of levels of metabolism and mitotic activity, the highest level being the **hypostome** or "head region", the lowest being the basal region, **peduncular region**. A transplanted head region has the ability to suppress the formation of another head region in its immediate vicinity, and there is evidence that it normally elaborates a **growth-inhibiting substance** that diffuses downward through the body column. The growth inhibitor is rendered impotent as it diffuses along the entire length of gastric region. Growth inhibitors have been demonstrated in *Hydra* and *Tubularia*. In *Tubularia* the growth inhibitor has been found to be thermostable, water and alcohol-soluble, dialyzable, resistant to weak hydrolysis and abosorbed by charcoal.

The investigations of **Burnett** and others (1966) have indicated that maintenance of body form in *Hydra* may be explainable on the basis of **single inhibitor** diffusing proximally from a growth region in the hypostome ; the hypostome is a distal-dominant centre and it may inhibit regeneration of the same structure within the piece but not inhibit regeneration of another organ, for example, basal disc or peduncle. This inhibitory diffusible substance is found to be a polypeptide which remains associated with **neurosecretory granules**. However, polar regeneration in hydroids may depend not only upon growth-inhibitor factors, but also upon the presence of **growth-stimulating principles**. Various transplanting experiments have suggested that the hypostomal region of *Hydra* and *Tubularia* contains a principle called **inducer** that stimulates cell proliferation in interstitial and other cells. Diffusion of this inducer proximally from the hypostome causes the cells just proximal to the hypostome to engage in more active proliferation than cells elsewhere in the body, and a **growth centre** is formed. An inhibitory principle is produced by the established growth centre. The inhibitor is formed only when the metabolism and mitotic activity of an area is high, and it may be a metabolic by-product. This growth-inhibitor does not affect cellular divisions in the area of its formation ; but the inhibitor diffuses downward and inhibits cell divisions to a level where it is either diluted below its threshold of activity is or rendered impotent. At this level a bud is formed, which, in turn produces its own growth inhibitor, preventing formation of additional buds there ; this last inhibitor diffuses downward into the peduncular area, at which level the inhibitor is no longer active and a basal disc is formed.

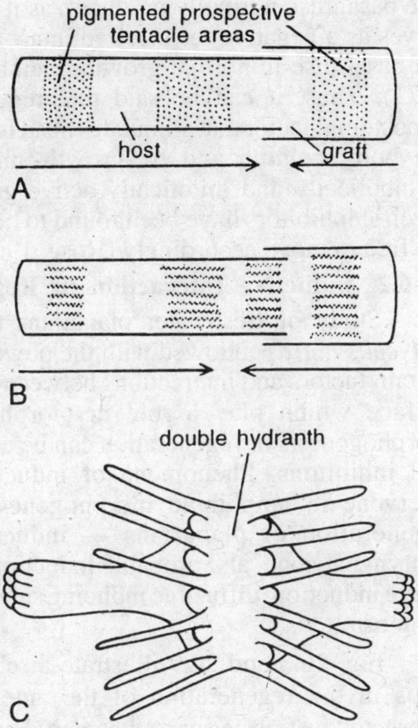

Fig. 26.7. Induction in *Tubularia*. A—Two portions of *Tubularia* stems showing regeneration and transformation when hydranth primordia are combined. The left component already has two pigmented prospective tentacle areas and represents a potential whole new hydranth; the right component posesses only one (anterior) pigmented area and gives rise only to the anterior part of a hydranth, if isolated; B—In combination the host develops two rows of posterior tentacles of which the second row, although in host territory, develops in conformity with the graft; C—The doubled hydranth with the secondary posterior tentacles oriented as part of the graft (after Berrill, 1970).

The basal disc normally produces, as it grows, an inhibitory substance which serves as a negative feedback in intact animal, thereby regulating the size of the basal disc in normal growth, maintenance and regeneration (see **Berrill** 1971). Thus, it can be said that regeneration in *Hydra* is controlled by hypostomal factor that stimulates head formation wherever it is introduced into the body column and by growth-inhibiting factor produced within the metabolically and mitotically active growth centre of the hypostome. The specific inhibitors have been found to be proteins having **positive charge** and are free to move accordingly (**Rose**, 1967) (see Chapter 25).

26-6.2. Inductive Interaction in Regeneration of Planarians

Study of freshwater planarians (*Euplanaria* (*Dugesia*) *lugubris* and *Polycelis nigra*) endowed with the power of total regeneration has elucidated certain factors and interactions between the regeneration blastema and the cut surface which play a role in morphogenesis during regeneration. The morphogenesis of regeneration can be explained by a succesion of inductions and inhibitons. Phenomena of induction, comparable with those which intervene in amphibian morphogenesis, have been demonstrated in the regeneration of planarians — induction of the eyes, induction of the pharyngeal zone, and probable induction of copulatory apparatus. In the case of eye induction, diffusible inducing substances, acting at a distance, have been demonstrated.

Fig. 26.8 and 26.9 illustrate a scheme interpreting the morphogenetic steps in the regeneration of the anterior structures in a planarian body transected just posterior to the genital orifice. It is assumed that tissues at the cut surface induce the formation of anterior blastema, in which the first differentiation is the brain. The regenerated brain (Fig. 26.9A) has an inducing action, through the intermediary of diffusible substances on the eyes (which may be 80 to 90 in *Polycelis nigra*). The prepharyngeal region (II, Fig. 26.8 A) is probably induced by the head ; the former in turn induces the pharynx (III, Fig.26.8 A). The copulatory organs are induced by the zone of the gential

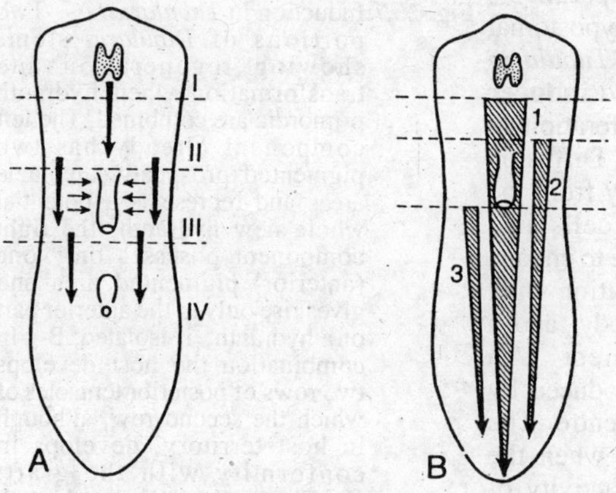

Fig. 26.8. Induction and inhibition during regeneration of planaria. A—Process of successive inductions in the course of regeneration anterior to a caudal fragment. Zones I, II, III, IV correspond to successive induction; B—Inhibitions exerted on anterior regions by the posterior regions. Arrows indicate gradients of inhibition due to the diffusion of inhibitory substances. The inhibitory substances in zones 2 and 3 are hypothetical (after Bodemer, 1968).

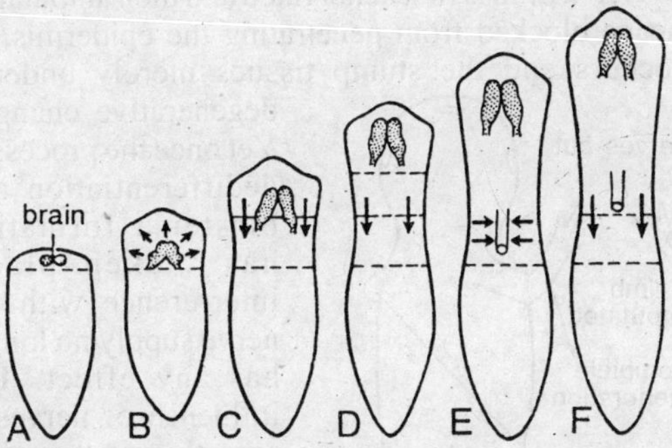

Fig. 26.9. Sequence of induction in regenerating flatworm. From A to F : brain first differentiates, which induces differentiation of eyes along anterior and antero-lateral margin, to constitute head structures; head induces prepharyngeal region and then the pharyngeal region posteriorly which induces reproductive organs posteriorly (after Berrill, 1971).

glands, located just anterior to them (IV, Fig. 26.8 A).

Inhibition mechanisms intervene parallel to the inductive processes, preventing repetitive regeneration of an organ. The production of a diffusible inhibitory substance has been demonstrated for the planarian brain, and it is believed that each zone of a regenerating planarian emits inhibiting substances (Fig. 26.8 B).

26-6.3. Induction in Regeneration of Annelids

In annelids, regeneration is controlled by neurosecretory hormones produced by the brain. It is also known that annelids nerve cord exerts a local trophic effect upon regeneration : not only is its presence necessary, but abnormal regeneration will occur if it is deflected from its usual position (**Highnam** and **Hill**, 1969). In this way, regeneration can be induced along the whole length of the body in *Myxicola aesthetica*, and *Nereis diversicolor*.

26-6.4. Inductive Interactions of Tissues in Limb Regeneration

The amphibian's regeneration of limb is found to be dependent upon epithelium and nerves. It is apparent that the wound epithelium that covers the amputation surface is necessary for blastema formation. The thickened epidermal cap that forms over the transected limb serves as a focal point for the gathering blastemal cells furnished from the tissues of stump. **C.S. Thornton** (1960) has shown that if the epidermal cap is placed eccentrically, an eccentric blastema will form ; if it is continually removed, blastema formation can be prevented. The epidermal thickening plays a rather prominent role in removal of debris from the amputation surface, and it is possible that it accounts for 5 kinds of inductive reaction with the underlying tissues which aid dedifferentiation and the appropriate arrangement of the migrating blastemal cells. The similarity of the epidermal cap and the apical cap of ectoderm in the embryonic limb is apparent.

Neural trophic factor. Regeneration of the limb blastema of vertebrates, particularly amphibians is normally dependent upon a critical supply of nerves at a very early stage. Many nerve endings in fact enter the epidermal cap that covers the wound, and make close, synapse-like contact with epidermal cells. However, if a limb is first denervated and then amputated, or if nerves are by any means blocked from penetrating the epidermis, no regenerative outgrowth occurs and the stump tissues merely undergo degenerative changes. Yet once the process of dedifferentiation and blastema formation has taken place, interference with the nerve supply no longer has any effect. The influence of nerves is something other than nerve function as such and is not due to acetylcholine. The agent is known as the **neural trophic factor**. It's exact function and nature are still unknown, but, it is known that neural-epidermal junctions are the essential conditions in such nerve-dependent regenerations.

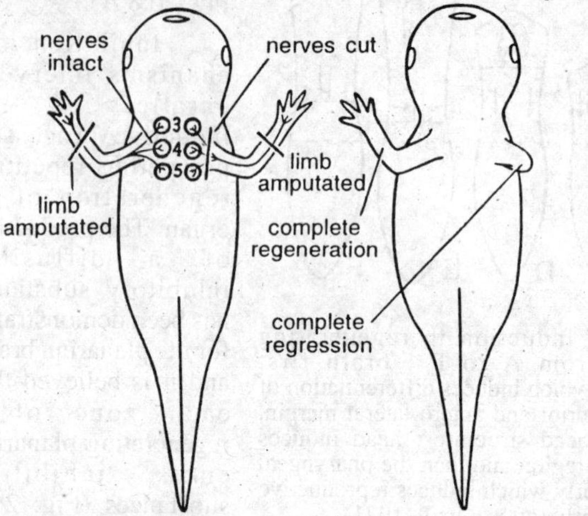

Fig. 26.10. The influence of nerves on limb regeneration. A—A larval salamander with both fore-limbs amputated through the radius and ulna. The nerves to the left fore-limb stump are intact, whereas those to the right fore-limb stump have been cut as indicated; B—Some weeks later the left fore-limb has regenerated whereas the right fore-limb has regressed (after Bodemer, 1968).

26-7. POLARITY IN REGENERATION

Generally, parts of an organism have a distinct polarity. An example is the hydroid, *Tubularia*, which has base for attachment, a long body, and a set of tentacles. Bisecting the stem of a hydroid into two equal segments and observing the development of each part reveals that the hypostomal, head end forms the tentacles and the other, penduncular end simply forms a base for attachment. This experiment clarifies a distinct polarity in the hydroid body, the upper end of each piece forms tentacles and the lower end forms base. This polarity can be compared with the animal-vegetal polarity of sea-urchin's egg and the anterior-posterior polarity in the amphibian limb. And, like the polarity in embryonic systems, the polarity of these regenerating parts can be altered by experimental methods.

For example, when a hydroid segment is placed in a divided chamber with a high concentration of O_2 in one chamber and a low O_2 concentration in the other chamber (Fig. 26.11) the polarity may be completely reversed, and on the end in high O_2 tentacles develop where a base would normally form. Obviously, the organization within the tissue is quite labile. This lability may be illustrated also by separation of a part of the hydroid body into two parts by means of a constricting ligature. When a piece of a hydroid is ligated, both the prospective base and tentacle- developing ends (Fig. 26.12) form tentacles.

Thus, two individuals are obtained from one mass of tissue. The result is similar to dividing an egg and obtaining two embryos or manipulating a limb bud to produce a duplicated limb.

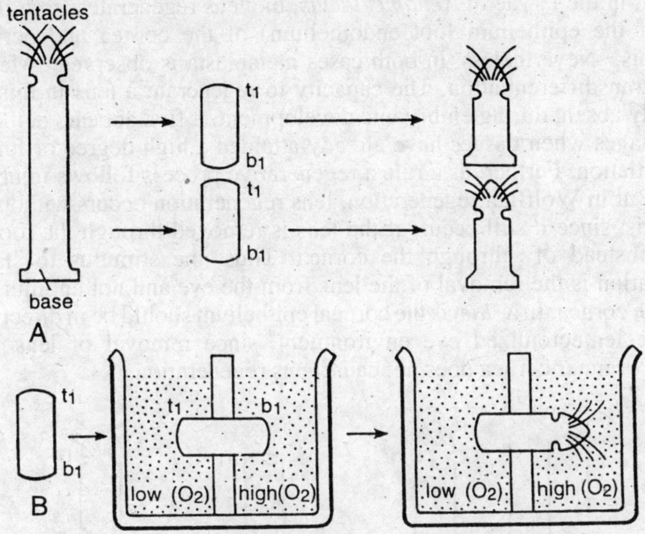

Fig. 26.11. The normal polarity of a hydroid and the reversal of polarity by experimental treatment. A—A hydroid is cut into equal parts. Each part retains the polarity of the original hyrdoid and regenerates a base and a set of tentacles; B—The end t, which would normally form a set of tentacles, is treated with a low concentration of oxygen and forms a base; the end b, which would normally form a base, is treated with a high concentration of oxygen and forms a set of tentacles. Polarity is reversed by other chemical treatments (after Bodemer; 1968).

26-8. METAPLASIA

The transformation of one cell type into another, usually into a cell type not normally associated with the anatomical structure or tissue concerned, is called **metaplasia**. Metaplasia has been observed in chordates as well as nonchordates.

26-8.1. Metaplasia in Chordates — Lens Regeneration in Amphibians (or Wolffian Regeneration)

The eye lens is readily regenerated in some vertebrates, notably in salamanders (*Salamandra*) and newts (*Triturus*) though rarely in frogs and toads. In fact, the regeneration of a new lens after the removal of the existing lens (called **lentectomy**) occurs in many urodele amphibians throughout post-embryonic life. The phenomenon is usually called **Wolffian regeneration** after **Gustav Wolff** who studied it in detail in 1894—95. However, it was discovered in 1885 by **Colucci**. Lens regeneration after lentectomy occurs also in anurans. However, in them, the lens regenerating capacity disappears at metamorphosis (*e.g.*, *Xenopus laevis*).

1. Which tissues form regenerating lens ? In urodeles, the lens, unlike the limb, regenerates from tissue different from that which give rise to it during embryonic development. Thus, the lens regenerates from the neural ectoderm (dorsal iris epithelium) and not the body ectoderm (outer layer of cornea). In contrast, in the larvae of *Xenopus laevis*, the lens regenerates from the basal layer of the epithelium (not endothelium) of the cornea and pericorneal epidermis. Nevertheless, in both cases metaplasia is observed. Metaplasia means transdifferentiation. The capacity to regenerate a lens in this way is generally absent during embryonic development, it first appears in the young larval stages when tissues have already attained a high degree of functional differentiation. Further, as a rule a regenerative process follows injury to the tissue. But in Wolffian regeneration, lens regeneration occurs without injury to the iris, since it still occurs if the lens is removed through the roof of the mouth instead of through the cornea. Thus, the stimulus to start lens regeneration is the removal of the lens from the eye and not an injury to the iris or the cornea. In *X. laevis* the corneal epithelium should be in direct contact with the lentectomized eye environment, since removal of lens without incision of inner cornea does not cause lens regeneration.

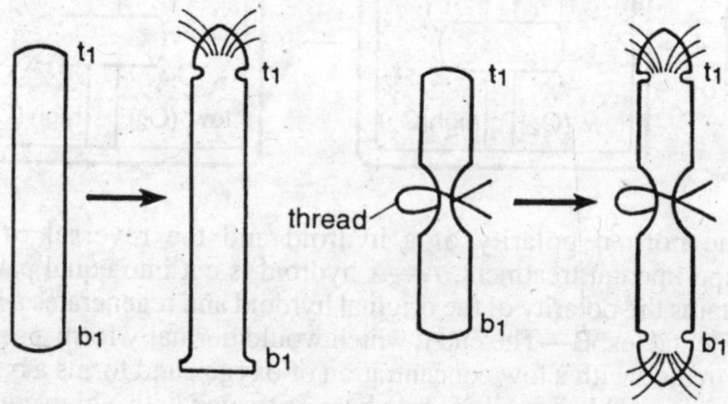

Fig. 26.12. Regeneration and polarity in coelenterate. A—A section of stem is shown with the upper end, t_1, giving rise to tentacles and the lower end, b_1, forming a base; B—A similar stem is ligatured in the middle and as a result tentacles develop at b_1 and as well at t_1 end (after Bodemer, 1968).

2. Events during lens regeneration. Immediately after the lentectomy, the iris encircles the space occupied by the lens, and the new lens arises from a restricted group of iris cells. The iris cells are normally non-dividing and heavily pigmented, but within 6 hours of lentectomy, iris cells that were previously in close proximity to the original lens begin to divide (Fig. 26.13). They also begin to lose pigment. Soon these de-pigmented cells begin to appear close to the mid-dorsal margin of the iris epithelium and these cells proliferate to form a distinct vesicle, the precursor of the new lens. In case of *Xenopus* the basal layer of the epithelium of the cornea forms a lens vesicle (Fig. 26.14). After some time, the cells on the internal layer of the lens vesicle cease division and begin synthesis of γ-crystallins. As these lens specific proteins concentrate, the internal cells begin to adapt to morphology of lens fibre cells and crystallin synthesis begins in the outer cells of the vesicle. By 25 days after the initial lentectomy a definitive lens is developed.

3. Interactions of tissue during lens regeneration. The iris or cornea does not regenerate a lens as long as the original lens is there ; or it is replaced by a lens from another individual. This inhibitory effect of the lens shows a constant interaction between the lens and the iris or cornea in the animal.

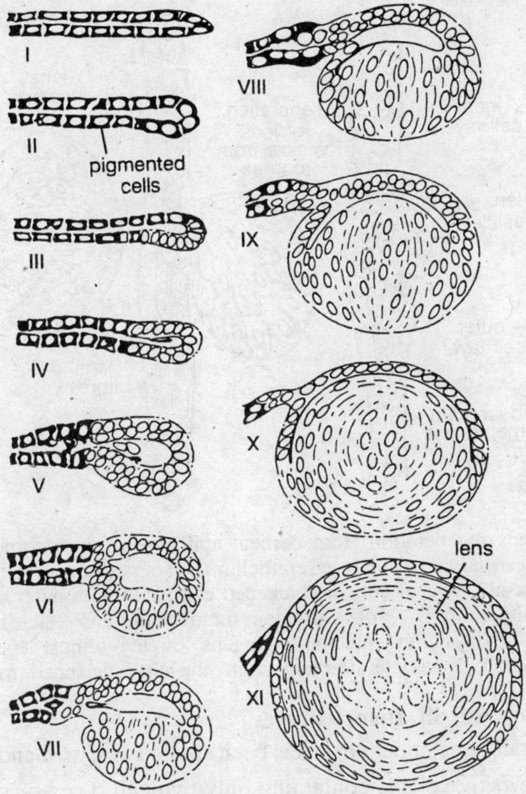

Fig. 26.13. Lens regeneration from dorsal iris in the newt, *Triturus viridescens* after lens removal or lentectomy (after Maclean and Hall, 1987).

The neural retina also appears to play a vital role in lens regeneration. The iris or cornea, when transplanted alone in a foreign site usually does not regenerate a lens but produces a lens if transplanted along with the retina. However, recently it is found that the iris (in *Notophthalmus*) or the cornea (in *Xenopus*) transplanted alone in the amputated limbs form the lenses. It seems probable that both neural retina and amputated limb provide the necessary neural trophic factor for lens regeneration.

Eguchi, **Abe** and **Watanable** (1971) have also obtained differentiation of **lentoids** — structures shown to be lens-like by electron microscopy and immunological techniques — from dissociated newt iris cells in a simple cell culture system (*i.e.*, *in vitro*). Here lentoid formed equally from cells of dorsal and ventral iris ; this difference may be related to the fact that, in the operation on the organism, the act of removal of the lens loosens cell association in the dorsal margin but not in the ventral margin of the iris, whereas in cultures all cells, from whichever source, are dispersed by enzymic dissociation ; this suggests that changes at the cell surface provide significant signals in both cases (**Ede**, 1978).

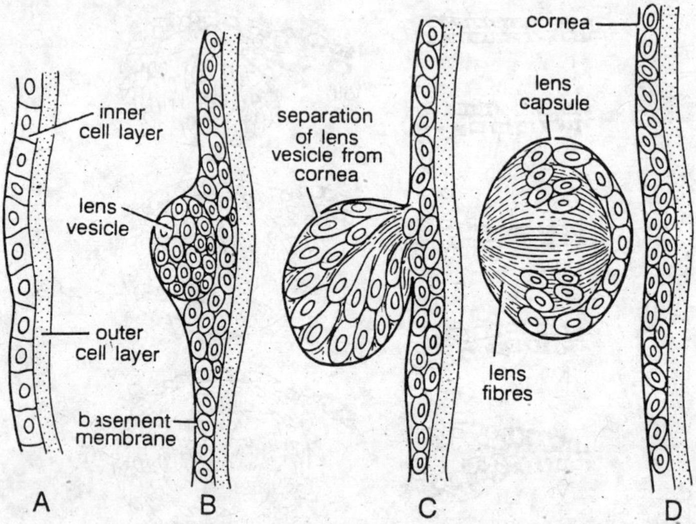

Fig. 26.14. Lens regeneration from corneal epithelium (outer cornea) in *Xenopus laevis* larvae. A—Healed epithelium 36 hours after lens removal. The inner (basal) layer of cells is thickened compared with outer cell layer of the epithelium; B—4 days after lens removal, the lens vesicle is being formed; C—6 days after lentectomy, the lens vesicle is almost separating from the cornea; D—8 days after lentectomy the lens is detached from cornea. Lens

26-8.2. Metaplasia in Invertebrates

In invertebrates, metaplasia has been studied in *Hydra* and *Podocoryne*.

1. In *Hydra*, explants containing only two cell types — digestive cells and gland cells, are claimed to be capable of regenerating a complete new hydra (**Davis**, 1973). But such regrowth involves the production of new interstitial cells from the gland cells, followed by development of nerve cells, cnidoblasts, and other cell types from these interstitial cells.

Perhaps a more interesting observation of a change in cell type in *Hydra*, made by **Burnett** (1973), is that, in gastrodermal explants, digestive cells transdifferentiate into muscle fibres. (This is called **direct metaplasia**). Since the digestive cells are characterized by symbiotic algae, which they lose in the process of the redifferentiation, and the longitudinal muscle fibres are readily distinguishable, there seems little room for confusion. The most interesting aspect of this experiment is that the cell type conversion apparently takes place without intervening cell division (**Maclean** ; 1976 ; **Bode** *et al.*, 1986).

2. Some impressive examples of metaplasia (transdifferentiation) have been provided by **Schmid** and **Alder** (1984) in anthomedusa *Podocoryne*

carnea. Pieces of striated muscle tissue were removed from medusae and cultured *in vitro* in the presence of either **collagenase** or **hyaluronidase** enzymes. Following 6–8 digestion with these enzymes, the muscle cells separate from the umbrellar endoderm of the medusa. The cultures were then washed and incubated at reduced tempeatures in a low concentration of collagenase for a period of up to 6 days. During this period, early transdifferentiation to smooth muscle is observed, followed later by the appearance of, amongst others, gland cells, nematocysts and nerve cells. Both smooth muscle and gland cells are conspicuously flagellated in contrast to the non-flagellated striated muscle cells. In addition, the chracteristic cross-striations of the original striated muscle cells are lost. Although most of these cell types appear only after cell proliferation, and, therefore, following mitosis and cell division, the transition to flagellated smooth muscle cells can occur without an intervening cell cycle and is independent of DNA synthesis (**Maclean** and **Hall**, 1987).

26-9. SUPER REGENERATION AND HETEROMORPHOSIS

Regeneration after extensive damage, but without loss of the limb or tail, results in the formation of an additional limb. This is called **super regeneration** and due to this two tails tend to regenerate (form) in place of one in the lizards.

Sometimes when an entire regeneration field is removed, an organ different from that which have been removed, may develop. For example, in the shrimp, *Palinurus*, if the eye along with its ganglion is removed (both of which forming a field), an antenna-like structure is formed in place of the eye. This biological phenomenon is called **heteromorphosis**. In planarian, regeneration of a second head in place of posterior end, if the cut is too close to the anterior end of the body forms another example of heteromorphosis.

REVISION QUESTIONS

1. What is regeneration ? Summarise the regenerative capacity in various animal groups.
2. Give a schematic account of mechanism of regeneration of salamander limb.
3. What do you understand by epimorphosis and morphallaxis ?
4. Comment upon the following statements :

   (i) Regeneration is a specialized embryogenesis ;

   (ii) Regeneration declines with age in higher vertebrates.
5. What is metaplasia ? Explain this phenomenon by citing examples from vertebrates and non-chordates.
6. Can a vital organ like the heart show regeneration ? Explain which types of organs are most likely to possess regenerative ability.
7. After autotomy in lizards, sometimes two tails are regenerated in place of one. Is this epimorphic, morphalactic or heteromorphic regeneration ? or is it super regeneration ? Explain.
8. Write short notes on the following :

   (i) Regenerative field ; (ii) Metaplasia ; (iii) Wolffian regeneration ;

   (iv) Heteromorphosis.

27

METAMORPHOSIS

The egg of an animal even if provided with an enormous amount of food, is certainly smaller than the sexually mature adult. The egg, therefore, does not directly develop into an organism of size of an adult. Nature has adopted several methods to overcome this difficulty. Some of these methods can be broadly classified into the following three categories : **1. Direct development.** In many animals, the hatched organism resembles an adult, except that it is small in size. It is often called a **juvenile**. In such cases only growth and sexual development occur in the juvenile after its hatching, *e.g.*, certain Turbellaria (*Planaria*), *Ascaris*, *Pheretima*, reptiles and birds. In the eggs with large amount of yolk (macrolecithal eggs of reptiles and birds), the development is direct with no larval stage and no metamorphosis, because the supply of food is sufficient for complete development. **2. Viviparity.** In a few animals, the hatched organism grows and develops inside the body of mother. It draws its food supply and oxygen from her by the help of a placenta. The remaining growth of infant takes place after birth. For example, in eggs with very little yolk (*i.e.*, microlecithal eggs of the placental mammals), the development is direct without larval stage. **3. Indirect development.** In many animals, the egg develops into a strikingly different organism from the adult. Such an organism leads an independent free life and is known as **larva**. Thus, these eggs are with yolk but that is not sufficient for the complete development, *e.g.*, *Herdmania*, *Amphioxus*, frog, sea urchin and insects. The larval stage is free-swimming or free-living and feeds voraciously to store food material for further development. Compared with their adults, the larvae usually have different habitat and a different way of life. Larvae can be so different from the adult that in the past some were mistakenly given a species or a generic status. For example, the first instar larva of *Meloidae* (entomophagous, oil or blister beetle) was originally given the generic status of *Triungulinus*. Likewise, ammocete larvae of lampreys were given the generic names such as *Ammocoetes branchioles* (see **Hickman** *et al.*, 1979). The process of the transformation of a larva into an adult is termed **mtamorphosis** (Gr., *metamorphoun* = to transform).

27-1. DEFINITION OF METAMORPHOSIS

Metamorphosis is a post-embryonic extension of the developmental potential and involves dramatic changes in habit, habitat, morphology, physiology and behaviour of larva so that it is transformed into the adult having entirely different habitat and structure.

Metamorphosis is a wide spread developmental phenomenon which is usually associated with a dramatic change in habitat and consequent way of life, such as the change from a planktonic to a benthic existence in the sea-urchin, from an aquatic to a terrestrial existence in frogs and toads, and from non-flying to a flying existence in insects. Such changes in environment and activities demand equally rapid transformations of the structure and function of the living machinery. Metamorphic change during the development cycle is an acceleration or condensation of essentially same basic process characteristic of most forms of development. Primarily, it consists of differential destruction of certain tissues, accompanied by an increase in growth and differentiation of other tissues.

27-2. OCCURRENCE OF METAMORPHOSIS

Metamorphosis occurs in most metazoan phyla, beginning with Porifera (amphiblastula) and ends with Amphibia (tadpole). Animals with indirect development exhibit a variety of larvae (Table 27-1) which undergo metamorphosis. The process of metamorphosis in different animal groups differ both in the nature of transformation and in the mode of causation of the whole sequence, so, it is not possible to describe their metamorphosis in common terms. Therefore, in this chapter, metamorphosis of amphibians and insects which serve as the representatives of Vertebrata and Non-chordata, respectively, will be considered.

Table 27-1. Representative types of larvae of different phyla which undergo the process of metamorphosis.

Phylum	Larval forms
1. Porifera	Amphiblastula
2. Coelenterata	Planula
3. Platyhelminthes –	
(a) Turbellaria (Polycladida)	Muller's larva
(b) Trematoda	Miracidium, sporocyst, redia, cercaria, metacercaria
(c) Cestoda	Onchosphere
4. Rhynchocoela (Nemertina)	Pilidium
5. Annelida	Trochophore
6. Mollusca	Trochophore and Veliger
7. Arthropoda –	
(a) Crustacea	Nauplius, metanuplius, zoaea, megalopa
(b) Insecta	Caterpillar, grub, maggot
8. Bryozoa (Ectoprocta)	Cyphonautes
9. Echinodermata –	
(a) Crinoidea	Doliolaria
(b) Holothuroidea	Auricularia
(c) Echinoidea	Echinopluteus
(d) Asteroidea	Bipinnaria, or brachiolaria
(e) Ophiuroidea	Pluteus
10. Hemichordata (Enteropneusta)	Tornaria
11. Ascidiacea (Urochordata)	Tadpole
12. Cyclostomata (*Petromyzon*)	Ammocoete
13. Amphibia	Tadpole

27-2.1. Significance of Larval Stage

In species with small eggs, a larval stage is interposed as feeding adaptation to support continued development of adult structures. As soon as adult tissues mature, the larva undergoes dramatic metamorphic changes as larval tissues are shed or transformed into an adult.

Once a larval stage is introduced, it is subject of **natural selection** and acquires adaptations for dispersal, site selection, and parasitism. Larvae possess anatomical and physiological adaptations for feeding, movement and attachment which successfully exploit new habitats and ecological niches, thereby avoiding competition with adults for limited resources. The larva evolves separately from the adult and may through natural selection and precocious maturity of reproductive function (neoteny) give rise to new group of organism.

27-2.2. Two Basic Types of Metamorphosis

Among chordates occur the following two types of metamorphosis :

1. Retrogressive metamorphosis. It occurs in ascidians where only the larvae (tadpoles) possess chordate features, the adults, adapted to a sessile existence, lose all larval locomotory organs and thereby give up all traces of chordate relationships (For details see Chapter 15).

2. Progressive metamorphosis. It occurs in amphibians (and insects) and includes change of simple larval organization into more complex organization of the adult (see Chapter 17).

27-3. METAMORPHOSIS IN AMPHIBIANS

Amphibians furnish the best example of metamorphosis in vertebrates. In them, metamorphosis incorporates following ecological, morphological, physiological and biochemical changes.

27-3.1. Ecological Metamorphic Changes

In amphibians, ecologically, metamorphosis is associated with a transition from an **aquatic** to a **terrestrial** mode of life. Superimposed on this change of environment, a change in feeding habit occurs in the **anuran amphibians** (*i.e.*, frogs and toads). For example, the tadpoles of most frogs and toads feed on vegetable matter—particles of plants, living and decaying—which they scrap off from submerged objects with the aid of the horny teeth surrounding their mouths. Some anurans are detritus feeders, and others, as the tadpoles of the clawed toad *Xenopus*, are plankton feeders. Adult frogs and toads are carnivorous, living on insects, worms and small vertebrate animals such as small frogs, birds, rodents, etc., the latter are catched, overpowered and swallowed by them. In the case of **urodele amphibians** (*i.e.*, salamanders, newts, etc.), however, there is no substantial change of diet, the larvae being as carnivorous as the adults though naturally they feed on smaller animals (mainly crustaceans and worms).

27-3.2. Morphological Metamorphic Changes

The changes in the organization or morphology of the animal during metamorphosis are in part progressive and in part regressive, and may be grouped into three categories : 1. The organs or structures necessary during larval life but redundant in the adults are reduced and may disappear completely. 2. Some organs develop and become functional only during and after metamorphosis. 3. A third group of structures, while present and functional both before and after metamorphosis, become changed so as to meet the requirements of the adult mode of life.

From morphological point of view, **anuran** amphibians undergo much extensive metamorphic changes because the degree of difference between their larvae and adults remain much profound. The metamorphosis of anuran amphibians includes following changes in organization :

1. Regressive metamorphic changes of Anura. Certain adaptive structures formed during embryonic development, namely, the ventral suckers, external gills and long tail along with fin folds of the tadpole larvae are resorbed during early functional life. Further, gill-clefts are closed, peribranchial cavities disappear, the horny teeth of the perioral disc are shed, as well as the horny lining of the jaws. Further, the shape of the mouth changes, the cloacal tube becomes shortened and reduced, some blood vessels are reduced, and also the lateral line organs of skin of tadpole disappear. These are merely resorptions of precociously formed structures, not a part of the true metamorphic event which occurs much later. They disappear when they have served their purpose.

2. Progressive metamorphic changes of Anura. The progressive or constructive metamorphic changes involve the progressive development of the limbs, which increase in size and differentiation. The **fore-limbs**, which in frogs develop under the cover of the opercular membrane break through to the exterior. The gill arches become modified into the **hyoid apparatus**. The **middle ear** develops in connection with the first pharyngeal pouch (which remains situated between the mandibular and the hyoid arch). The **tympanic membrane** develops and is supported by the circular tympanic cartilage. The eyes protrude on the dorsal surface of the head and develop eyelids. The **tongue** is developed from the floor of the mouth.

3. Organs which exist both in larva and adult of Anura. The organs which function both in the larva and the adult, but change their differentiation during metamorphosis, are primarily the skin, the intestine and the brain. The skin thickens, becomes more glandular by possessing multicellular mucous and serous glands ; attains an outer keratinized layer, and acquires a characteristic pattern of pigmentation. While, the intestine which is very long in tadpole, as in most herbivorous animals, becomes proportionately shorter and most of the coils which it forms in the tadpole become straightened out. The brain becomes more highly differentiated.

At the cellular level, cell modifications are evident in eyelids, limbs, lungs, tongue, eardrum, operculum, skin, liver, pancreas, and intestine. Probably no cell or tissue or organ of Anura remains unaffected during metamorphosis. Further in anuran amphibians the rate of metamorphosis remains very rapid and takes only a few days.

Urodele amhibians undergo less striking ecological and morphological metamorphic changes. For instance, in them the tail is retained, only the finfolds disappear. The branchial apparatus is reduced, the external gills become resorbed and the gill-clefts closed. The visceral skeleton becomes greatly reduced. The head changes its shape, becoming more oval. The progressive metamorphic changes occur chiefly in skin and the eyes. The skin becomes cornified and multicellular skin glands become differentiated. The pigmentation of the skin changes. The eyes bulge more on the dorsal surface of the head and develop lids. The legs and intestine suffer no change. In urodeles, thus, the metamorphosis is more gradual and may take up to several weeks.

It is worth mentioning that lungs do not undergo drastic changes during metamorphosis in both anurans and urodeles ; they develop very gradually and become fully funcational in the larval state. Long before metamorphosis, the larvae of frogs and salamanders start coming up to the surface and gulping air into their lungs and, thus, supplementing their aquatic respiration. This may be of considerable importance where the larvae develop in stagnant and polluted waters, as is often the case.

27-3.3. Physiological and Biochemical Metamorphic Changes

Hand in hand with ecological and morphological changes during metamorphosis go physiological and biochemical changes. The chief physiological changes of anuran amphibians are following : In frog tadpoles, the **endocrine function** of the pancreas (*i.e.*, productiron of insulin hormone by Islets of Langerhans of pancreas) starts at metamorphosis, and this is connected with the increased role of the liver in the turnover of carbohydrates (glycogen) (**Abeloos,** 1956). A profound change takes place in the **excretory mechanism**. In the tadpole, the end product of nitrogen metabolism is ammonia (**ammonotelism**) which is easily disposed off by diffusion in an aquatic medium, but which in a terrestrial animal might accumulate and become dangerous because of its high toxicity. Metamorphosed frogs, however, excrete most of their nitrogen in the form of urea (**ureotelism**) and only small amounts of ammonia. The change of ammonotelism to ureotelism occurs at the late stage of metamorphosis and is associated with changed function of the liver which performs the synthesis of urea (**Munro**, 1939). The biosynthesis of urea from bicarbonate and ammonia involves a number of enzymes, and their activity which is found to be related with the level of thyroxine hormone in blood, rises abruptly during metamorphic climax (see **Turner** and **Bagnara**, 1971).

Further, the visual pigments of tadpoles, are porphyropsin (retinene$_2$). During metamorphosis, there is a shift to the use of rhodopsin (retinene$_1$) as a visual pigment. Likewise, during metamorphosis larval haemoglobin is replaced by adult haemoglobin (see **Turner** and **Bagnara**, 1971).

The reduction of the gills and tail is affected by **autolysis** of the component tissues of these organs, with active participation of amoeboid macrophages which phagocytose the debris of the disintegrating cells. The same mechanism occurs in urodele amphibians when their external gills and fin folds are resorbed during metamorphosis.

Since destructive processes play such a considerable role in metamorphosis and since, in addition, food intake may be interrupted during the crucial part of the transformation especially in tadpoles, the mass of the body at the end of metamorphosis is smaller than at the beginning (**"degrowth"**). The reduction of body mass is not only due to loss of some parts (gill, tail, etc.), but the remaining parts, excepting the actively growing organs, also appear to shrink during metamorphosis ; the head and trunk of metamorphosed amphibians are smaller than in larvae just before the commencement of metamorphosis.

Lastly, according to **Freiden** (1961), the biochemical metamorphic alterations may have direct or indirect adaptive value relating to the transition from freshwater to land. Among the most important adaptive changes are the shift from ammonotelism to ureotelism, the increase in serum albumin and other serum proteins, the alteration in the properties and biosynthesis of haemoglobins. The development of certain digestive enzymes and augmentation of respiration also contribute to the success of the differentiation process. During metamorphosis, there are many additional important chemical developments which may be secondary to the primary morphological or cytological transformations which aid in the adjustment to the land. These include changes in carbohydrate, lipid, nucleic acid and nitrogen metabolism. Major modifications occur in water balance, visual pigments (vitamin A), pigmentation, and tail metabolism.

All the metamorphic changes of anuran amphibians have been summarized in Table 27-2.

According to **Grant** (1978) the sequence of events during anuran metamorphosis is divided into three distinct periods : 1. A **premetamorphic period**. This is first phase which is marked by tadpole growth with little morphogenetic change except for hindlimb growth. 2. **Prometamorphic phase**. During this second phase the body grows at a reduced rate as hindlimbs grow. The ratio of hindlimb length to body length increases rapidly. 3. **Metamorphic climax**. This is third and most dramatic phase and is marked by cessation of growth accompanied by rapid changes in limbs, tail, head and internal organs.

27-3.4. Time of Metamorphosis of Amphibians

Frogs metamorphose after various periods of growth according to the species, and it was early recognized that the attainment of a critical species-specific size, was much important than the duration of growth. Thus, in nature, American bullfrog tadpoles metamorphose at the end of the third summer-growth season in the north but at the end of the second summer-growth season in the south, after having attained a certain size, the time required depending on the mean environmental conditions. Further, it has also long been known that addition of iodine to the water or feeding with thyroid gland tissue causes metamorphosis to occur earlier at a smaller size, while elimination of iodine from the diet postpones it. From the first, therefore, the iodine-containing **thyroxine** of the thyroid gland has been studied intensively with regard to the timing of metamorphosis and the metabolic effect of the hormone.

27-3.5. Hormonal Control of Amphibian Metamorphosis

The concurrent changes in so many parts of the animal's body during metamorphosis suggest the existence of some common cause for all the transformations. It has been found that this common cause is a hormone released in large quantities from the thyroid gland of the animal which entering the stage of metamorphosis. The first indication of this was obtained when **Gundernatsch** (1912) fed some frog tadpoles on dried and powdered sheep thyroid gland and observed that they metamorphosed precociously. The proof that the thyroid hormone is actually the cause of metamorphosis in normal

Table 27-2. Summary of some metamorphic changes in anurans (**Turner** and **Bagnara**, 1971).

System	Larva	Adult
1. Locomotory	Aquatic; tail fins.	Terrestrial; tailless tetrapod.
2. Respiratory	Aortic arches; aorta; anterior, posterior, and common cardinal veins.	Carotid arch; systemic arch; pulmocutaneous arch; juglar veins, caval veins.
3. Nutritional	Vegetarian : long spiral gut–intestinal symbionts, carbohydrases; small mouth–horny jaws, labial teeth.	Carnivore : short gut—proteases; large mouth–long tongue.
4. Nervous	No corneal reflex, lack of nictitating membrane, porphyropsin; lateral-line system–Mauthner's neurons; branchial columella.	Corneal reflex— development of ocular muscles, nictitating membrane, rhodopsin; loss of lateral line system–degeneration of Mauthner's neurons, enlargement of mesencephalic V nucleus, enlargement of lateral motor columns, tympanic columella, tympanic membrane.
5. Excretory	Mesonephros : largely ammonia, some urea (ammonotelic).	Mesonephros : largely urea, high activity of enzymes or ornithine-urea cycle (ureotelic).

development, was further given by the following two experiments : In an experiment when the rudiment of thyroid gland was removed in frog embryos in the tail bud stage, then, it was found that the operated tadpoles were fully viable and showed normal growth but failed to metamorphose (**Allen**, 1918). The thyroidless tadpoles continued to grow and attained a much greater size than normal (*i.e.*, a total length up to 123 mm instead of about 60 mm as at the beginning of metamorphosis in normal tadpoles). It was, thus, proved that metamorphosis cannot set in without a stimulus coming from the thyroid gland. In final experiment, thyroidless tadpoles were provided with thyroid hormone either by feeding them on thyroid gland (**Allen**, 1918) or by immersing them in water containing soluble extracts from thyroid glands. The tadpoles treated in this way immediately proceeded to metamorphosis, thus, showing that their own thyroid glands are not necessary so long they are supplied with thyroid hormone (**Allen**, 1938). Similar experiments were carried out on urodele amphibians by **Marx** (1935).

Thyroid is found to contain a protein called **thyroglobulin** which has a molecular weight of about 675,000. The thyroglobulin has two tyrosine and iodine rich active compounds or factors namely **L-thyroxine** (or **T_4 hormone** or 3, 5, 3′, 5′- **tetraiodothyronine**) and **T_3 hormone** or 3, 3′ 5 - **triidothyronine** (Fig. 27.2 and 27.3). T_4 hormone is produced in much greater quantities than the T_3 hormone. In peripheral tissues T_4 hormone is converted to T_3 hormone which is the biologically active form of thyroid hormone (TH) (**Kent**, 1987).

It has been tested whether **iodine** alone can cause metamorphosis in amphibians, and positive results have been obtained either by keeping frog tadpoles or newt larvae in water containing the element iodine, or by injecting animals with iodine or by implanting iodine crystals into the body cavity. The element iodine caused metamorphosis even in thryroidectomized axolotl.

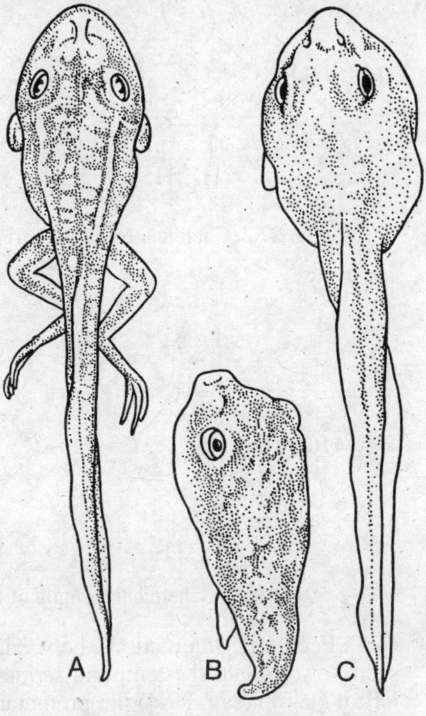

Fig. 27.1. Hormones and metamorphosis : A—Normal metamorphic stages; B—Premature metamor- phosis following exposure of young tadpoles to thyroxine; C—Inhibition of metamorphosis following removal of either the thyroid or the pituitary gland (after Berrill, 1971).

1. Mode of hormonal action. The amphibian metamorphosis is under neuroendocrine control, involving neurosecretory cells in the brain (the **hypothalamus**) and two endocrine glands, the **pituitary** (anterior pituitary or adenohypophysis) and the thyroid. The trigger to metamorphosis may be an environmental signal affecting the larval brain through the nervous system, or there may be an endogenous ‘clock’ in the hypothalamus. In a way, hypothalamus integrates the information received from body with the environmental information. **Neurosecretory cells** in the hypothalamus are stimulated to produce **TRF** or **thyroid-releasing-factor** (a brain hormone) which enters the pituitary portal circulation. The anterior pituitary, the target tissue for this brain hormone, is activated and releases **TSH** or **thyrotropic-stimulation-hormone** which stimulates the dormant thyroid gland to secrete two thyroid hormones, namely T_4 and T_3 hormones. Both of these hormones enter the circulation and gradually increae in titer.

Anterior pituitary of amphibians also secrete the **prolactin** hormone which similar to insectan "juvenile hormone" inhibits or delay thyroxine - induced metamorphosis, sustaining larval characteristics and promoting their growth.

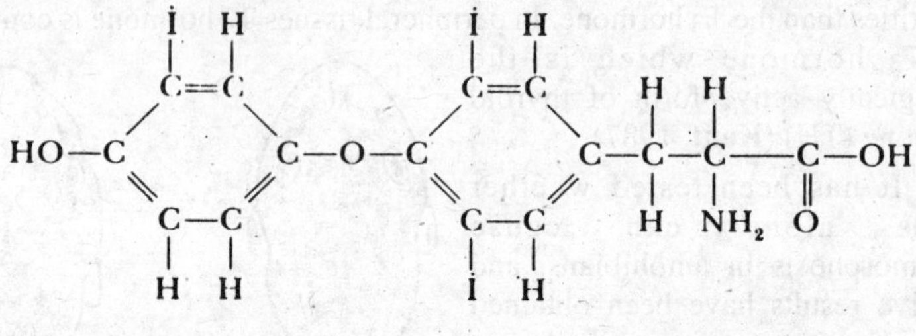

Fig. 27.2. Chemical formula of tri-iodothyronine (after Balinsky, 1970).

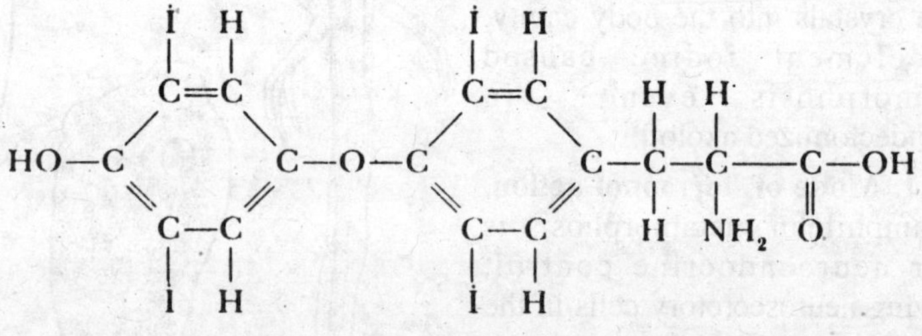

Fig. 27.3. Chemical formula of thyroxine (after Balinsky, 1970).

Feedback interactions between thyroid, pituitary and hypothalamus seem to regulate the temporal sequence of events. In a model proposed by **Etkin** (1967) (Fig. 27.4) the premetamorphic phase is marked by high titers of prolactin which favour growth and very low but detectable levels of thyroid hormones. Tissues are sensitive to thyroid hormone since they respond precociously if a tadpole is treated with thyroxine ; cells must have thyroxine receptors on their plasma membrane at this time. The pituitary gland is also able to secrete TSH if properly stimulated during this period. This means that onset of metamorphosis is limited by low levels of thyroid hormone. The hypothalamus is immature during the premetamorphic phase but seems to become active just before the prometamorphic phase. Activation by hypothalamus also depends on development of the portal circulation connecting hypothalamus and pituitary gland. Gradually, the hypothalamus become sensitive to low levels of thyroxine in the circulation and begins to secrete thyroid-releasing factor (TRF) which stimulates the pituitary gland to secrete TSH. The thyroid is activated, more thyroid hormone is secreted, and the prometamorphic phase is initiated. At the same time, a prolactin inhibitor is secreted by the hypothalamus, prolactin concentration decreases, and growth declines as metamorphic changes occur.

At metamorphic climax, there is a tenfold increase in thyroid hormone concentration and this initiates the rapid changes that follow. The tadpole tail is quickly resorbed as are the horny teeth and gills. The eyes shift from a lateral position to almost a dorso-lateral position on the head. Larval pigment in the retina is replaced by adult visual pigment, rhodopsin, better adapted to a terrestrial mode of vision. Meanwhile other tissues such as liver, kidney, red blood cells, etc., undergo constructive changes. Each tissue responds differently to the same hormonal signal ; its response pattern is programmed into its genome (**Grant**, 1978).

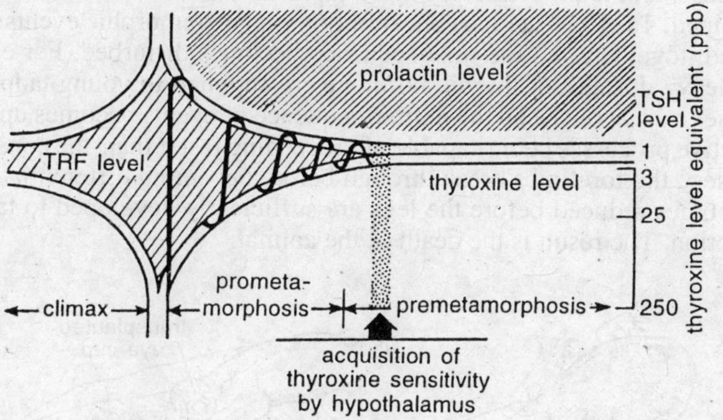

Fig. 27.4. Etkin's model illustrating the interaction of endocrine factors (or hormones) in determining the time and pattern of anuran metamorphosis (after Berrill, 1971).

In fact, a tadpole's thyroid gland undergoes developmental changes that correlate with metamorphic events. The thyroxine colloid begins to accumulate in the follicles prior to metamorphosis, and the follicular epithelium thickens at the time of metamorphic climax. Moreover, there are certain compounds which interfere with specific steps of thyroid function and inhibit metamorphosis. For instance, **thiourea** prevents metamorphosis when given to tadpoles. It does not, however, inhibit metamorphosis if given simultaneously with thyroxine to thyroidectomized tadpoles, which means it acts directly on the thyroid gland without antagonizing hormone action at the target site. It is not an anti-metamorphic agent ; rather it affects thyroid production of hormone.

2. Tissue reactivity in amphibian metamorphosis (or threshold and rate of response). The metamorphosis involves alteration of larval tissues which normally acquire sensitivity to thyroid hormones long before significant quantities of such hormones are released in to the circulation (**Kollros**, 1961). In general, structures that metamorphose early are more sensitive than those that undergo change later. However, by using greatly lowered hormone concentrations for long periods of time, up to a year, **Kollros** has shown that different tissues have different thresholds. Some tissues require much greater doses of hormones than others. For example, forelimbs require a stronger dose of thyroid hormone in order to break through than do hindlimbs and even greater dose is necessary for tail resorption. Thus, each tissue type is responsive to a certain threshold dose for reactivity ; once this threshold is reached, the tissue responds.

Successive metamorphic events have different and somewhat higher threshold from preceding ones. For example, the threshold for the onset of corneal reflex is lower than that for perforation of the operculum, and the threshold for perforation of the operculum is lower than the threshold for gill absorption. In the frog, however, as hormone concentrations increase, tissue responses become progressively more rapid until maximum rates of change are attained. In effect, at high concentrations, all metamorphic events become crowded, together, and the time sequence becomes disturbed. For example, when heavy dosages of thyroid hormone are supplied to young tadpoles, all processes start at once and the normal sequence of events becomes upset. The destructive processes being capable of proceeding faster than the constructive processes ; the forelimbs break through before becoming differentiated ; the tail becomes reduced before the legs are sufficiently developed to take over locomotion. The result is the death of the animal.

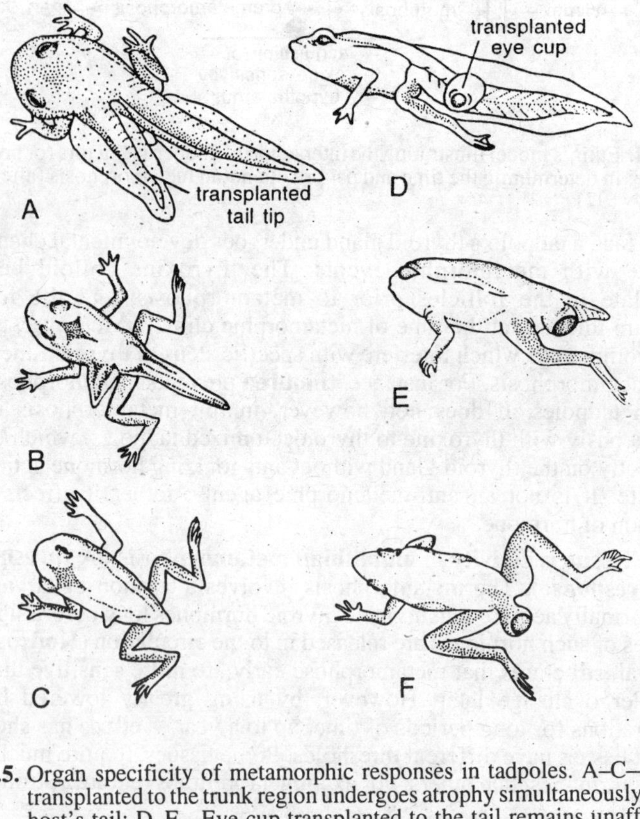

Fig. 27.5. Organ specificity of metamorphic responses in tadpoles. A–C—Tail tip transplanted to the trunk region undergoes atrophy simultaneously with the host's tail; D–F—Eye cup transplanted to the tail remains unaffected by the regressing tail tissue (after Berrill, 1971).

The concept of difference in tissue sensitivity is well-illustrated by an early experiment (Fig. 27.5), in which an eye cup was transplanted to a tadpole tail where it differentiated. During metamorphosis the eye moved forward as the tail reasorbed and finally come to rest in the sacral region when metamorphosis was complete. Transplanted limbs and transplanted kidney tumour likewise remain unaffected by the degenerative processes in the surrounding tail tissue.

Most important aspect of hormonal control of amphibian metamorphosis is that a single hormone, the low-molecular-weight compound **thyroxine**, evokes multiple responses in diversified tissues. Responses are specific although the inducer varies only quantitatively. Moreover, they are both constructive and destructive depending on the target tissue. Thus, in response to tri-iodothyronine biosynthesis of nucleic acids is decreased in tail but increased in the liver. Similarly, thyroxine induces rapid aging and destruction of the red blood cells of the metamorphosing tadpoles which carry tadpole-type haemoglobin, while simultaneously it stimulates the development of cells that synthesize the adult frog type of haemoglobin exclusively. The agent is the same, but the response is cell death and cell proliferation respectively, in the two erythrocytic cell populations. As thyroid hormone remains associated with the level of basal metabolism in higher vertebrates, so, it is suggested that an increase in oxidative metabolism may be discriminating factor which underlies the double mode of activities of thyroid hormones. However, there is no increase in oxygen uptake during spontaneous metamorphosis. On the other hand a flood of metabolites, resulting from tissue destruction, is immediately available for rapid protein and nucleic acid synthesis.

Experiments with tail resorption during metamorphosis of *Xenopus* suggested that the anterior pituitary of the tadpole releases a factor that interferes with the action of thyroxine on tail-disc resorption *in vitro*, suggesting that an **antagonism** exists between a pituitary growth factor, possibly a prolactin or a growth-hormone-like-molecule, and thyroid hormone, and that this interaction takes place peripherally through direct interactions at the site of target cells.

3. Regulation of molecular events by thyroid hormones. Thyroid hormones are known to affect the process of protein synthesis at the levels of transcription and translation and to have a role in cytodifferentiation.

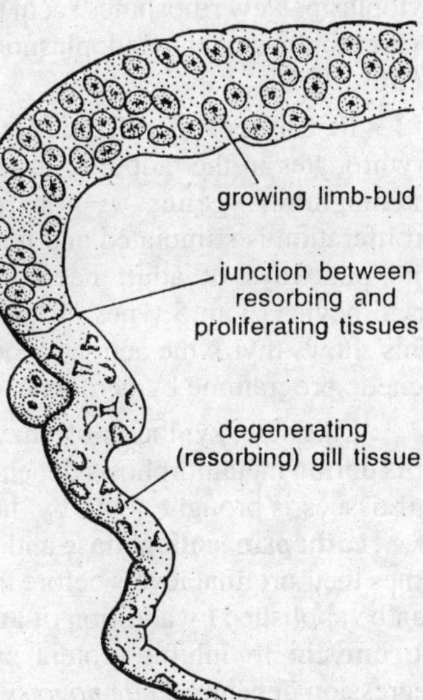

Fig. 27.6. Differential response of limb-bud and gill-tissue to thyroxine : the line separating the two districts. One growing, the other degenerating is very sharp (after Berrill, 1971).

A. Control of protein synthesis. Liver cells or hepatocytes experience important metabolic changes during metamorphosis. For example, the production of urea depends upon enzymes of the **arginine-ornithine cycle** (**Krebs urea cycle**) (**Brown** *et al.*, 1959) and these are low or absent in young tadpole liver but are actively synthesized during metamorphosis (Fig. 27.7). While urea cycle enzymes are synthesized, other liver enzymes either do not respond or decrease. Within the same hepatic cells, each enzyme system (and the encoding genes) respond to a common hormone signal.

Thyroxine stimulation of liver protein synthesis is preceded by a long latent period (almost 100 hours or about 5 days) a time when RNA synthesis is stimulated in nuclei (Fig. 27.8). The initial cellular response to the hormone is nuclear RNA synthesis, suggesting hormone action at the level of **transcription**. Increased RNA synthesis results from enhanced activity of **RNA polymerase enzymes**. Though all types of RNA are equally affected by thyroxine, the bulk of the newly synthesized RNA found in the cytoplasm is **ribosomal RNA (rRNA)**. **Messenger RNA (mRNA)** is also produced, but in smaller amounts.

The thyroxine hormone also acts at the **translational level** ; it stimulates **polysome** accumulation and changes the distribution of **ribosomes** in the cytoplasm. New ribosomes seem to bind more tenaciously than old ribosomes to membranes of the endoplasmic reticulum before protein synthesis begins (**Tata**, 1970).

B. Control of differentiation. Stem cells continuously generate the erythrocytes in the tadpole ; in low levels of thyroxine, they express **larval haemoglobin** genes. As the thyroxine level increases, stem cell's proliferation is stimulated and new population of **erythroblasts** are diverted into pathways of adult haemoglobin synthesis. Meanwhile, old tadpole erythrocytes (with 5 types of haemoglobin) persist for a time until they die. This shows thyroxine action at the transcription level, "selecting" a specific genetic programme by derepressing adult haemoglobin genes.

Likewise, synthesis of enzymes needed for regression of tail, gut and gills during metamorphosis depends upon thyroxine stimulation. Histolysis of tail tissues is brought about by the action of a variety of hydrolytic enzymes (*e.g.*, **cathepsin**, **collagenase** and **phosphatase**) which are synthesized to 200 times their original levels before the tail is resorbed. The effect of T_3 hormone can be abolished by addition of **actinomycin D** to inhibit RNA synthesis and **puromycin** to inhibit protein synthesis. These experiments indicate that regression depends on *de novo* synthesis of hydrolytic enzymes and does not result from hormone activation of lysosomes as was originally thought. Evidently metamorphic **cell death** as well as **cell proliferation** and **cell differentiation** require a genetically determined synthesis of specific hydrolytic enzymes (see **Grant**, 1978).

4. Acquisition of metabolic competence. The ability to respond to thyroxine is not always present in a cell. A cell acquires a **competence** to react to hormone during the course of its development. We find that the amphibian embryo and larva are not always responsive to thyroxine ; they develop metamorphic competence as they mature.

Studies on the action of oestrogens on uterine tissues have led to a concept of a target tissue where cells possess hormone **receptor sites** on their surfaces or in cytoplasm which recognize and "capture" circulating hormone. The receptor- hormone complex is conveyed to the nucleus where it regulate genome expression. An alternative model of **Sutherland** (1956) stressed the involvement of cAMP in hormonal regulation of cell metabolism (see Chapter 24). **Tata** (1970) tried to find out that whether the acquisition of metamorphic competence by amphibian larval cells (*i.e.*, *Xenopus* larva) require synthesis of specific thyroxine receptor protein ?

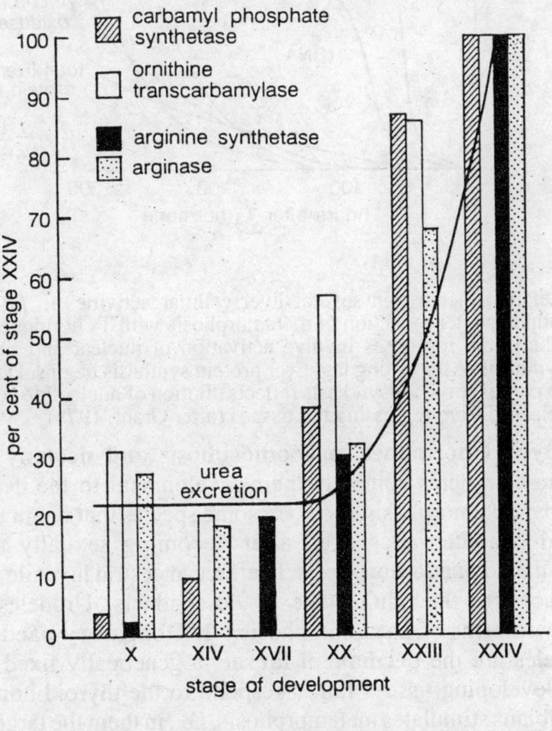

Fig. 27.7. Activities of Krebs-urea cycle enzymes and urea excretion during metamorphosis of the liver in *Rana catesbiana.* Assay for arginine synthetase reflects activity of rate-limiting component. All enzymes are coordinately synthesized during this period (after Grant, 1978).

To determine when embryonic cells become competent to respond to thyroid hormone, **J.R. Tata** exposed different developmental stages to radioactive thyroxine and followed its binding in the tissues. The embryo is insensitive to the hormone before hatching but soon shows an increase in binding capacity that increases slowly throughout larval life. There are evidences which indicate that embryonic tissues acquire thyroxine binding sites toward the end of embryogenesis.

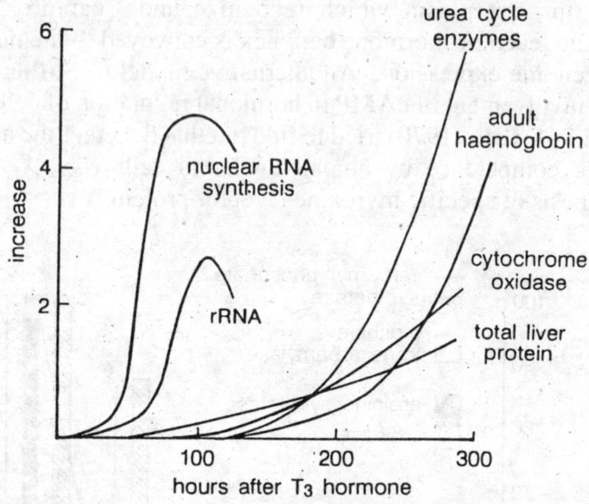

Fig. 27.8. Relative increase in several liver cellular activities in *Rana catesbiana* tadpoles after induction of metamorphosis with T_3 hormone (**Tata,** 1967). The initial responses involve activation of nuclear and ribosomal RNA synthesis. After a long lag, liver protein synthesis begins. Downward trend of curves for RNA synthesis reflects dilution of nucleotide pool as these are released from regressing tail tissues (after Grant, 1978).

5. Thyroid hormones, paedomorphosis and neoteny in Urodela. Whereas most urodeles complete their development to the definitive adult body form by metamorphosis, there are some species that retain their gills and other larval characteristics even after becoming sexually mature. This condition is called **paedomorphosis**. In effect, ancestral juvenile features have been displaced to the adult stage of descendants. Urodeles exhibit the following two forms of paedomorphosis : **1. Obligatory paedomorphosis.** Some urodeles are the **permanent larvae**, a genetically fixed condition in which the developing tissues fail to respond to the thyroid hormone that in other amphibians stimulates metamorphosis, *i.e.*, in them the target tissues lose competence to respond to TH. This condition is called **obligatory paedomorphosis**. Examples of such sexually mature permanent larvae are mud pupies of the genus *Necturus*, which live on bottoms of ponds and lakes and keep their external gills throughout the life, and *Proteus* (European blind cave salamander). **2. Facultative paedomorphosis.** Some urodeles such as *Ambystoma* and *Triturus*, become sexually mature and breed in the larval state, but, unlike the permanent larvae, they may metamorphose to adults if environmental conditions change. This is called **facultative paedomorphosis**. For example, the American axolotl *Ambystoma tigrinum* (tiger salamander) remains aquatic, gill-breathing, and fully reproductive larval form unless the water begins to dry up ; then it metamorphoses to an adult, loses its gills, develops lungs, and assumes the appearance of an ordinary salamander. Thus, in these forms, paedomorphosis is a handy adaptation for remaining aquatic in an environment where water is plentiful. This particular kind of paedomorphosis is called **neoteny** (Gr., *neos* = young + *teineu* = to extend) because it is achieved by delaying somatic development, but not sexual development. Neoteny is also exhibited by *Triturus vulgaris*, in which larval may not undergo metamorphosis for one or two winters. Axolotls can be made to metamorphose by treating them with the thyroid hormone, **thyroxine**. Recent findings suggest that the biological clock of hypothalamo-hypophyseal-thyroid axis is somehow affected, *i.e.*, for some reasons the pituitary gland fails to become fully active in neotenous forms and does not release the thyrotropin TSH, which is required to stimulate the production of thyroxine by the thyroid gland (**Hickman** *et al.*, 1979).

Lastly in some urodeles such as *Siren* (mud eel) and *Amphiuma means* (Congo eel), the metamorphosis is partial. This means that only some and not all tissues have lost their ability to respond to thyroxine.

6. Role of prolactin in Urodela. Etkin (1968) has suggested an evolutionary significance of prolactin hormone for urodele amphibians. The common newt after a typical first metamorphosis which produces the land form, and after a period of growth on land, undergoes a second metamorphosis for the return to water in order to breed. This second metamorphosis has been shown to be under the influence of prolactin, presumably resulting from a shift in the balance between pituitary factors, prolactin and TSH rather than from an activation of one of them. In other words, the first metamorphosis is induced by a shift in favour of TSH, the second metamorphosis by a return to the predominance of prolactin.

27.4. INSECT METAMORPHOSIS

Insect metamorphosis is a special form of moulting. Usually moulting comprises mainly the casting off of an old cuticle and the acquisition of a new one. This change of cuticle is necessary for **growth** and **morphogenesis** to occur because the fully sclerotized cuticle is rigid and does not expand. At each moult the epidermal mitosis allow the production and expansion of a new soft cuticle. The cuticle in the different stages has a different structure. For example cuticles of larva, pupa and adult are composed of different layers, each with a visibly different pattern of bristles and hairs. Besides moulting, insectan metamorphosis also involves major internal changes.

The interval between the two moults is known as a **stadium** ; and the form an insect takes in a stadium is an **instar**. Growth and morphogenesis tend to occur during the intermoult period. All insects undergo several moults after hatching from the egg and before emerging as adults (imagoes). The number of moults in an insect species is usually fixed but is not entirely constant. Metamorphosis normally refers to last moult, from which an imago (adult) emerges. Generally, insect metamorphosis is associated with the evolution of wings which are restricted to the reproductive stage where they can be of the most benefit. Like amphibian metamorphosis it involves both **cell death** and **differentiation** in many systems, including muscles, alimentary canal, Malpighian tubules, etc. The process of insect metamorphosis has been extensively studied in the blood-sucking bug *Rhodnius* (Hemimetabola; Hemiptera) and certain silkworms, *Bombyx mori* (silk moth ; Holometabola ; Lepidoptera).

Moulting cycles. Since the cuticle is a mosaic made up of the contributions of each individual epidermal cell, the growth activity of each cell at each stage of the whole development history, is firmly registered in the characters of the little patch of cuticle that it lays down. And since the cells comprising the single epidermal sheet not only lay down cuticle, a process incompatible with cell division but are also responsible for growth and change of form, which requires cell multiplication, these two processes do not occur at the same time. Accordingly, the growth of insects takes place in cycles, in which mitosis and cellular growth alternate with the deposition of a new cuticle and the shedding of the old, *i.e.*, in **moulting cycle.** Therefore, as growth proceeds insects periodically discard the old cuticular structures and produce a new set irrespective of whether, apart from growth, the new and the old need to be alike.

27-4.1. Types of Insect Metamorphosis

Not all insects undergo metamorphosis, but most of them do in some from or other :

1. Epimorphosis. The wingless insects such as silverfish and springtail (Order Apterygota) do not undergo the process of metamorphosis. They are said to be **ametabolous** or **ametabolic insects** (Note : **Schmidt - Neilsen**, 1983, has argued that term ametabolous is less desirable because it mixes a Greek prefix with a Latin ending, therefore, he preferred the term ametabolic). Ametabolic development is considered to be primitive, for the insect goes through moults without changing its appearance, merely getting bigger. In other words, such insects are said to undergo **epimorphosis**, *i.e.*, young or juvenile are similar to adults except in size. The stages are **egg**, **juvenile** and **adult**.

2. Gradual metamorphosis. Some insects (Order Exopterygota) undergo a type of **gradual** or **incomplete metamorphosis** and are called **hemimetabolous** or **hemimetabolic insects**. These include the grasshoppers, cockroaches, cicadas, and mantids, which have terrestrial youngs, called **nymphs** ; and mayflies, stoneflies, and dragonflies which lay their eggs in water and have youngs, called **naiads**. Nymphs look similar to adults but they lack the wings that gradually develop during the several moults. Their wings develop externally as bud-like outgrowths in the early instars and increase in size as the animal grows by successive moults and becomes winged adults. The aquatic naiads have tracheal gills or other modifications for aquatic life. The stages are **egg**, **nymph** (or **naiad**) (several instars), and **adult**.

3. Complete metamorphosis. Approximately 88 per cent insects undergo a complete metamorphosis. Insects undergoing complete metamorphosis are called **holometabolous** or **holometabolic insects** and they belong to order Endopterygota, *e.g.*, beetles, ants, bees, butterflies, moths, flies, etc. Complete metamorphosis separates the physiologic processes of **growth** (larva) from those of **differentiation** (pupa) and **reproduction** (adult). Each stage functions efficiently without competition with the other stages, for the larvae often live in entirely different surroundings and eat different foods from the adults. The worm-like larvae which usually have chewing mouthparts, are known as **caterpillars**, **maggots**, **bagworms**, **fuzzy worms**, **grubs**, and so on. After a series of instar stages during which the wings are developing internally, the larva forms a case or **cocoon** about itself and becomes a pupa or chrysalis (in butterfly), a non-feeding stage in which many insects pass the winter. When the final moult occurs over winter the full-grown adult emerges, pale with wings wrinkled. In a short time the wings expand and harden, and the insect is on its way. The stages are **egg**, **larva** (several instars), **pupa** and **adult**. The adults do not undergo further moulting.

Thus, in holometabolic insects body forms, mouth-parts and sense oragans (particularly in the absense of compound eyes) are all different from those in the adult, indicating the quite separate ways of life of the larva and adult. For example, the caterpillar larva of butterfly is wingless and eats leaves, while the adult drinks the nectar from flowers. The mosquito larva lives in ponds and eats algae and Protozoa, while adult sucks the blood of man and other mammals. The caterpillar crawls about, eat voraciously and moults several times, each time becoming a larger larva. The last larval moult forms a pupa, a form which neither moves nor feeds. Moth and butterfly larvae spin a **cocoon** and pupate within that. During pupation all the structures of the larva are broken down and are used as the raw materials in the development of the adult animal. Each part of the adult (legs, wings, eyes, etc.) develops from the groups of cells called **imaginal buds**, or **discs** which develop directly from the egg. They have never been a functional part of the larva but remain more or less **quiescent** during the larval period. During the pupal stage these discs grow and differentiate into the adult structures but remain collapsed and folded. When the adult **hatches** out of the pupa case, blood is pumped into these collapsed structures, they unfold and inflate, and chitin is deposited to make them hard.

27-4.2. Genetic Control of Metamorphosis

Complete metamorphosis requires three sets of genetic instructions : larval, pupal and adult. Larvae grow through a series of moulting periods and transform into quiescent but morphologically distinct pupae. During the pupal stage, larval gene sets are turned off and pupal genes are expressed. Later, all larval and pupal tissues degenerate, while adult primordia read the final set of genetic instructions and differentiate into the adult body plan.

The larval developmental programme is precocious, insuring rapid emergence of a functional feeding organism at the end of embryognesis. Pupal and adult programmes depend upon nutrients furnished by the larva. All programmes, however, are **determined** early in embryogenesis, usually when cleavage nuclei migrate to the periphery and form a cellular blastoderm.

Though larval and adult tissues are determined at the same time, they develop at different rates. Larval cells differentiate first, but adult primordia (*i.e.* imaginal discs) persist throughout larval life until they receive the metamorphic signal to differentiate.

27-4.3. Imaginal Discs

Imaginal discs are small pockets of determined but undifferentiated cells that are set aside within the body of the larval insect to differentiate into special parts of the adult insect at the time of metamorphosis. For example, in *Drosophila* there are many imaginal discs, namely **labial**, **clypeolabral**, **eye - antennal** (Fig. 27.9), **prothoracic**, **wing**, **halteres**, **leg** (three pairs) and **genital**. Special **histoblast cells** (or **nests** or **discs**) such as **spiracular**, **anterior dorsal**, **posterior dorsal** and **ventral histoblast** (7 pairs of each) produce the abdominal epidermis.

The mode of development of imaginal discs varies from species to species and also from organ to organ. Imaginal discs are sometimes formed during late embryonic development, and their cells are separate from the prospective larval cells, for example, in *Drosophila* and other diptera. In some insects, the imaginal discs are derived from larval cells later during development, for example, the **leg discs** of *Pieris* (Lepidoptera) are not distinguished from the larval epidermis till the IV larval instar. The discs originate as thickenings and later as infoldings of the embryonic or larval ectoderm.

Thus, in *Drosophila*, imaginal discs arise at the end of cleavage after several thousand nuclei migrate to the periphery to form a cellulated blastoderm (see Chapter 10). Most blastoderm cells develop into larval tissues but some are determined to become adult tissue after metamorphosis. These (usually small groups of 2 to 30 cells) are isolated as primordia for organ - specific imaginal discs and remain quiescent during embryogenesis. After hatching, these proliferate actively into visible imaginal discs, single or paired anlages (the primordia) distributed throughout the larval body. Disc cells proliferate during each moulting cycle (*e.g.*, three larval instars in *Drosophila*) until metamorphosis when they differentiate into specific adult structures.

Each imaginal discs consists of a single epithelial cell layer that is somewhat folded ; this, together with a narrow lumen, forms a hollow pocket in the larval integument. The outer surface of each disc is covered by a **basal lamina**. Each disc cell is about 5 μm in diameter and has a large nucleus and basophilic cytoplasm, with the possible exception of some cells in the eye discs, disc cells appear to be uniform and undifferentiated. These cells are, however, fully determined to follow their respective fates at the time of normal metamorphosis, or on transplantation in a III instar larva undergoing metamorphsis. The disc cells remain undifferentiated during the entire larval life. They do not acquire the competence to respond to the metamorphic stimuli till late in the II larval instar. If the disc cells are transplanted into the haemocoel of adult *Drosophila*, they undergo proliferation but not differentiation.

During pupal life, much of the larval integument is destroyed by autolysis while each disc becomes everted through its stalk so that the cells previously enclosed within the disc become outer cells. Dramatic changes in cell shape yield the resulting structures of essentially hollow legs, antennae, and so on. Thus, it is during this period of eversion that the long-standing commitment of these cells comes to be revealed, as they adopt characteristic shapes and go to form parts of wings, bristles, or simply surface architecture. All the products of the various discs finally fuse together, so that the elaborate exoskeleton of an adult insect is essentially a mosaic of cell population each of which has resulted from a distinct imaginal disc (**Maclean** and **Hall**, 1987).

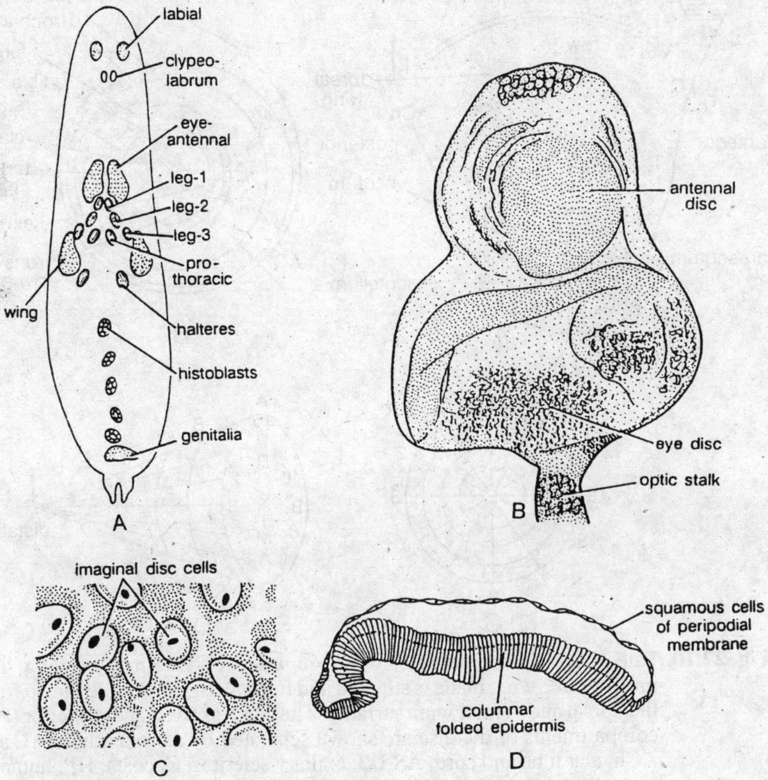

Fig. 27.9. Imaginal discs of *Drosophila melanogaster*. A—Location of different imaginal discs; B—Eye antennal disc in late third-instar larva (before pupation); C—Typical imaginal disc cells; D—Cellular structure of an imaginal disc (after Grant, 1978, Goel 1984).

As previously mentioned, cells of the imaginal discs are determined in the blastoderm, at least in terms of general fate, *i.e.*, wing, leg, or antenna. This determination is very stable since almost indefinite growth and culture does not change the state of commitment of the cells (**Hadorn**, 1978). But a progressive refinement of determination proceeds as each disc grows, so that distinct portions of any one disc come to have a fixed fate for a subsection of the appropriate organ. Thus, as shown in Fig. 27.10, **fate maps** can be constructed for leg (or wing) discs, for example, to produce tibia, cells in another portion produce femur, or trochanter, or coxa, and so on.

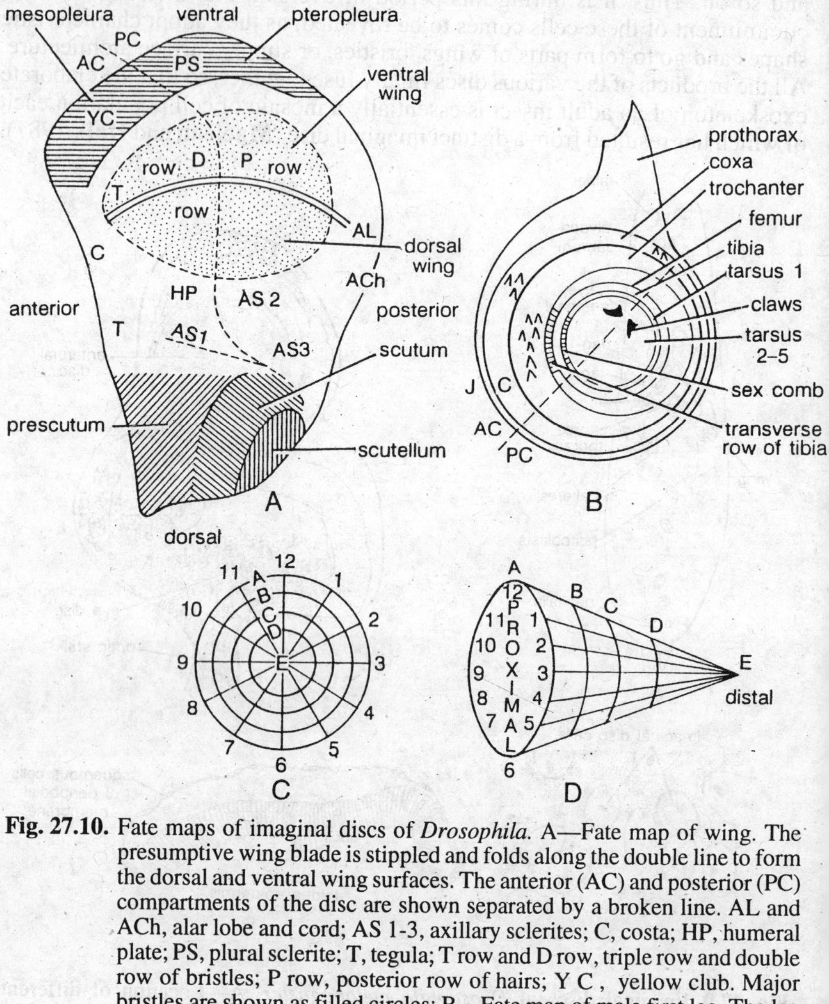

Fig. 27.10. Fate maps of imaginal discs of *Drosophila.* A—Fate map of wing. The presumptive wing blade is stippled and folds along the double line to form the dorsal and ventral wing surfaces. The anterior (AC) and posterior (PC) compartments of the disc are shown separated by a broken line. AL and ACh, alar lobe and cord; AS 1-3, axillary sclerites; C, costa; HP, humeral plate; PS, plural sclerite; T, tegula; T row and D row, triple row and double row of bristles; P row, posterior row of hairs; Y C , yellow club. Major bristles are shown as filled circles; B—Fate map of male first leg. The leg disc is telescoped segment by segment, and the anterior and posterior compartments are divided as shown by the broken line. C, costa; J, particular joints; S, sensilla; Tvr, transverse rows of bristles of tibia; C—The polar coordinates or clock-face model indicating positional information in an epimorphic field. The information is presumed to be both circumferential (0-12), where 0 and 12 are taken as identical and radial (A-E). Thus, each cell has precise information regarding its position in the surface area involved. The field of an imaginal disc could be seen as flat and such a model could account for the cell comitment evident in organs such as wings; D—Model of field of imaginal disc of legs, with the distal point of the cone being at the top of the limb. (A and B, **Sang**, 1984; C and D, **French**, 1984) (after Maclean and Hall, 1987).

27-4.4. Transdetermination of Imaginal Discs

Imaginal discs that have been continuously cultured (for months and years) in the adult body sometimes undergo alteration in their state of determination. For example, cells of the germinal disc become determined to form legs or antenna. This alteration from one determined state to another is known as **transdetermination**. Far from being a random process, transdetermination occurs in a definite number of fixed directions, each with a distinct probability, so that an antennal disc may become wing disc or leg disc, but not a haltere or balancer disc. Most discs show transdetermination in one particular direction, while some can alternate between two states, such as antenna and wing.

For many years it was assumed that all cells of a disc were equally capable of transdetermination and indeed, proceeded to change their collective fate together. But it is now clear from experiments of **Strub** (1977) that only certain cells within a disc are capable of transdetermination, and only when these are exposed to a cut surface or are in a dissociated fragment during manipulation of the discs. Cell division seems to be essential for transdetermination.

Maclean and **Hall** (1987) have suggested the following possible reasons of transdetermination :

1. The simplest explanation for transdetermination would be somatic mutation, but this has been ruled out by many observations, the most significant being that the frequency of transdetermination is much too high, and that it is a phenomenon of a cell population and not of a single cell. Are all the cells of a group transdetermine together ? Yes, there are known cases of **homeotic mutations** in *Drosophila* which may result in a change in the fate of a disc from one structure to another, frequently resulting in the duplication of an existing structure.

Thus, the phenomenon of transdetermination is too frequent for somatic mutation, unless some as yet unkown **transposable genes** or **elements** can raise the mutation rate to very high values, in some cases up to 100 per cent. The *Drosophila* genome is extremely rich in transposable elements. Some 5 per cent of the total genome is comprised of families of **nomadic sequences** that are some 5 kilobases (kb) in length, the best known family being the *Copia* elements, but it is hard to believe that even these abundant elements can account for this specific and directional change from one phenotypic state to another.

2. Discs will only transdetermine when a suitable age that is between 96 and 115 hours old, is attained; younger or older tissues are not competent to transdetermine.

3. There is sound experimental evidence that the phenomenon is not due to the presence of undertermined stem cells within the disc (**Gehring**, 1967).

4. Transdetermination requires growth and cell division.

5. Transdetermination only occurs when embryonic imaginal discs are cut and then grown *in vivo* and *in vitro*.

Lastly, imaginal discs are not necessary compliment of metamorphosis because no such disc occur in hemimetabolous insects. In them the tissues of the adult are derived from the general cells of larva.

27-4.5. Endocrine or Hormonal Regulation of Insect Metamorphosis

Hormonal control of metamorphosis in insects has been proved by numerous experiments : 1. implantation of tissues in hosts undergoing metamorphosis ; 2. implantation of endocrine glands ; 3. ligature of host body at different levels ; and 4. injection of synthetic hormones in very small quantities.

The sequence of events of moulting cycles and metmorphosis of insects has been found under precise hormonal control. In brief, certain stimuli associated with the state of nourishment cause the **neurosecretory cells** of the brain to discharge the "**brain hormone**",which is carried by neurosecretory cell axons to the **corpora cardiaca** (insect endocrine glands), from where it is released into the blood. This hormone then stimulates the **thoraic glands** (or their equivalents) to produce a hormone which causes the epithelial cells to begin the processes which lead to moulting. At the same time at all moults except the last, the **corpora allata** (insect endocrine glands which are located near brain) secrete a hormone which ensures the development of another larval instar. During the last moult, the corpora allata are inactive, and the adult is produced.

These hormones are named according to their function. The brain hormone because of its effect upon the thoracic gland is called **thoracotrophic hormone** ; that from the thoracic gland is called **ecdysone** or **moulting hormone** or **MH** because it initiates moulting or ecdysis ; and that from the corpora allata is called **juvenile hormone** (**JH** or **neotenin**) because it ensures the retention of juvenile characters. These insect hormones have following roles in the insect moulting and metamorphosis (Fig. 27.11) :

1. Ecdysone and its role in insect moulting and metamorphosis. Ecdysone is a cyclic compound of small molecular size (Fig. 27.12). Just as the same thyroid hormones are produced in all classes of vertebrates, the hormone ecdysone is produced by all insects and other arthropods that have been investigated. Moulting cycles are characteristic of all classes of arthropods, and ecdysone extracted from insect causes moulting in shrimps, while ecdysone from shrimps causes moulting in insects. Because, the ecdysone was found to be activator of a non-specific type, therefore, about the differential response of tissues, it was concluded that it was the property of the target tissues.

Under the impact of ecdysone, at each moult, the old cuticle is loosened and thrown off, the specialized epidermal cells are resorbed or discarded, and the ordinary epidermal cells which have been stimulated to grow and divide, give rise to new cuticle and two new groups of specialized cells. In fact, ecdysone acts directly upon those cells that are in a state of dormancy. Within the few hours, the nucleolus is enlarged, different species of RNA begin to accumulate in the cytoplasm and mitochondria enlarge and multiply by subdivision. By the time, the old cuticle is thrown off, the renewed epidermal growth has been virtually completed with locally expanded regions folded compactly while awaiting release from cuticular confinement. Remarkably, the muscles of the larva, *e.g.*, of *Rhodnius*, dedifferentiate between moults and redifferentiate muscle fibres shortly before the next moult occur. To a degree the organism is renewed on each occasion and at each time the volume of the growing animal doubles.

2. Juvenile hormone and its role in insect moulting and metamorphosis. The juvenile hormone (Fig. 27.13) controls the growth, moulting and

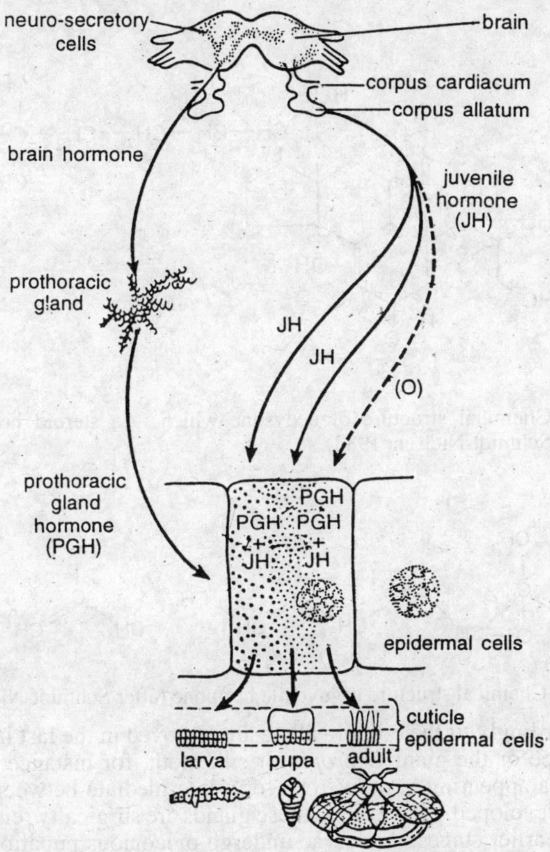

Fig. 27.11. The endocrine control of growth, moulting and metamorphosis in the insect *Cecropia,* silkworm (after Berrill, 1971).

onset of metamorphosis of insects. In the insect, whose life cycle consists of four larval moults, a larval-pupal moult and a pupal-adult moult, it has been observed that in every case the developing system, having been diverted to some degree during embryonic development, moves toward expression of the adult organization. The hormonal controls operate to determine when the major event occurs, to what extend development can be temprorarily arrested, and how long juvenile state may be prolonged. For example, in the presence of the ecdysone or prothoracic gland hormone (PGH), juvenile hormone (JH) promotes larval development or maintains the *status quo* and so prevents metamorphosis. The presence of JH in an immature insect, whether larva or pupa, ensures that when the immature insect moults it retains its larval or pupal characters and does not differentiate into an adult. In other words withdrawal of JH initiates metamorphosis. Conversely, implantation of the JH-secreting gland from a second-stage larva to a fourth-stage larva maintains the larval state, in effect inhibiting metamorphosis so that larval growth continue and a giant larva is produced.

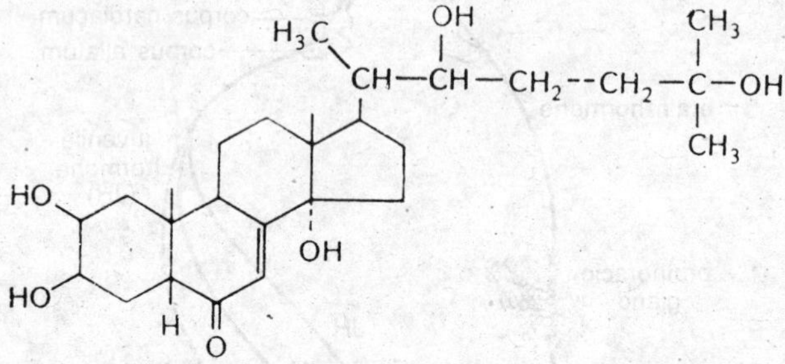

Fig. 27.12. Chemical structure of ecdysone which is a steroid hormone (after Schmidt-Nielsen, 1983).

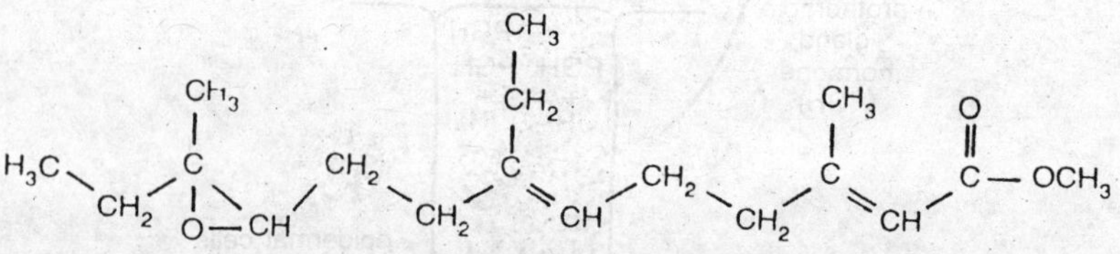

Fig. 27.13. Chemical structure of juvenile hormone (after Schmidt-Nielsen, 1983).

Accordingly, if the corpora allata are removed in the last larval stage of the honeybee or the giant *Cecropia* or silk moth, for instance, instead of a normal pupa appearing, a monstrous form intermediate between a pupa and an adult is developed. However, if these glands are surgically removed during one of the earlier stages, the larvae undergo precocious pupation at the very next stage and eventually develop into midget-sized adult insects. The larval form is developed in the presence of a high JH concentration, the pupa in the presence of a very low JH concentration and the adult when the JH is completely absent.

27-4.6. Hormones Affecting Gene Expression in Insects

The first report of a specific hormonal effect at the chromosomal level in any organism was made by **U. Clever** in the early 1950s for the dipteran insect, *Chironomus tentans*. He first studied carefully the patterns of polytene chromosome puffing at various stages of larval development and established that these patterns followed a fixed sequence, with certain bands puffings at one stage and regressing at a later stage. He then took intermoult larvae, in which puffing of a band called 1-19A was known to be particularly active, and injected the animal with ecdysone. He found that within 10 to 15 minutes puff 1-19A regressed, band 1-18C began to puff, and band IV-2B followed suit. This was exactly the puffing sequence known to take place just before moulting, a time at which the insect spontaneously releases ecdysone into the haemolymph. On the reasonably certain assumption that puffing corresponds to the transcription of one or several genes, these experiments at best interpreted to mean that ecdysone somehow turn on the transcription of particular genes.

Ecdysone induction of puffs at I-18C and IV-2B in *Chironomous* is inhibited by actinomycin D. As the inhibition is released and RNA synthesis recommences, these puffs reappear in normal sequence followed later by the four late puffs. If puromycin is injected 1 hour before ecdysone, then the first two puffs appear on time but late puffs fail to develop. **Clever** concluded that the activity of first puffs is essential for the activation of puffs at the subsequent sites. Presumably, the RNA product of the first puffs must be translated to induce late puffing, a type of cascading induction (see **Grant**, 1978).

Quite recently, the following two modes of hormone action have been suggested :

1. Ecdysone after conversion to **20-hydroxyecdysone**, enters the target cell and binds to a specific protein (**Ashburner** *et al.*, 1974). Then it associates with specific chromatin sites and activates puffing, itself a reflexion of localized transcriptional activity. These activated genes, the so-called **primary response genes**, produce various products, some of which shut off the primary response by a negative feedback mechanism, others of which proceed to turn on a larger set of genes.

2. Juvenile hormone (JH) on its own without ecdysone (MH) does not induce changes in puffing pattern of the salivary gland chromosomes of Diptera, and not on primary set puffs, produced by MH. Similarly, while the primary set puffs are the same in both the larval-larval and larval-pupal moults, the secondary set puffs are different. This suggests that *only the secondary set puffs are sensitive to JH*. It, thus, seems probable that JH somehow affects the **translation** process. Recent evidence from *Tenebrio* (Coleoptera) also suggest that JH influences the synthesis of adult and pupal cuticular protein at the translation level. So in *Tenebrio* pupae treated with JH the mRNA for the adult cuticular protein, though present, is not translated. The JH effect is mediated by the appearance of a new rate-limiting aminoacyl-$tRNA^{Leu}$ and its activating enzyme.

27-5. HORMONAL REGULATION OF METAMORPHOSIS IN OTHER ANIMALS

It is little known that whether in other chordates metamorphosis is regulated by hormones. Because in **cyclostomes** it has been found that thyroxine does not accelerate transformation of the ammocoete into a lamprey (**Gorbman** and **Bern**, 1962). However, in **ascidian tadpoles** the absorption of tail is in some way dependent on the anterior end of the body, since cutting off the anterior tip with adhesive papillae will prevent the necrotization of the tail (**Oka**, 1943). Treating ascidian tadpoles with thyroid hormone accelerated the metamorphosis (**Weiss**, 1928). Though, the larvae of various invertebrates (*Tubularia*, Bryozoa ; sea-urchins, molluscs) are metamorphosed by treatment with various chemicals (toxic agents particularly traces of metals or very dilute mercuric chloride, iodine, iron, copper and aluminium salts) but, it is still unknown that what factors cause metamorphosis in these animals under natural conditions (**Lynch**, 1961). However, in an annelid worm, *Platyneries* *dumerilli*, the asexual form is found to metamorphosed into sexual form by neurosecretory hormone released in the prostomium of the asexual individual (**Hauenschild**, 1964).

The pattern of endocrine regulation in crustaceans shows certain parallels to that in insects. Some crustanceans such as the crayfish, reach a definite adult size after a final moult ; others, such as the lobster, moult continuously and keep growing ever larger until death. In them moulting is controlled by a neurosecretory hormone, called **moult-inhibiting hormone**, produced by the **X-organ** in the eye stalk, and by a moulting hormone (ecdysone) produced by the **Y-organ** located in the thorax. During intermoult, the moult-inhibiting hormone inhibits activity of the Y-organ. But at intervals, just before the moult, the output of moult-inhibiting hormone drops, this permits the Y-organ to secrete moulting hormone, which stimulates moulting (**Hickman** *et al.*, 1979).

27-6. COMPARISON OF AMPHIBIAN AND INSECT METAMORPHOSIS

The metamorphosis of amphibians and insects has certain fundamental similarities. In both cases, the transformation is initiated by a secreting organ closely associated with the brain, the hypophysis in amphibians, the neurosecretory cells of the protocerebrum in insects. In both cases, the secretion of this primary centre does not act on the tissue directly but stimulates the activity of a second endocrine gland—the thyroid gland in amphibians, the thoracic gland in insects.

REVISION QUESTIONS

1. What is larva ? What is the advantage of having a larval stage in the life cycle ? How is a larva different from an embryo, a foetus, a nymph, a pupa or an imago ?
2. What is metamorphosis ? Describe the hormonal control of metamorphosis in amphibians.
3. What is the relationship between moulting and metamorphosis in insects ? Describe the process of metamorphosis in insects ?
4. Explain the significance of metamorphosis with reference to life history of an anuran/insect.
5. Discuss the morphological, anatomical and biochemical changes involved in the metamorphosis of tadpole of frog.
6. Discuss the role of various hormones that control metamorphosis in frog.
7. Discuss hormonal control of metamorphosis in insects.
8. Are imaginal discs necessary features of insect metamorphosis ? How are the discs are formed and what are their important properties ?
9. How do the hormones affect gene expression ?
10. Explain the role of juvenile hormone or JH in insect metamorphosis.
11. Write short notes on the following :

    (i) Indirect development ; (ii) Ecdysone ; (iii) Transdetermination ;

    (iv) Paedomorphosis ; (v) Imaginal discs ; (vi) T_3 and T_4 hormone.
12. What will happen if :

    (i) a young tadpole is put in water rich in iodine ; and

    (ii) a young tadpole is fed on thyroid gland ;

    (iii) thyroid gland of frog tadpole is removed or damaged; and

    (iv) a tadpole undergoing metamorphosis is put in a solution of thiourea.

28

TERATOGENESIS

(Occurrence of Congenital Malformations)

A **congenital malformation** is a gross structural deformity present at birth. Its incidence is about 2.5 per cent in all infants born. However, only half of these deformities are apparent at the time of delivery, most of the remainder coming to light during the first postnatal year. The term **congenital anomaly** is reserved for a minor congenital disorder such as a deformed finger or ear-lobe. Anomalies are found in a further 2.5 per cent of live-born infants.

Individuals of a species (including human and other animals) exhibiting minor deviation (variation) from each other are considered normal. Individuals that show gross deviation from the normal due to congenital malformations are known as **monsters** or **terata** (Gr., *teras* = monster). Abnormal development or formation of a terata is called **teratogenesis** and the science which is concerned with the investigation of terata and teratogenesis is called **teratology**. Since, the development of a normal phenotype requires both a normal genotype and a favourable environment, therefore, the teratogenesis can be due to either abnormal genotype or the environment.

28-1. GENETIC TERATOGENESIS

Genetic teratogenesis has been studied both in human beings and animals as has been described under :

28-1.1. Genetic Teratogenesis in Human Beings

Abnormal genes may be inherited from one or both parents and they may be dominant or recessive. In the majority of the afflicted individuals, however, a mendelian pattern of inheritance cannot be observed ; the abnormalities merely occur more frequently among relatives than in general population. In those occurring among relatives, there may be complex interplay between several genes. In this group of subjects occur the malformation listed in the data, tabulated in Table 28-1, of Birmingham study (**Mckeown** and **Record**, 1961).

Table 28-1. Percentage of congenital malformations in the human population of Birmingham, England (Source : **FitzGerald,** 1978).

Nature of Malformation	Percentage
1. Club foot	19
2. Cardiac abnormalities	18
3. Spina bifidia cystica (Fig. 1.5)	13
4. Hydrocephalus (greatly enlarged head)	11
5. Anencephaly (absence of the forebrain)	9
6. Cleft lip/cleft palate	8
7. Down's syndrome (mongolism)	7

Deformities due to abnormal dominant genes are rare. In most of these, the skeleton is affected and the deformities include **achondroplasia** (insufficient growth of long bones), **arachnodactyly** (abnormally long hand and foot bones), **craniocleidal dysostasis** (absence or rudimentary development of clavicle and abnormal shape of skull) and **osteogenesis imperfecta** (incomplete development or hypoplasia of osteoid tissue and collagen, resulting in bone fracture). Any of such conditions will appear in 50 per cent of the offspring of the affected parent.

Abnormal recessive genes do not find phenotypic expression unless inherited from both parents ; in such instances the abnormality may be found in 25 per cent of the offspring. Cystic fibrosis of the pancreas and Sprengel's deformity of the shoulder (*i.e.*, elevation of scapula or scapula elevata) are examples in human beings.

Genetic factors, when they are operative, are the **remote cause** (initiating agent) of malformations. Each gene locus is concerned with the synthesis of an enzyme or polypeptide ; and the **immediate cause**, which results from the genetic abnormality, is a biochemical disturbance at cell or tissue level.

28-1.2. Genetic Teratogenesis in Animals

It can be considered under the following two headings :

1. Gene-phene relationship. Several different genes can cause the same terata, though not necessarily by the same route. For example, there are more than twenty genes which affect eye colour in *Drosophila melanogaster*. The mutants causing the same defect may be either recessive or dominant. For example, in fowl the trait of rumplessness (absence of tail) is controlled by either a recessive or dominant gene. In some cases, the same mutation may behave as a recessive or a dominant depending on the genetic background. Thus, in mice the fused gene (for fusion or absence of ribs and/or absence of tail) is dominant in *Mus musculus musculus* but recessive in *Mus musculus bactrianus*.

The **penetrance** (*i.e.*, proportion of affected individuals in a population) and **expressivity** (*i.e.*, degree of effect) of mutant genes are dependent on both genetic and environmental factors. For example, fowl carrying **rumpless gene** can be selectively bred to produce a 'normal' tail phenotype. Similarly, in *Drosophila* carrying **Bar eye gene**, the size of the eye and the number of facets in the eye decreases by about 100 facets with an increase in temperature (15° to 31°C) during development.

2. Autophene, allophene and pleiotropy. Not all genetic terata are the result of intrinsic action of genes in the affected tissues. The following two examples will make the point clear : (1) **Creeper mutation** (cp/cp) in fowl affect the limbs (called **phocomelia** or abnormally short limbs) and the eyes (called **microphthalmia** or small eye) ; the embryo does not survive till hatching. Transplants of the cp/cp limb rudiments in the normal hosts produce phocomelia limbs. However, transplants of cp/cp eye rudiments in the normal hosts produce normal eyes. Therefore, the cp gene intrinsically affects limb development (it is called **autophene**) but only indirectly affects the eye development (it is called **allophene**). (2) The multiple effects of one gene (**pleiotropy**) are also now better understood. Any given gene mutation essentially affects the production or structure of one transcribed RNA molecule. The translation product of this RNA (*i.e.*, mRNA), the defective protein, may ultimately result in various defects due to correlation of various biochemical reactions in the body (*i.e.*, by a pedigree of causes), *e.g.*, death in rat due to grey lethal mutant gene (Fig. 28.1) and sickle cell anaemia in human beings.

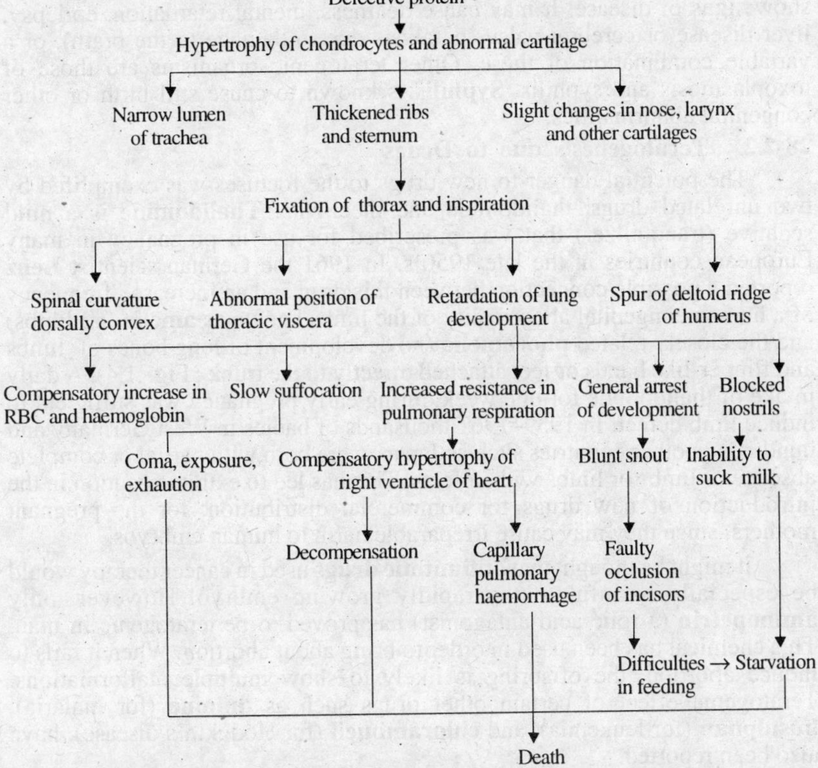

Fig. 28.1. Death due to pleiotropic grey lethal mutant gene of rat. The diagram shows pedigree of causes as well as an autophene for cartilage.

28-2. ENVIRONMENTAL TERATOGENESIS

Almost any environmental factor can be teratogenic. The environmental factor may be either **biological** (*e.g.*, viruses) or **non-biological** such as physical factors (*e.g.*, temperature, irradiation, mechanical disturbances) and chemical factors (*e.g.*, drugs, environmental chemicals and dietary imbalances or malnutrition).

28-2.1. Teratogenesis due to Infections

The most dangerous known teratogenic organism is the virus of **German measles** (**rubella**). Contraction of the disease by the mother during the first month of pregnancy appears to carry about a 50 per cent risk of producing a congenital abnormality ; during the second month the risk is about 25 per cent and during the third month about 7 per cent. The virus tends to affect the developing eyes, heart, ears and palate. The triad of congenital cataract (blindness), heart disease and deafness have become known as the **rubella syndrome**. Rubella also known to cause terata such as mental deficiency due to very small brain, cleft palate and hare-lip.

Cytomegalovirus (CMV) infection is found all over the world. Like the rubella virus, it passes through the placenta to infect the foetus. The CMV is found in 0.5 to 1 per cent of newborn babies and about 10 per cent of these show signs of disease. It may cause deafness, mental retardation, epilepsy, liver disease or cerebral palsy (paralysis due to damage to the brain), or a variable combination of these. Other teratogenic organisms are those of toxoplasmosis and syphilis. **Syphilis** is known to cause still birth or other congenital abnormalities.

28-2.2. Teratogenesis due to Drugs

The potential danger to new drugs to the foetuses was exemplifed by two unrelated drugs, thalidomide and meclizine. **Thalidomide** is a mild sedative (tranquilizer) that was prescribed for use in pregnancy in many European countries in the late 1950's. In 1961 the German scientist **Lenz** reported a possible connection between this drug and an increased frequency of a human congenital abnormality of the limbs known as **amelia** (no limbs) and the closely related **phocomelia** (no development of long bones of limbs and flipper-like hands or feet attached directly to the trunk ; Fig. 1.4). A daily intake of thalidomide for one week during early pregnancy was sufficient to induce limb defect. In 1959–1961 thousands of babies in West Germany and hundreds in other countries such as Japan, were born with partial or complete absence of limbs or limbs with defects. This has led to extreme caution in the introduction of new drugs for commercial distribution, for the pregnant mothers, since they may cause irreparable harm to human embryos.

It might be thought that **antimitotic drugs** used in cancer therapy would be especially harmful to the rapidly growing embryo. However, only **aminopetrin** (a folic acid antagonist) has proved to be teratogenic in man. This chemical has been used in order to bring about abortion. When it fails to induce abortion, the offspring is likely to show multiple malformations. Teratogenic effect of certain other drugs such as **quinine** (for malaria), **busulphan** (for leukemia), and **chlorambucil** (for Hodgkin's disease), have also been reported.

None of the **hormones** has been implicated in teratogenesis following their administration to the pregnant women. However, in mice and rabbits **cortisone** is known to induce cleft palate in the offspring (it also causes cleft palate in human foetuses but in less percentage).

Of the **antibiotics**, only **tetracyclines** may give rise to an anomaly. When these drugs are administrated during the period of enamel formation (for teeth), they may produce yellowing of deciduous teeth. Epileptic mothers taking **anticonvulsant drugs** such as **phenytoin** and **barbiturates**, are about three times as likely as normal mothers to give birth to malformed babies. And in some cases such harmless drugs such as **caffeine** and **aspirin** have teratogenic effects at very low doses. The teratogenic effect of **lysergic acid diethylamide (LSD)** is unproven. Some studies have implicated LSD in the appearance of defects in the hands and feet of offspring whose mothers have taken this drug either before or during pregnancy (see **FitzGerald**, 1978).

There is also some reason to believe that pregnant women working in operation theaters (such as anesthetists, operating nurses) have an increased risk of giving birth to malformed babies. The responsible factor could be one or more of the anesthetic gases.

28-2.3. Teratogenesis due to Radiations

It is well known that the use of **roentgen rays** or **radium** in the treatment of pregnant patients having pelvic tumours, is likely to produce skeletal abnormalities in the foetuses. The exposure of pregnant women to severe atomic radiation, as in the Hiroshima and Nagasaki explosions, led to foetal/neonatal death rate of about 40 per cent. The infants who survived tended to show evidence of brain-cell damage.

Irradiation of gonads, even at very low levels (50 r) may induce mutation of germ cells.

28-2.4. Teratogenesis due to Autoimmunization

Considerable interest was aroused by the observation that mothers of infants born without thyroid glands, *i.e.*, **athyroid cretins**, may have antithyroid antibodies in their blood. This observation suggests that products of embryonic organ primordia may cross the placenta, inducing maternal antibody production ; the antibodies may return to the embryo and interfere with organ differentiation.

28-2.5. Teratogenesis due to Malnutrition

It is a well known fact that deficiency of various vitamins causes a number of diseases in man ; the deficiency of vitamin A causes night blindness, the deficiency of vitamin D leads to abnormal development of bone and teeth and deficiency of vitamin K causes abnormalities in the clotting of blood. It has been studied that if rats, rabbits and other animals are given a diet deficient in vitamin A, B and D, they may produce various teratogenic effects such as hare-lip, cleft palate, spina bifida, skeletal defects, brain defect, etc. It is very likely that such vitmain malnutrition of pregnant mothers of human species may also cause similar congenital malformation.

28-2.6. Sensitive Period of the Teratogen

The developmental stage at which the embryo is treated by a teratogen has great relevance to teratogenesis. In general, the very early stages of development are not much affected by the teratogen, possibly due to the considerable regulatory capacity of the embryo. Similarly, the later stages of development, when maturation and growth of organs to occur, are also comparatively resistant to teratogens. The main **teratogenic period** starts with the formation of germ layers and continues up to organogenesis. For example, in rat **trypan blue** (20 mg intravenous injection to the mother) and **actinomyocin D** (0.3 mg/kg body weight to the mother) are maximally teratogenic on 8th and 9th day of gestation, respectively. Both chemicals are ineffective after 10th day and only minimally teratogenic before 6th day of gestation.

28-2.7. Specificity of the Teratogen

Almost any teratogen can produce almost any terata if applied at the right time in the experimental animal. However, each teratogen produces its own typical syndrome. For example, the **micromelia** (short limb) in fowl can be caused by a deficient or imbalanced diet, (*i.e.*, riboflavin (or biotin deficiency), insulin, thallium, boric acid, pilocarpine, propanediol-1,3, sulphanilamide or eserine sulphate. A similar situation occurs in the case of **tetraptera** (four wings) in *Drosophila melanogaster*, *i.e.*, phenocopies of tetraptera can be obtained by treating the early embryo with either ether or 40°C heat shock.

However, the types of micromelia produced in fowl by different teratogen are only superficially similar. For example, in the insulin induced micromelia all the long bones of the leg are shortened but there is little or no reduction in toes. In the boric acid induced terata the toes are absent or short and curled ; the tarsometatarsus bone is greatly reduced but the femur and tibia are relatively normal. Even more, the other defects that accompany micromelia differ with the type of the teratogen : boric acid causes high incidence (43.7 per cent) of colobomas (fissure of eye or eyelids), while insulin causes more eye defects (6.7 per cent). It is also remarkable that for maximum incidence and severity of micromelia, the different teratogens must be administrated at different times : insulin at 120 hours, boric acid at 96 hours, and sulphanilamide at 48 hours of development.

The type of malformation produced by a teratogen is also dependent on the developmental stage of the embryo at the time of treatment. For example, insulin (2 units in yolk sac), causes rumplessness in the young fowl embryo (24 hours) but micromelia and abnormal beak in the older embryos (70 to 170 hours) of fowl. Thus, each organ system appears to have its own sensitive period for a given teratogen.

28-2.8. Teratogenic Dose

A teratogen may be quite inefective at a very low dose, while at its high doses it may be lethal to all embryos. A **teratogenic dose range** comprises doses which produce at least some malformed and living embryos. It is also necessary to discriminate between the teratogenic and the toxic effects of an agent. By definition an agent is toxic, if it causes embryonic death by direct action and not through teratogenic interaction. For example, **trypan blue** (50 mg/kg) injected into 8.5 day pregnant rats affects 65 per cent of the embryos. But with advancing gestation, out of these 65 per cent of the embryos the number of dead one increases at the expense of live and malformed individuals. Moreover, when trypan blue injected outside the teratogenic period, it does not result in appreciable deaths. Trypan blue, therefore, is a teratogenic and not toxic chemical. On the other hand 0.05 ml **propanediol - 1, 3**, injected into the yolk sac of 4-day-old fowl embryo of white Leghorn strain 111, causes malformation in 5 per cent individuals ; and an increased dose (0.3 ml) is not more teratogenic but induces high mortality (80 per cent) ; so it is toxic not teratogenic.

28-2.9. Interaction of Teratogens with other Environmental Factors

The effect of any teratogen depends on other factors present in the environment of an organism. Many types of interactions between environmental factors do occur. For example, the teratogenic effects of insulin on a 96 or 120-hour-old fowl embryo are almost completely eliminated by the simultaneous injection of nicotinamide.

The synergistic effects of agents are also well known. The incidence of both the micromelia and abnormal beak in the insulin injected fowl embryos is significantly enhanced by **chlorpromazine**, a non-teratogenic but slightly toxic chemical. Likewise in another type of interaction the non-teratogenic doses of two teratogens, **sulphanilamide** and **6-aminonicotinamide**, can cause teratogenesis when injected simultaneously.

28-2.10. Phenocopies

An environmental teratogen can cause a normal genotype to produce a phenotype resembling an already known mutant, that is, a **phenocopy** of the mutant. **Landauer**, in fact, has proposed that a true phenocopy should satisfy two criteria : (1) The developmental stage affected by the mutant gene and the environmental teratogen must be same, that is, both should cause a terata by altering the same developmental pathway. (2) Modifications in the genetic background should have similar effects on the incidence and expressivity of both the mutant gene and the environmental teratogen. On these criteria, the insulin-induced rumplessness appeared to be a true phenocopy of the recessive rumpless mutant. In both cases, the abnormal tail structures are formed and later degenerate.

28-3. DEVELOPMENTAL MECHANISMS OF TERATOGENESIS

Every aspect of development is prone to a defect. The sequential nature of development is based on networks of interacting systems ; what happens at one stage is vital to all subsequent stages. Defects initiated at an early stage may be expressed at later stages because intermediate steps are abnormal. The magnitude of a defect will depend on the stage of development at which it originates and the developmental process that is the specific target. Lesions arising at early stages usually have more drastic effects since many more cells are affected (**Grant**, 1978).

Since in a terata any organ and any tissue may be affected, so the mode of production of terata are much varied. The developmental processes that may be affected by teratogenesis include competence, induction (evocation), determination, histogenesis and morphogenesis, cell growth, cell division, cell death and cell locomotion. The genetic and molecular studies on development indicate that each teratogen affects the developmental processes by essentially affecting cell metabolism and gene expression. This is brought about by several independent or inter-related mechanisms such as (1) production of defective, deficient or no proteins ; (2) production of proteins at an inappropriate developmental juncture ; (3) production of defective, deficient or no rRNA or tRNA of a particular kind ; or (4) alteration in membrane permeability. For example, in case of sickle cell anaemia, the production of a defective protein (β chain of haemoglobin) leads to sickle cell syndrome.

REVISION QUESTIONS

1. What is teratogenesis ? Write an essay on the teratogenesis.
2. Describe the genetic bases of the teratogenesis.
3. Relate the pedigree of causes to the pleiotropy concept and discuss its theoretical bases.
4. Enumerate the factors that influence the teratogenic effects of an environmental teratogen.
5. Distinguish between the teratogenic and toxic effects of a chemical.
6. An adult frog had no left forelimb. Is it a terata, if it is due to the amputation of the limb (a) in adult, (b) in tadpole, or (c) in embryonic stage ?

29

BIRTH CONTROL

The term **birth control** was first used by **Ms. Margaret Sanger** who started a movement in U.S.A. to popularize it. In 1916, for highlighting the worldwide need of birth control, she declared ''*the first right of every child is to be wanted, to be desired, to be planned with an intensity of love that gives it its title to being*''. Today, there is a pressing need for limiting the family size at a personal level and for the control of population at a national level.

29-1. REASONS OF BIRTH CONTROL

The main and compelling reason for birth control is the population explosion and resultant socio-economical global problems due to overpopulation. For example, the population in India has been growing rapidly. It has doubled from 344 million in 1947 to 685 million in 1981 and is likely to double in next thirty years. At the personal level, the following reasons stand for the need of the birth control : 1. To maintain the healthy family. 2. For better care and education of children in the present competitive world. 3. To fulfil a desire for an improved standard of living. 4. To help parents to rear the children without any difficulty. 5. For the interest of health of the mother. 6. To save the parents from worries and stress syndrome, since, when both parents are working then frequent child births affect the income of the mother. 7. For financial consideration of the parents. 8. To have only that number of children which couple wants to have. 9. For health and medical grounds. For example, a woman from low income group generally suffers from malnutrition and is predisposed to heart disease, prolapse, stress incontinence (*i.e.*, inability to prevent the discharge of urine or faeces) and cancer of cervix. Further, in India it is customary to offer sterilization operation at the time of two or third Ceasarean operations. The woman suffering from mental retardation and serious psychiatric disorders such as schizophrenia, also need birth control by sterilization. 10. To avoid birth of child having a genetic disorder such as Rhesus incompatibility.

Measures of birth Control. There are following two main types of birth control measures : 1. Contraception ; and (2) abortion.

29-2. CONTRACEPTION

Contraception is the prevention of fertilization without destroying fertility by natural, mechanical or chemical means. In other words, a method or system which allows intercourse and yet prevents conception is called **contraceptive method**. The contraception may be **temporary** when the effect of preventing pregnancy lasts while the couple uses the method, but the fertility returns immediately or within a few months of its discontinuation. The **permanent contraceptive methods** are surgical, *i.e.*, removal of gonads and uterus and sterilization (tubectomy in women and vasectomy in man).

29-2.1. Natural Contraception

1. Complete abstinence. When the couples refrain from the act of sexual intercourse, it is called **abstinence**. This is 100 per cent successful and has even religious sanction. However, it is very difficult to be followed by young people.

2. Periodic abstinence or rhythm method. This method is based on the study of the perishable nature of the sperm and ovum and also of the detection of approximate day of ovulation, *i.e.*, release of ovum from the ovary. Normally a sperm survives for about a day and an ovum for three or four days. The egg (ovum) undergoes many changes when it passes through the Fallopian tube and by the time it reaches the uterus it is over ripe and is not in a position to be fertilized. Therefore, it must be fertilized when it passes through the Fallopian tube. With these facts in hand, it is possible to avoid impregnation or fertilization, if the exact date of ovulation is known and the time when the egg will be in a fit condition to be fertilized.

To recall, the menstrual cycle of a woman involves 28 days. With the day on which menstruation begins as the first day (Fig. 29.1). In such a perfect cycle, ovulation occurs on the 14th day. This is a day that has to be avoided for the sexual intercourse, and so also a few days before and also three days after. This is because ovulation occurs in some women between 12th and 16th day. Thus, the couple refrains from intercourse from the 10th to the 17th day both days inclusive. This period of menstrual cycle is called the **danger** or **risk period**. The remaining days of the cycle are called **safe days**, since, sexual intercourse can take place during these days. Thus, the periods of 1st to 9th day and 18th to 28th day are called the **safe periods** when normally fertilization may not occur. However, the safe periods are not 100 per cent safe and there are cases where pregnancy has occurred during the so-called safe periods. An average of 25 pregnancies per 100 women per year has been reported by gynaecologists. This failure rate is found to be due to the following reasons :

1. Women (about 16 per cent) who have irregular menstrual cycle cannot use this method. 2. After childbirth, this method cannot be followed as menstruation may not occur for a few months and when it occurs it will be irregular. 3. Menstrual effect become irregular also due to mental upsets. 4. Women who are about to reach the menopause period (45 to 50 years of age; when menstruation/ovulation will stop) also have irregular menstrual cycle and cannot use this method. 5. In many women the menstrual cycle is irregular in their young age when they are normally at the height of their reproductive activities. They also cannot follow this method. 6. It is difficult for many women to keep track of their menstrual cycles and also of their safe periods and may commit mistakes. 7. Restricting love making for a few days and abstinence to other days is not practicable.

Effectiveness of rhythm method can be increased by recording oral or body temperature each morning before getting up (called **basal body temperature** or **BBT** ; see Chapter 30).

3. Withdrawal method (Coitus interruptus). This is a common practice and is one of the earliest birth control methods. In this method coitus takes place in a normal manner, but male withdraws his penis from the vagina during mating just before the ejaculation takes place. This involves an interruption in coitus or intercourse. This method is not always successful. There are two main drawbacks in this method : (1) some spermatozoa may be released even before the climax is reached and also some semen may ooze out without the awareness of the copulating male ; (2) it involves psychological and physiological problems to both partners (*e.g.*, it imposes a great strain upon the husband and is thought to be a causative factor in some anxiety states).

MENSTRUAL CALENDER

1 2 3 4 5 6 7 8 9 **10 11 12 13 14* 15 16 17** 18 19 20
← Menstruation →
21 22 23 24 25 26 27 28

* Ovulation

10 –17 Risky days

Fig. 29.1. Safe periods for love making in a typical 28-day menstrual cycle of human beings.

29-2.2. Barrier or Mechanical Contraceptives

1. Condom. In this method, the erect penis of copulating man is completely covered by a very thin, elastic, non-porous, tight-fitting sheath of rubber or of similar material, called **condom.** Condom is used only once and prevents deposition of sperm in the female reproductive tract. Condom should be used in conjuction with spermicidal (sperm killing) chemical agent to improve the efficiency of this method.

The main advantages of condoms include: (1) Condom is easily available, is cheap and requires no instruction for its use. 'Nirodh' brand of condom is distributed free of cost in the Government hospitals and Family Planning Centres in India. (2) Condom prevents transmission of sexually transmitted diseases from one partner to the other. The occurrence of cancer of the cervix is low amongst women whose partners use condom, because sexual transmission of the viral infection causing this disease is prevented. (3) Condom is strongly recommended worldwide to check the spread of the dreaded disease of AIDS. The method is only partially reliable, having a pregnancy rate of 10 to 14 per 100 women per year. Failures are caused due to bursting or slipping off of condom after climax or not putting it on soon enough.

2. Occlusive diaphragm. The diaphragm is in the form of a round thin membrane or flexible dome of rubber with an elastic ring at the rim. This is inserted into vagina to close the opening of the cervix. Though the diaphragm physically close the entrance to the uterus, sometimes sperm may find their way through the space between the wall of the vagina and the rim of diaphragm. To be double sure that no such entry takes place, spermicidal creams and jellies can be applied to the diaphragm before.

3. Intrauterine contraceptive device (IUCD) or Intrauterine device (IUD). IUD is a small object made of plastic, copper, or stainless steel and shaped like a loop, coil, T or 7. It is inserted into uterus by a gynaecologist. It may be left in place for a long period of time and does not require continued attention by the user. Various types of IUD include : **1. Biologically inert devices**, *e.g.*, Lippe loop and Saf - T - coil ; they can be left *in situ* for several years, provided they cause no side effects. **2. Copper carrying devices**, *e.g.*, Copper T 200, Copper 7, Copper 380, Soonawala Y and multi-load copper devices. In them copper wire is wrapped round the vertical stem of the plastic device. **3. Progestasert**, a T - shaped device carrying 38 mg of progesterone hormone in the vertical stem. Each of these IUDs has two nylon threads attached at the lower end and these threads protrude through the cervical canal into the vagina, where it can be felt by the patient herself and by the doctor. This is placed in position inside the uterus by an **inserter**. The inserter can be loaded with a loop by straightening it and then it is inserted into the uterus where the loop regains its original shape. The loop prevents conception by one or more of the following ways :

1. The presence of a foreign body in the uterine cavity renders the migration of spermatozoa difficult. 2. A foreign body within the uterus provokes uterine contractility and increases the peristalsis of the Fallopian tubes. Due to which, the fertilized egg is propelled down the Fallopian tube more rapidly than in normal and thereby prevent fertilization. 3. As the time taken by the egg to reach the uterine wall is very short, the wall (endometrium) of uterus remain unprepared to receive it. 4. The irritation caused by the loop makes an implanted zygote to move down the uterus. 5. The IUD *in situ* causes leucocytic infiltration in the endometrium. The macrophages engulf the fertilized egg if enters the endometrial tissue. 6. Copper T elutes about 50 μg of copper daily in the uterus which brings about certain enzymatic and metabolic changes in the endometrial tissue and these changes are hostile to the implantation of fertilized ovum. 7. Progestasert releases 60 μg of progesterone hormone per day which causes alteration in the cervical mucus which prevents penetration of sperm, in addition to its local action.

The contraceptive method of IUD is very cheap and efficient. Copper T is inserted free of cost in the Government hospitals in India. In a country like India where the women are not much educated and shun other methods which involves personal care and efficient handling, the loop is the answer. Its failure rate is very low being 1 to 5 per cent. However, IUD is contraindicated when pregnancy is suspected, or woman has pelvic infection, uterine fibroids, heart disease, ectopic pregnancy.

29-2.3. Chemical Methods of Contraception

1. Spermicidal agents. The sperm-killing chemicals kill the sperm before they gain access to the cervical canal. These chemical contraceptives are available in the form of soluble pessaries (*i.e.*, medicated devices worn in the vagina), jellies, creams and foaming tablets, which are inserted into the vagina prior to intercourse. They provide protection for about 1 hour. These agents are effective when used alone, but significantly more effective when used with a diaphragm or condom. The most recent spermicidal contraceptive available today is **nanoxynol - 9**. Failure rate is high, being 30 per 100 women a year.

2. Altering cervical mucus. Low-dose progestogen-only pill, called **minipill** is administered daily. (**Progestogen** = Any agent capable of producing biological effects similar to those of progesterone). Under the influence of progestogen, the cervical mucus becomes viscid and prevents penetration of sperms into the cervical canal. In addition, the minipill inhibits ovulation and reduces fertility.

3. Douching. This is one of the most oldest and common method. It involves the douching of the vagina with a spermicidal solution. An irrigator is used for this purpose. This has to be done immediately after the intercourse. The spermicidal chemicals immobilize the sperm which are washed only by the flow of liquid. The failure rate of this method is 40 per 100 woman per year.

4. Oral contraceptive (OC). The hormonal method, **oral contraception** (the pill) has found rapid and widespread use. This method causes suppression either of spermatogenesis in males and ovulation in females.

A. Suppression of spermatogenesis. The quest for an efficient male contraceptive has been disappointing until recently. However, an oral contraceptive, **gossypol**, which is derived from the cotton plant (*i.e.*, a yellow pigment isolated from cotton seed oil), has achieved an efficacy (power to produce effects) rate of 99.9 per cent in clinical tests in more than 4000 healthy males in China. Gossypol is administered orally 10 to 20 mg daily for 3 months and thereafter 20 mg twice weekly. The action of gossypol is to inhibit an enzyme required for spermatogenesis. Levels of blood luteinizing hormone (LH) and testosterone remained unchanged and potency was not impaired. After termination of therapy, the number and morphology of spermatozoa gradually recover and fertility is restored to normal within three months. However, some of its side effects such as weakness, hypokalemia (*i.e.*, deficiency of potassium in blood), and permanent sterility in 20 per cent cases has restricted its uses.

B. Suppression of ovulation. Although several types of pills are available, the one most commonly used contains a high concentration of **progesterone** and a low concentration of **estrogens**. These two hormones act on the anterior pituitary to decrease the secretion of FSH and LH by inhibiting the gonadotropin releasing factor (GnRF) by the hypothalamus. The low levels of FSH and LH are not adequate to initiate follicle maturation or ovulation. In the absence of mature ovum, pregnancy cannot occur.

Side effects of oral contraceptives of females include nausea, occasional light bleeding between periods, breast tenderness or enlargement, fluid retention and weight gain. Oral pills should not be used by the women who have cardiovascular conditions (such as thromboembolic disorders, cerebrovascular disease, heart diseases, hypertension, etc.), liver malfunction, cancer of breast or of reproductive organs or by women who smoke.

5. Immunological methods. Immunological approach to the family planning is in a developmental stage. The antigens which are being experimented upon are : 1. β subunit HCG evokes specific antibodies and thereby produces temporary sterility. 2. The antibodies for the zona pellucida can either prevent sperm penetration inside the ovum or prevent shedding of zona after fertilization so that implantation is impossible. 3. Antibodies to sperm antigens, namely **lactate dehydrogenase** and **hyaluronidase** and **acrosin** cause immobilization and destruction of sperms.

29-3. SURGICAL CONTRACEPTION

This includes the following permanent contraceptive methods :

29-3.1. Removal of Gonads and Uterus

Castration (removal of testes), **hysterectomy** (removal of the uterus) and **oophorectomy** (removal of ovaries) are absolute preventive methods. Once performed, these operations cannot be reversed and it is impossible to produce offspring. However, removal of testes or ovaries has adverse effects because of the importance of these organs in the endocrine system. Castration before puberty prevents the development of secondary sex characteristics.

29-3.2. Sterilization

The sterilization operation is undertaken with the primary objective of preventing further pregnancy permanently. It is suited to those couples who have completed their families and do not want to bear the inconvenience or cost of the other methods of contraception. It is of the following two types :

1. Male sterilization (Vasectomy). Vasectomy is a simple operation in which a portion of each ductus (vas) deferens is removed (Fig. 29.2). An incision is made in the scrotum under local anaesthesia. The tubes are located and each is tied into places. Then the portion between the ties is cut out. Sperm production can continue in the testes, but the sperm cannot reach the exterior.

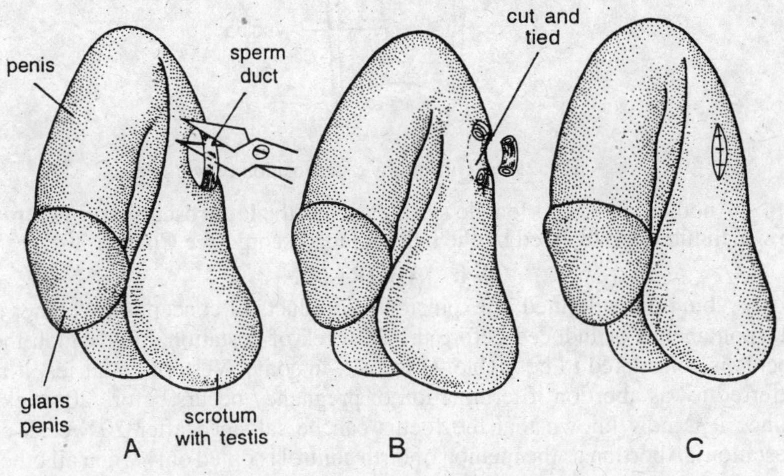

Fig. 29.2. Vasectomy : sterilization of man.

Sterility by vasectomy is not immediate. The sperms are stored in the reproductive tract for up to 3 months. The couple must, therefore, abstain from intercourse during this period or use other methods of contraception. Approximately ten ejaculates clear the semen of all sperms.

2. Female sterilization (Tubectomy or tubal ligation). Sterilization in females generally is achieved by performing a **tubal ligation**. An incision is made into the abdominal cavity, the uterine tubes (or Fallopian tubes) are squeezed and a small loop, called a **knuckle** is made. A suture is tied tightly at the base of the knuckle and the knuckle is then cut. After four or five days, the suture is digested by body fluids and two severed ends of the tubes separate. The ovum is, thus, prevented from passing to the uterus and the sperm cannot reach the ovum.

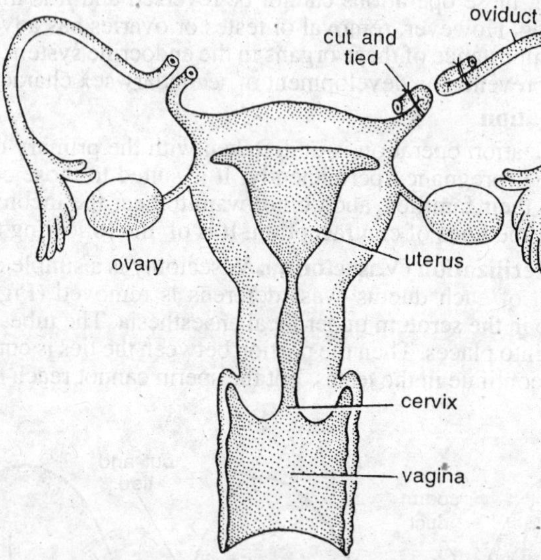

Fig. 29.3. Tubectomy : sterilization of woman.

Another method of female sterilization is the **laparoscopic tubectomy** in which tubes are blocked by the help of laparoscopy (see Chapter 30).

29-4. ABORTION

Abortion is defined as expulsion of products of conception whether it be spontaneous or induced before the 28 week of gestation, after which the foetus is considered to be viable. However, in some Western countries, it is referred to as abortion if termination of pregnancy occurs before 20 weeks (since it is now known that the foetus can be salvaged after 20 weeks of gestation). Abortion as the method of birth control is opted only when all other means of contraception fail and there is no alternative. In India, the **Medical Termination of Pregnancy** (**MTP**) **Act** was passed by the Parliament in 1971 and came into force on April 1, 1972. It liberalized the abortion law as a family welfare measure to ensure better health and avoid risks to the lives of women. The law specifices : (1) The conditions under which the pregnancy can be terminated, *i.e.*, **medical grounds** such as grave danger to mother's life ; **eugenic grounds** such as congenital physical or mental abnormalities ; **humanitarian grounds** such as rape ; or **social grounds** such as contraceptive failure. (2) The person who can perform such terminations, *i.e.*, Registered Medical Practitioners only can perform abortion. (3) The place where such terminations can be performed, *i.e.*, abortion can be performed in a Government Hospital. MTP is permitted up to 20 weeks of gestation.

Abortion or MTP is painful and repeated abortions are not conducive to the health of the mother. There are the following three methods of abortion :

1. Dilation and surgical curettage or D and C. In this method the cervix is gradually dilated with a **dilator** to enable the entry of a spoon-like apparatus called the **curette**. The embryo and the covering membranes (the endometrial tissue) are slowly scrapped and removed from its attachment to the uterus.

2. Dilation and suction curettage or vaccum suction. This method involves dilation of the cervix by instruments and aspiration with the suction curette, during which the embryo is sucked out. This operation is less painful than D and C and has many advantages such as it empties the uterus more rapidly, minimizes the blood loss and reduces the likelihood of perforation of the uterus.

3. Use of abortifacients. Various substances and solutions have been used to induce mid-trimester abortion by intra-amniotic instillation. These include saline, prostaglandin hormone and urea. Of all the three methods this is the least painful and now is in use.

REVISION QUESTIONS

1. What is contraception ? Describe various contraceptive methods of birth control.
2. Describe different methods of sterilizations.
3. What is abortion ? Describe various common methods of abortion.

30

IMPOTENCY, STERILITY, ARTIFICIAL INSEMINATION, TEST TUBE BABY AND GIFT

Impotency is defined as physical incapacity of accomplishing the sexual act. The word impotency implies "powerless" or "wanting in natural strength". Impotency in males is the persistent inability to achieve or maintain a penile erection sufficient to conclude coitus to orgasm and ejaculation. For impotency, sex therapists have coined a new term **erectile dysfunction** or **erection problem**. According to **Masters** and **Johnson** (1970), there are two different types of impotency : (1) **Primary impotency** where the male has failed every time he has attempted sex because of his inability to achieve an erection or to maintain it. (2) **Secondary impotency** which is a highly frequent complaint in males who have functioned adequately for some time but develop erection problem later. It usually begins with an initial erectile failure, resulting in anxiety and fear of failure in subsequent attempts at sex.

Sterility is inability to procreate or to have a child. Sterility and impotency are often confused by laymen and even by doctors. A sterile man can be potent ; that is, he can have normal intercourse, but he will not have children. An impotent man, on the other hand, may not necessarily be sterile, he can have children, as pregnancy can result by merely depositing the semen on the external genital organs of the female (**Franklin**, 1988). According to **Tortora** and **Anagnostakos** (1984) infertility or sterility is an inability to fertilize the ovum. Male fertility requires production of adequate amounts of viable, normal spermatozoa by the testes, unobstructed transportation of sperm through the seminal tract and satisfactory deposition in the vagina. However, **Padubidri** and **Daftary** (1989) have discriminated between two terms : infertility and sterility. Thus, infertility implies apparent failure of a couple to conceive ; while sterility indicates absolute inability to conceive for one or more reasons. Infertility is termed **primary** if conception has never occurred, and **secondary** if the patient fails to conceive after having produced a child or had an undoubted miscarriage. The incidence of infertility in any community varies between 5 per cent and 15 per cent. **Physiological sterility** is present before puberty and after the menopause.

30-1. CAUSES OF IMPOTENCY AND STERILITY IN THE MALE

The impotency and sterility in the male human beings may be due to the following factors :

1. Age. Impotency is generally observed at the extremes of age. Boys are considered to be sexually potent at the age of puberty which usally occurs at the 15th or 16th year. Sexual intercourse is, however, possible at about the

13th or 14th year, as the power of coitus begins earlier and ceases later than the power of procreation.

2. Malformation. Absence of the male organ (the penis) forms **absolute impotency** as the act of sexual intercourse is physically impossible and in case of partially developed penis, performance of sexual intercourse may be rendered impossible. Such a man is not sterile, if semen can be deposited into the vagina with such a ill-developed penis. Impotency may also be result of physical abnormalities of penis and systemic disorders such as vascular disturbances.

The congenital absence of the testicles produces sterility and impotency. **Monorchids**, *i.e.*, those who have one testicle only are physiologically quite potent ; while **cryptorchids**, those who have undescended testicles are usually impotent and sterile.

3. Local diseases or injuries. (1) A large **hydrocele** or **scrotal hernia**, **elephantiasis**, **phimosis** (elongation of the prepuce and constriction of the orifice, so that the foreskin cannot be retracted to uncover the glans penis), and **adherent prepuce** may cause temporary impotency by mechanical obstruction to coitus, as these conditions can be remedied by proper surgical treatment. (2) Marked diseases of penis or of testicles such as **orchitis** (inflammation of testicles) following mumps, syphilis, cancer and tuberculosis, may lead to impotency or sterility or both. (3) **Inflammatory afflictions** of the testicles, epididymis, prostatic gland and seminal vesicles of gonorrhoel origin are the frequent causes of impotency and sterility. (4) **Varicocele** (swelling of spermatic cord) is known to cause infertility. (5) The ejaculatory ducts may be obliterated by chronic gonorrhoea or tuberculosis, so that, the seminal discharge may be prevented from flowing into the urethera. (6) Atrophy of testicles occurring after mumps may produce impotency or sterility. (7) Accidental injury to genitals or any other parts of body (*e.g.*, head or brain and spinal cord or spine, *i.e.*, vertebral discs) may produce temporary and permanent impotency. (8) Individuals who have attended X-ray department without proper protection may suffer from temporary **azoospermia** (complete absence of spermatozoa in semen).

4. General diseases. Endocrine disturbances such as testosterone deficiency cause impotency. Certain general diseases such as diabetes mellitus, autonomic neuropathy, pulmonary tuberculosis, chronic nephritis, some forms of mental diseases (paralysis), etc., may produce temporary or permanent impotency. The excessive and continued use of some drugs such as alcohol, opium, *Cannabis indica*, tobacco, cocaine and bromides may render a man impotent.

5. Psychogenic origin. A temporary absence of desire for sexual intercourse may result from fear (of causing pregnancy or of sexually transmitted diseases), anxiety, guilt sense, timidity, aversion, hypochondriasis (a chronic condition in which the patient is morbidly concerned with his own health and believes himself suffering from grave bodily diseases), excessive passion, sexual over indulgence, and religious inhibitions and emotional immaturity. Persons with homosexual tendencies may be impotent. Sometimes, an individual may be impotent with one particular woman but not with another. Ignorance of sexual technique or infrequent intercourse in non-ovulating period may also cause sterility.

30-2. CAUSES OF STERILITY IN THE FEMALE

Female infertility or the inability to conceive may be caused due to the following factors :

1. Age. Puberty in the female usually begins at the age of thirteenth or fourteenth year in India. A state of functional or physiological sterility exists in most young girls because of the initial menstrual cycles being anovulatory. The age of onset of **menstruation** is very variable and factors like race, heredity, general health, environment, climate, diet and hygiene play a part. The average age of menarche (onset of menstruation) in India is 13th to 14th years with a range of 10th to 17th years (**Kerkar**, 1985). Luxurious living and early stimulation of the mental faculties tend to bring on menstruation at an earlier age, while feeble health and poor diet tend to retard it.

2. Malformation. Congenital malformations such as the total occlusion of the vagina, adhesion of the labia and the tough, imperforate hymen, are barriers to coitus and consequently lead to sterility, but these malformations can be remedied by surgery. The congenital absence of vagina will ordinarily render a female completely and permanently impotent and sterile. In such cases the uterus is commonly non-existent. Absence of ovaries and Fallopian tubes also produce sterility.

3. Local diseases or injuries. The inflammatory afflictions of gonorrhoel infection involving the cervix, uterus, and Fallopian tubes often produce sterility. Displacements and tumours of the uterus are possible causes of sterility. Rectovaginal fistula, disorders of menstruation, leucorrhoea and acid discharge from the vagina may contribute towards sterility.

4. Psychogenic origin. Hatred, fear, passion, neurotic temperment, etc. may produce a hysterical fit on an attempt at copulation and may, thus, render a women temporarily barren if she happens to be a virgin.

30-3. TESTS FOR STERILITY

30-3.1. Tests for Male Sterility

The first test is the microscopical examination of the semen of the man. This has to be done within an hour of its emission from the body. Sperms collected directly from man as well as from the neck of the cervix of the woman immediately after intercourse are examined. The fertilization potential of the human sperm is found to depend on the following six factors : 1. volume of the entire ejaculate ; 2. total number of sperms in the ejaculate ; 3. number of spermatozoa per millilitre of semen ; 4. percentage of motile spermatozoa; 5. quality of motility of the sperms ; and 6. percentage of spermatozoa with abnormal shape.

1. Effect of sperm count on fertility. The usual quantity of semen ejaculated at each coitus averages approximately 2.5 ml to 6 ml, and the average range of spermatozoa ejaculated is 50 to 100 million/ml. When the number of spermatozoa falls below 20 million/ml, the male is likely to be sterile (**Tortora** and **Anagnostakos**, 1984).

2. Effect of sperm morphology and motility on fertility. Occasionally a man has a completely normal number of sperm but is still infertile. When this occurs, often as many as half the sperm are found to be abnormal physically, such as having two heads, abnormally shaped heads, or abnormal tails. Whenever the majority of the sperms are abnormal in morphology or are found to be non-motile, the person is likely to be infertile. Lack or insufficient quantity of acrosomal enzymes such as hyaluronidases and nucleolytic enzymes too leads to male sterility.

30-3.2. Tests for Female Infertility (Sterility)

The main test for the female is that of finding out whether the Fallopian tube is open or not. This is done by the **Rubin test**. In this test carbon dioxide gas under pressure is sent into the uterus which will go through the Fallopian tubes and enter the abdomen. The presence of gas and its movements can be tested by stethoscope or by clinical symptoms. The Rubin test must be conducted twice.

Laparoscopy and diagnostic dilation and curettage (D and C) is also carried out for checking the defects of reproductive structures of sterile female. These tests are performed under anaesthesia. The technique of **laparoscopy** involves inflating the abdominal cavity with nitrous oxide or carbon dioxide. The inside structures are then visually examined with the help of a long flexible rope-like instrument (called **endoscope**) which is passed through a tiny hole made on the abdominal wall, below the naval. Abnormalities in the womb and pelvis, benign tumours, obstruction in the Fallopian tubes, activity of ovaries, etc., can be easily seen and identified. Visual examination of uterus, called **hysteroscopy** can also be done. As already described in previous chapter (Chapter 29), **D and C procedure** is nothing but dilating the mouth of the uterus and cleaning of the internal cavity. This helps in removing anything that prevents the fertilized egg from implanting on the uterine wall.

Tubal and ovarian disease are diagnosed by an injection of dye through the cervix during laparoscopy. The whole procedure called ''**dye spillage**'' is performed before the menstruation begins and the changes occurring in the lining of the uterus are followed closely which give definite information about ovulation and can also rule out any genital infection such as tuberculosis. Two days after the menstrual bleeding stops **hysterosalpingography** or **dye test** is performed. This procedure can detect abnormalities of the uterus and Fallopian tubes which may not have detected during laparoscopy. A radio-opaque 'dye' is inserted through the uterus and into the Fallopian tubes and X-ray photographs are taken.

In fact, the most common cause of female sterility is failure to ovulate. This can result from hyposecretion of gonadotropic hormones, in which case the intensity of the hormonal stimuli simply is not sufficient to cause ovulation. Because of the high incidence of anovulation in sterile women, special methods are often utilized to determine whether or not ovulation occur. These are all based on the effects of **progesterone** on the body, for the normal increase in progesterone secretion does not occur during the latter half of anovulatory cycles. In the absence of progesteronic effect, the cycle can be assumed to be anovulatory. One of these tests is simply to analyse the urine for a surge in **pregnanediol**, a end product of progesterone metabolism, during the latter half of the sexual cycle, the lack of which indicates the failure of ovulation.

Further, following laparascopic tests, cyclic evaluation is started. Daily measurement of basal body temperature (BBT) is done to detect when actually the ovulation is occurring. The temperature of the body slightly rises around ovulation time (secretion of progesterone hormone during the latter half of the cycle raises the body temperature about 0.5 °F and this indicates ovulation). The woman is asked to measure the oral temperature immediately after getting up from sleep, before leaving the bed every morning. The temperature is measure on a special fertility Farenheit thermometer. Along with this, the woman also has to undergo a cervical mucus study, especially during mid-cycle. The doctor studies the mucus taken from the cervix and a smear of fluid from the vagina is also taken for examination under the microscope. The thickness and viscosity of mucus and the vaginal cells undergo cyclical changes which help to know the hormonal status and to detect ovulation. The mucus is thin at ovulation time and it helps in sperm penetration.

To confirm the occurrence of ovulation, **ultrasonography** is done during the mid-cycle. Using high frequency sound waves and studying their reflection on a screen the ovarian follicular development can even be seen and the time of ovulation can be detected. In addition to these tests, **hormonal assays** are carried out. Measurements of serum (blood) levels of hormones such as FSH, LH, prolactin, progesterone, thyroid hormones (T_3, T_4) and thyroid stimulating hormones (TSH) are carried out using sophisticated techniques such as **radioimmunoassay** using a radioactive tracer. These tests can detect even minute changes in the levels of these hormones which occur in blood in trace amounts. Any abnormality in any of these hormonal levels, can be a cause for infertility.

30-4. TREATMENT OF STERILITY

In males if there is any blockage in the vas deferens, it can be removed by surgery. The sperm count can be improved by hormonal treatment such as injection of testosterone and oral administration of thyroxine sodium (or Elthroxin). Recently, following two techniques have also been employed to correct the erection problem of penis : (1) implantation of silicon rods in the penis ; (2) by-pass surgery to increase the flow of blood circulation in the penis (**Dastur**, 1984).

In females, most of the structural malformations can be corrected by surgery. The secretions (mucus) that form plug-like structure at the neck of the cervix can be removed. If the Rubin test shows positive reaction for blockage of Fallopian tubes, then inflation by carbon dioxide is tried. This may remove the blockage in the tubules. In case of faulty ovulation or infertility related to anovulation (no ovulation as in polysystic ovarian syndrome), ovulation can be induced by a synthetic drug, **clomiphene citrate**. Women who fail to respond to this drug, are treated with hormones such as **human menopausal gonadotropins** and **human chorionic gonadotropins** (**HCG**). **Bromocryptine** is used to induce ovulation when the prolactin level is raised in blood. Other drugs such as **Metrodin** or **Cyclofenil** may also be used for high prolactin level. In presence of **macroadenomas** (*i.e.*, tumour of the pituitary) removal by surgery of the tumour is essential before induction of ovulation (**Sheth**, 1991).

However, HCG, which is a hormone extracted from the human placenta, is found to have a disadvantage it can cause ovulation from many follicles simultaneously and this result in multiple births (up to 7). Dangerous enlargement of the ovaries and side effects may also occur. Therefore hormonal treatment should be given very carefully and by a competent physician. This must be resorted to only as a last resort.

Women with totally blocked or infected Fallopian tubes can resort either to *in vitro* fertilization (IVF) and embryo transfer (ET) (*i.e.*, **test-tube baby**) or **GIFT** (Gamete Intra Fallopian Transfer). Women whose husbands have low sperm count or no sperms are **artificially inseminated** with either husband's or donor sperm. Such varied latest techniques have struck a ray of hope in the infertile couples.

30-5. ARTIFICIAL INSEMINATION

When semen is deposited in the vagina, the cervical canal or the uterus by artificial means, such as instruments, this method of bringing about pregnancy in an otherwise healthy woman who is unable to conceive a child through sexual intercourse with her husband, is called **artificial insemination** (**A.I.**). Artificial insemination in women is not done in India mainly due to the moral and legal problems involved in it. In U.S.A. it has been a regular feature. The technique of artificial insemination has been used both for human beings and animals.

30-5.1. Artificial Insemination in Human Beings

For avoiding medico-legal complications, artificial insemination is done only when the woman's husband has the following defects : inability to deposit semen into the vagina ; inability of husbands spermatozoa to gain entry into the uterus ; occurrence of the urinary opening at the undersurface of the penis (**Hypospadias**) ; premature ejaculation ; azoospermia or painful sexual intercourse on the part of woman. AI is also done when there is (i) incurable defects in the husbands semen rendering him incapable of procreation ; (ii) mental disease in the husband ; (iii) hereditary disease in husband.

In AI, if the semen of women's husband is used, the procedure is called **artificial insemination homologus** (**AIH**) ; if the semen of some other unrelated donor is used, then it is known as **artificial insemination donor** (**AID**) or **heterologous artificial insemination.** In case of AID, the donor should be unkown to both the women and her husband and conversely the donor should not know either of them. Donor should not know the result of insemination. The doctor should choose a donor below the age of 40, having no relation to the couple, should have had children of his own, should be biologically and genetically satisfactory and healthy in all respects. The wife of the donor should also agree. Lastly, the donor should have resemblance to the husband and blood type of donor and that of husband should preferably be the same.

Further before the physician undertakes to inseminate a woman, it should have been established that there is no bar to her conceiving, that her husband is sterile, or that there is a risk, if he fathers a child, or that he may pass some hereditary disease to the child. The relevant record should be maintained by the doctor and while he or she artificially inseminating the patient, the presence of a nurse is essential.

Surrogate birth. This technique involves artificial insemination of another woman (called surrogate mother) with the sperms of the barren woman's husband. After surrogate birth the baby is handed over to its biological father and his wife. It is now also possible to remove a mature healthy ovum from the abdomen of the wife and fertilize it with the husband's sperm "in vitro" in the specialized laboratory and reimplant this fertilized ovum/embryo in the 'hired' womb of another woman.

30-5. 2. Artificial Insemination in Cattle

In cattle, artificial insemination is practiced for the following three reasons : (1) to obtain maximum utility of the semen of a superior breed of bull; (2) to bring about remote mating ; (3) to prevent sexually transmitted diseases such as Brucellosis (which causes abortion) and Vibriosis. The technique of artificial insemination of cattle involves the following steps :

1. Collection and processing of semen. Semen is collected only from a pedigreed bull, as the main aim is the propagation of superior germplasm. For example, for the milch cows, the breeds that are popularly selected are Jersy, Holstein, Fresion, and Red Dane ; while Murrah is the breed sought after for the buffaloes. Semen is collected at the rate of 2 or 3 collection per week per bull. An artificial vagina is used for this purpose. It is made up of rubber and has a latex lining, including a latex cone and a graduated **semen collecting tube**. The cone is surrounded by warm water at temperature of 4°C to simulate the normal vagina. The bull is made to mount on model of a cow or even males and the protruding penis is caught by the artificial vagina. Normally 5 to 10 ml of semen is ejaculated at a time.

The processing or extending of the semen involves dilution. The dilution material is called **egg yolk citrate dilutant**. The materials involved in it are yolk of hen's egg and sodium citrate. Yolk is very essential to prevent cold shock. Before the semen is processed, it has to be tested for quality. The factors taken for consideration are pH value, colour, consistency, motility, mass activity and density. The colour should be cream and consistency thick. Motility is described in terms of different grades. Grade 5, for example, has above 80 per cent spermatozoa with progressive movement. Grade 4 shows between 60 to 80 per cent progressive movement and grade 3 has between 40 to 60 per cent progressive movement of sperms. Normally semen with a motility above grade 3 is processed, while if it is grade 3 or less, it is discarded. The percentage of dilution depends on the need of the insemination material and also according to the concentration of the spermatozoa present. The presence of 10 to 15 million active spermatozoa per cubic millilitres is considered to be fit for fertilization.

2. Storage. The collected and processed semen can be stored at a temperature between 5 and 15°C in a refrigerator (ideal temperature is 5°C). A low temperature reduces metabolic activity and conserve energy. Maximum storage life at this temperature is only 72 hours. However, the spermatozoa with 24 to 48 hours of storage give ideal results.

Deep-freezing of semen. If the semen has to be stored for a long duration, then the deep-freezing method is adopted. Frozen semen can be preserved for years and can be transported to any part of the world and hybrids produced everywhere. Here the semen is stored under the ultra storage temperature of – 196°C in liquid nitrogen in a state of suspended animation. The semen is kept in PVC straws, each containing 0.5 to 0.25 ml., and frozen in liquid nitrogen vapour at — 145°C and then stored in liquid nitrogen at – 196°C. Semen banks can be established in this way. Whenever required for use the PVC straw is taken out, thawed and loaded in an artificial insemination gun for insemination purpose.

3. Insemination. In artificial insemination, normally the semen is introduced by means of a instrument called the **vaginal speculum**. This instrument dilates the vagina and helps in the introduction of the semen into the cervix directly. The latest method used for insemination is called the **recto-vaginal technique**. In this method, one hand is passed through the rectum and the opening of the cervix is grasped by it while the other hand is used for guiding the catheter into the vagina and for the deposition of the semen. Insemination is done to the cow only when it is in period of heat. About 1 ml of semen is injected in each insemination. In our country, the artificial insemination has helped in bringing about the much talked about **white revoluion**.

30-6. TEST TUBE BABY

In last two decades, tireless efforts of certain biologists have fulfilled the age-old dream of mankind to create human life in the laboratory. This phenomenon is called **extrauterine gestation** or **in vitro fertilization** (*in vitro* are Latin words which literally mean 'in glass', but meaning in any artificial medium). The child produced by such a method came to known as a **test tube baby**. In fact, the production of a test-tube baby involves two main steps : *in vitro* fertilization (IVF) and *embryo transfer* (ET) (**Menon** *et al.*, 1989).

Historically speaking, the first attempt to produce a test-tube baby was made by an Italian scientist **Dr. Petrucci** in 1959, when he began his epoch-making experiment. He removed an ovum from a patient and put it among swarming millions of spermatozoa, one of which met and fertilized the egg under rigidily monitored conditions in a glass dome. The embryo is developed. Although this human embryo survived for just 29 days, **Petrucci** experiment opened up new vistas that had potential to revolutionize the future of mankind (see **Sarin**, 1985). In 1974, **Dr. Douglas Bevis**, a gynaecologist from Leeds University in England, startled his scientific colleagues and rest of the world with the amazing announcement that he had *in vitro* fertilized human eggs that developed into embryos and implanted these in the uteri of three women ; each woman carried an embryo to term and gave birth to a baby. The children, from 12 to 18 months of age at the time of the announcement, were reported to be quite healthy. Many of Bevis' colleagues have doubted the veracity of his report, because, he has refused to publish the details of his research. Although Bevis has not described the means by which he brought about *in vitro* fertilization and coaxed the egg to develop into an embryo, the technique for doing so is fairly well established. In a book on the subject, the British researchers **Austin** and **Short** discussed the steps involved in the production of test tube babies (see **Goldsby**, 1976).

The best documented work of *in vitro* fertilization and embryo transfer has been done by two English scientists, the physiologist **Robert Edwards** and the gynaecologist **Patrick Steptoe** who had been working for over a decade to perfect a technique for fertilizing human eggs outside the human body. On July 25, 1978, the world's first 'test tube baby' – a baby girl named **Louise Joy Brown**, was born. The Brown's parents had been trying to have a child for some nine years. Though the experimental success of production of test tube baby is happened in this century, the very idea of transplanting an embryo has occurred in the mind of a Cambridge University biologist **Walter Heape** in 1890 when he successfully experimented with rabbits for the transfer of fertilized eggs from one animal to another.

30-6.1. Technique of Production of Test Tube Baby

According to **Chakravarthy**, 1989, the production of a test tube baby involves the following six steps : 1. Options [Ovulation by natural cycle or hormonal stimulation (or controlled ovarian hyperstimulation)]; 2. Ovarian functions ; 3. Follicular aspiration [Laparoscopic or ultrasonograph guided-USG-ovum pick up (OPU)] ; 4. In vitro fertilization (IVF) (Incubation and insemination) ; 5. Cleavage ; and 6. Embryo transfer (ET).

Thus, the first requisite to do the test tube baby operation is to get the ovum from a suitable woman. For this an inert gas such as carbon dioxide is pumped into a small incision made just below the navel of the female in order to distend the abdomen to allow for manipulation of instruments. A thin instrument called a **laparoscope**—which is a telescope having an external light source—is inserted through the incision to observe the ovaries. With the ovary clearly spotlighted, the physician inserts a syringe into a follicle on the surface of the ovary and sucks up any eggs present there. This operation is performed under general anesthia and the patient can usually leave the hospital the same day or shortly thereafter.

Some days before the surgery, the patient is injected with certain hormones (gonadotropin injection) which helps the eggs to ripen and assure that more than one egg is produced by the ovary. After the egg has been removed, it is placed in cultures containing calf serum or human serum and other chemicals, together with the spermatozoa ejaculated by the husband of the patient or of a donor. Within 12 hours after the introduction of sperm, the egg is fertilized. Within 38 hours, it cleaves into two çells, reaching the eight-celled stage within three days (72 hours) after sperm penetration and the 16-celled stage or blastocyst stage within five days. The developing blastocyst, now ready for implantation, is inserted into the uterus in one of two ways—by catheter through the vagina and cervical canal to the womb or through an incision made in the wall of the uterus.

However, prior to implantation of the blastocyst, the uterus of the recepient mother is made fit for receiving the 8-or 16-celled embryo by the hormonal treatment. After implantation, the developmental process follows the regular course.

The whole process of test tube baby formation requires a high level of technical skill and advanced facilities. It needs the following precautions :

1. The egg (ovum) must be taken out only in fully mature condition. 2. The egg as well as the culture medium must be kept at proper temperature. 3. Infection of any kind must be avoided at all levels. 4. Receipient mother's uterine wall must be properly prepared to receive the blastocyst. 5. Careful watch must be maintained throughout the gestation period to ensure that the embryo is not expelled.

30-6.2. Advantages and Disadvantages of Test Tube Baby

A. Advantages. 1. Chief advantage of *in vitro* fertilization is obvious : it enables a women with diseased or missing oviducts to give birth to her own child. Thus, it is a boon to infertile mothers. 2. Even mothers who have damaged ovaries can hope to get children, if another woman is willing to donate her ova. 3. For the woman whose uterus do not allow the spermatozoa to move up and men who have a low sperm count (oligospermia), this method can be ideally adopted. 4. Research with *in vitro* fertilization is yielding vital information in other areas, including birth control techniques and possibly the mechanisms involved in the growth of cancer.

B. Disadvantages. 1. This operation cannot be done in women who have damaged uterus wall. 2. The whole process requires high cost and technical skill and involves many risks. 3. Many women may not agree for the operation for the removal of eggs. 4. There may arise certain legal problems as well as religious and moral objections.

30-7. GAMETE-INTRA-FALLOPIAN TRANSFER (GIFT)

This is an alternative to *in vitro* fertilization and embryo transfrer for the women with prolonged sterility and at least one patent Fallopian tube. This is especially useful where the fimbriae (A fringe of finger-like projections that surround the infundibulum of a Fallopian tube) fails to capture the ovum or in those with cervical factors and sperm antibodies. Washed sperms and harvested ova are transferred via a laparoscope to the ampulla of the Fallopian tube. Fertilization and cleavage occur in the tube followed by transportation of the blastocyst to the endometrial cavity (*i.e.*, uterus). GIFT technique requires less sophisticated equipment and training. In a series of 500 GIFT operations, 192 (38.4 per cent) pregnancies occurred (**Rao**, 1989).

REVISION QUESTIONS

1. What is impotency ? How is it differ from sterility of man ?
2. Describe main causes of sterility in man and woman.
3. Summarise various tests for detecting women infertility.
4. Describe the treatment of male impotency and sterility and also of female infertility.
5. Describe the procedure of artificial insemination of cattle.
6. Describe the method of production of test tube babies.
7. Write short notes on the following :

   (i) GIFT ; (ii) Test tube baby ; (iii) Artificial insemination ; (iv) Laparoscopy ; (v) Male sterility ; and (vi) Female infertility.

GLOSSARY

A. (Gk., *an* = not). Prefix, negates the succeeding part of the word.

Ab. (L., away = off). Prefix, meaning ''away from'' or ''off''.

Abdomen. (L. *abdere* = to hide). In vertebrates that portion of the trunk containing visceral organs except for heart and the lungs ; in arthropods, the posterior portion of body composed of a group of similar segments and containing reproductive organs and posterior portion of digestive tract.

Absorption. The passage of water and dissolved substances into the cell or organism.

Acetylcholine. One of the known chemical transmitters of nervous impulse across synaptic junctions.

Acro. (*Gk., acron = tip*). Prefix, meaning ''pertaining to tip''.

Acrosome. (Gk., *acron*=tip, + *soma*=body). A membrane-bounded cap-like structure at the forwarding end of a spermatozoon. It contains proteolytic enzymes, has a derivation from Golgi-complex, and functions to create a clear passage through egg-envelops and egg-plasma membrane during fertilization process.

Actin. Force-generating cytoplasmic and muscle protein, present in both globular (G-actin) and fibrous (F-actin) forms.

Active transport. The energy-expending process by which a cell moves a substance across the cell plasma-membrane from a point of lower concentration to a point of higher concentration, against the diffusion gradient.

Ad. (L., toward to). Prefix, meaning ''toward'' or ''to''.

Adaptation. Adjustments in some way to the environment.

Adenine. A purine base occurring in DNA and RNA. Pairs normally with thymine in DNA.

Adenohypophysis. (Gk., *aden*=gland + *hypophysis*=an undergrowth). The anterior lobe of the hypophysis (pituitary).

Adenosine triphosphate (ATP). The major source of usable energy in cell metabolism; composed of adenine, ribose, and three phosphate groups. On hydrolysis, ATP loses one phosphate and one hydrogen to become adenosine diphosphate (ADP), releasing energy in the process.

Adenylate cyclase. Membrane-bound enzyme that catalyzes the production of cAMP from ATP.

Adhesion. (L., *adhaerere*=to stick to). A sticking together of substances, cells or blastomeres in a blastula.

Afferent. (L. *ad*=to + *ferre*=to bear). Bringing inward to a central part, applied to nerves and blood vessels.

Albumen. (L., *albumen*=white of egg). A proteinous substance which by absorption of water forms white of an egg such as in birds.

Allantois. (Gk., *allantoeides*=sausage-shaped). An extraembryonic sac-like extension of the hindgut of amniotes. It functions as an embryonic urinary bladder or as a respiratory extension of the hindgut in reptiles and birds, and in mammals it is modified to carry blood vessels to and from the placenta.

Amino-acids. (Gk., *Ammon*=referring to the Egyptian sun god, near whose temple ammonium salts were first prepared from camel dung). Organic molecules containing an amino (NH_2) group and a carboxyl (COOH) group, plus a side chain; the "building blocks" of protein molecules.

Amniocentesis. Puncture of the uterine wall with a needle for the purpose of obtaining amniotic fluid containing sloughed foetal cells, which can be analyzed to determine whether the foetus has a genetic abnormality.

Amnion. (Gk., *amnion*=foetal membrane). An extraembryonic, inner, fluid-filled sac composed of a thin double membrane that surrounds the embryo in reptiles, birds and mammals. It provides a kind of private aquarium to the embryo and protect it from mechanical shocks such as pressure, abrasion, irritation and desiccation.

Amniote egg. An egg which is isolated from the environment by a more or less impervious shell during the period of its development and which is completely self-sufficient, requiring only oxygen from the outside.

Amoeboid. (Gk., *amoibe*=change). Moving or eating by means of pseudopodia.

Amplification. The production of additional copies of a gene.

Anabolism. Constructive metabolic process during which protoplasm is produced from simple food molecules.

Anamniota. (Gk., *an*=without ; *amnion*=foetal membrane). Vertebrates that lack an amnion during development, *e.g.*, Cyclostomata, fishes and Amphibia.

Anaphase. (Gk., *ana*=up + *phasis*=form). A stage in karyokinesis of mitosis or meiosis in which chromatids of each chromosome separate and move from spindle-equator to opposite poles.

Andro. (Gk., *aner, andros*=man). Prefix, "man" or "male".

Androgen. A steroid male sex hormone secreted by interstitial and Sertoli cells of testes.

Angstrom unit (A°). A measurement of length or distance, often used in describing intra-or intermolecular dimensions ; equal to 1×10^{-10} metre and 1×10^{-1} nanometre.

Animal hemisphere. That half of the egg or zygote in which there is less yolk, which contains the nucleus and active cytoplasm, and which is most active in cell cleavage; contrasts with vegetal hemisphere.

Animal pole. Imaginary point on the surface of egg or zygote which lies in the centre of the animal hemisphere, opposite to the vegetal pole. The polar bodies are often given off at the animal pole of egg during maturation phase of oogenesis.

Antagonist. (Gk., *anti*=against, + *agonizomai*= I fight). That which opposes or resists the action of another.

Anterior. (L., *ante*=before, towards, in front of). The front end of an organism; in human anatomy, the ventral surface.

Antibiotic. An organic compound formed and secreted by an organism that is toxic to organisms of other species.

Antibody. A globular protein that is produced in the body fluids, in response to a foreign substance (antigen), with which it reacts specifically.

Antigen. A foreign substance, usually a protein or polysaccharide, that stimulates the formation of specific antibodies, when introduced into a living organism.

Antrum. (Gk., *antron*= a cavity). Any closed or nearly closed cavity, such as fluid-filled follicular cavity of mammalian follicle.

Anus. (L., *anus*=fundament). The rear opening of the digestive tract.

Appendicular. (L., *appendere*=to hang upon). Relating to appendages or limbs.

Archencephalon. (Gk., *arche*=beginning, + *enkephalos*=brain). The anterior division of the primitive brain.

Archenteron. (Gk., *arche*=beginning, + *enteron*=intestine). The principal cavity within the early embryo (gastrula) of many animals ; lined with endoderm, it opens to the outside by means of the blastopore and ultimately becomes the digestive tract.

Artery. A vessel carrying oxygenated blood from the heart to the tissues ; it is usually thick-walled, elastic and muscular.

Asexual reproduction. Any reproductive process, such as fission, budding or gemmule formation, that does not involve the union of gametes.

Atrium. (L., *yard* = court, hall). A thin-walled chamber of the heart which receives the blood from veins and passes it on to a ventricle.

Auditory. (L., *audire* = to hear). Relating to the perception of sound.

Autolysis. A process in which a cell is disintegrated by its lysosomal enzymes.

Autonomic nervous system. A special system of motor nerves and ganglia in vertebrates that is not under voluntary control and that innervates the heart, gland, visceral organs and smooth muscles. It is subdivided into the sympathetic and the parasympathetic nervous system.

Autoradiography. Technique by which emission from a radioactively labelled microscopic specimen can be visualized by exposure of photographic emulsion to the labelled tissue.

Autosome. Any chromosome other than the sex chromosomes. Man has 22 pairs of autosomes and 1 pair of sex chromosomes.

Axial. (L., *axis* = axle of a wheel). Relating to the central part of the body.

Axis. An imaginary line passing through an egg, body or organ around which parts are symmetrically aligned, *e.g.*, animal-vegetal axis of an egg.

Axoneme. Axial microtubular component of cilia and flagella.

B cells. White blood cells (lymphocytes) produced by the bone marrow which produce antibodies.

B DNA. Double helical DNA in which the helices are right handed.

Basal body. (L., *basis* = footing or base). A cytoplasmic organelle that organizes cilia or flagella, identical in form and structure to centriole.

Base pair. Two nitrogenous bases, one in each strand of double-helical DNA, that interact through hydrogen bonds. The base pairs are adenine and thymine (*i.e.*, A and T) and guanine and cytosine (*i.e.*, G and C).

Basibranchial. (L, *basis* = base, + *branchia* = gill). Median-ventral component of a gill-arch.

Bilateral symmetry. An anatomical arrangement in which the right and left halves of an organism, are approximate mirror images of each other.

Bile. (L. *bilis*=gall). A yellow secretion of the vertebrate liver, temporarily stored in the gall-bladder and composed of organic salts that emulsify fats in the small intestine.

Biological clock. An unidentified factor (or factors) in plants and animals that governs the innate biological rhythms (growth and activity patterns) of the organisms.

Biosynthesis. Formation of organic compounds from elements or simple compounds by living organisms.

Blast, Blasto. (Gk. *blastos* = germ, sprout). Suffix or prefix, meaning ''pertaining to an embryo''.

Blastema. (Gk., *blastema* = a sprout). A primitive cell aggregate (*e.g.*, regenerating bud, an asexual reproductive body) from which a new organ or new individual called **blastozooid** develops by regeneration or blastogenesis activity.

Blastocoele. (Gk., *blastos* = germ, + *koilos* = hollow). The fluid-filled cavity in the blastula.

Blastocyst. (Gk., *blastos* = germ, + *kystis* = bladder). The hollow spherical blastula stage of early mammalian development.

Blastoderm. (Gk., *blastos* = germ, + *derma* = skin). The primitive cellular plate of a blastula containing presumptive cells for ectoderm, mesoderm, and endoderm and surrounding the blastocoele.

Blastodisc. (Gk., *blastos* = germ, + L. *discus* = plate). Disc-like area of active cytoplasm that contains zygote nucleus and occurring atop yolk-mass of macrolecithal eggs of reptiles, birds, etc. It forms the animal pole of egg and undergoes cleavage, blastulation, gastrulation, etc., to give rise the embryo.

Blastomere. (Gk., *blastos* = germ, + *meros* = part). One of the cells into which the egg divides during the cleavage phase of embryogenesis.

Blastopore. (Gk., *blastos* = germ, + *poros* = opening). The opening connecting the archenteron of the gastrula stage of an embryo with the outside ; represents the future mouth in some animals (**protostomes**), the future anus in others (**deuterostomes**).

Blastula. (Gk., *blastos* = germ). An animal embryo after cleavage and before gastrulation ; usually consists of a hollow sphere the walls of which are composed of a single layer of cells called **blastoderm**.

Brachial. (L. *brachialis* = arm). Relating to the arms.

Branchial. (Gk., *branchia* = gill). Relating to the gills.

Bronchus, pl. bronchi. (Gk., *bronchos* = windpipe). One of a pair of respiratory tubes branching into either lung at the lower end of the trachea ; it subdivides into progressively finer passageways, the **bronchioles**, culminating in the **alveoli**.

Buccal. (L., *bucca* = cheek). Relating to the mouth ; the surface toward a cheek.

Bud. In animals, an asexually produced protuberance that develops into a new individual by going through blastogenesis.

C value. Characteristic DNA content of the cells of a given species.

Capsula. (L., *capsa* = a chest). A capsule in any sense.

Carcinogen. Any physical or chemical agent that cause cancer.

Carnivorous. (L. *caro, carnis*=flesh, + *vorare* = to devour). An organism that obtains its food energy by eating animals.

Cartilage. (L., *cartilago* = gristle). A skeletal connective tissue of vertebrates; forms much of the skeleton of adult lower vertebrates and immature higher vertebrates.

Caudal. (L. *cauda* = tail). Relating to the tail or rear.

Cavernous. (L. *caverna* = cave). Relating to a cavern or cavity.

Cell. (L. *cella* = a small chamber). The structural and functional unit of plant and animal organization or protoplasm. It is composed of **cytoplasm**, one or more **nuclei** and is surrounded by a lipoproteinous **plasma membrane**. Cytoplasm is composed of colloidal matrix and contains various cellular organelles such as mitochondria, endoplasmic reticulum, Golgi complex, lysosomes, centrioles, microtubules, microfilaments, etc.

Cell culture. A growth of cells *in vitro*.

Central nervous system. In vertebrates, the brain and spinal cord ; in invertebrates it usually consists of one or more cords of nervous tissue plus their associated ganglia.

Centriole. (Gk., *kentron* = center). A cytoplasmic organelle which is composed of microtubules and which forms mitotic or meiotic spindle from microtubules and microfilaments during mitosis or meiosis.

Centromere. (Gk., *kentron* = centre, + *meros* = part). Region of constriction of chromosome that holds sister chromatids together and to which the spindle fibres (microtubules) become attached by the help of the **kinetochore**.

Cephalo. (Gk., *kephale* = head). Prefix, meaning head.

Cerebellum. (L., diminutive of *cerebrum* = brain). An enlarged part of the dorsal side of the vertebrate hind-brain (*i.e.*, metencephalon) ; chief muscle-coordinating centre.

Cerebrum. (L. *cerebrum* = brain). The principal portion of vertebrate brain (forebrain) which is derived from the roof of telencephalon and consisting of two cerebral hemispheres united by the corpus callosum.

Chalaza. (Gk., *chalaza* = a sty). The suspensory ligament of the yolk in a bird's egg.

Chiasma, pl. Chiasmata. (Gk., *chiasma* = a cross). The X-shaped figure formed by the meeting of two non-sister chromatids of homologous chromosomes during prophase I of meiosis ; a site of crossing over.

Chondroblast. (Gk., *chondros* = cartilage, + *blastos* = germ). An embryonic cartilage forming cell.

Chondrocranium. (Gk., *chondros* = cartilage, + *kranion* = skull). A cartilaginous skull.

Chord, Chorda. (L., *chorda* = cord, string). Suffix or prefix, meaning "cord".

Chorion. (Gk., *chorion* = a skin). An outermost extra-embryonic membrane of reptiles, birds and mammals. It contains ectoderm and somatic layer of mesoderm and surrounds the whole embryonic systems of embryo. In placental mammals it contributes to the structure of placenta.

Choroid. (Gk., *chorion* = skin, + *eidos* = like). Relating to a coat of membrane.

Chrom. (Gk., *chroma* = colour). Prefix, meaning "colour".

Chromatid. One of the two daughter strands of a duplicated chromosome which are joined by a single centromere.

Chromatin. The deeply staining nucleoprotein complex (DNA + histone and non-histone chromosomal proteins) of the chromosome.

Chromosome. (Gk., *chroma* = colour, + *soma* = body). Nucleoprotein structures, generally more or less rod-like during mitotic or meiotic nuclear division, the physical sites of nuclear genes which are arranged in linear order. Each species has a characteristic number of chromosomes.

Cistron. A segment of DNA specifying one polypeptide chain in protein synthesis.

Cleavage. (Ger., *kleiben* = to cleave). Rapid cycles of DNA synthesis followed by cell division in which the cytoplasm is partitioned without growth, occurring very early in embyronic development.

Cleidoic. (Gk., *kleido* = lock up). Referring to eggs, such as those of birds, that are insulated from the environment by albumen, membranes and shell.

Clitoris. (Gk., *kleitoris* = small hill). A small erectile body at the anterior part of the vulva of mammalian female ; homologous to penis.

Cloaca. (L., *cloaca* = sewer). The exit chamber from the digestive system ; also may serve as exit for the reproductive system and urinary system.

Clone. A group of cells or organisms, derived from a single ancestral cell or individual and all are genetically alike.

Cloning. Asexual production of a line of cells or organisms or segments of DNA genetically identical to the original.

Coel-, Coela-, Coelo-. (Gk., from *koilos* = hollow). Suffix or prefix, meaning "cavity".

Coelom. (Gk., *koilos* = a hollow). A body cavity formed between layers of mesoderm.

Cohesion. (L., *cohaerere* = to stick together). The union or holding together of like molecules of like substances.

Collagen. (Gk., *kolla* = glue). A fibrous protein material in bones, tendons and connective tissues.

Cone. A type of light-sensitive neuron in the vertebrate retina, concerned with the perception of colour and with the most acute discrimination of detail.

Connective tissue. A type of tissue which lies between groups of nerves, glands and muscle cells and beneath epithelial cells, in which the cells are irregularly distributed through a relatively large amount of intercellular material, includes bones, cartilage, blood and lymph tissue.

Contact inhibition. The tendency of normal cultured cells to limit their own movement, cell division or growth.

Copula. (L., *copula* = yolk, joining). Relating to a part connecting two structures.

Coracoid. (Gk., *korax* = raven, + eidos = appearance) : Shaped like a crow's beak.

Cornea. (L., *corneus* = horny). The transparent front portion of the eye.

Corneum. (L., *corneus* = horny). Relating to a horny layer.

Corpus. (L., *corpus* = the body). Any body or mass.

Cortex. (L., *cortex* = bark, rind). The outer portion of an organ, part or egg. Egg cortex of frog, for example, includes ectoplasm containing hyaline cytoplasm, microfilaments, cortical granules and pigment granules plus plasma membrane and primary egg membrane.

Cotyledonary. (Gk., *kotyledon* = the hollow of a cup). Referring to aggregations of villi on the chorionic surface of a placenta.

Cranium (Gk., *kranion* = skull). The bones of the head collectively.

Crossing over. The exchange of corresponding segments of genetic material between non-sister-chromatids of homologous chromosomes at meiosis.

Cyclosis. (Gk., *kyklosis* = circulation). The circulation of protoplasm within a cell.

Cyte, Cyto. (Gk., *kytos* = cell, vessel, container). Suffix or prefix, meaning ''pertaining to cell''.

Cytokinesis. (Gk., *kytos* = cell, + *kinesis* = motion). Division of the cytoplasm of a cell, during cell division.

Cytoplasm. The living matter of a cell excluding nucleus.

Cytosine. A pyrimidine base occurring in DNA and RNA. Pairs with guanine in DNA.

Cytoskeleton. Complex cytoplasmic network of microtubules and microfilaments located predominantly near the cell membrane and thought to be responsible for cell shape and movement.

Cytotrophoblast. (Gk., *kytos* = cell + *trophe* = nourishment, + *blastos* = germ). The inner or cellular layer of the trophoblast.

Dalton. Unit of molecular weight ; one dalton equals one-twelfth the mass of carbon 12.

De. (L., away from, down, off). Prefix, meaning ''down'', ''away from'', or ''off'' : for example, dehydration, ''removal of water''.

Decidua. (L., *deciduus* = falling off). The endometrium of the pregnant uterus.

Dedifferentiation. A process in which the normally stable state of cell specialization to which a cell has been previously committed, is replaced by apparent backtracking to a less clearly specialized state.

Deferens. (L., *de* = away + *ferens* = carrying). Applying to a duct carrying away.

Deoxyribonucleic acid (DNA). A chemical macromolecule called nucleic acid, that acts as a carrier of genetic information in cells and which is composed of two chains or strands of phosphate, sugar molecules (deoxyribose) and purines (adenine and guanine) and pyrimidines (thymine and cytosine) wound in a double helix or duplex. It is capable of self-replication as well as of determining RNA synthesis and forms genetic material of all living organisms and many viruses.

Dermatome. (Gk., *derma* = skin, + *tome* = cut, section). One of the embryonic skin segments.

Dermis. (Gk., *derma* = skin). The inner layer of skin beneath the epidermis.

Desmosome. Mechanical attachment formed by the thickened plasma membrane of two adjacent cells, containing intermediate filaments.

Determination. A cell is said to be determined when its specialized fate is fixed but the overt demonstration and realization of that fate has not yet become apparent.

Development. The progressive production of the phenotypic characteristic of an organism.

Di. (Gk., *dis* = twice, double = two). Prefix, meaning "twice", "double".

Differentiation. The development process by which a relatively unspecialized cell or tissue undergoes a progressive change to become a more specialized cell or tissue.

Dioecious. (Gk., *di* = two, + *oikos* = house). Having the male and female reproductive systems in different individuals of the same species.

Diploid. (Gk., *diplous* = double, + *eidos* = resemblance). The full number of chromosomes in the fertilized ovum and in all cells, except the mature germ cells.

Discoidal. (Gk., *diskos* = disc, + *eidos* = appearance). Resembling a disc or plate.

Distal. (L., *distalis* = distant). Farthest from the centre or median line.

Dorsal. (L., *distalis* = back). Relating to the top or back.

Ductus. (L., *dustus* = to lead). A tubular structure giving exit to or conducting any fluid.

Ecdysone. Insect steroid hormone that can induce chromosomal puffing and regulates the process of moulting or ecdysis.

Ecto. (Gk., *ektos* = outside). Prefix, meaning "outside", "outer".

Ectoderm. (Gk., *ektos* = outside, + *derma* = skin). (1) The outermost layer of body tissue. (2) One of the three germinal layers of the gastrula that gives rise to the epithelial tissue of the skin, to sense organs, to the nerve cells, etc., of the adult animal.

Efferent. (L., *ex* = out, + *ferre* = to bear). Carrying away from a centre, applied to nerves and blood vessels.

Egg. A haploid large-sized, immotile, nutrient-filled and metabolically inert, female gamete which is manufactured by ovary of female through oogenesis.

Embryo. (Gk., *en* = in, + *bryei* = to swell). The early developmental stage of an organism produced from a fertilized egg ; a young organism before it is capable of leading a self-supporting existence and before it emerges from the seed, egg, or the body of its mother. In the mammal the later embryonic stage is called a foetus.

Embryogeny. (Gk., *embryon* = embryo, + *gennao* = I produce). The origin and growth of the embryo from a fertilized egg.

Embryology. (Gk., *embryon* = embryo, + *logia* = discourse). The study of the origin and development of the embryo.

Endocrine gland. (Gk., *endon* = within, + *krineni* = to separate). Ductless gland whose secretions (hormones) are released into the circulatory system ; in vertebrates, including pituitary, sex gland, adrenal, thyroid, and others.

Endocytosis. Collective term for phenomenon involving the surrounding and ingestion of various substances by the plasma membrane.

Endoderm. (Gk., *endon* = within, + *derma* = skin). (1) The innermost layer of body tissue. (2) One of the three germinal layers of the gastrula that gives rise to epithelium that lines certian internal structures, such as most of the digestive tract and its outgrowths, most of the respiratory tract and the urinary bladder, etc., of adult animal.

Endometrium. (Gk., *endo* = within, + *metra* = uterus). The glandular lining of the uterus in mammals ; thickens in response to progesterone secretion during ovulation and is sloughed off in menstruation.

Endotheliochorial. (Gk., *endon* = whtin, + *thele* = nipple). Placenta type involving contact between chorionic epithelium and uterine blood vessels.

Epitheliochorial. (Gk., *epi* = upon, + *thele* = nipple, + *chorion* = skin). Placenta type involving contact between chorionic epithelium and uterine epithelium.

Enzyme. A protein molecule that regulates the rate of (or catalyzes) a chemical reaction.

Epiblast. (Gk., *epi* = upon, + *blastos* = germ). The outer or upper layer of a double layered embryo, *e.g.*, chick.

Epidermis. (Gk., *epi* = upon, + *derma* = skin). The outermost epithelial portion of the skin.

Epimere. (Gk., *epi* = upon, + *meris* = part). The dorsal region of mesoderm on each side of the neural tube which gives origin to dermis, vertebral column, appendicular skeleton and skeletal muscles.

Epigenesis theory. This universally accepted theory proposes that development involves a gradual increase in organization and complexity from essentially simple initial structures. Presently this theory incorporates the processes whereby the genetic material, DNA, gives rise to the phenotype via transcription, translation and other aspects of gene expression.

Epithelial tissue. (Gk., *epi* = upon, + *thele* = nipple). A type of tissue that covers a body or structure, or lines a cavity ; epithelial cells from one or more regular layers with little intercellular material.

Erythrocyte. (Gk., *erythros* = red + *kytos* = cell). Red blood cells, the carrier of haemoglobin.

Estrogen. (Gk., *oistros* = mad desire, + *gennao* = produce). Female sex hormone that is secreted by ovaries and produce estrus.

Estrus. (Gk., *oistros* = mad desire, frenzy). Period of sexual excitement in the female.

Euchromatin. Uncondensed, genetically active regions of the chromatin.

Eukaryote. (Gk., *eu* = good, + *karyon* = nut, kernal). A cell having a membrane-bound nucleus, membrane-surrounded organelles, and multiple chromosomes in which the DNA is combined with special proteins (histone and non-histone proteins) ; an organism composed of such cells, *e.g.*, plants, animals.

Evolution. (L., *e* = out, + *volvere* = to roll). The doctrine that forms of animal or plant life have been derived by gradual changes from simpler forms.

Exocrine. (Gk., *exo* = out, + *krino* = I separate). Denoting external secretion of a gland or cell.

Exocytosis. Energy-requiring process by which secretory granules discharge their contents by fusing with the cell membrane.

Exon. Part of the gene that becomes part of the mature mRNA after the intervening sequences (introns) are spliced out.

Extra-embryonic membranes. Membranes formed of embryonic tissues that lie outside the embryo and are concerned with its protection and metabolism (respiration, excretion, food supply, etc.) ; includes amnion, chorion, allantois and yolk sac.

Fertilization. (L., *fertilis* = to bear). The fusion of two gametes (spermatozoa and ova) and so of their nuclei, to form a diploid or polypoid zygote ; it activates egg to form fertilization membrane outside egg plasma membrane, to start its metabolism and to start its cleavage.

Fibronectin. High-molecular weight glycoprotein component of the cell coat, thought to aid in cellular locomotion.

Fission. (L., *fissus* = split). Asexual reproduction in cells or organisms by division of body into two or more equal, or nearly equal parts.

Flagellum, pl. flagella (L., *flagellum* = whip). A fine, long, thread-like structure, which protrudes from a cell body and have a characteristic internal organization of nine pairs of microtubules encircling two central microtubules. Flagella are capable of vibratory motion and are used in locomotion (*e.g.*, sperm cell).

Fluid mosaic model. Model of plasma membrane structure postulating integral proteins embedded in the lipid bilayer but free to move laterally at one surface or the other.

Foetus. (L., *fetus* = pregnant). An unborn or unhatched vertebrate that has passed through the earliest development stages.

Follicle. (L., *folliculus* = a small bag). A vesicular body in the ovary, containing the ovum.

Frontal. (L., *frons* = brow). Relating to the front or forehead.

FSH (Follicle stimulating hormone). One of the gonadotropin hormones secreted in the anterior lobe of pituitary gland ; it stimulates the follicular development of ovaries.

Gamete (Gk., *gametes* = husband ; + *gamete* = wife). A mature functional haploid sex cell, spermatozoon or ovum whose nucleus fuses with that of another gamete of an opposite sex (fertilization), with the resulting cell (zygote) developing into a new individual.

Gametogenesis. (Gk., *gametes*, *gamete* = husband, wife, + *genesis* = origin). A process which is involved in the production of gametes inside the gonads.

Gap junction. Array of plaque-like connections between the plasma membrane of adjacent cells.

Gastrocoel. (Gk., *gaster* = stomach, + *koilos* = hollow). The cavity within the gastrula.

Gastrula. (Gk., *gaster* = stomach). An embryo in the process of gastrulation, the stage of development during which the blastula with its single layer of cells turns into a three-layered embryo, made up of ectoderm, mesoderm, and endoderm, often enclosing an archenteron.

Gene. (L., *gennao* = I produce). Hereditary unit specifying the production of a distinct protein (*e.g.*, an enzyme) or RNA.

Genital. (L., *genitalis* = pertaining to birth). Relation to reproduction or generation.

Genotype. The genetic constitution, latent or expressed, of an organism, as contrasted with the phenotype ; the sum total of all the genes present in an individual.

Germinativum. (L., *germen* = bud, germ). Relating to multiplication, development or growth.

Gill. The respiratory organ of aquatic animals, usually a thin-walled projection from some part of the digestive tract of a vertebrate.

Gland. A cell or organ producing one or more secretions which are discharged to the outside of the gland.

Glycogen. A polysaccharide that is used as stored food substance in most animals ; it is converted into glucose by hydrolysis.

Golgi complex. A cellular organelle present in eukaryotic cells, consisting of flat, disc-shaped sacs, tubules, and vesicles. It functions as a collecting and packaging centre for substances the cell manufactures.

Gonad. (Gk., *gone* = seed). Gamete-producing organ of multicellular animals ; ovary or testes.

Guanine. A purine base occurring in DNA and RNA ; pairs normally with cytosine in DNA.

Haploid. (Gk., *haploos* = single, + *eidos* = resemblance). An individual or cell (*e.g.*, gamete cells as a spermatozoon or ovum) containing a single set of chromosomes.

HeLa cell. Cell line originally (1952) derived from the uterine carcinoma cells of a patient named **Henrietta Lacks**. These cells have been maintained as a standard human cell line for histological and genetic experiments.

Hemochorial. (Gk., *haima* = blood, + *chorion* = skin). Placenta type involving contact between chorionic epithelium and maternal blood.

Herbivore. An organism that eats plants or other photosynthetic organisms to obtain its food energy.

Hermaphroditism. (Gk., *hermaphroditos* = the son of Hermes, Mercury, + *aphrodite* = venus). An organism possessing both male and female reproductive organs.

Histogenesis. (Gk., *histos* = tissue, + *genesis* = origin). The origin of a tissue ; the formation and development of the tissues of the body.

Histones. Group of five basic proteins that associate with DNA in the cell nucleus to form chromatin.

Holo. (Gk., *holos* = entire). A prefix, signifying entire or total.

Holoenzyme. The complete form of an enzyme (coenzyme + apoenzyme = holoenzyme).

Holoblastic. (Gk., *holos* = whole, + *blastos* = germ). Denoting the involvement of entire egg in cleavage.

Hormone. (Gk., *hormon* = I rouse or *hormain* = to excite). A chemical substance secreted, usually in minute amounts, in one part of an organism and transported via blood to another part of that organism where it has a specific effect on the target organ, the latter is stimulated to functional activity.

House keeping genes and proteins. A restricted set of genes whose products are required in every cell, whether specialized or not, since these products constitute the essential macromolecules (proteins) of cellular metabolism. Housekeeping proteins are found in all cells.

Hyaline. (Gk., *hyalos* = glass). Of a glassy, translucent appearance.

Hyperplasia. An increase in the amount of tissue as produced by cell division. Regeneration of damaged organs is normally accompanied by hyperplasia.

Hypertonic. (Gk., *hyper* = above, + *tonos* = tension). Having a concentration of solutes high enough to gain water (the solvent) across a selectively permeable membrane from another solution.

Hypoblast. (Gk., *hypo* = under, + *blastos* = germ). The inner or lower layer of a double-layered embryo.

Hypomere. (Gk., *hypo* = under, + *meris* = part). The most ventral sub-division of mesoderm which originates following parts of adult animal—parietal peritoneum, visceral peritoneum, mesenteries, heart, blood cells, blood vessels, gonads, muscles of viscera and coelom.

Hypothalamus. (Gk., *hypo* = under, + *thalamos* = inner room). The floor and sides of the vertebrate brain just below the cerebral hemispheres ; controls the autonomic nervous system and the pituitary gland and contains centers that regulate body temperature and appetite.

Hypothesis. (Gk., *hypo* = under, + *tithenai* = to put). A temporary working explanation or supposition based on accumulated facts and suggesting some general principle or relation of cause and effect ; a postulated solution to a scientific problem that must be tested by experimentation and, if not validated, discarded.

Hypotonic. (Gk., *hypo* = under, + *tonos* = tension). Having a concentration of solutes low enough to lose water across a membrane to another solution.

Immunoglobulin. One of the several types of antibodies secreted by plasma cells.

Induction. (L., *inducere* = to induce). The process in an animal embryo in which one tissue or body part causes the differentiation of another tissue or body part, *e.g.*, neural induction of neural ectoderm to form neural tube by chordamesodermal cells in amphibians.

Interstitial. (L., *inter* = between, + *sistere* = to set). Relating to space or structures within parts.

Interstitial cells (Leydig cells). The cells occurring in the connective tissues that remains filled in between seminiferous tubules of vertebrate testes and containing extensive smooth endoplasmic reticulum, a prominent Golgi complex, lipid droplets and numerous lysosomes and are the sites of steroidogenesis, *i.e.*, they synthesize steroid male sex hormones or androgens.

Intestine. (L., *intestinum*). The digestive tube passing from the stomach to the anus.

Intrinsic. (L., *intrinsecus* = on the inside). Belonging entirely to a part.

Intron. Intervening sequence of DNA, located within a gene, that is not included in the mature mRNA.

Invagination. (L., *in* = in + *vagina* = sheath). The local infolding of a layer of tissue, especially in animal embryos, so as to form a depression or pocket opening to the outside.

In vitro. (L., "in glass"). Experimentally induced biological processes outside the organism.

In vivo. (L., "in life"). Experimentally induced biological processes within the organism.

Involution. (L., *involvere* = to roll up). The turning of a part around a margin or edge.

Ion. An atom or molecule that has lost or gained one or more electrons. By this process, called ionization, the atom becomes electrically charged.

Iris. (L., *iris* = rainbow). The pigmented disc-like diaphragm of the eye.

Isolecithal. (Gk., *isos* = equal, + *lekithos* = yolk). Denoting a uniform distribution of yolk within an egg, *e.g.*, unfertilized eggs of *Amphioxus*.

Isotonic. Having the same solute concentration as that of the substance on the other side of a selectively permeable membrane.

Karyokinesis. (Gk., *karyon* = nucleus, + *kinesis* = motion). The division of the nucleus during cell division.

Karyotype. The general appearance of the chromosomes with regard to number, size, and shape.

Keratin. (Gk., *keras* = horn). One of a group of tough, fibrous proteins ; a horny substance formed by certain epidermal tissues, especially abundant in skin, claws, hair, feathers, and hooves.

Kidney. A vertebrate body organ that regulates the balance of water and solutes and the excretion of nitrogen wastes in the form of urine.

Kinetochore. Region of the centromere to which spindle microtubules attach during cell division.

Labium. (L., *labium* = a lip). Denoting any lip-shaped structure.

Lacuna. (L., *lacus* = a hollow). A small depression, gap or space.

Lagging strand. DNA strand formed by ligating the smaller fragments produced by discontinuous synthesis.

Lampbrush chromosomes. Chromosome with highly extended lateral loops of DNA; found in all animal oocytes at the diplonema stage of meiosis.

Lamella. (L., dim. of *lamina* = thin plate). A thin sheet, layer or scale.

Lateral. (L.,*latus* = side). On the side.

Larva. (L., ghost). An immature animal such as caterpillar or tadpole that is morphologically very different from the adult.

Leading strand. DNA strand formed by continuous synthesis.

Lipid. (Gk., *lipos* = fat). An organic compound containing C, H, and O and is made up of one molecule of glycerol and three molecules of one or more kind of fatty acids. Functions as a reserve supply of energy and also as building blocks for different cellular membranes, *e.g.*, fats, waxes, steroids, phospholipids, and carotenes.

Luteum. (L., *luteus* = yellow). Yellow in colour.

Luxury gene and proteins. Luxuary or cell-specific proteins are not found in all cells but only in certain specialized cell types. The genes coding them are called luxury genes.

Lysis, Lytic, Lyte. (Gk., *lysis* = a loosening). Suffix, meaning "pertaining to dissolving".

Lysis. Disintegration of cell.

Lysosome. A membrane-surrounded cellular oaganelle which contains many hydrolytic enzymes and originates from Golgi-complex ; they are used in the digestion of food contents of pinosome and phagosome and also in autolysis of cell.

Macrolecithal. (Gk., *makros* = large, + *lekythos* = yolk of egg). Referring to large amount of yolk stored in the egg, *e.g.*, eggs of insects, reptiles, birds and monotreme mammals.

Macromolecules. A molecule of very high molecular weight, refers specifically to proteins, nucleic acids, polysaccharides, and complexes of these.

Medulla. (L., *medulla* = marrow, pith). 1. The inner or central portion of an organ or part, as in the mammalian ovary. 2. The most posterior region of the vertebrate brain ; connects with the spinal cord.

Meiosis. (Gk., *meioum* = to make smaller). The two successive nuclear divisions in which the chromosome number is reduced from diploid (2n) to haploid (n) and various genetical events such as segregation and reassortment of the genes and synapsis and crossing-over occur ; gametes may be produced as a result of meiosis.

Menstrual cycle. (L., *mensis* = month, or *menstruare* = discharge). In certain primates (mammals), the periodic discharge of blood and disintegrated uterine lining through the vagina.

Meroblastic. (Gk., *meros* = part, + *blastos* = germ). Denoting the involvement of only restricted cytoplasmic area at the animal pole of the egg in cleavage, *e.g.*, birds.

Mesencephalon. (Gk., *mesos* = middle, + *enkephalos* = brain). The mid-brain, the second of the three primitive divisions of the brain.

Mesenchyme. (Gk., *mesos* = middle, + *enchyma* = infusion). Embryonic connective tissue.

Mesentery. (L., *mesenterium* = a membranous support). A double layer of peritoneum enclosing tissue.

Mesoderm. (Gk., *mesos* = middle, + *derma* = skin). The middle layer of the three primary germ layers that gives rise to muscle, connective tissue, the circulatory system, and most of the excretory and reproductive systems.

Mesolecithal. (Gk., *mesos* = middle, + *lekthyos* = yolk of egg). Referring to moderate amount of yolk stored in the egg.

Mesomere. (Gk., *mesos* = middle, + *meris* = segment). The middle region of mesoderm; it gives origin to excretory organs and reproductive tracts of adult animal.

Messenger RNA (mRNA). The RNA that carries genetic information from the gene to the ribosome, where it determines the order of the amino acids in the formation of a polypeptide.

Metabolism. (Gk., *metabole* = change). The sum of all chemical reactions occurring within a living unit.

Metamerism. (Gk., *meta* = after, + *meros* = part). Referring to segmentation resulting in a living unit.

Metamorphosis (Gk., *metamorphoun* = to transform). Abrupt transition from larval to adult form ; it includes certain morphological, anatomical, physiological, and behavioural, hormonally regulated changes in the larval form to transform it into the adult form.

Metaphase. (Gk., *meta* = middle, + *phasis* = form). The stage of mitosis or meiosis during which the chromosomes lie in the central plane of the spindle, prior to centromere division.

Metaplasia. Transformation of one cell type into another.

Metastasis. The separation of malignant tumour cells from a tumour followed by their spread to other tissues.

Metencephalon. (Gk., *meta* = after, + *enkephalos* = brain). The anterior of the two subdivisions of the primitive hind-brain.

Microfilament. Linear polymers of tensile proteins occurring in the egg cortex and play an active role in cell transformation, in morphogenetic movements, in the formation and orientation of mitotic spindles during cleavage, in the determination of ooplasm, and so on.

Microlecithal. (Gk., *mikros* = small, + *lekythos* = yolk of egg). Referring to small amount of yolk stored in the egg, *e.g., Amphioxus*, eutherian mammals.

Mitochondrion. (Gk., *mitos* = thread, + *chondros* = granule). A cellular organelle of eukaryotes which is bounded by double unit membranes, occurring in cytoplasm, containing enzymes and coenzymes for energy metabolism, and which stores ATP molecules during cellular respiration ; it contains a circular DNA molecule and in higher animals, has significant role in yolk- platelet formation.

Mitosis. (Gk., *mitos* = thread, + *osis* = state). Nuclear division characterized by exact chromosome duplications and the formation of two identical daughter cells ; the means by which somatic cells and/or zygotes (fertilized eggs) divide.

Morphogens. Substances that are deemed to induce morphogenesis or differentiation in target cells ; morphogens themselves being the products of the inductor cells.

Morphogenesis. (Gk., *morphe* = shape, + *genesis* = production). The origin and development of size, form and other structural features of organism.

Moulting. Shedding of all or part of outer covering ; in arthropods, periodic shedding of the exoskeleton to permit an increase in size.

Mucus. (L., *mucus* = juice). A clear, viscid secretion.

Myelencephalon. (Gk., *myelos* = marrow, + *enkephalos* = brain). The posterior of the two subdivisions of the primitive hind-brain.

Nanometre (nm). A unit of distance equal to 1×10^{-9} metre. Synonym, millimicron (mμ).

Neoplasia. A phenomenon in which a localized population of proliferating cells grows without reference to normal growth mechanism and, therefore, produces a tumour. Neoplasms are described as benign if they are non-invasive and do not produce metastasis, and malignant if they are invasive and produce secondary tumours. Malignant neoplasms are commonly referred to as cancers.

Neural groove. (Gk., *neuron* = nerve). Dorsal longitudinal groove which forms in a vertebrate embryo ; bordered by two neural folds ; preceded by neural plate stage and followed by neural tube stage.

Neural plate. Thickened strip of neural ectoderm in early vertebrate embryos ; forms along the dorsal side of the body and gives rise to the central nervous system.

Neural tube. Primitive hollow dorsal nervous system of the early vertebrate embryo formed by neural folds around neural groove.

Neuroblast. (Gk., *neuron* = nerve, + *blastos* = germ). An embryonic nerve cell.

Neuroglia. (Gk., *neuron* = nerve, + *glia* = glue). The supporting tissue of the brain and spinal cord.

Neurohypophysis. (Gk., *neuron* = nerve, + *hypophysis* = an undergrowth). The posterior lobe of the hypophysis (pituitary).

Neurosecretory cell. A neuron which produces one or more hormones.

Notochord. (Gk., *notos* = back, + *chorde* = cord, string). A longitudinal, solid, elastic rod-like structure serving as the endoskeleton in the embryos of all chordates ; in most adult chordates the notochord is replaced by a vertebral column which forms around (but not from) the notochord.

Nucleic acids. An organic acid consisting of nucleotides ; the principle types are deoxyribonucleic acid (DNA) and ribonucleic acid (RNA).

Nucleolus. (L., *nucleolus* = a small kernal). A conspicuous spherical intranuclear body which lacks a limiting membrane, remains associated with rDNA of nucleolar organizer region of chromosome, and involves in the storage and maturation of various ribosomal RNAs and ribosomal proteins; site of origin of ribosomes.

Nucleotide. A single unit of nucleic acid composed of phosphate, a five-carbon sugar (either ribose or deoxyribose), and a purine or a pyrimidine.

Nucleus. (L., *nucleus* = a kernal). The cellular structure that contains the genetic information in the form of DNA. In the eukaryotic cell, a specialized body bounded by a double membrane, containing the chromosomes.

Oblique. (L., *obliquus* = slanting). Deviating from the perpendicular or horizontal.

Olfactory. (L., *olfacere* = to smell). Pertaining to smell.

Oligo . (Gk., *oligos* = few, small). Prefix, meaning "few", "small".

Ontogeny. (Gk., *on* = being, + *genesis* = origin). The developmental history of an individual organism from zygote to maturity.

Oo. (Gk., *oion* = egg). Prefix, meaning "egg".

Oocyte. (Gk., *oion* = egg, + *kytos* = vessel). A cell that eventually gives rise by meiosis to an ovum.

Oogamy. Kind of sexual reproduction in which one of the gametes (the egg) is large and non-motile, while the other gamete (the sperm) is smaller and motile.

Oogenesis. (Gk., *oion* = egg, + *genesis* = origin). The process of formation and development of the ovum inside ovary.

Optic. (Gk., *opsis* = sight). Relating to the eye or vision.

Oral. (L., *os* = mouth). Relating to the mouth.

Organ. (Gk., *organon* = tool). A body part usually composed of several tissues grouped together in a structural and functional unit.

Organizer. Part of an embryo capable of inducing undifferentiated cells to follow a specific course of development ; in particular, the dorsal lip of the blastopore in the amphibian.

Organogenesis. (Gk., *organon* = organ, + *genesis* = production). The formation of an organ.

Osmosis. (Gk., *osmos* = impulse, thrust). The movement of water between two solutions separated by a membrane that permits the free passage of water and that prevents or retards the passage of the solute ; the water tends to move from the side containing a lesser concentration of solute to the side containing a greater concentration.

Otic. (Gk., *otikos* = belonging to the ear). Relating to the ear.

Ovi-, Ovo. (L., *ovum* = egg). Prefix, meaning "egg".

Ovary. (L., *ovarium* = egg-receptacle). The egg-producing reproductive organ of a female animal.

Oviduct. (L., *ductus* = duct). The tube serving to transport the eggs to the uterus or to the outside ; Fallopian tubes (in humans).

Ovulation. Release of an animal egg from the ovary.

Ovum. pl. **ova**. (L., *ovum* = egg). The female sex cell or gamete which carries female genotype to the offspring and contains stored food for the development of embryo up to feeding stage and which after fusion with sperm forms zygote that undergoes embryogenesis to produce new individual.

Oxidation. Loss of an electron by an atom ; an important energy transforming reaction of living systems.

Pancreas. (Gk., *pan* = all + *kreas* = meat, flesh). A small complex gland in vertebrates, located between the stomach and the duodenum; which produces digestive enzymes and the hormones insulin and glucagon.

Parthenogenesis. (Gk., *parthenos* = virgin, + *genesis* = origin). The development of an egg without fertilization.

Penis. (L., *penis*). The male copulatory or intromittent organ.

Peri. (Gk., around). Prefix, meaning "around".

Peritoneum. (Gk., *peritonos* = stretched over). A mesodermal epithelial membrane lining the body cavity and forming the external covering of the visceral organs.

pH. A symbol denoting the relative concentration of hydrogen ions in a solution ; pH values range from 0 to 14 ; the lower the value the more acid a solution, that is, the more hydrogen ions it contains ; pH 7 is neutral, less than 7 is acid, more than 7 is alkaline.

Phagocytosis. (Gk., *phagein* = to eat). Cell "eating" ; the intake of solid particles by a cell, by flowing over and engulfing them.

Pharynx. (Gk., *pharnyx* = the throat). Segment of gut between the mouth and oesophagus.

Phenotype. (Gk., *phainein* = to show). The observable properties of an organism resulting from interaction between its genetic constitution (genotype) and the environment.

Phylogeny. (Gk., *phylon* = race, tribe, + *genesis* = origin). Evolutionary relationships among organisms, developmental history of group of organisms.

Physiology. (Gk., *physio* = nature, + *logos* = a discourse). The study of functions in cells, organs, or entire organisms.

Pinocytosis. (Gk., *pinein* = to drink). Cell "drinking" ; the intake of fluid droplets by a cell.

Pituitary. (L., *pituita* = phlegm). Endocrine gland in vertebrates ; the anterior lobe (adenohypophysis) is the source of tropic hormones, growth hormones and prolactin and is stimulated by neurosecretory cells in the hypothalamus ; the posterior lobe stores and releases oxytocin and ADH produced by the hypothalamus.

Placenta. (Gk., *plax* = a flat object). A structure formed in part from the inner lining of the uterus and in part from the extra-embryonic membranes ; develops in most species of mammals and serves as the connection between the mother and the embryo during pregnancy through which exchanges occur between the blood of mother and of embryo.

Placode. (Gk., *plax* = plate + *eidos* = like). Denoting only thickened plate-like area.

Polar. Having parts or areas with opposed or contrasting properties, such as head and tail.

Polar-body (Polocyte). Minute non-functioning cell produced during meiotic divisions in egg cells ; contains a nucleus but very little cytoplasm.

Polymer. (Gk., *polys* = many, + *meros* = part). A large molecule composed of many identical molecular subunits.

Polypeptide. A molecule consisting of numerous amino acids linked together by peptide bonds.

Polysaccharide. A carbohydrate composed of many joined monosaccharide units in a long chain ; for example, glycogen, starch, cellulose.

Polyspermy. The entrance of more than one spermatozoon into the ovum.

Positional information. Signals from outside a cell that instruct the cell about its position relative to other cells in the same organism. Such information is important in the commitment of cells in early embryonic life.

Post-, Postero. (L., behind = after). Prefix, meaning "at" "near" or "toward the hind part" ; opposite of "pre-", and "anterior".

Posterior. Of or pertaining to the rear end of the animal.

Pre. (L., before, in front of). Prefix, meaning "before", or "in front of" ; opposite of post.

Preformation theory. This classical now obsolete theory maintains that a miniature adult organism (*e.g.*, the human homunculus) exists inside an egg or sperm and that development consists merely of the expansion of the miniature.

Primitive. (L., from *primus* = first). Not specialized ; at an early stage of evolution or development.

Primitive streak. The thickened dorsal longitudinal strip of ectoderm and mesoderm in early avian, reptilian, and mammalian embryos ; equivalent to blastopore in other forms.

Proestrus. The period immediately preceding heat (estrus) in female mammals.

Progesterone. (Gk., *pro* = in favour of, + L. *gestare* = to bear). In mammals, steroid hormone produced by corpus luteum that prepares the uterus for reception of the ovum.

Pronucleus. (Gk., *pro* = before, + L. dim. of *nux* = nut or kernel). The nucleus of the spermatozoon or of the ovum prior to fertilization.

Prophase. (Gk., *pro* = before, + *phasis* = form). An early stage in nuclear division, characterized by the condensing of the chromosomes and their movement to the equator of the spindle. During meiotic prophase-I certain important events such as synapsis, chromosome duplication, crossing-over and chiasma formation occur.

Prosencephalon. (Gk., *pros* = before, + *enkephalos* = brain). The most anterior of the three primitive divisions of the brain.

Prostaglandin. One of a group of fatty acid hormones discovered in semen, now found to be present in many other tissues. Believed to play a role in fertilization.

Prostate gland. (L., *pro* = before, + *stare* = to stand). A mass of muscle and glandular tissue surrounding the base of urethera in male mammals, through which the seminal vesicle passes and which secretes an alkaline fluid that has stimulating effect on sperm as they released.

Protamines. A class of proteins rich in the basic amino acid arginine. They are found complexed to the DNA of sperm in many invertebrates and fish.

Protein. (Gk., *proteios* = primary). A complex organic compound composed of one or more polypeptide chains, each made up of many (about 100 or more) amino acids joined by peptide bonds.

Protoplasm. (Gk., *protos* = first ; *plasma* = thing formed). The living substance of all cells.

Proximal. (L., *proximus* = next). Nearest the trunk or point of origin.

Pseudopod. (Gk., *pseudes* = false, + *pous, pod* = foot). A temporary cytoplasmic protrusion from an amoebid cell which functions in locomotion or in feeding by phagocytosis.

Puff. Local enlargement of a polytene chromosome in which active gene transcription can be visualized.

Pupa. (L., girl, doll). A developmental stage, non-feeding, immotile, and sometimes encapsulated or in a cocoon, between the larval and adult phases in insects.

Purine. A nitrogenous base such as adenine or guanine ; one of the components of nucleic acids.

Pulse-chase experiment. A radioactively labelled compound is added to living cells and a short time later an excess of unlabelled compound is added. Samples are then taken at periods after the pulse to follow the course of the label as the compound is metabolized (chase).

Pyrimidine. A nitrogenous base such as cytosine, thymine, or uracil ; one of the components of nucleic acids.

Radial symmetry. The regular arrangement of parts, around one longitudinal axis ; any line drawn through this oral-aboral axis will divide similar halves ; seen in coelenterates and adult echinoderms.

Renal. (L., *renes* =kidneys). Pertaining to kidney.

Replicon. ''Replication unit'' of eukaryotic DNA, having its own replication initiation site.

Reticulum. (L., *reticulum* = network). Any network of web-like structure, as in endoplasmic reticulum.

Retina. (L., *rete* = net). The innermost nervous-tissue layer of the eye-ball ; contains several layers of neurons and light-receptor cells (rods and cones) ; receives the image formed by the lens and is connected to the brain by the optic nerve.

Rhombencephalon. (Gk., *rhombos* = parallelogram, + *enkephalos* = brain). The posterior of the three primitive brain divisions.

Ribonucleic acid (RNA). A nucleic acid formed on chromosomal DNA and involved in protein synthesis ; similar in composition to DNA, except that the pyramidine uracil replace thymine. It may be of four main kinds—ribosomal RNA (rRNA), transfer RNA (tRNA), heterogenous nuclear RNA (hn RNA), and messenger RNA (mRNA).

Ribosome. A small cytoplasmic cellular organelle composed of protein and different kinds of ribosomal RNA (rRNA) ; the site of protein synthesis.

Sacculus. (L., *sacculus* = a little bag). Denoting any small pouch.

Schizocoel. (Gk., *schizo* = split, *koilos* = hollow). A coelom formed by the splitting of an originally solid layer of mesoderm.

Sclera. (Gk., *skleros* = hard). The tough outer coat of the eye.

Sclerotome. (Gk., *skleros* = hard, + *tomos* = cutting). A segment of the skeleton producing cells derived from a mesodermal somite.

Scrotum. (L., *scrotum* = hide). A musculo-cutaneous sac containing the testes.

Secondary sex characteristics. External characteristics that distinguish between the two sexes but which have no direct role in reproduction, such as the rooster's comb.

Self-fertilization. The union of egg and sperm produced by a single hermaphroditic organism.

Semen. (L., seed). Products of male reproductive systems ; includes sperm and sperm-carrying fluid.

Seminiferous. (L., *semen* = seed, + *ferre* = to carry). Carrying or conducting the semen.

Sexual reproduction. Reproduction involving meiosis and syngamy (union of haploid gametes, the sperm and ova).

Skeleton. (Gk., *skeletos* = dried). The supporting framework of the body.

Skull. (Early Eng., *skulle* = a bowl). The skeleton framework of head.

Somatic cells. (Gk., *somatikos* = bodily, + L., *cella* = a small chamber). The differentiated, usually diploid (2n) cells composing body tissues of multicellular animals ; all body cells except the germ cells.

Somatopleure. (Gk., *somatikos* = bodily, + *plura* = side). The embryonic layer formed by the union of the somatic layer of mesoderm with the ectoderm.

Specialized. (1) Of organisms, having special adaptations to a particular habitat or mode of life. (2) Of cells, having particular functions.

Species, pl. **species**. (L., kind, sort). A group of organisms that actually (or potentially) interbreed and are reproductively isolated from all other such groups.

Specificity. Uniqueness, as in proteins in a given organism and of enzymes in given reactions.

Sperm, Sperme-, Sperma-, Spermato-. (Gk., *sperma* = seed). Suffix or prefix, meaning "seed".

Sperm. (Gk., *sperma* = seed). A mature male sex cell or gamete ; usually motile flagellated, smaller than the female gamete and containing enzyme-containing body the acrosome at its forwarding end.

Spermatid. Each of four haploid cells resulting from the meiotic divisions of a spermatocyte, which becomes differentiated into sperm cell.

Spermatocytes. The diploid cells formed by the enlargement of the spermatogonia ; give rise by meiotic division to the spermatids.

Spermatogenesis. (Gk., *sperma* = seed, *genesis* = origin). The process by which spermatogonia develop into a mature sperm.

Spermatogonia. The unspecialized diploid (2n) germ cell on the walls of the testes (*i.e.*, seminiferous tubules) which by meiotic division becomes spermatocytes, then spermatids, and lastly spermatozoa or sperm cells.

Spermiogenesis. Transformation of a non-motile, unspecialized spermatid into a motile and specialized, mature sperm.

Spinal cord. The part of vertebrate central nervous system that consists of thick longitudinal bundle of nerve fibres extending from the brain posteriorly along the dorsal side.

Splanchnic. (Gk., *splanchnon* = viscus). Referring to the viscera.

Splanchnopleure. (Gk., *splanchnon* = viscus, + *plura* = side). The embryonic layer formed by the union of visceral layer of mesoderm with the endoderm.

Splicing. Precise excision of the intervening sequence from a RNA primary transcript, followed by ligation of the message to produce a functional molecule.

Stem cell. Undifferentiated cell capable of giving rise to specialized daughter cells.

Steric (Stereochemical). Relating to the arrangement in space of the atoms in molecules.

Steroid. (Gk., *steros* = solid, + L., *oleum* = oil, Gk., *eidos* = form). Resembling a sterol.

Stimulus. (L., god incentive). Any internal or external change which influences the activity of an organism or part of an organism.

Stomach. (L., *stomachus*). The portion of gut between the oesophagus and small intestine.

Stome,-Stoma,-Stomat (Gk., *stoma* = mouth). Suffix meaning "mouth".

Stomodaeum. (Gk., *stoma* = mouth ; *hodaios* = relating to a way). Ectodermal invagination forming the mouth cavity.

Stratum. (L., *stratus* = layer). Any given layer of differentiated tissues.

Sub-, Sus. (L., under below). Prefix, meaning "under" or "below".

sym-, Syn-. (Gk., *syn* = together, with). Prefix, meaning "together".

Synapsis. The pairing of homologous chromosomes that occurs prior to the first meiotic division ; crossing over occurs as a result of synapsis.

Syncytium. Large protoplasmic mass not divided into distinct cells.

Syndesmochorial (Gk., *syndesmos* = ligament, + *chorion* = skin). Placenta type involving contact between chorionic epithelium and connective tissue of uterine mucosa.

Synthesis. The formation of a more complex substances from simpler ones.

Telencephalon. (Gk., *teleos* = complete, + *eukephalos* = brain). The anterior division of fore-brain.

Telic. (Gk., *telikos* = final). Tending to a definite end.

Telolecithal. (Gk., *telos* = end, + *lekithos* = yolk). A condition of egg in which yolk remains accumulated at one pole.

Telophase. (Gk., *telos* = end, + *phasis* = form). The last stage in mitosis and meiosis, during which the chromosomes become reorganized into two new nuclei .

Template. A pattern or mould guiding the formation of a negative or complement.

Terminally differentiated cells. Non-dividing cells such as erythrocytes or neurons, which by dint of their extreme specialization, have become end cells and can no longer indulge in mitotic division.

Testis. pl. Testes. (L., witness). The male gamete-producing organ or gland ; also the source of male sex hormone.

Testosterone. A hormone secreted by the testes in higher vertebrates and stimulating the development and maintenance of male sex characteristics and the production of sperm ; an androgen.

Theca. (Gk., *theke* = a box). A sheath.

Theory. (Gk., *theorein* = to look at). A generalization which is based on observation and experiments that were conducted to test the validity of a hypothesis and found to support the hypothesis.

Thorax. (Gk., *thorax* = breast plate). (1) In vertebrates, that portion of the trunk containing the heart and lungs, (2) In insects, the three leg-bearing segments between head and abdomen.

Thymine. A pyrimidine base occurring in DNA ; pair normally with adenine.

Thyroid. (Gk., *thyreos* = an oblong shield, + *eidos* = form). Denoting a gland or cartilage.

Tight junction. Structures formed when adjacent cell membranes adhere to each other, creating a barrier to diffusion.

Tissue. (L., *texere* = to weave). A group of similar cells organized into a structural and functional unit.

Totipotent. Capable of giving rise to all of the various cell types that comprise the tissue of an organism.

Transcription. The enzymatic process, involving base pairing, by which the genetic information contained in DNA is used to specify a complementary sequence of bases in an RNA molecule.

Transfer RNA (tRNA). The type of RNA that recognizes specific amino acid, becomes attached to it, transports it to the site of protein synthesis and guides it to the correct position on the ribosome-mRNA complex for protein synthesis. There is at least one tRNA molecule for each amino acid.

Translation. The process by which the genetic information contained in a messenger RNA molecule dictates the sequence of amino acids in a protein molecule.

Transposon (Transposable element). A segment of DNA, which generally consists of more than 2,000 nucleotides, and is capable of moving into and out of a chromosome or plasmid, and/or from place to place within the chromosome in both prokaryotes and eukaryotes. It is capable of turning genes "on" and "off".

Troph, Tropho. (Gk., *trophos* = feeder). Suffix or prefix meaning "feeder", "feeding".

Trophoblast. (Gk., *trophe* = nourishment, + *blastos* = germ). The outer wall of the mammalian blastocyst, concerned with nutrition of embryo within.

Tunica. (L., *tunica* = a coat). An enveloping layer.

Uracil. A pyrimidine base occurring in RNA ; pairs normally with adenine.

Umbilical. (L., *umbilicus* = navel). Relating to the umbilicus.

Urea. (Gk., *ouron* = urine). An organic compound formed in the vertebrate liver from ammonia and carbon dioxide and excreted by the kidneys ; the principal form of disposal of nitrogenous wastes in mammals.

Ureter. (Gk., from *ourein* = to urinate). The tube carrying urine from the kidney to the cloaca (in reptiles and birds) or to the bladder (in amphibians and mammals).

Urethra. The tube carrying urine from the bladder to the exterior in mammals.

Uric acid. An insoluble nitrogeneous waste product that is the principal excretory product of birds, reptiles, and insects.

Uterus. (L., womb). The muscular, expanded portion of the female reproductive tract modified for the storage of eggs or for housing and nourishing the developing embryo. In it the impregnated ovum develops into the foetus.

Vacuole. (L., *vacuus* = empty). A space within a cell, bounded by a membrane and filled with water and solid materials in solution.

Vagina. (L., *vagina* = sheath). The genital canal in the female extending from the uterus to vulva, it receives the male penis during copulation.

Vas. (L., *vas* = vessel). Denoting a tube or vessel.

Vas deferens. In mammals, the tube carrying sperm from the testes to the urethra.

Vegetal. (L., *vegetare* = to animate). Vegetal pole, denoting the end of a telolecithal egg containing the yolk.

Vein. (L., *vena* = a blood vessel). A blood vessel carrying blood from the tissues to the heart.

Ventral. (L., *venter* = belly). Pertaining to the under surface of an animal that moves on all fours ; to the front surface of an animal that holds its body erect.

Ventricle. (L., *ventriculus* = small cavity). A chamber of the heart that receives blood from an atrium and pumps blood out of the heart.

Vertebral column. The backbone ; in nearly all vertebrates, it forms the supporting axis of the body and a protection for the spinal cord.

Vesicle. (L., *vesicula* = a little bladder). A small, intracellular membrane-bounded sac.

Viability. Ability to live.

Viscera. (L., *viscus* = an internal organ). The collective term for the internal organs of an animal.

Vitelline. (L., *vitellus* = yolk). Relating to the yolk of an egg.

Vitreous. (L., *vitreus* = glassy). Any structure having a glassy appearance.

Viviparous. (L., *vivus* = alive, + *parere* = to bear). Giving birth to living young.

Vulva. External genitalia of the human female ; includes the clitoris and the labia.

Yolk. The stored food material in egg cells that nourishes the embryos ; chemically it is composed of proteins and lipids.

Z - DNA. Double helical DNA in which the helices are left handed.

Zona. (L., *zona* = girdle). An encircling area of structure.

Zygote. (Gk., *zygon* = yolk, pair). The fertilized ovum or a diploid cell resulting from the fusion of male (sperm) and female (ovum) gametes.

SELECTED READING

Arey, L.B., 1965, *Developmental Anatomy,* W.B. Saunders Company, Philadelphia and London.

Austin, C.R. and R.V. Short, 1972, *Reproduction in Mammals 2. Embryonic and fetal development,* The University Press, Cambridge.

Balinsky, B.I., 1970, 1981, *An Introduction to Embryology,* W.B. Saunders Company, Philadelphia and London.

Berrill, N.J., 1971, *Developmental Biology,* McGraw Hill Book Company, New York.

Berrill, N.J., and Karp, G., 1987, *Development.* Tata McGraw Hill Publishing Company Ltd., New Delhi, India.

Bodemer, C.W., 1968, *Modern Embryology,* Holt. Reinehart and Winston Inc., New York, Chicago,

Brachet, J., 1974, *Introduction to Molecular Embryology,* The English University Press Ltd., London.

Browder, L.W., (Editor), 1988, *Developmental Biology (Vol. 5): The Molecular Biology of Cell Determination and Cell Differentiation,* Plenum Press, New York.

Burns, G.W. and Bottino, P.J., 1989, *The Science of Genetics,* Macmillan Publishing Company, New York.

Cockrum, E.L. and W.J. MacCauley, 1965, *Zoology,* W.B. Saunders Company, Philadelphia.

Cohen, J., 1967, *Living Embryos,* Pergamon Press, Oxford.

Craigmyle, M.B.L., and R. Presley, 1975, *Embryology,* Bailliere Tindall, London.

Dehaan, R.L. and H. Ursprung, 1965, *Organogenesis,* Holt, Rinehart and Winston, New York, Chicago.

De Robertis, E.D.P., and De Robertis Jr., E.M.F., 1987, *Cell and Molecular Biology,* Lea and Febiger, Philadelphia.

Dutta, D.C., 1987, *Text Book of Obstetrics,* New Central Book Agency, Calcutta.

Ebert, J.D. and I.M. Sussex, 1970, *Interacting System in Development,* Holt, Rinehart and Winston, Inc., New York.

Ede, D.A., 1978, *Developmental Biology,* Blackie, Glasgow and London.,

Epel, D., 1973, *Germ Cell and Early Embryonic Development.,* In—Cell biology in medicine by E.E. Bitnar.

FitzGerald, M.J. T., 1978, *Human Embryology,* Harper and Row, Publishers, Hagerstown, U.S.A.

Freeman, W.H. and B. Bracegirdle, 1963, *An Atlas of Embryology,* Heinemann, London.

Freifelder, D., 1985, *Essentials of Molecular Biology,* Jones and Bartlett Publishers, Inc., Boston.

Gilchrist, F.G., 1968, *A Survey of Embryology,* McGraw-Hill Book Company, New York.

Goel, S.C., 1984, *Principles of Animal Developmental Biology,* Himalaya Publishing House, Bombay.

Goldsby, R.A., 1976, *Biology,* Harper & Row, Publishers, New York

Grant, P., 1978, *Biology of Developing Systems,* Holt-Saunders International Editions, New York.

Guyton, A.C., 1986, *Medical Physiology,* W.B. Saunders Company, Philadelphia.

Hamburgh, M., 1971, *Theories of Differentiation,* Edward Arnold, London.

Herreid, II, C.F., 1977, *Biology,* Macmillan Publishing Co., Inc., New York.

Hickman, Sr., C.P., Hickman, Jr., C.P., and Hickman, M.F., 1979, *Zoology,* The C.V. Mosby Company, St. Louis, U.S.A.

Huettner, A.F., 1949, *Fundamentals of Comparative Embryology,* The Macmillian Company, New York.

Jordan, E.L. and Verma, P.S., 1994, *Chordate Zoology,* S. Chand & Co. Ltd., New Delhi.

Kent, G.C., 1987, *Comparative Anatomy of Vertebrates,* Times Mirror/Mosby, College Publishing, St. Louis, U.S.A.

Kuhn, A., 1971, *Lectures on Development Biology,* Springer Verlag, Berlin.

Lovtrup, Soren, 1974, *Epigenetics,* John-Wiley & Sons, London.

MacBride, E.W., 1914, *Text Book of Embryology,* (Vol I, Invertebrata), Macmillian and Co., Ltd. London.

Maclean, N. and Hall, B.K., 1987, *Cell Commitment and Differentiation,* Cambridge University Press, Cambridge.

Market, C.L. and H. Ursprung, 1974, *Developmental Genetics,* Prentice Hall of India, Pvt. Ltd., New Delhi.

McEwen, R.S., 1949, *Vertebrate Embryology,* Oxford and IBH Publishing Co., New Delhi.

Meglitsch, P.A., 1972, *Invertebrate Zoology,* Oxford University Press, Oxford, London.

Menon, Devi and Rao, 1989, *Obstetrics and Gynaecology,* Orient-Longman, Madras.

Nalbandov, A.V., 1964, *Reproductive Physiology,* W.H. Freeman and Company, San Francisco, London.

Nelsen, O.E., 1953, *Comparative Embryology of the Vertebrates,* The Blakistan Company, Inc., New York.

Newth, D.R., 1970, *Animal Growth and Development,* Edward Arnold (Publishers) Ltd., London.

Patten, B.M., 1958, *Foundation of Embryology.* McGraw-Hill Book Company, Inc., New York.

Phillips, J.B., 1975, *Development of Vertebrate Anatomy,* The C.V. Mosby Company, Saint Louis, U.S.A.

Ribchester, R.R., 1986, *Molecule, Nerve, and Embryo,* Blackie, Glasgow.

Pollak, J.K. and G.E. Sullivan, 1972, *Developmental Biology,* Van Nostrand Reinhold, Melbourne.

Saunders, J.W., 1968, *Animal Morphogenesis,* The Macmillan Company, New York.

Saunders, J.W., 1970, *Pattern and Principles of Animal Development,* The Macmillan Company, London.

Scandalios, J.G., 1987, *Molecular Genetics of Development,* Academic Press, Inc., San Diego, U.S.A.

Schmidt-Nielsen, K., 1983, *Animal Physiology,* Cambridge University Press, Cambridge.

Sheeler, P, and Bianchi, 1987, *Cell and Molecular Biology,* John Wiley & Sons, Inc., New York.

Sussman, M., 1958, *Amimal Growth and Development,* Prentice Hall of India, (Pvt.) Ltd., New Delhi.

Torrey, T.W., 1971, *Morphogenesis of Vertebrates,* John Wiley and Sons. Inc., New York.

Tortora, G.J. and Anagnostakos, N.P., 1984, *Principles of Anatomy and Physiology,* Harper & Row Publishers, New York.

Trinkaus, J.P., 1969, 1984, *Cells into Organs,* Prentice-Hall, Inc., Engelewood Cliff, New Jersey.

Verma P.S. and Agarwal, V.K., 1994, *Cytology,* S. Chand and Company Ltd., New Delhi.

Wheater, P.R., Burkitt, H.G., and Daniels, V.G., 1979, *Functional Histology,* English Language Book Society, Churchill Livingstone.

Waddington, C.H., 1966, *Principles of Development and Differentiation,* The Macmillan Company, New York.

Willier, B.H., Paul A., Weiss and V. Hamburger, 1955, *Analysis of Development,* Hafner Publishing Company, New York.

INDEX

B

F

N

P

Q

R

S

T

SOME SELECTED BOOKS FOR DEGREE STUDENTS

BOTANY

B.R. Vashishta

BOTANY FOR DEGREE STUDENTS : ALGAE

The new edition includes life histories of a few more types such as Ulva, Enteromorpha, Protsiphen, Pleurococcus and Mougeotia.

03 011 8/e 1995 pp 463
03 229 Hard Bound Edition

B.R. Vashishta

BOTANY FOR DEGREE STUDENTS : FUNGI

Reflects recent advances in the field of Mycology. Gives a vivid account of the various aspects of Botany.

03 012 9/e 1995 pp 518
03 230 Hard Bound Edition

B.R. Vashishta

BOTANY FOR DEGREE STUDENTS : BRYOPHYTA

Contains the latest research material and includes the life history of Notothylus. Gives a clear picture of the pattern of Alternation of Generations.

03 013 8/e 1994 pp. 368

P.C. Vashishta

BOTANY FOR DEGREE STUDENTS : PTERIDOPHYTA

Present edition incorporates new chapters on life cycles of Isoetes, Salvinia and Azolla. Good many diagrams are given to illustrate important aspects of morphology and development.

03 014 5/e Rev. 1995 pp 522
03 232 Hard Bound Edition

P.C. Vashishta

BOTANY FOR DEGREE STUDENTS : GYMNOSPERMS

It makes a detailed study of morphology and reproduction of both living and fossil forms.

03 015 3/e 1995 pp. 496

B.P. Pandey

COLLEGE BOTANY (Vol.I & II)

These two volumes are compilation work which embodies a fairly comprehensive treatment of the fundamental facts of the subject to the beginners in this field. Simple language, self-explanatory labelled diagrams with detailed legends are some of the salient features of this elementary text for the Degree students of Botany.

Vol. I — Algae, Fungi & Bryophyta

03 151 5/e 1994 pp.936

Vol. II — Pteridophyta, Gymnosperms & Palaeobotany

03 152 5/e1994 pp. 396

Also available in parts by B.P. Pandey :

A TEXT-BOOK OF BOTANY : ALGAE

03 212 5/e 1994 pp 286

A TEXT-BOOK OF BOTANY : FUNGI

03 213 5/e1994 pp. 424

A TEXT-BOOK OF BOTANY : BRYOPHYTA

03 214 5/e 1994 pp. 232

A TEXT-BOOK OF BOTANY : PTERIDOPHYTA

03 215 5/e 1994 PP. 184

A TEXT-BOOK OF BOTANY : GYMNOSPERMS & PALAEOBOTANY

03 216 5/e 1994 PP. 213

(All these parts are also available in Hindi)

B.P. Pandey

PLANT ANATOMY

This book provides the fundamental facts of plant Anatomy for the graduate and post-graduate students of Botany with detailed facts, terminology and internal structures of common plant. This book may safely be used as laboratory guide also.

03 016 5 /e 1994 pp 656

B.P. Pandey

INTRODUCTION TO PLANT ANATOMY

This book embodies a comprehensive treatment of the various aspects of plant Anatomy. The text deals with various aspects of plant Anatomy. The text deals mainly with the anatomy of vascular plants. Emphasis has been given on the anatomical study of the Angiosperms.

03 072 1/e (Rep)1993 pp. 318
03 161 1/e Hindi 1993 pp. 390

B.P. Pandey

A TEXT-BOOK OF BOTANY: ANGIOSPERMS

Dealing primarily with Taxonomy, Anatomy, Economic Botany & Embryology of Angiospermic plants, this book also covers Pharmacognosy Morphogenesis and is profusely illustrated with self-explanatory and well-labelled diagrams.

03 118 4 /e 1995 pp. 830

B.P. Pandey

TAXONOMY OF ANGIOSPERMS (SYSTEMATIC BOTANY)

In all sixty-five families have been discussed in detail, forty-nine from Dicots and sixteen from Monocots. The families have been arranged according to the phylogenetic system of classification propounded by John Hutchinson.

03 149 5/e 1995 pp. 650

B.P. Pandey

A TEXTBOOK OF PLANT PATHOLOGY

(Pathogen and Plant Diseases)

This text embodies a fairly comprehensive treatment of the fundamental facts and aspects of plant pathology, and also serves as a laboratory manual. This book will also serve as an introduction to plant pathology to the beginners in this field.

03 091 2/e 1994 pp. 532

Purekar, Singh & Deshmukh

PLANT PHYSIOLOGY AND ECOLOGY

In a compact size, this book present a very comprehensive and up to date knowledge regarding Plant Physiology, Ecology and Phytogeography and cover the syllabus of many Indian universities.

03 054 2/e 1995 pp. 390

R.S. Shukla & P.S. Chandel

PLANT ECOLOGY

In this book the chapter on Environment has been upgraded. The chapter on Ecosystem contains more information and better diagrams. Chapters on Autecology and Ecological Genetics of Population has been re-written. Similarly there are better diagrams and additional matter in the chapter on Synecology.

03017 8/e 1994 pp. 328

R.S. Shukla & P.S. Chandel

PLANT ECOLOGY AND SOIL SCIENCE

Since the Ecology courses at Undergraduate and Post-graduate levels have been substantially revised in all the Indian Universities, this book has been revised keeping in view the requirements of the students.

03 018 8/e 1994 pp. 421

B.P. Pandey

ECONOMIC BOTANY

Several new topics and photographs have been added in this thoroughly revised edition.

03 021 4/e 1995 pp. 520

V.K. Jain

FUNDAMENTALS OF PLANT PHYSIOLOGY

Great emphasis has been laid in this book on basic fundamentals with simultaneous incorporation of most up-to-date information in sufficient details. The subject-matter has been splitted into 'typical questions' to make it more meaningful to the students.

03 020 4/e 1995 pp. 416

R.S. Shukla & P.S. Chandel

CYTOGENETICS, EVOLUTION AND PLANT BREEDING

This book is suitable for biology and agriculture students at the under-graduate and graduate level. The subject matter is systematically arranged and illustrated with easy diagrams. Each topic is discussed in the light of recent researches.

03 019 2/e 1995 (In Press)

CYTOGENETICS AND EVOLUTION

03 073 3/e 1995 (in Press)

J.L. Jain

FUNDAMENTALS OF BIOCHEMISTRY

There have been several new discoveries in biochemistry which had necessitated revision of almost all chapters of this book. Five new chapters concerning 'Bio-energetics and Metabolism' have been added, incorporating the latest information on the subject.

03 115 4/e 1994 pp. 673

J.L. Jain

ELEMENTARY BIOCHEMISTRY

Contains the basic concept of biochemistry written in a manner suited to the broad spectrum of the college students and providing them an insight at a molecular level all those materials which are considered to be hidden secrets of life.

03 095 **2/e 1990** **pp. 372**

R.C. Dubey

TEXT-BOOK OF BIOTECHNOLOGY

Bringing Forth the Current informations on different areas of Biotechnology, this book also provides the details of the researches being carried out in all the aspects in India and abroad. In seventeen chapters which forms the five part this petite book contains illustrations, diagrams, figures and tables.

03 217 **2/e 1995** **pp. 290**

B.P. Pandey

TANNINS AND DYES

This work presents a comprehensive account dealing with plants yielding tannins and dyes from their different parts. Both Tannins and Dyes have been classified according to the morphological nature, distribution and usage.

03 119 **1/e 1981** **pp. 141**

J.F. Southey, Ed.

PLANT NEMATOLOGY

Reveals the various important aspects of Nematology. Conveys the subject-matter in a lucid and comprehensive manner.

03 121 **2/e** **pp. 282**

ZOOLOGY

E.L. Jordan & P.S. Verma

INVERTEBRATE ZOOLOGY

The new edition has a separate section on 'Minor Phyla'. Provides good many illustrations on Ctenophora, Endoprocta, Nemertinea, Acanthocephala, Rotifera, Ectoprocta Brachiopoda and Phoronida. It contains latest material obtained from the recent researches.

03 027 **14/e 1995** **pp. 856**
03 227 **Hard Bound Edition**
03 048 **2/e Hindi 1993** **pp. 1160**

E.L. Jordan & P.S. Verma

CHORDATE ZOOLOGY AND ANIMAL PHYSIOLOGY

Incorporates the Morphology and Physiology of Chordate types. Has new figures at appropriate places.

03 028 **10/e 1995** **pp. 1163**
03 228 **Hard Bound Edition**
03 088 **2/e Hindi 1993** **pp.1247**

P.S. Verma, V.K. Agarwal

CHORDATE EMBRYOLOGY

Gives a lucid, well-illustrated and up-dated account of the subject. Contains new chapters such as Gradient Theory, Embryonic Inductions, Metamorphosis, regeneration and differentiation.

03 026 **9/e 1995** **pp. 688**

Verma, Tyagi & Agarwal

ANIMAL PHYSIOLOGY AND ECOLOGY

This book provides the students of B.Sc. and M.Sc., a better understanding of the functions of different organ systems of animals, the inter-relationship & co-existence between animals, plants and mirobes and interactions occurring between organisms and their environments

03 031 **4/e 1995** **pp. 718**

Verma, Tyagi & Agarwal

ANIMAL PHYSIOLOGY

03 030 **6/e 1995** **pp. 448**

Agarwal, Srivastava & Kumar

ANIMAL PHYSIOLOGY AND BIOCHEMISTRY

The book emphasizes in depth about the knowledge of physiology which have assumed gigantic proportions in modern times. However, in this compact book only the central concept such as various biological molecules and physiology of different organs-systems has been dealt in detail.

03 087 **3/e 1994** **pp. 452**

P.S. Verma & V.K. Agarwal

ENVIRONMENTAL BIOLOGY

The thoughtless exploitation of various natural resources by man has resulted in a very critical state of existence. To forestall this, the science of Ecology has come forward to human rescue to save them from this state of self-destruction, dilemma and Paradox. Visualising this, the book has been especially prepared for the B.Sc. & M.Sc. students of different Indian universities of Biology as well Environmental science.

03 171 **4/e 1995** **pp. 599**

P.S. Verma & V.K. Agarwal

TEXTBOOK OF CYTOLOGY

The book provides a lucid, simple, detailed and well illustrated account of cell biology and molecular biology.

03 022 **8/e 1994** **pp. 570**

P.S. Verma & V.K. Agarwal

CELL BIOLOGY, GENETICS, EVOLUTION & ECOLOGY

The present edition contains several new illustrations and new chapters such as inbreeding, outbreeding, hybrid vigour in Genetics.

03 024 12/e 1994 pp. 860

P.S. Verma & V.K. Agarwal

CELL BIOLOGY, GENETICS AND ECOLOGY

03 173 7/e 1994 pp. 737

CELL BIOLOGY, GENETICS, MOLECULAR BIOLOGY AND EVOLUTION

03 025 9/e 1994 pp. 612

CELL BIOLOGY, MOLECULAR BIOLOGY AND GENETICS

03 093 7/e 1994 pp. 488

P.J. Bentley

COMPARATIVE VERTEBRATE ENDOCRINOLOGY

This book is a suitable text for senior undergraduate/graduate courses in vertebrate endocrinology and comparative endocrinology and as a background reading for courses in comparative animal physiology. It is also useful for First-year medical students and endocrinology researchers.

03 129 2/e 1990 pp. 415

P.S. Verma & V.K. Agarwal

GENETICS

All the chapters of this book have been thoroughly revised in keeping pace with the rapid progress in the field of life sciences.

03 023 8/e 1994 pp. 589

L.R. Patki, B.L. Bhalchandra & I.H. Jeevaji

AN INTRODUCTION TO MICROTECHNIQUE

This small and compact book contains all the aspects of mircotechnique in a fair Quantum in the right sequence, which makes it very useful and important for the students working in Biological and Pathological laboratories, as well as the practitioners in any laboratory using Mircotechnique.

03 047 2/e 1992 pp. 167

Jawaid Ahsan & Subhas Prasad Sinha

A HANDBOOK ON ECONOMIC ZOOLOGY

This hand-book is a practical guide to applied Zoology in which science of Fisheries, Sericulture, Lac culture Apiculture, Sponge culture and Pearl formation have been dealt with an economic point of view.

03 128 3/e 1995 pp. 186

S.N. Prasad & Vasantika Kashyap

AN INTRODUCTION TO TOXICOLOGY

This book provides a useful detail of toxic chemicals which is widely used for different purposes especially in agriculture where the benefits are manifold which outweigh the risks.

03 203 1/e 1991 pp. 200

OTHER IMPORTANT BOOKS

BOTANY

Mandahar, C.L.

03 102 Introduction to Plant Viruses 2/e 1987

Pandey, B.P.

Modern Practical Botany Vol. I & II

ZOOLOGY

Verma, P.S. & Srivastava, P.C.

(B.Sc. F.Y. Zoology Unified Course M.P.)

03 190 Non-chordate & Cell Biology Paper I 2/e 1993

03 193 Chordate & General Biology Paper II 2/e 1993

03 222 Advanced Practical Zoology 2/e 1995

Jajurley, Puranik & Thakur

03 080 Invertebrate Zoology 2/e 1992

Verma, P.S.

03 032 A Manual of Practical Zoology : Invertebrates 12/e 1995

03 033 A Manual of Practical Zoology Chordates 10/e 1995